OXFORD SERIES ON MATERIALS MODELLING

Oxford Series on Materials Modelling

Materials modelling is one of the fastest growing areas in the science and engineering of materials, both in academe and in industry. It is a very wide field covering materials phenomena and processes that span ten orders of magnitude in length and more than twenty in time. A broad range of models and computational techniques has been developed to model separately atomistic, microstructural and continuum processes. A new field of multi-scale modelling has also emerged in which two or more length scales are modelled sequentially or concurrently. The aim of this series is to provide a pedagogical set of texts spanning the atomistic and microstructural scales of materials modelling, written by acknowledged experts. Each book will assume at most a rudimentary knowledge of the field it covers and it will bring the reader to the frontiers of current research. It is hoped that the series will be useful for teaching materials modelling at the postgraduate level.

APS, London
RER, Livermore, California

1. M.W. Finnis: *Interatomic forces in condensed matter*
2. K. Bhattacharya: *Microstructure of martensite – Why it forms and how it gives rise to the shape-memory effects*
3. V.V. Bulatov, W. Cai: *Computer simulations of dislocations*
4. A.S. Argon: *Strengthening mechanisms in crystal plasticity*
5. L.P. Kubin: *Dislocations, mesoscale simulations and plastic flow*
6. A. P. Sutton: *Physics of elasticity and crystal defects, Second edition*
7. D. Steigmann: *A course on plasticity theory*
8. J.A. Moriarty: *Theory and application of quantum-based interatomic potentials in metals and alloys*
9. D.N. Theodorou, V.G. Mavrantzas: *Multiscale modelling of polymers*

Multiscale Modelling of Polymers

Doros N. Theodorou

School of Chemical Engineering, National Technical University of Athens and Academy of Athens

Vlasis G. Mavrantzas

Department of Chemical Engineering, University of Patras and Department of Mechanical and Process Engineering, ETH Zürich

OXFORD
UNIVERSITY PRESS

OXFORD
UNIVERSITY PRESS

Great Clarendon Street, Oxford, OX2 6DP,
United Kingdom

Oxford University Press is a department of the University of Oxford.
It furthers the University's objective of excellence in research, scholarship,
and education by publishing worldwide. Oxford is a registered trade mark of
Oxford University Press in the UK and in certain other countries.

Published in the United States of America by Oxford University Press
198 Madison Avenue, New York, NY 10016, United States of America

British Library Cataloguing in Publication Data
Data available

Library of Congress Control Number: 2025936910

ISBN 9780199226191

DOI: 10.1093/oso/9780199226191.001.0001

Printed and bound by
CPI Group (UK) Ltd, Croydon, CR0 4YY

The manufacturer's authorized representative in the EU for product safety is
Oxford University Press España S.A. of Parque Empresarial San Fernando de Henares,
Avenida de Castilla, 2 – 28830 Madrid (www.oup.es/en or product.safety@oup.com).
OUP España S.A. also acts as importer into Spain of products made by the manufacturer.

to Ueli Suter, who showed us the way
and to our beloved families
for their unconditional support

Preface

Understanding and predicting the relations between structure, properties, processing, and performance of materials is a central goal of materials science and engineering, since it can form a basis for more efficient and economical design of products and processes for specific applications. Materials modelling and simulation play an increasingly important role in developing this understanding and predictive ability. Firmly founded on the molecular and continuum engineering sciences, making use of the unprecedented computational capabilities available today, and complementing the constellation of powerful contemporary experimental techniques for probing matter at a variety of length and time scales, modelling and simulation have become indispensable to the materials researcher, in both industrial and academic settings. With the current trends towards miniaturization of devices, nanoengineering, modularization, and redesign of processes for minimal environmental impact and a circular economy, understanding materials down to the atomic and electronic levels becomes imperative.

A challenge we face in modelling materials, which is especially serious in the case of polymeric materials, is that structure and dynamics are characterized by extremely broad spectra of length and time scales. Intramolecular correlations and local packing of chains in a bulk polymer exhibit features on the length scale of bond lengths and atomic radii, i.e. 0.1 nm. The length of a Kuhn (statistical) segment, defined such that the real chain can be considered as a random flight of such segments, is on the order of 1 nm for a typical synthetic randomly coiled polymer and can be more than an order of magnitude higher for macromolecules with stiff backbones. The radius of gyration of entire chains in the amorphous bulk scales with the chain length N as $N^{1/2}$ and is on the order of 10 nm for typical molecular weights. Domain sizes of different phases in semicrystalline polymers and in immiscible polymer blends may well be on the order of 1 μm. Even broader is the range of time scales characterizing molecular motion in polymers. While localized vibrational modes of chains have periods on the order of 10^{-14} s, conformational transitions of individual bonds over torsional energy barriers in the melt state have waiting times in excess of 10^{-11} s. Longer and longer sequences of segments along the backbone exhibit longer and longer correlation times. The longest relaxation time, required for a chain to diffuse by a length commensurate to its size and thus 'forget' its previous conformation, controls the viscoelastic response of polymer melts to flow. This time scales as N^2 for low-molecular-weight melts and as $N^{3.4}$ for N exceeding a critical chain length sufficient for the development of entanglements, in the reptation regime; for the molecular weights encountered in typical processing operations, it easily exceeds the millisecond range. Microphase and phase separation phenomena in molten copolymers and polymer mixtures may have characteristic times longer than seconds, whereas structural relaxation in the glassy state, 20°C below the glass temperature, has characteristic times on the order of a year (10^7 s).

This tremendous multiplicity of characteristic times and length scales, to which polymeric materials largely owe their widespread technological use, makes it clear that a single modelling or simulation approach cannot address all the properties of these materials. The development of hierarchical or multiscale modelling approaches is imperative. Such approaches comprise many levels, each level addressing phenomena over a window of specific length and time scales; links between different levels can be established through systematic *coarse-graining* of the model used to represent the material.

It is the aim of this book to introduce some key methods for modelling polymeric materials computationally at the atomistic and mesoscopic levels. We have done our best to present the methods in a coherent fashion, emphasizing their theoretical underpinnings and explaining how bridges between different levels of modelling can be built with minimal loss of predictive ability.

Chapter 1 presents some basic elements of statistical mechanics and highlights the hierarchy of models, from chemically detailed assemblages of atoms to single soft ellipsoid particles, that can be used to represent polymer chains. In addition to providing background for the following chapters, this chapter serves to introduce the concepts of coarse-graining and multiscale modelling. How to build force fields for atomistic modelling based on electronic structure calculations is briefly discussed in Chapter 2. Chapters 3, 4, and 5 introduce molecular mechanics (energy minimization), molecular dynamics, and Monte Carlo methods, respectively. Links between molecular mechanics and continuum thermoelasticity are made in Chapter 3, where the statistical mechanics of condensed phases formed by polymer chains in the absence and presence of mechanical deformation is treated in some detail. Chapter 4 includes a discussion of multiple time step methods for the integration of the dynamical equations of motion, of fluctuation relations for extracting equilibrium properties, and of ways to predict segmental dynamics for comparison with and interpretation of experimental measurements. State-of-the-art Monte Carlo moves, which have greatly enhanced our ability to equilibrate long-chain polymer systems, are discussed at some length in Chapter 5, as are free energy calculations, potentials of mean force, and their role in coarse-graining. Chapter 6 illustrates techniques for the analysis and simulation of infrequent events and their utility in addressing dynamical phenomena over time scales by many orders of magnitude longer than those accessible by molecular dynamics, with applications in small molecule diffusion through polymers at low temperatures and subglass relaxation transitions. The problem of coarse-graining from atomistic to mesoscopic models involving fewer degrees of freedom is discussed in Chapter 7. This chapter includes an overview of methods for extracting effective interaction models for use at the coarse-grained level from detailed atomistic force fields. It also incorporates an introduction to the projection operation formalism and its use in deriving Langevin dynamics and Brownian dynamics descriptions, relating the latter to the Rouse and reptation theories of polymer melt dynamics. Mesoscopic entanglement-based simulations of the rheological behaviour of polymer melts and of the large-deformation mechanical response of solid polymers are briefly presented in Chapter 8. Dissipative particle dynamics simulations of polymer fluids are discussed in Chapter 9. Chapter 10 introduces the generalized bracket and the GENERIC approaches as two thermodynamically consistent frameworks for bridging to the macroscopic level and discusses how they can be used as guides for the design of self-consistent beyond-equilibrium atomistic simulations.

Emphasis is placed on physical principles and on the fundamental theoretical underpinnings of simulation methods. Our intention is to show how a hierarchy of methods for multiscale modelling and simulations can be developed *bottom up*, based on widely accepted scientific principles and well-defined approximations. Mathematical derivations are included in considerable detail to enforce rigorous quantitative physical understanding. Although the book is structured by methods, it can be read by applications. We have expended every effort to include examples from a variety of problems in the prediction of properties of polymers. Properties covered include equilibrium volumetric behaviour and chain conformation in melts (Chapters 1, 2, 5, and 7); phase equilibria of polymer–solvent systems (Chapter 5); sorption and diffusion of gases and solvents in molten and glassy polymers (Chapters 5 and 6); segmental dynamics and interpretation of relaxation measurements above the glass temperature (Chapter 4); elastic constants, plasticity, and structural relaxation in polymer glasses (Chapters 3, 4, and 6); rheological properties of entangled polymers (Chapters 7, 8, 10); microphase separation and morphology development in block copolymers and nanocomposites (Chapter 9); and polymers at interfaces (Chapters 4, 5, 8). Problems for solution by the reader are included at the end of each chapter, based on the material presented in the text. They are intended to involve the reader actively in some mathematical derivations and thereby enforce quantitative understanding.

The book draws heavily on our research and teaching in the field of polymer modelling over the past forty years. It is intended as a graduate textbook for chemical engineering, materials science, physics, chemistry, and applied mathematics students interested in polymers and as a reference book for practitioners of polymer modelling and simulation. Undergraduate courses in calculus, physical chemistry, statistical mechanics, properties of materials, and polymers would be helpful as background, although we have done our best to make the material presented self-consistent and understandable with minimal use of external literature sources.

We hope that readers will find this book of some use in embarking upon the fascinating and far-reaching world of multiscale modelling of polymers.

Vlasis G. Mavrantzas, Patras & Zürich
Doros N. Theodorou, Athens
August 2024

Contents

1

Models for Polymer Chains

1.1 Basic concepts

1.1.1 Chain conformation

A polymeric chain is a collection of bonded atoms whose spatial arrangement, defining what is known as the *chain configuration*,[1–4] can be specified in terms of their coordinates. For a polymer molecule consisting of N such atoms, $3N$ coordinates must be specified, six of which are needed to fix the absolute location of the molecule and its orientation in space. Figure 1.1, for example, shows the configuration of the skeletal atoms of a hypothetical linear chain consisting of $N+1$ atoms (N bonds), specifying their location by means of the position vectors, $\boldsymbol{R}_0, \boldsymbol{R}_1, \ldots, \boldsymbol{R}_N$ with respect to some fixed coordinate frame. The vector of the Cartesian coordinates of all $N+1$ atoms is written as

$$\{\boldsymbol{R}\} = \{\boldsymbol{R}_0, \boldsymbol{R}_1, \boldsymbol{R}_2 \ldots, \boldsymbol{R}_N\}. \tag{1.1.1}$$

Alternatively, instead of the position vectors $\boldsymbol{R}_0, \boldsymbol{R}_1, \ldots, \boldsymbol{R}_N$ of the $N+1$ skeletal atoms, the configuration of the chain can de described by $\boldsymbol{R}_0$ and the set

$$\{\boldsymbol{l}\} = \{\boldsymbol{l}_1, \boldsymbol{l}_2, \ldots, \boldsymbol{l}_N\} \tag{1.1.2}$$

of the corresponding N bond vectors $\boldsymbol{l}_1, \boldsymbol{l}_2, \ldots, \boldsymbol{l}_N$, defined through

$$\boldsymbol{l}_i = \boldsymbol{R}_i - \boldsymbol{R}_{i-1}, \quad 1 \leq i \leq N. \tag{1.1.3}$$

The bonds may be either true covalent bonds or equivalent bonds (each representing a section of chain containing a number of chemical bonds in a coarse-grained, statistical way). The $3N$-long vector $\boldsymbol{l}$ defines the overall orientation and the conformation of the chain. By *conformation* we mean the exact shape of the chain, as specified, for example, by all internal distances between its constituent atoms. To fix the configuration of a chain we need to specify, in addition to the conformational degrees of freedom that determine its shape (e.g. bond lengths, angles between pairs of successive bonds, torsion angles formed by triplets of successive bonds), three translational and three orientational degrees of freedom. The *translational* degrees of freedom specify the position of the chain as a whole object; they can be chosen here e.g. as the coordinates of the position vector of the chain start, $\boldsymbol{R}_0$. The *orientational* degrees of freedom specify the overall orientation of the chain in space;

Multiscale Modelling of Polymers. Doros N. Theodorou and Vlasis G. Mavrantzas, Oxford University Press.
 DOI: 10.1093/oso/9780199226191.003.0001

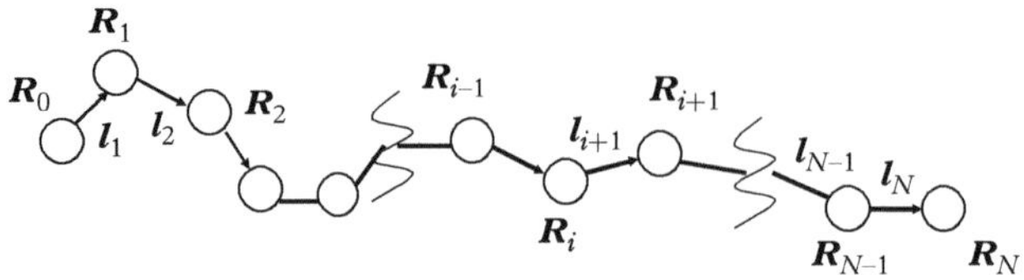

Figure 1.1 The configuration of the skeletal atoms of a hypothetical linear chain consisting of $N + 1$ atoms (N bonds) is depicted by specifying the locations of all atoms by means of their positions vectors, $\boldsymbol{R}_0, \boldsymbol{R}_1, \ldots, \boldsymbol{R}_N$, with respect to some fixed coordinate system.

they can be chosen here e.g. as the polar and azimuthal angles of bond $\boldsymbol{l}_1$, along with a dihedral angle around bond $\boldsymbol{l}_1$ that fixes the orientation of the plane of the chain's first two bonds. The choice of configurational degrees of freedom is further discussed in Section 1.1.6.

We define the probability density function $\Psi(\boldsymbol{l}_1, \boldsymbol{l}_2, \ldots, \boldsymbol{l}_N) = \Psi(\{\boldsymbol{l}\})$ such that $\Psi(\boldsymbol{l}_1, \boldsymbol{l}_2, \ldots, \boldsymbol{l}_N)$ $\mathrm{d}\boldsymbol{l}_1 \mathrm{d}\boldsymbol{l}_2 \ldots \mathrm{d}\boldsymbol{l}_N$ is equal to the probability of finding the chain in a given overall orientation and conformation where the vector of the first bond is between $\boldsymbol{l}_1$ and $\boldsymbol{l}_1 + \mathrm{d}\boldsymbol{l}_1$, the vector of the second bond between $\boldsymbol{l}_2$ and $\boldsymbol{l}_2 + \mathrm{d}\boldsymbol{l}_2$, . . ., and the vector of the Nth bond between $\boldsymbol{l}_N$ and $\boldsymbol{l}_N + \mathrm{d}\boldsymbol{l}_N$, with $\mathrm{d}\boldsymbol{l}_i = \mathrm{d}l_{xi}\, \mathrm{d}l_{yi}\, \mathrm{d}l_{zi} \equiv \mathrm{d}^3 l_i$ symbolizing a small volume in three-dimensional space. The function $\Psi(\boldsymbol{l}_1, \boldsymbol{l}_2, \ldots, \boldsymbol{l}_N)$ governs the mean value of any property A depending on the conformation of the chain through

$$\langle A(\boldsymbol{l}_1, \boldsymbol{l}_2, \ldots, \boldsymbol{l}_N)\rangle = \frac{\int A(\boldsymbol{l}_1, \boldsymbol{l}_2, \ldots, \boldsymbol{l}_N)\, \Psi(\boldsymbol{l}_1, \boldsymbol{l}_2, \ldots, \boldsymbol{l}_N) \mathrm{d}\boldsymbol{l}_1 \mathrm{d}\boldsymbol{l}_2 \ldots \mathrm{d}\boldsymbol{l}_N}{\int \Psi(\boldsymbol{l}_1, \boldsymbol{l}_2, \ldots, \boldsymbol{l}_N) \mathrm{d}\boldsymbol{l}_1 \mathrm{d}\boldsymbol{l}_2 \ldots \mathrm{d}\boldsymbol{l}_N}. \tag{1.1.4}$$

The function $\Psi(\boldsymbol{l}_1, \boldsymbol{l}_2, \ldots, \boldsymbol{l}_N)$ is usually normalized, i.e.

$$\int \Psi(\boldsymbol{l}_1, \boldsymbol{l}_2, \ldots, \boldsymbol{l}_N) \mathrm{d}\boldsymbol{l}_1 \mathrm{d}\boldsymbol{l}_2 \ldots \mathrm{d}\boldsymbol{l}_N = 1. \tag{1.1.5}$$

Since the information included in the set $\{\boldsymbol{l}\}$ is enormous, the need immediately arises for some more concise measures of the chain configuration. One such measure (see Figure 1.2) is the vector that connects the two ends of the chain, i.e. the *chain end-to-end vector*

$$\boldsymbol{r} \equiv \boldsymbol{R}_N - \boldsymbol{R}_0 = \sum_{i=1}^{N} \boldsymbol{l}_i. \tag{1.1.6}$$

Although under equilibrium conditions in an isotropic medium the mean value $\langle \boldsymbol{r} \rangle$ of the vector $\boldsymbol{r}$ is zero, the average $\langle r^2 \rangle$ of its square magnitude

$$r^2 \equiv \boldsymbol{r} \cdot \boldsymbol{r} = \sum_{i=1}^{N} \sum_{j=1}^{N} \boldsymbol{l}_i \cdot \boldsymbol{l}_j \tag{1.1.7}$$

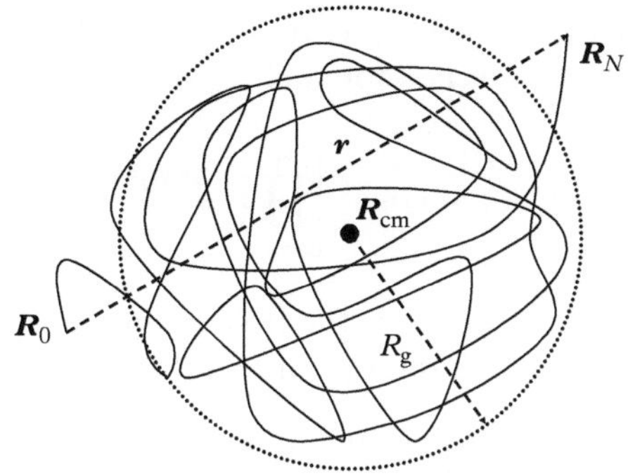

Figure 1.2 Definition of the chain end-to-end vector $\boldsymbol{r}$ and of the chain radius of gyration R_g.

is in general nonzero; it can therefore be used to quantify the size of the chain. A measure of the conformational stiffness of the chain is the *persistence length*, defined as the mean value of the projection of the end-to-end vector $\boldsymbol{r}$ on the direction of the first bond of the chain

$$a = \frac{\langle \boldsymbol{r} \cdot \boldsymbol{l}_1 \rangle}{l_1}, \tag{1.1.8}$$

where a provides a measure of the chain stiffness. A third useful measure of chain size is the *radius of gyration.* In general, for any collection of $N + 1$ particles of equal mass, the square radius of gyration (see Figure 1.2) is defined through

$$R_g^2 = \frac{1}{N+1} \sum_{i=0}^{N} (\boldsymbol{R}_i - \boldsymbol{R}_{cm})^2, \tag{1.1.9}$$

where

$$\boldsymbol{R}_{cm} = \frac{1}{N+1} \sum_{i=0}^{N} \boldsymbol{R}_i \tag{1.1.10}$$

defines the position vector of the chain centre of mass. According to Lagrange's theorem [5] (see the Appendix), R_g^2 is directly related to the sum of the squares of the distances between all pairs of atoms making up the chain:

$$R_g^2 = \frac{1}{(N+1)^2} \sum_{i=0}^{N-1} \sum_{j=i}^{N} R_{ij}^2. \tag{1.1.11}$$

1.1.2 Local and non-local interactions. The unperturbed state

We consider a system containing N_c polymeric chains, each of length N bonds, and possibly also solvent molecules, so that the total number of atoms is n, within a region of volume V at temperature T.

Clearly, $n \geq N_c(N+1)$, the equality holding where no solvent molecules are present. In the *classical* approximation, the *canonical partition function* for such a system is[6]

$$Q(N_c, N, V, T) = \text{const.}\left[\prod_{i=1}^{n}\left(\frac{2\pi m_i k_B T}{h}\right)^{\frac{3}{2}}\right] Z, \tag{1.1.12}$$

where m_i denotes the mass of the ith atom, k_B is the Boltzmann constant, T the temperature, h the Planck constant, and Z the classical configurational integral

$$Z = \int \exp\left[-\frac{\mathcal{V}(\boldsymbol{R})}{k_B T}\right] d\boldsymbol{R} \tag{1.1.13}$$

with the function $\mathcal{V}(\boldsymbol{R})$ denoting the *total potential energy* of the system being in configuration $\boldsymbol{R}$ (the symbol $\boldsymbol{R}$ is a shorthand for the set of vectors $\{\boldsymbol{R}\}$ of atomic positions defined in Eq. (1.1.1) for all chains and solvent molecules present) due to all interactions among all atoms of all (macro)molecules in the space of volume V. In a melt of N_c identical chains at equilibrium, the constant prefactor in Eq. (1.1.12) is $1/N_c!$

In general,[1] the interactions defining the potential energy function $\mathcal{V}(\boldsymbol{R})$ of Eq. (1.1.13) in a system of polymer chains can be classified as *intramolecular* and *intermolecular*. The latter are exerted on the atoms of a given chain by its environment, which consists of the atoms of all other chains and the molecules of the solvent (if we deal with a polymer solution—see e.g. Figure 1.3). The former are those exerted among the atoms of the same chain; as shown in Figure 1.4, they can be further classified as *local* and *nonlocal*. Local interactions have to do with the bonded structure and are exerted among atoms and groups which are neighbours in sequence along the chain (e.g. less than about five chemical bonds apart in the case of a polyethylene melt). Nonlocal interactions, on the other hand, involve pairs of atoms which are *topologically* remote in the chain sequence (separated by many bonds), although they may be spatially near one another. For a chain in dilute solution we often resort to a representation wherein solvent degrees of freedom are integrated out. The function $\mathcal{V}(\boldsymbol{R})$ is then a free energy (potential of mean force) incorporating solvent-mediated interactions among the chain segments in configuration $\boldsymbol{R}$.

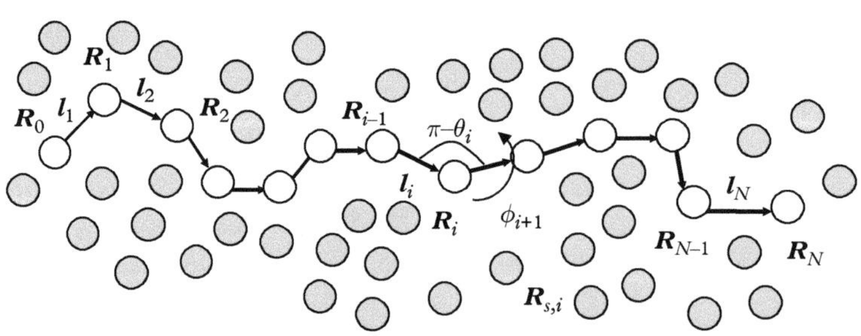

Figure 1.3 A macromolecule in the environment of solvent molecules or other macromolecules. l_i denotes the length of bond i, θ_i is the bond angle at chain atom i, and ϕ_i is the torsion angle measuring the dihedral angle between the two planes defined by bond pairs $(i-1, i)$ and $(i, i+1)$, respectively.

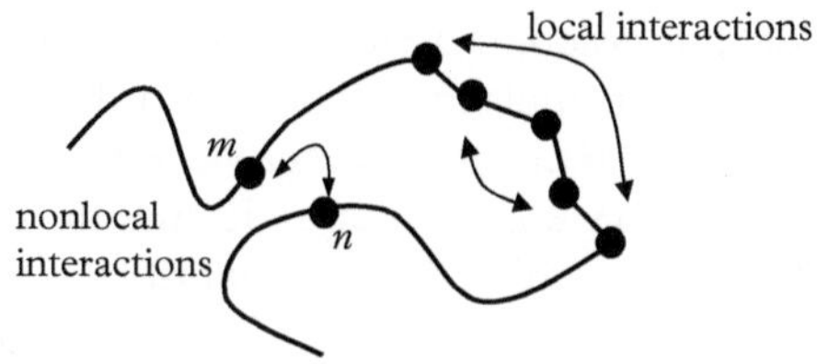

Figure 1.4 Local and nonlocal interactions between two segments *n* and *m* along a chain.

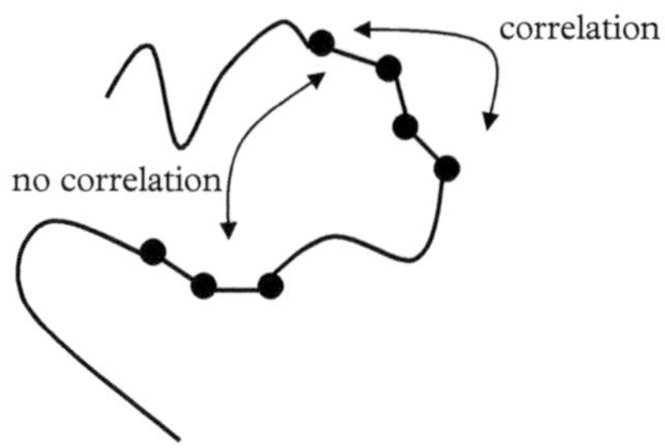

Figure 1.5 Sketch of an unperturbed chain.

Under certain conditions (e.g. for some specific solvent at a specific temperature),[1–2,7] a chain can behave as if it did not feel the nonlocal interactions at all. Such an *ideal* chain, whose conformation is solely determined by local interactions, is termed *unperturbed* (by nonlocal interactions). The fact that only local interactions dictate the conformation of an unperturbed chain considerably facilitates its statistical description. In particular, for an unperturbed chain (see Figure 1.5) we anticipate a probability density function of the form[1]

$$\Psi(\{\boldsymbol{l}_i\}) = \prod_i \psi_i(\boldsymbol{l}_i, \boldsymbol{l}_{i+1}, \ldots, \boldsymbol{l}_{i+n_c}), \tag{1.1.14}$$

where n_c is an integer (considerably smaller than the chain length N) which defines the range of local interactions. Equation (1.1.14) is the starting point for the development of some simple models for the statistical description of the shape and conformational properties of flexible polymers in their equilibrium unperturbed state. The most important of these models in which the interaction among the polymer segments is limited to within a few neighbours along the chain are reviewed in later sections of this chapter.

In reality, however, segments that are topologically quite distant along the chain also interact if they come close to one another in space (see Figure 1.4). This nonlocal interaction is often mainly *steric* in nature: since a segment has a finite volume, other segments cannot come into its own region. Such an interaction, which is known as the *excluded volume effect* (see Section 1.1.3), swells the polymer, since the coil size of a flexible chain in the presence of such an interaction is larger than that of the corresponding ideal chain (which is subject to the same local interactions but) whose units do not feel any nonlocal steric hindrance. The chain will swell even in the presence of attractive forces among its atoms, as long as the repulsive forces due to the excluded volume effect dominate.

In a polymer solution, the nonlocal interaction depends not only on the direct interaction between the chain segments involved, but also on their interaction with the surrounding solvent molecules; it is best described through a *solvent-mediated potential of mean force.*

According to Kuhn[8] and Flory,[9] who were the first to discuss the excluded volume effect, nonlocal interactions perturb the configuration which the chain would have assumed if it had been subject solely to local interactions, to a degree which is associated with the *effective covolume*[1] (analogous to a second virial coefficient) for a pair of segments immersed in the given solvent, rather than with the actual volume of the chain segments. The value adopted by the mean-square end-to-end distance $\langle r^2 \rangle$ over all conformations, which in the absence of any perturbation of the distribution of $\boldsymbol{r}$ by nonlocal interactions is proportional to chain length N (see Section 1.1.4), is calculated to vary with chain length N as

$$\langle r^2 \rangle \propto N^{2\nu}, \tag{1.1.15}$$

where ν is about 3/5 in a good solvent. If present, the excluded volume effect is very important for long chains. In general, a good solvent will increase the covolume of a chain segment, whereas a bad solvent will reduce it.

For the particular set of conditions (solvent and temperature) for which excluded volume interactions are exactly compensated by solvent-mediated attractions between chain segments, the covolume is zero and the excluded volume effect vanishes. In this case, the chain behaves as an ideal chain, i.e. as if it were unperturbed by nonlocal interactions. The temperature at which the covolume effect disappears for the given solvent is called the theta (ϑ) temperature or the *Flory temperature.*[2]

The nonlocal interactions can be either repulsive or attractive and depend on a number of parameters, such as the chemical details of the chain, the solvent conditions, and the temperature. They include steric effects, van der Waals attraction, and specific interactions mediated by solvent molecules, that is, complicated effects which, when introduced in the analysis,[8–11] render the exact analytical calculation of the statistical properties of the polymer impossible; they can be accurately represented only through a detailed atomistic model for the description of the system, such as those discussed in Sections 1.1.6 and 2.6 of this book. Here we restrict ourselves to a simple representation of these interactions with emphasis on their effect on the chain properties on longer length scales (mainly chain dimensions).

1.1.3 The excluded volume effect

By assuming that the nonlocal interaction affects two chain segments m and n only when they occupy the same point in space, we can express it in terms of a delta function:[3,12–13]

$$\upsilon k_{\mathrm{B}} T \delta(\boldsymbol{R}_n - \boldsymbol{R}_m), \tag{1.1.16}$$

where υ is a constant with dimensions of volume and is called the *excluded volume*; it depends on the nature (chemical constitution) of the polymer and solvent molecules, and on the temperature. Having introduced the parameter υ, we can write down an expression for the total interaction energy of the chain due to nonlocal forces by employing a continuous representation for the segment indices n and m and retaining the notation $\boldsymbol{R}_n \equiv \boldsymbol{R}(n)$ to indicate that $\boldsymbol{R}$ is a function of n:

$$\mathcal{V}_1 = \frac{1}{2} v k_B T \int_0^N dn \int_0^N \delta(\boldsymbol{R}_n - \boldsymbol{R}_m) dm. \tag{1.1.17}$$

By defining the local concentration of segments through

$$C(\boldsymbol{R}') = \left\langle \int_0^N \delta(\boldsymbol{R}' - \boldsymbol{R}_n) dn \right\rangle, \tag{1.1.18}$$

we expect Eq. (1.1.17) to lead to a contribution to the total free energy of the chain of the form

$$A_1 = \int \frac{1}{2} v k_B T \left[C(\boldsymbol{R}')\right]^2 d\boldsymbol{R}', \tag{1.1.19}$$

i.e. similar to the expression defining the Helmholtz free energy of a dilute gas in reference to that of an ideal gas at the same density;[6,14] the latter is calculated from the virial equation of state truncated at the second virial term as

$$A_1 = NB(T) k_B T \rho = \int B(T) k_B T \left[\rho(\boldsymbol{R}')\right]^2 d\boldsymbol{R}' \tag{1.1.20}$$

with N denoting now the number of molecules and ρ the molecular density. There exists, therefore, an analogy between the excluded volume parameter v and the quantity $2B(T)$, where $B(T)$ is the *second virial coefficient* characterizing the dilute gas of polymer segments in the solvent. Guided by the microscopic expression for the second virial coefficient of an imperfect gas,[6,14] we can thus write that

$$v = \int \left[1 - \exp\left(-\frac{u(\boldsymbol{r})}{k_B T}\right)\right] d\boldsymbol{r}, \tag{1.1.21}$$

where $u(\boldsymbol{r})$ is the effective potential describing the environment-mediated interaction between polymer segments and the integration takes place over all the space surrounding the position of a polymer segment; $u(\boldsymbol{r})$ is in essence a potential of mean force (see Eq. (1.2.2) below) accumulated as a function only of segment separation r, i.e. $u = u(r)$.

If the potential $u(r)$ consisted only of a hard sphere repulsion with diameter d_{HS}, Eq. (1.1.21) would give $v_{HS} = \int_0^{d_{HS}} r^2 dr \int_0^{\pi} \sin\theta d\theta \int_0^{2\pi} d\phi = \frac{4}{3}\pi d_{HS}^3$, which justifies the term *excluded volume* for the parameter v. In general, we anticipate the potential $u(r)$ to consist of a hard repulsive part $u_{HS}(r)$ and a soft attractive tail $u_{attr}(r)$ (see e.g. Figure 1.6) so that Eq. (1.1.21) gives

$$
\begin{aligned}
v &= \int d\boldsymbol{r}\left[1-\exp\left(-\frac{u_{\text{HS}}(r)}{k_{\text{B}}T}\right)\exp\left(-\frac{u_{\text{attr}}(r)}{k_{\text{B}}T}\right)\right] \\
&= \int_{|r|\leq d_{\text{HS}}} d\boldsymbol{r}\left[1-\exp\left(-\frac{u_{\text{HS}}(r)}{k_{\text{B}}T}\right)\exp\left(-\frac{u_{\text{attr}}(r)}{k_{\text{B}}T}\right)\right] \\
&\quad + \int_{|r|>d_{\text{HS}}} d\boldsymbol{r}\left[1-\exp\left(-\frac{u_{\text{HS}}(r)}{k_{\text{B}}T}\right)\exp\left(-\frac{u_{\text{attr}}(r)}{k_{\text{B}}T}\right)\right] \\
&= \int_{|r|\leq d_{\text{HS}}} d\boldsymbol{r} + \int_{|r|>d_{\text{HS}}} d\boldsymbol{r}\left[1-\exp\left(-\frac{u_{\text{attr}}(r)}{k_{\text{B}}T}\right)\right] = \frac{4\pi}{3}d_{\text{HS}}^3 + \int_{|r|>d_{\text{HS}}} d\boldsymbol{r}\left[1-\exp\left(-\frac{u_{\text{attr}}(r)}{k_{\text{B}}T}\right)\right] \\
&\simeq \frac{4\pi}{3}d_{\text{HS}}^3 + \int_{|r|>d_{\text{HS}}} d\boldsymbol{r}\left[1-\left(1-\frac{u_{\text{attr}}(r)}{k_{\text{B}}T}\right)\right] = \frac{4\pi}{3}d_{\text{HS}}^3 + \frac{1}{k_{\text{B}}T}\int_{|r|>d_{\text{HS}}} d\boldsymbol{r}\, u_{\text{attr}}(r)
\end{aligned}
\tag{1.1.22}
$$

where the assumption $|u_{\text{attr}}(r)| << k_{\text{B}}T$ for $r \equiv |\boldsymbol{r}| > d_{\text{HS}}$ has been invoked. This implies[2] that

$$
v \cong A' - \frac{B'}{T} = v_{\text{HS}}\left(1-\frac{\vartheta}{T}\right), \tag{1.1.23}
$$

where the parameters $v_{\text{HS}} = A'$ and $\vartheta = B'/v_{\text{HS}}$ are insensitive to temperature and depend on the nature of the segment and solvent molecules. In a good solvent, the polymer segments will prefer to be surrounded by solvent molecules rather than by other polymer segments. In this case, the interaction $u(r)$ will be strongly repulsive and the excluded volume parameter v large and positive, close in value to v_{HS}. In a bad solvent, on the other hand, the polymer segments will tend to cling together, avoiding unfavourable contacts with the solvent molecules. In this case, the attractive component in the interaction $u(r)$ will be significant, and this will result in a small positive value for the excluded volume parameter v. (If the solvent is very bad, then it is also possible that the excluded volume parameter v may become negative.) There will be, therefore, a particular case of bad solvent for which attractive effective interactions between segments will exactly compensate the repulsive ones, so that $v = 0$. According to Eq. (1.1.23), the condition for this to happen is $T = \vartheta$. In this case, the chains are said to be under ϑ conditions, the solvent is called ϑ-solvent (for the given temperature T), and the chains adopt unperturbed dimensions. The ϑ temperature of polymer solutions is analogous to the Boyle temperature of dilute gases, where the second virial coefficient vanishes as a result of a compensation between repulsive and attractive interactions.

The factor $(1 - \vartheta/T)$ in Eq. (1.1.23) is usually set proportional to $(1 - 2\chi)$, where χ, a dimensionless interchange energy between polymer segments and solvent molecules, is the celebrated Flory's interaction parameter or chi factor.[2] A value $\chi = 0$ corresponds to an *athermal solvent* (segment–solvent interactions very similar to segment–segment and solvent–solvent interactions, leading to self-avoiding walk conformational statistics for the chain). Higher values of χ correspond to more unfavourable segment–solvent interactions, up to $\chi = 1/2$ corresponding to a ϑ-solvent.

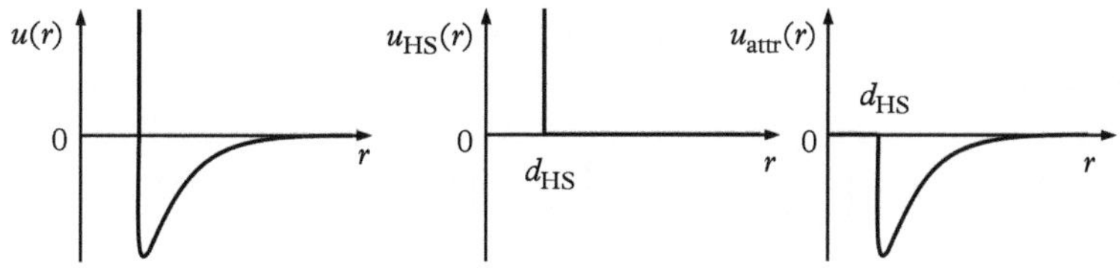

Figure 1.6 The excluded volume potential $u(r)$ and its decomposition into a hard repulsive $u_{\text{HS}}(r)$ and a soft attractive $u_{\text{attr}}(r)$ part.

1.1.4 Excluded volume effect on chain size

The above simple model of the excluded volume effect can be used to study (at least semi-analytically) the statistical properties of a randomly coiled chain for any value of υ.[6,12,15] Here we will limit the discussion to a scaling analysis of the effects of the excluded volume interaction on the mean-square value of the chain end-to-end vector for the case $\upsilon > 0$.[3] The analysis is inspired by the work of Flory,[9] who envisioned the size of a polymer chain as resulting from the balance of two effects: a repulsive interaction among polymer segments which tends to swell the chain, and an entropic (elastic) contribution due to chain connectivity which opposes the swelling, since the number of conformations accessible to an expanded chain is reduced. To estimate this elastic contribution, Flory considered the free energy of a chain whose end-to-end vector is *fixed* at $\boldsymbol{r}$. We expect the equilibrium distribution function of the end-to-end distance $\Phi\,(\boldsymbol{r}, N)$ to be proportional to $\exp\left(-\dfrac{A_{\text{el}}(\boldsymbol{r})}{k_{\text{B}}T}\right)$; thus, $A_{\text{el}}(r)$ should be of the form

$$A_{\text{el}}(r) = -k_{\text{B}}T\ln\Phi\,(\boldsymbol{r}, N) + \text{terms independent of } \boldsymbol{r}. \tag{1.1.24}$$

For the case $\upsilon = 0$ and long enough chains, $\Phi\,(\boldsymbol{r}, N)$ is Gaussian (see Section 1.3.4):

$$\Phi\,(\boldsymbol{r}, N) = \left(\frac{3}{2\pi N l^2}\right)^{\frac{3}{2}} \exp\left[-\frac{3r^2}{2Nl^2}\right], \text{ so that}$$

$$A_{\text{el}}(\boldsymbol{r}) = k_{\text{B}}T\frac{3r^2}{2Nl^2} + \text{terms independent of } \boldsymbol{r}. \tag{1.1.25}$$

As far as the repulsive interaction is concerned, we neglect chain connectivity and consider the polymer units as behaving as a dilute gas inside the sea of solvent molecules. The Helmholtz free energy of such a system is given by Eq. (1.1.19), where, for the purposes of this analysis, we assume a uniform distribution of chain segments with a mean concentration equal to (omitting any numerical prefactors), $C(\boldsymbol{R}) = \overline{C} \cong \dfrac{N}{r^3}$. Consequently,

$$A_{\text{excl}}(r) \cong r^3 \upsilon k_{\text{B}}T\overline{C}^2 = \upsilon k_{\text{B}}T\frac{N^2}{r^3}. \tag{1.1.26}$$

Combining Eqs. (1.1.25) and (1.1.26), we get the following expression for the total Helmholtz free energy of the chain:

$$A(r) = k_B T \left(\frac{3r^2}{2Nl^2} + \upsilon \frac{N^2}{r^3} \right) + \text{terms independent of } r. \tag{1.1.27}$$

The chain will attain that conformation which minimizes the Helmholtz free energy $A(r)$. The result is

$$r \cong \left(Nl^2\right)^{1/2} \left(\frac{\upsilon N^{1/2}}{l^3} \right)^{1/5} \propto N^{\frac{3}{5}}. \tag{1.1.28}$$

This gives the exponent $\nu = 3/5$ quoted earlier, which is close to the exact value $\nu = 0.588 \pm 0.001$ calculated with more accurate methods, and to the values measured experimentally or computed through molecular simulations.[16] Although the value of ν as calculated through such a mean-field approach comes out to be very close to the exact one, other predictions (e.g. that for the free energy) are not consistent with the results of detailed simulations, where the excluded volume effect is accounted for exactly (within computer experimental error).[17]

At finite concentrations, solvent-mediated interactions between segments belonging to *different chains* come into play (semidilute and concentrated regimes). On large length scales, this limits the swelling of chains in a good solvent, leading to a scaling exponent ν smaller than the self-avoiding walk value discussed in connection with Eq. (1.1.28). The reader is referred to de Gennes for a discussion of the *blob* picture he developed to deal with the semidilute regime.[10] In the limit of very high concentrations, where the solvent disappears and the solution becomes a melt, intra- and intermolecular excluded volume interactions compensate each other, and a chain behaves as if unperturbed by nonlocal interactions ($\nu = 1/2$). This is Flory's *random coil hypothesis* for polymer melts,[2] for which ample experimental and simulation evidence exists today, as discussed in Chapters 2 and 3.

1.1.5 Classical equations of motion

We return now to the canonical partition function (Eq. (1.1.12)) of the classical system (N_{ch} polymeric chains, each of length N, so that the total number of atoms is n, in volume V and temperature T) that we began to discuss in Section 1.1.2. Since each atom has three degrees of freedom (i.e. each atom requires three coordinates for a complete description of its position), the number of coordinates necessary to describe the positions of all n atoms is $3n$. These $3n$ coordinates, $q_1, q_2, \ldots, q_{3n}$, completely describe the configuration of the entire n-body system. To each of these $3n$ coordinates corresponds a conjugate momentum p_j, defined by

$$p_j = \frac{\partial \mathcal{L}}{\partial \dot{q}_j}, \quad j = 1, 2, \ldots, 3n, \tag{1.1.29}$$

where $\mathcal{L}(\boldsymbol{q}, \dot{\boldsymbol{q}}, t)$ is the *Lagrangian function*, that is, the difference

$$\mathcal{L}(\boldsymbol{q}, \dot{\boldsymbol{q}}, t) \equiv \mathcal{K}(\boldsymbol{q}, \dot{\boldsymbol{q}}, t) - \mathcal{V}(\boldsymbol{q}, t) \tag{1.1.30}$$

between the *kinetic energy* $\mathcal{K}$ and the *potential energy* $\mathcal{V}$ of the system, and $\dot{q}_j$ denotes the time derivative of q_j. The set $\boldsymbol{p} \equiv (p_1, p_2, \ldots, p_{3n})$ of the *generalized momenta* is said to be *conjugate* to the set $\boldsymbol{q} \equiv (q_1, q_2, \ldots, q_{3n})$ of the *generalized coordinates*. In classical mechanics[6] the state of the system is completely specified in terms of these two sets of generalized coordinates $\boldsymbol{q}$ and generalized momenta $\boldsymbol{p}$. The $3n$-dimensional space from which the generalized coordinates $\boldsymbol{q}$ take on values is referred to as the *configuration space*. The $3n$-dimensional space from which the generalized momenta $\boldsymbol{p}$ take on values is referred to as the *momentum space*. The $6n$-dimensional vector $\boldsymbol{X} \equiv (\boldsymbol{p}, \boldsymbol{q}) \equiv (p_1, p_2, \ldots, p_{3n}; q_1, q_2, \ldots, q_{3n})$ is referred to as the *state vector* or *representative point* of the system; this takes on values from the *phase space* of the system, symbolized by $\boldsymbol{\Gamma}$. A point in this space represents a *microstate* of the entire system.

For a monatomic fluid one can use the Cartesian position vectors of all molecules to specify the system. For systems consisting of polyatomic molecules, however, other (than Cartesian) coordinates, e.g. bond lengths, bond angles, and dihedral angles, can be used to specify their configuration. In this case, the components of $\boldsymbol{q}$ are also termed *configurational degrees of freedom* of the system.

The *Hamiltonian* of the system is defined by

$$\mathcal{H}(\boldsymbol{p}, \boldsymbol{q}, t) = \dot{\boldsymbol{q}} \cdot \boldsymbol{p} - \mathcal{L}(\boldsymbol{q}, \dot{\boldsymbol{q}}, t), \tag{1.1.31}$$

where it is understood that the set $\dot{\boldsymbol{q}}$ is eliminated in favour of the set $\boldsymbol{p}$ by means of Eq. (1.1.29). This is an important difference between the Lagrangian approach and the Hamiltonian approach: The Lagrangian is considered to be a function of the *generalized velocities* $\dot{\boldsymbol{q}}$ and the *generalized coordinates* $\boldsymbol{q}$, whereas the Hamiltonian is considered to be a function of the *generalized momenta* $\boldsymbol{p}$ and the *generalized coordinates* $\boldsymbol{q}$. Clearly, $\mathcal{H}$ is a Legendre transform of $-\mathcal{L}$ with respect to the variables $\dot{\boldsymbol{q}}$. In the Lagrangian approach, the equations of motion are written in the form

$$\frac{\mathrm{d}}{\mathrm{d}t}\left(\frac{\partial \mathcal{L}(\boldsymbol{q}, \dot{\boldsymbol{q}}, t)}{\partial \dot{\boldsymbol{q}}}\right) = \frac{\partial \mathcal{L}(\boldsymbol{q}, \dot{\boldsymbol{q}}, t)}{\partial \boldsymbol{q}}, \tag{1.1.32}$$

whereas in the Hamiltonian approach, the time evolution of the system in phase space is dictated by *Hamilton's equations*:

$$\dot{p}_j \equiv \frac{\mathrm{d}p_j}{\mathrm{d}t} = -\frac{\partial \mathcal{H}(\boldsymbol{p}, \boldsymbol{q}, t)}{\partial q_j} \tag{1.1.33}$$

$$\dot{q}_j \equiv \frac{\mathrm{d}q_j}{\mathrm{d}t} = \frac{\partial \mathcal{H}(\boldsymbol{p}, \boldsymbol{q}, t)}{\partial p_j}. \tag{1.1.34}$$

In a description cast in terms of Cartesian coordinates, Eq. (1.1.33), which states that the rate of change of momentum for particle j equals the total force on that particle, is a statement of Newton's second law of motion. Eq. (1.1.34), on the other hand, simply states that the rate of change of the system position vector, i.e. the velocity, equals the momentum divided by the mass.

For the majority of classical systems the kinetic energy is a quadratic function of the generalized velocities, i.e. of the form

$$\mathcal{K}(\boldsymbol{q}, \dot{\boldsymbol{q}}, t) = \sum_{j=1}^{3n} a_j (q_1, q_2, \ldots, q_{3n}) \dot{q}_j^2, \qquad (1.1.35)$$

where the coefficients a_j are in general functions of generalized coordinates but do not depend explicitly on time. If, in addition, the potential energy is a function only of the generalized coordinates, then it can be shown that

$$\mathcal{H} = \mathcal{K} + \mathcal{V}; \qquad (1.1.36)$$

that is, the Hamiltonian equals the *total energy* of the system. If the Lagrangian is not an explicit function of time (as e.g. for conservative systems, where the potential energy is a function of the particle positions only), then it is straightforward to prove that $\frac{d\mathcal{H}}{dt} = 0$, which means that $\mathcal{H}$ is a constant of the motion

$$\mathcal{H}(\boldsymbol{X}) \equiv \mathcal{H}(\boldsymbol{p}, \boldsymbol{q}) = E, \text{ a constant.} \qquad (1.1.37)$$

Given the microstate $\boldsymbol{X}(t_1)$ of the system at time t_1, one can uniquely determine the microstate $\boldsymbol{X}(t_2)$ at a later time t_2, where $t_2 > t_1$, by integrating Hamilton's equations of motion. Furthermore, given $\boldsymbol{X}(t_2)$, one can uniquely determine $\boldsymbol{X}(t_1)$ by integrating Hamilton's equations backwards in time. Hamiltonian dynamics is therefore completely *time-reversible*.

1.1.6 Potential energy function

In general, as mentioned in Section 1.1.2, the potential energy of a system comprising polymeric chains can be distinguished into *intramolecular* and *intermolecular* contributions. Intramolecular contributions are associated with nuclear, electronic, and the bonded vibrational and torsional degrees of freedom; they also include nonbonded contributions arising from the potential energy of interaction between atoms belonging to the same chain but separated by several bonds (in most models, at least four). Intermolecular contributions arise from the potential energy of interaction between atoms belonging to different chains. The nonbonded component of the intramolecular energy and the intermolecular energy are functions of the relative positions of the constituent atoms.

In the most general case,[1,18] a complete microscopic description of a polymer chain in configuration space can be conveniently given by specifying the following (see Figure 1.7): (a) the location of the chain start, $\boldsymbol{R}_0$; (b) the Euler angles $\boldsymbol{\psi} = (\psi_1, \psi_2, \psi_3)$ specifying the overall orientation of the chain with respect to a fixed frame of reference; (c) all bond lengths of the chain, which are collectively denoted by $\{\boldsymbol{l}\}$ or simply $\boldsymbol{l}$ (see Eq. (1.1.2)); (d) all bond angles in the chain, which are collectively denoted by $\{\boldsymbol{\theta}\}$ or simply $\boldsymbol{\theta}$; and (e) all torsion angles in the chain, which are collectively denoted by $\{\boldsymbol{\phi}\}$ or simply $\boldsymbol{\phi}$. Variables $\boldsymbol{R}_0$ and $\boldsymbol{\psi}$ are *external*, while variables $\boldsymbol{l}$, $\boldsymbol{\theta}$, and $\boldsymbol{\phi}$ are *internal*. Furthermore, variables $\boldsymbol{R}_0$, $\boldsymbol{\psi}$, and $\boldsymbol{\phi}$ are considered to be *soft*, whereas variables $\boldsymbol{l}$ and $\boldsymbol{\theta}$ are considered to be *hard*, their variation being associated with large changes in potential energy.

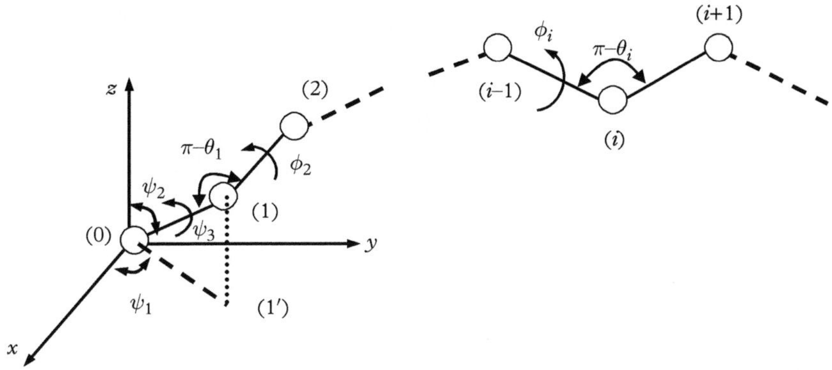

Figure 1.7 Generalized coordinates along a chain. Numbering of skeletal atoms in parentheses. (1′) is the projection of atom 1 on the xy plane.

For a multi-chain system, $\boldsymbol{R}_0$, $\boldsymbol{\psi}$, $\boldsymbol{l}$, $\boldsymbol{\theta}$, and $\boldsymbol{\phi}$ are extended variables spanning the configurational space of all N_{ch} chains.

The total potential energy function of the system, $\mathcal{V}$, is the sum of contributions arising from all bonded and nonbonded interactions. For a stiff model where all bond lengths and bond angles are assumed to be fixed at their equilibrium values, $\mathcal{V}$ is considerably simplified, since it contains contributions only from the soft degrees of freedom, evaluated by setting all bond lengths and bond angles equal to their equilibrium values. In such a case, $\mathcal{V}$ can be considered as a sum of torsional potentials associated with the dihedral angles, usually modelled by simple expressions such as

$$\mathcal{V}_\phi(\phi) = \mathcal{V}_0^\phi + \frac{k_\phi}{2}\left(1 - \cos\left(3\phi\right)\right) \tag{1.1.38}$$

for torsion angles formed by four consecutive bonded atoms, and

$$\mathcal{V}_\chi(\chi) = \mathcal{V}_0^\chi + k_\chi(\chi - \chi_0)^2 \tag{1.1.39}$$

for dihedral angles corresponding to four atoms not successively bonded as e.g. at junction or branch points (see Figure 1.8) and serving to conserve the configuration of chiral molecules; these are typically called improper dihedrals. In the above two expressions, $\mathcal{V}_0^\phi$ and $\mathcal{V}_0^\chi$ denote the torsional and improper torsion energies at angles $\phi = 0$ and $\chi = 0$, respectively. Nonbonded interactions, on the other hand, are usually modelled by a Lennard-Jones 6–12 potential expression,

$$\mathcal{V}_{\mathrm{LJ}}(r) = 4\varepsilon\left\{\left(\frac{\sigma}{r}\right)^{12} - \left(\frac{\sigma}{r}\right)^{6}\right\}. \tag{1.1.40}$$

When polar groups characterized by explicit separation of charges leading to dipole–dipole interactions, hydrogen bonding, or other types of specific forces are present, additional electrostatic terms

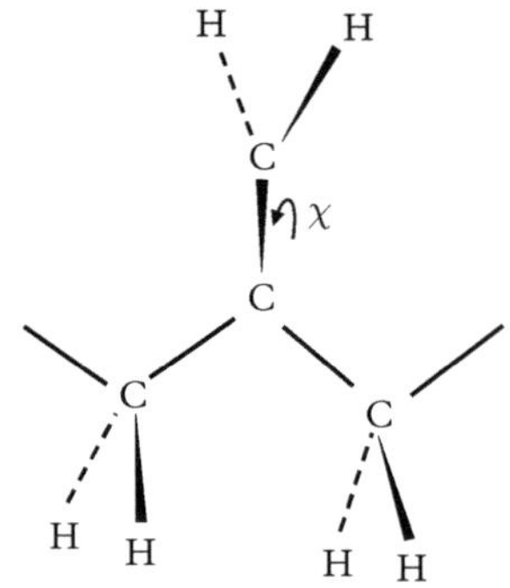

Figure 1.8 Example of an improper torsion angle defined by four bonded carbon atoms in a polymer chain. The three consecutive skeletal carbons lie in the plane of the page. Improper torsion angle χ is formed between this plane and pendant bond C-C, which sticks out of the plane towards the reader.

must be included in the nonbonded force field. A simple way to model these terms is to invoke Coulomb potentials,

$$\mathcal{V}_C(r) = \frac{1}{4\pi\varepsilon\varepsilon_0}\frac{q_i q_j}{r}, \tag{1.1.41}$$

with $\varepsilon_0 = 8.85 \times 10^{-12}\mathrm{C/(Vm)}$ being the electric permittivity of free space and ε being the dielectric constant of the polarizable medium between interacting charges q_i and q_j. In fully atomistic representations one typically takes $\varepsilon = 1$ and sums electrostatic interactions out to infinite distance. The partial charges q_i are often based on electronic structure calculations (see Chapter 2).

For a flexible model (variable bond lengths and angles), the total potential energy function $\mathcal{V}$ should include additional terms accounting for bond stretching and bond bending contributions. Gō and Scheraga have shown that, although motion associated with the soft degrees of freedom can be treated classically, (vibrational) motion associated with the hard degrees of freedom must be treated quantum mechanically.[19] Simple models have been proposed in the literature for the potential energy terms associated with hard degrees of freedom, usually in the form of simple harmonic functions of the type

$$\mathcal{V}_l(l) = \mathcal{V}_0^l + \frac{k_l}{2}(l - l_0)^2 \tag{1.1.42}$$

and

$$\mathcal{V}_\theta(\theta) = \mathcal{V}_0^\theta + \frac{k_\theta}{2}(\theta - \theta_0)^2, \tag{1.1.43}$$

where $\mathcal{V}_0^l$ and $\mathcal{V}_0^\theta$ are the (chemical) bond energies at the equilibrium length l_0 and equilibrium angle θ_0, respectively. In the limit of infinite stiffness $k_l \to \infty$, $k_\theta \to \infty$, bond lengths and bond angles cannot depart from their equilibrium values.

To improve the predictive capability of the model, either more complicated expressions are used for the various components of the potential energy function or cross terms are added, which couple the various degrees of freedom. Today, sophisticated potential energy functions are available as part of commercial software packages and are known under various names, such as the COMPASS forcefield, the PCFF forcefield, or the Dreiding forcefield (see Chapter 2). To reduce computational requirements, today extensive use is made also of the so-called united-atom models for polymers. *United-atom* models differ from *explicit-atom* ones in that certain atoms (such as the hydrogen and the carbon atoms in the chemical groups CH_3, CH_2, or CH) are lumped together to be considered as a single, spherically symmetric or spherically asymmetric site; this results in a significant reduction in the total number of the configurational degrees of freedom. While the description appears to be oversimplified, a number of reliable united-atom force fields have been proposed for many polymers, providing quite accurate estimates of their volumetric and conformational properties. We will review many of these models in the next chapters (see e.g. Section 2.6). In atomistic simulations of polymers, the parameters contained in this section's expressions for the components of potential energy ($k_\phi, k_\chi, k_\theta, k_l, \chi_0, \theta_0, l_0, \sigma, \varepsilon, q$) are fixed a priori (usually they are derived by fitting quantum-mechanically derived potential energy data to simple algebraic equations, such as those mentioned earlier) and are not adjustable. Chapter 2 outlines typical methods for deriving these expressions and their parameters from first principles employing accurate quantum-mechanical calculations.

1.2 Rotational isomeric state model

1.2.1 Conformational partition function

A chain molecule is not a mere collection of atoms. The existence of chemical bonds connecting atoms together considerably restricts the number of configurations that the atoms of the chain can assume. Thus, the coordinates of the various atoms cannot be assigned in a completely arbitrary way; certain geometric constraints should be respected, associated with the fact that certain bond lengths and bond angles should be restricted within very narrow limits. For molecules consisting of a small number of atoms, these constraints or requirements are enough to fully determine the atomic configurations given their chemical structures. For chain-like molecules, however, torsion angles about bonds along the chain bring up a diversity of shapes, or *conformations*. The rotations about chemical bonds are governed by potentials which are less restrictive than those governing bond lengths and bond angles; therefore, they play a dominant role in the analysis of the spatial configurations of various chain molecules.

For a hypothetical linear polymer molecule consisting of N bonds, whose configuration is specified by means of the position vectors $\boldsymbol{R}_0, \boldsymbol{R}_1, \ldots,$ and $\boldsymbol{R}_N$ of its $N+1$ skeletal atoms with respect to some fixed coordinate system, it is therefore more convenient to transform (see Section 1.1.1) from the space $\{\boldsymbol{R}\}$ of Cartesian coordinates of all atoms (or, equivalently, from the space of bond vectors $\{\boldsymbol{l}\}$) to the space of generalized or internal coordinates

$$\{\boldsymbol{R}_0, \boldsymbol{R}_1, \boldsymbol{R}_2, \ldots, \boldsymbol{R}_N\} \rightarrow \{\boldsymbol{R}_0, \psi_1, \psi_2, \psi_3, l_1, \theta_1, l_2, \theta_2, \phi_2, \ldots, l_{N-1}, \theta_{N-1}, \phi_{N-1}, l_N\}, \qquad (1.2.1)$$

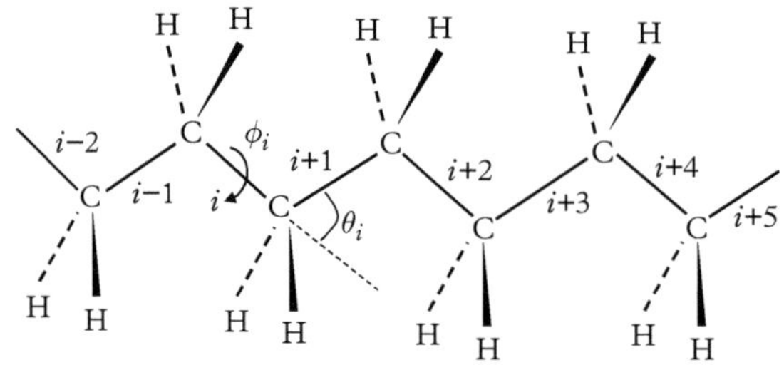

Figure 1.9 Definition of the *trans* state along a segment of a polyethylene chain including bonds $i-1$ to $i+4$, indicating also bond angle θ_i and torsion (bond rotation) angle ϕ_i.

where l_i denotes the length of bond i, $\pi-\theta_i$ is the bond angle at chain atom i, and ϕ_i is the torsion angle measuring the dihedral angle between the two planes defined by bond pairs $(i-1, i)$ and $(i, i+1)$, and $(i, i+1)$ and $(i+1, i+2)$ (see Figure 1.7). Following Flory's convention, the sign of angle ϕ_i is measured here with respect to the *trans* conformation and taken positive if it would lead to unscrewing of a right-handed screw directed along bond i (see Figure 1.9 for polyethylene). The values of bond lengths l_i and bond angles θ_i are practically fixed (*hard degrees of freedom*), so that the conformation of the chain is primarily differentiated by the bond rotation angles ϕ_i.[1] The chain conformation is therefore determined by the set $\{\phi\}=\{\phi_2, \phi_3, \ldots, \phi_{N-1}\}$ of its dihedral angles.

By integrating out all hard degrees of freedom in Eq. (1.1.13) for the configurational integral, as well as those associated with the chain start and overall chain orientation in space, we obtain that[1]

$$Z = C_{\{\phi\}} \int \exp\left[-\frac{\widehat{\mathcal{V}}(\phi_2, \phi_3, \ldots, \phi_{N-1})}{k_B T}\right] d\phi_2 d\phi_3 \ldots d\phi_{N-1}, \tag{1.2.2}$$

where $C_{\{\phi\}}$ is a constant with dimensions V^{N+1} and $\widehat{\mathcal{V}}(\phi_2, \phi_3, \ldots, \phi_{N-1})$ is the potential of mean force associated with internal rotations. Z is then known as the *conformational partition function* of the chain. Specification of the chain configuration by the set of variables $\{\phi\}$, along with $\boldsymbol{R}_0, \psi_1, \psi_2, \psi_3$, directly defines the positions of all atoms along the chain skeleton in the approximation of infinitely stiff bond lengths and angles.

1.2.2 The form of function $\widehat{\mathcal{V}}(\phi_2, \phi_3, \ldots, \phi_{N-1})$ for some simple molecules

Treatment of the conformational properties of chain molecules is greatly simplified if one assumes that the function $\widehat{\mathcal{V}}(\phi_2, \phi_3, \ldots, \phi_{N-1})$ is separable into a sum of energies

$$\widehat{\mathcal{V}}(\phi_2, \phi_3, \ldots, \phi_{N-1}) = \sum_{i=2}^{N-1} \widehat{\mathcal{V}}_i(\phi_i), \tag{1.2.3}$$

where $\widehat{\mathcal{V}}_i(\phi_i)$ represents the rotational potential for bond i. Although instances where such a separation can be justified are the exception (rather than the rule),[1] Eq. (1.2.3) permits the exact calculation of a number of chain properties, such as the mean-square end-to-end distance and the mean-square radius of gyration. Assuming that Eq. (1.2.3) is valid, the probability that each rotation angle ϕ_i is found within the range ϕ_i to $\phi_i + \mathrm{d}\phi_i$ is given by

$$p(\phi_i)\,\mathrm{d}\phi_i = \frac{1}{z_i}\exp\left[-\frac{\widehat{\mathcal{V}}_i(\phi_i)}{k_\mathrm{B}T}\right]\mathrm{d}\phi_i, \tag{1.2.4}$$

where

$$z_i = \int_0^{2\pi} \exp\left[-\frac{\widehat{\mathcal{V}}_i(\phi_i)}{k_\mathrm{B}T}\right]\mathrm{d}\phi_i \equiv \sum_k \exp\left[-\frac{E_k}{k_\mathrm{B}T}\right] \tag{1.2.5}$$

is the *bond rotational partition function*. On the right-hand side of Eq. (1.2.5) the domain of ϕ_i is assumed to be separable into subdomains corresponding to discrete conformational states k. E_k is a free energy for conformational state k.

Equations (1.2.3)–(1.2.5) are, of course, an excellent approximation for a molecule with only one rotatable skeletal bond, namely n-butane. For the central bond in an n-butane molecule, a form used to represent the torsional potential $\widehat{\mathcal{V}}_2(\phi_2)$ in simulations is shown in Figure 1.10. Its most important features are the dihedral angle values of the three minima for the *trans* (at 0°) and the two *gauche* conformations (*c.*±120°, respectively). The energy of the latter minima is *c.*3 kJ mol^{-1} higher than that of the former. The energy barrier between the *trans* and each one of the two *gauche* states is *c.*14.5 kJ mol^{-1}, while that between the two *gauche* states is much higher. Molecules confined to the regions of the three potential energy minima are well differentiated at room temperature, so we can speak of three clearly distinguishable conformational states of n-butane. These are designated as t (*trans*), g^+ (*gauche*$^+$), and g^- (*gauche*$^-$). On the same basis, in n-pentane (whose internal configuration is determined by two dihedral angles ϕ_2 and ϕ_3), we expect nine rotational isomers or conformers:[1] $tt, tg^-, tg^+, g^+t, g^-t, g^+g^+, g^-g^-, g^+g^-, g^-g^+$. However, the

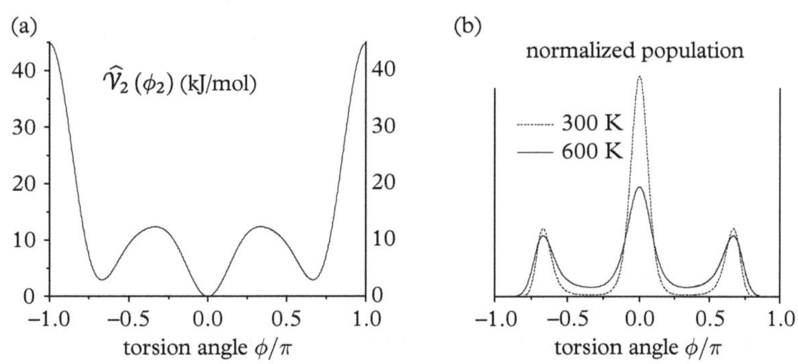

Figure 1.10 (a) The form of the torsional potential for n-butane and (b) the corresponding graph of the normalized population of torsion angles at two temperatures.

conformers g^+t, g^-t, and g^-g^+ are identical with tg^+, tg^-, and g^+g^-, respectively, which reduces the total number of independent conformers to six, including two pairs of mirror images, namely (tg^+, tg^-) and (g^+g^+, g^-g^-). Thus, one is left with only four principal conformations: tt, tg^-, g^+g^+, and g^+g^-. If tt is defined to be the reference state, then we can write for the free energies of the conformations $E_{tg^+} = E_{tg^-} = E_{g^+t} = E_{g^-t} \equiv E_\sigma$, where E_σ is a *first-order* interaction term depending only on the corresponding dihedral angle. Due to the additivity of the energetic contributions for the *gauche* bonds, we also have $E_{g^+g^+} = E_{g^-g^-} \cong 2E_\sigma$. However, $E_{g^+g^-} = E_{g^-g^+} >> 2E_\sigma$ due to the *pentane effect*, indicating that a separation into independent terms of the type of Eq. (1.2.3) is not appropriate for pentane. In fact, we can set $E_{g^+g^-} = E_{g^-g^+} = 2E_\sigma + E_\omega$, with the $2E_\sigma$ term accounting for the two first-order interactions (each one depending on the corresponding dihedral angle) and the E_ω term for the second-order interactions (which depend on both dihedral angles); Flory[1] and colleagues have used $E_\sigma \cong 2.1\,\text{kJ mol}^{-1}$ and $E_\omega \cong 7.5\,\text{kJ mol}^{-1}$. A successful representation of the conformations of n-pentane and higher members of the alkane series in terms of discrete states is provided by the *rotational isomeric state* model.

1.2.3 The rotational isomeric state approximation

In the rotational isomeric state (RIS) approximation,[1,20] the *continuous* space (−180°, 180°) spanned by each torsion angle ϕ of a macromolecule (e.g. polyethylene) is replaced by a set of *discrete* rotational states (e.g. t, g^+, g^-), which are ordinarily constructed around the minima of the energy function $\widehat{\mathcal{V}}_i(\phi_i)$ for a molecule with only one torsion angle of the type considered (e.g. n-butane). Each bond of the macromolecule is treated as occurring in one or another of these discrete rotational states; the conformations adopted by the entire macromolecule are thus discretized into sequences of states adopted by the individual bonds. For a molecule with only one torsion angle, the bond rotational partition function (compare Eq. (1.2.5)) is written as

$$z = u_1 + u_2 + \cdots + u_\nu, \tag{1.2.6}$$

where $u_1 + u_2 + \cdots + u_\nu$, etc., are the ν statistical weights applicable to the respective ν isomeric states. In practice, it is convenient to assign a value of unity to one of these weights and express the others relative to the one chosen. Thus, if $u_1 = 1$, then

$$u_k = \exp\left[-\frac{E_k - E_1}{k_B T}\right]. \tag{1.2.7}$$

In n-butane, for example, we can take the energy of the *trans* state as the reference level. In this case, we denote by E_σ the (free) energy of the *gauche* state relative to *trans* and define the quantity σ through

$$\sigma = \exp\left(-\frac{E_\sigma}{k_B T}\right). \tag{1.2.8}$$

With these choices, the conformational partition function for n-butane becomes $1 + 2\sigma$. We can thus form Table 1.1 for the three states t, g^+, and g^- in n-butane.

Table 1.1 *Rotational Isomeric States for* n-*Butane*

State	*trans*	*gauche*$^+$	*gauche*$^-$
Statistical weight	1	σ	σ
Population	$(1+2\sigma)^{-1}$	$\sigma(1+2\sigma)^{-1}$	$\sigma(1+2\sigma)^{-1}$
Population at 300K	0.54	0.23	0.23
Mean ϕ value at 300K*	0°	~110°	~−110°

* This is calculated through $\langle \phi_i \rangle = \dfrac{\displaystyle\int_{i-\text{domain}} \phi \exp\left[-\dfrac{\widehat{\mathcal{V}}(\phi)}{k_{\mathrm{B}}T}\right] \mathrm{d}\phi}{\displaystyle\int_{i-\text{domain}} \exp\left[-\dfrac{\widehat{\mathcal{V}}(\phi)}{k_{\mathrm{B}}T}\right] \mathrm{d}\phi}$ and is not necessarily equal to the value of ϕ at the minimum. The convention wherein ϕ is 0° at the minimum is adopted here. Another convention places the minimum of *trans*, ϕ_{trans}, at 180°. To cover both conventions, values reported in the last row of the table must be interpreted as values of $\phi - \phi_{\text{trans}}$.

Similarly, in *n*-pentane, whose internal configuration is determined by two rotational angles, we can define two quantities, σ and ω, through Eq. (1.2.8) and

$$\omega = \exp\left(-\frac{E_\omega}{k_{\mathrm{B}}T}\right). \tag{1.2.9}$$

At $T = 300\mathrm{K}$, $\sigma \cong 0.43$, and $\omega \cong 0.049$. This allows us to create Table 1.2 for the energies and statistical weights of the various conformers of *n*-pentane.

The conformational partition function for n-pentane is then

$$z = \sum_i u_i = 1 + 4\sigma + 2\sigma^2 + 2\sigma^2\omega, \tag{1.2.10}$$

with the corresponding probability for each possible conformational state being

$$p_i = \frac{u_i}{z}. \tag{1.2.11}$$

Table 1.2 *Rotational Isomeric States for* n-*Pentane*

State i	tt	tg^+	tg^-	g^+t	g^+g^+	g^+g^-	g^-t	g^-g^+	g^-g^-
Energy, E_i	0	E_σ	E_σ	E_σ	$2E_\sigma$	$2E_\sigma + E_w$	E_σ	$2E_\sigma + E_w$	$2E_\sigma$
Statistical weight, u_i	1	σ	σ	σ	σ^2	$\sigma^2\omega$	σ	$\sigma^2\omega$	σ^2

1.2.4 The conformational partition function of a macromolecule written as a product of statistical weight matrices

The statistical weights for the nine possible combinations of the three rotational isomeric states t, g^+, and g^-, adopted by bonds 2 and 3 of n-pentane, can be arranged in a matrix-like form known as the *statistical weight matrix*, as follows:

$$U = \begin{bmatrix} 1 & \sigma & \sigma \\ 1 & \sigma & \sigma\omega \\ 1 & \sigma\omega & \sigma \end{bmatrix}. \tag{1.2.12}$$

Here, each column refers to the t, g^+, or g^- conformational state of bond 3 and each row to the t, g^+, or g^- state, respectively, of bond 2. Thus, an element $u_{\alpha\beta}$ of the matrix $\boldsymbol{U}$ denotes the statistical weight of that conformation where bond 2 is in state α and bond 3 in state β, divided by the statistical weight of state α for bond 2. With the help of such a statistical weight matrix, one can verify that the conformational partition function for n-butane can be written as the matrix product

$$z = \text{const.} \begin{bmatrix} 1 & \sigma & \sigma \end{bmatrix} \cdot \begin{bmatrix} 1 \\ 1 \\ 1 \end{bmatrix} = \text{const.} \begin{bmatrix} 1 & 0 & 0 \end{bmatrix} \cdot \begin{bmatrix} 1 & \sigma & \sigma \\ 1 & \sigma & \sigma\omega \\ 1 & \sigma\omega & \sigma \end{bmatrix}^1 \cdot \begin{bmatrix} 1 \\ 1 \\ 1 \end{bmatrix}, \tag{1.2.13}$$

where the exponent 1 accounts for the fact that in n-butane there is only one rotatable bond. Similarly, for n-pentane,

$$\begin{aligned} z &= \text{const.} \begin{bmatrix} 1 & \sigma & \sigma \end{bmatrix} \cdot \begin{bmatrix} 1 & \sigma & \sigma \\ 1 & \sigma & \sigma\omega \\ 1 & \sigma\omega & \sigma \end{bmatrix} \begin{bmatrix} 1 \\ 1 \\ 1 \end{bmatrix} \\ &= \text{const.} \begin{bmatrix} 1 & 0 & 0 \end{bmatrix} \cdot \begin{bmatrix} 1 & \sigma & \sigma \\ 1 & \sigma & \sigma\omega \\ 1 & \sigma\omega & \sigma \end{bmatrix}^2 \cdot \begin{bmatrix} 1 \\ 1 \\ 1 \end{bmatrix}, \end{aligned} \tag{1.2.14}$$

where the exponent 2 accounts for the fact that we have two rotatable bonds.

Based on the above observation, the RIS model was developed to express the conformational partition function as the product of statistical weight matrices for any polymeric chain (i.e. for any chain consisting of bonds adopting any kind of rotational state). Overall, the RIS model rests on the following assumptions:

- the conformational space can be satisfactorily described as a set of discrete conformational states for the chain, each one of which is a sequence of rotational isomeric states for the various bonds. That is, for the conformation $\{\eta_2, \eta_3, \ldots, \eta_{N-1}\}$ (with, for example, $\eta_i \in \{t, g^+, g^-\}$ for polyethylene),

$$z = \text{const.} \sum_{\eta_2} \sum_{\eta_3} \cdots \sum_{\eta_{N-1}} \exp[-\beta E(\eta_2, \eta_3, \ldots, \eta_{N-1})], \tag{1.2.15}$$

where we have set $\beta = \frac{1}{k_B T}$.

- the free energy $E(\eta_2, \eta_3, \ldots, \eta_{N-1})$ for the considered conformation can be written as a sum of contributions from pairs of consecutive bonds along the chain,

$$\begin{aligned} &E(\eta_2, \eta_3, \ldots, \eta_{N-1}) \\ &= E_2(\eta_2) + E_3(\eta_2, \eta_3) + E_4(\eta_3, \eta_4) + \cdots + E_{n-1}(\eta_{N-2}, \eta_{N-1}), \end{aligned} \tag{1.2.16}$$

  where (see Figure 1.11)
  - $E_2(\eta_2)$ is the first-order (free) energy of interaction contributed by bond 2 when it is in state η_2, and
  - $E_i(\eta_{i-1}, \eta_i)$ is the sum of the first-order (free) energy of interaction contributed by bond i when it is in state η_i plus the second-order energy of interaction contributed by bonds $i-1$ and i when they are in states η_{i-1} and η_i, respectively.
- All nonlocal interactions are neglected. Also, second-order are the highest order terms taken into account. One should note, however, that incorporation of third- or higher-order interactions is also possible and sometimes necessary for several structures.[1,21]

The energies $E_2(\eta_2)$ and $E_i(\eta_{i-1}, \eta_i)$ are written with respect to a reference rotational isomeric state, e.g. the trans state, t. For example, for the (t, t, g^+, g^-) conformation in n-heptane, $E(t, t, g^+, g^-) = E_2(t) + E_3(t, t) + E_4(t, g^+) + E_5(g^+, g^-) = 0 + 0 + E_\sigma + E_\sigma + E_\omega = 2E_\sigma + E_\omega$.

The corresponding statistical weight for the given rotational isomeric state is then expressed as

$$\begin{aligned} &\exp[-\beta E(\eta_2, \eta_3, \ldots, \eta_{N-1})] \\ &= \exp[-\beta E_2(\eta_2)] \exp[-\beta E_3(\eta_2, \eta_3)] \ldots \exp[-\beta E_{N-1}(\eta_{N-2}, \eta_{N-1})] \\ &= u_2(\eta_2) u_3(\eta_2, \eta_3) \ldots u_{N-1}(\eta_{N-2}, \eta_{N-1}), \end{aligned} \tag{1.2.17}$$

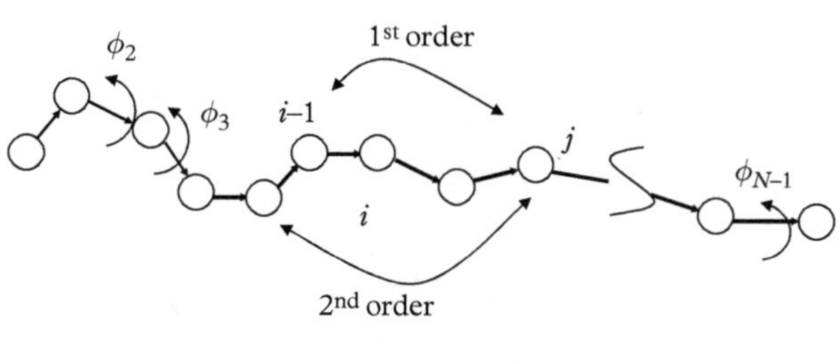

Figure 1.11 First- and second-order interactions along a chain considered in the RIS model.

where we have defined that

$$u_2(\eta_2) = \exp[-\beta E_2(\eta_2)] \tag{1.2.18}$$

and

$$u_i(\eta_{i-1}, \eta_i) = \exp[-\beta E_i(\eta_{i-1}, \eta_i)]. \tag{1.2.19}$$

For n-heptane, for example, the previous expression is equivalent to $\exp[-\beta E(t, t, g^+, g^-)]$
$= u_2(t)u_3(t, t)u_4(t, g^+)u_5(g^+, g^-)$
$= 1.1.\sigma.\sigma\omega = \sigma^2\omega$.

The corresponding conformational partition function is then expressed as

$$z = \text{const.} \sum_{\eta_2} \sum_{\eta_3} \cdots \sum_{\eta_{N-1}} u_2(\eta_2) u_3(\eta_2, \eta_3) \cdots u_{N-1}(\eta_{N-2}, \eta_{N-1}) \tag{1.2.20}$$

or, equivalently, as the matrix product

$$z = \text{const.}\ \boldsymbol{R}_2 \cdot \boldsymbol{U}_3 \cdot \boldsymbol{U}_4 \cdot \ldots \cdot \boldsymbol{U}_{N-1} \cdot \boldsymbol{U}_N, \tag{1.2.21}$$

where

$$\boldsymbol{R}_2 = [u_2(\text{state 1})\ u_2(\text{state 2})\ \ldots u_2(\text{state } \nu_2)] \tag{1.2.22}$$

is a $(1 \times \nu_2)$ matrix with elements equal to the statistical weights of the ν_2 rotational isomeric states of the second bond in the chain;

$$\boldsymbol{U}_i = \begin{bmatrix} u_i(\text{state 1, state 1}) & \ldots & u_i(\text{state 1, state } \nu_i) \\ \cdot & \ldots & \cdot \\ \cdot & \ldots & \cdot \\ \cdot & \ldots & \cdot \\ u_i(\text{state } \nu_{i-1}, \text{state 1}) & \ldots & u_i(\text{state } \nu_{i-1}, \text{state } \nu_i) \end{bmatrix} \quad 3 \leq i \leq N-1 \tag{1.2.23}$$

is a $(\nu_{i-1} \times \nu_i)$ matrix with elements equal to the statistical weights of all possible conformations for bonds $i - 1$ and i (the symbol ν_i denoting how many different conformational states correspond to bond i, $i = 2, \ldots, N - 1$); and

$$\boldsymbol{U}_N = \begin{bmatrix} 1 \\ 1 \\ \cdot \\ \cdot \\ \cdot \\ 1 \end{bmatrix} \tag{1.2.24}$$

is a $(\nu_{N-1} \times 1)$ matrix with all of its elements equal to 1.

In the special case where all bonds are similar, in the sense that they adopt exactly the same isomeric states (as e.g. in linear PE or in PTFE), let their number be ν (for polyethylene, e.g. $\nu = 3$). Then

$$z = \text{const.}\ \boldsymbol{R}_2 \cdot \underbrace{\boldsymbol{U} \cdot \boldsymbol{U} \cdot \ldots \cdot \boldsymbol{U}}_{3,4,\ldots,N-1} \cdot \boldsymbol{U}_N, \tag{1.2.25}$$

where $\boldsymbol{U}$ is a $(\nu \times \nu)$ matrix and U_N a $(\nu \times 1)$ matrix. Usually,

$$\boldsymbol{R}_2 = \underbrace{\begin{bmatrix} 1 & 0 & 0 & \ldots & 0 \end{bmatrix}}_{\text{dimension } (1\times\nu)} \cdot \boldsymbol{U} \tag{1.2.26}$$

since we can isolate the first row of $\boldsymbol{U}$, as the reference state t does not introduce any first- or second-order interactions. For PE, for example:

$$\boldsymbol{R}_2 = \begin{bmatrix} 1 & \sigma & \sigma \end{bmatrix} = \begin{bmatrix} 1 & 0 & 0 \end{bmatrix} \cdot \begin{bmatrix} 1 & \sigma & \sigma \\ 1 & \sigma & \sigma\omega \\ 1 & \sigma\omega & \sigma \end{bmatrix}. \tag{1.2.27}$$

This means that we can formally write z as

$$z = \text{const.}\ \boldsymbol{U}_1 \cdot \boldsymbol{U}_2 \cdot \boldsymbol{U}_3 \cdot \ldots \cdot \boldsymbol{U}_{N-1} \cdot \boldsymbol{U}_N = \text{const.}\ \boldsymbol{U}_1 \cdot \boldsymbol{U}^{N-2} \cdot \boldsymbol{U}_N \tag{1.2.28}$$

with

$$\boldsymbol{U}_1 = \begin{bmatrix} 1 & 0 & 0 & \ldots & 0 \end{bmatrix}, \text{ dimension } 1 \times \nu, \tag{1.2.29}$$

$$\boldsymbol{U} = \left.\begin{bmatrix} u\,(\text{state }1, \text{state }1) & \ldots & u\,(\text{state }1, \text{state }\nu) \\ \cdot & \ldots & \cdot \\ \cdot & \ldots & \cdot \\ \cdot & \ldots & \cdot \\ u\,(\text{state }\nu, \text{state }1) & \ldots & u\,(\text{state }\nu, \text{state }\nu) \end{bmatrix}\right\} \text{dimension } \nu \times \nu, \tag{1.2.30}$$

and

$$\boldsymbol{U}_N = \left.\begin{bmatrix} 1 \\ 1 \\ \cdot \\ \cdot \\ \cdot \\ 1 \end{bmatrix}\right\} \text{dimension } \nu \times 1. \tag{1.2.31}$$

This technique for calculating z is inspired by the solution of the one-dimensional Ising model with the *transfer matrix method.*[1,20]

1.2.5 State probabilities for individual bonds and pairs of bonds

Given the statistical weight matrices for all bonds of a polymeric chain, the RIS model allows one to calculate state probabilities for specified bonds quite easily. For example,

(A) the probability $p_{\eta;i}$ for bond i to be in the rotational isomeric state η is simply

$$p_{\eta;i} = \frac{\sum\limits_{\eta_2}\sum\limits_{\eta_3}\cdots\sum\limits_{\eta_{i-1}}\sum\limits_{\eta_{i+1}}\cdots\sum\limits_{\eta_{N-1}} \exp[-\beta E(\eta_2,\eta_3,\ldots\eta_{i-1},\eta,\eta_{i+1},\ldots,\eta_{N-1})]}{\sum\limits_{\eta_2}\sum\limits_{\eta_3}\cdots\sum\limits_{\eta_{i-1}}\sum\limits_{\eta_i}\sum\limits_{\eta_{i+1}}\cdots\sum\limits_{\eta_{N-1}} \exp[-\beta E(\eta_2,\eta_3,\ldots\eta_{i-1},\eta_i,\eta_{i+1},\ldots,\eta_{N-1})]}$$

$$= \frac{\begin{array}{l}\sum\limits_{\eta_2}\sum\limits_{\eta_3}\cdots\sum\limits_{\eta_{i-1}}\sum\limits_{\eta_{i+1}}\cdots\sum\limits_{\eta_{N-1}} u_2(\eta_2)u_3(\eta_2,\eta_3)\ldots u_{i-1}(\eta_{i-2},\eta_{i-1})u_i(\eta_{i-1},\eta)\\ u_{i+1}(\eta,\eta_{i+1})\ldots u_{N-1}(\eta_{N-2},\eta_{N-1})\end{array}}{\begin{array}{l}\sum\limits_{\eta_2}\sum\limits_{\eta_3}\cdots\sum\limits_{\eta_{i-1}}\sum\limits_{\eta_i}\sum\limits_{\eta_{i+1}}\cdots\sum\limits_{\eta_{N-1}} u_2(\eta_2)u_3(\eta_2,\eta_3)\ldots u_{i-1}(\eta_{i-2},\eta_{i-1})u_i(\eta_{i-1},\eta_i)\\ u_{i+1}(\eta_i,\eta_{i+1})\ldots u_{N-1}(\eta_{N-2},\eta_{N-1})\end{array}} \tag{1.2.32}$$

i.e.

$$p_{\eta;\,i} = \frac{1}{z}\boldsymbol{U}_1\cdot\boldsymbol{U}_2\cdot\ldots\cdot\boldsymbol{U}_{i-1}\cdot\boldsymbol{U}'_{\eta;i}\cdot\boldsymbol{U}_{i+1}\cdot\ldots\cdot\boldsymbol{U}_N, \tag{1.2.33}$$

where $\boldsymbol{U}'_{\eta;i}$ is the $(\nu_{i-1}\times\nu_i)$ matrix that results from $\boldsymbol{U}_i$ (see Eq. (1.2.23)) if we set all its columns except column η equal to 0:

$$\boldsymbol{U}'_{\eta;\,i} = \begin{bmatrix} 0 & \ldots & 0 & u_i(\text{state }1,\text{state }\eta) & 0 & \ldots & 0\\ 0 & \ldots & 0 & u_i(\text{state }2,\text{state }\eta) & 0 & \ldots & 0\\ \cdot & \ldots & \cdot & \cdot & \cdot & \ldots & \cdot\\ \cdot & \ldots & \cdot & \cdot & \cdot & \ldots & \cdot\\ \cdot & \ldots & \cdot & \cdot & \cdot & \ldots & \cdot\\ 0 & \ldots & 0 & u_i(\text{state }\nu_{i-1},\text{state }\eta) & 0 & \ldots & 0 \end{bmatrix}. \tag{1.2.34}$$

(B) the joint probability $p_{\xi;j,\eta;i}$ for bond j to be in the rotational isomeric state ξ and for bond i to be in rotational isomeric state η is given as follows:
For $j\le i-2$,

$$p_{\xi;j,\,\eta;\,i} = \frac{1}{z}\boldsymbol{U}_1\cdot\boldsymbol{U}_2\cdot\ldots\cdot\boldsymbol{U}_{j-1}\cdot\boldsymbol{U}'_{\xi;j}\cdot\boldsymbol{U}_{j+1}\cdot\ldots\cdot\boldsymbol{U}_{i-1}\cdot\boldsymbol{U}'_{\eta;\,i}\cdot\boldsymbol{U}_{i+1}\cdot\ldots\cdot\boldsymbol{U}_N. \tag{1.2.35}$$

For $i-j=1$ (i.e. when bonds i and j are successive),

$$p_{\xi\eta;\,i} = p_{\xi;i-1,\eta;i} = \frac{1}{z}\boldsymbol{U}_1\cdot\boldsymbol{U}_2\cdot\ldots\cdot\boldsymbol{U}_{i-1}\cdot\boldsymbol{U}''_{\xi\eta;i}\cdot\boldsymbol{U}_{i+1}\cdot\ldots\cdot\boldsymbol{U}_{N-1}\cdot\boldsymbol{U}_N, \tag{1.2.36}$$

where $\boldsymbol{U}''_{\xi\eta;i}$ is the $(\nu_{i-1} \times \nu_i)$ matrix that results from $\boldsymbol{U}_i$ if we set all its elements except the element belonging to row ξ and column η equal to 0,

$$\boldsymbol{U}''_{\xi\eta;i} = \begin{bmatrix} 0 & \dots & 0 & 0 & 0 & 0 & \dots & 0 \\ \cdot & \dots & \cdot & \cdot & \cdot & \cdot & \dots & \cdot \\ \cdot & \dots & \cdot & \cdot & \cdot & \cdot & \dots & \cdot \\ \cdot & \dots & \cdot & \cdot & \cdot & \cdot & \dots & \cdot \\ 0 & \dots & 0 & u_i\,(\text{state } \xi,\ \text{state } \eta) & 0 & 0 & \dots & 0 \\ \cdot & \dots & \cdot & \cdot & \cdot & \cdot & \dots & \cdot \\ \cdot & \dots & \cdot & \cdot & \cdot & \cdot & \dots & \cdot \\ \cdot & \dots & \cdot & \cdot & \cdot & \cdot & \dots & \cdot \\ 0 & \dots & 0 & 0 & 0 & 0 & \dots & 0 \end{bmatrix}. \tag{1.2.37}$$

In general, $\boldsymbol{U}'_{\xi;i-1} \cdot \boldsymbol{U}'_{\eta;\,i} = \boldsymbol{U}_{i-1} \cdot \boldsymbol{U}''_{\xi\eta;i}$.

(C) the conditional probability $q_{\xi\eta;i}$ for bond i to be in the rotational isomeric state η given that bond $i-1$ is in rotational isomeric state ξ is

$$q_{\xi\eta;i} = \frac{p_{\xi\eta;i}}{p_{\xi;i-1}} = \frac{\boldsymbol{U}_1 \cdot \boldsymbol{U}_2 \cdot \ldots \cdot \boldsymbol{U}_{i-1} \cdot \boldsymbol{U}''_{\xi\eta;i} \cdot \boldsymbol{U}_{i+1} \cdot \ldots \cdot \boldsymbol{U}_{N-1} \cdot \boldsymbol{U}_N}{\boldsymbol{U}_1 \cdot \boldsymbol{U}_2 \cdot \ldots \cdot \boldsymbol{U}_{i-2} \cdot \boldsymbol{U}'_{\xi;i-1} \cdot \boldsymbol{U}_i \cdot \boldsymbol{U}_{i+1} \cdot \ldots \cdot \boldsymbol{U}_{N-1} \cdot \boldsymbol{U}_N}. \tag{1.2.38}$$

The probabilities $p_{\eta;2}$ and $p_{\xi\eta;i}$ define an equivalent Markov scheme for the generation of a representative sample of conformations of an unperturbed chain. These probabilities can be used to generate unperturbed chain conformations bond by bond with the Monte Carlo method and the use of pseudo-random numbers (see Figure 1.12). The procedure is an *equivalent* and not a *true* stochastic Markov process, because the state that will be chosen for bond i depends on the statistical weights of all bonds of the chain and not only on the statistical weights of the bond that precedes bond i. The formulation of the RIS model is such that, in the process of selecting a state for bond i, the effect of chain segments following bond i is taken into account.

When one uses the RIS model in order to generate conformations of unperturbed chains, the following normalization conditions apply:

$$\begin{aligned} &\sum_\eta p_{\eta;i} = 1 \\ &\sum_\xi \sum_\eta p_{\xi\eta;i} = 1, \ \sum_\eta p_{\xi\eta;i} = p_{\xi,i-1}, \ \sum_\xi p_{\xi\eta;i} = p_{\eta,i} \\ &\sum_\eta q_{\xi\eta;i} = 1, \ \sum_\xi \sum_\eta q_{\xi\eta,i} = \nu_{i-1}. \end{aligned} \tag{1.2.39}$$

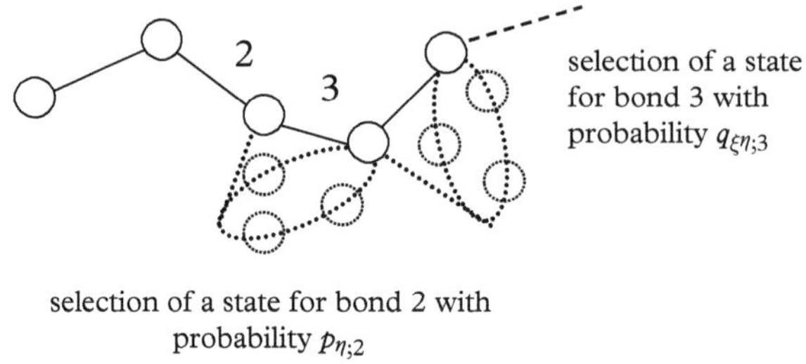

Figure 1.12 Generation of chain configurations bond by bond with a Monte Carlo method and use of pseudo-random numbers based on probabilities $p_{\eta;2}$ and $q_{\xi\eta;i}$.

1.2.6 Calculation of mean values over all conformations

The major utility of the RIS model is that it provides a route to the accurate (subject only to the assumptions introduced during the formulation of the model) and efficient calculation, through simple algebraic operations, of average values for unperturbed chain properties that depend on the conformation. Typical examples of such properties include the mean-square end-to-end distance $\langle r^2 \rangle$, the mean-square radius of gyration $\langle R_g^2 \rangle$, the mean radius of gyration tensor $\langle \boldsymbol{S} \rangle$, the mean dipole moment vector $\langle \boldsymbol{\mu} \rangle$ (projected on a specific direction), the mean polarizability anisotropy $\langle \boldsymbol{r}^{\mathrm{T}} \cdot \hat{\boldsymbol{a}} \cdot \boldsymbol{r} \rangle$ (which is associated with the stress optical law coefficient describing strain birefringence), and the mean Kerr constant $\langle \boldsymbol{\mu}^{T} \cdot \hat{\boldsymbol{a}} \cdot \boldsymbol{\mu} \rangle$ (describing electric birefringence). These calculations are facilitated by employing the *generator matrix technique*. This technique is based on the use of *local (i.e. bond) coordinate frames* and of certain *transformation matrices*. The local coordinate frame of bond i is defined[1] as follows (see Figure 1.13):

- The origin of the frame lies on atom $i-1$.
- The x_i axis is defined to be along bond i with direction from atom $i-1$ to atom i.
- The y_i axis is defined to be perpendicular to the x_i axis, lying in the plane of bonds $i-1$ and i, but on the other side of the x_i axis relative to atom $i-1$.
- The z_i axis is defined to be perpendicular to the x_i and y_i axes, such that it forms a right-handed coordinate frame with them.

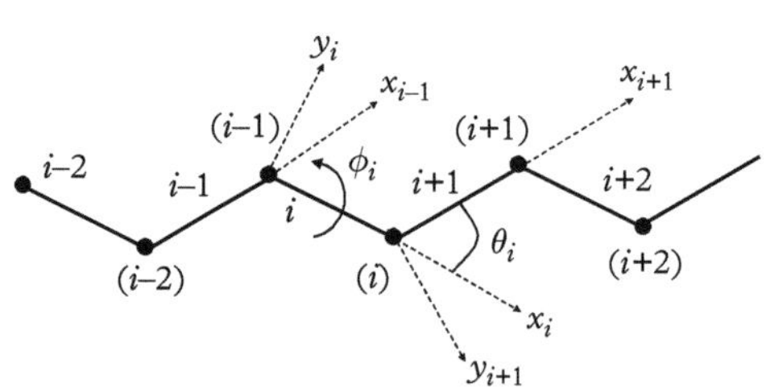

Figure 1.13 Local Cartesian reference frames for consecutive bonds along a chain.

In this coordinate frame, bond vector $\boldsymbol{l}_i$ is simply represented as

$$\boldsymbol{l}_i \equiv \boldsymbol{l}_i^i = \begin{bmatrix} l_i \\ 0 \\ 0 \end{bmatrix}, \tag{1.2.40}$$

where l_i is its length. If $\boldsymbol{\nu}^i$ is the representation of a vector in the local coordinate frame of bond i and $\boldsymbol{\nu}^{i+1}$ its representation in the local coordinate frame of bond $i+1$, the two representations are connected to one another through the transformation equation

$$\boldsymbol{\nu}^i = \boldsymbol{T}_i \cdot \boldsymbol{\nu}^{i+1}, \tag{1.2.41}$$

where

$$\boldsymbol{T}_i = \begin{bmatrix} \cos\theta_i & \sin\theta_i & 0 \\ \cos\phi_i \sin\theta_i & -\cos\phi_i \cos\theta_i & \sin\phi_i \\ \sin\phi_i \sin\theta_i & -\sin\phi_i \quad \cos\theta_i & -\cos\phi_i \end{bmatrix}$$
$$= \begin{bmatrix} 1 & 0 & 0 \\ 0 & \cos\phi_i & -\sin\phi_i \\ 0 & \sin\phi_i & \cos\phi_i \end{bmatrix} \cdot \begin{bmatrix} \cos\theta_i & \sin\theta_i & 0 \\ \sin\theta_i & -\cos\theta_i & 0 \\ 0 & 0 & -1 \end{bmatrix}. \tag{1.2.42}$$

Note that this is the original form of the transformation matrix given by Flory,[1] corresponding to the definitions of angles ϕ_i and θ_i given in Section 1.2.1 and Figure 1.9. Mattice and Suter use the supplement of θ_i in place of θ_i and define ϕ_i to be π, rather than 0, in the *trans* (antiperiplanar) conformational state, with ϕ_i increasing in the same right-handed screw sense introduced by Flory.[21] As a consequence, the $\boldsymbol{T}_i$ matrix of Mattice and Suter is different and can be obtained from Eq. (1.2.42) by using $\pi - \theta_i$ in place of θ_i and $\pi + \phi_i$ in place of ϕ_i.

Applying Eq. (1.2.41) recursively, we obtain the following simple rule for the transformation from the local coordinate frame of bond i to that of bond j with $j < i$:

$$\boldsymbol{\nu}^j = \boldsymbol{T}_j \cdot \boldsymbol{T}_{j+1} \cdot \ldots \cdot \boldsymbol{T}_{i-1} \cdot \boldsymbol{\nu}^i. \tag{1.2.43}$$

In the local coordinate frame of the first bond, the end-to-end vector of the chain is written as

$$\boldsymbol{r} \equiv \boldsymbol{r}^1 = \boldsymbol{l}_1 + \boldsymbol{T}_1 \cdot \boldsymbol{l}_2 + \boldsymbol{T}_1 \cdot \boldsymbol{T}_2 \cdot \boldsymbol{l}_3 + \ldots + \boldsymbol{T}_1 \cdot \boldsymbol{T}_2 \cdot \ldots \cdot \boldsymbol{T}_{N-1} \cdot \boldsymbol{l}_N$$
$$= \sum_{i=1}^{N} \left(\prod_{j=0}^{i-1} \boldsymbol{T}_j \right) \cdot \boldsymbol{l}_i, \quad \boldsymbol{l}_i = \begin{bmatrix} l_i \\ 0 \\ 0 \end{bmatrix}, \tag{1.2.44}$$

where $\boldsymbol{T}_0$ is defined as the unit 3×3 matrix and ϕ_1 is taken as 0 for the definition of $\boldsymbol{T}_1$.

Equations (1.2.20)–(1.2.31) allow the calculation of the mean value $\langle f \rangle$ of any property f that depends on the chain conformation and can be expressed as a product of *generator* matrices that depend on the rotational states of individual bonds. To see this, we note that, according to the RIS model, a conformation is a succession of rotational isomeric states of all rotational bonds, so that

$$\langle f \rangle = \sum_k p_k f_k, \tag{1.2.45}$$

where

- k denotes the conformation, i.e. $k = (\eta_2, \eta_3, \ldots, \eta_{N-1})$,
- p_k is the probability of conformation k, which, according to Section 1.2.3, is written as a product of statistical weights divided by a normalization factor,

$$p_k = \frac{1}{Z} u_2(\eta_2)\, u_3(\eta_2, \eta_3) \ldots u_{N-1}(\eta_{N-2}, \eta_{N-1}), \tag{1.2.46}$$

  and
- f is assumed to be expressible as a product of generator matrices:

$$f_k = (\boldsymbol{G}_1 \cdot \boldsymbol{G}_2 \cdot \ldots \cdot \boldsymbol{G}_{N-1} \cdot \boldsymbol{G}_N)_k = \boldsymbol{G}_1 \cdot \boldsymbol{G}_2(\eta_2) \cdot \boldsymbol{G}_3(\eta_3) \cdot \ldots \cdot \boldsymbol{G}_{N-1}(\eta_{N-1}) \cdot \boldsymbol{G}_N. \tag{1.2.47}$$

Then,

$$\langle f \rangle = \frac{1}{Z} \sum_{\eta_2} \sum_{\eta_3} \cdots \sum_{\eta_{N-1}} u_2(\eta_2) u_3(\eta_2, \eta_3) \ldots u_{N-1}(\eta_{N-2}, \eta_{N-1}) \boldsymbol{G}_1 \cdot \boldsymbol{G}_2(\eta_2) \cdot \ldots \cdot \boldsymbol{G}_{N-1}(\eta_{N-1}) \cdot \boldsymbol{G}_N. \tag{1.2.48}$$

The identification of generator matrices for the squared end-to-end vector, $f = r^2$, is given as Problem 1.4. The generator matrices for this problem are readily constructed from the transformation matrices $\boldsymbol{T}_i$ introduced above and the skeletal bond lengths. $\boldsymbol{G}_1$ is a 1×5 matrix, the $\boldsymbol{G}_i$ for $2 \le i \le N-1$ are 5×5 matrices, and $\boldsymbol{G}_N$ is a 5×1 matrix.

Expression (1.2.48) for $\langle f \rangle$ can also be written in a more compact form by defining the matrices

$$\mathfrak{S}_1 = \boldsymbol{G}_1 \tag{1.2.49}$$

$$\mathfrak{S}_i = \begin{bmatrix} u_i(1,1)\,\boldsymbol{G}_i(1) & u_i(1,2)\,\boldsymbol{G}_i(2) & \ldots & u_i(1,\nu_i)\,\boldsymbol{G}_i(\nu_i) \\ u_i(2,1)\,\boldsymbol{G}_i(1) & u_i(2,2)\,\boldsymbol{G}_i(2) & \ldots & u_i(2,\nu_i)\,\boldsymbol{G}_i(\nu_i) \\ \cdot & \cdot & \ldots & \cdot \\ \cdot & \cdot & \ldots & \cdot \\ \cdot & \cdot & \ldots & \cdot \\ u_i(\nu_{i-1},1)\,\boldsymbol{G}_i(1) & u_i(\nu_{i-1},2)\,\boldsymbol{G}_i(2) & \ldots & u_i(\nu_{i-1},\nu_i)\,\boldsymbol{G}_i(\nu_i) \end{bmatrix}, \quad 1 < i < N \tag{1.2.50}$$

$$\Im_N = \begin{bmatrix} \boldsymbol{G}_N \\ \boldsymbol{G}_N \\ \cdot \\ \cdot \\ \cdot \\ \boldsymbol{G}_N \end{bmatrix}. \tag{1.2.51}$$

In the calculation of the mean-square end-to-end distance ($f = r^2$), $\Im_1$ is a $(1 \times 5\nu_1)$ matrix with $\nu_1 = 1$, $\Im_i$ a $(5\nu_{i-1} \times 5\nu_i)$ matrix, and $\Im_N$ a $(5\nu_{N-1} \times 1)$ matrix. The final result is

$$\langle f \rangle = \frac{1}{Z}\Im_1 \cdot \Im_2 \cdot \ldots \cdot \Im_{N-1} \cdot \Im_N = \frac{\Im_1 \cdot \Im_2 \cdot \ldots \cdot \Im_{N-1} \cdot \Im_N}{\boldsymbol{U}_1 \cdot \boldsymbol{U}_2 \cdot \ldots \cdot \boldsymbol{U}_{N-1} \cdot \boldsymbol{U}_N}. \tag{1.2.52}$$

According to the latter equation, the calculation of the mean value $\langle f \rangle$ involves the following:

- formation of the statistical weight matrices U_i, $1 \le i \le N$
- formation of the generator matrices $\boldsymbol{G}_i$, $1 \le i \le N$, for every state that bond i can attain
- formation of the extended matrices $\Im_i$, $1 \le i \le N$, and
- multiplication of matrices $\boldsymbol{U}_i$ and $\Im_i$, $1 \le i \le N$, and calculation of $\langle f \rangle$.

To this end, it is often very convenient to make use of the following symbolic operations: given a matrix $\boldsymbol{A}$ with dimensions $m_1 \times n_1$ and a matrix $\boldsymbol{B}$ with dimensions $m_2 \times n_2$, we can define their direct (or Kronecker) product $\boldsymbol{A} \otimes \boldsymbol{B}$, which is a $m_1 m_2 \times n_1 n_2$ matrix, as follows:

$$\boldsymbol{A} \otimes \boldsymbol{B} = \begin{bmatrix} a_{11}\boldsymbol{B} & a_{12}\boldsymbol{B} & \ldots & a_{1n_1}\boldsymbol{B} \\ \cdot & & & \\ \cdot & & & \\ \cdot & & & \\ a_{m_1 1}\boldsymbol{B} & a_{m_1 2}\boldsymbol{B} & \ldots & a_{m_1 n_1}\boldsymbol{B} \end{bmatrix}. \tag{1.2.53}$$

Thus, if we define the following block-diagonal matrix, with its blocks being the generator matrices for the various states that bond i can attain,

$$\|\boldsymbol{G}\|_i = \begin{bmatrix} \boldsymbol{G}_i(1) & 0 & \ldots & 0 \\ 0 & \boldsymbol{G}_i(2) & \ldots & 0 \\ \cdot & & & \\ \cdot & & & \\ 0 & 0 & \ldots & \boldsymbol{G}_i(\nu_i) \end{bmatrix}. \tag{1.2.54}$$

then

$$\Im_1 = \boldsymbol{U}_1 \otimes \boldsymbol{G}_1 \quad (1.2.55)$$

$$\Im_i = (\boldsymbol{U}_i \otimes \boldsymbol{I_G}) \cdot \|\boldsymbol{G}\|_i, 1 < i < N \quad (1.2.56)$$

$$\Im_N = \boldsymbol{U}_1 \otimes \boldsymbol{G}_N, \quad (1.2.57)$$

where $\boldsymbol{I_G}$ denotes the unit matrix of the same rank as matrix $\boldsymbol{G}$.

The above methodology allows for the easy and quick calculation of the mean value (with respect to all conformations that an unperturbed chain can attain) of any property that can be expressed as the product of generator matrices $\boldsymbol{G}_1, \boldsymbol{G}_2, \ldots, \boldsymbol{G}_N$. In the literature,[1,21] generator matrices have been proposed for r^2, $a = (\boldsymbol{r} \cdot \boldsymbol{l}_1) \frac{1}{l_1}$ (the persistence length), R_g^2, the radius of gyration tensor $\boldsymbol{S}$, the projection of the chain dipole moment on its first bond $\boldsymbol{\mu} \cdot \frac{\boldsymbol{l}_1}{l_1}, \mu^2, \gamma^2 \left(= \frac{3}{2} tr\hat{\boldsymbol{a}}^2\right), \boldsymbol{r}^T \cdot \hat{\boldsymbol{a}} \cdot \boldsymbol{r}, \boldsymbol{\mu}^T \cdot \hat{\boldsymbol{a}} \cdot \boldsymbol{\mu}$ with $\hat{\boldsymbol{a}}$ being the traceless part of the molecular polarizability tensor of the chain, $\boldsymbol{r} \cdot \boldsymbol{\mu}$, and a few other properties.

1.3 Freely jointed (random flight) and freely rotating chain models

1.3.1 Freely jointed chain

In the hypothetical freely jointed (or random flight) chain model (see Figure 1.14),[1,3,21–23], all (effective) bonds along the chain are considered to have the same length. Using the symbol $\boldsymbol{l}_i$ to denote the *bond vector i*, irrespective of the coordinate system used for its representation,

$$|\boldsymbol{l}_i| = l, \quad \forall i. \quad (1.3.1)$$

In addition, the orientation of bond vectors is assumed to be completely random. In other words, all directions are equally probable for a given bond, irrespective of the directions of neighbouring bonds. As a consequence,

$$\langle \boldsymbol{l}_i \cdot \boldsymbol{l}_j \rangle = 0, \quad \forall i \neq j. \quad (1.3.2)$$

This complete absence of directional correlation between bonds implies that the conformation of a freely jointed chain is a random flight in 3D space of step length $\boldsymbol{l}$, with a probability density function given by the equation

$$\Psi(\{\boldsymbol{l}_n\}) = \prod_{n=1}^{N} \psi(\boldsymbol{l}_n) \text{ with } \psi(\boldsymbol{l}_n) = \frac{1}{4\pi l^2} \delta(|\boldsymbol{l}_n| - l.) \quad (1.3.3)$$

The mean-square end-to-end distance for the freely jointed chain is then calculated to be

$$\langle r^2 \rangle = \langle \boldsymbol{r} \cdot \boldsymbol{r} \rangle = \left\langle \left(\sum_{n=1}^{N} \boldsymbol{l}_n \right) \cdot \left(\sum_{m=1}^{N} \boldsymbol{l}_m \right) \right\rangle = \left\langle \sum_{n=1}^{N} \sum_{m=1}^{N} \boldsymbol{l}_n \cdot \boldsymbol{l}_m \right\rangle$$

$$= \sum_{n=1}^{N} l_n^2 + 2 \sum_{n=1}^{N-1} \sum_{m=n+1}^{N} \langle \boldsymbol{l}_n \cdot \boldsymbol{l}_m \rangle = \sum_{n=1}^{N} l^2 = Nl^2, \tag{1.3.4}$$

where we have used that $l_n^2 = l^2$ and $\langle \boldsymbol{l}_n \cdot \boldsymbol{l}_m \rangle = 0$, according to Eqs. (1.3.1)–(1.3.2). As a result, for the freely jointed chain model

$$\langle r^2 \rangle = Nl^2. \tag{1.3.5}$$

Having obtained an expression for $\langle r^2 \rangle$, we can calculate the value of the *characteristic ratio*, defined by

$$C_N \equiv \frac{\langle r^2 \rangle}{Nl^2} \tag{1.3.6}$$

for the freely jointed chain model; the result is unity for all values of N. The random flight scaling $\langle r^2 \rangle \propto N^1$ arises from the absence of nonlocal interactions and it is typical of long, unperturbed chains.

The mean-square radius of gyration for the freely jointed chain model is readily derived from the Lagrange theorem,[5] Eq. (1.1.11) (see Appendix),

$$\langle R_g^2 \rangle = \frac{1}{(N+1)^2} \sum_{i=0}^{N-1} \sum_{j=i}^{N} \langle R_{ij}^2 \rangle \tag{1.3.7}$$

by considering that all subchains $i \ldots j$ along the chain also obey random flight statistics, so that

$$\langle R_{ij}^2 \rangle = (j - i)\, l^2. \tag{1.3.8}$$

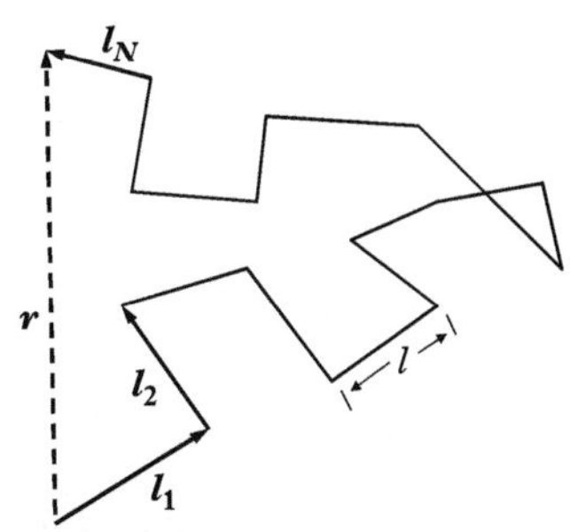

Figure 1.14 The freely jointed (or random flight) chain model.

By substituting Eq. (1.3.8) in Eq. (1.3.7) and carrying out the summations, we obtain that

$$\begin{aligned}\langle R_g^2\rangle &= \frac{l^2}{(N+1)^2}\sum_{i=0}^{N-1}\sum_{j=i}^{N}(j-i) = \frac{l^2}{(N+1)^2}\sum_{i=0}^{N-1}\sum_{j'=0}^{N-i}j' \\ &= \frac{l^2}{(N+1)^2}\sum_{i=0}^{N-1}\frac{(N-i)(N-i+1)}{2} = \frac{l^2}{(N+1)^2}\left[\sum_{i'=1}^{N}\frac{i'(i'+1)}{2}\right] \\ &= \frac{l^2}{2(N+1)^2}\left[\sum_{i'=1}^{N}i'^2+\sum_{i'=1}^{N}i'\right] = \frac{l^2}{2(N+1)^2}\left[\frac{N(N+1)(2N+1)}{6}+\frac{N(N+1)}{2}\right], \end{aligned} \tag{1.3.9}$$

where we have used the transformations $j' = j - i$ and $i' = N - i$ and, after rearrangement,

$$\langle R_g^2\rangle = \frac{1}{6}\frac{N+2}{N+1}Nl^2 = \frac{1}{6}\frac{N+2}{N+1}\langle r^2\rangle. \tag{1.3.10}$$

Equation (1.3.10) implies that

$$\frac{\langle R_g^2\rangle}{Nl^2} = \frac{1}{6}\frac{N+2}{N+1}, \tag{1.3.11}$$

which, unlike $\frac{\langle r^2\rangle}{Nl^2}$, depends on N. For long chains, however, it is expected that[24]

$$\langle R_g^2\rangle \underset{N\to\infty}{\to} \frac{1}{6}\langle r^2\rangle. \tag{1.3.12}$$

This result is also generally applicable for any unperturbed chain.

1.3.2 Freely rotating chain

In the hypothetical freely rotating chain model,[1] in addition to constant bond lengths, $|\boldsymbol{l}_n| = l$, one also assumes constant bond angles π–θ (see Figure 1.15):

$$\boldsymbol{l}_n \cdot \boldsymbol{l}_{n+1} = l^2\cos\theta, \quad n = 1, 2, \ldots, N-1. \tag{1.3.13}$$

No energy barriers are assumed for rotation around the bonds. Then,

$$\langle \boldsymbol{l}_n\rangle_{\boldsymbol{l}_{n-1}\text{ constant}} = \boldsymbol{l}_{n-1}(\cos\theta). \tag{1.3.14}$$

Premultiplying both sides with the vector $\boldsymbol{l}_m$ with $m < n$ and averaging over all conformations, we obtain that

$$\langle \boldsymbol{l}_m\cdot\boldsymbol{l}_n\rangle = (\cos\theta)\,\langle \boldsymbol{l}_m\cdot\boldsymbol{l}_{n-1}\rangle, \tag{1.3.15}$$

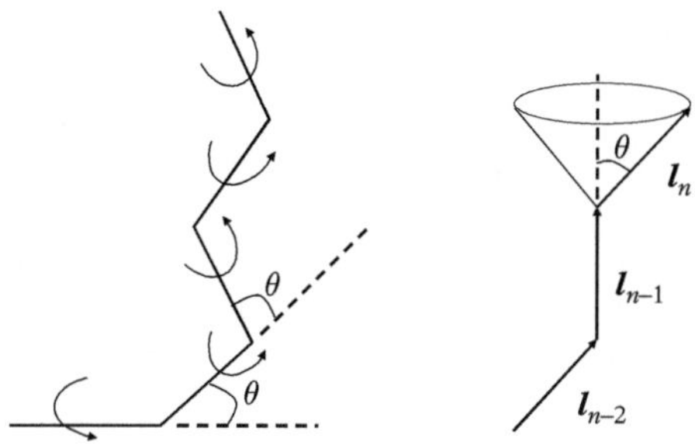

Figure 1.15 The freely rotating chain model.

which, if applied recursively down to $n = m + 1$, results in

$$\langle \boldsymbol{l}_m \cdot \boldsymbol{l}_n \rangle = l^2 (\cos\theta)^{|n-m|}. \tag{1.3.16}$$

Equation (1.3.16) means that the directional correlation of bond vectors decays exponentially with the separation of the bonds along the contour of a freely rotating chain.

The mean-square chain end-to-end distance can be obtained from Eq. (1.1.7) through the use of Eq. (1.3.16). Thus,

$$\begin{aligned}
\langle r^2 \rangle &= \sum_{m=1}^{N}\sum_{n=1}^{N} \langle \boldsymbol{l}_m \cdot \boldsymbol{l}_n \rangle = l^2 \sum_{m=1}^{N}\sum_{n=1}^{N} (\cos\theta)^{|n-m|} = l^2 \left[N + 2\sum_{m=1}^{N-1}\sum_{n=m+1}^{N} (\cos\theta)^{n-m} \right] \\
&= l^2 \left[N + 2\sum_{m=1}^{N-1}\sum_{k=1}^{N-m} (\cos\theta)^{k} \right] = l^2 \left[N + 2\sum_{k=1}^{N-1}\sum_{m=1}^{N-k} (\cos\theta)^{k} \right] = l^2 \left[N + 2\sum_{k=1}^{N-1} (N-k)(\cos\theta)^{k} \right] \\
&= l^2 \left[N + 2N\sum_{k=1}^{N-1} (\cos\theta)^{k} - 2\sum_{k=1}^{N-1} k(\cos\theta)^{k} \right] \\
&= l^2 \left[N + 2N\frac{(\cos\theta)^N - (\cos\theta)}{(\cos\theta) - 1} - 2(\cos\theta)\frac{\mathrm{d}}{\mathrm{d}(\cos\theta)}\left\{ \frac{(\cos\theta)^N - (\cos\theta)}{(\cos\theta) - 1} \right\} \right] \\
&= Nl^2 \left[\frac{1+\cos\theta}{1-\cos\theta} - 2\cos\theta \frac{1-(\cos\theta)^N}{N(1-\cos\theta)^2} \right].
\end{aligned} \tag{1.3.17}$$

The expression derived by the freely rotating chain model for the characteristic ratio is therefore

$$C_N = \frac{1+\cos\theta}{1-\cos\theta} - 2\cos\theta \frac{1-(\cos\theta)^N}{N(1-\cos\theta)^2}. \tag{1.3.18}$$

For $N >> 1$, the last two equations become[8,25]

$$\langle r^2 \rangle = \frac{1+\cos\theta}{1-\cos\theta} Nl^2 \quad \text{and} \quad C_\infty = \frac{1+\cos\theta}{1-\cos\theta}. \tag{1.3.19}$$

C_∞ is called the *characteristic ratio at infinite chain length.* We note that, for $\theta < 90°$, $\langle r^2 \rangle$ comes out to be larger than the corresponding value for the freely jointed chain model with the same N and l values. In fact, in the particular case of a tetrahedrally bonded chain (i.e. for $\pi - \theta = 109.5°$ so that $\cos\theta = \frac{1}{3}$),[1] one finds that $\langle r^2 \rangle = 2Nl^2$, giving a value of 2 for C_∞. The larger value of C_∞ in relation to the freely jointed chain is, of course, a consequence of the directional correlations present in the freely rotating chain. Although still unperturbed, the freely rotating chain has greater conformational stiffness than a random flight of the same bond length.

We can also obtain an expression for the mean-square radius of gyration of the freely rotating chain. The result is[1]

$$\frac{\langle R_g^2 \rangle}{Nl^2} = \frac{(N+2)(1+\cos\theta)}{6(N+1)(1-\cos\theta)} - \frac{\cos\theta}{(N+1)(1-\cos\theta)^2} + \frac{2\cos^2\theta}{(N+1)^2(1-\cos\theta)^3} - \frac{2\cos^3\theta\left(1-(\cos\theta)^N\right)}{N(N+1)^2(1-\cos\theta)^4}, \tag{1.3.20}$$

which, for large N, gives

$$\frac{\langle R_g^2 \rangle}{Nl^2} \underset{N\to\infty}{\to} \frac{1}{6}\frac{1+\cos\theta}{1-\cos\theta}, \tag{1.3.21}$$

leading again to Eq. (1.3.12).

1.3.3 The Kuhn segment or statistical segment length

In the freely jointed chain model the directions of all bonds are completely uncorrelated (see Eq. (1.3.2)), whereas in the freely rotating chain model neighbouring bonds are correlated according to Eq. (1.3.16). In general, the presence of local interactions, which depend on the chemical details of a polymeric chain, results in correlations among the directions of neighbouring bonds. The magnitude of these correlations decreases quickly with increasing separation between chemical bonds along the chain contour. Kuhn was the first to propose that the correlation between bonds i and $i+k$ in any real unperturbed chain with some finite degree of flexibility must vanish as k increases.[8] For chains with large molecular lengths N, the correlations will always vanish for some finite value of k which is much less than N, $k << N$. This ensures a scaling $\langle r^2 \rangle \propto N^1$ for any real, flexible chain. More precisely, for a real chain of N chemical bonds under unperturbed conditions,

$$\langle r^2 \rangle = C_\infty N l^2 \propto N^1 \tag{1.3.22}$$

for $N >> 1$, where N is the number of (chemical) skeletal bonds and l is the length of each such (chemical) bond. The proportionality constant C_∞ is the characteristic ratio. It is a coefficient that depends on the nature of the local interactions, i.e. on the molecular details of the chain (chemical constitution, tacticity, etc.), and its value will be in general different from those derived by the freely jointed (for which $C_\infty = 1$) and freely rotating models (for which $C_\infty = \dfrac{1+\cos\theta}{1-\cos\theta}$).

For real chains under unperturbed conditions, the rotational isomeric state model provides an excellent avenue for obtaining C_∞ from chemical constitution (compare Section 1.2.5, which discusses how $\langle r^2 \rangle$ can be obtained analytically from the RIS model). Sole input to this calculation is the bonded geometry of the chain and the statistical weight matrices, obtained through conformational analysis of chemically realistic models of oligomers of the chain.[1,21] Monte Carlo sampling of single unperturbed chains in continuous space offers another way for calculating C_∞ from chemical constitution.

The complete absence of directional correlations in a real unperturbed chain after some distance along its contour allows one to represent it by an *equivalent chain* (an equivalent freely jointed chain model) by grouping the real (chemical) skeletal segments into equivalent statistical segments, which are totally uncorrelated with each other. The equivalent chain will comprise N_K such hypothetical bonds, each of length b_K, connected by free joints as in the freely jointed chain model. The mapping (see Figure 1.16)

$$N \text{ real bonds of length } l \Rightarrow N_K \text{ statistical segments (Kuhn segments) of length } b_K$$

is made a one-to-one correspondence by requiring that that the mean-square chain end-to-end distance $\langle r^2 \rangle$ and the maximum contour length of the fully extended real chain, L, equal those of the equivalent chain:

$$\begin{aligned} \langle r^2 \rangle &= N_K b_K^2 \\ L &= N_K b_K. \end{aligned} \qquad (1.3.23)$$

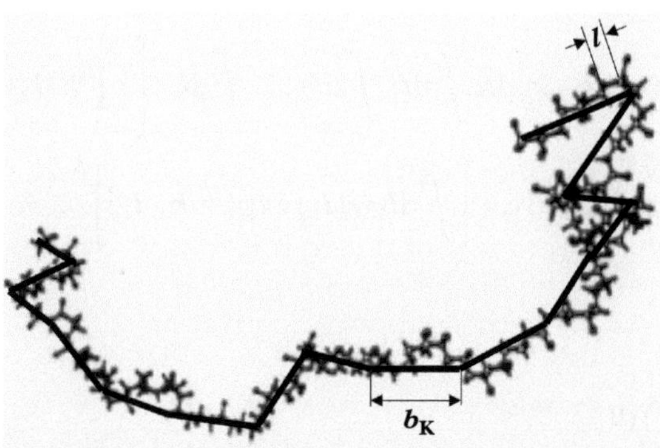

Figure 1.16 Mapping of a real chain to an *equivalent* freely jointed chain (bold zig-zag line) by grouping the real (chemical) segments into equivalent statistical segments. The chemical bond length and the Kuhn length are denoted by $\boldsymbol{l}$ and b_K, respectively.

Solving for N_K and b_K, then, gives

$$N_K = \frac{L^2}{\langle r^2 \rangle}$$
$$b_K = \frac{\langle r^2 \rangle}{L}. \qquad (1.3.24)$$

1.3.4 Distribution of the end-to-end vector for the freely jointed chain model

Here we are interested not only in estimating the mean $\langle \boldsymbol{r} \rangle$ and $\langle r^2 \rangle$ values, but also the entire probability density function followed by the chain end-to-end vector $\boldsymbol{r}$. Given the conformational probability density function $\Psi(\{\boldsymbol{l}_n\})$, the distribution of $\boldsymbol{r}$ can be found as

$$\Phi(\boldsymbol{r}, N) = \int d\boldsymbol{l}_1 \int d\boldsymbol{l}_2 \ldots \int d\boldsymbol{l}_N \delta\left(\boldsymbol{r} - \sum_{n=1}^{N} \boldsymbol{l}_n\right). \qquad (1.3.25)$$

Using the following identity (from the theory of Fourier transforms)[3]

$$\delta(\boldsymbol{r}) = \frac{1}{(2\pi)^3} \int \exp(i\boldsymbol{k} \cdot \boldsymbol{r})\, d\boldsymbol{k}, \qquad (1.3.26)$$

with i being the imaginary unit, the equation for $\Phi(\boldsymbol{r}, N)$ is also written as

$$\begin{aligned}
\Phi(\boldsymbol{r}, N) &= \frac{1}{(2\pi)^3} \int d\boldsymbol{k} \int d\boldsymbol{l}_1 \int d\boldsymbol{l}_2 \ldots \int d\boldsymbol{l}_N \Psi(\{\boldsymbol{l}_n\}) \exp\left[i\boldsymbol{k} \cdot \left(\boldsymbol{r} - \sum_{n=1}^{N} \boldsymbol{l}_n\right)\right] \\
&= \frac{1}{(2\pi)^3} \int \exp(i\boldsymbol{k} \cdot \boldsymbol{r})\, d\boldsymbol{k} \int d\boldsymbol{l}_1 \int d\boldsymbol{l}_2 \ldots \int d\boldsymbol{l}_N \left[\prod_{n=1}^{N} \psi(\boldsymbol{l}_n)\right] \left[\prod_{n=1}^{N} \exp(-i\boldsymbol{k} \cdot \boldsymbol{l}_n)\right] \\
&= \frac{1}{(2\pi)^3} \int \exp(i\boldsymbol{k} \cdot \boldsymbol{r})\, d\boldsymbol{k} \left[\int d\boldsymbol{l}_1 \psi(\boldsymbol{l}_1) \exp(-i\boldsymbol{k} \cdot \boldsymbol{l}_1)\right]^N. \qquad (1.3.27)
\end{aligned}$$

But

$$\begin{aligned}
&\int \psi(\boldsymbol{l}_1) \exp(-i\boldsymbol{k} \cdot \boldsymbol{l}_1)\, d\boldsymbol{l}_1 \\
&= \frac{1}{4\pi l_1^2} \int_0^\infty l_1^2 dl_1 \int_0^{2\pi} d\phi \int_0^{\pi} d\theta \sin\theta \exp(-ikl_1 \cos\theta)\, \delta(l_1 - l) = \frac{\sin kl}{kl}, \qquad (1.3.28)
\end{aligned}$$

where $l_1 = |\boldsymbol{l}_1|$ and $k = |\boldsymbol{k}|$. Thus,

$$\Phi(\boldsymbol{r}, N) = \frac{1}{(2\pi)^3} \int \exp(i\boldsymbol{k} \cdot \boldsymbol{r}) \left(\frac{\sin kl}{kl} \right)^N \mathrm{d}\boldsymbol{k}. \tag{1.3.29}$$

This equation was derived for the first time by Lord Rayleigh[22] in the context of his classical work on random flights. Of particular interest is the asymptotic form of $\Phi(\boldsymbol{r}, N)$ in the limit of very long chains. If $N >> 1$, the quantity $\left(\frac{\sin kl}{kl} \right)^N$ becomes very small, unless kl is very close to 0. Under these conditions, we can write that

$$\left(\frac{\sin kl}{kl} \right)^N \cong \left(1 - \frac{k^2 l^2}{6} \right)^N \cong \exp\left(-\frac{Nk^2 l^2}{6} \right) \tag{1.3.30}$$

and by replacing back in Eq. (1.3.31), we obtain that

$$\Phi(\boldsymbol{r}, N) = \frac{1}{(2\pi)^3} \int \exp(i\boldsymbol{k} \cdot \boldsymbol{r}) \exp\left(-\frac{Nk^2 l^2}{6} \right) \mathrm{d}\boldsymbol{k}, \tag{1.3.31}$$

that is, a Gauss integral. If $\boldsymbol{r} = (x, y, z)$ are the coordinates of the chain end-to-end vector, this Gauss integral can be decomposed into three factors

$$\begin{aligned}\Phi(\boldsymbol{r}, N) =& \frac{1}{(2\pi)^3} \int_{-\infty}^{+\infty} \exp\left(ik_x x - \frac{Nk_x^2 l^2}{6} \right) \mathrm{d}k_x \int_{-\infty}^{+\infty} \exp\left(ik_y y - \frac{Nk_y^2 l^2}{6} \right) \mathrm{d}k_y \\ & \int_{-\infty}^{+\infty} \exp\left(ik_z z - \frac{Nk_z^2 l^2}{6} \right) \mathrm{d}k_z = \left(\frac{3}{2\pi N l^2} \right)^{\frac{3}{2}} \exp\left[-\frac{3}{2Nl^2} \left(x^2 + y^2 + z^2 \right) \right],\end{aligned} \tag{1.3.32}$$

or, equivalently,

$$\begin{aligned}\Phi(\boldsymbol{r}, N) =& \left(\frac{3}{2\pi N l^2} \right)^{\frac{3}{2}} \exp\left[-\frac{3\left(x^2 + y^2 + z^2 \right)}{2Nl^2} \right] \\ =& \left(\frac{3}{2\pi N l^2} \right)^{\frac{3}{2}} \exp\left[-\frac{3r^2}{2Nl^2} \right].\end{aligned} \tag{1.3.33}$$

According to Eq. (1.3.33), $\Phi(\boldsymbol{r}, N)$ obeys a 3D Gaussian distribution (see Figure 1.17) in which the local structure of the chain appears only through the effective bond length l. From Eq. (1.3.33) for $\Phi(\boldsymbol{r}, N)$, we further find then that $\langle \boldsymbol{r} \rangle = 0$ and $\langle r^2 \rangle = Nl^2$. We can also calculate the probability density function for the length of the chain end-to-end vector through

$$W(r,N)\,dr = \iiint\limits_{r<(x^2+y^2+z^2)^{1/2}<r+dr} \Phi(x,\quad y,z)\mathrm{d}x\mathrm{d}y\mathrm{d}z, \qquad (1.3.34)$$

which gives that

$$W(r,N) = \left(\frac{3}{2\pi Nl^2}\right)^{\frac{3}{2}} 4\pi r^2 \exp\left[-\frac{3r^2}{2Nl^2}\right]. \qquad (1.3.35)$$

This is very similar in form to the Maxwell–Boltzmann distribution of molecular speeds.[6] Note that the exact $\Phi(\boldsymbol{r},N)$ and $W(r,N)$ should attain nonzero values only for $r < L = Nl$. This is not satisfied by the approximate Eqs. (1.3.33) and (1.3.35).

The Gaussian model does not describe correctly the local conformation of a real polymer chain, but it does describe chain properties on the larger (b_K and up) length scales, provided the chain is not strongly extended. In the case of extended configurations we can use a more accurate equation, the Kuhn–Grün–Flory distribution,[2,26–27]

$$\Phi(\boldsymbol{r},N) = \text{const.}\exp\left[-\frac{1}{l}\int_0^r \mathscr{L}^{-1}\left(\frac{r'}{Nl}\right)dr'\right] \qquad (1.3.36)$$

Equation (1.3.36) is preferred to the Gaussian distribution of Eq. (1.3.35) only when we are interested in the behaviour of the chain for values $\frac{r}{L} = \frac{r}{Nl} \cong 1$. The function $\mathscr{L}^{-1}$ appearing in Eq. (1.3.36) is the inverse Langevin function,[2,26–27] defined through

$$\mathscr{L}^{-1}(\nu) = u, \text{ where } \nu = \mathscr{L}(u) \equiv \coth(u) - \frac{1}{u} \qquad (1.3.37)$$

with $\mathscr{L}(u)$ being the (direct) Langevin function. Practical approximations to the inverse Langevin function have been discussed by Kröger.[28]

In general,[1–4] it can be shown that any unperturbed chain with a distribution of conformations of the form (1.1.14) is characterized by an asymptotic distribution of its chain end-to-end vector which is Gaussian,

$$\Phi(\boldsymbol{r},N) = \left(\frac{3}{2\pi\langle r^2\rangle}\right)^{\frac{3}{2}} \exp\left(-\frac{3r^2}{2\langle r^2\rangle}\right), \quad N >> 1, \qquad (1.3.38)$$

where $\langle r^2\rangle$ is the mean-square end-to-end distance. Such a proposition results from the central limit theorem in statistics and probability.[29]

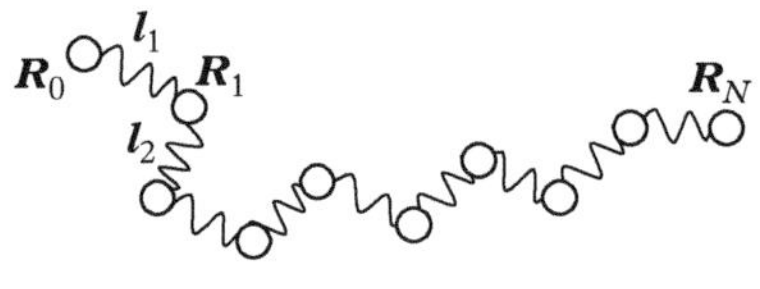

Figure 1.17 The Gaussian chain, also known as the bead–spring chain with Hookean springs.

1.4 Bead-and-spring models

1.4.1 Gaussian model

Inspired by the asymptotic form of the distribution for the chain end-to-end vector derived in Section 1.3, one can express the conformational density distribution function for an unperturbed chain as a product of individual density distribution functions for the N bonds,

$$\Psi\left(\boldsymbol{l}_1, \boldsymbol{l}_2, \ldots, \boldsymbol{l}_N\right) = \psi\left(\boldsymbol{l}_1\right)\psi\left(\boldsymbol{l}_2\right)\ldots\psi\left(\boldsymbol{l}_N\right), \tag{1.4.1}$$

each one of which follows a Gaussian distribution,

$$\psi\left(\boldsymbol{l}_n\right) = \left[\frac{3}{2\pi l^2}\right]^{\frac{3}{2}} \exp\left(-\frac{3l_n^2}{2l^2}\right), \tag{1.4.2}$$

with $\langle l_n \rangle = 0$ and $\langle l_n^2 \rangle = l^2$. Consequently,

$$\Psi\left(\{\boldsymbol{l}_n\}\right) = \left[\frac{3}{2\pi l^2}\right]^{\frac{3N}{2}} \exp\left[-\sum_{n=1}^{N} \frac{3(\boldsymbol{R}_n - \boldsymbol{R}_{n-1})^2}{2l^2}\right]. \tag{1.4.3}$$

The mechanical analogue to which such a probability density function corresponds is that of a chain of beads connected to each other with harmonic springs, the so-called *bead-and-spring chain model* (see Figure 1.17). According to Eq. (1.4.3), the value of each spring constant is equal to $\frac{3k_B T}{l^2}$, so that the total *potential energy* function for the chain is given by

$$U(\{\boldsymbol{l}_n\}) = \frac{3}{2l^2} k_B T \sum_{n=1}^{N} (\boldsymbol{R}_n - \boldsymbol{R}_{n-1})^2. \tag{1.4.4}$$

Of course, as already pointed out, at this level of description $U(\{\boldsymbol{l}_n\})$ is in reality a free energy of entirely entropic origin, as evidenced by the fact that it is proportional to T. It is important here to realize that the vector $\boldsymbol{R}_n - \boldsymbol{R}_m$ between any monomers n and m along the chain (see Figure 1.18) also follows a Gaussian distribution:

$$\Phi\left(\boldsymbol{R}_n - \boldsymbol{R}_m, n - m\right) = \left[\frac{3}{2\pi l^2 |n-m|}\right]^{\frac{3}{2}} \exp\left[-\sum_{n=1}^{N} \frac{3(\boldsymbol{R}_n - \boldsymbol{R}_m)^2}{2|n-m| l^2}\right]. \tag{1.4.5}$$

Consequently,

$$\left\langle (\boldsymbol{R}_n - \boldsymbol{R}_m)^2 \right\rangle = |n-m|\, l^2. \tag{1.4.6}$$

Often, the index n in the distribution $\Psi\left(\{\boldsymbol{l}_n\}\right)$ of a Gaussian chain is taken as a continuous variable (see Figure 1.19). In this consideration, the contour of the chain becomes a continuous curve $\boldsymbol{R}(n)$

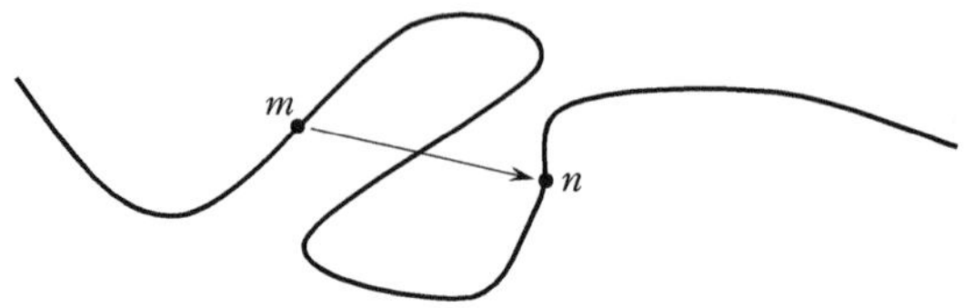

Figure 1.18 The vector $\boldsymbol{R}_n - \boldsymbol{R}_m$ between any monomers n and m along a Gaussian chain follows also a Gaussian distribution.

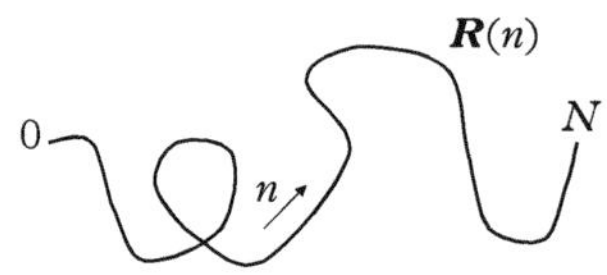

Figure 1.19 A chain in the continuous model representation.

(*Gaussian thread*) and the quantity $\boldsymbol{R}_n - \boldsymbol{R}_{n-1}$ is replaced by the partial derivative $\frac{\partial \boldsymbol{R}}{\partial n}$.[3] In this case, the probability density distribution function becomes a Wiener distribution:[30]

$$\Psi\left[\boldsymbol{R}(n)\right] = \text{const. } \exp\left[-\frac{3}{2l^2}\int_0^N \left(\frac{\partial \boldsymbol{R}}{\partial n}\right)^2 \mathrm{d}n\right]. \qquad (1.4.7)$$

Such a consideration allows for the use of methods of functional analysis for the description of polymer physics (path integral formulation, field theoretic approaches; see also Section 1.5.2).[31]

1.4.2 FENE model

According to Eq. (1.3.33) for the (equilibrium) distribution of the end-to-end vector of a freely jointed chain, if one end of the molecule is located at the origin, the most probable location of the other end is also at the origin. (Of course, this does not mean that the most likely value of the end-to-end distance $r = |\boldsymbol{r}|$ is 0, since we have shown in Sections 1.3 and 1.4.1 that the average value of the square of the end-to-end distance $\langle r^2 \rangle$ for both the freely jointed chain and the Gaussian model is equal to Nl^2.) Expression (1.3.33) for $\Phi(\boldsymbol{r}, N)$ is of importance, however, for another reason: it can be used to derive the force or tension $\boldsymbol{F}$ needed to keep the two ends of the chain a vector $\boldsymbol{r}$ apart. There are two ways to proceed: the first[1,32] makes use of the configurational partition function for a chain subject to a constant force $\boldsymbol{F}$ on its ends. The second[4] makes use of the relation between the Helmholtz free energy of a polymer chain and its distribution function for the end-to-end vector (partition function at given $\boldsymbol{r}$):

$$A(\boldsymbol{r}) = A_0 - k_B T \ln \Phi(\boldsymbol{r}, N), \qquad (1.4.8)$$

where A_0 does not depend on $\boldsymbol{r}$. For a Gaussian chain, $\Phi(\boldsymbol{r}, N)$ is given by Eq. (1.3.33). When the end-to-end vector of the chain is changed by a small amount $\mathrm{d}\boldsymbol{r}$ at constant temperature T, the change in its Helmholtz free energy is

$$\mathrm{d}A = -k_\mathrm{B}T\mathrm{d}\ln\Phi(\boldsymbol{r}, N) = \frac{3k_\mathrm{B}T}{Nl^2}\left(\boldsymbol{r}\cdot\mathrm{d}\boldsymbol{r}\right). \tag{1.4.9}$$

But for a reversible isothermal process, the change in the Helmholtz free energy is related to the applied tension $\boldsymbol{F}$ in the chain by

$$\mathrm{d}A = \boldsymbol{F}\cdot\mathrm{d}\boldsymbol{r}. \tag{1.4.10}$$

By comparing Eqs. (1.4.9) and (1.4.10), we get

$$\boldsymbol{F}(\boldsymbol{r}) = \frac{3k_\mathrm{B}T}{Nl^2}\boldsymbol{r}. \tag{1.4.11}$$

Equation (1.4.11) implies that a freely jointed or a freely rotating chain behaves like a spring with a force constant given by

$$k_0 = \frac{3k_\mathrm{B}T}{Nl^2} \tag{1.4.12}$$

and zero equilibrium length. Here we have used the subscript 0 to emphasize that this is a constant independent of the spring extension. The force law of Eq. (1.4.11) permits infinite extension of the chain, provided sufficient force is applied. We know, however, that a polymer molecule cannot be extended beyond its maximum contour length $L(= Nl)$. We also anticipate that the force should change more rapidly than linearly as the chain becomes more and more extended, i.e. the spring will stiffen up, since the number of available conformations declines sharply as the end-to-end distance increases, approaching full extension. It seems, therefore, more appropriate to describe the flexible macromolecule with a force law of the form

$$\boldsymbol{F}(\boldsymbol{r}) = \frac{k_0}{1-\left(\frac{r}{L}\right)^2}\boldsymbol{r}, \quad r \leq L. \tag{1.4.13}$$

A spring with such a force law will be linear (Hookean) for small extensions but will get stiffer as the spring is extended; furthermore, it can never be extended beyond a separation L. This is the Warner *finitely extensible nonlinear elastic* (FENE) mechanical model for a polymer molecule.[4]

A second model that has been proposed in the literature on molecular arguments to account for the non-Hookean character of a chain modelled as a dumbbell is the *inverse Langevin force*,[1,4,26–27,33–35]

$$F(r) = \frac{k_\mathrm{B}T}{l}\mathscr{L}^{-1}\left[\frac{r}{L}\right], \quad r \leq L, \tag{1.4.14}$$

where $\mathscr{L}^{-1}$ is the inverse of the Langevin function $\mathscr{L}$, defined in Eq. (1.3.37). The various spring force laws are compared in Figure 1.20.[36]

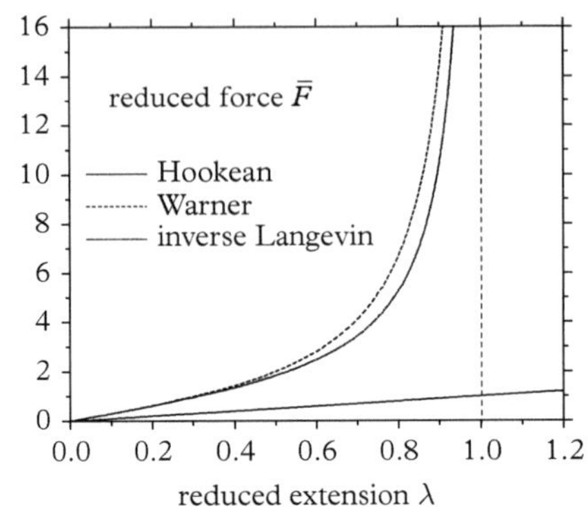

Figure 1.20 Comparison of the three spring force laws presented by Eqs. (1.4.11), (1.4.13), and (1.4.14). The reduced force is defined as $\bar{F} = FL/(3Nk_BT)$ and the reduced extension is $\lambda = r/L$.

1.4.3 A note on the Rouse model

The bead–spring chain with Hookean springs is also called a *Rouse chain* (see also Section 7.8). The Rouse chain model has been very popular because, although it has a large number of internal degrees of freedom, it affords analytical solutions to many problems concerning the equilibrium and nonequilibrium properties of polymer melts. It was first proposed by Rouse,[37] who represented a polymer chain as a string of N beads connected together with Hookean springs of equilibrium length 0 (compare Eqs. (1.4.3) and (1.4.4)). The beads move as tethered Brownian particles, experiencing forces from the springs, as well as Langevin and frictional forces from their surroundings (see Figure 1.21). The dynamical properties of the system (self-diffusivity, viscosity, stress relaxation modulus, and associated spectrum of relaxation times) are derived from the motion of the beads. In the Rouse model, the excluded volume interaction is disregarded. Also disregarded are hydrodynamic interactions, namely that the motion of a bead relative to its surroundings may induce a flow field that ultimately exerts a force on another bead. The central assumption of the Rouse model is therefore that the chain dynamics is solely governed by localized interactions along the chain. The Rouse model is particularly successful in capturing the long length- and time-scale dynamics of short-chain, unentangled polymer melts.

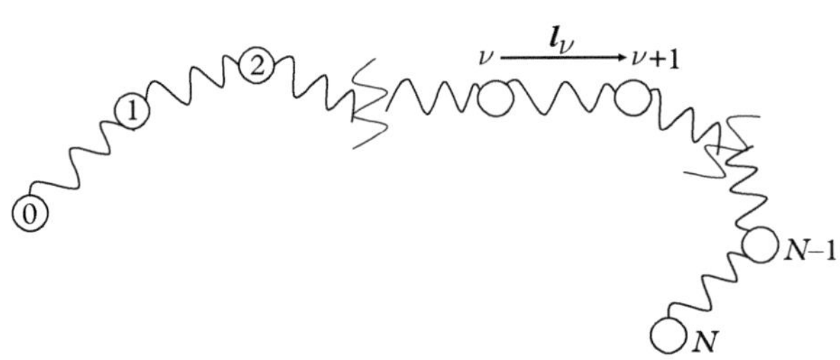

Figure 1.21 The bead–spring or Rouse chain.

There exist two versions of the Rouse model, the discrete (following Rouse's original formulation) and the continuous.[3,4,37] Their results agree with each other for properties on the long time scale, but differ for short times. Both models describe the dynamics of the chain in the sea of the surrounding solvent molecules and/or the segments of other chains as a Brownian motion of coupled harmonic oscillators. A standard way to solve such a system is to find the normal coordinates, each capable of independent motion. By analysing the time autocorrelation functions of these normal modes, the Rouse model leads to a set of very significant results for the dynamics of the chain. For example, it predicts that the *self-diffusion coefficient* of the chain centre of mass, D, defined through

$$D = \lim_{t\to\infty} \frac{\left\langle \left(\boldsymbol{R}_{\mathrm{G}}(t) - \boldsymbol{R}_{\mathrm{G}}(0)^2 \right) \right\rangle}{6t} \tag{1.4.15}$$

where $\boldsymbol{R}_{\mathrm{G}}$ denotes the center-of-mass vector, is given by

$$D = \frac{k_{\mathrm{B}} T}{N\zeta}, \tag{1.4.16}$$

where ζ is the *friction constant* for each bead and N the number of beads. It also predicts the decay of the chain end-to-end vector *time autocorrelation function* (*TAF*) as

$$\langle \boldsymbol{r}(t) \cdot \boldsymbol{r}(0) \rangle = Nl^2 \sum_{p=1,3,5,\ldots} \frac{8}{p^2 \pi^2} \exp\left(-\frac{t}{\tau_p} \right), \tag{1.4.17}$$

where

$$\tau_p = \frac{\tau_1}{p^2}, \quad p = 1, 2, 3, \ldots \tag{1.4.18}$$

defines the *spectrum of relaxation times* describing the dynamics of each mode, with

$$\tau_1 = \tau_{\mathrm{R}} = \frac{\zeta N^2 l^2}{3\pi^2 k_{\mathrm{B}} T} = \frac{1}{3\pi^2} \frac{\langle r^2 \rangle}{D} \tag{1.4.19}$$

being the longest relaxation time, also known as the *rotational* or Rouse relaxation time τ_{R} of the polymer. The normal mode of order p describes collective motions involving a fraction of roughly $1/p$ of the chain segments; higher modes thus describe more localized motion. According to Eqs. (1.4.16)–(1.4.19), the Rouse model indicates that D and τ_{R} depend on the molar mass M as

$$D \propto M^{-1} \quad \text{and} \quad \tau_{\mathrm{R}} \propto M^2. \tag{1.4.20}$$

We refer the reader to Refs. 3 and 4 and to Chapter 7 of this book for more details on the Rouse model.

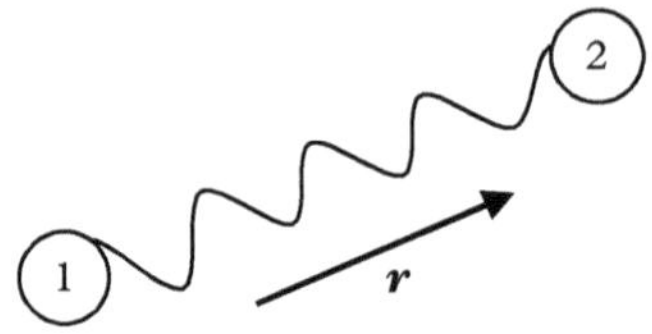

Figure 1.22 The dumbbell model.

1.4.4 Dumbbell models

As mentioned in Section 1.3.4, the Gaussian distribution for the end-to-end vector of a freely jointed or a freely rotating chain suggests that such a chain behaves like a spring with constant k:

$$\Phi(\boldsymbol{r}) = \left(\frac{k}{2\pi k_B T}\right)^{\frac{3}{2}} \exp\left[-\frac{k}{2k_B T}r^2\right]. \tag{1.4.21}$$

Guided by this, an idealization often made is to forget the internal configuration of the chain and, instead, consider it as an elastic dumbbell consisting of two beads labelled as '1' and '2', which are connected by a nonbendable spring (see Figure 1.22).[4] The dumbbell is a very simplified model of a polymer chain consisting of N beads connected together with rigid bonds or springs for the special case $N = 2$. If the instantaneous locations of the two beads are $\boldsymbol{R}_1$ and $\boldsymbol{R}_2$, then the overall orientation and internal configuration of the dumbbell is described by the connector vector $\boldsymbol{r} = \boldsymbol{R}_2 - \boldsymbol{R}_1$. If the two beads are connected with a Hookean spring, we speak of a *Hookean* or elastic *dumbbell.* If the connector vector obeys Eq. (1.4.11), then we speak of a *FENE dumbbell.* Apart from these dumbbell models, other known models of elastic dumbbells are the encapsulated dumbbell, the Fraenkel dumbbell, the modified Fraenkel dumbbell, and the dumbbell with internal viscosity. Their spring force laws are discussed in detail in Ref. 4. Also discussed in Ref. 4 are a number of other mechanical models for macromolecules, such as the rigid dumbbell and the multibead–rod chain models.

1.5 Continuous-curve models: wormlike chain and Edwards diffusion equation for a Gaussian string

1.5.1 Porod–Kratky chain

The limit of the freely jointed chain model when $l \to 0$, $N \to \infty$, and $\theta \to 0$, such that the persistence length $a = \dfrac{l}{1 - \cos\theta}$ and the contour length $L = Nl$ remain constant, is a continuously curved chain whose radius of curvature is everywhere the same, but the direction towards which the chain curves is random (see Figure 1.19). This is called the continuously curved or *wormlike* chain; it is also known as the *Porod–Kratky* chain.[38–39] Such a chain is described by two parameters: the stiffness a and the length at maximal extension, L. For this chain, it can be shown[40–42] that the mean-square end-to-end distance $\langle r^2 \rangle$ and the mean-square radius of gyration $\langle R_g^2 \rangle$ are given by the equations[1]

$$\langle r^2\rangle = 2aL\left\{1-\frac{a}{L}\left[1-\exp\left(-\frac{L}{a}\right)\right]\right\} \tag{1.5.1}$$

and

$$\langle R_g^2\rangle = \frac{aL}{3}\left\{1-\frac{3a}{L}\left[1-\frac{2a}{L}+2\left(\frac{a}{L}\right)^2\left(1-\exp\left(-\frac{L}{a}\right)\right)\right]\right\}. \tag{1.5.2}$$

In the limit $\frac{a}{L}\to 0$ we recover the following results for a long, flexible polymeric chain: $\frac{\langle r^2\rangle}{L} = 2a\left(1-\frac{a}{L}\right)\cong 2a$ and $\frac{\langle R_g^2\rangle}{L} = \frac{a}{3}\left(1-\frac{3a}{L}\right)\cong\frac{a}{3}$. On the other hand, in the limit $\frac{L}{a}\to 0$, we recover the rigid rod results: $\langle r^2\rangle\cong L^2$ and $\langle R_g^2\rangle\cong\frac{L^2}{12}$. It is clear that the *continuously curved chain model* can describe very different types of macromolecules.

1.5.2 Edwards diffusion equation

In the presence of an (external) field $U_e(\boldsymbol{r})$ acting on the segments, the equilibrium distribution of the Gaussian thread is modified[43] by the Boltzmann factor

$$\exp\left[-\frac{1}{k_B T}\int_0^N U_e\left(\boldsymbol{R}(n)\right)\mathrm{d}n\right] \tag{1.5.3}$$

so that the conformational distribution function becomes (compare Eq. (1.4.7))

$$\Psi\left[\boldsymbol{R}(n)\right] = \text{const.}\ \exp\left[-\frac{3}{2l^2}\int_0^N\left(\frac{\partial\boldsymbol{R}}{\partial n}\right)^2\mathrm{d}n - \frac{1}{k_B T}\int_0^N U_e\left(\boldsymbol{R}(n)\right)\mathrm{d}n\right]. \tag{1.5.4}$$

The statistical properties of such a system are more conveniently discussed[44–47] by considering the *Green function* or chain propagator, $G(\boldsymbol{r},\boldsymbol{r}_0;n)$, representing the statistical weight (unnormalized probability density) for a chain that has started at $\boldsymbol{r}_0$ to have segment n along its contour at position $\boldsymbol{r}$ (see Figure 1.23). It is convenient to define $G(\boldsymbol{r},\boldsymbol{r}_0;n)$ relative to the corresponding propagator of a field-free chain integrated over all possible positions $\boldsymbol{r}$ of its nth segment:[3]

$$G(\boldsymbol{r},\boldsymbol{r}_0;n) = \frac{\displaystyle\int_{R(0)=r_0}^{R(n)=r}\mathcal{D}[\boldsymbol{R}(n_1)]\exp\left[-\frac{3}{2l^2}\int_0^n\left(\frac{\partial\boldsymbol{R}}{\partial n_1}\right)^2\mathrm{d}n_1 - \frac{1}{k_B T}\int_0^n\mathrm{d}n_1\,U_e(\boldsymbol{R}(n_1))\right]}{\displaystyle\int\mathrm{d}^3r\int_{\boldsymbol{R}(0)=\boldsymbol{r}_0}^{\boldsymbol{R}(n)=\boldsymbol{r}}\mathcal{D}[\boldsymbol{R}(n_1)]\exp\left[-\frac{3}{2l^2}\int_0^n\left(\frac{\partial\boldsymbol{R}}{\partial n_1}\right)^2\mathrm{d}n_1\right]}. \tag{1.5.5}$$

In the definition of the Green function, Eq. (1.5.5), the numerator and the denominator contain functional integrals or path integrals over all possible paths $\boldsymbol{R}(n_1)$, $0\le n_1\le n$ that can be followed

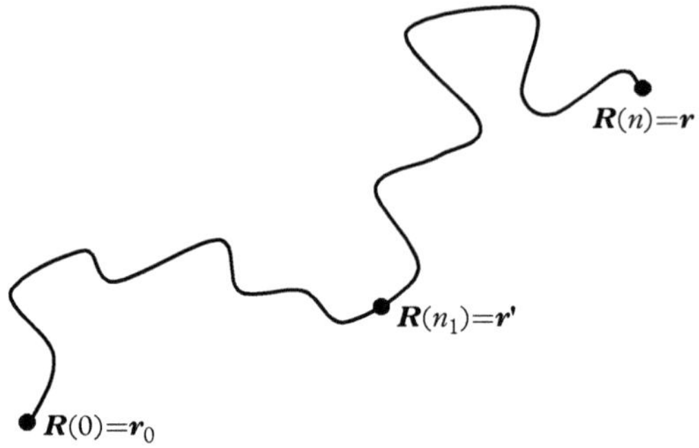

Figure 1.23 The propagator $G(\boldsymbol{r}_0, \boldsymbol{r}; n)$ represents the (unnormalized) probability that a chain which starts from $\boldsymbol{r}_0$ ends at $\boldsymbol{r}$ at contour length $n\,(0 \leq n \leq N)$.

by an n segment-long chain with its start fixed at point $\boldsymbol{R}(0) = \boldsymbol{r}_0$ and its end fixed at point $\boldsymbol{R}(n) = \boldsymbol{r}$ in 3D space. Defined in this way, $G(\boldsymbol{r}, \boldsymbol{r}_0; n)$ has dimensions of inverse volume and satisfies the conditions

$$G(\boldsymbol{r}, \boldsymbol{r}_0; n) = 0 \text{ for } n < 0 \quad \text{and} \quad G(\boldsymbol{r}, \boldsymbol{r}_0; n) = \delta(\boldsymbol{r} - \boldsymbol{r}_0) \quad \text{for } n = 0 \tag{1.5.6}$$

with δ being the Dirac delta function.

It can be shown that $G(\boldsymbol{r}, \boldsymbol{r}_0; n)$ satisfies the partial differential equation[48]

$$\left(\frac{\partial}{\partial n} - \frac{l^2}{6}\nabla_{\boldsymbol{r}}^2 + \frac{1}{k_B T}U_e(\boldsymbol{r})\right) G(\boldsymbol{r}, \boldsymbol{r}_0; n) = \delta(\boldsymbol{r} - \boldsymbol{r}_0)\,\delta(n) \tag{1.5.7}$$

with Nl^2 being the mean-square end-to-end distance of the Gaussian chain. Equation (1.5.7) is known as the Edwards diffusion equation.

Closely related to the propagator is the restricted partition function $q(\boldsymbol{r}, n)$. This is proportional to the probability that segment n of a chain, which may have started anywhere in the considered system, finds itself at position $\boldsymbol{r}$. It is defined as

$$\begin{aligned} q(\boldsymbol{r}, n) &= \int d^3 r_0\; G(\boldsymbol{r}, \boldsymbol{r}_0; n) \\ &= \frac{\displaystyle\int d^3 r_0 \int_{\boldsymbol{R}(0)=\boldsymbol{r}_0}^{\boldsymbol{R}(n)=\boldsymbol{r}} \mathcal{D}[\boldsymbol{R}(n_1)] \exp\left[-\frac{3}{2l^2}\int_0^n \left(\frac{\partial R}{\partial n_1}\right)^2 dn_1 - \frac{1}{k_B T}\int_0^n dn_1\, U_e(\boldsymbol{R}(n_1))\right]}{\displaystyle\int d^3 r \int_{\boldsymbol{R}(0)=\boldsymbol{r}_0}^{\boldsymbol{R}(n)=\boldsymbol{r}} \mathcal{D}[\boldsymbol{R}(n_1)] \exp\left[-\frac{3}{2l^2}\int_0^n \left(\frac{\partial \boldsymbol{R}}{\partial n_1}\right)^2 dn_1\right]}. \end{aligned} \tag{1.5.8}$$

It is important to note that the denominator, which refers to a field-free chain, is independent of $\boldsymbol{r}_0$. Integrating Eq. (1.5.7) with respect to $\boldsymbol{r}_0$, one obtains a diffusion equation for $q(\boldsymbol{r}, n)$. For $n > 0$ this reads as

$$\left[\frac{\partial}{\partial n} - \frac{l^2}{6}\nabla_{\boldsymbol{r}}^2 + \frac{1}{k_{\mathrm{B}}T}U_{\mathrm{e}}(\boldsymbol{r})\right] q(\boldsymbol{r}, n) = 0. \tag{1.5.9}$$

By spatial averaging of $q(\boldsymbol{r}, N)$ over a region of volume V one obtains a partition function $Q[U_{\mathrm{e}}]$ for a chain of N segments in that volume subject to the field $U_{\mathrm{e}}(\boldsymbol{r})$, normalized by the corresponding partition function of the field-free chain:

$$\begin{aligned} Q[U_{\mathrm{e}}] &= \frac{1}{V}\int \mathrm{d}^3 r\, q(\mathbf{r}, N) \\ &= \frac{\displaystyle\int \mathrm{d}^3 r_0 \int \mathrm{d}^3 r \int_{\boldsymbol{R}(0)=\boldsymbol{r}_0}^{\boldsymbol{R}(N)=\boldsymbol{r}} \mathcal{D}\left[\boldsymbol{R}(n)\right] \exp\left[-\frac{3}{2l^2}\int_0^N \left(\frac{\partial \boldsymbol{R}}{\partial n}\right)^2 \mathrm{d}n - \frac{1}{k_{\mathrm{B}}T}\int_0^N \mathrm{d}n\, U_{\mathrm{e}}(\boldsymbol{R}(n))\right]}{\displaystyle\int \mathrm{d}^3 r_0 \int \mathrm{d}^3 r \int_{\boldsymbol{R}(0)=\boldsymbol{r}_0}^{\boldsymbol{R}(N)=\boldsymbol{r}} \mathcal{D}\left[\boldsymbol{R}(n)\right] \exp\left[-\frac{3}{2l^2}\int_0^N \left(\frac{\partial \boldsymbol{R}}{\partial n}\right)^2 dn\right]}. \end{aligned} \tag{1.5.10}$$

Both $q(\boldsymbol{r}, n)$ and $Q\left[U_{\mathrm{e}}\right]$ are dimensionless quantities.

The field theoretic formulation of Eqs. (1.5.5) to (1.5.10) is very useful in the study of inhomogeneous polymer systems. For certain coarse-grained models of polymers and copolymers it is possible to recast the problem of many interacting chains into an equivalent problem of individual chains interacting with a fluctuating field. If the latter field is replaced by its average value at each point (saddle point approximation), one obtains an approximate formulation that must be solved self-consistently. The field dictates the spatial distribution and conformation of chains, as is obvious from the Edwards diffusion equation, but is in turn shaped by the same spatial distribution and conformation. This is the basis of the self-consistent field method,[49–54] which is valuable in the study of microphase-separated copolymer systems and polymers at interfaces.

1.6 Soft ellipsoid and soft sphere models for entire chains

In addressing the phase diagrams and long-time (e.g. phase separation) dynamics of multicomponent polymer systems, it is desirable to develop very coarse-grained models, where entire polymer chains are represented as single particles. What are the effective potentials required at this level of description? They will necessarily have to be soft, as entire polymer chains can interpenetrate.

The instantaneous shape of a polymer chain can be characterized through its radius of gyration tensor $\boldsymbol{S}$, with elements (compare with notation of Section 1.1)

$$S_{\alpha\beta} = \frac{1}{N+1}\sum_{k=0}^{N}(R_{k,\alpha} - R_{\mathrm{cm},\alpha})(R_{k,\beta} - R_{\mathrm{cm},\beta}), \tag{1.6.1}$$

where index k runs over all segments of the chain and $\alpha, \beta \in \{1, 2, 3\}$ designate the three coordinate directions. The mean eigenvalues of the radius of gyration tensor, R_1^2, R_2^2, R_3^2, define the lengths

R_1, R_2, R_3 of its principal axes and the corresponding eigenvectors characterize the orientation of the chain in space. Clearly, $R_g^2 = Tr(\boldsymbol{S}) = R_1^2 + R_2^2 + R_3^2$. The average chain shape can be characterized through the average values $\langle R_1^2 \rangle, \langle R_2^2 \rangle, \langle R_3^2 \rangle$ over an ensemble of chains. Furthermore, if one orients all chains in the ensemble such that their longest, intermediate, and shortest principal axes coincide with prescribed directions and superposes them such that their centres of mass coincide, one can define the segment density distribution $\rho(\boldsymbol{r})$ of the chains in their principal axis system.

As early as 1934, Kuhn realized that the instantaneous shape of randomly coiled linear chains is not spherical. Generating random walks on a simple cubic lattice, Šolc and Stockmayer[55] found $\langle R_1^2 \rangle : \langle R_2^2 \rangle : \langle R_3^2 \rangle = 11.7 : 2.7 : 1$ in the limit of very long N. Later simulations of unperturbed chains grown using realistic representations with the help of the RIS model (compare Eqs. (1.2.33) and (1.2.38)) led to very similar results for shape in the long chain limit, also allowing a close examination of finite length and tacticity effects for short unperturbed chains.[56] As Šolc and Stockmayer put it,[55] the instantaneous shape of unperturbed chains is that of a 'cake of soap', with very different principal axes. The segment density distribution $\rho(\boldsymbol{r})$ in the principal axis system has also been visualized; an example is given in Figure 1.24.[56] The soap bar-like shape of the overall segment distribution is evident from the low-density contour plots. Remarkably, the high-density core of the distribution is bimodal along the longest principal axis, lending support to dumbbell model representations (see Section 1.4.3).

Murat and Kremer were the first to invoke a soft ellipsoid representation for entire chains to simulate polymers in the melt.[57] Each chain was represented in terms of the position of its centre of mass, the lengths of its principal axes R_1, R_2, and R_3, and its overall orientation in space. The starting point was a coarse-grained melt model consisting of bead-and-spring chains with nonbonded interactions described by a truncated and shifted Lennard-Jones potential with energy parameter ε. At the soft ellipsoid level of description, the free energy of a melt consisting of N_{ch} chains was assumed to be of the form

$$A = \sum_{i=1}^{N_{ch}} A_i^{intra} + \sum_{i=1}^{N_{ch}-1} \sum_{j=i+1}^{N_{ch}} A_{ij}^{inter}. \tag{1.6.2}$$

The intramolecular part A_i^{intra} contributed by each chain i was expressed as a function of its radius of gyration tensor, by inverting the distribution of radius of gyration tensors obtained by sampling single (self-avoiding walk) chains in vacuo,

$$A_i^{intra} = -k_B T \ln [P(\boldsymbol{S}_i)], \tag{1.6.3}$$

where $P(\boldsymbol{S}_i)$ is the 3D probability density for the squared principal axes R_1^2, R_2^2, R_3^2. Intermolecular interactions, assumed to be pairwise additive in Eq. (1.6.2), were described by

$$A_{ij}^{inter} = \hat{\varepsilon}(N) \int \rho_i(\boldsymbol{r}', \boldsymbol{S}_i)\rho_j(\boldsymbol{r}', \boldsymbol{S}_j)\mathrm{d}\boldsymbol{r}', \tag{1.6.4}$$

which involves an overlap integral in 3D space of the segment density distributions of the interacting chains i and j. For the purpose of calculating the overlap integral, $\rho(\boldsymbol{r}', \boldsymbol{S})$ was assumed to be a

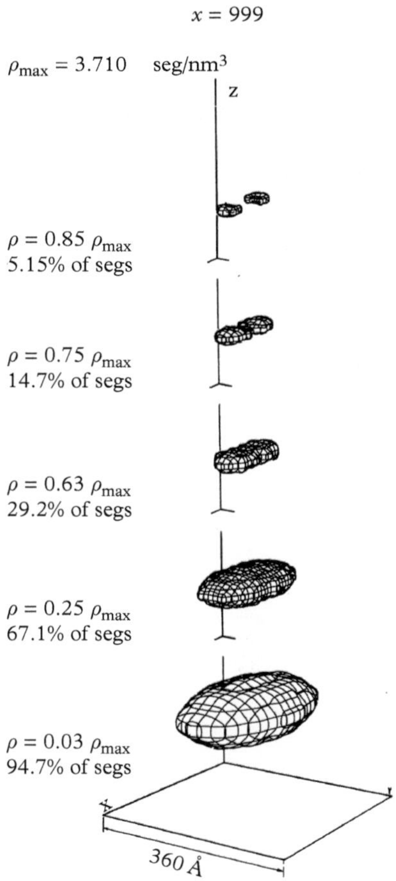

Figure 1.24 Segment density distribution of a syndiotactic polypropylene chain of degree of polymerization $x = 999$ (1999 segments) under unperturbed conditions in its principal axis frame. Isodensity surfaces at various values of ρ, indicated as fractions of its maximum value, are shown. The percentage of segments enclosed by each isodensity surface is also indicated.

Reprinted (adapted) with permission from Theodorou, D.N., and Suter, U.W., *Macromolecules* **18**, 1206 (1985).

function of coordinates r_1, r_2, r_3 in the principal axis system (centred at the centre of mass), scaled by the principal axis lengths:

$$\rho(\boldsymbol{r}', \boldsymbol{S}) = \rho(r_1/R_1, r_2/R_2, r_3/R_3). \tag{1.6.5}$$

This approximation becomes exact for the average density distribution of Gaussian chains. The prefactor $\hat{\varepsilon}(N)$ in Eq. (1.6.4) was adjusted so that the radius of gyration in the melt obtained from simulations in the soft ellipsoid representation matched that of the original bead-and-spring model. Murat and Kremer used the soft ellipsoid model in homopolymer melt simulations to explore correlation hole effects.[57] Furthermore, they extended it to binary blends to study spinodal decomposition resulting from unfavourable interactions between the component chains.

An even simpler model, representing entire chains as single spherical particles, was developed by J. P. Hansen and collaborators.[58] Here one develops an orientationally averaged potential of mean force $U_2^{\text{eff}}(r)$ as a function of distance r between the chain centres of mass (compare Figure 1.25a).

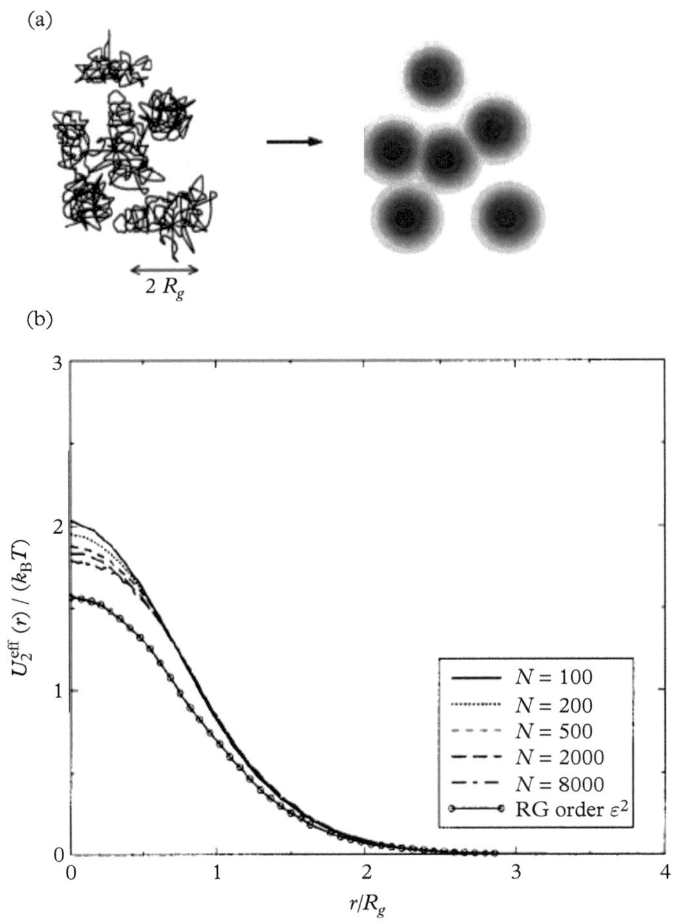

Figure 1.25 (a) Modelling entire polymer coils by effective soft particles. (b) Effective potential between two isolated self-avoiding walk chains, as obtained from the lattice-based Monte Carlo simulations of Bolhuis et al.[58] The distance between the two centres of mass has been scaled by the radius of gyration. Potential curves, all approximately Gaussian in shape, are shown for various values of the chain length $N+1$. The result from an approximate renormalization group (RG) theoretical treatment is also shown.

Reprinted from Bolhuis, P.G., Louis, A.A., Hansen, J.-P., and Meijer, E.J. *J. Chem. Phys.* **114**, 4296 (2001), with the permission of AIP Publishing.

At infinite dilution in a good solvent, using a self-avoiding walk lattice model, they obtained $U_2^{\text{eff}}(r)$ by Monte Carlo sampling of the configurations of two chains and calculating the probability of overlap between their segments at various distances between their centres of mass. The resulting $U_2^{\text{eff}}(r)$ is shown in Figure 1.25b as a function of r/R_g for a number of chain lengths. The potential of mean force is repulsive and its form is Gaussian, to a good approximation. The interaction strength at full overlap between the centres of mass, $U_2^{\text{eff}}(0)$, is finite: there is a finite probability that chains may interpenetrate without any overlap between their segments, such that their centres of mass coincide. The Gaussian form was proposed for U_2^{eff} a long time ago by Flory and Krigbaum,[59] but their mean-field analysis predicted an incorrect scaling of the interaction strength at full centre of mass overlap with the chain length. From Figure 1.25b we see that the range of U_2^{eff} between two chains in a good solvent at infinite dilution is commensurate with R_g. The strength of the interaction is remarkably insensitive to the chain length in the limit of very long chains. As a consequence, the second virial coefficient between chains will scale as $B_2 \propto R_g^3$ for very long chains in a good solvent. Bolhuis et al. conducted lattice-based simulations of polymers in a good solvent also at finite concentrations.[58] From the pair distribution function of chain centres of mass they extracted $U_2^{\text{eff}}(r)$ for various concentrations using integral equation theory. Potentials of mean force extracted in this way have been used in mesoscopic simulations to calculate phase diagrams for colloid–polymer mixtures.[60]

Appendix

The theorem of Lagrange: The theorem of Lagrange[5,61,62] relates the radius of gyration of a system of point masses to the distances between them taken pairwise. In particular, the theorem states that for any collection of $N+1$ particles of equal mass, the square of their radius of gyration is given by Eq. (1.1.11).

Proof:

$$
\begin{aligned}
R_g^2 &= \frac{1}{N+1}\sum_{i=0}^{N}(\boldsymbol{R}_i - \boldsymbol{R}_{\text{cm}})^2 = \frac{1}{N+1}\sum_{i=0}^{N}\left(\boldsymbol{R}_i - \frac{1}{N+1}\sum_{j=0}^{N}\boldsymbol{R}_j\right)^2 = \frac{1}{(N+1)^3}\sum_{i=0}^{N}\left[(N+1)\,\boldsymbol{R}_i - \sum_{j=0}^{N}\boldsymbol{R}_j\right]^2 \\
&= \frac{1}{(N+1)^3}\sum_{i=0}^{N}\left[\sum_{j=0}^{N}(\boldsymbol{R}_i - \boldsymbol{R}_j)\right]^2 = \frac{1}{(N+1)^3}\sum_{i=0}^{N}\left[\sum_{j=0}^{N}(\boldsymbol{R}_i - \boldsymbol{R}_j)\right]\cdot\left[\sum_{k=0}^{N}(\boldsymbol{R}_i - \boldsymbol{R}_k)\right] \\
&= \frac{1}{(N+1)^3}\sum_{i=0}^{N}\left[\sum_{j=0}^{N}\sum_{k=0}^{N}(\boldsymbol{R}_i - \boldsymbol{R}_j)\cdot(\boldsymbol{R}_i - \boldsymbol{R}_k)\right] \\
&= \frac{1}{(N+1)^3}\sum_{i=0}^{N}\left[\sum_{j=0}^{N}\sum_{k=0}^{N}\frac{(\boldsymbol{R}_i - \boldsymbol{R}_j)^2 + (\boldsymbol{R}_i - \boldsymbol{R}_k)^2 - (\boldsymbol{R}_j - \boldsymbol{R}_k)^2}{2}\right] \\
&= \frac{1}{2(N+1)^3}\sum_{i=0}^{N}\left[\sum_{j=0}^{N}\sum_{k=0}^{N}\left(R_{ij}^2 + R_{ik}^2 - R_{jk}^2\right)\right]
\end{aligned}
$$

$$= \frac{1}{2(N+1)^3}\sum_{i=0}^{N}\sum_{j=0}^{N}\sum_{k=0}^{N}R_{ij}^2 + \frac{1}{2(N+1)^3}\sum_{i=0}^{N}\sum_{j=0}^{N}\sum_{k=0}^{N}R_{ik}^2 - \frac{1}{2(N+1)^3}\sum_{i=0}^{N}\sum_{j=0}^{N}\sum_{k=0}^{N}R_{jk}^2$$

$$= \frac{1}{2(N+1)^2}\sum_{i=0}^{N}\sum_{j=0}^{N}R_{ij}^2 + \frac{1}{2(N+1)^2}\sum_{i=0}^{N}\sum_{k=0}^{N}R_{ik}^2 - \frac{1}{2(N+1)^2}\sum_{j=0}^{N}\sum_{k=0}^{N}R_{jk}^2$$

$$= \frac{1}{2(N+1)^2}\sum_{i=0}^{N}\sum_{j=0}^{N}R_{ij}^2 = \frac{1}{(N+1)^2}\sum_{i=0}^{N-1}\sum_{j=i}^{N}R_{ij}^2, \qquad \text{(A.1.1)}$$

The Lagrange theorem is very useful not only in calculating R_g^2 and its average $\langle R_g^2 \rangle$ over all chain configurations, but also in scattering theory.[63]

Problems

Problem 1.1: Convince yourself that Eqs. (1.1.16)–(1.1.18) lead indeed to Eq. (1.1.19).

Problem 1.2: For *n*-hexane we know for the rotational isomeric state model statistical weight matrices of bonds 2 to 4 that $\boldsymbol{U}_2 = \boldsymbol{U}_3 = \boldsymbol{U}_4 = \boldsymbol{U}$, where $\boldsymbol{U}$ is given by Eq. (1.2.12). Calculate (a) the probability for bond 2 to be in state g^+, and (b) the probability for bond 3 to be in state t.

Problem 1.3: For *n*-hexane, calculate the probability for bond 3 to be in state t given that bond 2 is in state g^+.

Problem 1.4: Cast the squared end-to-end distance of an unperturbed chain in the form of Eq. (1.2.47).

Hint:

$$\boldsymbol{G}_i = \begin{bmatrix} 1 & 2\boldsymbol{l}_i^T \cdot \boldsymbol{T}_i & l_i^2 \\ 0 & \boldsymbol{T}_i & \boldsymbol{l}_i \\ 0 & 0 & 1 \end{bmatrix}, \quad 2 < i < N-1, \quad \boldsymbol{G}_1 = \begin{bmatrix} 1 & 2\boldsymbol{l}_1^T \cdot \boldsymbol{T}_1 & l_1^2 \end{bmatrix}, \quad \boldsymbol{G}_N = \begin{bmatrix} l_N^2 \\ \boldsymbol{l}_N \\ 1 \end{bmatrix}$$

Problem 1.5: Derive an analytic expression for the persistence length of the freely rotating chain.

Problem 1.6: Calculate the values of the Kuhn length b_K and the number of Kuhn lengths per chain N_K for (a) a freely jointed chain of N steps; (b) a freely rotating chain of N bonds; and (c) a real unperturbed chain of N bonds with fixed bond lengths and bond angles.

Problem 1.7: Show the Edwards diffusion equation for the propagator $G(\boldsymbol{r}, \boldsymbol{r}_0; n)$.

References

1. Flory, P.J. *Statistical Mechanics of Chain Molecules*. Interscience, New York (1969).
2. Flory, P.J. *Principles of Polymer Chemistry*. Cornell University Press, Ithaca, NY (1953).
3. Doi, M., and Edwards, S.F. *The Theory of Polymer Dynamics*. Oxford Science Publications, Oxford (1986).
4. Bird, B.R., Curtiss, C.F., Armstrong, R.C., and Hassager, O. *Dynamics of Polymeric Liquids* (2nd ed.), Volume 2: *Kinetic Theory*. John Wiley & Sons, New York (1987).

5. Lagrange, J.L. *Oeuvres (Paris)* **5**, *535* (1870).
6. McQuarrie, D.A. *Statistical Mechanics*. Harper & Row, New York (1973).
7. Yamakawa, H. *Modern Theory of Polymer Solutions*. Harper & Row, New York (1971).
8. Kuhn, W. *Kolloid Z.* **68**, 2 (1934); **76**, 258 (1936); **87**, 3 (1939).
9. Flory, P.J. *J. Chem. Phys.* **17**, 303 (1949).
10. de Gennes, P.G., *Scaling Concepts in Polymer Physics*. Cornell University Press, Ithaca, NY (1979).
11. Oono, Y. *Adv. Chem. Phys.* **61**, 301 (1985).
12. Zimm, B.H., Stockmayer, W.H., and Fixman, M. *J. Chem. Phys.* **21**, 1716 (1953).
13. Stockmayer, W.H. *Macromol. Chem.* **35**, 54 (1960).
14. Reichl, L.E. *A Modern Course in Statistical Physics*. University of Texas Press, Austin (1980).
15. Tanaka, G. *Macromolecules* **13**, 1513 (1980).
16. A complete list of references for the calculation of excluded volume effects on chain dimensions is given in Ref. 3, Section 2.5.
17. Domb, C., Gillis, J., and Wilmers, G. *Proc. Phys. Soc.* **85**, 625 (1965).
18. Theodorou, D.N. *Molecular Modelling of Polymeric Glasses*. Ph.D. Thesis, MIT, Department of Chemical Engineering, Boston (1985).
19. Gō, N., and Scheraga, H.A. *Macromolecules* **9**, 535 (1976).
20. Volkenstein, M.V. *Configurational Statistics of Polymeric Chains*, translated from the Russian ed., S.N. Timasheff and M.J. Timasheff. Interscience, New York (1963).
21. Mattice, W.L., and Suter, U.W. *Conformational Theory of Large Molecules*. Wiley, New York (1994).
22. Lord Rayleigh. *Phil. Mag.* **37**, 321 (1919).
23. Chandrasekhar, S. *Rev. Mod. Phys.* **15**, 1 (1943).
24. Debye, P. *J. Chem. Phys.* **14**, 636 (1946).
25. Eyring, H. *Phys. Rev.* **39**, 746 (1932).
26. Kuhn, W., and Grün, F. *Kolloid Z.* **101**, 248 (1942).
27. Kuhn, W., and Kuhn, H. *Helv. Chim. Acta* **26**, 1394 (1943); Kuhn, W., *Helv. Chim. Acta* **29**, 1095 (1946).
28. Kröger, M. *J. non-Newt. Fluid Mech.* **223**, 77 (2015).
29. Feller, W. *An Introduction to Probability Theory and its Applications* (3rd ed.). Wiley, New York (1968).
30. Van Kampen, N.G. *Stochastic Processes in Physics and Chemistry*. North-Holland, Amsterdam (1987).
31. Fredrickson, G.H., Ganesan, V., and Drolet, F. *Macromolecules* **35**, 16 (2002).
32. Weiner, J.H. *Statistical Mechanics of Elasticity*. Wiley, New York (1983).
33. James, H.M., and Guth, E. *J. Chem. Phys.* **11**, 470 (1943).
34. Treloar, L.R.G. *Trans. Farad. Soc.* **42**, 77 (1946).
35. Treloar, L.R.G. *The Physics of Rubber Elasticity* (2nd ed.). Oxford University Press, Oxford (1958).
36. See also Ref. 4, p. 17.
37. Rouse, P.E. *J. Chem. Phys.* **21**, 1272 (1953).
38. Porod, G. *Monatsh. Chem.* **80**, 251 (1949).
39. Kratky, O., and Porod, G. *Rec. Trav. Chim.* **68**, 1106 (1949).
40. Hermans, J.J., and Ullman, R. *Physica* **18**, 951 (1952).
41. Heine, S., Kratky, O, and Porod, G. *Makromol. Chem.* **44**, 682 (1961).
42. Benoit, H., and Doty, P. *J. Phys. Chem.* **57**, 958 (1953).
43. See e.g. Ref. 3, p. 17.
44. Edwards, S.F. *Proc. Phys. Soc.* **85**, 613 (1965).
45. Edwards, S.F., and Freed, K.F. *J. Phys. A* **2**, 145 (1969).
46. Freed, K.F. *Adv. Chem. Phys.* **22**, 1 (1972).
47. Edwards, S.F. The configurations and dynamics of polymer chains. In Balian, R., and Weill, G. (eds.), *Molecular Fluids*. Gordon & Breach, London (1976).

48. See e.g. Ref. 3, p. 40–41.
49. Helfand, G. In Šolc, K. (ed.), *Polymer Compatibility and Incompatibility*, p. 143. Harwood Academic Publishers, Reading, UK (1982).
50. Napper, D.H. *Polymeric Stabilization of Colloidal Dispersions*. Academic Press, London (1983).
51. Fleer, G.J., Cohen Stuart, M.A., Scheutjens, J.M.H.M., Cosgrove, T., and Vincent, B. *Polymers at Interfaces*. Chapman and Hall, Cambridge, UK (1993).
52. Jones, R.A.L., and Richards, R.W. *Polymers at Surfaces and Interfaces*. Cambridge University Press, Cambridge, UK (1999).
53. Wu, S. *Polymer Interface and Adhesion*. Marcel Dekker, New York (1982).
54. Fredrickson, G.H. *The Equilibrium Theory of Inhomogeneous Polymers*. Oxford Science, Oxford (2006).
55. Šolc, K., and Stockmayer, W.H. *J. Chem. Phys.* **54**, 2756 (1971).
56. Theodorou, D.N., and Suter, U.W. *Macromolecules* **18**, 1206 (1985).
57. Murat, M., and Kremer, K. *J. Chem. Phys.* **108**, 4340 (1998).
58. Bolhuis, P.G., Louis, A.A., Hansen, J.-P., and Meijer, E.J. *J. Chem. Phys.* **114**, 4296 (2001).
59. Flory, P.J., and Krigbaum, W.R. *J. Chem. Phys.* **18**, 1086 (1950).
60. Bolhuis, P.G., Louis, A.A., and Hansen, J.-P. *Phys. Rev. Lett.* **89**, 128302 (2002).
61. Lamb, H. *Statics* (3rd edn), p. 166. Cambridge University Press, Cambridge, UK (1928).
62. Sack, R.A. *Nature* **171**, 310 (1953).
63. See e.g. Ref. 3, Section 2.4.

2

From Electronic Structure Calculations to Classical Force Fields

2.1 Role of quantum mechanical calculations in polymer modelling

The hierarchy of modelling methods for the prediction of polymer properties starts with quantum mechanics. Today we accept the quantum mechanical description as the only fundamentally true description of matter. Classical atomistic or coarse-grained descriptions are merely approximations to which we resort because of computer time limitations. Although most of the strategies described in this book are rooted in classical mechanics and statistical mechanics, we should not forget the connection to quantum mechanical foundations. The following aspects of this connection are especially important:

(a) *Force fields*. Quantum mechanical electronic structure calculations are essential in deriving force fields for the accurate description of the potential energy $\mathcal{V}$ as a function of the nuclear position vectors $\boldsymbol{r}_1, \boldsymbol{r}_2, \ldots, \boldsymbol{r}_n$ of all atoms comprising the system under study. As will be evident from Chapters 3–5, the potential energy hypersurface $\mathcal{V}\,(\boldsymbol{r}_1, \boldsymbol{r}_2, \ldots, \boldsymbol{r}_n)$ dictates the equations of motion of the atoms and the configuration-space distribution of the system under given external constraints; all properties are averages with respect to this distribution. While ab initio molecular dynamics methods are available for solving the electronic problem *on the fly* as nuclear degrees of freedom evolve in time,[1] most often a specific expression is postulated for $\mathcal{V}\,(\boldsymbol{r}_1, \boldsymbol{r}_2, \ldots, \boldsymbol{r}_n)$ and the parameters appearing in this expression are fitted to electronic structure calculations conducted on appropriately defined oligomers or macromolecular fragments. The expression and parameters for $\mathcal{V}\,(\boldsymbol{r}_1, \boldsymbol{r}_2, \ldots, \boldsymbol{r}_n)$ are referred to as a *force field*. The force field serves as input to classical formulations for the calculation of properties.

(b) *Chemical reactions*. Phenomena involving the formation and/or breakage of bonds naturally call for electronic structure calculations. Reaction path and transition state theory-based calculations of kinetic constants for polymerization reactions,[2] chemical ageing, wear, and fracture of chains under load fall in this category. First-principles molecular dynamics calculations have been invoked with the same objectives.[3,4] Reactive force fields,[5] usually

Multiscale Modelling of Polymers. Doros N. Theodorou and Vlasis G. Mavrantzas, Oxford University Press.
 DOI: 10.1093/oso/9780199226191.003.0002

based on the *bond order* concept,[6,7] have been developed by fitting quantum mechanical calculations. These force fields allow for bond formation, bond breaking, and changes in atomic hybridization in atomistic simulations.

(c) *Prediction of electrical and optical properties.* In electronically conductive or semiconductive oligomers and polymers, electronic structure calculations are valuable for predicting how conductivities, band gaps, and related electronic properties depend on chemical constitution. Work on charge transfer and energy transfer processes in π-conjugated polymers has been reviewed.[8] The interplay between molecular interactions, molecular packing, and electronic properties can be investigated through multiscale simulation approaches where electronic structure calculations play a prominent role. Results from such calculations can provide insights for molecular design. Examples include prediction of charge carrier mobilities in self-organizing discotic liquid crystals.[9] On the other hand, high-level ab initio calculations can provide molecular polarizabilities and hyperpolarizabilities and, in the context of multiscale simulations, linear and nonlinear electric susceptibilities useful in predicting the dielectric, optical, and second harmonic generation properties of organic materials.[10] In polymeric materials that are proton conductors, quantum mechanics provides the safest way to deal with proton hopping, as exemplified by calculations on morphology and conductivity of hydrated Nafion membranes.[11]

(d) *Prediction of thermal properties.* Apart from the motion of light particles, such as electrons, protons, and hydrogen atoms, it is well known from solid state physics[12] that vibrational motions of angular frequency $\omega >> k_B T/h$, with k_B, h, and T being the Boltzmann constant, Planck's constant, and absolute temperature, respectively, must be treated quantum mechanically. As early as 1976, Gō and Scheraga[13] considered the quantum mechanical partition function for a polymer, and its *flexible* and *rigid* classical approximations. Thermal properties which depend on fast vibrations, such as heat capacity, thermal conductivity, elastic constants at low temperatures, and miscibility of hydrogenous and deuterated polymers, require quantum statistical mechanics for their proper treatment. Such treatments have been invoked for crystalline[14] and glassy polymers.

This chapter focuses on connection (a) between electronic structure calculations and force fields for simulations of polymers. Connection (d), the treatment of vibrational modes with quantum statistical mechanics, is briefly addressed in Chapters 3 and 6.

2.2 Fundamentals of quantum mechanics

2.2.1 Feynman's path integral formulation of quantum mechanics

In classical mechanics, an isolated system described in terms of the generalized coordinates $\boldsymbol{q}$ and generalized momenta $\dot{\boldsymbol{q}}$ evolving between two specific times t_a and t_b under the influence of its own internal interactions will follow that dynamical trajectory $\boldsymbol{q}(t)$ which minimizes the integral

$$S_{\text{cl}}[\boldsymbol{q}(t)] = \int_{t_a}^{t_b} \mathcal{L}(\boldsymbol{q}, \dot{\boldsymbol{q}}, t)\, \mathrm{d}t \qquad (2.2.1)$$

with $\mathcal{L}(\boldsymbol{q}, \dot{\boldsymbol{q}}, t)$ being the Lagrangian, defined in Eq. (1.1.30).[15] Equation (2.2.1) defines the classical action S_{cl} as a functional of the trajectory or path $\boldsymbol{q}(t)$. The requirement that S_{cl} be minimal along the actual path followed by the system is the celebrated *Hamilton's principle of least action.*[16] The Euler–Lagrange conditions[17] associated with the problem of minimizing the functional $S_{\text{cl}}[\boldsymbol{q}(t)]$ defined by Eq. (2.2.1) are the Lagrangian equations, Eq. (1.1.32), which constitute a set of ordinary differential equations in time (initial value problem) for the calculation of the classical trajectory.

Feynman put forth a formulation for quantum mechanics that constitutes a generalization of the principle of least action.[18] Feynman considered the *probability* $P(\boldsymbol{r}_b, t_b; \boldsymbol{r}_a, t_a)$ for a system to evolve from configuration $\boldsymbol{r}_a$ at time t_a to configuration $\boldsymbol{r}_b$ at time t_b. The configuration is here expressed by specifying the $3n$ coordinates $\boldsymbol{r}$ for all n particles constituting the system. The probability $P(\boldsymbol{r}_b, t_b; \boldsymbol{r}_a, t_a)$ can be expressed as the squared modulus of an, in general complex, *probability kernel* or *probability amplitude* $K(\boldsymbol{r}_b, t_b; \boldsymbol{r}_a, t_a)$:

$$P(\boldsymbol{r}_b, t_b; \boldsymbol{r}_a, t_a) = |K(\boldsymbol{r}_b, t_b; \boldsymbol{r}_a, t_a)|^2. \tag{2.2.2}$$

Feynman proposed that $K(\boldsymbol{r}_b, t_b; \boldsymbol{r}_a, t_a)$ has the form of a functional integral over all possible paths $\boldsymbol{r}(t)$ leading from $\boldsymbol{r}_a$ at time t_a to $\boldsymbol{r}_b$ at time t_b:

$$K(\boldsymbol{r}_b, t_b; \boldsymbol{r}_a, t_a) = \int_{\boldsymbol{r}_a(t_a)}^{\boldsymbol{r}_b(t_b)} \mathcal{D}[\boldsymbol{r}(t)] \exp\left(\frac{i}{\hbar} S_{\text{cl}}[\boldsymbol{r}(t)]\right). \tag{2.2.3}$$

In Eq. (2.2.3), $S_{\text{cl}}[\boldsymbol{r}(t)]$ is the classical action for each path $\boldsymbol{r}(t)$, defined by Eq. (2.2.1) with $\boldsymbol{r}$ in place of $\boldsymbol{q}$. The number $i = \sqrt{-1}$ is the imaginary unit, while $\hbar = h/(2\pi)$, where $h = 6.6261 \times 10^{-34}$ Js is Planck's constant, with units of action.

It is worth thinking about the physical content of Eqs. (2.2.2) and (2.2.3). In a quantum mechanical description there is not a single trajectory taken by the system as it goes from $\boldsymbol{r}_a$ to $\boldsymbol{r}_b$ within time $t_b - t_a$. Rather, there is an infinite number of trajectories (paths) $\boldsymbol{r}(t)$ that can be followed. Each trajectory contributes to the probability amplitude according to the weight

$$\Phi[\boldsymbol{r}(t)] = \text{const. } \exp\left(\frac{i}{\hbar} S_{\text{cl}}[\boldsymbol{r}(t)]\right), \tag{2.2.4}$$

where $S_{\text{cl}}[\boldsymbol{r}(t)]$ is the classical action of the path. Remembering that $\exp(ix) = \cos x + i \sin x$ for $x \in \mathbb{R}$, we see that the probability amplitude of a path is expressed as a *wave* of phase $S_{\text{cl}}[\boldsymbol{r}(t)]/\hbar$. If the considered system is made up of massive particles, small deviations in the path will lead to large changes in the phase. Thus, there will be destructive interference between waves describing the probability amplitudes of neighbouring paths and only one path's probability amplitude will overwhelmingly dominate the integral of Eq. (2.2.3). This will be the path which minimizes $S_{\text{cl}}[\boldsymbol{r}(t)]$, i.e. the classical path. For systems composed of light particles, such as electrons, however, a multitude of paths can be found whose actions differ from that of the path which minimizes $S_{\text{cl}}[\boldsymbol{r}(t)]$ to within order $\hbar$. All these paths contribute to the probability amplitude $K(\boldsymbol{r}_b, t_b; \boldsymbol{r}_a, t_a)$, introducing an element of uncertainty or delocalization in the description of the evolution of the system. The reader may have noticed an analogy between the probability amplitude $K(\boldsymbol{r}_b, t_b; \boldsymbol{r}_a, t_a)$ and the propagator

$G(\boldsymbol{R}, \boldsymbol{R}_0; n)$ introduced in Section 1.5 in the context of Edwards' field theoretic description of a chain in a field; length along the chain contour in the latter description is analogous to time, and the weight of a path $\Phi[\boldsymbol{r}(t)]$ is analogous to the conformational distribution function $\Psi[\boldsymbol{R}(n)]$ given in Eq. (1.5.4).

Let us consider an intermediate point in space-time $(\boldsymbol{r}_c, t_c)$ through which the system passes on its way from $(\boldsymbol{r}_a, t_a)$ to $(\boldsymbol{r}_b, t_b)$. By the form of the probability amplitude, Eq. (2.2.3), the following relationship is true:

$$K(\boldsymbol{r}_b, t_b; \boldsymbol{r}_a, t_a) = \int \mathrm{d}\boldsymbol{r}_c K(\boldsymbol{r}_b, t_b; \boldsymbol{r}_c, t_c) K(\boldsymbol{r}_c, t_c; \boldsymbol{r}_a, t_a), \tag{2.2.5}$$

where the position of the intermediate point c is integrated over all space. This relation can be extended to an arbitrary number of intermediate points:

$$K(\boldsymbol{r}_b, t_b; \boldsymbol{r}_a, t_a) = \int \mathrm{d}\boldsymbol{r}_c \int \mathrm{d}\boldsymbol{r}_d \ldots \int \mathrm{d}\boldsymbol{r}_q K(\boldsymbol{r}_b, t_b; \boldsymbol{r}_q, t_q) K(\boldsymbol{r}_q, t_q; \ldots) K(\ldots; \boldsymbol{r}_d, t_d) K(\boldsymbol{r}_d, t_d; \boldsymbol{r}_c, t_c) K(\boldsymbol{r}_c, t_c; \boldsymbol{r}_a, t_a). \tag{2.2.6}$$

2.2.2 Wave function and the Schrödinger equation

Let us now consider the probability $P(\boldsymbol{r}, t)$ for our system to arrive at a point $(\boldsymbol{r}, t)$ in space-time without any consideration of its previous motion. We define the probability amplitude for the system to arrive at point $(\boldsymbol{r}, t)$ from the past as $\Psi(\boldsymbol{r}, t)$. By definition,

$$P(\boldsymbol{r}, t) = |\Psi(\boldsymbol{r}, t)|^2 = \Psi^*(\boldsymbol{r}, t)\Psi(\boldsymbol{r}, t), \tag{2.2.7}$$

where we use the asterisk to denote the complex conjugate. The probability amplitude $\Psi(\boldsymbol{r}_b, t_b)$ is very similar to the amplitude $K(\boldsymbol{r}_b, t_b; \boldsymbol{r}_a, t_a)$ we have introduced previously. The difference is that, whereas $K(\boldsymbol{r}_b, t_b; \boldsymbol{r}_a, t_a)$ tells us about the system arriving at $(\boldsymbol{r}_b, t_b)$ for the special case wherein it came from point $(\boldsymbol{r}_a, t_a)$ in space-time, $\Psi(\boldsymbol{r}_b, t_b)$ is not concerned with the system's past history and only cares about the probability for arrival at $(\boldsymbol{r}_b, t_b)$ from all possible histories. (A similar relation exists between the restricted partition function $q(\boldsymbol{r}, n)$ and the propagator $G(\boldsymbol{r}, \boldsymbol{r}_0; n)$ in the field theoretic formulation for polymers, Section 1.5.) We will call $\Psi(\boldsymbol{r}, t)$ the *wave function*.

If we know the wave function at $(\boldsymbol{r}_a, t_a)$, we can calculate the wave function at $(\boldsymbol{r}_b, t_b)$ via

$$\Psi(\boldsymbol{r}_b, t_b) = \int \mathrm{d}\boldsymbol{r}_a K(\boldsymbol{r}_b, t_b; \boldsymbol{r}_a, t_a) \Psi(\boldsymbol{r}_a, t_a). \tag{2.2.8}$$

Let us now consider that $(\boldsymbol{r}_a, t_a)$ and $(\boldsymbol{r}_b, t_b)$ are infinitesimally far apart, with $t_b = t_a + \varepsilon$ and ε being a very small time interval. By Eq. (2.2.3) we can assume that

$$K(\boldsymbol{r}_b, t_b; \boldsymbol{r}_a, t_a) = K(\boldsymbol{r}_b, t_a + \varepsilon; \boldsymbol{r}_a, t_a) = \frac{1}{A} \exp\left(\frac{i}{\hbar} S_{\mathrm{cl}}[\boldsymbol{r}_b, \boldsymbol{r}_a]\right), \tag{2.2.9}$$

where A is a normalization constant (related to ε) to be determined (see below). By definition [Eq. (2.2.1)], we can approximate the classical action $S_{\rm cl}[\boldsymbol{r}_b, \boldsymbol{r}_a]$ for going from point $\boldsymbol{r}_a$ to point $\boldsymbol{r}_b$, which is infinitesimally close to it, within time ε, by using the average value of the Lagrangian between the two points during time ε, as

$$S_{\rm cl}[\boldsymbol{r}_b, \boldsymbol{r}_a] = \mathcal{L}_{\rm avg} \cdot \varepsilon = \mathcal{L}\left(\frac{\boldsymbol{r}_b + \boldsymbol{r}_a}{2}, \frac{\boldsymbol{r}_b - \boldsymbol{r}_a}{\varepsilon}, t_a + \frac{\varepsilon}{2}\right) \cdot \varepsilon. \tag{2.2.10}$$

Now, $\mathcal{L}(\boldsymbol{r}, \dot{\boldsymbol{r}}, t) = \mathcal{K}(\dot{\boldsymbol{r}}) - \mathcal{V}(\boldsymbol{r}) = \frac{1}{2} m \dot{\boldsymbol{r}}^2 - \mathcal{V}(\boldsymbol{r}, t)$, where we have assumed for simplicity that the masses of all n particles constituting the system are equal to m. Then,

$$\mathcal{L}\left(\frac{\boldsymbol{r}_b + \boldsymbol{r}_a}{2}, \frac{\boldsymbol{r}_b - \boldsymbol{r}_a}{\varepsilon}, t_a + \frac{\varepsilon}{2}\right) = \frac{1}{2} m \left(\frac{\boldsymbol{r}_b - \boldsymbol{r}_a}{\varepsilon}\right)^2 - \mathcal{V}\left(\frac{\boldsymbol{r}_b + \boldsymbol{r}_a}{2}, t_a + \frac{\varepsilon}{2}\right). \tag{2.2.11}$$

Combining Eqs. (2.2.8)–(2.2.11) we obtain

$$\Psi(\boldsymbol{r}_b, t_a + \varepsilon) = \int \mathrm{d}\boldsymbol{r}_a \frac{1}{A} \exp\left\{\frac{i\varepsilon}{\hbar}\left[\frac{1}{2} m \left(\frac{\boldsymbol{r}_b - \boldsymbol{r}_a}{\varepsilon}\right)^2 - \mathcal{V}\left(\frac{\boldsymbol{r}_b + \boldsymbol{r}_a}{2}, t_a + \frac{\varepsilon}{2}\right)\right]\right\} \Psi(\boldsymbol{r}_a, t_a)$$

or

$$\Psi(\boldsymbol{r}_b, t_a + \varepsilon) = \int \mathrm{d}\boldsymbol{\eta} \frac{1}{A} \exp\left[\left(\frac{i}{\hbar}\right)\left(\frac{m}{2\varepsilon}\right)\boldsymbol{\eta}^2\right] \exp\left\{-\frac{i\varepsilon}{\hbar}\mathcal{V}\left(\boldsymbol{r}_b + \frac{\boldsymbol{\eta}}{2}, t_a + \frac{\varepsilon}{2}\right)\right\} \Psi(\boldsymbol{r}_b + \boldsymbol{\eta}, t_a), \tag{2.2.12}$$

where we have set $\boldsymbol{\eta} = \boldsymbol{r}_a - \boldsymbol{r}_b$. Due to the exponential, only the domain with $|\boldsymbol{\eta}|$ of order $(\varepsilon\hbar/m)^{1/2}$ will contribute significantly to the integral of Eq. (2.2.12). Thus, we Taylor-expand terms on both sides of Eq. (2.2.12) up to second order in $\boldsymbol{\eta}$ and up to first order in ε. Equation (2.2.12) thus leads to

$$\Psi(\boldsymbol{r}_b, t_a) + \varepsilon \frac{\partial \Psi(\boldsymbol{r}_b, t_a)}{\partial t_a} = \int \mathrm{d}\boldsymbol{\eta} \frac{1}{A} \exp\left[\left(\frac{i}{\hbar}\right)\left(\frac{m}{2\varepsilon}\right)\boldsymbol{\eta}^2\right]\left[1 - \frac{i\varepsilon}{\hbar}\mathcal{V}(\boldsymbol{r}_b, t_a)\right]$$
$$\left[\Psi(\boldsymbol{r}_b, t_a) + \boldsymbol{\eta}^T \left.\frac{\partial \Psi(\boldsymbol{r}, t_a)}{\partial \boldsymbol{r}}\right|_{\boldsymbol{r}=\boldsymbol{r}_b} + \frac{1}{2}\boldsymbol{\eta}^T \left.\frac{\partial^2 \Psi(\boldsymbol{r}, t_a)}{\partial \boldsymbol{r}\partial \boldsymbol{r}^T}\right|_{\boldsymbol{r}=\boldsymbol{r}_b} \boldsymbol{\eta}\right], \tag{2.2.13}$$

where $\left.\frac{\partial \Psi(\boldsymbol{r}, t_a)}{\partial \boldsymbol{r}}\right|_{\boldsymbol{r}=\boldsymbol{r}_b}$ and $\left.\frac{\partial^2 \Psi(\boldsymbol{r}, t_a)}{\partial \boldsymbol{r}\partial \boldsymbol{r}^T}\right|_{\boldsymbol{r}=\boldsymbol{r}_b}$ stand for the gradient vector and the Hessian matrix of second derivatives of Ψ with respect to $\boldsymbol{r}$ evaluated at $\boldsymbol{r} = \boldsymbol{r}_b$ (at time t_a).

Before performing the integrations on the right-hand side of Eq. (2.2.13), we can establish the value of the normalization constant A. The leading term on the right-hand side of Eq. (2.2.13) is $\int \mathrm{d}\boldsymbol{\eta} \frac{1}{A} \exp\left[\left(\frac{i}{\hbar}\right)\left(\frac{m}{2\varepsilon}\right)\boldsymbol{\eta}^2\right] \Psi(\boldsymbol{r}_b, t_a)$, which, after performing the integration, yields

$\frac{1}{A}(2\pi i\hbar\varepsilon/m)^{3n/2}\Psi(\boldsymbol{r}_b, t_a)$. The leading term on the left-hand side is $\Psi(\boldsymbol{r}_b, t_a)$. For the two terms of the equation to be equal in the limit $\varepsilon \to 0$, the leading terms must be equal. Therefore,

$$A = (2\pi i\hbar\varepsilon/m)^{3n/2}. \tag{2.2.14}$$

Now, returning to the integrals which appear on the right-hand side of Eq. (2.2.13), we see that

$$\int \mathrm{d}\boldsymbol{\eta}\frac{1}{A}\exp\left[\left(\frac{i}{\hbar}\right)\left(\frac{m}{2\varepsilon}\right)\boldsymbol{\eta}^2\right] = \frac{1}{A}(2\pi i\hbar\varepsilon/m)^{3n/2},$$

$$\int \mathrm{d}\boldsymbol{\eta}\frac{1}{A}\exp\left[\left(\frac{i}{\hbar}\right)\left(\frac{m}{2\varepsilon}\right)\boldsymbol{\eta}^2\right]\boldsymbol{\eta}^{\mathrm{T}} = 0,$$

and

$$\int \mathrm{d}\boldsymbol{\eta}\frac{1}{A}\exp\left[\left(\frac{i}{\hbar}\right)\left(\frac{m}{2\varepsilon}\right)\boldsymbol{\eta}^2\right]\boldsymbol{\eta}^{\mathrm{T}}\left.\frac{\partial^2\Psi(\boldsymbol{r}, t_a)}{\partial\boldsymbol{r}\partial\boldsymbol{r}^T}\right|_{\boldsymbol{r}=\boldsymbol{r}_b}\boldsymbol{\eta} = \frac{1}{A}\left(\frac{2i\hbar\varepsilon}{m}\right)^{3n/2+1}\frac{\pi^{3n/2}}{2}Tr\left[\left.\frac{\partial^2\Psi(\boldsymbol{r}, t_a)}{\partial\boldsymbol{r}\partial\boldsymbol{r}^T}\right|_{\boldsymbol{r}=\boldsymbol{r}_b}\right]$$

$$= \frac{1}{A}\left(\frac{2\pi i\hbar\varepsilon}{m}\right)^{3n/2}\left(\frac{i\hbar\varepsilon}{m}\right)\nabla_{\boldsymbol{r}}^2\Psi(\boldsymbol{r}, t_a)\Big|_{\boldsymbol{r}=\boldsymbol{r}_b}$$

Using the value of A from Eq. (2.2.14), Eq. (2.2.13) gives

$$\Psi(\boldsymbol{r}_b, t_a) + \varepsilon\frac{\partial\Psi(\boldsymbol{r}_b, t_a)}{\partial t_a} = \Psi(\boldsymbol{r}_b, t_a) - \frac{i\varepsilon}{\hbar}\mathcal{V}(\boldsymbol{r}_b, t_a)\Psi(\boldsymbol{r}_b, t_a) + \frac{1}{2}\left(\frac{i\hbar\varepsilon}{m}\right)\nabla_{\boldsymbol{r}}^2\Psi(\boldsymbol{r}, t_a)\big|_{\boldsymbol{r}=\boldsymbol{r}_b}$$

or

$$-\frac{\hbar}{i}\frac{\partial\Psi(\boldsymbol{r}_b, t_a)}{\partial t_a} = -\frac{\hbar^2}{2m}\nabla_{\boldsymbol{r}}^2\Psi(\boldsymbol{r}, t_a)\big|_{\boldsymbol{r}=\boldsymbol{r}_b} + \mathcal{V}(\boldsymbol{r}_b, t_a)\Psi(\boldsymbol{r}_b, t_a)$$

or, using t in place of t_a and $\boldsymbol{r}$ in place of $\boldsymbol{r}_b$,

$$\left[-\frac{\hbar^2}{2m}\nabla_{\boldsymbol{r}}^2 + \mathcal{V}(\boldsymbol{r}, t)\right]\Psi(\boldsymbol{r}, t) = i\hbar\frac{\partial\Psi(\boldsymbol{r}, t)}{\partial t}. \tag{2.2.15}$$

Equation (2.2.15) is a partial differential equation in the wave function $\Psi(\boldsymbol{r}, t)$ that is equivalent to the integral Eq. (2.2.8). It is none else than the celebrated *time-dependent Schrödinger equation* published by Schrödinger in 1926.[19] The operator in brackets on the left-hand side is called the Hamiltonian operator and is symbolized by $\mathcal{H}$:

$$\mathcal{H} \equiv -\frac{\hbar^2}{2m}\nabla_{\boldsymbol{r}}^2 + \mathcal{V}(\boldsymbol{r}, t). \tag{2.2.16}$$

The derivation of the time-dependent Schrödinger equation from Feynman's path integral formulation, Eqs. (2.2.2) and (2.2.3), is analogous to the derivation of the Edwards diffusion equation from the path integral formulation for a polymer in a field, discussed in Section 1.5.

2.2.3 Time-independent Schrödinger equation, Heisenberg's uncertainty principle, and variational theorem

If the potential energy $\mathcal{V}(\boldsymbol{r}, t)$ is not a function of time, solutions to the time-dependent Schrödinger equation, Eq. (2.2.15), are separated into time- and configuration-dependent parts. That is to say, if $\mathcal{V}(\boldsymbol{r}, t) = \mathcal{V}(\boldsymbol{r})$, then

$$\Psi(\boldsymbol{r}, t) = f(t)\psi(\boldsymbol{r}). \tag{2.2.17}$$

Substituting Eq. (2.2.17) into Eq. (2.2.15) (method of separation of variables), we obtain

$$f(t) = \exp(-iEt/\hbar), \tag{2.2.18}$$

where E is a constant that corresponds to the total energy of the system. Note that, by Eq. (2.2.7), the probability of finding the system in configuration $\boldsymbol{r}$ at time t is

$$P(\boldsymbol{r}, t) = |\Psi(\boldsymbol{r}, t)|^2 = |\exp(-iEt/\hbar)|^2|\psi(\boldsymbol{r})|^2 = |\psi(\boldsymbol{r})|^2 = \psi^*(\boldsymbol{r})\psi(\boldsymbol{r}) \tag{2.2.19}$$

and is therefore independent of t and expressible in terms of the *configuration-dependent* part of the wave function, $\psi(\boldsymbol{r})$, only. From now on we will be concerned with this part only, which we will simply call the *wave function*. Furthermore, we will interpret $|\psi(\boldsymbol{r})|^2 = \psi^*(\boldsymbol{r})\psi(\boldsymbol{r})$ as a *probability density*, i.e. as a probability per unit volume in configuration space. By the method of separation of variables applied to Eq. (2.2.15), $\psi(\boldsymbol{r})$ satisfies the second-order differential equation:

$$\left[-\frac{\hbar^2}{2m}\nabla_{\boldsymbol{r}}^2 + \mathcal{V}(\boldsymbol{r})\right]\psi(\boldsymbol{r}) = E\psi(\boldsymbol{r}), \text{ or } \mathcal{H}\psi(\boldsymbol{r}) = E\psi(\boldsymbol{r}). \tag{2.2.20}$$

This is the *time-independent* Schrödinger equation, which will be the focus of our attention from now on.

From the mathematical point of view, Eq. (2.2.20) is a Sturm–Liouville[17] or eigenvalue-eigenfunction problem. It can have solutions, under given boundary conditions, only if the parameter E (the total energy) assumes specific, discrete values, which are called *eigenvalues* of the equation. The solutions $\psi(\boldsymbol{r})$ obtained for a given eigenvalue (energy level) E are called *eigenfunctions* corresponding to that energy level. When there are more than one distinct eigenfunctions corresponding to the same energy level, we say that the energy level is degenerate. Eigenfunctions are normalized:

$$\int |\psi(\boldsymbol{r})|^2 \mathrm{d}\boldsymbol{r} = \int \psi^*(\boldsymbol{r})\psi(\boldsymbol{r})\mathrm{d}\boldsymbol{r} = 1. \tag{2.2.21}$$

Equation (2.2.21) expresses the simple physical fact that the total probability of finding the system in some configuration must equal 1.

In quantum mechanics, every physical observable A has a corresponding operator $\mathcal{A}$. The value of A that is measured in an experiment is obtained as an expectation value over the probability density defined by $\psi^*(\boldsymbol{r})\psi(\boldsymbol{r})$:

$$A \equiv \langle \mathcal{A} \rangle = \frac{\int \psi^*(\boldsymbol{r})\mathcal{A}\psi(\boldsymbol{r})\mathrm{d}\boldsymbol{r}}{\int \psi^*(\boldsymbol{r})\psi(\boldsymbol{r})\mathrm{d}\boldsymbol{r}}. \tag{2.2.22}$$

The operator for the total energy is $\mathcal{H}$ (see Eq. (2.2.20)). Indeed, the expectation value of $\mathcal{H}$ is the energy E:

$$\langle \mathcal{H} \rangle = \frac{\int \psi^*(\boldsymbol{r})\mathcal{H}\psi(\boldsymbol{r})\mathrm{d}\boldsymbol{r}}{\int \psi^*(\boldsymbol{r})\psi(\boldsymbol{r})\mathrm{d}\boldsymbol{r}} = \frac{\int \psi^*(\boldsymbol{r})E\psi(\boldsymbol{r})\mathrm{d}\boldsymbol{r}}{\int \psi^*(\boldsymbol{r})\psi(\boldsymbol{r})\mathrm{d}\boldsymbol{r}} = E\frac{\int \psi^*(\boldsymbol{r})\psi(\boldsymbol{r})\mathrm{d}\boldsymbol{r}}{\int \psi^*(\boldsymbol{r})\psi(\boldsymbol{r})\mathrm{d}\boldsymbol{r}} = E \tag{2.2.23}$$

The operator for the momentum vector of the system is $-i\hbar\nabla_{\boldsymbol{r}}$. The operator for $p_{x,k}$, the x-component of the momentum of a particle k in the system, is $-i\hbar\frac{\partial}{\partial x_k}$, while the operator for the x-component of the position of particle k is simply x_k.

All operators we deal with in quantum mechanics are linear. Furthermore, they are Hermitian. Operator $\mathcal{A}$ is Hermitian if, for any well-behaved functions $\varphi_i(\boldsymbol{r})$ and $\varphi_j(\boldsymbol{r})$, the following equality is true:

$$\int \varphi_i{}^*(\boldsymbol{r})\mathcal{A}\varphi_j(\boldsymbol{r})\mathrm{d}\boldsymbol{r} = \int \varphi_j(\boldsymbol{r})(\mathcal{A}\varphi_i(\boldsymbol{r}))^*\mathrm{d}\boldsymbol{r}. \tag{2.2.24}$$

If, for operator $\mathcal{A}$, there is a number λ_i and a function $\varphi_i(\boldsymbol{r})$ that is not identically equal to 0 such that

$$\mathcal{A}\varphi_i(\boldsymbol{r}) = \lambda_i\varphi_i(\boldsymbol{r}), \tag{2.2.25}$$

then we say that λ_i is an eigenvalue of $\mathcal{A}$ with corresponding eigenfunction $\varphi_i(\boldsymbol{r})$. We have seen a special case of this definition in connection with the time-independent Schrödinger equation, Eq. (2.2.20), where the eigenvalues of the Hamiltonian operator $\mathcal{H}$ have the physical meaning of the energy levels that can be assumed by a system, and the eigenfunctions are the corresponding wave functions, i.e. probability amplitudes defining the distribution of the system in configuration space. For a Hermitian operator, the eigenvalues are real and the eigenfunctions corresponding to distinct eigenvalues are orthogonal (see Problem 2.1).

If, on the other hand, one has a set of degenerate eigenfunctions corresponding to the same eigenvalue, one can always construct a set of new eigenfunctions as linear combinations of the original eigenfunctions such that the new eigenfunctions are orthogonal to each other. One way to accomplish this is through Gram–Schmidt orthogonalization.[17]

The eigenfunctions of a linear Hermitian operator form a complete set. This means that, if any function of configuration $f(\boldsymbol{r})$ is expanded as a linear combination of eigenfunctions of a linear Hermitian operator,

$$f(\boldsymbol{r}) = \sum_{n=0}^{m} a_n\varphi_n(\boldsymbol{r}), \tag{2.2.26}$$

where the φ_n are eigenfunctions, then coefficients a_n can be found, such that

$$\lim_{m\to\infty}\int\left[f(\boldsymbol{r})-\sum_{n=0}^{m}a_n\varphi_n(\boldsymbol{r})\right]^2\mathrm{d}\boldsymbol{r}=0. \tag{2.2.27}$$

Practically speaking, if we choose a large enough set of eigenfunctions of a linear Hermitian operator, we can represent any function of configuration well as a linear combination of the eigenfunctions. As discussed earlier, the set of eigenfunctions can be chosen such that $\int\varphi_i{}^*(\boldsymbol{r})\,\varphi_j(\boldsymbol{r})\mathrm{d}\boldsymbol{r}=\delta_{ij}$ with δ_{ij} being the Kronecker delta. Thus, we say that the eigenfunctions of the Hamiltonian form a complete set of normalized and orthonormal functions associated with real eigenvalues. By Eq. (2.2.26), this set can be used as a basis for expressing any configuration-dependent function $f(\boldsymbol{r})$.

We now return to the question of expressing experimentally measurable observables via Eq. (2.2.22). We will assume that the wave function $\psi(\boldsymbol{r})$ is normalized, i.e. satisfies Eq. (2.2.21). The denominator can then be omitted in expressing the expectation value of $\mathcal{A}$. By analogy to the expectation value, we can also define a variance σ_A^2 for observable A as

$$\sigma_{\mathcal{A}}^2=\int\psi^*(\boldsymbol{r})\left(\mathcal{A}-\langle\mathcal{A}\rangle\right)^2\psi(\boldsymbol{r})\mathrm{d}\boldsymbol{r}. \tag{2.2.28}$$

σ_A is a measure of the uncertainty in knowing A due to the intrinsic quantum mechanical nature of the system.

The uncertainties of conjugate variables are related to each other. In particular, the uncertainties σ_x in measuring the x-component of the position vector of a particle in the system and σ_{p_x} in measuring the corresponding component of the particle's momentum in the x-direction satisfy the relation

$$\sigma_x\sigma_{p_x}\geq\frac{\hbar}{2}. \tag{2.2.29}$$

Equation (2.2.29) is a statement of Heisenberg's famous *uncertainty principle* (1927).[20] In a quantum mechanical description, which we accept as the only true description of dynamics, simultaneously specifying the position and the momentum of a particle is impossible. If we want to know position precisely, we must give up on the precise specification of momentum, and vice versa. The product of uncertainties in the two quantities must always be larger than or equal to $\hbar/2$. For a single particle moving in a one-dimensional potential energy field $\mathcal{V}(x)$, Heisenberg's uncertainty principle can be derived by invoking the *Schwarz inequality* (see Problem 2.2), which states that, for any complex functions $f(x)$ and $g(x)$ of a real variable x,

$$\left|\int_a^b f^*(x)\,g(x)\,\mathrm{d}x\right|^2\leq\int_a^b|f(x)|^2\mathrm{d}x\int_a^b|g(x)|^2\mathrm{d}x. \tag{2.2.30}$$

Moreover, for the position and momentum operators $\mathcal{A}$ and $\mathcal{B}$ of a one-dimensional system, we find that (see Problem 2.3)

$$\mathcal{AB} - \mathcal{BA} = i\hbar; \tag{2.2.31}$$

i.e. operating with $\mathcal{AB} - \mathcal{BA}$ on a function amounts to premultiplying the function with $i\hbar$. We note that $\mathcal{AB} \neq \mathcal{BA}$; i.e. operators $\mathcal{A}$ and $\mathcal{B}$ do not commute. In general, we define the commutator as $[\mathcal{A}, \mathcal{B}] = \mathcal{AB} - \mathcal{BA}$. For the two operators defined in this example, $[\mathcal{A}, \mathcal{B}] = i\hbar$. We also find that $\langle \mathcal{BA} \rangle = \langle \mathcal{AB} \rangle^*$.

Because the time-independent Schrödinger equation, Eq. (2.2.20), is a Sturm–Liouville problem, it satisfies a variational principle that is very useful for determining its lowest eigenvalue (ground state energy) and associated eigenfunction(s). Let $f(\boldsymbol{r})$ be an *arbitrary* normalized function of configuration. As discussed earlier, this can be expanded in terms of the complete set of eigenfunctions of the Hamiltonian operator,[21]

$$f(\boldsymbol{r}) = \sum_{n=0}^{\infty} a_n \psi_n(\boldsymbol{r}), \tag{2.2.32}$$

where $\psi_n(\boldsymbol{r})$ satisfy $\mathcal{H}\psi_n(\boldsymbol{r}) = E_n \psi_n(\boldsymbol{r})$ and constitute a complete normalized and orthonormal set with E_n being real numbers. Since $f(\boldsymbol{r})$ is normalized, the following condition holds for the coefficients of the expansion:

$$\sum_{n=0}^{\infty} |a_n|^2 = 1. \tag{2.2.33}$$

Now, we consider the quantity $\int f^*(\boldsymbol{r}) \mathcal{H} f(\boldsymbol{r})\, \mathrm{d}\boldsymbol{r}$. By virtue of the expansion, Eq. (2.2.32),

$$\begin{aligned}
&\int f^*(\boldsymbol{r}) \mathcal{H} f(\boldsymbol{r})\, \mathrm{d}\boldsymbol{r} \\
&= \sum_{n=0}^{\infty} \sum_{n'=0}^{\infty} a_n^* a_{n'} \int \psi_n^*(\boldsymbol{r}) \mathcal{H} \psi_{n'}(\boldsymbol{r}) \mathrm{d}\boldsymbol{r} \\
&= \sum_{n=0}^{\infty} \sum_{n'=0}^{\infty} a_n^* a_{n'} E_n \delta_{nn'} \\
&= \sum_{n=0}^{\infty} |a_n|^2 E_n.
\end{aligned} \tag{2.2.34}$$

If we now replace E_n by the lowest eigenvalue of the Hamiltonian operator (ground state energy) E_0 on the right-hand side of Eq. (2.2.34), we obtain

$$\int f^*(\boldsymbol{r}) \mathcal{H} f(\boldsymbol{r})\, \mathrm{d}\boldsymbol{r} \geq \sum_{n=0}^{\infty} |a_n|^2 E_0 = E_0 \sum_{n=0}^{\infty} |a_n|^2 = E_0,$$

i.e.

$$E_0 \leq \int f^*(\boldsymbol{r}) \mathcal{H} f(\boldsymbol{r})\, \mathrm{d}\boldsymbol{r}, \tag{2.2.35}$$

where the normalization condition for $f(\boldsymbol{r})$, Eq. (2.2.33), has been used. If $f(\boldsymbol{r})$ is not normalized, Eq. (2.2.3) can be rewritten as

$$E_0 \leq \frac{\int f^*(\boldsymbol{r}) \mathcal{H} f(\boldsymbol{r}) \mathrm{d}\boldsymbol{r}}{\int f^*(\boldsymbol{r}) f(\boldsymbol{r}) \mathrm{d}\boldsymbol{r}}. \tag{2.2.36}$$

From Eq. (2.2.36) we conclude that, in order to determine the ground state energy E_0 and wave function $\psi_0(\boldsymbol{r})$, we could merely minimize the quantity $\dfrac{\int f^*(\boldsymbol{r}) \mathcal{H} f(\boldsymbol{r}) \mathrm{d}\boldsymbol{r}}{\int f^*(\boldsymbol{r}) f(\boldsymbol{r}) \mathrm{d}\boldsymbol{r}}$ with respect to the function $f(\boldsymbol{r})$. The function which minimizes this quantity will be $\psi_0(\boldsymbol{r})$ and the minimum value attained will be E_0. This *variational theorem* has formed a basis for practical schemes that determine the ground-state energy.[21]

2.3 London dispersion forces

Between nonpolar molecules (or parts thereof) attractive interactions develop as a consequence of instantaneous fluctuations of the electronic charge around the nuclei. These interactions, known as *London dispersion forces*, are weak (on the order of a few kJ mol^{-1}, or less than $k_B T$, for a pair of small molecules such as methane at room temperature) but ubiquitous, and therefore play a profound role in shaping the properties of matter. They are intimately related to the size and deformability of electron clouds, hence to the polarizability of the interacting species. Although empirical functionals for taking into account dispersion interactions in the context of density functional theory (DFT) calculations have been developed (see Section 2.5), our current understanding of their nature and distance dependence still owes a lot to their first successful analysis by London in 1930.[22] London conducted a simple quantum mechanical calculation of dispersion forces, which we reconstruct here.

For an approximate analysis of the dispersion energy we will invoke a very simplified model of a molecule, originally proposed by Drude.[23] We will postulate that a molecule is composed of two charges, $+Q$ and $-Q$, not interacting with each other and connected by a harmonic spring of spring constant k. $+Q$ loosely represents the charge of the nuclei in the molecule and $-Q$ the charge of the electronic cloud around them. In the absence of an external field, the equilibrium length of the spring is 0 and the spatial positions of the two charges coincide, on average (see Figure 2.1a).

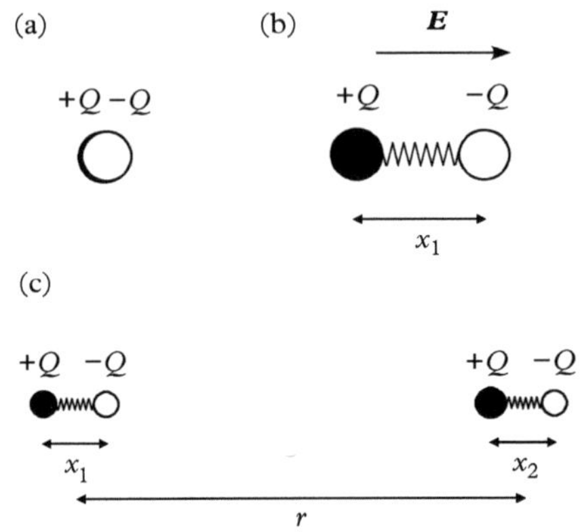

Figure 2.1 Model considered for the derivation of the London dispersion energy. (a) Representation of a molecule by a Drude oscillator. (b) The same molecule under an external electric field $\boldsymbol{E}$. The charges are now separated by x_1, giving rise to a dipole. (c) Two molecules at distance r. The conditions $r >> x_1$ and $r >> x_2$ are assumed in the derivation.

When an electric field $\boldsymbol{E}$ is applied, the spring will elongate to distance x (Figure 2.1b) and the molecule will develop a dipole moment along the direction of the field, with magnitude

$$\mu = Qx. \tag{2.3.1}$$

On the other hand, mechanical equilibrium in the presence of the field dictates that

$$QE = kx. \tag{2.3.2}$$

By combining Eqs. (2.3.1) and (2.3.2) one obtains

$$\mu = \frac{Q^2}{k}E, \tag{2.3.3}$$

which allows one to express the polarizability of the molecule, α, in terms of Q and k:

$$\alpha = \frac{\mu}{E} = \frac{Q^2}{k} \Rightarrow k = \frac{Q^2}{\alpha}. \tag{2.3.4}$$

For the simple Drude model, the molecular polarizability, which in general is a second-order tensor, is isotropic and can be fully represented by one-third its trace.

Consider two identical molecules represented by the Drude model, 1 and 2, at a distance r from each other (Figure 2.1c). Let x_1 and x_2 be the separation of the charges in each molecule along the direction of $\boldsymbol{r}$. When $r \to \infty$, molecule 1 is isolated. Its wave function $\psi_1(x_1)$ will obey the Schrödinger equation

$$-\frac{\hbar^2}{2m}\frac{\partial^2\psi_1(x_1)}{\partial x_1^2}+\frac{1}{2}kx_1^2\psi_1(x_1)=E_1\psi_1(x_1). \tag{2.3.5}$$

This is the equation of a harmonic oscillator. It describes the oscillation of the electronic charge around the nuclei, with m being an effective mass of the electrons in a molecule. The frequency of such an isolated Drude oscillator is

$$\nu=\frac{1}{2\pi}\sqrt{\frac{k}{m}} \tag{2.3.6}$$

and its ground state energy is

$$E=\frac{1}{2}h\nu. \tag{2.3.7}$$

The frequency of oscillation ν and ground state energy E of the second molecule when the two molecules are infinitely far apart is given by expressions identical to Eqs. (2.3.6) and (2.3.7). Thus, the total ground state energy of the two molecules at infinite separation is

$$E(\infty)=\frac{1}{2}h\nu+\frac{1}{2}h\nu=h\nu. \tag{2.3.8}$$

Consider, now, the two molecules at a fixed finite distance r. To describe the system, we need two degrees of freedom, x_1 and x_2. The total potential energy of the system at configuration (x_1, x_2) is given by the sum of the potential energies of the two Drude oscillators plus the electrostatic interaction energy between the two molecules. Assuming that $r >> x_1$ and $r >> x_2$, the latter can be estimated as the electrostatic energy of two point dipoles with dipole moments Qx_1 and Qx_2, at distance r. This can readily be estimated by classical electrostatics[24] and is proportional to the two dipole moments and inversely proportional to the third power of the separation. For the one-dimensional model of Figure 2.1c with the two dipoles collinear and arranged head to tail, it is negative (attractive) with a coefficient of -2,

$$\mathcal{V}(x_1,x_2)=\frac{1}{2}kx_1^2+\frac{1}{2}kx_2^2-\frac{2}{4\pi\varepsilon_0}\frac{(Qx_1)(Qx_2)}{r^3}, \tag{2.3.9}$$

where $\varepsilon_0 = 8.85419 \times 10^{-12}\mathrm{C/(Vm)}$ is the permittivity of free space. The Schrödinger equation obeyed by the wave function $\psi(x_1, x_2)$ of the two dipoles is

$$-\frac{\hbar^2}{2m}\left(\frac{\partial^2\psi(x_1,x_2)}{\partial x_1^2}+\frac{\partial^2\psi(x_1,x_2)}{\partial x_2^2}\right)+\left(\frac{1}{2}kx_1^2+\frac{1}{2}kx_2^2-\frac{2}{4\pi\varepsilon_0}\frac{Q^2x_1x_2}{r^3}\right)\psi(x_1,x_2)=E\psi(x_1,x_2). \tag{2.3.10}$$

To solve Eq. (2.3.10) it is convenient to diagonalize the quadratic form multiplying ψ on the left-hand side, making the transformation

$$z_1=\frac{x_1+x_2}{\sqrt{2}},\quad z_2=\frac{x_1-x_2}{\sqrt{2}}. \tag{2.3.11}$$

The Jacobian of the transformation from (x_1, x_2) to (z_1, z_2) is 1. In terms of the new variables z_1 and z_2, the equation becomes

$$-\frac{\hbar^2}{2m}\left(\frac{\partial^2\psi(z_1,z_2)}{\partial z_1^2}+\frac{\partial^2\psi(z_1,z_2)}{\partial z_2^2}\right)+\left(\frac{1}{2}k_1\,z_1^2+\frac{1}{2}k_2\,z_2^2\right)\psi\,(z_1,z_2)=E\psi\,(z_1,z_2) \tag{2.3.12}$$

with

$$k_1=k-\frac{2Q^2}{4\pi\varepsilon_0 r^3},\, k_2=k+\frac{2Q^2}{4\pi\varepsilon_0 r^3}. \tag{2.3.13}$$

Equation (2.3.12) is the wave equation for two *independent* harmonic oscillators of mass m in the coordinates z_1 and z_2. Thus, the ground state energy of the system at separation r is

$$E\,(r)=\frac{1}{2}h(\nu_1+\nu_2)=\frac{1}{2}h\frac{1}{2\pi}\left(\sqrt{\frac{k_1}{m}}+\sqrt{\frac{k_2}{m}}\right). \tag{2.3.14}$$

The interaction energy between the two molecules at internuclear separation r is obtained directly from Eqs. (2.3.8) and (2.3.14):

$$\mathcal{V}\,(r)=E\,(r)-E\,(\infty)=\frac{1}{2}h\,(\nu_1+\nu_2-2\nu)=\frac{h}{4\pi}\left(\sqrt{\frac{k_1}{m}}+\sqrt{\frac{k_2}{m}}-2\sqrt{\frac{k}{m}}\right)\Rightarrow$$

$$\mathcal{V}\,(r)=\frac{h}{4\pi}\sqrt{\frac{k}{m}}\left(\sqrt{1-2\frac{Q^2}{4\pi\varepsilon_0 r^3 k}}+\sqrt{1+2\frac{Q^2}{4\pi\varepsilon_0 r^3 k}}-2\right). \tag{2.3.15}$$

Taylor-expanding the square roots in parentheses to second order, we obtain

$$\mathcal{V}\,(r)=-\frac{1}{2}h\nu\left(\frac{Q^2}{4\pi\varepsilon_0 r^3 k}\right)^2. \tag{2.3.16}$$

A better, but still not exact, model can be constructed by assuming that the motion of the oscillating charge in each molecule can be resolved into three oscillations of identical frequency along three Cartesian coordinates. In this case, Eq. (2.3.16) is modified to read

$$\mathcal{V}\,(r)=-\frac{3}{4}h\nu\left(\frac{Q^2}{4\pi\varepsilon_0 r^3 k}\right)^2. \tag{2.3.17}$$

Recalling Eq. (2.3.4) for the polarizability, we can rewrite Eq. (2.3.17) as

$$\mathcal{V}\,(r)=-\frac{3}{4}\left(\frac{1}{4\pi\varepsilon_0}\right)^2 h\nu\frac{\alpha^2}{r^6}. \tag{2.3.18}$$

A generalization of Eq. (2.3.18) to the case of dissimilar molecules 1 and 2 can be shown to be

$$\mathcal{V}(r) = -\frac{3}{2}\left(\frac{1}{4\pi\varepsilon_0}\right)^2 \frac{(h\nu_1)(h\nu_2)}{h\nu_1 + h\nu_2}\frac{\alpha_1\alpha_2}{r^6}. \tag{2.3.19}$$

Equation (2.3.19) establishes the very important fact that the attractive interaction energy between two nonpolar molecular species due to correlated fluctuations of their electronic clouds scales as the inverse sixth power of their separation and is proportional to the polarizabilities of the two species. The term *dispersion interaction* arises from the fact that, in the Drude model of propagation of electromagnetic radiation through matter,[24] the frequency ν of oscillation of the electronic cloud around the nuclei appears in the relation expressing the speed of light in a gas of the considered molecules as a function of the light frequency; such a relation is commonly referred to as a dispersion relation.

The frequencies ν_1 and ν_2 appearing in Eq. (2.3.19) are not easy to estimate. One approach is to interpret the charge $-Q$ in the Drude model as being due to an effective number N_e of outer shell electrons which contribute to the molecule's polarizability, while the balancing charge $+Q$ is due to the nuclei and the remaining (mainly inner core) electrons. One then sets $m = N_e m_e$ and $Q = N_e\, e$ in the Drude model, with m_e and e being the mass of an electron and the elementary electronic charge, respectively. From Eq. (2.3.4) the spring constant k is then estimated as $k = N_e^2 e^2/\alpha$, and consequently from Eq. (2.3.6) the frequency ν is estimated as $\nu = e/\left(2\pi m_e^{1/2}\right)(N_e/\alpha)^{1/2}$. Replacing ν_1 and ν_2 in Eq. (2.3.19) by this expression, we obtain

$$\mathcal{V}(r) = -\frac{3}{2}\left(\frac{1}{4\pi\varepsilon_0}\right)^2 \frac{he}{2\pi m_e^{1/2}} \frac{\alpha_1\,\alpha_2}{\left(\frac{\alpha_1}{N_{e,1}}\right)^{1/2} + \left(\frac{\alpha_2}{N_{e,2}}\right)^{1/2}} \frac{1}{r^6}. \tag{2.3.20}$$

Equation (2.3.20) is known as the *Slater–Kirkwood formula.* Values of polarizabilities and effective numbers of electrons to be used with this formula for various groups have been proposed by Ketelaar[25] and have been invoked by Flory and collaborators in developing rotational isomeric state (RIS) models for various polymers (see Section 1.2.3).

Another approach is to replace the energies $h\,\nu_i$ appearing in Eq. (2.3.19) by measurable quantities. One such quantity is the ionization potential I_i of molecule i. If one uses I_i as an estimate of $h\,\nu_i$, the equation for the dispersion energy becomes

$$\mathcal{V}(r) = -\frac{3}{2}\left(\frac{1}{4\pi\varepsilon_0}\right)^2 \frac{I_1 I_2}{I_1 + I_2}\frac{\alpha_1\,\alpha_2}{r^6}. \tag{2.3.21}$$

Equations such as (2.3.20) and (2.3.21) are valuable in pointing out how dispersion interactions should depend on molecular characteristics. One should bear in mind that they are only approximate, however.

Interactions between bodies of colloidal dimensions due to dispersion forces between their constituent atoms can be calculated by spatial integration of $\mathcal{V}(r)$ between the atoms, considering the bodies as continua of given atomic densities. This has led to the well-known

Hamaker potentials,[26] which are very useful in understanding colloidal phenomena. For a didactic treatment of these forces and their consequences, the reader is referred to the textbook by Hiemenz.[27]

2.4 Hartree–Fock calculations

2.4.1 Born–Oppenheimer approximation and Schrödinger equation for the electrons

We now start to address how the Schrödinger description of quantum mechanics, Eq. (2.2.20), can be applied for the investigation of the electronic properties of atoms and molecules. We consider a molecule made up of M nuclei and N electrons. The study of this system is greatly aided by invoking several reasonable approximations.

The first approximation we will invoke is the so-called *Born–Oppenheimer approximation*. Since the nuclei are more than three orders of magnitude heavier than the electrons (the ratio of the rest mass of a proton to the rest mass of an electron being 1836), nuclei move much more slowly than electrons. We will thus assume that the electrons adjust to changes in the nuclear configuration instantaneously. We will solve the electronic problem at various fixed nuclear configurations, the nuclei merely acting as sources of an external potential for the electrons. This amounts to a *quasi-steady state* approximation, as practiced in engineering. Invoking the Born–Oppenheimer approximation, the Hamiltonian for the electrons becomes

$$\mathcal{H} = -\frac{\hbar^2}{2m_\mathrm{e}}\sum_{i=1}^{N}\nabla_{\boldsymbol{r}_i}^2 - \sum_{i=1}^{N}\sum_{l=1}^{M}\frac{1}{4\pi\varepsilon_0}\frac{Z_l e^2}{|\boldsymbol{r}_i - \boldsymbol{R}_l|} + \sum_{i=1}^{N-1}\sum_{j=i+1}^{N}\frac{1}{4\pi\varepsilon_0}\frac{e^2}{|\boldsymbol{r}_i - \boldsymbol{r}_j|}, \tag{2.4.1}$$

where m_e and e are the mass and the charge of an electron, ε_0 is the permittivity of free space, and Z_l is the charge on nucleus l divided by the elementary electronic charge. We denote the position vectors of electrons and nuclei by $\boldsymbol{r}_i$ and $\boldsymbol{R}_l$, respectively. The three terms on the right-hand side of Eq. (2.4.1) express the kinetic energy of the electrons, the potential energy of (attractive) electrostatic interactions between nuclei and electrons, and the potential energy of (repulsive) electrostatic interactions between electrons, respectively. Note that the energy of electrostatic interactions among the nuclei is not included in $\mathcal{H}$. It can be added a posteriori to find the total energy of the system.

We will adopt the atomic units customary in electronic structure calculations: lengths will be measured in units of

$$a_0 = \frac{4\pi\hbar^2\varepsilon_0}{m_\mathrm{e}e^2} = 5.29177 \times 10^{-11}\mathrm{m} \tag{2.4.2}$$

and energies in units of

$$E_\mathrm{a} = \frac{\hbar^2}{m_\mathrm{e}a_0^2} = m_\mathrm{e}\left(\frac{e^2}{4\pi\varepsilon_0\hbar}\right)^2 = 4.35974 \times 10^{-18}\,\mathrm{J}. \tag{2.4.3}$$

The quantities a_0 and E_a are called the Bohr radius and the Hartree, respectively. If all lengths are reduced by a_0 and all energies are reduced by E_a, the Hamiltonian of Eq. (2.4.1) assumes the simpler expression:

$$\mathcal{H} = -\frac{1}{2}\sum_{i=1}^{N}\nabla_{\boldsymbol{r}_i}^2 - \sum_{i=1}^{N}\sum_{l=1}^{M}\frac{Z_l}{|\boldsymbol{r}_i - \boldsymbol{R}_l|} + \sum_{i=1}^{N-1}\sum_{j=i+1}^{N}\frac{1}{|\boldsymbol{r}_i - \boldsymbol{r}_j|}. \tag{2.4.4}$$

We have to solve the Schrödinger equation $\mathcal{H}\psi = E\psi$ with $\mathcal{H}$ given by Eq. (2.4.4) to determine its eigenfunctions $\psi(\boldsymbol{r}_1, \boldsymbol{r}_2, \ldots, \boldsymbol{r}_N)$ as functions of the positions of all electrons in the system at fixed nuclear positions $\boldsymbol{R}_l, l = 1, 2, \ldots, M$, all positions being measured in Bohr radii, and the corresponding eigenvalues E for the total energy of the electrons being measured in Hartrees. If the last term of Eq. (2.4.4), involving interelectronic repulsions, were absent, then the problem would be separable and ψ could be expressed as a product of one-electron wave functions. This last term, however, complicates the solution.

2.4.2 Electronic spin, Pauli exclusion principle, and antisymmetry of electronic wave functions

In order to explain the doubling of spectral lines in the emission spectra of atoms, Uhlenbeck and Goudsmit proposed in 1925 that the electron has an intrinsic angular momentum associated with it, in addition to that due to its motion about the nucleus.[28] This additional angular momentum is called *spin angular momentum* and arises naturally from the relativistic description of electronic motion developed by Dirac in 1928.[29] Thus, in addition to the spatial coordinate $\boldsymbol{r}_i$, we will need a spin coordinate s_i to describe each electron. This can have values of $+1/2$ (spin up) or $-1/2$ (spin down). Let us use the symbol

$$\boldsymbol{q}_i = (\boldsymbol{r}_i, s_i) = (x_i, y_i, z_i, s_i) \tag{2.4.5}$$

to denote the spatial and spin coordinates of an electron i. The wave function for N electrons should then be specified as $\psi(\boldsymbol{q}_1, \boldsymbol{q}_2, \ldots, \boldsymbol{q}_N)$. In this book, this is one of the very few cases where the symbols $\boldsymbol{q}_i$ and $\boldsymbol{r}_i$ are not equivalent.

Quantum mechanics tells us that in a system of several *identical* particles, such as electrons, different particles are *indistinguishable* from each other. This means that the probability density $|\psi(\boldsymbol{q}_1, \boldsymbol{q}_2, \ldots, \boldsymbol{q}_N)|^2$ must remain invariant with respect to interchange of the coordinates of two particles. For example, if the coordinates of particles 1 and 2 are interchanged, the following condition must hold:

$$|\psi(\boldsymbol{q}_2, \boldsymbol{q}_1, \ldots, \boldsymbol{q}_N)|^2 = |\psi(\boldsymbol{q}_1, \boldsymbol{q}_2, \ldots, \boldsymbol{q}_N)|^2, \text{or}$$
$$\psi^*(\boldsymbol{q}_2, \boldsymbol{q}_1, \ldots, \boldsymbol{q}_N)\psi(\boldsymbol{q}_2, \boldsymbol{q}_1, \ldots, \boldsymbol{q}_N) = \psi^*(\boldsymbol{q}_1, \boldsymbol{q}_2, \ldots, \boldsymbol{q}_N)\psi(\boldsymbol{q}_1, \boldsymbol{q}_2, \ldots, \boldsymbol{q}_N)$$

There are two ways in which the wave function may transform under interchange of coordinates so as to ensure that this invariance is satisfied:

$$\psi(\boldsymbol{q}_2, \boldsymbol{q}_1, \ldots, \boldsymbol{q}_N) = \psi(\boldsymbol{q}_1, \boldsymbol{q}_2, \ldots, \boldsymbol{q}_N) \text{ (Bosons), or}$$
$$\psi(\boldsymbol{q}_2, \boldsymbol{q}_1, \ldots, \boldsymbol{q}_N) = -\psi(\boldsymbol{q}_1, \boldsymbol{q}_2, \ldots, \boldsymbol{q}_N) \text{ (Fermions).} \tag{2.4.6}$$

Electrons are fermions. The wave function of a many-electron system is antisymmetric with respect to interchange of the electrons, Eq. (2.4.6).

Application of Eq. (2.4.6) to the special case when $\boldsymbol{q}_1 = \boldsymbol{q}_2$ leads to

$$\psi(\boldsymbol{q}_1, \boldsymbol{q}_1, \ldots, \boldsymbol{q}_N) = -\psi(\boldsymbol{q}_1, \boldsymbol{q}_1, \ldots, \boldsymbol{q}_N) \Rightarrow 2\psi(\boldsymbol{q}_1, \boldsymbol{q}_1, \ldots, \boldsymbol{q}_N) = 0 \Rightarrow \psi(\boldsymbol{q}_1, \boldsymbol{q}_1, \ldots, \boldsymbol{q}_N) = 0. \tag{2.4.7}$$

Equation (2.4.7) tells us that the probability of finding two electrons with the same spin coordinate at the same spatial location is 0. This is the celebrated *Pauli exclusion principle* (1925).[30] Since the wave function is continuous, the probability of finding two electrons with the same spin coordinate very close to each other is small; thus, one often speaks of a *Pauli repulsion* between electrons of the same spin. Of course, irrespective of spin, two electrons cannot approach arbitrarily close to each other due to *Coulomb repulsion*, since they carry the same charge. Thus, electrons are kept apart from each other via two mechanisms: Coulomb repulsion, which is active irrespective of whether the spins are the same or opposite, and Pauli repulsion, which is active only when the spins are the same.

2.4.3 Hartree–Fock equations

In the Hartree–Fock method,[31,32] the wave function of the N-electron system is written as a determinant of a $N \times N$ matrix,

$$\psi_{\mathrm{HF}} = \frac{1}{\sqrt{N!}} \begin{vmatrix} \phi_1(\boldsymbol{q}_1) & \phi_2(\boldsymbol{q}_1) & \ldots & \phi_N(\boldsymbol{q}_1) \\ \phi_1(\boldsymbol{q}_2) & \phi_2(\boldsymbol{q}_2) & \ldots & \phi_N(\boldsymbol{q}_2) \\ \cdot & \cdot & \cdot & \cdot \\ \cdot & \cdot & \cdot & \cdot \\ \phi_1(\boldsymbol{q}_N) & \phi_2(\boldsymbol{q}_N) & \ldots & \phi_N(\boldsymbol{q}_N) \end{vmatrix} \tag{2.4.8}$$

with $\phi_i(\boldsymbol{q})$ being one-electron spin orbitals. Such a determinantal wave function was proposed by Slater in 1930.[33] It is an ingenious construction, for it guarantees that ψ_{HF} satisfies automatically the antisymmetry condition, Eq. (2.4.6). If the one-electron spin orbitals are orthonormal, then ψ_{HF} is normalized. This is ensured by the prefactor $1/\sqrt{N!}$.

The Hartree–Fock method attempts to find the optimal set of one-electron spin orbitals $\phi_1(\boldsymbol{q}), \phi_2(\boldsymbol{q}), \ldots, \phi_N(\boldsymbol{q})$ that minimizes the energy functional,

$$E_{\mathrm{HF}}[\psi_{\mathrm{HF}}] = \int \psi_{\mathrm{HF}}^* \mathcal{H} \psi_{\mathrm{HF}} \mathrm{d}\boldsymbol{q}_1 \mathrm{d}\boldsymbol{q}_2 \ldots \mathrm{d}\boldsymbol{q}_N \tag{2.4.9}$$

as dictated by the variational theorem, Eq. (2.2.36). The minimum value of the functional $E_{HF}[\psi_{HF}]$ will be the ground state energy of the M-atom, N-electron system. The functional minimization must take place under the constraints imposed by the orthonormality of the one-electron spin orbitals:

$$\int \phi_i^*(\boldsymbol{q})\phi_j(\boldsymbol{q})\,\mathrm{d}\boldsymbol{q} = \delta_{ij}. \tag{2.4.10}$$

Introducing a Lagrange multiplier ε_{ij} for each constraint,[17] we have to find an unconstrained stationary point of the functional

$$L = E_{HF} - \sum_i \sum_j \varepsilon_{ij}\left[\int \phi_i^*(\boldsymbol{q})\phi_j(\boldsymbol{q})\,\mathrm{d}\boldsymbol{q} - \delta_{ij}\right] \tag{2.4.11}$$

with respect to the functions $\phi_1(\boldsymbol{q})$, $\phi_2(\boldsymbol{q}), \ldots, \phi_N(\boldsymbol{q})$ and the multipliers ε_{ij}.

Substituting ψ_{HF} from Eq. (2.4.8) into Eq. (2.4.9) and explicitly considering the Hamiltonian operator leads to the following expression for E_{HF},

$$E_{HF} = \sum_{i=1}^{N} H_i + \frac{1}{2}\sum_{i=1}^{N}\sum_{j=1}^{N}(\mathcal{J}_{ij} - K_{ij}) \tag{2.4.12}$$

where

$$H_i = \int \phi_i^*(\boldsymbol{q})\left[-\frac{1}{2}\nabla_{\boldsymbol{r}}^2 + \mathcal{V}(\boldsymbol{q})\right]\phi_i(\boldsymbol{q})\,\mathrm{d}\boldsymbol{q} \tag{2.4.13}$$

$$\mathcal{J}_{ij} = \int \phi_i^*(\boldsymbol{q})\phi_i(\boldsymbol{q})\,\frac{1}{|\boldsymbol{q}-\boldsymbol{q}'|}\,\phi_j^*(\boldsymbol{q}')\,\phi_j(\boldsymbol{q}')\,\mathrm{d}\boldsymbol{q}\mathrm{d}\boldsymbol{q}' \tag{2.4.14}$$

$$K_{ij} = \int \phi_i^*(\boldsymbol{q})\phi_j(\boldsymbol{q})\,\frac{1}{|\boldsymbol{q}-\boldsymbol{q}'|}\,\phi_i(\boldsymbol{q}')\,\phi_j^*(\boldsymbol{q}')\,\mathrm{d}\boldsymbol{q}\mathrm{d}\boldsymbol{q}'. \tag{2.4.15}$$

In Eq. (2.4.13), $\mathcal{V}(\boldsymbol{q})$ stands for the external electrostatic field exerted on an electron due to the nuclei (compare Eq. (2.4.4)):

$$\mathcal{V}(\boldsymbol{q}) = -\sum_{l=1}^{M}\frac{Z_l}{|\boldsymbol{r}-\boldsymbol{R}_l|}. \tag{2.4.16}$$

The integrals $\mathcal{J}_{ij}$ and K_{ij}, on the other hand, have to do with interelectronic repulsions.

Differentiating Eq. (2.4.11) with respect to $\phi_i(\boldsymbol{q})$ and invoking Eqs. (2.4.12)–(2.4.15), one obtains

$$\mathcal{F}\phi_i(\boldsymbol{q}) = \sum_{j=1}^{N}\varepsilon_{ij}\,\phi_j(\boldsymbol{q}), \tag{2.4.17}$$

where the Fock operator $\mathcal{F}$ is defined as

$$\mathcal{F} = -\frac{1}{2}\nabla_{\boldsymbol{r}}^2 + \mathcal{V}(\boldsymbol{q}) + j - k \tag{2.4.18}$$

with

$$j\phi_i(\boldsymbol{q}) = \int \sum_{j=1}^{N} \frac{\phi_j^*(\boldsymbol{q}')\phi_j(\boldsymbol{q}')}{|\boldsymbol{q}-\boldsymbol{q}'|}\,\mathrm{d}\boldsymbol{q}' \tag{2.4.19}$$

and

$$k\phi_i(\boldsymbol{q}) = \int \sum_{j=1}^{N} \frac{\phi_j^*(\boldsymbol{q}')\phi_j(\boldsymbol{q})\,\phi_i(\boldsymbol{q}')}{|\boldsymbol{q}-\boldsymbol{q}'|}\,\mathrm{d}\boldsymbol{q}' \equiv \int K(\boldsymbol{q},\boldsymbol{q}')\,\phi_i(\boldsymbol{q}')\mathrm{d}\boldsymbol{q}', \tag{2.4.20}$$

the kernel $K(\boldsymbol{q},\boldsymbol{q}')$ being defined by

$$K(\boldsymbol{q},\boldsymbol{q}') = \sum_{j=1}^{N} \frac{\phi_j^*(\boldsymbol{q}')\,\phi_j(\boldsymbol{q})}{|\boldsymbol{q}-\boldsymbol{q}'|}. \tag{2.4.21}$$

The set of Eq. (2.4.17) with the definitions, Eqs. (2.4.16) and (2.4.18)–(2.4.21), are the Hartree-Fock equations. The first term in Eq. (2.4.18) for the Fock operator, $-\frac{1}{2}\nabla_{\boldsymbol{r}}^2$, represents the kinetic energy operator for the electrons; the second term, $\mathcal{V}(\boldsymbol{q})$, corresponds to the potential energy of the electrons due to the nuclei; the third term, j, given by Eq. (2.4.19), is the potential energy due to interelectronic repulsions; the fourth term, k, given by Eq. (2.4.20), is a result of incorporating Pauli correlations in the theory.

In a closed-shell system containing N electrons in $N/2$ orbitals, there are two spin orbitals associated with each spatial orbital $\phi_i(\boldsymbol{r})$. Each electron in spatial orbital $\phi_i(\boldsymbol{r})$ contributes an energy H_i associated with its motion and its interaction with the bare nuclei. For the two electrons in an orbital, this energy becomes $2H_i$. On the other hand, the interaction between each pair of spatial orbitals $\phi_i(\boldsymbol{r})$ and $\phi_j(\boldsymbol{r})$ involves a total of four electrons. There are four ways in which two electrons in one orbital can interact in a Coulomb sense with two electrons in a second orbital, thus giving an energy of $4\mathcal{J}_{ij}$ for interelectronic Coulombic repulsions, with $\mathcal{J}_{ij}$ being given by Eq. (2.4.14) with $\boldsymbol{q}$ and $\boldsymbol{q}'$ replaced by $\boldsymbol{r}$ and $\boldsymbol{r}'$, respectively. However, there are only two ways of obtaining paired electrons from this arrangement, giving a total exchange contribution of $-2K_{ij}$, with K_{ij} being given by Eq. (2.4.15) with $\boldsymbol{q}$ and $\boldsymbol{q}'$ replaced by $\boldsymbol{r}$ and $\boldsymbol{r}'$, respectively. Finally, the Coulomb interaction $\mathcal{J}_{ii}$ between each pair of electrons in the same orbital must be included; there is no exchange interaction between the two electrons in an orbital, because the two electrons have paired spins. The total energy for a closed-shell system of N electrons in $N/2$ orbitals is thus given by

$$E_{\mathrm{HF}} = 2\sum_{i=1}^{N/2} H_i + \sum_{i=1}^{N/2-1}\sum_{j=i+1}^{N/2} (4\mathcal{J}_{ij} - 2K_{ij}) + \sum_{i=1}^{N/2} \mathcal{J}_{ii}.$$

Recognizing that $\mathcal{J}_{ii} = K_{ii}$ (compare Eqs. (2.4.14) and (2.4.15)), we can rewrite the above expression more concisely as

$$E_{\text{HF}} = 2\sum_{i=1}^{N/2} H_i + \sum_{i=1}^{N/2}\sum_{j=1}^{N/2}(2\mathcal{J}_{ij} - K_{ij}). \tag{2.4.22}$$

For such a closed-shell system of N electrons in $N/2$ orbitals, the Fock operator can be written [compare Eq. (2.4.22) with Eq. (2.4.12)] as

$$\mathcal{F} = -\frac{1}{2}\nabla_{\boldsymbol{r}}^2 + \mathcal{V}(\boldsymbol{r}) + 2j - k \tag{2.4.23}$$

with

$$j\phi_i(\boldsymbol{r}) = \int \sum_{j=1}^{N/2} \frac{\phi_j^*(\boldsymbol{r}')\phi_j(\boldsymbol{r}')}{|\boldsymbol{r} - \boldsymbol{r}'|} \, d\boldsymbol{r}' \tag{2.4.24}$$

and

$$k\phi_i(\boldsymbol{r}) = \int \sum_{j=1}^{N/2} \frac{\phi_j^*(\boldsymbol{r}')\phi_j(\boldsymbol{r})\phi_i(\boldsymbol{r}')}{|\boldsymbol{r} - \boldsymbol{r}'|} \, d\boldsymbol{r}'. \tag{2.4.25}$$

By analogy to the definition, Eq. (2.2.24), of a Hermitian operator, we will say that a square matrix $\boldsymbol{A}$ is Hermitian (or self-adjoint) when it is equal to its own conjugate transpose. We will use the superscript † to denote the conjugate transpose of a matrix. Furthermore, we will say that a square matrix $\boldsymbol{U}$ is unitary, when its inverse is equal to its conjugate transpose:

$$\begin{aligned} &\boldsymbol{A} \text{ Hermitian} : \boldsymbol{A} = \boldsymbol{A}^{\dagger} \equiv (\boldsymbol{A}^*)^{\mathrm{T}} \\ &\boldsymbol{U} \text{ unitary: } \boldsymbol{U}^{-1} = \boldsymbol{U}^{\dagger}. \end{aligned} \tag{2.4.26}$$

One can readily show that all the eigenvalues of a Hermitian matrix must be real. One can also show that the squared modulus of the determinant of a unitary matrix must be 1; i.e. the most general form for the determinant of a unitary matrix is $e^{i\gamma}$ with γ real. As a consequence, if we generate a new set of one-electron wave functions for Eq. (2.4.8) by premultiplying the original matrix appearing in the Slater determinant by a unitary matrix, the resulting wave function will differ from the original one by at most a phase factor. This will leave the probability density $\psi_{\text{HF}}^* \psi_{\text{HF}}$ unaffected; hence, from a physical point of view, the unitary transformation will leave the wave function unchanged. That is, there is no change in the Hartree–Fock wave function upon changing from one set of basis one-electron functions to another, provided this change in basis set is the result of a unitary transformation.

From Eq. (2.4.11), given that the functional L is real and E_{HF} is real, one can show that the matrix ϵ of Lagrange multipliers is Hermitian. For every Hermitian matrix there exists a unitary matrix that diagonalizes it. Given that the Fock operator is unchanged upon any unitary transformation and that the set of one-electron wave functions obtained via unitary transformation does not change the physical content of the determinantal Hartree–Fock wave function, one can conclude that there exists a unitary transformation that will convert Eq. (2.4.17) to the diagonal form:

$$\mathcal{F}\phi'_i(\boldsymbol{q}) = \varepsilon_i\phi'_i(\boldsymbol{q}). \tag{2.4.27}$$

The set of N one-electron equations represented by Eq. (2.4.27) is called the set of *canonical* Hartree–Fock equations. In the following we will be concerned only with those and will drop the primes on the ϕ's for convenience.

2.4.4 The Roothaan–Hall method

We now examine some practical methods for solving the Hartree–Fock equations, Eq. (2.4.27), with the Fock operator given by Eqs. (2.4.23)–(2.4.25). A method introduced by Roothaan[34] and Hall[35] in 1951 expands the one-electron functions $\phi_i(\boldsymbol{q})$, often called *molecular orbitals*, as linear combinations of a complete set of basis functions $\chi_k(\boldsymbol{q})$:

$$\phi_i = \sum_k C_{ki}\chi_k. \tag{2.4.28}$$

Substituting Eq. (2.4.28) into Eq. (2.4.27) with the primes removed, multiplying both sides of the resulting equation by χ_j^* and integrating over all space leads to

$$\sum_k C_{ki}\int \chi_j^*\mathcal{F}\chi_k \mathrm{d}\boldsymbol{q} = \varepsilon_i \sum_k C_{ki}\int \chi_j^*\chi_k \mathrm{d}\boldsymbol{q}. \tag{2.4.29}$$

Setting $\int \chi_j^*\mathcal{F}\chi_k \mathrm{d}\boldsymbol{q} = F_{jk}$, an element of the *Fock matrix*, and $\int \chi_j^*\chi_k \mathrm{d}\boldsymbol{q} = S_{jk}$, an element of the *overlap matrix*, we rewrite Eq. (2.4.29) in the form of the *Roothaan–Hall equations*:

$$\sum_k C_{ki}\left(F_{jk} - \varepsilon_i S_{jk}\right) = 0 \quad \text{for } j = 1, 2, \ldots K, \tag{2.4.30}$$

where K is the total number of basis functions used.

If we consider element F_{jk} of the Fock matrix in more detail, we find (see Problem 2.4) that for the closed-shell system of N electrons in $N/2$ orbitals it takes the form

$$F_{jk} = H_{jk}^{\text{core}} + \sum_\lambda \sum_\sigma P_{\lambda\sigma}\left[(jk|\lambda\sigma) - \frac{1}{2}(j\sigma|\lambda k)\right], \tag{2.4.31}$$

where the following shorthand notations have been introduced:

$$\int d\boldsymbol{r}\chi_j^*(\boldsymbol{r})\left[-\frac{1}{2}\nabla_{\boldsymbol{r}}^2 + \mathcal{V}(\boldsymbol{r})\right]\chi_k(\boldsymbol{r}) = H_{jk}^{\text{core}} \tag{2.4.32}$$

$$\int d\boldsymbol{r}\int d\boldsymbol{r}'\chi_j^*(\boldsymbol{r})\chi_k(\boldsymbol{r})\frac{1}{|\boldsymbol{r}-\boldsymbol{r}'|}\chi_\lambda^*(\boldsymbol{r}')\chi_\sigma(\boldsymbol{r}') = (jk|\lambda\sigma) \tag{2.4.33}$$

$$\sum_{l=1}^{N} C_{\lambda l}^* C_{\sigma l} = 2\sum_{l=1}^{N/2} C_{\lambda l}^* C_{\sigma l} = P_{\lambda\sigma}. \tag{2.4.34}$$

As we will explain in Section 2.5.4, $P_{\lambda\sigma}$ is an element of the so-called *density matrix*.

For a closed-shell system, the matrix $\boldsymbol{C}$ of expansion coefficients in the Roothaan–Hall equations, Eq. (2.4.30), has dimensions $K \times N/2$, where, as already mentioned, K is the total number of basis functions used. To solve the Roothaan–Hall equations, it is implicitly assumed that $K = N/2$; formulations with $K > N/2$ are also possible, by extending the number of spatial orbitals sought to K, having in mind that only $N/2$ of these orbitals will be occupied. To see how Eq. (2.4.30) can be solved, we note that the Fock matrix $\boldsymbol{F}$ with elements F_{jk} and the overlap matrix $\boldsymbol{S}$ with elements S_{jk} are both of dimension $K \times K$. Also, if real basis functions are used, $\boldsymbol{F}$ and $\boldsymbol{S}$ are symmetric. We also consider the matrix $\boldsymbol{E}$, a diagonal $N/2 \times N/2$ matrix with diagonal elements ε_i. Both $\boldsymbol{C}$ and $\boldsymbol{E}$ are square $K \times K$ matrices. Then, the Roothaan–Hall equations which can be written as

$$\boldsymbol{F}\cdot\boldsymbol{C} = \boldsymbol{S}\cdot\boldsymbol{C}\cdot\boldsymbol{E} \tag{2.4.35a}$$

can be brought to the form[36]

$$\boldsymbol{F}'\cdot\boldsymbol{C}' = \boldsymbol{C}'\cdot\boldsymbol{E}, \tag{2.4.35b}$$

where $\boldsymbol{F}' = \boldsymbol{S}^{-1/2}\cdot\boldsymbol{F}\cdot\boldsymbol{S}^{-1/2}$ and $\boldsymbol{C}' = \boldsymbol{S}^{1/2}\cdot\boldsymbol{C}$. This is a common matrix diagonalization problem which can be solved for the determination of the expansion coefficients C_{ki} in a closed-shell system using known linear algebra tools.

Having solved for $\boldsymbol{C}$, the density of electrons (number of electrons per unit volume) at position $\boldsymbol{r}$, $n(\boldsymbol{r})$, can be expressed in terms of the wave functions of the molecular orbitals $\phi_i(\boldsymbol{r})$ as $n(\boldsymbol{r}) = \sum_i n_i|\phi_i(\boldsymbol{r})|^2$, where n_i is the number of electrons in orbital i. For a closed-shell system of N electrons, $n_i = 2$ for each orbital. Invoking Eq. (2.4.28) to express the wave functions in terms of the basis functions, we obtain that

$$\begin{aligned} n(\boldsymbol{r}) &= 2\sum_{i=1}^{N/2}|\phi_i(\boldsymbol{r})|^2 = 2\sum_{i=1}^{N/2}\phi_i^*(\boldsymbol{r})\phi_i(\boldsymbol{r}) = 2\sum_{i=1}^{N/2}\left(\sum_{\lambda=1}^{K}C_{\lambda i}^*\chi_\lambda^*(\boldsymbol{r})\right)\left(\sum_{\sigma=1}^{K}C_{\sigma i}\chi_\sigma(\boldsymbol{r})\right) \\ &= 2\sum_{\lambda=1}^{K}\sum_{\sigma=1}^{K}\left(\sum_{i=1}^{N/2}C_{\lambda i}^*C_{\sigma i}\right)\chi_\lambda^*(\boldsymbol{r})\chi_\sigma(\boldsymbol{r}) = \sum_{\lambda=1}^{K}\sum_{\sigma=1}^{K}P_{\lambda\sigma}\chi_\lambda^*(\boldsymbol{r})\chi_\sigma(\boldsymbol{r}). \end{aligned} \tag{2.4.36}$$

Similarly, for the total Hartree–Fock energy, which is given by Eq. (2.4.22), we find (see Problem 2.5) the very useful expressions

$$E_{\mathrm{HF}} = \sum_{i=1}^{N/2} \varepsilon_i + \frac{1}{2} \sum_{\lambda=1}^{K} \sum_{\sigma=1}^{K} P_{\lambda\sigma} H_{\lambda\sigma}^{\mathrm{core}} \tag{2.4.37a}$$

and

$$E_{\mathrm{HF}} = \frac{1}{2} \sum_{\lambda=1}^{K} \sum_{\sigma=1}^{K} P_{\lambda\sigma} (F_{\lambda\sigma} + H_{\lambda\sigma}^{\mathrm{core}}). \tag{2.4.37b}$$

2.4.5 Basis set expansions

What functions χ_k should be used as basis functions in expanding the molecular orbitals ϕ_i according to Eq. (2.4.28)? For most Hartree–Fock calculations, the basis functions chosen are the atomic orbitals of the constituent atoms. In other words, molecular orbitals are expanded as linear combinations of the atomic orbitals of the constituent atoms. This method is often referred to as the Linear Combination of Atomic Orbitals–Molecular Orbital (LCAO-MO) method.

Atomic orbitals are eigenfunctions of a Sturm–Liouville problem. A complete set of atomic orbitals would comprise an infinite number of such functions. In practice, a finite number must be used, as the computational cost of a Hartree–Fock calculation generally scales with the fourth power of the number of basis functions (K^4). This is because each two-electron integral $(j\sigma|k\lambda)$ involves four basis functions, so the number of two-electron integrals is expected to increase in proportion to the fourth power of the number of basis functions. In fact, use of symmetry conditions and careful choice of the basis functions may lead to a weaker scaling of the computation time with K.[36] In any case, however, the choice of an appropriate basis set results from a compromise between accuracy and efficiency.

Two types of atomic basis functions have found widespread use: Slater-type orbitals (STOs) and Gaussian basis functions. STOs have exponential radial parts and are labelled like hydrogen atomic orbitals. Some examples of STOs are

$$\begin{aligned} \chi_{1s} &= \left(\frac{\zeta_1^3}{\pi}\right)^{0.5} \exp(-\zeta_1 r) \\ \chi_{2s} &= \left(\frac{\zeta_2^5}{96\pi}\right)^{0.5} r \exp\left(-\frac{\zeta_2 r}{2}\right) \end{aligned} \tag{2.4.38}$$

with r being the distance from the centre of the nucleus. The reader can verify that the probability densities $\chi^*\chi$ defined by these functions are normalized in 3D space. The constants ζ_1 and ζ_2 define the spatial extent (size) of the atomic basis functions. Although STOs provide accurate descriptions of atomic orbitals, they necessitate numerical evaluation of the various integrals involved in the Hartree–Fock calculation, and are thus computationally costly.

Gaussian basis functions offer the advantage that, if these functions are used, the integrals defining the elements of the Fock and overlap matrices can be determined analytically. The following are examples of Gaussian basis functions characterized by the angular symmetry of s and p atomic orbitals:

$$g_s(\alpha, r) = \left(\frac{2\alpha}{\pi}\right)^{3/4} \exp(-\alpha r^2)$$
$$g_x(\alpha, r) = \left(\frac{128\alpha^5}{\pi^3}\right)^{1/4} x \exp(-\alpha r^2)$$
$$g_y(\alpha, r) = \left(\frac{128\alpha^5}{\pi^3}\right)^{1/4} y \exp(-\alpha r^2)$$
$$g_z(\alpha, r) = \left(\frac{128\alpha^5}{\pi^3}\right)^{1/4} z \exp(-\alpha r^2). \tag{2.4.39}$$

Functions such as those of Eq. (2.4.39) are called *primitive Gaussians* and serve as building blocks for the construction of atomic functions termed *contracted Gaussians* by linear combination,

$$\chi_\mu = \sum_{k=1}^{n} d_{k\mu} g_k(\alpha_k, r), \tag{2.4.40}$$

with n being the *expansion length* of the contraction and $d_{k\mu}$ the *expansion coefficients*.

A *minimal basis* set representation contains exactly as many functions as are needed to accommodate all the electrons of an atom. For example, a minimal basis set representation for hydrogen would contain only a $1s$ function. The most common minimal set representations are the so-called STO-nG basis sets. These are Slater-type functions χ_μ that are constructed as linear combinations of primitive Gaussians with an expansion length n. The Gaussian exponents α_k and the expansion coefficients $d_{k\mu}$ are obtained by minimizing the least-squares error between the actual Slater atomic function and the minimal basis set representation. Atomic functions that have the same principal quantum number (e.g. χ_{2s} and χ_{2p}) are fitted simultaneously, so as to share common Gaussian exponents. The absolute energies of atoms and molecules obtained via Hartree–Fock calculations that use the STO-nG basis set exhibit a strong dependence on n. However, energy differences, minimum energy geometries, charge distributions, and electric dipole moments are much less sensitive to the expansion length. A commonly used minimal basis set representation is STO-3G.

Clearly, an exhaustive discussion of basis functions, their applicability, and their computational requirements is outside the scope of this book. The interested reader is referred to Huzinaga's compendium.[37]

2.4.6 Mulliken population analysis

The nuclei in a molecule are surrounded by a continuous electronic charge distribution, $-e\,n(\boldsymbol{r})$, where e is the elementary electronic charge and $n(\boldsymbol{r})$, the spatial density of electrons, can be computed from the solution of the Hartree–Fock equations (see Eq. (2.4.36)). In actual applications it is useful to apportion the electronic charge density among the different atoms in a molecule. Such a partitioning is clearly artificial; nevertheless, it helps to guide intuitive thinking. Furthermore, it is useful in developing classical force fields for describing intra- and intermolecular interactions in molecular systems, wherein atomic coordinates are typically used as degrees of freedom. There are many ways to partition electronic charge density among nuclear centres. Here we briefly outline one of the most traditional ones, developed by Mulliken in 1955, which is known as Mulliken population analysis.[38]

We have shown [Eq. (2.4.36)] that, within the Roothaan–Hall treatment, the electronic density can be expressed in terms of the basis functions as

$$n(\boldsymbol{r}) = 2\sum_{i=1}^{N/2} |\phi_i(\boldsymbol{r})|^2 = \sum_{\lambda=1}^{K}\sum_{\sigma=1}^{K} P_{\lambda\sigma}\chi_\lambda^*(\boldsymbol{r})\chi_\sigma(\boldsymbol{r}).$$

Integrating the electronic density over all space, one obtains that

$$\int n(\boldsymbol{r})\mathrm{d}\boldsymbol{r} = \sum_{\lambda=1}^{K}\sum_{\sigma=1}^{K} P_{\lambda\sigma}\int \mathrm{d}\boldsymbol{r}\chi_\lambda^*(\boldsymbol{r})\chi_\sigma(\boldsymbol{r}) = \sum_{\lambda=1}^{K}\sum_{\sigma=1}^{K} P_{\lambda\sigma}S_{\lambda\sigma} = N. \tag{2.4.41}$$

Given that the basis functions are normalized, the diagonal elements in the sum of Eq. (2.4.41) equal $P_{\lambda\lambda}$. $P_{\lambda\lambda}$ is known as the net population of the basis function χ_λ. The off-diagonal terms $P_{\lambda\sigma}S_{\lambda\sigma}$ and $P_{\sigma\lambda}S_{\sigma\lambda}$ in the same sum are equal to each other, by definition of the density and overlap matrices. The contribution for each pair of indices,

$$Q_{\lambda\sigma} = P_{\lambda\sigma}S_{\lambda\sigma} + P_{\sigma\lambda}S_{\sigma\lambda} = 2P_{\lambda\sigma}S_{\lambda\sigma}, \tag{2.4.42}$$

is called the overlap population between basis functions χ_λ and χ_σ, which may be on the same or different atoms. In order to define the number of electrons associated with each basis function, we need to partition the overlap population between the two basis functions. *One way* to do this is to divide the overlap population equally between the two basis functions involved. With this ad hoc recipe, the number of electrons that belong to a particular basis function is defined as

$$q_\lambda = P_{\lambda\lambda} + \sum_{\substack{\sigma=1 \\ \sigma\neq\lambda}}^{K} P_{\sigma\lambda}S_{\sigma\lambda}. \tag{2.4.43}$$

From Eqs. (2.4.41) and (2.4.43),

$$\sum_{\lambda=1}^{K} q_\lambda = N. \tag{2.4.44}$$

In the LCAO-MO approach, basis functions are centred at individual atoms. The electronic population ascribed to an atom is thus simply estimated by summing up q_λ over all basis functions that are centred on the atom in question. Thus, for an atom A,

$$q_A = \sum_{\lambda \in A} q_\lambda. \tag{2.4.45}$$

This is the Mulliken population analysis recipe for attributing partial electronic charges to atoms. One can also define an overlap population between atoms A and B:

$$q_{AB} = \sum_{\lambda \in A} \sum_{\sigma \in B} Q_{\lambda\sigma}. \tag{2.4.46}$$

Large positive values of q_{AB} are indicative of a large electron density in the region between atoms A and B, and therefore of strong bonding between the two atoms.

In many practical applications, one needs to optimize the geometry of a molecule, in other words perform Hartree–Fock calculations for different sets of nuclear coordinates and locate that set which minimizes the total energy (minimum on the Born–Oppenheimer energy hypersurface). In such calculations, the gradient of the electronic energy with respect to nuclear coordinates is a highly useful quantity. There are many methods for estimating the derivative of the electronic energy with respect to a parameter λ. In the calculation of the gradient, i.e. the contribution of the electrons to the forces felt by atoms, λ is a nuclear coordinate. An ingeniously simple expression for the calculation of this derivative was derived independently in 1939 by Feynman (at the time an undergraduate student at MIT)[39] and Hellman[40]—apparently the theorem had been proved even earlier by Güttinger (1932)[41] and Pauli (1933)[42]—and is known as the Hellman–Feynman theorem (see Problem 2.6),

$$\frac{\partial E}{\partial \lambda} = \left\langle \frac{\partial \mathcal{H}}{\partial \lambda} \right\rangle, \tag{2.4.47}$$

where the expectation value has been defined in Eq. (2.2.22).

For an electronic structure calculation, with λ being a nuclear coordinate, the kinetic energy part of $\mathcal{H}$ is independent of λ, while the potential energy $\mathcal{V}$ is merely a multiplicative operator. Equation (2.4.47) then reduces to

$$\frac{\partial E}{\partial \lambda} = \int \psi^* \psi \frac{\partial \mathcal{V}}{\partial \lambda} \mathrm{d}\boldsymbol{r} = \int n(\boldsymbol{r}) \frac{\partial \mathcal{V}(\boldsymbol{r})}{\partial \lambda} \mathrm{d}\boldsymbol{r}. \tag{2.4.48}$$

For all its generality and elegance, the Hellman–Feynman theorem is of limited utility in Hartree–Fock calculations. Forces calculated from Eq. (2.4.48) are very inaccurate, unless very large basis set expansions are employed. Small errors in the electron density $n(\boldsymbol{r})$, arising from the approximations inherent in the Hartree–Fock calculation, lead to large errors in the forces. This is particularly true in the core regions, where the electronic density is high and varies rapidly, yet the small basis sets typically used lead to large errors. More practical ways of calculating derivatives of the Hartree–Fock energy are discussed in the literature.[36]

2.5 Density functional theory

2.5.1 Limitations of Hartree–Fock theory

In the simple formulation of Hartree–Fock theory, discussed in Section 2.4.3, exchange correlations between electrons, associated with the Pauli exclusion principle, are incorporated exactly by use of the Slater determinant. Coulomb correlations associated with the electrostatic repulsion between electrons, however, are not incorporated; each electron is treated as moving in the average electrostatic potential that is generated by the other electrons. This neglect of Coulomb correlations may not lead to serious error in systems with high electron density, such as small organic molecules; it does lead to serious error in extended systems with many low-energy valence electrons, such as metals, however. Coulomb correlations can be treated using multideterminantal wave functions within the Hartree–Fock framework (Hartree–Fock configuration interaction method), at substantial computational cost. These methods, or other methods for incorporating Coulomb correlations in Hartree–Fock calculations (e.g. generalized valence bond theory), will not be discussed here.

Another difficulty associated with Hartree–Fock calculations for relatively large systems is associated with their computational cost. As briefly discussed in connection with the Roothaan–Hall scheme (Section 2.4.4), this scales roughly with the fourth power of the number of basis functions. Memory requirements also increase steeply with the number of electrons in the system, making it impractical to perform accurate Hartree–Fock calculations for large systems.

An alternative avenue to electronic structure calculations, which alleviates these problems, is offered by density functional theory (DFT) of the inhomogeneous electron gas. The basic idea in DFT is to replace the complicated N-electron wave function $\psi(\boldsymbol{q}_1, \boldsymbol{q}_2, \ldots, \boldsymbol{q}_N)$ by the 3D density distribution of electrons, $n(\boldsymbol{r})$, as an unknown function. Note that, with the notation introduced in Eq. (2.4.5), $n(\boldsymbol{r})$, measured in electrons per unit volume, is related to the wave function as

$$n(\boldsymbol{r}) = \int \mathrm{d}\boldsymbol{r}_1 \mathrm{d}\boldsymbol{r}_2 \ldots \mathrm{d}\boldsymbol{r}_N \left(\sum_{i=1}^{N} \sum_{s_i} \delta(\boldsymbol{r} - \boldsymbol{r}_i) \right) \psi^* (\boldsymbol{q}_1, \boldsymbol{q}_2, \ldots, \boldsymbol{q}_N) \psi (\boldsymbol{q}_1, \boldsymbol{q}_2, \ldots, \boldsymbol{q}_N). \quad (2.5.1)$$

In the following we briefly discuss its foundations and some aspects of its practical implementation.

2.5.2 Hohenberg–Kohn theorems

DFT was put on a firm theoretical basis by two theorems proved by Hohenberg and Kohn in 1964.[43] Hohenberg and Kohn considered a system of N electrons under the influence of an external scalar potential $\mathcal{V}(\boldsymbol{r})$, and their mutual Coulomb repulsions. In molecular systems treated in the Born–Oppenheimer approximation, $\mathcal{V}(\boldsymbol{r})$ is the electrostatic potential generated by the nuclei, but the theorems are more general than that. The Hamiltonian operator for the electrons in a molecular system can be written as shown in Eq. (2.4.4).

Let us restrict ourselves to systems with a nondegenerate ground state. Clearly, N and $\mathcal{V}(\boldsymbol{r})$, via the Schrödinger equation, determine the wave function and all the electronic properties of the ground state corresponding to the Hamiltonian of Eq. (2.4.4), hence also the spatial density of electrons $n(\boldsymbol{r})$. But is the inverse true? That is to say, does $n(\boldsymbol{r})$ uniquely define the number of electrons N and the external potential $\mathcal{V}(\boldsymbol{r})$? That $n(\boldsymbol{r})$ defines the number of electrons is clear from the balance equation

$$\int n(\boldsymbol{r})\mathrm{d}\boldsymbol{r} = N. \tag{2.5.2}$$

But does $n(\boldsymbol{r})$ also uniquely define the external potential? The answer is given by the first theorem (see Problem 2.7) of Hohenberg and Kohn (HK):

Theorem HK1: For a nondegenerate ground state, the external potential $\mathcal{V}(\boldsymbol{r})$, up to an additive constant, is a unique functional of the electron density $n(\boldsymbol{r})$. Hence, the total energy E is also uniquely determined by the electron density.

According to Theorem HK1, there exists a universal functional $F[n(\boldsymbol{r})]$ independent of the external potential that gives the kinetic and interaction energy of an arbitrary collection of electrons. For example, let us consider a system of N electrons subject to the external potential due to nuclei $\mathcal{V}(\boldsymbol{r})$. Let $n(\boldsymbol{r})$ be the ground state electron density distribution and $\psi(\boldsymbol{r})$ the ground state wave function of the system. The Hamiltonian of the system is related to $\mathcal{V}(\boldsymbol{r})$ via Eq. (2.2.16). According to Theorem HK1, the density distribution uniquely determines its corresponding external potential and wave function. If we consider another density distribution $n'(\boldsymbol{r})$, that will necessarily correspond to a different external potential $\mathcal{V}'(\boldsymbol{r})$ and wave function $\psi'(\boldsymbol{r})$. This is a remarkable result: if we know $n(\boldsymbol{r})$, we know all the electronic properties of the ground state. In particular, the ground state energy E is a functional of $n(\boldsymbol{r})$. We may write this as

$$E[n(\boldsymbol{r})] = F[n(\boldsymbol{r})] + \int n(\boldsymbol{r})\mathcal{V}(\boldsymbol{r})\mathrm{d}\boldsymbol{r}, \tag{2.5.3}$$

where the first term on the right-hand side incorporates the kinetic energy and the interelectronic interaction energy (compare Eq. (2.4.4)).

The next question is how to obtain the ground state electron density distribution. This is addressed by the second theorem of Hohenberg and Kohn.

Theorem HK2: The ground state electron density distribution can be obtained variationally, as the density distribution $n(\boldsymbol{r})$ that minimizes the energy functional of Eq. (2.5.3) under the constraint of Eq. (2.5.2).

The ground state energy of the considered system will be (compare Eqs. (2.2.23), (2.5.3))

$$E[\psi] = \int \mathrm{d}\boldsymbol{r}\,\psi^*\mathcal{H}\psi = E[n(\boldsymbol{r})] = F[n(\boldsymbol{r})] + \int n(\boldsymbol{r})\mathcal{V}(\boldsymbol{r})\mathrm{d}\boldsymbol{r}. \tag{2.5.4}$$

By the variational principle for the ground state energy in terms of the wave function, Eq. (2.2.36),

$$E[\psi'] = \int \mathrm{d}\boldsymbol{r}\,\psi'^*\mathcal{H}\,\psi' = E[n'(\boldsymbol{r})] = F[n'(\boldsymbol{r})] + \int n'(\boldsymbol{r})\mathcal{V}(\boldsymbol{r})\mathrm{d}\boldsymbol{r} > E[\psi] \tag{2.5.5}$$

or $E[n(\boldsymbol{r})] < E[n'(\boldsymbol{r})]$. This tells us that the energy functional defined by Eq. (2.5.3) assumes its minimal value for the correct density distribution $n(\boldsymbol{r})$, which proves the variational principle stated by Theorem HK2.

Assuming differentiability of the energy functional $E[n(\boldsymbol{r})]$, the variational principle requires that the ground state density $n(\boldsymbol{r})$ satisfy the condition

$$\frac{\delta}{\delta n(\boldsymbol{r})}\left[E[n(\boldsymbol{r})] - \mu\left(\int n(\boldsymbol{r})\mathrm{d}\boldsymbol{r} - N\right)\right] = 0, \tag{2.5.6}$$

where μ, the chemical potential of the electrons, is a Lagrange multiplier associated with the constraint on the total number of electrons, Eq. (2.5.2). From Eqs. (2.5.3) and (2.5.6) one obtains that

$$\mu = \mathcal{V}(\boldsymbol{r}) + \frac{\delta F[n(\boldsymbol{r})]}{\delta n(\boldsymbol{r})}. \tag{2.5.7}$$

If the universal functional $F[n(\boldsymbol{r})]$ were known, Eq. (2.5.7) could be used directly to determine the correct ground state distribution for any given $\mathcal{V}(\boldsymbol{r})$. Unfortunately, however, an exact expression for the functional $F[n(\boldsymbol{r})]$ is not available, and judicious approximations have to be used.

2.5.3 Local density approximation: kinetic energy and exchange–correlation energy functionals

We have seen that the functional $F[n(\boldsymbol{r})]$ appearing in Eqs. (2.5.3) and (2.5.7) contains contributions from the kinetic energy of the electrons and the interaction energy among the electrons. We write

$$F[n(\boldsymbol{r})] = T[n(\boldsymbol{r})] + \frac{1}{2}\int \mathrm{d}\boldsymbol{r}\int \mathrm{d}\boldsymbol{r}'\frac{n(\boldsymbol{r})n(\boldsymbol{r}')}{|\boldsymbol{r}-\boldsymbol{r}'|} + E_{\mathrm{xc}}[n(\boldsymbol{r})], \tag{2.5.8}$$

where $T[n(\boldsymbol{r})]$ is the kinetic energy functional of electrons, $\frac{1}{2}\int \mathrm{d}\boldsymbol{r}\int \mathrm{d}\boldsymbol{r}'\frac{n(\boldsymbol{r})n(\boldsymbol{r}')}{|\boldsymbol{r}-\boldsymbol{r}'|}$ is the energy of Coulombic repulsions among electrons estimated based on a mean field approximation, and $E_{\mathrm{xc}}[n(\boldsymbol{r})]$ embodies all the many-body effects in the interaction among electrons not contained in the mean field approximation; it is called the exchange-correlation energy functional. With these definitions, the Euler–Lagrange condition, Eq. (2.5.7), arising from the variational principle for the total energy becomes

$$\mu = \mathcal{V}(\boldsymbol{r}) + \frac{\delta T[n(\boldsymbol{r})]}{\delta n(\boldsymbol{r})} + \int \mathrm{d}\boldsymbol{r}'\frac{n(\boldsymbol{r}')}{|\boldsymbol{r}-\boldsymbol{r}'|} + \frac{\delta E_{\mathrm{xc}}[n(\boldsymbol{r})]}{\delta n(\boldsymbol{r})}. \tag{2.5.9}$$

To proceed we need expressions for the kinetic energy $T[n(\boldsymbol{r})]$ and exchange–correlation energy functional $E_{\mathrm{xc}}[n(\boldsymbol{r})]$. The simplest approximation that can be used for expressing these functionals is the *local density approximation* (LDA).[44] This assumes that the contribution to each type of energy (kinetic energy, exchange–correlation energy) from each point $\boldsymbol{r}$ in the space occupied by the *inhomogeneous* gas of electrons is the same as that of a *homogeneous* electron gas with density equal to the local density $n(\boldsymbol{r})$ prevailing at the considered point.

A simple LDA form for the kinetic energy functional $T[n(\boldsymbol{r})]$ can be based on the theory of Thomas and Fermi (1927)[45,46] for a gas of noninteracting electrons (ideal Fermi gas). This leads to the expression (in the dimensionless units defined on page 70)

$$T[n(\boldsymbol{r})] = C_{\mathrm{TF}} \int [n(\boldsymbol{r})]^{5/3} \mathrm{d}\boldsymbol{r}, \tag{2.5.10}$$

where

$$C_{\mathrm{TF}} = \frac{3}{10}\left(3\pi^2\right)^{2/3}. \tag{2.5.11}$$

Starting from the solution of the Schrödinger equation for a particle in a box, one can show (see Problem 2.8) that the kinetic energy density of a *homogeneous* electron gas of density n is (in absolute units) $C_{\mathrm{TF}} E_{\mathrm{a}} a_0^2 n^{5/3}$, thus leading to Eq. (2.5.10) for an *inhomogeneous* electron gas in the LDA.

The exchange–correlation energy $E_{\mathrm{xc}}[n(\boldsymbol{r})]$ appearing in Eqs. (2.5.8) and (2.5.9) is expressed in the LDA as

$$E_{\mathrm{xc}}[n(\boldsymbol{r})] = \int \varepsilon_{\mathrm{xc}}(n(\boldsymbol{r}))\, n(\boldsymbol{r}) \mathrm{d}\boldsymbol{r}, \tag{2.5.12}$$

where $\varepsilon_{\mathrm{xc}}(n)$ is the exchange–correlation energy density of a *homogeneous* electron gas of density n. To proceed, we need explicit functional expressions for $\varepsilon_{\mathrm{xc}}(n)$. Typically, $\varepsilon_{\mathrm{xc}}(n)$ is divided into two parts, one due to exchange (the Pauli exclusion principle) and another due to Coulomb correlations:

$$\varepsilon_{\mathrm{xc}}(n) = \varepsilon_{\mathrm{x}}(n) + \varepsilon_{\mathrm{c}}(n). \tag{2.5.13}$$

An expression for $\varepsilon_{\mathrm{x}}(n)$ has been derived by Dirac (1928).[29,47] For a closed-shell system containing $s = +1/2$ and $s = -1/2$ electrons in equal numbers, it reads, in atomic units,

$$\varepsilon_{\mathrm{x}}(n) = C_{\mathrm{x}} n^{1/3}, \tag{2.5.14}$$

where $C_{\mathrm{x}} = -\dfrac{3}{4}\left(\dfrac{3}{\pi}\right)^{1/3}$, a known constant.

A practically useful expression for $\varepsilon_{\mathrm{c}}(n)$ has been derived from the results of quantum Monte Carlo calculations on a homogeneous electron gas performed by Ceperley and Alder (1980),[48] as parametrized by Perdew and Zunger (1981).[49]

2.5.4 Kohn–Sham equations and their numerical solution

With Eq. (2.5.10) for the kinetic energy and Eqs. (2.5.12)–(2.5.14), or analogues thereof, for the exchange–correlation energy as functionals of the local electron density $n(\boldsymbol{r})$, we can return to

Eqs. (2.5.4) and (2.5.8) and cast density functional theory of the inhomogeneous electron gas as a problem of minimizing the functional

$$E[n(\boldsymbol{r})] = T[n(\boldsymbol{r})] + \frac{1}{2}\int \mathrm{d}\boldsymbol{r} \int \mathrm{d}\boldsymbol{r}' \frac{n(\boldsymbol{r})n(\boldsymbol{r}')}{|\boldsymbol{r}-\boldsymbol{r}'|} + E_{\mathrm{xc}}[n(\boldsymbol{r})] + \int n(\boldsymbol{r})\mathcal{V}(\boldsymbol{r})\mathrm{d}\boldsymbol{r} \tag{2.5.15}$$

with respect to $n(\boldsymbol{r})$, subject to the constraint, Eq. (2.5.2).

In common electronic structure calculations, the external potential $\mathcal{V}(\boldsymbol{r})$ is the electrostatic potential exerted on an electron from all nuclei in the system, Eq. (2.4.16). Such a formulation, based on the Thomas–Fermi–Dirac model for the kinetic and exchange energies, offers a nice simplicity, as the equations involve the electron density alone. Unfortunately, however, it does not produce quantitatively satisfactory results.[42]

Kohn and Sham[50] traded simplicity for accuracy, inventing an ingenious indirect approach to the kinetic energy functional $T[n(\boldsymbol{r})]$, the Kohn–Sham (KS) method. This turned DFT into a practical tool for rigorous calculations, explaining why Kohn was awarded the Nobel Prize for chemistry in 1998, jointly with Pople. Kohn and Sham proposed introducing orbitals into the problem in such a way that the kinetic energy can be computed simply to good accuracy, leaving a small residual correction that is handled separately.[44]

In the KS method the electron density is written as the sum of the square moduli of a set of one-electron orthonormal orbitals,

$$n(\boldsymbol{r}) = \sum_{i=1}^{N}\sum_{s} |\phi_i(\boldsymbol{r}, s)|^2, \tag{2.5.16}$$

which automatically satisfies the normalization requirement, Eq. (2.5.2). And in place of Eq. (2.5.15) one has the requirement of minimizing the energy:

$$\begin{aligned} E[n(\boldsymbol{r})] =& \sum_{i=1}^{N}\sum_{s} \int \phi_i^*(\boldsymbol{r}) \left(-\frac{\nabla_{\boldsymbol{r}}^2}{2}\right) \phi_i(\boldsymbol{r})\mathrm{d}\boldsymbol{r} + \frac{1}{2}\int \mathrm{d}\boldsymbol{r} \int \mathrm{d}\boldsymbol{r}' \frac{n(\boldsymbol{r})n(\boldsymbol{r}')}{|\boldsymbol{r}-\boldsymbol{r}'|} \\ &+ E_{\mathrm{xc}}[n(\boldsymbol{r})] + \int n(\boldsymbol{r})\mathcal{V}(\boldsymbol{r})\mathrm{d}\boldsymbol{r}. \end{aligned} \tag{2.5.17}$$

The Euler–Lagrange conditions corresponding to Eq. (2.5.17) can be reduced to

$$\begin{aligned} &\left[-\frac{1}{2}\nabla_{\boldsymbol{r}}^2 + \mathcal{V}_{\mathrm{eff}}(\boldsymbol{r})\right] \phi_i(\boldsymbol{r}) = \varepsilon_i \phi_i(\boldsymbol{r}) \\ &\mathcal{V}_{\mathrm{eff}}(\boldsymbol{r}) = \mathcal{V}(\boldsymbol{r}) + \int \frac{n(\boldsymbol{r}')}{|\boldsymbol{r}-\boldsymbol{r}'|}\mathrm{d}\boldsymbol{r}' + \varepsilon_{\mathrm{xc}}[n(\boldsymbol{r})]. \end{aligned} \tag{2.5.18}$$

Equations (2.5.16) and (2.5.18) form the Kohn–Sham orbital equations in their canonical form. This system is then solved iteratively, until self-consistency is reached between the assumed $n(\boldsymbol{r})$ and the computed $n(\boldsymbol{r})$. The eigenvalues ε_i have no physical meaning, but their total sum at the optimum corresponds to the total energy through

$$E = \sum_{i=1}^{N} \varepsilon_i - \frac{1}{2} \int \mathrm{d}\boldsymbol{r} \int \mathrm{d}\boldsymbol{r}' \frac{n(\boldsymbol{r})n(\boldsymbol{r}')}{|\boldsymbol{r}-\boldsymbol{r}'|} + E_{\mathrm{xc}}[n(\boldsymbol{r})] - \int \varepsilon_{\mathrm{xc}}[n(\boldsymbol{r})]\, n(\boldsymbol{r})\mathrm{d}\boldsymbol{r}. \tag{2.5.19}$$

With the reintroduction of electronic orbitals ϕ_i, most of the numerical machinery that has been developed for Hartree–Fock calculations can be ported to the DFT problem. The Kohn–Sham orbitals are usually expressed as linear combinations of atomic-centred basis functions $\chi_k(\boldsymbol{r})$ as in Eq. (2.4.28), Gaussians and Slater-type orbitals being popular choices. If such a basis set expansion is substituted into the KS equations, then a set of matrix equations identical in form to the Roothaan–Hall equations Eq. (2.4.35) is obtained:

$$\boldsymbol{H} \cdot \boldsymbol{C} = \boldsymbol{S} \cdot \boldsymbol{C} \cdot \boldsymbol{E}. \tag{2.5.20}$$

The elements of the Kohn–Sham matrix $\boldsymbol{H}$ are given by

$$H_{\mu\nu} = \int \chi_\mu^*(\boldsymbol{r}) \left\{ -\frac{\nabla_{\boldsymbol{r}}^2}{2} + \mathcal{V}(\boldsymbol{r}) + \int \frac{n(\boldsymbol{r}')}{|\boldsymbol{r}-\boldsymbol{r}'|} \mathrm{d}\boldsymbol{r}' + \varepsilon_{\mathrm{xc}}[n(\boldsymbol{r})] \right\} \chi_\nu(\boldsymbol{r})\mathrm{d}\boldsymbol{r}. \tag{2.5.21}$$

The first two terms constitute the core contribution. The third term is the Coulomb repulsion term, which can be expanded in terms of the basis functions and the density matrix $\boldsymbol{P}$,

$$\iint \frac{\chi_\mu^*(\boldsymbol{r})n(\boldsymbol{r}')\chi_\nu(\boldsymbol{r})}{|\boldsymbol{r}-\boldsymbol{r}'|} \mathrm{d}\boldsymbol{r}\,\mathrm{d}\boldsymbol{r}' = \sum_{\lambda=1}^{K}\sum_{\sigma=1}^{K} P_{\lambda\sigma} \iint \frac{\chi_\mu^*(\boldsymbol{r})\chi_\lambda^*(\boldsymbol{r}')\chi_\sigma(\boldsymbol{r}')\chi_\nu(\boldsymbol{r})}{|\boldsymbol{r}-\boldsymbol{r}'|} \mathrm{d}\boldsymbol{r}\,\mathrm{d}\boldsymbol{r}', \tag{2.5.22}$$

where the density matrix is defined in terms of the expansion coefficients as in Eq. (2.4.36). Likewise, the overlap matrix $\boldsymbol{S}$ is defined as in Eq. (2.4.30).

Dealing with the integrals appearing in Eq. (2.5.22) is easier than with those in Hartree–Fock theory. One way to do this is to expand the electron density $n(\boldsymbol{r}')$ in a set of basis functions that has the same functional form as that used for the orbital expansion, leading to three-centre, two-electron integrals. Another way is to use Poisson's equation in order to compute the electrostatic potential generated by $n(\boldsymbol{r}')$ on a grid running through $\boldsymbol{r}$.[27] These techniques cannot be used in Hartree–Fock theory, for there the exchange contribution is a nonlocal functional requiring the computation of four-centre, two-electron integrals of the form of Eq. (2.4.33).

The Kohn–Sham equations can be solved numerically by spatial discretization (on a grid) through *basis-set free* approaches.[51] For spatially periodic bulk systems, plane waves are the most obvious basis set to use, providing a Fourier series representation of the orbital wave functions. These are very important in solid-state calculations.[52]

The local density approximation to the exchange–correlation energy $\varepsilon_{\mathrm{xc}}$ is surprisingly accurate, given its simplicity. Nevertheless, it has been shown to be inadequate for some applications. For this reason, *nonlocal* functionals have been developed, which express $\varepsilon_{\mathrm{xc}}[n(\boldsymbol{r})]$ in terms of not only $n(\boldsymbol{r})$, but also its gradient, $|\nabla n(\boldsymbol{r})|$, at each point in space. A variety of gradient-corrected expressions have been proposed for both the exchange and the correlation contributions. A combination of the standard expression, Eq. (2.5.14), for the exchange energy with Becke's[53] gradient correction for the exchange energy and the Lee–Yang–Parr correlation energy functional[54] has become quite popular and is commonly abbreviated to BLYP.

2.6 Empirical force fields and their parameterization

Most of the problems encountered in molecular modelling of polymers are too computationally demanding to be tackled by direct quantum mechanical calculations. To make progress, empirical force fields are invoked to express the total potential energy $\mathcal{V}(\boldsymbol{r})$ as a function of the *nuclear* degrees of freedom $\boldsymbol{r}$. When the form of $\mathcal{V}(\boldsymbol{r})$ and the chemical constitution and composition-related parameters employed in it are judiciously chosen, its use in well-designed and correctly executed simulations can yield very good estimates of structure, dynamics, volumetric, thermal, phase equilibrium, mechanical, rheological, transport, and interfacial properties. Thanks to painstaking efforts by many investigators, many force fields, which vary widely in complexity and predictive ability, exist today for modelling synthetic and biological macromolecules and materials made therefrom. It is not our intention to provide an overview of these force fields here. The reader is referred to more extensive treatments on force fields, for example Chapter 4 of Leach's book on molecular modelling.[36]

In most empirical force fields, $\mathcal{V}(\boldsymbol{r})$ is expressed as a sum of contributions from (a) the deformation (stretching–compression) of chemical bonds between pairs of atoms; (b) the deformation (bending) of bond angles formed by triplets of atoms bonded in pairs; (c) torsion of dihedral angles formed by quartets of bonded atoms connected by three consecutive bonds; (d) out-of-plane bending contributions (e.g. associated with quartets of atoms that are kept in a planar geometry by their chemical bonding departing from that geometry); (e) nonbonded contributions between pairs of atoms, incorporating both excluded volume interactions at short distances and dispersion interactions at larger distances; and (f) electrostatic interactions between partial charges on the atoms.

Let us discuss these contributions using as an example a relatively simple generic force field that was proposed in 1990, the DREIDING force field.[55] One form of the DREIDING expression for the total potential energy as a function of the nuclear degrees of freedom reads (see Figure 2.2)

$$
\begin{aligned}
\mathcal{V}(\boldsymbol{r}) = & \sum_{\text{bonds } ij} \frac{1}{2} K_{ij}\left(l_{ij} - l_{ij}^0\right)^2 + \sum_{\text{bond angles } ijk} \frac{1}{2} K_{ijk}\left(\cos\theta_{ijk} - \cos\theta_{ijk}^0\right)^2 \\
& + \sum_{\text{torsion angles } ijkm} \frac{1}{2} V_{jk}\left\{1 - \cos\left[n_{jk}\left(\phi_{ijkm} - \phi_{jk}^0\right)\right]\right\} \\
& + \sum_{\text{inversions } ijkm} \frac{1}{2} C_{ijkm}\left(\cos\psi_{ijkm} - \cos\psi_{ijkm}^0\right)^2 + \sum_{\text{planar } ijkm} \frac{1}{2} K_1\left(1 - \cos\psi_{ijkm}\right)^2 \\
& + \frac{1}{2}\sum_i \sum_{j \neq i} \left\{4\varepsilon_{ij}\left[\left(\frac{\sigma_{ij}}{r_{ij}}\right)^{12} - \left(\frac{\sigma_{ij}}{r_{ij}}\right)^6\right] + \frac{q_i q_j}{4\pi\varepsilon_0 r_{ij}}\right\}.
\end{aligned}
\qquad (2.6.1)
$$

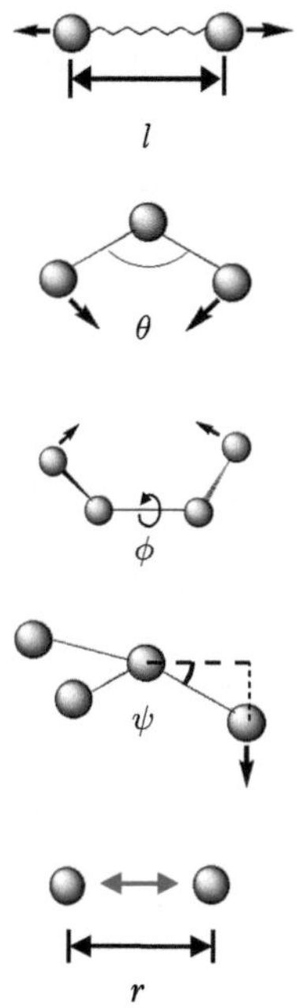

Figure 2.2 Schematic of various contributions to an empirical force field.

2.6.1 Bond stretching

In Eq. (2.6.1) the stretching contribution is expressed as a parabolic or harmonic or Hooke's law term which incurs an energetic penalty whenever the length of a bond departs from its *equilibrium* value l^0. The value of K_{ij} is typically on the order of 10^3 kJ mol^{-1} Å^{-2}, reflecting the small deformability of bonds. In reality, a better representation of the interaction contributed by a bonded pair of atoms is provided by the Morse potential,

$$\mathcal{V}_l(l) = D_e\{1 - \exp[-a(l - l^0)]\}^2, \tag{2.6.2}$$

where D_e is the depth of the potential energy minimum and $a = \omega\sqrt{\mu/2D_e}$ with ω being the natural frequency of vibration of the bond and μ the reduced mass of the two atoms connected by the bond. Morse potential parameters for specific bonds have been estimated by DFT electronic structure calculations and even found to be in good agreement with experimental measurements of the force-lifetime relationship of covalent bonds between single chains and solid substrates subjected to tension in a surface force apparatus.[56]

2.6.2 Bond angle bending

In Eq. (2.6.1) the bending contribution of a bond angle departing from the preferred value θ^0 dictated by chemical bonding is expressed as a quadratic function in $\cos\theta$. Alternative expressions of the form $\frac{1}{2}K_{ijk}\left(\theta_{ijk} - \theta^0_{ijk}\right)^2$ are also widespread. Generally, less energy is required to distort an angle than to stretch or compress a bond, K_{ijk} typically being on the order of 10^{-2} kJ mol^{-1}rad^{-2}. Values of K_{ijk} and θ^0_{ijk} are obtained by fitting electronic structure calculations.

2.6.3 Torsion

As already pointed out in Chapter 1, torsion angles are extremely important for the flexibility leading to the multitude of conformations adopted by macromolecules and hence to the remarkable properties of soft matter. The torsion angle ϕ formed by a linear sequence of bonded atoms $ijkm$ is the dihedral angle between planes ijk and jkm. Care should be taken concerning the conventions for measuring ϕ. As explained in Chapter 1, a convention used by Flory and collaborators[57] is to take $\phi = 0$ in the *trans* (antiperiplanar) conformation, with i, j, k, m coplanar and bonds ji and km pointing to different sides of the central bond jk. A newer convention adopted by IUPAC, on the other hand, is to take $\phi = 0$ in the *cis* (synperiplanar) conformation, with i, j, k, m coplanar and bonds ji and km pointing to the same side of the central bond jk. Changes in ϕ are generally taken as positive when they would lead to unscrewing if the central bond jk were a right-handed screw.

The torsional potential $\mathcal{V}_\phi(\phi)$ is clearly a periodic function of ϕ, with period 2π. It is thus natural to express it as a Fourier series in ϕ. The expression in Eq. (2.6.1) is equivalent to a simple Fourier series, with n_{jk} local minima equal in depth, one being at ϕ^0_{jk}. For torsion around the skeletal bond of a normal alkane, for example, a natural choice would be $n_{jk} = 3$, corresponding to the *trans*, *gauche*$^+$, and *gauche*$^-$ conformational states. V_{jk} is the conformational energy barrier separating the states; it is typically on the order of 10 kJ mol^{-1} and can therefore be overcome at ordinary temperatures. Torsional potentials involving minima and barriers of unequal depth can be described through a generalization of the term appearing in Eq. (2.6.1) to

$$\mathcal{V}_\phi(\phi) = \sum_{n=1}^{N} \frac{\mathcal{V}^{(n)}}{2} \left\{1 - \cos\left[n(\phi - \phi^0)\right]\right\}. \tag{2.6.3}$$

Expansion of the cosine term in Eq. (2.6.3) leads to the equivalent expression

$$\mathcal{V}_\phi(\phi) = \sum_{n=0}^{N} C^{(n)} \cos^n(\phi - \phi^0). \tag{2.6.4}$$

2.6.4 Out-of-plane bending

The term marked *planar ijkm* in Eq. (2.6.1) is intended to keep four atoms connected by bonds in a nonlinear manner, with a central atom being bonded to three peripheral atoms (see Figure 2.2) close to a planar configuration. To define the angle Ψ one can project one of the peripheral atoms on the plane defined by the remaining three. Such a term, for example, could be used to keep a C-H bond in benzene coplanar with the two C-C bonds formed by the aromatic carbon from which it emanates.

The term marked *inversions ijkm* in Eq. (2.6.1) plays a similar role for quartets of nonlinearly connected atoms that must be kept near a fixed nonplanar geometry. Ψ^0 is the value near which Ψ

should be kept at all times. Such terms are often used to prevent the inversion of chiral centres in the course of molecular dynamics simulations of organic molecules.

An alternative to using out-of-plane bending expressions such as those of Eq. (2.6.1) is to define an *improper* torsion angle from the positions of the four centres i, j, k, m (improper in the sense that these centres are not linearly connected) and to use a term of the form of Eq. (1.1.39) to maintain the improper torsion angle near prescribed values (e.g. 0 and π for a planar arrangement).[36]

2.6.5 Excluded volume and dispersion interactions

The double sum in Eq. (2.6.1) corresponds to nonbonded interactions between pairs of sites i and j. The two sites in the pair (i,j) may belong to different molecules, or to different images of the same molecule in a spatially periodic model system subjected to simulation, in which case we speak of intermolecular interactions; or they may belong to the same image of the same molecule, their distance varying as the conformation of the molecule varies, in which case we speak of intramolecular nonbonded interactions. The first term in the sum represents excluded volume and dispersion interactions between sites i and j. The second term in the sum represents Coulomb interactions between partial charges located at sites i and j.

In a fully atomistic model, i and j in the excluded volume and dispersion terms are taken as the nuclear centre positions of atoms i and j, the distance between them being r_{ij}. In the attractive London dispersion term one can recognize the r_{ij}^{-6} dependence predicted by the London analysis (compare Eqs. (2.3.19) to (2.3.21)).

At small internuclear distances a steeply repulsive interaction potential develops between neutral atoms with a closed-shell electronic configuration. This harsh *excluded volume* repulsion, also referred to as *exchange force* or *overlap force*, can be understood in terms of the Pauli exclusion principle, Eq. (2.4.7). Placing two atoms on top of each other would result in electrons of the same spin occupying the same position in space, which is forbidden by the antisymmetry of the many-electron wave function. Pairs of electrons with the same quantum numbers thus move away from the internuclear region. The reduced electron density in the internuclear region leads to repulsion between the incompletely shielded nuclei. At very short internuclear separations the repulsive interaction energy should scale as $1/r_{ij}$ due to the nuclear repulsion. At larger separations the repulsive energy is predicted to vary as $\exp(-r_{ij}/a_0)$, with a_0 being the Bohr radius, Eq. (2.4.2). Combining this exponentially decaying repulsive term with the r_{ij}^{-6} term of the dispersion attraction, one would obtain the following expression for the nonbonded interaction between a pair of neutral atoms:[58]

$$\mathcal{V}^{\mathrm{NB}}(r_{ij}) = A_{ij}\exp(-C_{ij}r_{ij}) - \frac{B_{ij}}{r_{ij}^6}. \tag{2.6.5}$$

In practice, use of an empirical r_{ij}^{-12} repulsive term in place of the theoretically more justifiable exponential has prevailed, leading to the celebrated potential first proposed by Lennard-Jones[59] in 1924, also referred to as '12-6' potential

$$\mathcal{V}_{\mathrm{LJ}}(r_{ij}) = 4\varepsilon_{ij}\left[\left(\frac{\sigma_{ij}}{r_{ij}}\right)^{12} - \left(\frac{\sigma_{ij}}{r_{ij}}\right)^{6}\right] = \varepsilon_{ij}\left[\left(\frac{r_{\mathrm{m},ij}}{r_{ij}}\right)^{12} - 2\left(\frac{r_{\mathrm{m},ij}}{r_{ij}}\right)^{6}\right]. \tag{2.6.6}$$

The r_{ij}^{-12} repulsive term offers the computational advantage that it can be formed directly by squaring the r_{ij}^{-6} term. The r_{ij}^{-12} repulsive energy is quite satisfactory in describing interactions between noble gas atoms. It is too steep for other systems, such as hydrocarbons, however.[36]

The Lennard-Jones potential (see Figure 2.2) requires two parameters per pair, the collision diameter σ_{ij} and the well depth ε_{ij}. Alternatively, the distance at the minimum, $r_{\mathrm{m},ij} = 2^{1/6}\sigma_{ij}$, can be used in place of σ_{ij}, as shown in Eq. (2.6.6).

As the number of pairs of atomic species is much greater than the number of atomic species, it makes sense to tabulate the parameters ε_{ii} and σ_{ii} for homonuclear pairs and then estimate from those ε_{ij} and σ_{ij} for heteronuclear pairs by suitable combining rules. For this purpose, the Lorentz–Berthelot rules are commonly invoked:

$$\sigma_{ij} = \frac{1}{2}(\sigma_{ii} + \sigma_{jj}) \tag{2.6.7}$$

$$\varepsilon_{ij} = (\varepsilon_{ii}\varepsilon_{jj})^{1/2}. \tag{2.6.8}$$

The arithmetic mean combining rule, Eq. (2.6.7), is based on a picture of atoms at the potential energy minimum as tangent spheres; the centre-to-centre distance of dissimilar spheres at contact equals the average of their diameters. The geometric mean combining rule, Eq. (2.6.8), is inspired by the proportionality between the dispersion energy and the product of polarizabilities $\alpha_i\alpha_j$ in Eq. (2.3.19). With the interaction energy at a given separation distance between two i centres being proportional to α_i^2 and that between two j centres being proportional to α_j^2, the interaction energy between an i and a j will be proportional to $\alpha_i\alpha_j = \left(\alpha_i^2\alpha_j^2\right)^{1/2}$. This rule is certainly not universally applicable. In fact, the reader may verify that the Slater–Kirkwood Eq. (2.3.20) would lead to a more complex combining rule.

2.6.6 Coulomb interactions

The last nonbonded term in the force field expression, Eq. (2.6.1), describes nonbonded interactions between point charges q_i permanently affixed to specific sites i on the molecules. These are typically smaller than the elementary electronic charge e and are therefore referred to as *partial charges*. Their positions do not necessarily coincide with those of nuclear centres and, in fact, in many cases it is advantageous that they do not. For example, the substantial quadrupole moment of nitrogen, N_2, cannot be reproduced by assigning charges to its two atoms, for these would necessarily have to be 0 for the molecule to be electroneutral. Three or five partial charges distributed along the axis of the molecule are required for this.[36]

Partial charges reflect the spatial distribution of nuclear and electronic charge in a molecule, as shaped by chemical bonding. They should therefore be derivable from quantum mechanical electronic structure calculations (see e.g. Section 2.4.6). A more promising approach is to fit the values and positions of partial charges in such a way that they reproduce the electrostatic potential field generated by a molecule. In terms of the quantities we considered in our discussion of quantum mechanical electronic structure calculations, the electrostatic potential of a molecule, in atomic units, is

$$\Phi(\boldsymbol{r}) = \Phi_{\text{nuclei}}(\boldsymbol{r}) + \Phi_{\text{electrons}}(\boldsymbol{r}) = \sum_{l=1}^{M} \frac{Z_l}{|\boldsymbol{r} - \boldsymbol{R}_l|} - \int \frac{n(\boldsymbol{r}')}{|\boldsymbol{r} - \boldsymbol{r}'|} \mathrm{d}\boldsymbol{r}'. \tag{2.6.9}$$

A common strategy is to construct a 3D grid of points around the molecule. The electrostatic potential $\Phi(\boldsymbol{r})$ at each one of these points is computed according to Eq. (2.6.9) from the quantum mechanical solution to the electronic problem. The partial charges that best reproduce this electrostatic potential are then computed through a least-squares regression procedure. Partial charges derived in this way are referred to as electrostatic potential-based (ESP) charges.

Coulombic interactions between partial charges should not be truncated, as this may cause spurious peaks in the pair distribution functions describing the structure of a material and distort estimates of its thermodynamic properties. Techniques for the summation of Coulombic interactions out to infinite distance, which exploit the spatial periodicity of simulation model systems, have been developed. Such techniques include the Ewald summation method for model systems of modest size and the Particle Particle Particle Mesh (P^3M) for large model systems. The reader is referred to standard simulation texts for more details.[60,61]

In reality the charge distribution in a molecule responds both to changes in the instantaneous configuration of the molecule and to changes in its environment, through polarization effects. Such changes are not captured by the fixed charges envisioned in a force field such that of Eq. (2.6.1). One approach for capturing these effects is to distribute throughout the molecule not only point charges (monopoles), but also point dipoles, quadrupoles, and even higher multipoles, computed from a quantum mechanical wave function. Another approach is to represent a polarizable centre in the spirit of the Drude model as a collection of closely spaced charges which are allowed to vary as the system evolves, but whose total sums to 0. Polarizable models are significantly more expensive computationally than fixed charge models, as they require a self-consistent calculation of the charges and electrostatic field in each iteration. They are used only when polarization effects are likely to be significant, e.g. in simulations of ionic solutions.

In developing molecular mechanics-type force fields, intramolecular nonbonded interactions, whether excluded volume, dispersion, or electrostatic, are typically not taken into account between centres whose distance is kept nearly constant due to chemical bonding in a molecule. For example, 1–3 nonbonded interactions (i.e. interactions between atoms separated by two bonds) are seldom included in a force field; their effect is effectively incorporated into the bending potential of the intervening bond angle. The user of a force field should always be cautious about the range of nonbonded interactions that should be used in conjunction with a particular force field. For example, the intramolecular potential energy of butane can be represented using a threefold symmetric torsional potential of the type appearing in Eq. (2.6.1) if 1–4 interactions (i.e. nonbonded interactions between atoms separated by three bonds) are included in the model. The difference in the Lennard–Jones parameters σ and ε between carbon and hydrogen atoms, along with the threefold symmetric torsional potential, can generate the required energetic difference between *trans* and *gauche* conformational states. Alternatively, however, one can employ a more complex torsional potential expression with unequal local minima, of the type of Eq. (2.6.3) or (2.6.4) and omit 1–4 nonbonded interactions altogether. This is the strategy adopted in most *united-atom* representations of alkanes, where hydrogens are collapsed onto the carbon atoms with which they are connected, to save computer time. One should note that relatively simple united atom models can be remarkably successful, if parameterized judiciously through a combination of quantum mechanical calculations and fitting to experiment. An example is the TraPPE-UA (Transferable Potentials for Phase Equilibria—United Atom) force field developed by Siepmann and collaborators.[62]

At the other end of the spectrum, one has force fields that are much more sophisticated than that of Eq. (2.6.1). For example, the Condensed-Phase Optimized Molecular Potentials for Atomistic Simulation Studies (COMPASS) force field[63] uses anharmonic stretching and bending terms, as well as coupling terms for stretching between adjacent bonds, between stretching and bending, between stretching and torsion, between bending and torsion, between bending of adjacent bond angles, as well as bending–bending–torsion coupling terms:

$$\begin{aligned}
\mathcal{V}(\boldsymbol{r}) = & \sum_{l} \left[k_2^l (l - l^0)^2 + k_3^l (l - l^0)^3 + k_4^l (l - l^0)^4\right] \\
& + \sum_{\theta} \left[k_2^\theta (\theta - \theta^0)^2 + k_3^\theta (\theta - \theta^0)^3 + k_4^\theta (\theta - \theta^0)^4\right] \\
& + \sum_{\phi} \left[k_1^\phi (1 - \cos\phi) + k_2^\phi (1 - \cos 2\phi) + k_3^\phi (1 - \cos 3\phi)\right] + \sum_{\chi} k_2^\chi \chi^2 \\
& + \sum_{l} \sum_{l'} k^{ll'} (l - l^0) \left(l' - l^{0'}\right) + \sum_{l} \sum_{\theta} k^{l\theta} (l - l^0)(\theta - \theta^0) \\
& + \sum_{l} \sum_{\phi} (l - l^0) \left[k_1^{l\phi} \cos\phi + k_2^{l\phi} \cos 2\phi + k_3^{l\phi} \cos 3\phi\right] + \sum_{\theta} \sum_{\theta'} k^{\theta\theta'} (\theta - \theta^0) \left(\theta' - \theta^{0'}\right) \\
& + \sum_{\theta} \sum_{\phi} (\theta - \theta^0) \left[k_1^{\theta\phi} \cos\phi + k_2^{\theta\phi} \cos 2\phi + k_3^{\theta\phi} \cos 3\phi\right] \\
& + \sum_{\theta} \sum_{\theta'} \sum_{\phi} k^{\theta\theta'\phi} (\theta - \theta^0) \left(\theta' - \theta^{0'}\right) \cos\phi + \frac{1}{4\pi\varepsilon_0} \sum_{i} \sum_{j<i} \frac{q_i q_j}{r_{ij}} \\
& + \sum_{i} \sum_{j<i} \varepsilon_{ij} \left[2\left(\frac{r_{ij}^0}{r_{ij}}\right)^9 - 3\left(\frac{r_{ij}^0}{r_{ij}}\right)^6\right].
\end{aligned} \tag{2.6.10}$$

Thanks to extensive efforts in parameterization, COMPASS performs very well, capturing very accurately the volumetric and thermal properties of pure liquids. Its performance in predicting mixture properties may be less satisfactory, as these have not been included in the parameterization.

2.7 Example: development of a force field for poly(dimethylsilylenemethylene)

As an example of how quantum mechanical electronic structure calculations can be used for the development of a molecular mechanics force field for classical mechanical simulations we briefly discuss work by Raptis and Melissas[64] on poly(dimethylsilylenemethylene) -[$Si(CH_3)_2$-CH_2]$_n$-, abbreviated as PDMSM in the following.

Quantum mechanical electronic structure calculations were conducted on the dimeric analogue di(trimethylsilyl) methylene, to be referred to as DTMSM in the following, with the objective of developing a united-atom (UA) model wherein methyls and methylenes are represented as single quasiatomic sites. Thus, three types of sites were used: Si, CH_3, and CH_2. The molecular mechanics model postulated was of the form of Eq. (2.6.1) with bond angle bending terms proportional to $\left(\theta_{ijk} - \theta^0_{ijk}\right)^2$, with $n_{jk} = 3$ and $\phi^0_{ij} = \pi$ in all torsional terms, and with no improper torsion terms. Coulombic interactions were neglected, on the basis of the argument that no significant dipole moments or ions are present in the system of interest; therefore, nonbonded interactions can be adequately described by Lennard–Jones potentials.

In the molecular mechanics model invoked by Raptis and Melissas[64] for PDMSM, two sets of σ and ε parameters were employed: one for *local* nonbonded interactions between sites separated by three or four bonds along the same chain, and a second one for *nonlocal* nonbonded interactions, i.e. intramolecular nonbonded interactions between sites separated by five or more bonds or intermolecular interactions. Local interactions, along with all bonded (stretching, bending, and torsional) terms in the molecular mechanics force field, are very important for chain conformational properties, while nonlocal interactions are very important for thermodynamic properties in the melt state. Distinguishing between local and nonlocal nonbonded parameters allows for separate refinement of each parameter class and is thus a promising strategy for developing portable force fields for polymers. In the case of PDMSM, local and bonded parameters were determined by fitting to quantum mechanical electronic structure calculations on DTMSM, while nonlocal parameters were fit to monomer and dimer thermodynamic data.[65]

Quantum mechanical calculations on DTMSM were conducted with both DFT and second-order Møller-Plesset perturbation theory (MP2), using the Gaussian 98 suite of programs.[64] To map out the potential energy hypersurface of DTMSM with quantum mechanical calculations, the two main dihedral angles $\phi_1 = \phi_{1234}$ and $\phi_2 = \phi_{2345}$ were varied systematically in steps of 30°. For each combination of (ϕ_1, ϕ_2) values, a constrained minimization of the energy was undertaken with respect to all other degrees of freedom. Because of the high symmetry characterizing the molecule, ten (ϕ_1, ϕ_2) combinations, not including the global energy minimum, sufficed for probing its full conformational space. Additional constrained minimizations were undertaken with quantum mechanics, in which ϕ_1 and ϕ_2 were fixed at 180° (*trans* states) and selected bond angles θ were varied by $\pm1°$, $\pm2°$, and $\pm3°$ from their values at the *trans–trans* energy minimum.

To determine the molecular mechanics force field parameters on the basis of these quantum mechanical calculations, a two-stage procedure was adopted. The first stage involved a minimization, with respect to the force field parameters $\boldsymbol{p}$, of the sum, over all studied configurations i, of the squared deviation between the quantum mechanically computed energy $\mathcal{V}_i^{\mathrm{MP2}}$ relative to the global minimum and the corresponding force-field-based potential energy $\mathcal{V}_i^{\mathrm{FF}}(\boldsymbol{p}; \boldsymbol{r}_i)$ by minimizing the following function F with respect to $\boldsymbol{p}$:

$$F(\boldsymbol{p}) = \frac{\sum_{i=1}^{N_{\mathrm{conf}}} \left(\mathcal{V}_i^{\mathrm{MP2}} - \mathcal{V}_i^{\mathrm{FF}}(\boldsymbol{p}; \boldsymbol{r}_i)\right)^2}{\sum_{i=1}^{N_{\mathrm{conf}}} \left(\mathcal{V}_i^{\mathrm{MP2}}\right)^2}. \tag{2.7.1}$$

Here N_{conf} is the total number of configurations for which a quantum mechanical electronic structure calculation was undertaken and $\boldsymbol{r}_i$ is the quantum mechanically determined set of nuclear coordinates at configuration i. In a second stage, molecular mechanics energy minimizations on the DTMSM molecule were conducted, both in the absence of constraints and in the presence of constraints, setting the (ϕ_1, ϕ_2) angles equal to the values used in the quantum mechanical calculations. The objective of this stage was to determine how close the molecular mechanics-based minimum energy configurations would be to their quantum mechanically computed analogues, $\boldsymbol{r}_i$.

The final parameter set of the molecular mechanics force field[64] was used to quantify the conformational stiffness of PDMSM. To this end single chains of various lengths were sampled with Monte Carlo simulation, systematically varying the range of local interactions until an asymptotic value of the mean square end-to-end distance was reached. A characteristic ratio of 5.4 ± 0.2 was estimated at 300 K, in excellent agreement with the available experimental value.

Problems

Problem 2.1: Prove that eigenfunctions corresponding to distinct eigenvalues of a Hermitian operator are orthogonal.

Problem 2.2: Derive Heisenberg's uncertainty principle, Eq. (2.2.29), for a single particle moving in a one-dimensional potential energy field, $\mathcal{V}(x)$, by invoking the Schwarz inequality, Eq. (2.2.30).

Problem 2.3: For the position and momentum operators $\mathcal{A}$ and B of a one-dimensional system calculate the operator $\mathcal{AB} - \mathcal{BA}$, Eq. (2.2.31). Also calculate the expectation value $\langle \mathcal{BA} \rangle$ and show that it is equal to $\langle \mathcal{AB} \rangle^*$.

Problem 2.4: Prove Eq. (2.4.31) for the elements of the Fock matrix.

Problem 2.5: Derive Eqs. (2.4.37a) and (2.4.37b) for the Hartree–Fock energy E_{HF}.

Problem 2.6: Prove the Hellman–Feynman theorem.

Problem 2.7: Prove the first theorem of Hohenberg and Kohn (HK1).

Problem 2.8: Starting from the solution of the Schrödinger equation for a particle in a box, prove that the kinetic energy density of a homogeneous electron gas of density n is $C_{\text{TF}} E_{\text{a}} a_0^2 n^{5/3}$.

References

1. Car, R., and Parrinello, M. *Phys. Rev. Lett.* **55**, 2471 (1985).
2. Heuts, J.P.A., Gilbert, R.G., and Radom, L. *Macromolecules* **28**, 8771 (1995).
3. Saitta, A.M., and Klein, M.L. *J. Chem. Phys.* **111**, 9434 (1999).
4. Röthlisberger, U., Sprik, M., and Klein, M.L. *J. Chem. Soc. Faraday Trans.* **94**, 501 (1998).
5. Van Duin, A.C.T., Dasgupta, S., Lorant, F., and Goddard, W.A. *J. Phys. Chem. A* **105**, 9396 (2001).
6. Tersoff, J. *Phys. Rev. B* **37**, 6991 (1988).
7. Stuart, S.J., Tutain, A.B., and Harrison, J.A., *J. Chem. Phys.* **112**, 6472 (2000).
8. Brédas, J.-L., Coropceanu, V., and Cornil, J. *Chem. Rev.* **104**, 4971 (2004).

9. Feng, X.L., Marcon, V., Pisula, W., Hansen, M.R., Kirkpatrick, J., Grozema, F., Andrienko, D., Kremer, K., and Müllen, K. *Nat. Mater.* **8**, 421 (2009).
10. Janssen, R.H.C., Bomont, J.-M., Theodorou, D.N., Raptis, S., and Papadopoulos, M.G. *J. Chem. Phys.* **110**, 6463 (1999).
11. Devanathan, R., Venkatnathan, A., Rousseau, R., Dupuis, M., Frigato, T., Gu, W., and Helms, V. *J. Phys. Chem. B* **114**, 13681 (2010).
12. Kittel, C. *Introduction to Solid State Physics* (7th ed.). Wiley, New York (1996).
13. Gō, N., and Scheraga, H.A., *Macromolecules* **9**, 535 (1976).
14. Rutledge, G.C., in Kotelyanskii, M., and Theodorou, D.N. (eds.), *Simulation Methods for Polymers.* Marcel Dekker, New York (2004), pp 359–388.
15. Much of the development in this chapter is based on the class notes developed by Arup Chakraborty for the graduate class *Molecular Theory for Chemical Engineers* co-taught by Arup Chakraborty and Doros Theodorou in the Chemical Engineering Department at the University of California, Berkeley in 1993.
16. Goldstein, H. *Classical Mechanics* (2nd ed.). Addison-Wesley, New York (1980).
17. Wylie, C.R., and Barrett, L.C. *Advanced Engineering Mathematics* (5th ed.). McGraw-Hill, New York (1982).
18. Feynman, R.P., and Hibbs, A.R. *Quantum Mechanics and Path Integrals.* McGraw-Hill, New York (1965).
19. Schrödinger, E. *Phys. Rev.* **28**, 1049 (1926).
20. Heisenberg, W. *Z. Phys.* **43**, 172 (1927).
21. Schiff, L.I. *Quantum Mechanics.* McGraw-Hill, New York (1968).
22. London, F.Z. *Phys. Chem. (B)* **11**, 222 (1930).
23. Maitland, G.C., Rigby, M., Smith, E.B., and Wakeham, W.A. *Intermolecular Forces: Their Origin and Determination.* Clarendon Press, Oxford (1981).
24. Hirschfelder, J.O., Curtiss, C.F., and Bird, R.B. *Molecular Theory of Gases and Liquids.* Wiley, New York (1954).
25. Ketelaar, J.A.A. *Chemical Constitution: An Introduction to the Theory of the Chemical Bond* (2nd ed.). Elsevier: Amsterdam (1958).
26. Hamaker, H. *Physica (Amsterdam)* **4**, 1058 (1937).
27. Hiemenz, P.C. *Principles of Colloid and Surface Chemistry* (2nd ed.). Marcel Dekker, New York (1977).
28. Uhlenbeck, G.E., and Goudsmit, S. *Naturwissenschaften* **47**, 953 (1925).
29. Dirac, P.A.M. *Proc. Royal Soc. London A* **117**, 610 (1928).
30. Pauli, W. *Z. Phys.* **31**, 765 (1925).
31. Hartree, D.R. *Proc. Camb. Phil. Soc.* **24**, 89 (1928).
32. Fock, V.Z. *Phys.* **61**, 126 (1930).
33. Slater, J. *Phys. Rev.* **36**, 5764 (1930).
34. Roothaan, C.C.J. *Rev. Mod. Phys.* **23**, 69 (1951).
35. Hall, G.G. *Proc. Roy. Soc. A* **205**, 541 (1951).
36. Leach, A.R. *Molecular Modelling: Principles and Applications* (2nd ed.). Pearson Education, Edinburgh (2001).
37. Huzinaga, S. (ed.) *Gaussian Basis Sets for Molecular Calculations.* Elsevier, St Louis, MO (1984).
38. Mulliken, R.S. *J. Chem. Phys.* **23**, 1833 (1955).
39. Feynman, R.P. *Phys. Rev.* **56**, 340 (1939).
40. Hellmann, H. *Einführung in die Quantenchemie.* Franz Deutike, Leipzig (1937).
41. Güttinger, P.Z. Physik **73**, 169 (1932).
42. Pauli, W. *Die allgemeine Prinzipien der Wellenmechanik.* In Geiger, H., and Scheel, K. (eds.), *Handbuch der Physik*, vol. 24. Springer, Berlin (1933).
43. Hohenberg, P., and Kohn, W. *Phys. Rev. B* **136**, 864 (1964).

44. Parr, R.G., and Yang, W. *Density-Functional Theory of Atoms and Molecules*. Oxford University Press, New York (1989).
45. Thomas, L.H. *Proc. Camb. Phil. Soc.* **23**, 542 (1927).
46. Fermi, E. *Rend. Acc. Naz. Lincei* **6**, 602 (1927).
47. Dirac, P.A.M. *Proc. Royal Soc. London A* **118**, 351 (1928).
48. Ceperley, D., and Alder, B. *Phys. Rev. Lett.* **45**, 566 (1980).
49. Perdew, J.P., and Zunger, A. *Phys. Rev. B* **23**, 5048 (1981).
50. Kohn, W., and Sham, L.J. *Physical Review* **140**, A1133 (1965).
51. Becke, A.D., and Dickson, R.M. *J. Chem. Phys.* **92**, 3610 (1990).
52. Ashcroft, N.W., and Mermin, N.D. *Solid State Physics*. Saunders College, Philadelphia (1976).
53. Becke, A.D. *Phys. Rev. A* **38**, 2098 (1988).
54. Lee, C., Yang, W., and Parr, R.G. *Phys. Rev. B* **37**, 785 (1988).
55. Mayo, S.L., Olafson, B.D., and Goddard, W.A. *J. Phys. Chem.* **94**, 8897 (1990).
56. Beyer, M. *J. Chem. Phys.* **112**, 7307 (2000).
57. Mattice, W.L., and Suter, U.W. *Conformational Theory of Large Molecules*. Wiley, New York (1994).
58. Buckingham, R.A. *Proc. Roy. Soc. London A* **168**, 264 (1938).
59. Lennard-Jones, J.E. *Proc. Roy. Soc. London A* **106**, 463 (1924).
60. Allen, M.P., and Tildesley, D.J. *Computer Simulations of Liquids*. Clarendon Press, Oxford (1989).
61. Frenkel, D., and Smit, B. *Understanding Molecular Simulation: From Algorithms to Applications*. Academic Press, New York (2002).
62. See http://trappe.oit.umn.edu/ and references therein.
63. Sun, H. *J. Phys. Chem. B* **102**, 7338 (1998).
64. Raptis, V.E., and Melissas, V.S. *J. Phys. Chem.* **110**, 14929 (2006).
65. Economou, I.G., Raptis, V.E., Melissas, V.S., Theodorou, D.N., Petrou, J., and Petropoulos, J.H. *Fluid Phase Equilib.* **228–229**, 15 (2005).

3

Molecular Mechanics

The simplest calculation one can attempt on a molecular model is to minimize its potential energy $\mathcal{V}(\boldsymbol{r})$ with respect to all microscopic degrees of freedom (the set $\{\boldsymbol{r}\}$). Local energy minima created in this way are very useful in computational materials science for several reasons.

First, they constitute excellent starting points for conducting Monte Carlo (MC) or molecular dynamics simulations (MD) which sample thermal fluctuations at finite temperature. Equilibrium statistical mechanics teaches us that the probability of a microscopic configuration $\{\boldsymbol{r}\}$ adopted by a system of given mass, spatial extent, and temperature T is proportional to the Boltzmann factor $\exp\left[-\frac{\mathcal{V}(\boldsymbol{r})}{k_B T}\right]$ of its potential energy. Thus, macroscopic properties are shaped primarily by low-energy configurations. Starting off a molecular simulation at or near a potential energy minimum makes sense, as one is already within a probable region of configuration space and does not need to expend valuable computation time to reach the part of configuration space that mainly contributes to properties; in technical terms, the *equilibration part* of the simulation is reduced.

Especially in the crystalline state, potential energy minima and how they respond to strain are key to understanding cohesion and elastic behaviour. Thermodynamic equilibrium under given temperature T and pressure P requires that the Gibbs energy $G = U + PV - TS$, with U, V, S symbolizing the internal energy, volume, and entropy, respectively, be minimal with respect to the imposition of any internal constraint in the system.[1] In condensed phases at ordinary pressures the PV term is small relative to the other two, and equilibrium largely entails a tug of war between reducing energy and increasing entropy. At low temperature, the role of the TS term becomes less important. The tug of war is won by ordered phases of low U and low S, and energy dominates properties.

Minima of the potential energy play a significant role not only in crystalline, but also in amorphous solids, i.e. in the glassy state, which is extremely important from the point of view of applications of polymers. As we shall see, a useful molecular picture of a glass is to consider that the configuration of a glassy region of given mass and spatial extent is locked in the vicinity of a local minimum of the potential energy $\mathcal{V}(\boldsymbol{r})$ with respect to the set $\{\boldsymbol{r}\}$. The multidimensional potential energy hypersurface $\mathcal{V}(\boldsymbol{r})$ is very rugged, characterized by the presence of a multitude of such minima. The probability distribution of minima (or *inherent structures*[2]) around which a macroscopic glass is locked in depends on its formation history. Good estimates of the macroscopic properties of the glass can be obtained as averages of the corresponding properties of the configuration-space

Multiscale Modelling of Polymers. Doros N. Theodorou and Vlasis G. Mavrantzas, Oxford University Press.
 DOI: 10.1093/oso/9780199226191.003.0003

regions (*basins*[2]) around the different inherent structures near which it is locked in. Being able to generate representative minimum energy configurations is thus of paramount importance for modelling polymer glasses.

Potential energy minimization with respect to all microscopic degrees of freedom is often referred to as *molecular mechanics*. In this chapter we discuss how to generate dense amorphous multichain minimum energy configurations from atomistic models of polymers, with emphasis on modelling glassy polymers, and how to use these configurations in order to predict structural, volumetric, and thermodynamic properties. To fully appreciate the conditions under which molecular mechanics can be useful, we will discuss in considerable detail the statistical mechanics of condensed multichain polymer phases. The statistical mechanical foundations we will introduce here will also form a basis for our discussion of simulation approaches incorporating thermal motion in Chapters 4 and 5.

3.1 Molecular models for polymer configurations

We consider a system of N_c chains, each containing $N+1$ skeletal atoms, so that the total number of atomic coordinates is $n=(3N+3)N_c$. In a classical, fully flexible description, the microscopic state of the system is defined by specifying the configuration space $\{\boldsymbol{r}\} \equiv \{\boldsymbol{r}_{0(1)}, \boldsymbol{r}_{1(1)}, \ldots, \boldsymbol{r}_{N(1)}, \boldsymbol{r}_{0(2)}, \boldsymbol{r}_{1(2)}, \ldots, \boldsymbol{r}_{N(2)}, \ldots, \boldsymbol{r}_{0(N_c)}, \boldsymbol{r}_{1(N_c)} \ldots, \boldsymbol{r}_{N(N_c)}\}$ of the Cartesian coordinates of all atoms and the space $\{\boldsymbol{p}\} \equiv \{\boldsymbol{p}_{0(1)}, \boldsymbol{p}_{1(1)}, \ldots, \boldsymbol{p}_{N(1)}, \boldsymbol{p}_{0(2)}, \boldsymbol{p}_{1(2)}, \ldots, \boldsymbol{p}_{N(2)}, \ldots, \boldsymbol{p}_{0(N_c)}, \boldsymbol{p}_{1(N_c)}, \ldots, \boldsymbol{p}_{N(N_c)}\}$ of their conjugate momenta, where $\boldsymbol{r}_{i(k)}$ denotes the position vector and $\boldsymbol{p}_{i(k)}$ the momentum vector of the ith atom in the kth chain, $i=0,1,2,\ldots,N$ and $k=1,2,\ldots,N_c$ (compare Section 1.1.1 and Figure 1.1). If atom i has mass m_i, $\boldsymbol{p}_{i(k)}=m_i\dot{\boldsymbol{r}}_{i(k)}$. Much of the analysis in this and the subsequent chapters makes use of the canonical partition function $Q(N_c,N,V,T)$ of such a system. To calculate $Q(N_c,N,V,T)$, we should choose an appropriate molecular model for polymer configurations. Traditionally, a number of such models have been proposed:[3–8] (i) the flexible model, (ii) the flexible model in the limit of infinite (bond length and/or bond angle) stiffness, and (iii) the rigid model. We discuss all three of them in this section.

3.1.1 The flexible model

In the flexible model, no geometric constraints are imposed on the coordinates of atoms along the polymer chain. Thus, the kinetic energy of the system is the sum of the individual uncoupled kinetic energies of all atoms. For this model, then, according to classical statistical mechanics,

$$Q=\frac{1}{h^n N_c!}\int \exp[-\beta\mathcal{H}(\boldsymbol{r},\boldsymbol{p})]\mathrm{d}\boldsymbol{r}\,\mathrm{d}\boldsymbol{p}, \tag{3.1.1}$$

where the Hamiltonian $\mathcal{H}(\boldsymbol{r},\boldsymbol{p})$ denotes the total energy of the system,

$$\mathcal{H}(\boldsymbol{r},\boldsymbol{p})=\mathcal{K}(\boldsymbol{p})+\mathcal{V}(\boldsymbol{r})=\sum_{i=1}^{n}\frac{\boldsymbol{p}_i^2}{2m_i}+\mathcal{V}(\boldsymbol{r}), \tag{3.1.2}$$

i.e. the sum of the kinetic energy $\mathcal{K}$ and of the potential energy $\mathcal{V}$, and $\beta = \frac{1}{k_B T}$. By substituting Eq. (3.1.2) into (3.1.1) and carrying out the integrations with respect to the particle momenta, we obtain

$$Q = \frac{1}{N_c!}\left[\prod_{i=0}^{N}\left(\frac{2\pi m_i k_B T}{h^2}\right)^{3/2}\right]^{N_c} \int \exp[-\beta\mathcal{V}(\boldsymbol{r})]d\boldsymbol{r}. \tag{3.1.3}$$

Equation (3.1.3) is the partition function for the flexible model cast in terms of Cartesian atomic coordinates. We wish to recast it in terms of the internal coordinates discussed briefly also in Section 1.1.6. Since we are addressing a multichain system (as opposed to a single-chain system discussed there), the configuration space consists of the space $\{\boldsymbol{r}_0\} \equiv \{\boldsymbol{r}_{0(1)}, \boldsymbol{r}_{0(2)}, \ldots, \boldsymbol{r}_{0(N_c)}\}$ of chain starts, of the space $\{\psi\} \equiv \{\psi_{1(1)}, \psi_{2(1)}, \psi_{3(1)}, \psi_{1(2)}, \psi_{2(2)}, \psi_{3(2)}, \ldots, \psi_{1(N_c)}, \psi_{2(N_c)}, \psi_{3(N_c)}\}$ of Euler angles, of the space $\{\boldsymbol{\theta}\} \equiv \{\theta_{1(1)}, \theta_{2(1)}, \ldots, \theta_{(N-1)(1)}, \theta_{1(2)}, \theta_{2(2)}, \ldots, \theta_{(N-1)(2)}, \ldots, \theta_{1(N_c)}, \theta_{2(N_c)}, \ldots, \theta_{(N-1)(N_c)}\}$ of bond angle supplements (compare Figure 1.7), of the space of skeletal bond lengths $\{\boldsymbol{l}\} \equiv \{l_{1(1)}, l_{2(1)}, \ldots, l_{N(1)}, l_{1(2)}, l_{2(2)}, \ldots, l_{N(2)}, \ldots, l_{1(N_c)}, l_{2(N_c)}, \ldots, l_{N(N_c)}\}$, and of the space $\{\boldsymbol{\phi}\} \equiv \{\phi_{2(1)}, \phi_{3(1)}, \ldots, \phi_{(N-1)(1)}, \phi_{2(2)}, \phi_{3(2)}, \ldots, \phi_{(N-1)(2)}, \ldots, \phi_{2(N_c)}, \phi_{3(N_c)}, \ldots, \phi_{(N-1)(N_c)}\}$ of torsion angles. By denoting this space as $\{\boldsymbol{r}_0, \psi, \boldsymbol{l}, \boldsymbol{\theta}, \boldsymbol{\phi}\}$, the transformation from Cartesian atomic coordinates to internal coordinates is described through

$$\{\boldsymbol{r}\} \to \{\boldsymbol{r}_0, \psi, \boldsymbol{l}, \boldsymbol{\theta}, \boldsymbol{\phi}\}. \tag{3.1.4}$$

If

$$\boldsymbol{J}(\boldsymbol{r}_0, \psi, \boldsymbol{l}, \boldsymbol{\theta}, \boldsymbol{\phi}) = \frac{\partial\{\boldsymbol{r}\}}{\partial\{\boldsymbol{r}_0, \psi, \boldsymbol{l}, \boldsymbol{\theta}, \boldsymbol{\phi}\}} \tag{3.1.5}$$

denotes the Jacobian matrix of this transformation, then the partition function of the system in the phase space of generalized coordinates is written as

$$Q = \frac{1}{N_c!}\left[\prod_{i=0}^{N}\left(\frac{2\pi m_i k_B T}{h^2}\right)^{3/2}\right]^{N_c} \int \det(\boldsymbol{J}) \exp[-\beta\mathcal{V}(\boldsymbol{r}_0, \psi, \boldsymbol{l}, \boldsymbol{\theta}, \boldsymbol{\phi})]d\boldsymbol{r}_0\, d\psi\, d\boldsymbol{l}\, d\boldsymbol{\theta}\, d\boldsymbol{\phi}. \tag{3.1.6}$$

Further progress requires that we calculate the determinant of the Jacobian $\boldsymbol{J}(\boldsymbol{r}_0, \boldsymbol{l}, \boldsymbol{\theta}, \boldsymbol{\phi})$. For linear polymers, this comes out to be (see also Problem 3.1)

$$\det(\boldsymbol{J}) = \prod_{k=1}^{N_c}\left[\sin\psi_{2(k)}\left(\prod_{b=1}^{N} l_{b(k)}^2\right)\left(\prod_{c=1}^{N-1}\sin\theta_{c(k)}\right)\right]. \tag{3.1.7}$$

A remarkable attribute of $\det(\boldsymbol{J})$ is that it does not depend on the torsion angles $\boldsymbol{\phi}$. The reader is asked to prove Eq. (3.1.7) in Problem 3.1. The Jacobian of the transformation from generalized

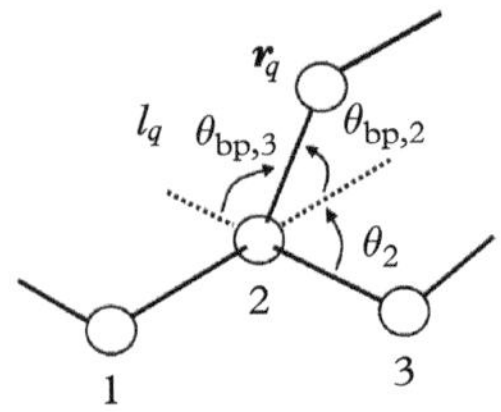

Figure 3.1 Local geometry around a branch point.[9] The position of the branch point can be specified either through its Cartesian coordinates $\boldsymbol{r}_q$ or through the length l_q of the pendant bond connecting it to skeletal atom 2 and the bond angles $\pi - \theta_{bp,2}$, $\pi - \theta_{bp,3}$ formed between the pendant bond and skeletal bonds 2 and 3, respectively. The bond angle between skeletal bonds 2 (connecting atoms 1 and 2) and 3 (connecting atoms 2 and 3) along the chain is $\pi - \theta_2$.

Reprinted (adapted) with permission from Peristeras, L.D., Economou, I.G., and Theodorou, D.N., *Macromolecules* **38**, 386 (2005). Copyright 2005 American Chemical Society.

to Cartesian coordinates at a branch point (see Figure 3.1) can be calculated in a similar fashion (please see Problem 3.2).

By substituting Eq. (3.1.7) into (3.1.6), we get

$$Q = \frac{1}{N_c!}\left[\prod_{i=0}^{N}\left(\frac{2\pi m_i k_B T}{h^2}\right)^{3/2}\right]^{N_c}$$
$$\times \int \prod_{k=1}^{N_c}\left[\sin\psi_{2(k)}\left(\prod_{b=1}^{N} l_{b(k)}^2\right)\left(\prod_{c=1}^{N-1}\sin\theta_{c(k)}\right)\right]\exp[-\beta\mathcal{V}(\boldsymbol{r}_0,\boldsymbol{\psi},\boldsymbol{l},\boldsymbol{\theta},\boldsymbol{\phi})]\mathrm{d}\boldsymbol{r}_0\,\mathrm{d}\boldsymbol{\psi}\,\mathrm{d}\boldsymbol{l}\,\mathrm{d}\boldsymbol{\theta}\,\mathrm{d}\boldsymbol{\phi}. \tag{3.1.8}$$

This is the partition function for the flexible model of polymer configurations.

3.1.2 The flexible model in the limit of infinitely stiff vibrational degrees of freedom

As mentioned in Sections 1.1.6 and 2.6, successive atoms along a polymer chain interact with each other through strong, highly directional covalent bonds; therefore, it is natural to model such molecules as having nearly constant bond lengths. To deal with this, we assume, without loss of generality, that each bond contributes an energy which has the form of a quadratic function,

$$\mathcal{V}_l(l) = \mathcal{V}_0^l + \frac{k_l}{2}(l - l_0)^2, \tag{3.1.9}$$

where $\mathcal{V}_0^l$ is the (chemical) bond energy at the equilibrium length $l = l_0$ and $\frac{1}{2}k_l(l - l_0)^2$ the energetic cost of changing the value of the bond length with the understanding that $k_l \to \infty$ implies that $l \to l_0$. By substituting expression (3.1.9) for $\mathcal{V}_l$ in (3.1.8), carrying out the integrations with respect to the set $\{\boldsymbol{l}\}$ in the limit $k_l \to \infty$ so that[6,8] $\int_0^\infty l^2 \exp\left[-\frac{1}{2}\beta k_l (l-l_0)^2\right] \mathrm{d}l \xrightarrow[k_l\to\infty]{} \left(\frac{2\pi}{\beta k_l}\right)^{1/2} l_0^2$, we obtain that

$$Q = \frac{1}{N_c!}\left\{\left[\prod_{i=0}^{N}\left(\frac{2\pi m_i k_B T}{h^2}\right)^{3/2}\right]\left[\prod_{l=1}^{N}\left(\frac{2\pi k_B T}{k_l}\right)^{1/2} l_0^2\right]\right\}^{N_c}$$
$$\int \prod_{k=1}^{N_c}\left[\sin\psi_{2(k)}\left(\prod_{c=1}^{N-1}\sin\theta_{c(k)}\right)\right]\exp[-\beta\mathcal{V}(\boldsymbol{r}_0,\boldsymbol{\psi},\boldsymbol{l}_0,\boldsymbol{\theta},\boldsymbol{\phi})]\mathrm{d}\boldsymbol{r}_0\,\mathrm{d}\boldsymbol{\psi}\,\mathrm{d}\boldsymbol{\theta}\,\mathrm{d}\boldsymbol{\phi}, \qquad (3.1.10)$$

where $\mathcal{V}$ is evaluated now by setting $l = l_0$ in the expression for the potential energy (i.e. it does not include any bond stretching contributions). Equation (3.1.10) is the partition function for the flexible model in the limit of infinitely stiff bond lengths with stiff degrees of freedom (bond lengths) treated classically.

For a more general formulation, let N_f be the number of soft degrees of freedom (chain start coordinates, Euler angles, and torsion angles, $N_f = (N+4)N_c$ in the system of linear chains considered above) and N_h be the total number of hard degrees of freedom (bond lengths and bond angles, $N_h = (2N-1)N_c$ in the system of linear chains considered above). We will use the symbol $\boldsymbol{q}_f = (\boldsymbol{q}_{f1}, \boldsymbol{q}_{f2}, \ldots, \boldsymbol{q}_{fN_f})$ to denote the soft degrees of freedom of our molecular system (in our example, $\boldsymbol{q}_f = \{\boldsymbol{r}_0, \boldsymbol{\psi}, \boldsymbol{\phi}\}$) and the symbol $\boldsymbol{q}_h = (\boldsymbol{q}_{h1}, \boldsymbol{q}_{h2}, \ldots, \boldsymbol{q}_{hN_h})$ to denote the vector of hard degrees of freedom (in our example, $\boldsymbol{q}_h = \{\boldsymbol{l}, \boldsymbol{\theta}\}$). As above, we will symbolize the Jacobian matrix of transformation from Cartesian to generalized coordinates as

$$\boldsymbol{J} = \frac{\partial \boldsymbol{r}}{\partial \boldsymbol{q}^T} = \frac{\partial \boldsymbol{r}}{\partial(\boldsymbol{q}_f^T, \boldsymbol{q}_h^T)} = \begin{bmatrix} \frac{\partial \boldsymbol{r}_1}{\partial \boldsymbol{q}_{f1}} & \cdots & \frac{\partial \boldsymbol{r}_1}{\partial \boldsymbol{q}_{fN_f}} & \frac{\partial \boldsymbol{r}_1}{\partial \boldsymbol{q}_{h1}} & \cdots & \frac{\partial \boldsymbol{r}_1}{\partial \boldsymbol{q}_{hN_h}} \\ \vdots & & \vdots & \vdots & & \vdots \\ \frac{\partial \boldsymbol{r}_{3N}}{\partial \boldsymbol{q}_{f1}} & \cdots & \frac{\partial \boldsymbol{r}_{3N}}{\partial \boldsymbol{q}_{fN_f}} & \frac{\partial \boldsymbol{r}_{3N}}{\partial \boldsymbol{q}_{h1}} & \cdots & \frac{\partial \boldsymbol{r}_{3N}}{\partial \boldsymbol{q}_{hN_h}} \end{bmatrix} = [\boldsymbol{J}_f, \boldsymbol{J}_h]. \qquad (3.1.11)$$

In the above, superscript T symbolizes the transpose of a matrix or vector; a vector such as $\boldsymbol{q}$ is a column vector, while $\boldsymbol{q}^T$ is a row vector. How $\det(\boldsymbol{J})$ can be calculated for linear and branched chains is outlined above and in Problems 3.1 and 3.2. It has been shown that $\det(\boldsymbol{J})$ is a function of only $\boldsymbol{q}_h$ and the Euler angles for each chain. The partition function of the classical flexible model can be written as

$$Q = \frac{1}{N_c!}\left[\prod_{i=1}^{N_f+N_h}\left(\frac{2\pi m_i k_B T}{h^2}\right)^{1/2}\right]\int \det(\boldsymbol{J})\exp[-\beta\mathcal{V}(\boldsymbol{q}_f, \boldsymbol{q}_h)]\mathrm{d}\boldsymbol{q}_f\,\mathrm{d}\boldsymbol{q}_h, \qquad (3.1.12)$$

where m_i is the mass associated with atomic coordinate vector $\boldsymbol{r}_i$. Equation (3.1.8) is a special case of Eq. (3.1.12).

The hard degrees of freedom can be assumed to be governed by a harmonic force field. That is, one can set[4]

$$\begin{aligned}\mathcal{V}(\boldsymbol{q}_\mathrm{f},\boldsymbol{q}_\mathrm{h}) &= \mathcal{V}(\boldsymbol{q}_\mathrm{f},\boldsymbol{q}_{\mathrm{h},0}) + \frac{1}{2}\sum_{i=1}^{N_\mathrm{h}}\sum_{j=1}^{N_\mathrm{h}} \left.\frac{\partial^2\mathcal{V}}{\partial q_{\mathrm{h}i}\partial q_{\mathrm{h}j}}\right|_{q_\mathrm{f},q_{\mathrm{h}0}} (q_{\mathrm{h}i}-q_{\mathrm{h}i,0})(q_{\mathrm{h}j}-q_{\mathrm{h}j,0}) \\ &= \mathcal{V}(\boldsymbol{q}_\mathrm{f},\boldsymbol{q}_{\mathrm{h},0}) + \frac{1}{2}(\boldsymbol{q}_\mathrm{h}-\boldsymbol{q}_{\mathrm{h},0})^\mathrm{T} \left.\frac{\partial^2\mathcal{V}}{\partial \boldsymbol{q}_\mathrm{h}\partial \boldsymbol{q}_\mathrm{h}^\mathrm{T}}\right|_{\boldsymbol{q}_\mathrm{f},\boldsymbol{q}_{\mathrm{h}0}} (\boldsymbol{q}_\mathrm{h}-\boldsymbol{q}_{\mathrm{h},0}) \\ &= \mathcal{V}(\boldsymbol{q}_\mathrm{f},\boldsymbol{q}_{\mathrm{h},0}) + \frac{1}{2}(\boldsymbol{q}_\mathrm{h}-\boldsymbol{q}_{\mathrm{h},0})^\mathrm{T}\boldsymbol{H}_{\boldsymbol{qq},\mathrm{hh}}(\boldsymbol{q}_\mathrm{h}-\boldsymbol{q}_{\mathrm{h},0}),\end{aligned} \tag{3.1.13}$$

where $\boldsymbol{q}_{\mathrm{h},0}$ are the values of the hard degrees of freedom at which the potential energy function $\mathcal{V}(\boldsymbol{q}_\mathrm{f},\boldsymbol{q}_\mathrm{h})$ attains a minimum under given values $\boldsymbol{q}_\mathrm{f}$ of the soft degrees of freedom and $\mathcal{V}(\boldsymbol{q}_\mathrm{f},\boldsymbol{q}_{\mathrm{h},0})$ is the value of that minimum. In general, $\boldsymbol{q}_{\mathrm{h},0}$ may depend on $\boldsymbol{q}_\mathrm{f}$, but here this dependence will be neglected. This neglect is legitimate for simple force fields incorporating independent harmonic contributions from individual bond lengths and bond angles, as considered above. We will adopt the symbolism $H_{qq,\mathrm{hh},ij} = \left.\frac{\partial^2 V}{\partial q_{\mathrm{h}i}\partial q_{\mathrm{h}j}}\right|_{\boldsymbol{q}_\mathrm{f},\boldsymbol{q}_{\mathrm{h},0}}$ and $\boldsymbol{H}_{\boldsymbol{qq},\mathrm{hh}}$ for the $N_\mathrm{h}\times N_\mathrm{h}$ Hessian matrix of second derivatives of the potential energy with respect to the hard degrees of freedom. Substituting Eq. (3.1.13) within Eq. (3.1.12) and performing the integration with respect to the hard degrees of freedom in the limit where the force constants associated with the hard degrees of freedom become very large, we obtain that

$$Q = \frac{1}{N_\mathrm{c}!}\left[\prod_{a=1}^{N_\mathrm{f}+N_\mathrm{h}}\left(\frac{2\pi m_a k_B T}{h^2}\right)^{1/2}\right]\int \frac{(2\pi k_\mathrm{B}T)^{N_\mathrm{h}/2}}{\sqrt{\det(\boldsymbol{H}_{\boldsymbol{qq},\mathrm{hh}})}}\det(\boldsymbol{J}|_{\boldsymbol{q}_{\mathrm{h},0}})\exp[-\beta\mathcal{V}(\boldsymbol{q}_\mathrm{f},\boldsymbol{q}_{\mathrm{h},0})]\mathrm{d}\boldsymbol{q}_\mathrm{f}. \tag{3.1.14}$$

Within the integral of Eq. (3.1.14), the Jacobian determinant matrix $\det(\boldsymbol{J})$ is evaluated at $\boldsymbol{q}_{\mathrm{h},0}$. As a consequence (compare Eq. (3.1.7)) most of it can be factored out of the integration,[4] leaving only a term $\prod_{k=1}^{N_\mathrm{c}}\sin\psi_{2(k)}$ having to do with the polar angles of the first bonds of the chains within the integral. Equation (3.1.14) gives the partition function of the classical flexible model in the limit where the force constants of all hard variables become very large. It is a generalization of Eq. (3.1.10) for the case where both bond lengths and bond angles are treated classically and the force constants governing their contributions to the potential energy are considered as approaching infinity.

As pointed out by Gō and Scheraga,[4] the vibrational frequencies associated with hard degrees of freedom are high (in the infrared region between 400 and 3000 cm^{-1}) and therefore a correct treatment of these vibrational motions demands the use of quantum statistical mechanics. For the hard degrees of freedom, the separate integration over momentum and configurational space which led to Eq. (3.1.14) is not legitimate. To proceed with a more rigorous formulation of the partition function for the flexible model, we start by defining the mass-weighted coordinates for the atoms,[7]

$$x_i = m_i^{1/2} r_i \quad (1 \leq i \leq (3N+3)N_c). \tag{3.1.15}$$

The kinetic energy of the system can be written as

$$\begin{aligned} \mathcal{K} &= \frac{1}{2}\sum_{k=1}^{(3N+3)N_c} m_k \dot{r}_k^2 = \frac{1}{2}\sum_{k=1}^{N_f+N_h} \dot{x}_k^2 = \frac{1}{2}\sum_{k=1}^{N_f+N_h}\left(\sum_{i=1}^{N_f+N_h}\frac{\partial x_k}{\partial q_i}\dot{q}_i\right)^2 \\ &= \frac{1}{2}\sum_{i=1}^{N_f+N_h}\sum_{j=1}^{N_f+N_h}\left(\sum_{k=1}^{N_f+N_h}\frac{\partial x_k}{\partial q_i}\frac{\partial x_k}{\partial q_j}\right)\dot{q}_i\cdot\dot{q}_j = \frac{1}{2}\dot{\boldsymbol{q}}^{\mathrm{T}}\cdot\boldsymbol{a}\cdot\dot{\boldsymbol{q}} \end{aligned} \tag{3.1.16}$$

with $\dot{\boldsymbol{q}}$ symbolizing the column vector of time derivatives of all $N_f + N_h$ generalized coordinates, and the $(N_f + N_h) \times (N_f + N_h)$ matrix $\boldsymbol{a}$ being the covariant metric tensor of the transformation from mass-weighted Cartesian coordinates $\{\boldsymbol{x}\}$ to generalized coordinates $\{\boldsymbol{q}\} = \{\boldsymbol{q}_f, \boldsymbol{q}_h\}$:

$$a_{ij} = \sum_{k=1}^{N_f+N_h}\frac{\partial x_k}{\partial q_i}\frac{\partial x_k}{\partial q_j}. \tag{3.1.17}$$

If we define the mass matrix $\boldsymbol{M}$ through

$$\boldsymbol{M} = \begin{bmatrix} m_1 & 0 & \cdots & 0 & 0 \\ 0 & m_2 & \cdots & 0 & 0 \\ \vdots & \vdots & \ddots & \vdots & \vdots \\ 0 & 0 & \cdots & m_{N_f+N_h-1} & 0 \\ 0 & 0 & \cdots & 0 & m_{N_f+N_h} \end{bmatrix}, \tag{3.1.18}$$

then the matrix $\boldsymbol{a}$ can be written in terms of the Jacobian $\boldsymbol{J}$ of Eq. (3.1.11) as

$$\boldsymbol{a} = \frac{\partial \boldsymbol{x}^{\mathrm{T}}}{\partial \boldsymbol{q}}\frac{\partial \boldsymbol{x}}{\partial \boldsymbol{q}^{\mathrm{T}}} = \frac{\partial \boldsymbol{r}^{\mathrm{T}}}{\partial \boldsymbol{q}}\cdot\boldsymbol{M}\cdot\frac{\partial \boldsymbol{r}}{\partial \boldsymbol{q}^{\mathrm{T}}} = \boldsymbol{J}^{\mathrm{T}}\cdot\boldsymbol{M}\cdot\boldsymbol{J}. \tag{3.1.19}$$

Clearly, the matrix $\boldsymbol{a}$ is symmetric, of the form[7]

$$\boldsymbol{a} = \left(\begin{array}{c|c} \boldsymbol{a}_{ff} & \boldsymbol{a}_{fh} \\ \hline \boldsymbol{a}_{hf} & \boldsymbol{a}_{hh} \end{array}\right) \tag{3.1.20}$$

with $\boldsymbol{a}_{hf} = \boldsymbol{a}_{fh}^{\mathrm{T}}$. Submatrix $\boldsymbol{a}_{ff}$ is of dimension $N_f \times N_f$ and contains elements of the form of Eq. (3.1.17) with both indices i and j corresponding to soft generalized coordinates, submatrix $\boldsymbol{a}_{fh}$ is of dimension $N_f \times N_h$ and contains elements of the form of Eq. (3.1.17) with index i corresponding to a soft and index j to a hard generalized coordinate, submatrix $\boldsymbol{a}_{\text{hf}}$ is of dimension $N_\text{h} \times N_\text{f}$ and contains elements of the form of Eq. (3.1.17) with index i corresponding to a hard and index j to a soft generalized coordinate, and submatrix $\boldsymbol{a}_{\text{hh}}$ is of dimension $N_\text{h} \times N_\text{h}$ and contains elements of the form of Eq. (3.1.17) with both indices i and j corresponding to hard generalized coordinates. The $N_\text{f} \times N_\text{f}$ matrix $\boldsymbol{a}_{\text{ff}}$ is the covariant metric tensor in the soft generalized coordinates. It is related to the $3N \times N_\text{f}$ submatrix $\boldsymbol{J}_\text{f}$ of transformation from Cartesian to generalized coordinates, appearing in Eq. (3.1.11), via

$$\boldsymbol{a}_{\text{ff}} = \boldsymbol{J}_\text{f}^\text{T} \cdot \boldsymbol{M} \cdot \boldsymbol{J}_\text{f}. \tag{3.1.21}$$

Correspondingly, one can define a contravariant metric tensor, $\boldsymbol{a}^{-1}$, through

$$(a^{-1})_{ij} = \sum_{k=1}^{N_\text{f}+N_\text{h}} \frac{\partial q_i}{\partial x_k} \frac{\partial q_j}{\partial x_k} \tag{3.1.22}$$

and partition it into soft–soft, soft–hard, hard–soft, and hard–hard submatrices as

$$\boldsymbol{a}^{-1} = \begin{pmatrix} (\boldsymbol{a}^{-1})_{\text{ff}} & | & (\boldsymbol{a}^{-1})_{\text{fh}} \\ -- & + & -- \\ (\boldsymbol{a}^{-1})_{\text{hf}} & | & (\boldsymbol{a}^{-1})_{\text{hh}} \end{pmatrix}, \tag{3.1.23}$$

where, again, $\boldsymbol{a}_{\text{hf}}^{-1} = (\boldsymbol{a}_{\text{fh}}^{-1})^\text{T}$. Note that $(\boldsymbol{a}^{-1})_{\text{ff}} \neq (\boldsymbol{a}_{\text{ff}})^{-1}$ and similarly for the other submatrices appearing in Eqs. (3.1.20) and (3.1.23).

Consider the system residing at a local minimum $\{\boldsymbol{x}_0\}$ of its potential energy $\mathcal{V}$, written as a function of all mass-weighted atomic coordinates $\{\boldsymbol{x}\}$. If the configuration of the system is displaced slightly from the minimum, its potential energy can be approximated through a Taylor expansion around $\{\boldsymbol{x}_0\}$ truncated at the second-order term:

$$\begin{aligned} \mathcal{V}(\boldsymbol{x}) &= \mathcal{V}(\boldsymbol{x}_0) + \frac{1}{2} \sum_{i,j=1}^{N_\text{f}+N_\text{h}} \left.\frac{\partial^2 \mathcal{V}}{\partial x_i \partial x_j}\right|_{\boldsymbol{x}_0} (x_i - x_{i,0})(x_j - x_{j,0}) \\ &= \mathcal{V}(\boldsymbol{x}_0) + \frac{1}{2} (\boldsymbol{x} - \boldsymbol{x}_0)^\text{T} \cdot \boldsymbol{H} \cdot (\boldsymbol{x} - \boldsymbol{x}_0). \end{aligned} \tag{3.1.24}$$

In Eq. (3.1.24), $\boldsymbol{H}$ is the $(N_\text{f} + N_\text{h}) \times (N_\text{f} + N_\text{h})$ Hessian matrix of second derivatives of the potential energy with respect to all mass-weighted coordinates of the atoms, evaluated at the energy minimum

$$H_{ij} = \left.\frac{\partial^2 \mathcal{V}}{\partial x_i \partial x_j}\right|_{\boldsymbol{x}_0}. \tag{3.1.25}$$

This matrix is positive definite, since $\{\boldsymbol{x}_0\}$ is a minimum of the potential energy. We will refer to Eq. (3.1.24) as a harmonic approximation for the energy with respect to the mass-weighted coordinates of the atoms.

With the potential energy function of Eq. (3.1.24), the equations of motion take the form

$$\boldsymbol{M} \cdot \ddot{\boldsymbol{r}} = -\nabla_{\boldsymbol{r}} \mathcal{V}$$

or, equivalently,

$$\ddot{\boldsymbol{x}} = -\nabla_{\boldsymbol{x}} \mathcal{V} = -\boldsymbol{H} \cdot (\boldsymbol{x} - \boldsymbol{x}_0). \tag{3.1.26}$$

From Eq. (3.1.26) we see that the mass-weighted atomic displacements $(\boldsymbol{x} - \boldsymbol{x}_0)$ move as a set of coupled harmonic oscillators. We can define normal modes for the motion as linear combinations of the atomic displacements which behave as *independent* harmonic oscillators. Mathematically, the normal modes and their frequencies are determined as eigenvectors and eigenvalues of the matrix diagonalization problem[7]

$$\boldsymbol{H} \cdot \boldsymbol{x}' = \lambda \boldsymbol{x}'. \tag{3.1.27}$$

In general, Eq. (3.1.27) yields $n = N_\mathrm{f} + N_\mathrm{h}$ eigenvectors $\boldsymbol{x}_i'$ and corresponding eigenvalues λ_i. The reader can verify that the eigenvalues have dimensions of squared inverse time. The vibrational frequency, ν_i, of mode i is related to λ_i by

$$\lambda_i = (2\pi\nu_i)^2. \tag{3.1.28}$$

From the fact that the product of eigenvalues of a matrix equals the determinant of the matrix, we obtain the condition

$$\prod_{i=1}^{N_\mathrm{f}+N_\mathrm{h}} (2\pi\nu_i) = \sqrt{\det(\boldsymbol{H})}. \tag{3.1.29}$$

The gradient of the potential energy with respect to mass-weighted atomic coordinates, $\nabla_{\boldsymbol{x}} \mathcal{V}$, is related to the gradient with respect to generalized coordinates, $\nabla_q \mathcal{V}$, via the equation

$$\nabla_{\boldsymbol{q}} \mathcal{V} = \boldsymbol{J}^\mathrm{T} \cdot \boldsymbol{M}^{1/2} \cdot \nabla_{\boldsymbol{x}} \mathcal{V}, \tag{3.1.30}$$

which, by inversion, gives

$$\nabla_{\boldsymbol{x}} \mathcal{V} = \boldsymbol{M}^{-1/2} \cdot (\boldsymbol{J}^{-1})^\mathrm{T} \cdot \nabla_{\boldsymbol{q}} \mathcal{V}. \tag{3.1.31}$$

From this we conclude that any stationary point of $\mathcal{V}$ with respect to $\boldsymbol{x}$ will also be a stationary point of $\mathcal{V}$ with respect to $\boldsymbol{q}$. At the considered minimum energy configuration, the set of coordinates $\{\boldsymbol{x}_0\}$ will correspond to a set of values of the generalized coordinates $\{\boldsymbol{q}_0\}$ at which the gradient $\nabla_{\boldsymbol{q}} \mathcal{V}$ is 0.

Differentiating Eq. (3.1.31) once more with respect to $\boldsymbol{x}$, we obtain

$$\left(\frac{\partial^2 \mathcal{V}}{\partial \boldsymbol{x}\, \partial \boldsymbol{x}^\mathrm{T}}\right) = \boldsymbol{M}^{-1/2} \cdot (\boldsymbol{J}^{-1})^\mathrm{T} \cdot \left(\frac{\partial^2 \mathcal{V}}{\partial \boldsymbol{q}\, \partial \boldsymbol{q}^\mathrm{T}}\right) \cdot \boldsymbol{J}^{-1} \cdot \boldsymbol{M}^{-1/2} + \left(\frac{\partial^2 \boldsymbol{q}^\mathrm{T}}{\partial \boldsymbol{x}\, \partial \boldsymbol{x}^\mathrm{T}}\right) \nabla_{\boldsymbol{q}} \mathcal{V}. \tag{3.1.32}$$

The third rank tensor $\left(\dfrac{\partial^2 \boldsymbol{q}^{\mathrm{T}}}{\partial \boldsymbol{x}\, \partial \boldsymbol{x}^{\mathrm{T}}}\right)$ can be thought of as a $n \times n$ matrix[7] in which each component, itself a vector, contains the coordinate transformation second derivatives of $\boldsymbol{q}$ with respect to x_i and x_j. Applying Eq. (3.1.32) at the minimum, where the gradient $\nabla_{\boldsymbol{q}}\mathcal{V}$ is 0, we obtain a relation between the Hessian matrix $\boldsymbol{H}$ with respect to mass-weighted atomic coordinates at the minimum and the corresponding Hessian matrix $\boldsymbol{H}_{qq}$ with respect to generalized coordinates at the minimum, defined by

$$H_{qq,ij} = \left.\frac{\partial^2 \mathcal{V}}{\partial q_i \partial q_j}\right|_{\boldsymbol{q}_0} \tag{3.1.33}$$

as

$$\boldsymbol{H} = \boldsymbol{M}^{-1/2} \cdot (\boldsymbol{J}^{-1})^{\mathrm{T}} \cdot \boldsymbol{H}_{\boldsymbol{qq}} \cdot \boldsymbol{J}^{-1} \cdot \boldsymbol{M}^{-1/2}. \tag{3.1.34}$$

How is the diagonalization problem, Eq. (3.1.27), for determining the vibrational modes translated into generalized degrees of freedom? To each eigenvector $\boldsymbol{x}'$ in the mass-weighted coordinates corresponds an eigenvector $\boldsymbol{q}'$ in generalized coordinates, defined such that (compare Eqs. (3.1.11) and (3.1.19)):

$$\boldsymbol{x}' = \boldsymbol{M}^{1/2} \cdot \boldsymbol{J} \cdot \boldsymbol{q}'. \tag{3.1.35}$$

Using Eqs. (3.1.35) and (3.1.34) in Eq. (3.1.27) leads to

$$\boldsymbol{H}_{\boldsymbol{qq}} \cdot \boldsymbol{q}' = \lambda \boldsymbol{J}^{\mathrm{T}} \cdot \boldsymbol{M} \cdot \boldsymbol{J} \cdot \boldsymbol{q}'$$

or, recalling Eq. (3.1.19) for the covariant metric tensor,

$$\boldsymbol{H}_{\boldsymbol{qq}} \cdot \boldsymbol{q}' = \lambda\, \boldsymbol{a} \cdot \boldsymbol{q}'. \tag{3.1.36}$$

Solving the generalized eigenvalue problem of Eq. (3.1.36) allows us to determine the normal modes and their frequencies by working in the space of generalized coordinates.[7]

To explore the behaviour of the normal modes and their frequencies in the limit where the force constants associated with hard degrees of freedom become very large, we rewrite the generalized eigenvalue problem, Eq. (3.1.36), using Eq. (3.1.23), as

$$\boldsymbol{a}^{-1} \cdot \boldsymbol{H}_{\boldsymbol{qq}} \cdot \boldsymbol{q}' = \lambda \boldsymbol{q}' \Leftrightarrow \begin{pmatrix} (\boldsymbol{a}^{-1})_{\mathrm{ff}} & | & (\boldsymbol{a}^{-1})_{\mathrm{fh}} \\ -- & + & -- \\ (\boldsymbol{a}^{-1})_{\mathrm{hf}} & | & (\boldsymbol{a}^{-1})_{\mathrm{hh}} \end{pmatrix} \cdot \begin{pmatrix} \boldsymbol{H}_{\boldsymbol{qq},\mathrm{ff}} & \boldsymbol{H}_{\boldsymbol{qq},\mathrm{fh}} \\ \boldsymbol{H}_{\boldsymbol{qq},\mathrm{hf}} & \boldsymbol{H}_{\boldsymbol{qq},\mathrm{hh}} \end{pmatrix} \cdot \begin{pmatrix} \boldsymbol{q}'_{\mathrm{f}} \\ \boldsymbol{q}'_{\mathrm{h}} \end{pmatrix} = \lambda \begin{pmatrix} \boldsymbol{q}'_{\mathrm{f}} \\ \boldsymbol{q}'_{\mathrm{h}} \end{pmatrix}, \tag{3.1.37}$$

where the eigenvector $\boldsymbol{q}'$ has been partitioned into its soft and hard components and the Hessian matrix into its soft–soft, soft–hard, hard–soft, and hard–hard submatrices, all evaluated at the energy minimum with respect to all generalized coordinates. Note also that $\boldsymbol{H}_{\boldsymbol{qq},\mathrm{hf}} = \boldsymbol{H}^{\mathrm{T}}_{\boldsymbol{qq},\mathrm{fh}}$. Equation (3.1.37) can be rewritten[7] as

$$[(\boldsymbol{a}^{-1})_{\mathrm{ff}} \cdot \boldsymbol{H}_{\boldsymbol{qq},\mathrm{ff}} + (\boldsymbol{a}^{-1})_{\mathrm{fh}} \cdot \boldsymbol{H}^{\mathrm{T}}_{\boldsymbol{qq},\mathrm{fh}}] \cdot \boldsymbol{q}'_{\mathrm{f}} + [(\boldsymbol{a}^{-1})_{\mathrm{ff}} \cdot \boldsymbol{H}_{\boldsymbol{qq},\mathrm{fh}} + (\boldsymbol{a}^{-1})_{\mathrm{fh}} \cdot \boldsymbol{H}_{\boldsymbol{qq},\mathrm{hh}}] \cdot \boldsymbol{q}'_{\mathrm{h}} = \lambda\, \boldsymbol{q}'_{\mathrm{f}} \quad (3.1.38\mathrm{a})$$

and

$$[(\boldsymbol{a}^{-1})_{\mathrm{hf}} \cdot \boldsymbol{H}_{\boldsymbol{qq},\mathrm{ff}} + (\boldsymbol{a}^{-1})_{\mathrm{hh}} \cdot \boldsymbol{H}^{\mathrm{T}}_{\boldsymbol{qq},\mathrm{fh}}] \cdot \boldsymbol{q}'_{\mathrm{f}} + [(\boldsymbol{a}^{-1})_{\mathrm{hf}} \cdot \boldsymbol{H}_{\boldsymbol{qq},\mathrm{fh}} + (\boldsymbol{a}^{-1})_{\mathrm{hh}} \cdot \boldsymbol{H}_{\boldsymbol{qq},\mathrm{hh}}] \cdot \boldsymbol{q}'_{\mathrm{h}} = \lambda\, \boldsymbol{q}'_{\mathrm{h}}. \quad (3.1.38\mathrm{b})$$

The matrix $\boldsymbol{H}_{\boldsymbol{qq},\mathrm{hh}}$ is the sum of the second derivative matrices, with respect to two hard generalized coordinates, of the nonbonded interactions and the harmonic stiffness interactions. The diagonal terms of $\boldsymbol{H}_{\boldsymbol{qq},\mathrm{hh}}$ are the stiffness force constants, which are allowed to approach infinity. The off-diagonal terms of $\boldsymbol{H}_{\boldsymbol{qq},\mathrm{hh}}$, on the other hand, remain finite. The elements of $\boldsymbol{H}_{\boldsymbol{qq},\mathrm{ff}}$, $\boldsymbol{H}_{\boldsymbol{qq},\mathrm{fh}}$, and its transpose $\boldsymbol{H}^{\mathrm{T}}_{\boldsymbol{qq},\mathrm{fh}}$ remain finite as the force constants associated with the hard degrees of freedom approach infinity, and so do the elements of the $\boldsymbol{a}^{-1}$ submatrices. From Eqs. (3.1.38a, b), if an eigenvalue λ is to be finite, then necessarily one must have $\boldsymbol{q}'_{\mathrm{h}} \to 0$; i.e. all hard components of the eigenvector must go to 0 and therefore the left-hand side of Eq. (3.1.38b) must approach 0 also. From this we deduce that $(\boldsymbol{a}^{-1})_{\mathrm{hh}} \cdot \boldsymbol{H}^{\mathrm{T}}_{\boldsymbol{qq},\mathrm{fh}} = -(\boldsymbol{a}^{-1})_{\mathrm{hf}} \cdot \boldsymbol{H}_{\boldsymbol{qq},\mathrm{ff}}$. Solving for $\boldsymbol{H}^{\mathrm{T}}_{\boldsymbol{qq},\mathrm{fh}}$ in this limit and substituting in Eq. (3.1.38a), we obtain that

$$\left[(\boldsymbol{a}^{-1})_{\mathrm{ff}} - (\boldsymbol{a}^{-1})_{\mathrm{fh}} \cdot ((\boldsymbol{a}^{-1})_{\mathrm{hh}})^{-1} \cdot (\boldsymbol{a}^{-1})_{\mathrm{hf}}\right] \cdot \boldsymbol{H}_{\boldsymbol{qq},\mathrm{ff}} \cdot \boldsymbol{q}'_{\mathrm{f}} = \lambda \boldsymbol{q}'_{\mathrm{f}}. \quad (3.1.39)$$

The factor in Eq. (3.1.39) premultiplying $\boldsymbol{H}_{\boldsymbol{qq},\mathrm{ff}}$ equals[4] the inverse of $\boldsymbol{a}_{\mathrm{ff}}$. Thus, the N_{f} finite-frequency modes in the limit of infinitely stiff force constants associated with the hard degrees of freedom satisfy

$$(\boldsymbol{a}_{\mathrm{ff}})^{-1} \cdot \boldsymbol{H}_{\boldsymbol{qq},\mathrm{ff}} \cdot \boldsymbol{q}'_{\mathrm{f}} = \lambda \boldsymbol{q}'_{\mathrm{f}} \Leftrightarrow \boldsymbol{H}_{\boldsymbol{qq},\mathrm{ff}} \cdot \boldsymbol{q}'_{\mathrm{f}} = \lambda \boldsymbol{a}_{\mathrm{ff}} \cdot \boldsymbol{q}'_{\mathrm{f}}. \quad (3.1.40)$$

The diagonalization problem, Eq. (3.1.40), yields N_{f} finite vibrational frequencies $\nu_{\mathrm{f},i}$ $(i = 1, 2, \ldots, N_{\mathrm{f}})$. The product of these frequencies satisfies

$$\prod_{i=1}^{N_{\mathrm{f}}} (2\pi\nu_{\mathrm{f}i}) = \sqrt{\det((\boldsymbol{a}_{\mathrm{ff}})^{-1} \cdot \boldsymbol{H}_{\boldsymbol{qq},\mathrm{ff}})}. \quad (3.1.41)$$

On the other hand, Eq. (3.1.38b) gives an additional N_{h} modes whose frequencies tend to infinity. These are solutions to the generalized eigenvalue problem

$$(\boldsymbol{a}^{-1})_{\mathrm{hh}} \cdot \boldsymbol{H}_{\boldsymbol{qq},\mathrm{hh}} \cdot \boldsymbol{q}'_{\mathrm{h}} = \lambda\, \boldsymbol{q}'_{\mathrm{h}} \Leftrightarrow \boldsymbol{H}_{\boldsymbol{qq},\mathrm{hh}} \cdot \boldsymbol{q}'_{\mathrm{h}} = \lambda (\boldsymbol{a}^{-1})^{-1}_{\mathrm{hh}} \cdot \boldsymbol{q}'_{\mathrm{h}}. \quad (3.1.42)$$

The eigenvectors corresponding to these eigenvalues are vibrational modes associated with the stiff degrees of freedom. The diagonalization problem, Eq. (3.1.42), yields N_{h} vibrational frequencies

$\nu_{\mathrm{h},i}$ $(i = 1, 2, \ldots, N_{\mathrm{h}})$, which tend to infinity. The product of these frequencies satisfies

$$\prod_{i=1}^{N_{\mathrm{h}}} (2\pi\nu_{\mathrm{h}i}) = \sqrt{\det((\boldsymbol{a}^{-1})_{\mathrm{hh}} \cdot \boldsymbol{H}_{\boldsymbol{qq},\mathrm{hh}})}. \tag{3.1.43}$$

Of course, $\nu_{\mathrm{f},i}$ and the $\nu_{\mathrm{h},i}$ taken together are none else than the vibrational frequencies ν_i appearing in Eq. (3.1.29). Thus, the product of the right-hand sides of Eqs. (3.1.41) and (3.1.43) must equal the right-hand side of Eq. (3.1.29). Together with Eqs. (3.1.19) and (3.1.34), this leads to the *Fixman relation*:

$$\det(\boldsymbol{a}) \det((\boldsymbol{a}^{-1})_{\mathrm{hh}}) = \det(\boldsymbol{a}_{\mathrm{ff}}). \tag{3.1.44}$$

Let us now turn to the partition function of the flexible model in the classical approximation. From Eq. (3.1.19) we obtain that

$$\det(\boldsymbol{a}) = [\det(\boldsymbol{J})]^2 \prod_{i=1}^{N_{\mathrm{f}}+N_{\mathrm{h}}} m_i. \tag{3.1.45}$$

Solving Eq. (3.1.45) for $\det(\boldsymbol{J})$ and substituting in Eq. (3.1.12) leads to

$$\begin{aligned} Q &= \frac{1}{N_{\mathrm{c}}!} \left[\prod_{i=1}^{N_{\mathrm{f}}+N_{\mathrm{h}}} \left(\frac{2\pi\, m_i k_{\mathrm{B}} T}{h^2} \right)^{1/2} \right] \int \frac{\sqrt{\det(\boldsymbol{a})}}{\sqrt{\prod_{a=1}^{N_{\mathrm{f}}+N_{\mathrm{h}}} m_a}} \exp[-\beta \mathcal{V}(\boldsymbol{q})] \mathrm{d}\boldsymbol{q} \\ &= \frac{1}{N_{\mathrm{c}}!} \left(\frac{2\pi\, k_{\mathrm{B}} T}{h^2} \right)^{(N_{\mathrm{f}}+N_{\mathrm{h}})/2} \int \sqrt{\det(\boldsymbol{a})} \exp[-\beta \mathcal{V}(\boldsymbol{q}_{\mathrm{f}}, \boldsymbol{q}_{\mathrm{h}})] \mathrm{d}\boldsymbol{q}_{\mathrm{f}}\, \mathrm{d}\boldsymbol{q}_{\mathrm{h}}. \end{aligned} \tag{3.1.46}$$

Invoking the harmonic approximation, Eq. (3.1.13), for the contribution of the hard degrees of freedom to the potential energy and performing all integrations with respect to $\boldsymbol{q}_{\mathrm{h}}$ we obtain that, as in Eq. (3.1.14),

$$Q = \frac{1}{N_{\mathrm{c}}!} \left[\prod_{i=1}^{N_{\mathrm{f}}+N_{\mathrm{h}}} \left(\frac{2\pi\, k_{\mathrm{B}} T}{h^2} \right)^{1/2} \right] \int \sqrt{\det(\boldsymbol{a})} \frac{(2\pi\, k_{\mathrm{B}} T)^{N_{\mathrm{h}}/2}}{\sqrt{\det(\boldsymbol{H}_{\boldsymbol{qq},\mathrm{hh}})}} \exp[-\beta \mathcal{V}(\boldsymbol{q}_{\mathrm{f}}, \boldsymbol{q}_{\mathrm{h}0})] \mathrm{d}\boldsymbol{q}_{\mathrm{f}}. \tag{3.1.47}$$

Combining Eqs. (3.1.44) and (3.1.47) yields

$$\begin{aligned} Q &= \frac{1}{N_{\mathrm{c}}!} \left(\frac{2\pi\, k_{\mathrm{B}} T}{h^2} \right)^{(N_{\mathrm{f}}+N_{\mathrm{h}})/2} \int \frac{\sqrt{\det(\boldsymbol{a}_{\mathrm{ff}})}}{\sqrt{\det((\boldsymbol{a}^{-1})_{\mathrm{hh}})}} \frac{(2\pi\, k_{\mathrm{B}} T)^{N_{\mathrm{h}}/2}}{\sqrt{\det(\boldsymbol{H}_{\boldsymbol{qq},\mathrm{hh}})}} \exp[-\beta \mathcal{V}(\boldsymbol{q}_{\mathrm{f}}, \boldsymbol{q}_{\mathrm{h},0})] \mathrm{d}\boldsymbol{q}_{\mathrm{f}} \\ &= \frac{1}{N_{\mathrm{c}}!} \left(\frac{2\pi\, k_{\mathrm{B}} T}{h^2} \right)^{N_{\mathrm{f}}/2} \int \sqrt{\det(\boldsymbol{a}_{\mathrm{ff}})} \frac{(2\pi k_{\mathrm{B}} T)^{N_{\mathrm{h}}}}{h^{N_{\mathrm{h}}} \sqrt{\det((\boldsymbol{a}^{-1})_{\mathrm{hh}} \cdot \boldsymbol{H}_{\boldsymbol{qq},\mathrm{hh}})}} \exp[-\beta \mathcal{V}(\boldsymbol{q}_{\mathrm{f}}, \boldsymbol{q}_{\mathrm{h},0})] \mathrm{d}\boldsymbol{q}_{\mathrm{f}}. \end{aligned} \tag{3.1.48}$$

Using Eq. (3.1.43) within Eq. (3.1.48),

$$Q = \frac{1}{N_c!}\left(\frac{2\pi k_B T}{h^2}\right)^{N_f/2} \int \sqrt{\det(\boldsymbol{a}_{ff})}\frac{(2\pi k_B T)^{N_h}}{h^{N_h}\prod_{i=1}^{N_h}(2\pi\nu_{hi})} \exp[-\beta\mathcal{V}(\boldsymbol{q}_f, \boldsymbol{q}_{h,0})]d\boldsymbol{q}_f$$

$$= \frac{1}{N_c!}\left(\frac{2\pi k_B T}{h^2}\right)^{N_f/2} \int \sqrt{\det(\boldsymbol{a}_{ff})}\left[\prod_{i=1}^{N_h}\left(\frac{k_B T}{h\nu_{hi}}\right)\right] \exp[-\beta\mathcal{V}(\boldsymbol{q}_f, \boldsymbol{q}_{h,0})]d\boldsymbol{q}_f. \qquad (3.1.49)$$

Equation (3.1.49) is an alternative form to Eq. (3.1.14) for the partition function of the classical flexible model in the limit of very large force constants governing the hard degrees of freedom. The contribution of the vibrational modes associated with the hard degrees of freedom appears in Eq. (3.1.49) explicitly. For each mode i, the quantity $\frac{k_B T}{h\nu_{hi}}$ is the partition function for that mode, treated classically.[10]

As pointed out earlier, the vibrational frequencies associated with the hard degrees of freedom are typically high, such that $\frac{k_B T}{h\nu_{hi}} < 1$ at the temperatures of interest. Thus, a more correct form of Eq. (3.1.49) is obtained if, for each hard mode, we use the quantum mechanical expression for its partition function instead of the classical one. The quantum mechanical partition function of a harmonic oscillator of frequency ν_{hi}, obtained by summation of the Boltzmann factor of the energy over its ground state and all its excited states, is[10]

$$\frac{\exp\left(-\frac{h\nu_{hi}}{2k_B T}\right)}{1-\exp\left(-\frac{h\nu_{hi}}{k_B T}\right)} = \frac{1}{\exp\left(\frac{h\nu_{hi}}{2k_B T}\right)-\exp\left(-\frac{h\nu_{hi}}{2k_B T}\right)} = \frac{1}{2\sinh\left(\frac{h\nu_{hi}}{2k_B T}\right)}. \qquad (3.1.50)$$

The reader can verify that Eq. (3.1.50) approaches the classical limit $\frac{k_B T}{h\nu_{hi}}$ when $\frac{k_B T}{h\nu_{hi}} >> 1$, i.e. when the spacing $h\nu_{hi}$ between successive energy levels of the oscillator becomes very small relative to the thermal energy, and therefore the set of discrete states of the quantum mechanical description can be approximated as a continuum.

Use of Eq. (3.1.50) instead of its classical counterpart within Eq. (3.1.49) leads to the partition function

$$Q_{QM} = \frac{1}{N_c!}\left(\frac{2\pi k_B T}{h^2}\right)^{N_f/2} \int \sqrt{\det \boldsymbol{a}_{ff}}\left[\prod_{i=1}^{N_h}\left(\frac{1}{2\sinh\left(\frac{h\nu_{hi}}{2k_B T}\right)}\right)\right] \exp\{-\beta\mathcal{V}(\boldsymbol{q}_f, \boldsymbol{q}_{h,0})\}d\boldsymbol{q}_f. \qquad (3.1.51)$$

This is the *quantum mechanical* partition function of the flexible model, discussed by Gō and Scheraga.[4] It constitutes the most appropriate starting point for statistical mechanical analyses of polymer systems.

3.1.3 The rigid model

In Section 3.1.1 we considered the flexible model for chain systems, where all bonded interactions are governed by stiff harmonic potentials. We developed the Jacobian $\boldsymbol{J}$ for transforming the description of this model from (mass-weighted) coordinates of all atoms to generalized coordinates (chain start coordinates, Euler angles, bond lengths, bond angles, and torsion angles for each chain molecule). In Section 3.1.2 we considered the behaviour of this model as the force constants governing the hard degrees of freedom (bond lengths and bond angles) become very large. The quantum mechanical partition function governing the flexible model in the limit of very large force constants is given by Eq. (3.1.51). If a classical partition function is used to describe vibrational modes associated with the hard degrees of freedom, the partition function is given by Eq. (3.1.49) or, equivalently, by Eq. (3.1.47) or Eq. (3.1.14).

There is an alternative model one can use to describe systems of chain molecules. In that model, all hard degrees of freedom are considered fixed at the values $\boldsymbol{q}_{\mathrm{h},0}$ at the outset. The phase space of the model is thus spanned by the soft generalized coordinates $\boldsymbol{q}_{\mathrm{f}}$ and their conjugate momenta. This is called the *rigid* model in the statistical mechanics of polymers. To calculate the partition function of the rigid model, the Boltzmann factor of its Hamiltonian must be integrated over the soft generalized coordinates and their conjugate momenta only, without any consideration of the hard generalized coordinates. Equivalently, to obtain the partition function of the rigid model, one can start from the quantum mechanical partition function of Eq. (3.1.51) and set all vibrational partition functions of the hard modes equal to 1. (This is equivalent to requiring that all hard modes be confined to their ground states.) The partition function of the rigid model thus reads

$$Q_{\mathrm{R}} = \frac{1}{N_{\mathrm{c}}!}\left(\frac{2\pi k_B T}{h^2}\right)^{N_{\mathrm{f}}/2} \int \sqrt{\det(\boldsymbol{a}_{\mathrm{ff}})}\exp\{-\beta\mathcal{V}(\boldsymbol{q}_{\mathrm{f}}, \boldsymbol{q}_{\mathrm{h},0})\}\mathrm{d}\boldsymbol{q}_{\mathrm{f}}. \tag{3.1.52}$$

The rigid model, Eq. (3.1.52), and the classical flexible model in the limit of infinitely stiff vibrational force constants, Eq. (3.1.14) or, equivalently, Eq. (3.1.48), produce different conformational distributions. This is because the factors multiplying the Boltzmann factor of the energy $\mathcal{V}(\boldsymbol{q}_{\mathrm{f}}, \boldsymbol{q}_{\mathrm{h},0})$ exhibit different dependences on the torsion angles $\boldsymbol{\phi}$. It has already been mentioned that the Jacobian factor $\det(\boldsymbol{J}|_{\boldsymbol{q}_{\mathrm{h},0}})$ appearing in the partition function of the classical flexible model in the limit of infinite stiffness is independent of $\boldsymbol{\phi}$; torsion angles in this model are thus distributed according to the Boltzmann factor $\exp[-\beta\mathcal{V}(\boldsymbol{q}_{\mathrm{f}}, \boldsymbol{q}_{\mathrm{h},0})]$. Such a distribution was invoked in developing the rotational isomeric state model in Chapter 1. In contrast, the factor $\sqrt{\det(\boldsymbol{a}_{\mathrm{ff}})}$ premultiplying the Boltzmann factor in the partition function of the rigid model, Eq. (3.1.52), depends on $\boldsymbol{\phi}$.

The nonequivalence of the rigid model and the classical flexible model in the limit of infinite stiffness is perhaps counterintuitive and perplexed chemical physicists in the 1970s. It was resolved by Fixman.[3] As Fixman showed, and as one can deduce by comparing Eqs. (3.1.52) and (3.1.48), if the conformation-dependent term $k_B T \ln(\sqrt{\det(\boldsymbol{a}^{-1})_{\mathrm{hh}}})$ is added to the potential energy of the rigid model, that model can be made to sample the conformational energy distribution of the flexible model in the limit of infinite stiffness. This correction term is known as the *Fixman potential.*

For an actual derivation of the Fixman potential in the case of a polypropylene model with fixed bond lengths, see Figure 3.2 and Problem 3.5. This Fixman potential for polypropylene, originally derived by Greenfield,[11] was used by Antoniadis, Samara, and Theodorou[12] in molecular dynamics simulations of atactic polypropylene melts.

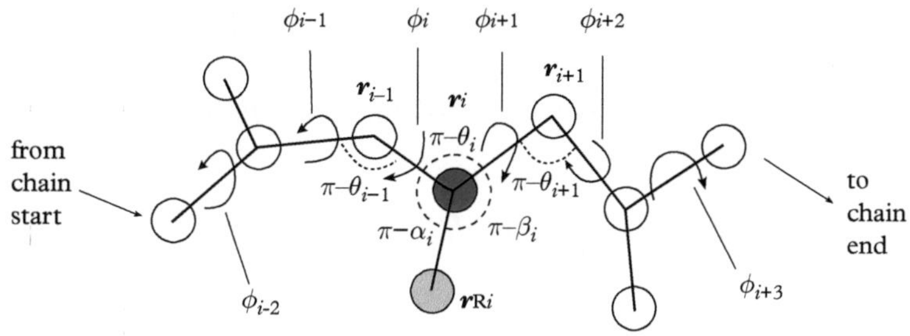

Figure 3.2 Geometry of a united-atom polypropylene chain. The position vectors of consecutive skeletal carbons are denoted as $\boldsymbol{r}_{i-1}, \boldsymbol{r}_i, \boldsymbol{r}_{i+1}$. Chiral carbon at $\boldsymbol{r}_i$ is bonded to a united-atom methyl group at $\boldsymbol{r}_{R_i}$. The masses of the united atoms are referred to as $m_{i-1}, m_i, m_{i+1}, m_{R_i}$, respectively. Skeletal bond vectors are denoted as $\boldsymbol{l}_i = \boldsymbol{r}_i - \boldsymbol{r}_{i-1}$, $\boldsymbol{l}_{i+1} = \boldsymbol{r}_{i+1} - \boldsymbol{r}_i$, their lengths being l_i and l_{i+1}, respectively. Pendant bond vector $\boldsymbol{l}_{R_i} = \boldsymbol{r}_{R_i} - \boldsymbol{r}_i$ connects skeletal united atom at $\boldsymbol{r}_i$ to the methyl group at $\boldsymbol{r}_{R_i}$. The supplement of a skeletal bond angle with its apex at $\boldsymbol{r}_i$ is denoted by θ_i. The bond angle between skeletal bond i and the pendant carbon-methyl bond at i is denoted by $\pi - \alpha_i$. The bond angle between skeletal bond $i + 1$ and the pendant carbon-methyl bond at i is denoted by $\pi - \beta_i$.

Gō and Scheraga[4] analysed both the rigid model and the classical flexible model in the limit of infinitely stiff vibrational force constants in connection with the quantum mechanical flexible model, Eq. (3.1.51). They showed that the classical flexible model in the limit of infinitely stiff vibrational force constants constitutes a better approximation to the quantum mechanical model than does the rigid model. Between the rigid model and the classical flexible model in the limit of infinite stiffness, the latter is thus physically more justifiable.

3.2 Forces, instantaneous and thermodynamic stress tensors, and potential energy minimization in Cartesian coordinates

3.2.1 Potential energy as a function of atomic coordinates

In our discussion in Section 3.1 we have focused on a model system containing N_c chain molecules in a simulation box, each molecule consisting of $N + 1$ interaction sites. The total number of sites in the primary box is therefore $n = N_c(N + 1)$. A typical expression for the potential energy function describing interactions in such a polymeric system has been discussed in Sections 1.1.6 and 2.6. In general, neighbouring sites within a molecule exert on each other bonded forces, which are usually expressed as functions of bond lengths (two-site), valence bond angles (three-site), and torsion angles (four-site potentials). In addition, there are nonbonded interactions both between sites in a given chain (intramolecular) and between sites belonging to different chains (intermolecular). Without loss of generality, we can assume that both types of nonbonded interactions are described by the same pairwise additive potential $\mathcal{V}_{\mathrm{NB}}(|\boldsymbol{r}_i - \boldsymbol{r}_j|)$, which depends only on the distance $r_{ij} = |\boldsymbol{r}_i - \boldsymbol{r}_j|$ between sites. When $\mathcal{V}_{\mathrm{NB}}$ decays faster than the inverse third power of intersite distance, it is convenient to separate $\mathcal{V}_{\mathrm{NB}}$ into an explicit contribution $\mathcal{V}_{\mathrm{SR}}(r)$ and a tail contribution $\mathcal{V}_{\mathrm{tail}}(r)$, where $\mathcal{V}_{\mathrm{SR}}(r)$ and $\mathcal{V}_{\mathrm{tail}}(r)$ are 0 for $r > r_c$ and $r < r_c$, respectively, and r_c is an appropriately selected cutoff distance. The bonded potentials governing the variation of bond lengths and bond angles (hard degrees of freedom) are typically stronger by five to two orders of magnitude than torsional and attractive nonbonded interaction potentials.

If a fully flexible model is used, then for a system of united-atom molecules, a frequently employed potential energy model is

$$\mathcal{V}(\boldsymbol{r}_0, \boldsymbol{r}_1, \ldots, \boldsymbol{r}_{n-1}) = \sum_{\substack{\text{all skeletal}\\ \text{bonds } l_i}} \mathcal{V}_l(|\boldsymbol{l}_i|) + \sum_{\substack{\text{all valence}\\ \text{angles } \theta_i}} \mathcal{V}_\theta(\cos\theta_i)$$
$$+ \sum_{\substack{\text{all torsion}\\ \text{angles } \phi_i}} \mathcal{V}_\phi(\cos\phi_i) + \sum_{i<j}\sum_{j} \mathcal{V}_{\text{NB}}(|\boldsymbol{r}_i - \boldsymbol{r}_j|) \qquad (3.2.1)$$

where

- $\boldsymbol{l}_i = \boldsymbol{r}_i - \boldsymbol{r}_{i-1}$ is the bond vector i connecting skeletal sites $i-1$ and i in a chain and $\hat{\boldsymbol{l}}_i = \frac{\boldsymbol{l}_i}{|\boldsymbol{l}_i|}$ the unit vector along bond $\boldsymbol{l}_i$,
- $\cos\theta_i = \hat{\boldsymbol{l}}_i \cdot \hat{\boldsymbol{l}}_{i+1}$, where θ_i is the supplement of the valence angle between bonds i and $i+1$ with its apex at site i, and
- $\cos\phi_i = -\left(\frac{\hat{\boldsymbol{l}}_{i-1} \times \hat{\boldsymbol{l}}_i}{|\hat{\boldsymbol{l}}_{i-1} \times \hat{\boldsymbol{l}}_i|}\right) \cdot \left(\frac{\hat{\boldsymbol{l}}_i \times \hat{\boldsymbol{l}}_{i+1}}{|\hat{\boldsymbol{l}}_i \times \hat{\boldsymbol{l}}_{i+1}|}\right) = \frac{(\hat{\boldsymbol{l}}_{i-1} \cdot \hat{\boldsymbol{l}}_{i+1}) - (\hat{\boldsymbol{l}}_{i-1} \cdot \hat{\boldsymbol{l}}_i)(\hat{\boldsymbol{l}}_i \cdot \hat{\boldsymbol{l}}_{i+1})}{|\hat{\boldsymbol{l}}_{i-1} \times \hat{\boldsymbol{l}}_i||\hat{\boldsymbol{l}}_i \times \hat{\boldsymbol{l}}_{i+1}|}$, where ϕ_i is the torsion angle of skeletal bond i, i.e. the dihedral angle between the plane of bonds $\boldsymbol{l}_{i-1}$ and bond $\boldsymbol{l}_i$, and the plane of bonds bond $\boldsymbol{l}_i$ and bond $\boldsymbol{l}_{i+1}$; it is defined so that bond $\phi_i = 0$ corresponds to the *trans* torsional state. ϕ_i takes values from the interval $[-\pi, \pi)$, its sign being the same as that of the triple product $\hat{\boldsymbol{l}}_i \cdot (\hat{\boldsymbol{l}}_{i-1} \times \hat{\boldsymbol{l}}_{i+1})$.

The three first terms in Eq. (3.2.1) constitute the bonded contributions ($\mathcal{V}_l$, $\mathcal{V}_\theta$, and $\mathcal{V}_\phi$ are the bond stretching, bond angle bending, and torsional potentials, respectively). All of them depend on vectorial differences of the type $\boldsymbol{r}_i - \boldsymbol{r}_{i-1}$ between adjacent skeletal atoms that are connected directly by a bond. The fourth term is a nonbonded contribution between sites on the same chain or on different chains, and is a function of the separation $|\boldsymbol{r}_i - \boldsymbol{r}_j|$ between nonadjacent sites; it is assumed to be pairwise additive here, for simplicity. Typical expressions for the four types are shown in Figure 3.3.

3.2.2 Gradient of the potential energy and interatomic forces

Given that $\mathcal{V}$ is a function of intersite separation vectors only, we can define the gradient vector in a system of n interaction sites numbered from 0 to $n-1$ as

$$\boldsymbol{F}_{ij} = -\nabla_{\boldsymbol{r}_i - \boldsymbol{r}_j} \mathcal{V}(\boldsymbol{r}_1 - \boldsymbol{r}_0, \boldsymbol{r}_2 - \boldsymbol{r}_0, \ldots, \boldsymbol{r}_{n-1} - \boldsymbol{r}_0, \boldsymbol{r}_2 - \boldsymbol{r}_1, \ldots, \boldsymbol{r}_{n-1} - \boldsymbol{r}_1, \ldots, \boldsymbol{r}_{n-2} - \boldsymbol{r}_{n-1}), \quad (3.2.2)$$

defining also the interatomic force between atomistic units i and j. In taking the gradient in this equation, all intersite separation vectors but $\boldsymbol{r}_i - \boldsymbol{r}_j$ are kept constant. We call $\boldsymbol{F}_{ij}$ the force on site i due to site j. The value of $\boldsymbol{F}_{ij}$ may depend on position vectors other than $\boldsymbol{r}_i$ and $\boldsymbol{r}_j$; thus, its definition does not presuppose pairwise additivity of the potential energy function. $\boldsymbol{F}_{ij}$ can be resolved into

Bond stretching energy: $\mathcal{V}_l(l) = \frac{1}{2}k_l(l - l_0)^2$

Bond bending energy: $\mathcal{V}_\theta(\theta) = \frac{1}{2}k_\theta(\cos\theta - \cos\theta_0)^2$

Bond torsional energy: $\mathcal{V}_\phi(\phi) = \sum_{i=0}^{n} a_i \cos^i(\phi)$

Nonbonded LJ energy: $\mathcal{V}^{\mathrm{NB}}(r) = 4\varepsilon\left[\left(\frac{\sigma}{r}\right)^{12} - \left(\frac{\sigma}{r}\right)^{6}\right]$

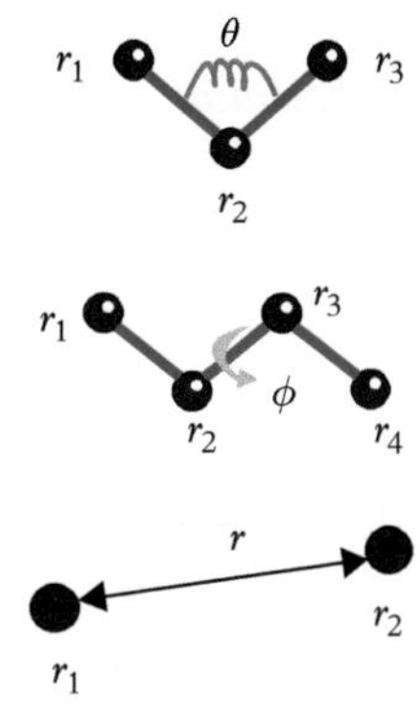

Figure 3.3 Typical mathematical expressions $\mathcal{V}$ of the different contributions to the potential energy in a model polymeric system.

bonded $\boldsymbol{F}_{ij}^{\mathrm{B}}$ and nonbonded $\boldsymbol{F}_{ij}^{\mathrm{NB}}$ contributions, which are derived from the bonded and nonbonded parts of $\mathcal{V}$, respectively. The bonded and nonbonded contributions, as well as the total force $\boldsymbol{F}_{ij}$, satisfy the reciprocity condition, namely $\boldsymbol{F}_{ij} = -\boldsymbol{F}_{ji}$. We can obtain the total force $\boldsymbol{F}_i$ on site i as $\boldsymbol{F}_i = -\nabla_{\boldsymbol{r}_i}\mathcal{V} = \sum_{j \neq i} \boldsymbol{F}_{ij}$ as can be seen through simple differentiation by parts using the definition, Eq. (3.2.2).

To compute the gradients for each of the three potential energy terms in Eq. (3.2.1) as functions of the set of atomic coordinates, we use that[13]

$$\boldsymbol{F}_{ij}^{\mathrm{NB}} = -\nabla_{\boldsymbol{r}_i-\boldsymbol{r}_j}\mathcal{V}_{\mathrm{NB}} = -\left.\frac{\mathrm{d}\mathcal{V}_{\mathrm{NB}}(r)}{\mathrm{d}r}\right|_{r=|\boldsymbol{r}_i-\boldsymbol{r}_j|} \nabla_{\boldsymbol{r}_i-\boldsymbol{r}_j}|\boldsymbol{r}_i - \boldsymbol{r}_j| \tag{3.2.3}$$

$$\boldsymbol{F}_{ij}^{l} = -\nabla_{\boldsymbol{r}_i-\boldsymbol{r}_j}\mathcal{V}_l = -\left.\frac{\mathrm{d}\mathcal{V}_l(l)}{\mathrm{d}l}\right|_{l=|\boldsymbol{r}_i-\boldsymbol{r}_j|} \nabla_{\boldsymbol{r}_i-\boldsymbol{r}_j}|\boldsymbol{r}_i - \boldsymbol{r}_j| \tag{3.2.4}$$

$$\begin{aligned}\boldsymbol{F}_{ij}^{\theta} = -\nabla_{\boldsymbol{r}_i-\boldsymbol{r}_j}\mathcal{V}_\theta(\cos\theta) &= -\left.\frac{\mathrm{d}\mathcal{V}_\theta(\cos\theta)}{\mathrm{d}\cos\theta}\right|_{\theta=\theta_{i-1}} \nabla_{\boldsymbol{r}_i-\boldsymbol{r}_j}\cos\theta_{i-1} \\ &\quad -\left.\frac{\mathrm{d}\mathcal{V}_\theta(\cos\theta)}{\mathrm{d}\cos\theta}\right|_{\theta=\theta_i} \nabla_{\boldsymbol{r}_i-\boldsymbol{r}_j}\cos\theta_i\end{aligned} \tag{3.2.5}$$

and

$$\begin{aligned}\boldsymbol{F}_{ij}^{\phi} = -\nabla_{\boldsymbol{r}_i-\boldsymbol{r}_j}\mathcal{V}_\phi &= -\left.\frac{\mathrm{d}\mathcal{V}_\phi(\cos\phi)}{\mathrm{d}\cos\phi}\right|_{\phi=\phi_{i-1}} \nabla_{\boldsymbol{r}_i-\boldsymbol{r}_j}\cos\phi_{i-1} \\ &\quad -\left.\frac{\mathrm{d}\mathcal{V}_\phi(\cos\phi)}{\mathrm{d}\cos\phi}\right|_{\phi=\phi_i} \nabla_{\boldsymbol{r}_i-\boldsymbol{r}_j}\cos\phi_i - \left.\frac{\mathrm{d}\mathcal{V}_\phi(\cos\phi)}{\mathrm{d}\cos\phi}\right|_{\phi=\phi_{i+1}} \nabla_{\boldsymbol{r}_i-\boldsymbol{r}_j}\cos\phi_{i+1}.\end{aligned} \tag{3.2.6}$$

The gradients appearing in these equations are computed with the help of the following expressions (see Figure 3.3).

For the nonbonded potential:

$$\begin{aligned}\nabla_{\boldsymbol{r}_1}(r_{12}) &= -\hat{\boldsymbol{r}}_{12}\\ \nabla_{\boldsymbol{r}_2}(r_{12}) &= +\hat{\boldsymbol{r}}_{12}\\ \nabla_{\boldsymbol{r}_{12}}(r_{12}) &= +\hat{\boldsymbol{r}}_{12},\end{aligned} \tag{3.2.7}$$

where $r_{12} = |\boldsymbol{r}_{12}|$, $\boldsymbol{r}_{12} = \boldsymbol{r}_2 - \boldsymbol{r}_1$, $\hat{\boldsymbol{r}}_{12} = \dfrac{\boldsymbol{r}_{12}}{r_{12}} = \dfrac{\boldsymbol{r}_2 - \boldsymbol{r}_1}{|\boldsymbol{r}_2 - \boldsymbol{r}_1|}$.

For the bond stretching potential:

$$\begin{aligned}\nabla_{\boldsymbol{r}_1}(l_2) &= -\hat{\boldsymbol{l}}_2\\ \nabla_{\boldsymbol{r}_2}(l_2) &= +\hat{\boldsymbol{l}}_2.\end{aligned} \tag{3.2.8}$$

For the bending potential:

$$\begin{aligned}\nabla_{\boldsymbol{r}_1}(\cos\theta_2) &= -\frac{1}{l_2}\hat{\boldsymbol{l}}_3 + \frac{\cos\theta_2}{l_2}\hat{\boldsymbol{l}}_2\\ \nabla_{\boldsymbol{r}_2}(\cos\theta_2) &= \frac{1}{l_2}\hat{\boldsymbol{l}}_3 - \frac{1}{l_3}\hat{\boldsymbol{l}}_2 + \frac{\cos\theta_2}{l_3}\hat{\boldsymbol{l}}_3 - \frac{\cos\theta_2}{l_2}\hat{\boldsymbol{l}}_2\\ \nabla_{\boldsymbol{r}_3}(\cos\theta_2) &= \frac{1}{l_3}\hat{\boldsymbol{l}}_2 - \frac{\cos\theta_2}{l_3}\hat{\boldsymbol{l}}_3.\end{aligned} \tag{3.2.9}$$

For the torsional potential:

$$\begin{aligned}\nabla_{\boldsymbol{r}_1}(\cos\phi_2) &= \frac{1}{l_2\sin\theta_2\sin\theta_3}\hat{\boldsymbol{l}}_4 - \frac{\cos\theta_3}{l_2\sin\theta_2\sin\theta_3}\hat{\boldsymbol{l}}_3 + \frac{\cos\theta_3\cos\theta_2 - \hat{\boldsymbol{l}}_4\cdot\hat{\boldsymbol{l}}_2}{\sin\theta_2\sin\theta_3}\left[\frac{1}{l_2\sin^2\theta_2}\hat{\boldsymbol{l}}_2 - \frac{\cos\theta_2}{l_2\sin^2\theta_2}\hat{\boldsymbol{l}}_3\right]\\
\nabla_{\boldsymbol{r}_2}(\cos\phi_2) &= -\frac{\cos\theta_2}{l_3\sin\theta_2\sin\theta_3}\hat{\boldsymbol{l}}_4 + \frac{\cos\theta_3}{l_2\sin\theta_2\sin\theta_3}\hat{\boldsymbol{l}}_3 - \frac{\cos\theta_3}{l_3\sin\theta_2\sin\theta_3}\hat{\boldsymbol{l}}_2\\
&\quad - \frac{1}{l_2\sin\theta_2\sin\theta_3}\hat{\boldsymbol{l}}_4 + \frac{2\hat{\boldsymbol{l}}_4\cdot\hat{\boldsymbol{l}}_2}{l_3\sin\theta_2\sin\theta_3}\hat{\boldsymbol{l}}_3 + \left[\frac{\cos\theta_3\cos\theta_2 - \hat{\boldsymbol{l}}_4\cdot\hat{\boldsymbol{l}}_2}{\sin\theta_2\sin\theta_3}\right]\\
&\quad \times\left[\frac{1}{l_3\sin^2\theta_3}\hat{\boldsymbol{l}}_3 - \frac{\cos\theta_3}{l_3\sin^2\theta_3}\hat{\boldsymbol{l}}_4 + \frac{1}{l_3\sin^2\theta_2}\hat{\boldsymbol{l}}_3 - \frac{1}{l_2\sin^2\theta_2}\hat{\boldsymbol{l}}_2 + \frac{\cos\theta_2}{l_2\sin^2\theta_2}\hat{\boldsymbol{l}}_3 - \frac{\cos\theta_2}{l_3\sin^2\theta_2}\hat{\boldsymbol{l}}_2\right]\\
\nabla_{\boldsymbol{r}_3}(\cos\phi_2) &= \frac{\cos\theta_2}{l_3\sin\theta_2\sin\theta_3}\hat{\boldsymbol{l}}_4 - \frac{\cos\theta_2}{l_4\sin\theta_2\sin\theta_3}\hat{\boldsymbol{l}}_3 + \frac{\cos\theta_3}{l_3\sin\theta_2\sin\theta_3}\hat{\boldsymbol{l}}_2\\
&\quad + \frac{1}{l_4\sin\theta_2\sin\theta_3}\hat{\boldsymbol{l}}_2 - \frac{2\hat{\boldsymbol{l}}_4\cdot\hat{\boldsymbol{l}}_2}{l_3\sin\theta_2\sin\theta_3}\hat{\boldsymbol{l}}_3 + \left[\frac{\cos\theta_3\cos\theta_2 - \hat{\boldsymbol{l}}_4\cdot\hat{\boldsymbol{l}}_2}{\sin\theta_2\sin\theta_3}\right]\end{aligned}$$

$$\times\left[\frac{1}{l_4\sin^2\theta_3}\hat{\boldsymbol{l}}_4-\frac{1}{l_3\sin^2\theta_3}\hat{\boldsymbol{l}}_3+\frac{\cos\theta_3}{l_3\sin^2\theta_3}\hat{\boldsymbol{l}}_4-\frac{\cos\theta_3}{l_4\sin^2\theta_3}\hat{\boldsymbol{l}}_3-\frac{1}{l_3\sin^2\theta_2}\hat{\boldsymbol{l}}_3+\frac{\cos\theta_2}{l_3\sin^2\theta_2}\hat{\boldsymbol{l}}_2\right]$$

$$\nabla_{\boldsymbol{r}_4}(\cos\phi_2)=\frac{\cos\theta_2}{l_4\sin\theta_2\sin\theta_3}\hat{\boldsymbol{l}}_3-\frac{1}{l_4\sin\theta_2\sin\theta_3}\hat{\boldsymbol{l}}_2+\frac{\cos\theta_3\cos\theta_2-\hat{\boldsymbol{l}}_4\cdot\hat{\boldsymbol{l}}_2}{\sin\theta_2\sin\theta_3}\left[\frac{\cos\theta_3}{l_4\sin^2\theta_3}\hat{\boldsymbol{l}}_3-\frac{1}{l_4\sin^2\theta_3}\hat{\boldsymbol{l}}_4\right],\tag{3.2.10}$$

where we have used that $|\hat{\boldsymbol{l}}_i\times\hat{\boldsymbol{l}}_{i+1}|=\sin\theta_i$.

The following results are then obtained:

$$\boldsymbol{F}_{ij}^{\mathrm{NB}}=-\frac{\mathrm{d}\mathcal{V}_{NB}(|\boldsymbol{r}_i-\boldsymbol{r}_j|)}{\mathrm{d}|\boldsymbol{r}_i-\boldsymbol{r}_j|}\frac{\boldsymbol{r}_i-\boldsymbol{r}_j}{|\boldsymbol{r}_i-\boldsymbol{r}_j|}=-\left.\frac{\mathrm{d}\mathcal{V}_{\mathrm{NB}}}{\mathrm{d}r}\right|_{r=|\boldsymbol{r}_i-\boldsymbol{r}_j|}\frac{\boldsymbol{r}_i-\boldsymbol{r}_j}{|\boldsymbol{r}_i-\boldsymbol{r}_j|}\tag{3.2.11}$$

$$\boldsymbol{F}_{i,i-1}^{l}=-\left.\frac{\mathrm{d}\mathcal{V}_l(l)}{\mathrm{d}l}\right|_{l=l_i}\hat{\boldsymbol{l}}_i\tag{3.2.12}$$

$$\boldsymbol{F}_{i,i-1}^{\theta}=\left.\frac{\mathrm{d}\mathcal{V}_\theta(\cos\theta)}{\mathrm{d}\cos\theta}\right|_{\theta=\theta_{i-1}}\frac{(\hat{\boldsymbol{l}}_{i-1}\cdot\hat{\boldsymbol{l}}_i)\hat{\boldsymbol{l}}_i-\hat{\boldsymbol{l}}_{i-1}}{|\hat{\boldsymbol{l}}_i|}+\left.\frac{\mathrm{d}\mathcal{V}_\theta(\cos\theta)}{\mathrm{d}\cos\theta}\right|_{\theta=\theta_i}\frac{(\hat{\boldsymbol{l}}_i\cdot\hat{\boldsymbol{l}}_{i+1})\hat{\boldsymbol{l}}_i-\hat{\boldsymbol{l}}_{i+1}}{|\hat{\boldsymbol{l}}_i|}\tag{3.2.13}$$

and

$$\begin{aligned}\boldsymbol{F}_{i,i-1}^{\phi}=&\left.\frac{\mathrm{d}\mathcal{V}_\phi(\cos\phi)}{\mathrm{d}\cos\phi}\right|_{\phi=\phi_{i-1}}\frac{1}{|\boldsymbol{l}_i||\hat{\boldsymbol{l}}_{i-2}\times\hat{\boldsymbol{l}}_{i-1}||\hat{\boldsymbol{l}}_{i-1}\times\hat{\boldsymbol{l}}_i|}\\&\times\left[\frac{(\hat{\boldsymbol{l}}_{i-2}\cdot\hat{\boldsymbol{l}}_i)-(\hat{\boldsymbol{l}}_{i-2}\cdot\hat{\boldsymbol{l}}_{i-1})(\hat{\boldsymbol{l}}_{i-1}\cdot\hat{\boldsymbol{l}}_i)}{|\hat{\boldsymbol{l}}_{i-1}\times\hat{\boldsymbol{l}}_i|^2}\left\{\hat{\boldsymbol{l}}_i-(\hat{\boldsymbol{l}}_{i-1}\cdot\hat{\boldsymbol{l}}_i)\hat{\boldsymbol{l}}_{i-1}\right\}-\left\{\hat{\boldsymbol{l}}_{i-2}-(\hat{\boldsymbol{l}}_{i-2}\cdot\hat{\boldsymbol{l}}_{i-1})\hat{\boldsymbol{l}}_{i-1}\right\}\right]\\&+\left.\frac{\mathrm{d}\mathcal{V}_\phi(\cos\phi)}{\mathrm{d}\cos\phi}\right|_{\phi=\phi_i}\frac{1}{|\boldsymbol{l}_i||\hat{\mathbf{l}}_{i-1}\times\hat{\mathbf{l}}_i||\hat{\mathbf{l}}_i\times\hat{\mathbf{l}}_{i+1}|}\\&\times\left[\frac{(\hat{\boldsymbol{l}}_i\cdot\hat{\boldsymbol{l}}_{i+1})-(\hat{\boldsymbol{l}}_{i-1}\cdot\hat{\boldsymbol{l}}_{i+1})(\hat{\boldsymbol{l}}_{i-1}\cdot\hat{\boldsymbol{l}}_i)}{|\hat{\boldsymbol{l}}_{i-1}\times\hat{\boldsymbol{l}}_i|^2}\left\{\hat{\boldsymbol{l}}_{i-1}-(\hat{\boldsymbol{l}}_{i-1}\cdot\hat{\boldsymbol{l}}_i)\hat{\boldsymbol{l}}_i\right\}\right.\\&\left.+\frac{(\hat{\boldsymbol{l}}_{i-1}\cdot\hat{\boldsymbol{l}}_i)-(\hat{\boldsymbol{l}}_{i-1}\cdot\hat{\boldsymbol{l}}_{i+1})(\hat{\boldsymbol{l}}_i\cdot\hat{\boldsymbol{l}}_{i+1})}{|\hat{\boldsymbol{l}}_i\times\hat{\boldsymbol{l}}_{i+1}|^2}\left\{\hat{\boldsymbol{l}}_{i+1}-(\hat{\boldsymbol{l}}_i\cdot\hat{\boldsymbol{l}}_{i+1})\hat{\boldsymbol{l}}_i\right\}\right]\\&+\left.\frac{\mathrm{d}\mathcal{V}_\phi(\cos\phi)}{\mathrm{d}\cos\phi}\right|_{\phi=\phi_{i+1}}\frac{1}{|\boldsymbol{l}_i||\hat{\boldsymbol{l}}_i\times\hat{\boldsymbol{l}}_{i+1}||\hat{\boldsymbol{l}}_{i+1}\times\hat{\boldsymbol{l}}_{i+2}|}\\&\times\left[\frac{(\hat{\boldsymbol{l}}_i\cdot\hat{\boldsymbol{l}}_{i+2})-(\hat{\boldsymbol{l}}_i\cdot\hat{\boldsymbol{l}}_{i+1})(\hat{\boldsymbol{l}}_{i+1}\cdot\hat{\boldsymbol{l}}_{i+2})}{|\hat{\boldsymbol{l}}_i\times\hat{\boldsymbol{l}}_{i+1}|^2}\left\{\hat{\boldsymbol{l}}_i-(\hat{\boldsymbol{l}}_i\cdot\hat{\boldsymbol{l}}_{i+1})\hat{\boldsymbol{l}}_{i+1}\right\}-\left\{\hat{\boldsymbol{l}}_{i+2}-(\hat{\boldsymbol{l}}_{i+1}\cdot\hat{\boldsymbol{l}}_{i+2})\hat{\boldsymbol{l}}_{i+1}\right\}\right].\end{aligned}\tag{3.2.14}$$

The contributions $\boldsymbol{F}^{l}_{i,i+1} = -\boldsymbol{F}^{l}_{i+1,i}$, $\boldsymbol{F}^{\theta}_{i,i+1} = -\boldsymbol{F}^{\theta}_{i+1,i}$, $\boldsymbol{F}^{\phi}_{i,i+1} = -\boldsymbol{F}^{\phi}_{i+1,i}$ are obtainable from Eqs. (3.2.12)–(3.2.14) by increasing index i by 1. A point to make in Eqs. (3.2.11)–(3.2.14) is that the valence angle and torsional contributions to the bonded force $\boldsymbol{F}^{\mathrm{B}}_{i,i-1}$

$$\boldsymbol{F}^{\mathrm{B}}_{i,i-1} = \boldsymbol{F}^{l}_{i,i-1} + \boldsymbol{F}^{\theta}_{i,i-1} + \boldsymbol{F}^{\phi}_{i,i-1} \tag{3.2.15}$$

on site i from site $i-1$ (and thus to $\boldsymbol{F}_{i,i-1}$) depend on the positions of sites other than i and $i-1$ and are not necessarily directed along $\boldsymbol{r}_i - \boldsymbol{r}_{i-1}$.

3.2.3 The microscopic or instantaneous stress tensor

Once the total force $\boldsymbol{F}_i$ on site i due to its interaction with other sites has been calculated, one can extract an expression for the stress tensor in the system in terms of an ensemble average of a mechanical quantity involving atomic positions and momenta. Such an expression constitutes an important link between molecular constitution and the macroscopic thermodynamic and mechanical behaviour of liquid or solid polymers. For a *bounded* atomic or molecular system, the instantaneous equilibrium stress tensor $\boldsymbol{\sigma}$ can be obtained by the general equation

$$\boldsymbol{F}_f = \boldsymbol{\sigma}^{\mathrm{T}} \cdot \boldsymbol{n}_f S_f, \tag{3.2.16}$$

where $\boldsymbol{F}_f$ is the face force exerted on the interior of the material element from its surroundings through the plane of face f with area S_f and normal unit vector $\boldsymbol{n}_f$, including wall forces. Following the convention shown in Figure 3.4, the resulting expression for $\boldsymbol{\sigma}$ from Eq. (3.2.16) is [13]

$$\sigma = -\frac{1}{V}[\boldsymbol{h}_1 \boldsymbol{h}_2 \boldsymbol{h}_3] \begin{bmatrix} \boldsymbol{F}_1^{\mathrm{T}} \\ \boldsymbol{F}_2^{\mathrm{T}} \\ \boldsymbol{F}_3^{\mathrm{T}} \end{bmatrix} \tag{3.2.17}$$

or, in component form,

$$\sigma^{\alpha\beta} = -\frac{1}{V}[h_1^{\alpha} F_1^{\beta} + h_1^{\alpha} F_3^{\beta} + h_3^{\alpha} F_5^{\beta}], \tag{3.2.18}$$

where $\boldsymbol{h}_1$, $\boldsymbol{h}_2$, and $\boldsymbol{h}_3$ denote the edge vectors of the orthogonal material element. By invoking then the equations of motion, we can derive the following expression for the *instantaneous* stress tensor, which is known as the virial theorem of Clausius:[13]

$$\sigma^{\alpha\beta} = -\frac{1}{V}\sum_i \frac{p_i^{\alpha} p_i^{\beta}}{m_i} - \frac{1}{V}\sum_i r_i^{\alpha} F_i^{\beta}, \tag{3.2.19}$$

In Eq. (3.2.19), p_i^{α} is the α-component of the momentum of site i with mass m_i and $\boldsymbol{F}_i = -\nabla_{\boldsymbol{r}_i}\mathcal{V}$ stands for the total force on site i due to interactions with all other sites in the bounded system. The

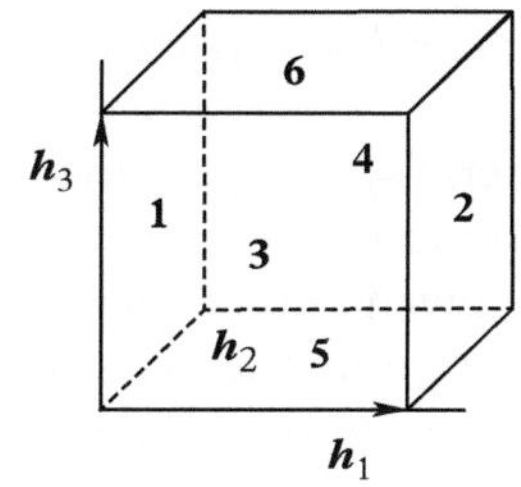

Figure 3.4 Numbering of the six faces and the edge vectors of the simulation box.
Reprinted (adapted) with permission from Theodorou, D.N., Boone, T.D., Dodd, L.R., and Mansfield, K.F. *Macromolecul. Chem., Theory Simul.* **2**, 191 (1993) by John Wiley and Sons.

corresponding thermodynamic stress tensor $\boldsymbol{\tau}$ is obtained by taking the ensemble average of $\boldsymbol{\sigma}$:

$$\boldsymbol{\tau} = \left\langle -\frac{1}{V}\sum_i \frac{\boldsymbol{p}_i\boldsymbol{p}_i}{m_i} - \frac{1}{V}\sum_i \boldsymbol{r}_i\boldsymbol{F}_i \right\rangle. \tag{3.2.20}$$

Equation (3.2.20) is an expression for the $\alpha\beta$-component of $\boldsymbol{\tau}$ in a macroscopic system containing a fixed number of atoms (interaction sites) and occupying a total volume V. The momentum term on the right-hand side of Eq. (3.2.20) can be written in terms of the system temperature as $\left\langle -\frac{n}{V}\right\rangle k_B T\boldsymbol{I}$, where the unit tensor $\boldsymbol{I}$ selects only normal component contributions. Thus, Eq. (3.2.20) is also written as

$$\boldsymbol{\tau} = \left\langle -\frac{n}{V}\right\rangle k_B T\boldsymbol{I} + \left\langle -\frac{1}{V}\sum_i \boldsymbol{r}_i\boldsymbol{F}_i \right\rangle. \tag{3.2.21}$$

It is also instructive to note that, for a system in equilibrium, the instantaneous stress (or pressure) tensor defines the scalar pressure P through the following reduced form of the virial theorem,

$$P = -\frac{1}{3}\mathrm{tr}(\langle\boldsymbol{\sigma}\rangle) = -\frac{1}{3}\mathrm{tr}(\boldsymbol{\tau}), \tag{3.2.22}$$

implying also that

$$P = \left\langle \frac{n}{V}\right\rangle k_B T + \frac{1}{3V}\left\langle \sum_i \boldsymbol{r}_i \cdot \boldsymbol{F}_i \right\rangle. \tag{3.2.23a}$$

For pairwise additive forces this can be rewritten as

$$P = \langle P_{\mathrm{int}}\rangle = \frac{1}{3V}\left\langle \sum_i \frac{\boldsymbol{p}_i^2}{m_i} + \sum_i\sum_{j>i} (\boldsymbol{r}_i - \boldsymbol{r}_j)\cdot\boldsymbol{F}_{ij} \right\rangle = \frac{1}{3V}\left\langle \sum_i \frac{\boldsymbol{p}_i^2}{m_i} \right\rangle + \frac{1}{V}\langle W\rangle \tag{3.2.23b}$$

with $\boldsymbol{F}_{ij}$ being the force on i due to j. The quantity $W = \frac{1}{3}\sum_i\sum_{j>i} (\boldsymbol{r}_i - \boldsymbol{r}_j)\cdot\boldsymbol{F}_{ij}$ is referred to as the *internal virial*. The quantity $P_{\mathrm{int}} = \frac{1}{3V}\sum_i \frac{\boldsymbol{p}_i^2}{m_i} + \frac{W}{V}$ is sometimes referred to as the *instantaneous internal pressure*, not to be confused with the thermodynamic *internal pressure* defined in Section 3.5.1.4.

Molecular simulations of bulk materials, on the other hand, employ *infinite periodic* model systems (systems subject to periodic boundary conditions) on the basis of the minimum image convention for site–site interactions. A given site in such a periodic system moves under the sole influence of the force $\boldsymbol{F}_i$ generated through interactions with other sites and their images. There are no wall forces, and this constitutes an important difference between the model systems used in these simulations and macroscopic bounded systems. It turns out that the single summation form, Eq. (3.2.19), is inapplicable to periodic model systems. For such systems, the correct *atomic virial* expression is derived[13] by calculating the total momentum flux through the boundaries of the periodic simulation box:

$$\boldsymbol{\sigma} = -\frac{1}{V}\sum_i \frac{\boldsymbol{p}_i\boldsymbol{p}_i}{m_i} - \frac{1}{2V}\sum_i\sum_j (\boldsymbol{r}_i - \boldsymbol{r}_{j_{\min}(i)})\boldsymbol{F}_{ij}^{\min} + \boldsymbol{\tau}_{\text{tail}}. \tag{3.2.24}$$

All summations in this equation take place over all sites in the primary simulation box. In addition, the symbol $j_{\min}(i)$ stands for the minimum image of site j with respect to site i and $\boldsymbol{F}_{ij}^{\min}$ is the force on site i due to $j_{\min}(i)$. In general, the force $\boldsymbol{F}_{ij}^{\min}$ between two sites a and b in the infinite periodic model system is defined as $\boldsymbol{F}_{ij} = -\boldsymbol{\nabla}_{\boldsymbol{r}_i-\boldsymbol{r}_j}\mathcal{V}$, where the gradient is taken keeping all intersite separation vectors other than $\boldsymbol{r}_i - \boldsymbol{r}_j$ constant; $\boldsymbol{F}_{ij}$ may depend on separation vectors other than $\mathbf{r}_i - \mathbf{r}_j$; thus, it is not necessarily central. Equation (3.2.24) therefore does not presuppose pairwise additivity. Also, $\boldsymbol{\tau}_{\text{tail}}$ is the *tail* contribution from the interactions $\mathcal{V}_{\text{tail}}$ beyond the cutoff distance of the nonbonded potential. For an isotropic melt or amorphous solid,

$$\boldsymbol{\tau}_{\text{tail}} = \boldsymbol{I}\frac{2\pi}{3}\left\langle\frac{n}{V}\right\rangle^2 \int_{r_c}^{\infty} g(r)\frac{\mathrm{d}\mathcal{V}_{\text{tail}}}{\mathrm{d}r} r^3 \mathrm{d}r, \tag{3.2.25}$$

where $g(r)$ is the pair distribution function of sites. Equation (3.2.25) arises by integrating the *intermolecular virial* $r\frac{\mathrm{d}\mathcal{V}_{\text{tail}}}{\mathrm{d}r}$ around a reference atom in spherical coordinates for all nonbonded pairs (as estimated by the pair distribution function at the given distance and the average number density of atoms n/V) from the cutoff distance r_c out to infinity, multiplying by the number of all atoms and dividing by the volume, and finally dividing also by 2 to correct for double counting of interactions.

In Eq. (3.2.24), the symbol $\boldsymbol{\sigma}$ has been used to denote the instantaneous mechanical quantity whose ensemble average gives the mean momentum flux through the faces of the periodic simulation box. An estimate of the thermodynamic stress $\boldsymbol{\tau}$ at equilibrium, including tail contributions, can be obtained as

$$\boldsymbol{\tau} = \langle \boldsymbol{\sigma}_{\text{a.v.}} \rangle = \left\langle -\frac{1}{V}\sum_{i \text{ in box}} \frac{\boldsymbol{p}_i^{\text{P}}\boldsymbol{p}_i^{\text{P}}}{m_i} - \frac{1}{2V}\sum_{i \text{ in box}}\sum_{j \text{ in box}} (\boldsymbol{r}_i - \boldsymbol{r}_{j_{\min}(i)})\boldsymbol{F}_{ij}^{\min} \right\rangle + \boldsymbol{\tau}_{\text{tail}}. \tag{3.2.26}$$

The subscript a.v. in Eq. (3.2.26) emphasizes stress calculation on the basis of the atomic virial. As in (3.2.24), all summations in Eq. (3.2.31) take place over all sites in the primary box. The symbol $j_{\min}(i)$ indicates the minimum image of site j with respect to site i, and $\boldsymbol{F}_{ij}^{\min}$ is the force on i due to $j_{\min}(i)$. As we saw earlier, in general, the force $\boldsymbol{F}_{ij}$ between two sites i and j in the infinite periodic system is defined as $\boldsymbol{F}_{ij} = -\boldsymbol{\nabla}_{\boldsymbol{r}_i-\boldsymbol{r}_j}\mathcal{V}$, where the gradient is taken keeping all intersite separation vectors other than $\boldsymbol{r}_i - \boldsymbol{r}_j$ constant. Also, the superscript P on the momenta denotes *parent image*, i.e. the image of i belonging to a chain which starts in the primary simulation box. $\boldsymbol{\tau}_{\text{tail}}$ is a the potential tail contribution from Eq. (3.2.25). For a proof of Eq. (3.2.26) starting from continuum thermoelasticity the reader is referred to Section 3.5.1 of this chapter. The atomic virial expression, Eq. (3.2.26), is widely used for the calculation of stress and pressure in equilibrium simulations of both fluids and solids. Solid mechanics, on the other hand, focuses on the interatomic force term and the tail contribution, excluding the part of Eq. (3.2.26) which depends on momenta. The part of $\boldsymbol{\tau}$ which is associated with interatomic forces is often called *internal stress* and is sometimes identified with the Cauchy stress,[14] although the Cauchy stress is more correctly defined as the full tensor $\boldsymbol{\tau}$, including the momentum contribution.

Equations (3.2.24) and (3.2.26) require knowledge of bonded forces. When a flexible model is used for the polymer chains, these are readily available at every step along the simulation. The situation is a bit more subtle when the simulation employs the flexible model in the limit of infinitely stiff force constants, or the rigid model. When bonds and/or bond angles are infinitely stiff or rigid, they can give rise to forces of arbitrary magnitude on the interaction sites. Knowledge of the system configuration alone does not suffice for calculating these forces. This is not a problem in MD simulations, which track the temporal evolution of the system in both the configuration and momentum space. In MD, atomic accelerations are known at all times, and the bonded forces generated by infinitely stiff or rigid bonds and bond angles can be backed out from the dynamical equations of motion of the interaction sites; we discuss this in Chapter 4. Molecular mechanics or energy minimization is another simulation technique where the calculation of bonded forces is possible despite the use of infinitely stiff or rigid bonds and bond angles based on the condition of detailed mechanical equilibrium, namely $\sum_{j \neq i} \boldsymbol{F}_{ij} = \mathbf{0}, \forall i$ in the simulation box; we discuss this in Section 3.4 of the current chapter. Then, Eq. (3.2.24) without the momentum term can be used to estimate the internal stress tensor from minimum potential energy configurations. However, there is at least one case where the use of Eq. (3.2.24) is impossible. As we will see in Chapter 5, a very powerful method for simulating polymers nowadays is Monte Carlo; by making use of a number of revolutionary moves, this method can induce thermal equilibration very rapidly. Typically, however, the models sampled with MC employ infinitely stiff bond lengths, to keep the molecular description as economical as possible. As a result, the evaluation of the bonded forces in these simulations is not possible since MC is a simulation in configuration space only, so accelerations are not known, and conditions of mechanical equilibrium are not satisfied on a configuration-by-configuration basis. Consequently, expressions for the stress such as Eq. (3.2.24) which involve bonded interatomic forces cannot be used in MC simulations of models with infinitely stiff or rigid bonds and bond angles. The only option is to have available an expression for the stress cast in terms of nonbonded interactions only. This issue has been addressed[13] and the resulting expression for the instantaneous stress as a flux of momentum through the faces of the simulation box has been found to have the form

$$\boldsymbol{\sigma}_{\text{face}}^{\alpha\beta} = \frac{1}{V} \lim_{\Delta t \to 0} \sum_{i \text{ in box}} \frac{\Delta}{\Delta t} \left[r_i^{\alpha} p_i^{\text{P},\beta} - r_i^{\text{P},\alpha} p_i^{\text{P},\beta} \right] + \frac{1}{V} \sum_{i \text{ in box}} r_i^{\text{P},\alpha} \sum_{\substack{j \text{ in box} \\ i, j_{\min}(i) \text{ belong to different} \\ \text{images of the same parent chain} \\ \text{or to different parent chains}}} F_{ij}^{\text{NB},\min,\beta}$$

$$- \frac{1}{2V} \sum_{i \text{ in box}} \sum_{\substack{j \text{ in box} \\ i, j_{\min}(i) \text{ belong to different} \\ \text{images of the same parent chain} \\ \text{or to different parent chains}}} \left(r_i^{\alpha} - r_{j_{\min}(i)}^{\beta} \right) F_{ij}^{\text{NB},\min,\beta} + \boldsymbol{\tau}_{\text{tail}}^{\alpha\beta} \tag{3.2.27}$$

where the subscript *face* has been used as reminder that Eq. (3.2.27) is based directly on the consideration of momentum flux through the faces of the simulation box and a tail contribution is appended. We see that, indeed, all forces appearing in Eq. (3.2.27) are nonbonded. However, not all nonbonded interactions are included. Nonbonded interactions between sites on the same image of the same parent chain do not contribute; only true intermolecular (i.e. true interchain) interactions play a role.

By ensemble averaging Eq. (3.2.27), one arrives at two alternative expressions for the thermodynamic stress $\boldsymbol{\tau}$ cast in terms of intermolecular forces only. The first corresponds to the so-called *intermolecular force-based instantaneous stress*, and reads[13]

$$\begin{aligned}\boldsymbol{\tau} &= \langle \boldsymbol{\sigma}_{\text{inter}} \rangle \\ &= \left\langle \frac{1}{V} \sum_{i \text{ in box}} \boldsymbol{r}_i^{\text{P}} \left(\sum_{\substack{j \text{ in box} \\ i, j_{\min}(i) \text{ belong to different} \\ \text{images of the same parent chain} \\ \text{or to different parent chains}}} \boldsymbol{F}_{ij}^{\text{NB,min}} \right) \right\rangle \\ &\quad - \left\langle \frac{1}{2V} \sum_{i \text{ in box}} \sum_{\substack{j \text{ in box} \\ i, j_{\min}(i) \text{ belong to different} \\ \text{images of the same parent chain} \\ \text{or to different parent chains}}} (\boldsymbol{r}_i - \boldsymbol{r}_{j_{\min}(i)}) \boldsymbol{F}_{ij}^{\text{NB,min}} \right\rangle + \boldsymbol{\tau}_{\text{tail}} \end{aligned} \qquad (3.2.28)$$

where $\boldsymbol{\sigma}_{\text{inter}}$ stands for the mechanical property being averaged, including the tail contribution. The atomic virial expression, Eq. (3.2.26), falls out as a special case of Eq. (3.2.28) for the case of a monatomic fluid. The second expression corresponds to the *molecular virial* and reads

$$\begin{aligned}\boldsymbol{\tau} = \langle \boldsymbol{\sigma}_{\text{m.v.}} \rangle = \Bigg\langle -\frac{1}{V} \sum_{c \text{ chains in box}} \frac{\boldsymbol{p}_{\text{cm},c} \boldsymbol{p}_{\text{cm},c}}{m_c} - \frac{1}{2V} \sum_{i \text{ in box}} \\ \sum_{\substack{j \text{ in box} \\ i, j_{\min}(i) \text{ belong to different} \\ \text{images of the same parent chain} \\ \text{or to different parent chains}}} (\boldsymbol{r}_{\text{cm},c(i)} - \boldsymbol{r}_{\text{cm},c(j_{\min}(i))}) \boldsymbol{F}_{ij}^{\text{NB,min}} \Bigg\rangle + \boldsymbol{\tau}_{\text{tail}}, \end{aligned} \qquad (3.2.29)$$

where the index c runs over all parent chains, $c(i)$ denotes the chain image to which site i belongs, and cm, $c(i)$ is the centre of mass of that image. Also, the centre-of-mass momentum term on the right-hand side of Eq. (3.2.29) can be substituted by $-\left\langle \frac{N_c}{V} \right\rangle k_B T \boldsymbol{I}$. Equations (3.2.28) and (3.2.29) provide a means of calculating the internal stress tensor from knowledge of the nonbonded interactions only. Since only minimum–image pairs are involved, it is evident that both are invariant to translations of the chain origins.

In summary, there are three different approaches to computing the thermodynamic stress tensor $\boldsymbol{\tau}$. The first and most conventional is the atomic virial expression, Eq. (3.2.26), cast in terms of total (bonded and nonbonded) forces on sites. The second is the intermolecular force-based, Eq. (3.2.28), cast exclusively in terms of intermolecular forces on sites. The third is the molecular virial expression, Eq. (3.2.29), cast in terms of interchain forces and chain centre-of-mass coordinates and momenta, with provisions for handling periodic boundary conditions when the molecular size becomes comparable to the simulation box. The instantaneous mechanical quantities which, upon averaging, give the thermodynamic stress have been labelled $\boldsymbol{\sigma}_{\text{a.v.}}$, $\boldsymbol{\sigma}_{\text{inter}}$, and $\boldsymbol{\sigma}_{\text{m.v.}}$, respectively. $\boldsymbol{\sigma}_{\text{a.v.}}$ is cast in terms of total forces, intersite distances, and parent image momenta. $\boldsymbol{\sigma}_{\text{inter}}$ is cast in terms of intermolecular forces on individual sites. $\boldsymbol{\sigma}_{\text{m.v.}}$ is cast in terms of chain centre-of-mass positions and momenta, and overall forces between chains. In a sense, $\boldsymbol{\sigma}_{\text{inter}}$ is intermediate between the atomic and molecular virial expressions. In addition, Eq. (3.2.27) for $\boldsymbol{\sigma}_{\text{face}}$ provides an expression for the instantaneous stress as a momentum flux through the box faces. The four quantities $\boldsymbol{\sigma}_{\text{a.v.}}$, $\boldsymbol{\sigma}_{\text{inter}}$, $\boldsymbol{\sigma}_{\text{m.v.}}$, and $\boldsymbol{\sigma}_{\text{face}}$ must all have the same average value, but may very well exhibit different fluctuations.

We should also note here that a rigid model governed by a potential function $\mathcal{V}$ can be thought of as a *flexible model in the limit of infinite force constants* governed by a potential function equal to $\mathcal{V}$ plus the Fixman compensating potential. The Fixman potential is a function of θ_i's and ϕ_i's, and therefore depends exclusively on separation vectors between sites connected directly by bonds. Upon differentiation, therefore, it will generate forces analogous to the bonded contributions $\boldsymbol{F}^{\theta}_{i,i-1}$ and $\boldsymbol{F}^{\phi}_{i,i-1}$, which are thus purely intramolecular. For a given configuration of the multichain system, the intermolecular nonbonded forces are identical between the flexible model with infinite force constants and the rigid model. Therefore, Eqs. (3.2.28) and (3.2.29) for $\boldsymbol{\sigma}_{\text{inter}}$ are valid for the rigid model as well.

An interesting observation[13] to make from Eqs. (3.2.13) and (3.2.14) is that

$$(\boldsymbol{r}_i - \boldsymbol{r}_{i-1}) \cdot \boldsymbol{F}^{\theta}_{i,i-1} = \boldsymbol{l}_i \cdot \boldsymbol{F}^{\theta}_{i,i-1} = 0 \tag{3.2.30a}$$

and

$$(\boldsymbol{r}_i - \boldsymbol{r}_{i-1}) \cdot \boldsymbol{F}^{\phi}_{i,i-1} = \boldsymbol{l}_i \cdot \boldsymbol{F}^{\phi}_{i,i-1} = 0, \tag{3.2.30b}$$

implying that the contribution of valence angle and torsional forces to the trace of the instantaneous value of the stress tensor is 0. In other words, valence angles and torsional forces do not contribute to the atomic virial pressure, which is shaped by the nonbonded forces and the strictly two-body bond length distortion part of the bonded forces $\boldsymbol{F}^{\text{B}}_{ij}$. On the other hand, valence angle and torsional forces do affect the instantaneous individual values of the diagonal components $\sigma^{11}_{\text{a.v.}}$, $\sigma^{22}_{\text{a.v.}}$, and $\sigma^{33}_{\text{a.v.}}$ of the atomic virial stresses as well as the shear stresses in the system.

3.3 Potential energy minimization in generalized coordinates

3.3.1 Matrix notation for the position vector and its partial derivatives with respect to internal coordinates

For simplicity, in what follows we will limit our discussion to linear chains consisting of $N + 1$ skeletal atoms or groups numbered as 0, 1, 2, . . ., N. Skeletal bonds are numbered from 1 to N so that bond i connects atoms $i - 1$ and i. Skeletal bond angles are numbered from 1 to $N - 1$, bond angle i having its apex on skeletal atom i. In this section we will use a model wherein bond lengths l_i and bond angles θ_i are fixed. The torsion angle ϕ_i is the dihedral angle between the planes of bonds $(i - 1,\ i)$ and bonds $(i,\ i + 1)$ measured in the right-hand sense relative to trans. There are $(N - 2)$ skeletal torsion angles in a chain, their indices running from 2 to $N - 1$. There are N_c parent chains in the system. The development presented in the following is an extension to multiparent chain systems of the formulation of Theodorou and Suter.[15]

Local reference frames are introduced at skeletal bonds following the conventions of Flory. The simulation box, characterized by periodic boundary conditions in all three directions, has fixed spatial extent in an *external frame* of reference. The Cartesian coordinates of atoms with respect to that frame are denoted by $\boldsymbol{r} = (x, y, z)$. For each chain in the system, there is an image of that chain which starts inside the simulation box; this is referred to as the *parent image* of the chain. We use three Eulerian angles to specify the overall orientation of the chain section comprising the first three skeletal atoms of the chain. Eulerian angle ψ_1 is taken as the angle between the x-direction and the projection of bond 1 on the x-y plane; Eulerian angle ψ_2 is taken as the angle between the z-direction and bond 1; and Eulerian angle ψ_3 is taken as the dihedral angle between the planes (z-direction, bond 1) and (bond 1, bond 2). See Figure 1.7 for the definition of these degrees of freedom. The start coordinates of the parent chain are denoted by $\boldsymbol{r}_0$. For an atom in the cube, let j symbolize the index of the skeletal atom to which the parent image of this atom is connected. The configuration of the system is then completely defined if one specifies: (i) the positions $\boldsymbol{r}_{0(k)}, k = 1, 2, \ldots, N_c$ of all chain starts in the simulation box; (ii) the three Eulerian angles $\psi_{1(k)}, \psi_{2(k)}, \psi_{3(k)}, k = 1, 2, \ldots, N_c$ for all chains describing the orientation of each parent chain with respect to the frame of reference of the cubic simulation box; and (iii) the rotation angles $\phi_{i(k)},\quad i = 2, \ldots, N - 1,\ k = 1, 2, \ldots, N_c$ of all skeletal bonds but the first and the last of all parent chains. The vector

$$\{\boldsymbol{r}_{0(k)}, \psi_{1(k)}, \psi_{2(k)}, \psi_{3(k)}, \phi_{2(k)}, \phi_{3(k)}, \ldots, \phi_{N-1(k)}\}; \quad k = 1, 2, \ldots, N_c \tag{3.3.1a}$$

is the vector of degrees of freedom of the system. However, we note that as far as the start coordinates are concerned, $\boldsymbol{r}_{0(1)}$, the start of the first chain, is defined with respect to the box frame of reference, and chain start coordinates $\boldsymbol{r}_{0(k)}$ of subsequent chains are defined relative to the first one. In our periodic model system, a displacement of the first chain start will not change any of the atom–atom separations; thus, in the absence of walls, none of the coordinates $x_{0(1)}, y_{0(1)}, z_{0(1)}$ is a *true* degree of freedom. In this case, the vector of degrees of freedom of the system is actually

$$\{\boldsymbol{r}_{0(k')} - \boldsymbol{r}_{0(1)}, \psi_{1(k)}, \psi_{2(k)}, \psi_{3(k)}, \phi_{2(k)}, \phi_{3(k)}, \ldots, \phi_{N-1(k)}\}; \quad k' = 2, 3, \ldots, N_c, k = 1, 2, \ldots, N_c. \tag{3.3.1b}$$

Assuming that bond lengths and bond angles are held fixed at their equilibrium values, the potential energy function of the system will be the sum

$$\mathcal{V}(\boldsymbol{r}_0, \psi, \phi) = \sum_{\substack{k,\ \text{all parent chains;} \\ k \geq 1}} \left\{ \begin{array}{l} \displaystyle\sum_{\substack{i_k \\ \text{all bonds } i_k \\ \text{of parent} \\ \text{chain } k; \\ 2 \leq i_k \leq N-1}} \mathcal{V}_\phi(\phi_{i_k}) \\ + \displaystyle\sum_{\substack{i(k) \\ i(k) > 0}} \sum_{\substack{j(k) \\ 0 \leq j(k) < i(k)}} \mathcal{V}_{\text{NB}}^{\text{intra}} \left(\left| \boldsymbol{r}_{i(k)} - \boldsymbol{r}_{j(k)}^{\min[i(k)]} \right| \right) \\ + \displaystyle\sum_{\substack{l,\ \text{all parent chains;} \\ l < k}} \sum_{\substack{i(k) \\ i(k) \geq 0}} \sum_{\substack{j(l) \\ j(l) \geq 0}} \mathcal{V}_{\text{NB}}^{\text{inter}} \left(\left| \boldsymbol{r}_{i(k)} - \boldsymbol{r}_{j(l)}^{\min[i(k)]} \right| \right) \end{array} \right\}, \tag{3.3.2}$$

where $\mathcal{V}_\phi$ is an intrinsic torsional potential associated with skeletal bonds, and the last two terms correspond to self-chain and cross-chain nonbonded van der Waals interactions, respectively. The notation $i(k)$ is used to indicate atom i on parent chain k. $\boldsymbol{r}_{j(l)}^{\min[i(k)]}$ stands for the position of that image of atom j belonging to chain l that lies closest to atom i of chain k.

To calculate the gradient and the Hessian of the function $\mathcal{V}(\boldsymbol{r}_0, \psi, \phi)$ wrt the set $\{\boldsymbol{r}_0, \psi, \phi\}$ of the flexible degrees of freedom, we need first to express the position vector of an atom in the system in terms of this set. In matrix notation, the position vector $\boldsymbol{r}_{i(k)}$ of atom i ($i = 0, 1, \ldots, N$) in the simulation box belonging to chain k ($k = 1, 2, \ldots, N_c$) is expressed in terms of the set $\{\mathbf{r}_{0(k)}, \psi_{1(k)}, \psi_{2(k)}, \psi_{3(k)}, \phi_{2(k)}, \ldots, \phi_{N-1(k)}\}$ of flexible variables according to

$$\begin{aligned} \boldsymbol{r}_{i(k)} &= \boldsymbol{r}_{0(k)} + \boldsymbol{T}_{0(k)} \cdot \boldsymbol{l}_{1(k)} + \boldsymbol{T}_{0(k)} \cdot \boldsymbol{T}_{1(k)} \cdot \boldsymbol{l}_{2(k)} + \ldots + \boldsymbol{T}_{0(k)} \cdot \boldsymbol{T}_{1(k)} \cdot \ldots \cdot \boldsymbol{T}_{i-1(k)} \cdot \boldsymbol{l}_{i(k)} \\ &\quad + \lambda_{i(k)} \boldsymbol{h}_1 + \mu_{i(k)} \boldsymbol{h}_2 + \nu_{i(k)} \boldsymbol{h}_3, \end{aligned} \tag{3.3.3}$$

where

$$\mathbf{T}_{0(k)} = \begin{bmatrix} \cos\psi_{1(k)} \cos\psi_{2(k)} & -\cos\psi_{1(k)} \cos\psi_{2(k)} \cos\psi_{3(k)} - \sin\psi_{1(k)} \sin\psi_{3(k)} & -\cos\psi_{1(k)} \cos\psi_{2(k)} \sin\psi_{3(k)} + \sin\psi_{1(k)} \sin\psi_{3(k)} \\ \sin\psi_{1(k)} \sin\psi_{2(k)} & -\sin\psi_{1(k)} \cos\psi_{2(k)} \cos\psi_{3(k)} + \cos\psi_{1(k)} \sin\psi_{3(k)} & -\sin\psi_{1(k)} \cos\psi_{2(k)} \sin\psi_{3(k)} - \cos\psi_{1(k)} \cos\psi_{3(k)} \\ \cos\psi_{2(k)} & \sin\psi_{2(k)} \cos\psi_{3(k)} & \sin\psi_{2(k)} \sin\psi_{3(k)} \end{bmatrix} \tag{3.3.4}$$

is the orthogonal transformation matrix of *internal* to *external* coordinates,

$$\mathbf{T}_{i(k)} = \begin{bmatrix} \cos\theta_{i(k)} & \sin\theta_{i(k)} & 0 \\ \sin\theta_{i(k)} \cos\phi_{i(k)} & -\cos\theta_{i(k)} \cos\phi_{i(k)} & \sin\phi_{i(k)} \\ \sin\theta_{i(k)} \sin\phi_{i(k)} & -\cos\theta_{i(k)} \sin\phi_{i(k)} & -\cos\phi_{i(k)} \end{bmatrix} \tag{3.3.5}$$

is the orthogonal transformation matrix from the reference frame of bond $(i+1)$ to the reference frame of bond i (compare Eq. (1.2.42); by definition, $\phi_1 = 0$), and $\boldsymbol{l}_{i(k)}$ is the column vector

$$\boldsymbol{l}_{i(k)} = l_{i(k)}(1, 0, 0)^{\mathrm{T}}. \tag{3.3.6}$$

The last three terms in Eq. (3.3.3) arise from the periodicity of the system: $\boldsymbol{h}_1$, $\boldsymbol{h}_2$, $\boldsymbol{h}_3$ are the edge vectors of the simulation box, also referred to as continuation vectors. On the other hand, the integers $\lambda_{i(k)}, \mu_{i(k)}, \nu_{i(k)}$ measure the number of translations along $\boldsymbol{h}_1, \boldsymbol{h}_2, \boldsymbol{h}_3$, respectively, that are required in order to reach the considered atom of the cube from its image belonging to parent chain k. The integers $\lambda_{i(k)}, \mu_{i(k)}, \nu_{i(k)}$ will be referred to as periodic continuation coefficients of atom $i(k)$.

The gradient of $\mathcal{V}(\boldsymbol{r}_0, \psi, \phi)$ is obtained by differentiation of the potential energy with respect to each of the degrees of freedom. The corresponding derivatives are calculated as shown in Sections 3.3.2 through 3.3.8. If atoms i and j belong to the same chain, we assume that $i < j$.

3.3.2 Partial derivative of the position vector with respect to chain start coordinates

As explained in Section 3.3.1, the start coordinates of the first parent chain, $\boldsymbol{r}_{0(1)}$, are not degrees of freedom in the presence of 3D periodic boundary conditions. The start coordinates of the remaining parent chains, $\boldsymbol{r}_{0(k')} - \boldsymbol{r}_{0(1)}$, $k' = 2, 3, \ldots, N_{\mathrm{c}}$, defined here relative to the start coordinates of the first chain, are degrees of freedom, however (see Eq. (3.3.1b)). Changing a component of $\boldsymbol{r}_{0(k')} - \boldsymbol{r}_{0(1)}$ while keeping all other degrees of freedom constant amounts to a rigid displacement of the entire chain k', while all other chains are kept in place. Thus,

$$\frac{\partial r^{\alpha}_{i(k)}}{\partial\left(\boldsymbol{r}^{\beta}_{0(k')} - r^{\beta}_{0(1)}\right)} = \delta_{kk'}\delta_{\alpha\beta}, \quad k' = 2, 3, \ldots, N_{\mathrm{c}} \tag{3.3.7}$$

with $r^{\alpha}_{i(k)}$, $\alpha \in \{1, 2, 3\}$ symbolizing the Cartesian coordinates of position vector $\boldsymbol{r}_{i(k)}$ of atom i of chain k and $\delta_{\alpha\beta}$ being the Kronecker delta. Note that the derivative of Eq. (3.3.7) is independent of the periodic continuation coefficients of atom $i(k)$. Upon displacing the start of a parent chain, all atoms on all images of the chain are displaced by the same vector.

3.3.3 Partial derivative of the position vector with respect to Euler angles

$$\frac{\partial \boldsymbol{r}_{i(k)}}{\partial \psi_{a(l)}} = \delta_{kl}\boldsymbol{v}_{\psi_{a(l)}} \times (\boldsymbol{r}_{i(k)} - \boldsymbol{r}_{0(k)} - \lambda_{i(k)}\boldsymbol{h}_1 - \mu_{i(k)}\boldsymbol{h}_2 - \nu_{i(k)}\boldsymbol{h}_3), \tag{3.3.8}$$

where

$$\begin{aligned}
\boldsymbol{\nu}_{\psi_{1(k)}} &= (0, 0, 1) \\
\boldsymbol{\nu}_{\psi_{2(k)}} &= (-\sin\psi_{1(k)}, \cos\psi_{1(k)}, 0) \\
\boldsymbol{\nu}_{\psi_{3(k)}} &= (-\cos\psi_{1(k)}\sin\psi_{2(k)}, -\sin\psi_{1(k)}\sin\psi_{2(k)}, -\cos\psi_{2(k)}).
\end{aligned} \tag{3.3.9}$$

3.3.4 Partial derivative of the position vector with respect to torsional angles

$$\frac{\partial \boldsymbol{r}_{i(k)}}{\partial \phi_{a(l)}} = 0 \quad \forall i, a \quad \text{if } k \neq l$$

$$\frac{\partial \boldsymbol{r}_{i(k)}}{\partial \phi_{a(k)}} = 0, \quad \text{if } a \geq i$$

$$\frac{\partial \boldsymbol{r}_{i(k)}}{\partial \phi_{a(k)}} = \frac{1}{l_{a(k)}}\left[(\boldsymbol{r}_{a(k)} - \boldsymbol{r}_{a-1(k)}) - (\lambda_{a(k)} - \lambda_{a-1(k)})\boldsymbol{h}_1 - (\mu_{a(k)} - \mu_{a-1(k)})\boldsymbol{h}_2 - (\nu_{a(k)} - \nu_{a-1(k)})\boldsymbol{h}_3\right]$$
$$\times \left[(\boldsymbol{r}_{i(k)} - \boldsymbol{r}_{a(k)}) - (\lambda_{i(k)} - \lambda_{a(k)})\boldsymbol{h}_1 - (\mu_{i(k)} - \mu_{a(k)})\boldsymbol{h}_2 - (\nu_{i(k)} - \nu_{a(k)})\boldsymbol{h}_3\right], \quad \text{if } a < i. \tag{3.3.10}$$

3.3.5 Partial derivative of the interatomic distances with respect to chain start coordinates

$$\frac{\partial \left|\boldsymbol{r}_{i(k)} - \boldsymbol{r}_{j(l)}^{\min[i(k)]}\right|}{\partial \left(r_{0(k')}^{\beta} - r_{0(1)}^{\beta}\right)} = (\delta_{kk'} - \delta_{lk'})(1 - \delta_{kl}) \frac{\boldsymbol{r}_{i(k)}^{\beta} - \boldsymbol{r}_{j(l)}^{\min[i(k)],\beta}}{\left|\boldsymbol{r}_{i(k)} - \boldsymbol{r}_{j(l)}^{\min[i(k)]}\right|}$$
$$= \left[\delta_{kk'} \frac{\boldsymbol{r}_{i(k)}^{\beta} - \boldsymbol{r}_{j(l)}^{\min[i(k)],\beta}}{\left|\boldsymbol{r}_{i(k)} - \boldsymbol{r}_{j(l)}^{\min[i(k)]}\right|} + \delta_{lk'} \frac{\boldsymbol{r}_{j(l)}^{\beta} - \boldsymbol{r}_{i(k)}^{\min[j(l)],\beta}}{\left|\boldsymbol{r}_{i(k)} - \boldsymbol{r}_{j(l)}^{\min[i(k)]}\right|}\right](1 - \delta_{kl}). \tag{3.3.11}$$

Note that pairs of atoms corresponding to the same parent chain have zero contribution to the derivative of Eq. (3.3.11), regardless of their periodic continuation coefficients.

3.3.6 Partial derivative of the interatomic distances with respect to Euler angles

$$\frac{\partial \left|\boldsymbol{r}_{i(k)} - \boldsymbol{r}_{j(l)}^{\min[i(k)]}\right|}{\partial \psi_{a(k)}} = \begin{cases} \boldsymbol{\nu}_{\psi_{a(k)}} \cdot \left\{[\boldsymbol{r}_{i(k)} - \boldsymbol{r}_{0(k)} - \lambda_{i(k)}\boldsymbol{h}_1 - \mu_{i(k)}\boldsymbol{h}_2 - \nu_{i(k)}\boldsymbol{h}_3] \times \frac{\boldsymbol{r}_{i(k)} - \boldsymbol{r}_{j(l)}^{\min[i(k)]}}{\left|\boldsymbol{r}_{i(k)} - \boldsymbol{r}_{j(l)}^{\min[i(k)]}\right|}\right\}, k \neq l \\ -\boldsymbol{\nu}_{\psi_{a(k)}} \cdot \left\{\left[\left(\lambda_{i(k)} - \lambda_{j(l)}^{\min[i(k)]}\right)\boldsymbol{h}_1 + \left(\mu_{i(k)} - \mu_{j(l)}^{\min[i(k)]}\right)\boldsymbol{h}_2 \right.\right. \\ \left.\left. + \left(\nu_{i(k)} - \nu_{j(l)}^{\min[i(k)]}\right)\boldsymbol{h}_3\right] \times \frac{\boldsymbol{r}_{i(k)} - \boldsymbol{r}_{j(l)}^{\min[i(k)]}}{\left|\boldsymbol{r}_{i(k)} - \boldsymbol{r}_{j(l)}^{\min[i(k)]}\right|}\right\}, \ k = l. \end{cases} \tag{3.3.12}$$

3.3.7 Partial derivative of the interatomic distances with respect to torsional angles

$$\frac{\partial \left| \boldsymbol{r}_{i(k)} - \boldsymbol{r}_{j(l)}^{\min[i(k)]} \right|}{\partial \phi_{a(k)}} = 0, \quad \text{if } k \neq l, i < a \text{ or } k = l, j < i < a$$

$$\frac{\partial \left| \boldsymbol{r}_{i(k)} - \boldsymbol{r}_{j(l)}^{\min[i(k)]} \right|}{\partial \phi_{a(k)}} = -\frac{1}{l_{a(k)}} \Big[(\boldsymbol{r}_{a(k)} - \boldsymbol{r}_{a-1(k)}) - (\lambda_{a(k)} - \lambda_{a-1(k)}) \boldsymbol{h}_1 - (\mu_{a(k)} - \mu_{a-1(k)}) \boldsymbol{h}_2 - (\nu_{a(k)} - \nu_{a-1(k)}) \boldsymbol{h}_3 \Big] \cdot \frac{\boldsymbol{r}_{i(k)} - \boldsymbol{r}_{j(l)}^{\min[i(k)]}}{\left| \boldsymbol{r}_{i(k)} - \boldsymbol{r}_{j(l)}^{\min[i(k)]} \right|}$$
$$\times \Big[(\boldsymbol{r}_{i(k)} - \boldsymbol{r}_{a(k)}) - (\lambda_{i(k)} - \lambda_{a(k)}) \boldsymbol{h}_1 - (\mu_{i(k)} - \mu_{a(k)}) \boldsymbol{h}_2 - (\nu_{i(k)} - \nu_{a(k)}) \boldsymbol{h}_3 \Big], \quad \text{if } k \neq l, a < i \text{ or } k = l, j < a < i$$

$$\frac{\partial \left| \boldsymbol{r}_{i(k)} - \boldsymbol{r}_{j(l)}^{\min[i(k)]} \right|}{\partial \phi_{a(k)}} = -\frac{1}{l_{a(k)}} \Big[(\boldsymbol{r}_{a(k)} - \boldsymbol{r}_{a-1(k)}) - (\lambda_{a(k)} - \lambda_{a-1(k)}) \boldsymbol{h}_1 - (\mu_{a(k)} - \mu_{a-1(k)}) \boldsymbol{h}_2 - (\nu_{a(k)} - \nu_{a-1(k)}) \boldsymbol{h}_3 \Big] \cdot \frac{\mathbf{r}_{i(k)} - \mathbf{r}_{j(l)}^{\min[i(k)]}}{\left| \mathbf{r}_{i(k)} - \mathbf{r}_{j(l)}^{\min[i(k)]} \right|}$$
$$\times \Big[-\left(\lambda_{i(k)} - \lambda_{j(l)}^{\min[i(k)]}\right) \boldsymbol{h}_1 - \left(\mu_{i(k)} - \mu_{j(kl)}^{\min[i(k)]}\right) \boldsymbol{h}_2 - \left(\nu_{i(k)} - \nu_{j(l)}^{\min[i(k)]}\right) \boldsymbol{h}_3 \Big], \quad \text{if } k = l, a < j < i. \tag{3.3.13}$$

3.3.8 The gradient of the potential energy function with respect to generalized coordinates

Having calculated the partial derivatives of the interatomic distances with respect to the Euler and torsional angles, we can now report the corresponding elements of the gradient of the potential energy function with respect to the flexible degrees of freedom of the system. To this end, and in order to cast the resulting equations in a more compact form, we introduce the abbreviated notations

$$\boldsymbol{b}_{a(k)} \equiv (\boldsymbol{r}_{a(k)} - \boldsymbol{r}_{a-1(k)}) - (\lambda_{a(k)} - \lambda_{a-1(k)}) \boldsymbol{h}_1 - (\mu_{a(k)} - \mu_{a-1(k)}) \boldsymbol{h}_2 - (\nu_{a(k)} - \nu_{a-1(k)}) \boldsymbol{h}_3 \tag{3.3.14a}$$

$$\boldsymbol{r}_{i(k)a(k)} \equiv (\boldsymbol{r}_{i(k)} - \boldsymbol{r}_{a(k)}) - (\lambda_{i(k)} - \lambda_{a(k)}) \boldsymbol{h}_1 - (\mu_{i(k)} - \mu_{a(k)}) \boldsymbol{h}_2 - (\nu_{i(k)} - \nu_{a(k)}) \boldsymbol{h}_3 \tag{3.3.14b}$$

$$\boldsymbol{r}_{i(k)j(l)} \equiv (\boldsymbol{r}_{i(k)} - \boldsymbol{r}_{j(l)}) - (\lambda_{i(k)} - \lambda_{j(l)}) \boldsymbol{h}_1 - (\mu_{i(k)} - \mu_{j(l)}) \boldsymbol{h}_2 - (\nu_{i(k)} - \nu_{j(l)}) \boldsymbol{h}_3 \tag{3.3.14c}$$

$$\boldsymbol{r}_{i(k)j(l)}^{\min} \equiv \left(\boldsymbol{r}_{i(k)} - \boldsymbol{r}_{j(l)}^{\min[i(k)]} \right) \tag{3.3.14d}$$

$$r_{i(k)j(l)}^{\min} = \left| \boldsymbol{r}_{i(k)j(l)}^{\min} \right| \tag{3.3.14e}$$

$$\boldsymbol{h}_{i(k)j(l)} \equiv -\left(\lambda_{i(k)} - \lambda_{j(l)}^{\min[i(k)]}\right)\boldsymbol{h}_1 - \left(\mu_{i(k)} - \mu_{j(l)}^{\min[i(k)]}\right)\boldsymbol{h}_2 - \left(\nu_{i(k)} - \nu_{j(l)}^{\min[i(k)]}\right)\boldsymbol{h}_3 \equiv \boldsymbol{r}_{i(k)j(l)} - \boldsymbol{r}_{i(k)j(l)}^{\min}. \tag{3.3.14f}$$

We also introduce the symbols

$$\mathcal{V}_\phi' = \frac{\partial \mathcal{V}_\phi}{\partial \phi}, \mathcal{V}_\phi'' = \frac{\partial^2 \mathcal{V}_\phi}{\partial \phi^2}, \mathcal{V}_r = \frac{\partial \mathcal{V}_{\mathrm{NB}}}{\partial r}, \mathcal{V}_r'' = \frac{\partial^2 \mathcal{V}_{\mathrm{NB}}}{\partial r^2}. \tag{3.3.15}$$

Then, the following equations provide the gradient of the potential energy function with respect to Euler and torsional angles, respectively:

$$\frac{\partial \mathcal{V}}{\partial(\boldsymbol{r}_{0(k')} - \boldsymbol{r}_{0(1)})} = \sum_{\substack{\text{all atoms } i\\ \text{in chain } k'}} \sum_{\substack{\text{all atoms } j\\ \text{in chains } k \neq k'}} \mathcal{V}_r'\left(r = r_{i(k')j(k)}^{\min}\right) \frac{\boldsymbol{r}_{i(k')j(k)}^{\min}}{r_{i(k')j(k)}^{\min}} \tag{3.3.16}$$

$$\begin{aligned}\frac{\partial \mathcal{V}}{\partial \psi_{a(k)}} = &\sum_{\substack{\text{all atoms } i\\ \text{in chain } k}} \sum_{\substack{\text{all atoms } j\\ \text{in chain } k\\ \text{with } j<i}} \frac{\mathcal{V}_r'\left(r = r_{i(k)j(k)}^{\min}\right)}{r_{i(k)j(k)}^{\min}} \boldsymbol{\nu}_{\psi_{a(k)}} \cdot \left(\boldsymbol{h}_{i(k)j(k)} \times \boldsymbol{r}_{i(k)j(k)}^{\min}\right)\\ &+ \sum_{\substack{\text{all atoms } i\\ \text{in chain } k}} \sum_{\substack{\text{all atoms } j\\ \text{in chains } l\\ \text{with } k \neq l}} \frac{\mathcal{V}_r'\left(r = r_{i(k)j(l)}^{\min}\right)}{r_{i(k)j(l)}^{\min}} \boldsymbol{\nu}_{\psi_{a(k)}} \cdot \left(\boldsymbol{r}_{i(k)0(k)} \times \boldsymbol{r}_{i(k)j(l)}^{\min}\right)\end{aligned} \tag{3.3.17}$$

$$\begin{aligned}\frac{\partial \mathcal{V}}{\partial \phi_{a(k)}} = &\mathcal{V}_\phi'(\phi = \phi_{a(k)})\\ &+ \frac{1}{l_{a(k)}}\left[\sum_{\substack{\text{all atoms } i\\ \text{in chain } k\\ \text{with } i \geq a}} \sum_{\substack{\text{all atoms } j\\ \text{in chain } k\\ \text{with } j<a}} \frac{\mathcal{V}_r'\left(r = r_{i(k)j(k)}^{\min}\right)}{r_{i(k)j(k)}^{\min}} \boldsymbol{b}_{i(k)} \cdot \left(\boldsymbol{r}_{i(k)a(k)} \times \boldsymbol{r}_{i(k)j(k)}^{\min}\right)\right.\\ &\left.+ \sum_{\substack{\text{all atoms } i\\ \text{in chain } k\\ \text{with } i>a}} \sum_{\substack{\text{all atoms } j\\ \text{in chain } k\\ \text{with } a \leq j<i}} \frac{\mathcal{V}_r'\left(r = r_{i(k)j(k)}^{\min}\right)}{r_{i(k)j(k)}^{\min}} \boldsymbol{b}_{i(k)} \cdot \left(\boldsymbol{h}_{i(k)j(k)} \times \boldsymbol{r}_{i(k)j(k)}^{\min}\right)\right]\\ &+ \frac{1}{l_{a(k)}} \sum_{\substack{\text{all atoms } i\\ \text{in chain } k\\ \text{with } i \geq a}} \sum_{\substack{\text{all atoms } j\\ \text{in chains } l\\ \text{with } k \neq l}} \frac{\mathcal{V}_r'\left(r = r_{i(k)j(l)}^{\min}\right)}{r_{i(k)j(l)}^{\min}} \boldsymbol{b}_{i(k)} \cdot \left(\boldsymbol{r}_{i(k)a(k)} \times \boldsymbol{r}_{i(k)j(l)}^{\min}\right).\end{aligned} \tag{3.3.18}$$

The Hessian matrix of second derivatives of the potential energy with respect to the generalized coordinates can be obtained by similar differentiations.[15]

3.4 Modelling glassy polymers as collections of energy minima

From a statistical mechanics point of view, minimizing the potential energy with respect to the vector of internal degrees of freedom of the system leads to a microstate which is at a local minimum of the total potential energy, i.e. in mechanical equilibrium. This, of course, should not be interpreted as implying thermodynamic equilibrium (which would require the system to be at a minimum of the Helmholtz free energy, as obtained from an ensemble of microstates at given mass and spatial extent). In molecular mechanics, the local configuration of the polymer is regarded as residing in a local minimum of the potential energy. Different regions in a macroscopic sample may be trapped in different minima, so a collection of minima, each generated by potential energy minimization of a different initial configuration, is used to represent the bulk polymer. Estimates of properties are obtained by arithmetic (not Boltzmann) averaging of the properties of these different minima. This is a static model for the polymer, in the sense that it does not incorporate thermal fluctuations. Such a molecular mechanics approach has been quite successful in modelling polymer glasses at relatively small computational cost. It was introduced for the explicit purpose of modelling polymer glasses with the following assumptions:[15]

A. The model does not incorporate thermal motion; i.e. it is static. Temperature enters only indirectly, through specification of the density.

B. The model is represented as a collection of microscopic structures whose parts (atoms, groups, and bonds) satisfy the requirements of detailed mechanical equilibrium. The glassy polymer is envisioned as locally consisting of solid-like microstates, each *locked* in the vicinity of a local minimum of total potential energy in configuration space. The microstates are mutually inaccessible; transitions between neighbouring minima in configuration space are inhibited by their high energy barriers.

C. Estimates of the macroscopic properties of the entire glass are obtained by arithmetically averaging the corresponding properties of the individual microstates.

D. Bond lengths and bond angles are held fixed. Molecular movements occur exclusively through rotation around skeletal bonds and overall chain translation and rotation.

Assumption (A) reflects that we focus on atomic positions at static mechanical equilibrium, as one would do in a crystal. Although stripped of thermal motion, the minimum energy microstates will not be 0 K structures; their density will be prespecified at a value equal to the experimental density of a well-relaxed polymer glass at the temperature and pressure of interest. Alternatively, a full simulation of the system, in both configuration and momentum space, could be attempted by a dynamic method (e.g. molecular dynamics), at much higher computational cost. When molecular mechanics was first applied to study polymer glasses, it was anticipated that, in the course of such a dynamical simulation, the configuration of the glass would not depart significantly from that of the local energy minimum. Nowadays dynamical simulations starting at energy-minimized configurations are conducted routinely, allowing us to explore the evolution of the model multichain system over microsecond time scales and to test the assumptions invoked in strategies for property prediction that are based on local energy minima.[16,17]

Assumption (B) implies that local regions of the polymeric glass are viewed as fluctuating around configurations satisfying the conditions of detailed mechanical equilibrium, i.e. constituting local minima of the total potential energy in configuration space. These minimum energy configurations are called *inherent structures*[2] or *static model microstates*.[15] Transitions from the vicinity of one local minimum to the vicinity of another are infrequent due to the presence of high (relative to $k_B T$) potential energy barriers between minima. The probability distribution, in configuration space, of a glass among the local energy minima used to represent it depends on its formation history. The minima do not comprise an equilibrium ensemble; the glass is not in thermodynamic equilibrium. In fact, its configuration-space distribution would gradually change over very long times through physical ageing processes involving infrequent transitions between minima. These processes are not considered in the simple molecular mechanics-based picture presented here; they will be briefly discussed in Chapter 6.

Regarding assumption (C), arithmetic averaging of properties over individual static microstates can be viewed as spatial averaging over different regions of a macroscopic glass, each of which may be trapped in the vicinity of a different static model microstate. Assumption (C) implies that the method of generating liquid-like initial-guess configurations and subsequently relaxing them by energy minimization yields a sufficiently accurate representation of the actual distribution of local environments within a real polymer glass formed by slowly cooling a melt.

Assumption (D) implies that the bond lengths and bond angles (hard degrees of freedom) of the model chains are treated as classical springs of very high stiffness. As we will see in Section 3.5.3 from an analysis of vibrational motion associated with the hard degrees of freedom, thermal motion does not significantly affect the values of thermodynamic properties related to the spatial extent of the system, such as stress, elastic constants, and surface tension, at the temperatures of interest here.

Based on the above assumptions, one can simulate glassy polymers by subjecting an initial guess structure to straightforward total potential energy minimization, to reach a microscopic structure that satisfies the conditions of detailed mechanical equilibrium. Given that the total potential energy is the sum of all bond intrinsic torsional potentials and all nonbonded interaction potentials computed according to the minimum image convention, the structures in mechanical equilibrium will be defined by local minima of the total potential energy with respect to all degrees of freedom. The minimum energy microstates become therefore the focus of attention; generating them is the central objective of a molecular mechanics simulation.

Because of its high dimensionality and nonlinearity, this problem may demand a very carefully planned method of attack to facilitate the minimization, even if the gradient and Hessian are analytically known. For example, since the computation time for a potential energy and gradient evaluation increases steeply with the range of nonbonded interactions, and to avoid the risk of being trapped in high-lying local energy minima, one can resort to a multistage optimization strategy that progressively augments the interatomic potential range, a different form of the total energy function being used at each stage. Such a strategy, consisting of three stages, was proposed in Ref. 15. It goes as follows: in the first stage one makes the system *feel* and alleviate only the most severe repulsive interactions, while keeping computational time small, by including no torsional potential and using a modified, purely repulsive, soft-sphere nonbonded potential with all atomic radii reduced to half their actual size. The second stage absorbs the bulk of the optimization iterations. It is executed by introducing the full torsional barriers and using again a repulsive soft-sphere potential, but with the atomic radii now at their actual size. It is only in the third and final stage of the minimization that the attractive part of the potential is switched on in order for the system to reach the final minimum subject to the full potential.

From a numerical point of view, the potential energy minimization can be carried out by a number of minimization algorithms: steepest descent, conjugate gradient, or even quasi-Newton matrix-updating algorithms can be implemented.[18] That the optimization routine employs an analytically calculated gradient is a crucial aspect for the success of the method, making it possible in practice to obtain minimum energy structures characterized by a Euclidean norm of the gradient smaller than 10^{-6} kcal/(mol K) in model microstructures of glassy atactic polypropylene formed from a single parent chain of $N = 152$ skeletal bonds. One should also verify that the exact Hessian calculated for the final structures is always positive definite, indicating true minima.

Once the potential energy has been minimized, one can compute a number of descriptors of the system's structure and its thermodynamic properties. These include the cohesive energy E_{coh}, the Hildebrand solubility parameter δ, the orientational correlation between bonds S, the average bond direction correlation function $S(r)$, shape and size characteristics of the chains, and the pair distribution functions for all species pairs in the system. By definition, the cohesive energy E_{coh} is the change in internal energy per amount of substance if all intermolecular forces are eliminated. It can thus be obtained by comparing the total potential energy of the resulting model microstructures with the corresponding energy for the isolated parent chains. In the model system, each chain is surrounded by other chains which are displaced images of itself, formed from the same parent chain, or are images of different parent chains. The cohesive energy is the energy of interaction between all these images. The N_{c} parent chains contain the same amount of matter as a model cube of the bulk polymer, and an estimate of E_{coh} can thus be obtained as an average, over all model microstructures, of the difference:

$$E_{\mathrm{coh}} = \langle \mathcal{V}_{\mathrm{parent}} - \mathcal{V} \rangle. \qquad (3.4.1)$$

Hildebrand's solubility parameter is simply the square root of the cohesive energy density, usually measured in $(\mathrm{J/cm^3})^{1/2}$:

$$\delta = (E_{\mathrm{coh}}/V)^{1/2}. \qquad (3.4.2)$$

In Eq. (3.4.2) it is important that E_{coh} and V refer to the same amount of material, e.g. a mole of simulation boxes.

The method illustrated above has been successfully applied for the detailed molecular modelling of a number of vinyl polymers, the first being glassy atactic polypropylene[15] at −40°C. For each static model microstate an initial configuration was generated by growing a single parent chain using a modified Markov process, based on rotational isomeric state theory discussed in Chapter 1 and incorporating long-range interactions. This configuration was then *relaxed* by a stagewise potential energy minimization method, as outlined above. Fifteen static model microstates were obtained.

The geometric parameters of the chain employed in these studies are listed in Table 3.1. Nonbonded interatomic interactions were modelled by a finite range modification of the Lennard-Jones potential function in which the potential tail was substituted by a quintic spline,

$$\mathcal{V}^{\mathrm{NB}}(r)=\begin{cases}\mathcal{V}_{\mathrm{LJ}}(r)=4\varepsilon\left\{\left(\frac{\sigma}{r}\right)^{12}-\left(\frac{\sigma}{r}\right)^{6}\right\}, & r<R_1\\ \varepsilon(1-\xi)^3\left\{\frac{\mathcal{V}_1^{\mathrm{LJ}}}{\varepsilon}+\left(3\frac{\mathcal{V}_1^{\mathrm{LJ}}}{\varepsilon}+\Delta\frac{\mathcal{V}_1^{\prime\mathrm{LJ}}}{\varepsilon/\sigma}\right)\xi+\left(6\frac{\mathcal{V}_1^{\mathrm{LJ}}}{\varepsilon}+3\Delta\frac{\mathcal{V}_1^{\prime\mathrm{LJ}}}{\varepsilon/\sigma}+\frac{\Delta^2}{2}\frac{\mathcal{V}_1^{\prime\prime\mathrm{LJ}}}{\varepsilon/\sigma^2}\right)\xi^2\right\}, & R_1<r<R\\ 0, \quad R<r\end{cases}$$

(3.4.3)

where $\mathcal{V}_1^{\mathrm{LJ}}$, $\mathcal{V}_1^{\prime\mathrm{LJ}}$, and $\mathcal{V}_1^{\prime\prime\mathrm{LJ}}$ stand for the value, the first derivative, and the second derivative of the Lennard-Jones potential at the junction point $r=R_1$, $\Delta=(R-R_1)/\sigma$, and $\xi=(r-R_1)/(R-R_1)=(r/\sigma-R_1/\sigma)/\Delta$. Expression (3.4.3) is designed so that the functions $\mathcal{V}_{\mathrm{NB}}$, $\mathrm{d}\mathcal{V}_{\mathrm{NB}}/\mathrm{d}r$, and $\mathrm{d}^2\mathcal{V}_{\mathrm{NB}}/\mathrm{d}r^2$ are continuous in the entire range of r and assume values of 0 for $r\geq R$. In the simulations of Ref. 15, two versions of the nonbonded potential energy function $\mathcal{V}_{\mathrm{NB}}(r)$ from Eq. (3.4.3) were employed: (i) the full potential, with $R_1=1.45\sigma$ and $R=2.30\sigma$, which reproduces the shape of the Lennard-Jones curve ($\mathcal{V}_{\mathrm{LJ}}(r)$) well, but brings the attractive tail smoothly to 0 at large distances, and (ii) a purely repulsive approximation to the Lennard-Jones function, referred to as the soft-sphere potential, which is obtained from Eq. (3.4.3) by employing the parameter values $R_1=0.94\sigma$ and $R=1.04\sigma$. The parameters σ and ε were obtained from literature values for atomic polarizabilities, effective numbers of electrons, and var der Waals radii, all listed in Table 3.1. A tail correction to the potential energy was estimated by integrating the difference $\mathcal{V}_{\mathrm{LJ}}(r)-\mathcal{V}_{\mathrm{NB}}(r)$ from R_1 out to infinite distance, taking into account the pair distribution functions for all pairs of interaction sites present.[15] As far as the contribution from dihedral angles is concerned, an intrinsic torsional potential associated with skeletal bond angle rotation was assumed of the form

$$\mathcal{V}_\phi(\phi)=\frac{k_\phi}{2}(1-\cos 3\phi) \tag{3.4.4}$$

with a barrier height of $k_\phi=2.8$ kcal mol^{-1}. It is important to emphasize here that the above model for the potential energy function employed by Theodorou and Suter[15] for the atomistic simulation of atactic PP (as well as similar models employed in other simulation studies of polymers) contained no adjustable parameters, so the results of the energy minimization process obtained constitute a true test of its predictive power.

The total potential energy averaged over the fifteen model structures, including the tail correction for nonbonded interactions, was

$$\mathcal{V}=28(\pm 20)\frac{\text{kcal}}{\text{mol of structures}}.$$

The estimate of the cohesive energy density obtained from Eq. (3.4.1) from a collection of fifteen static model microstates of glassy atactic polypropylene was

$$E_{\mathrm{coh}}=174(\pm 20)\frac{\text{kcal}}{\text{mol of structures}}=227(\pm 26)\frac{\text{kJ}}{\text{kg of polymer}}$$

Table 3.1 *Molecular Geometry and Potential Parameter Values for Atactic PP*[15].

			Bond lengths (Å)			
C-C			$l = 1.53$			
C-H			$l_H = 1.10$			
C-R			$l_R = 1.53$			
			Bond angle supplements (degrees)			
Intradyad C-C-C			$\theta' = 66$			
Interdyad C-C-C			$\theta'' = 68$			
Intradyad C-C-H			$\theta'_H = 71$			
Interdyad C-C-R			$\theta''_R = 68$			
Interdyad C-C-H			$\theta''_H = 73.2 = \cos^{-1}\left\{[(1-2\cos\theta'')/3]^{1/2}\right\}$			
Lennard-Jones parameters*						
I	**Atom or group**	**Van der Waals radius, r_i^0 (Å)**	**Polarizability α_i (Å^3)**	**Effective number of electrons, $N_{e,i}$**	LJ size parameter σ (Å)	**LJ energy parameter ε (kcal/mol)**
1	H	1.3	0.42	0.9	2.32	0.076
2	C	1.8	0.93	5.0	3.21	0.084
3	R (CH_3)	2.0	1.77	7.0	3.56	0.139

Lennard-Jones parameters σ for both like and unlike pairs of interaction sites were calculated from r_i^0, a_i, and $N_{e,i}$ via the Slater–Kirkwood formula, Eq. (2.3.20).

and of the Hildebrand solubility parameter

$$\delta = 14.2(\pm 0.8)\ (\mathrm{J/cm^3})^{1/2}.$$

The corresponding experimental value is

$$\delta_{\mathrm{exp}} = 16.8\ (\mathrm{J/cm^3})^{1/2}.$$

As a measure of orientational correlation between bonds, *the bond order parameter* or the *bond direction correlation function* S was used, defined by

$$S = \frac{1}{2}\left[3\left\langle \cos^2\theta \right\rangle - 1\right], \tag{3.4.5}$$

where θ is the angle between two skeletal bond chords of the chain (chord i is the vector connecting the midpoints of the skeletal bonds flanking carbon atom i). We usually compute the order

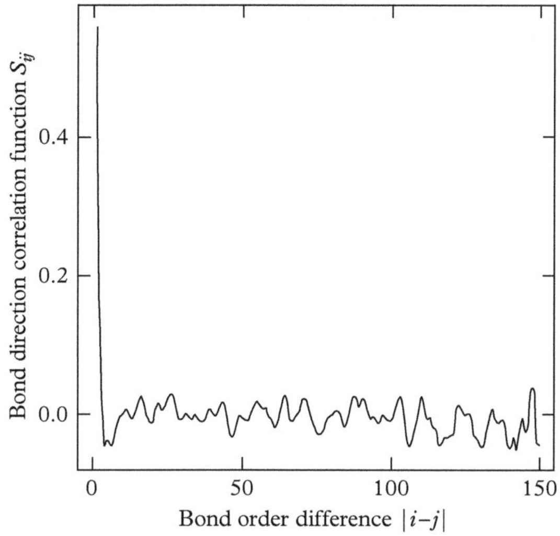

Figure 3.5 Average bond direction correlation function S_{ij} between skeletal chords versus the index difference $|i-j|$ in glassy atactic PP at $-40°$C. Computed as averages over fifteen energy minimized structures.

Reprinted (adapted) with permission from Theodorou, D.N., and Suter, U.W. *Macromolecules* 18, 1467 (1985). Copyright (1985) American Chemical Society.

parameter S_{ij} that is obtained by taking the average in Eq. (3.4.5) over all chord pairs with given index difference $|i-j|$ and over all structures, that is by averaging the squared cosine of the angle between all chords separated by a fixed number of bonds. A typical plot of S_{ij} as a function of $|i-j|$ is shown in Figure 3.5. The strong directional correlation between neighbouring chords for very small values of $|i-j|$ vanishes rapidly with increasing distance along the chain. The absence of any long-range correlation effects indicates that the procedure used to generate the minimum potential energy structures of atactic PP leads indeed to parent chains that are *random coils*.

Bond orientational correlation can be analysed also through the bond direction correlation function $S(r)$, derived from Eq. (3.4.5) by averaging over chord pairs with given distance r between the chord centres. The resulting curve is plotted in Figure 3.6. The fact that $S(r)$ goes to 0 at large distances proves that there is no long-range orientational correlation; i.e. the model cubes are truly amorphous systems. The sharp correlation pattern observed at short distances is completely intramolecular, i.e. caused by connectivity and conformation along the parent chains. For example, the three maxima at 1.3, 2.5, and 3.8 Å are attributable to t, tt, and ttt conformations. At the shortest possible intermolecular distances, there is a tendency for perpendicular arrangements (perhaps due to the *rough* surface of the chains); at slightly larger distances, there is a very weak trend towards parallelism.

To explore the size and shape of the parent chains, one can compute the ensemble-averaged end-to-end distance $\langle r^2\rangle^{1/2}$, the radius of gyration $\langle R_g^2\rangle^{1/2}$, and the eigenvalues $\overline{X^2}$, $\overline{Y^2}$, and $\overline{Z^2}$ of the radius of gyration tensor, and from these the shape asphericity $b=\overline{X^2}-\frac{1}{2}(\overline{Y^2}+\overline{Z^2})$, the

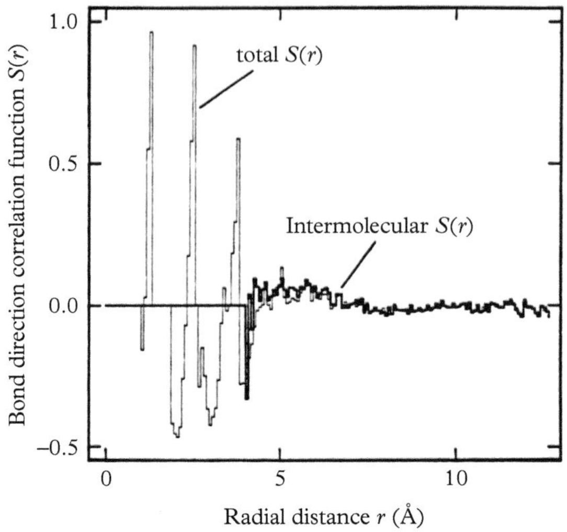

Figure 3.6 Average bond direction correlation function $S(r)$ between skeletal chords as a function of distance r between their centres in glassy atactic PP at −40°C. Results are shown for both the total correlation and its intermolecular component, accumulated from chords belonging to different chains. They were computed as averages over fifteen energy minimized structures.

shape acylindricity $c = \overline{Y^2} - \overline{Z^2}$, and the relative anisotropy $\kappa^2 = \dfrac{b^2 + \frac{3}{4}c^2}{\left\langle R_g^2 \right\rangle^2}$. Values of the reduced measures $\overline{X^2}/\overline{Z^2}$, $\overline{Y^2}/\overline{Z^2}$, $\langle b \rangle / \langle R_g^2 \rangle$, $\langle c \rangle / \langle s^2 \rangle$, and $\langle \kappa^2 \rangle$, obtained by averaging over the collection of fifteen static microstructures, are shown in Table 3.2.

Information on intermolecular structure of the minimum energy configurations is provided by the pair distribution functions $g_{\alpha\beta}(\alpha, \beta \in \{R,C,H\})$ for all six pairs in the minimum energy systems. If V is the volume of a model cube, the function $g_{\alpha\beta}$ is defined so that the quantity $(1/V)(4\pi r^2) g_{\alpha\beta}(r)dr$ is equal to the probability of finding an α centre and a β centre in the distance interval r to $r + dr$. The method of obtaining $g_{\alpha\beta}$ from the model involves[15] dividing first the r space into small intervals (bins) $\Delta r_i = r_i - r_{i-1}$ with $r_0 = 0$. Then, all $\alpha\beta$ pairs of the system are classified in the appropriate intervals, according to the distance between the atoms in each pair. If $N_{\alpha\beta}^{(i)}$ is the number of $\alpha\beta$ pairs at a distance lying in the interval (r_{i-1}, r_i), $N_{\alpha\beta}$ the total number of $\alpha\beta$ pairs, and V the total volume of the system, then the value of $g_{\alpha\beta}$ within the interval (r_{i-1}, r_i) is estimated by

$$g_{\alpha\beta}^{(i)} = \frac{V}{V(r_i) - V(r_{i-1})} \frac{N_{\alpha\beta}^{(i)}}{N_{\alpha\beta}}, \tag{3.4.6}$$

where $V(r)$ is the volume accessible to a second atom under the condition that its distance from a first atom is less than or equal to r. If we limit our discussion to a system with cubic boundaries

Table 3.2 *Shape and Size Characteristics of Chains in Bulk Glassy Atactic PP of Degree of Polymerization 76, as Obtained from Molecular Mechanics Simulations of a Model System Formed from a Single Parent Chain*[15]

Measure	**Average value from 15 minimum energy structures**	**RIS (rotational isomeric state) model value either by exact generator matrix method or by Monte Carlo estimation**
Root mean square end-to-end distance $\langle r^2 \rangle^{1/2}$ (Å)	40.12 ± 3.18	47.00 ± 0.10
Root mean square radius of gyration $\langle R_g^2 \rangle^{1/2}$ (Å)	17.20 ± 1.06	18.76 ± 0.05
$\langle r^2 \rangle / \langle R_g^2 \rangle$	5.44 ± 1.09	6.28 ± 0.04
$\overline{X^2}/\overline{Z^2}$	15.24 ± 2.55	14.84 ± 0.25
$\overline{Y^2}/\overline{Z^2}$	3.01 ± 0.47	3.13 ± 0.05
$\langle b \rangle / \langle R_g^2 \rangle$	0.687 ± 0.141	0.674 ± 0.013
$\langle c \rangle / \langle R_g^2 \rangle$	0.104 ± 0.093	0.112 ± 0.002
$\langle \kappa^2 \rangle$	0.444 ± 0.057	0.431 ± 0.004

of edge length a, the maximum possible value of the minimum image distance is simply the largest semi-diagonal of the minimum image locus, i.e. $r_{\max} = \frac{a\sqrt{3}}{2}$. Then, $V(r)$, the volume of that part of the minimum image locus around an atom that lies within a sphere of radius r centred at the atom, is given as[19]

$$V(r) = \begin{cases} \frac{4}{3}\pi r^3, \quad \text{for } r \leq \frac{a}{2} \\ 2\pi\left\{\frac{2}{3}r^3 - \left(r - \frac{a}{2}\right)^2\left(2r + \frac{a}{2}\right)\right\}, \quad \text{for } \frac{a}{2} \leq r \leq \frac{a}{\sqrt{2}} \\ a^3\left\{(\lambda^2 - 2)^{1/2} + \left(\lambda^2 - \frac{1}{3}\right)\left[2\sin^{-1}\frac{1}{(\lambda^2 - 1)^{1/2}} - \sin^{-1}\left(\frac{\lambda^2 - 2}{\lambda^2 - 1}\right)^{1/2} - \frac{\pi}{4}\right]\right. \\ \quad - \frac{\lambda^3}{3}\left[\sin^{-1}\left(\frac{\lambda - (\lambda^2 - 2)^{1/2}}{2(\lambda^2 - 1)^{1/2}}\right) - \sin^{-1}\left(\frac{\lambda + (\lambda^2 - 2)^{1/2}}{2(\lambda^2 - 1)^{1/2}}\right)\right. \\ \quad \left.\left. - \sin^{-1}\left(\frac{\lambda^2 - \lambda - 1}{(\lambda - 1)(\lambda^2 - 1)^{1/2}}\right) + \sin^{-1}\left(\frac{\lambda^2 + \lambda - 1}{(\lambda + 1)(\lambda^2 - 1)^{1/2}}\right)\right]\right\}, \\ \qquad \text{for } \frac{a}{\sqrt{2}} \leq r \leq \frac{a\sqrt{3}}{2}, \end{cases} \tag{3.4.7}$$

where $\lambda = \frac{2r}{a}$.

In the plots, contributions from pairs whose distance is fixed due to connectivity appear as *Dirac pulses*. More interesting is the intermolecular part of the pair distribution functions, $g_{\alpha\beta}^{\text{inter}}(r)$. The function $g_{\alpha\beta}^{\text{inter}}(r)$ is defined such that the quantity $(1/V)(4\pi r^2)g_{\alpha\beta}^{\text{inter}}(r)dr$ equals the probability of finding an α centre and a β centre *not* belonging to the same chain and separated by a distance in the interval r to $r + dr$. Typical plots of $g_{\alpha\beta}(r)$ and $g_{\alpha\beta}^{\text{inter}}(r)$ are reported in Figure 3.7(the intermolecular parts are displayed as thin lines). At long distances, all pair distribution functions approach unity, indicating that no long-range spatial correlation exists (as one would expect from a purely amorphous system). The intermolecular functions $g_{\alpha\beta}^{\text{inter}}(r)$ rise more or less smoothly with increasing distance towards an asymptotic value of 1. The sharp correlations observed at small distances in $g_{\alpha\beta}(r)$ are almost entirely intramolecular. The intermolecular pair distribution functions display some subtle structural features, but beyond 10 Å there are essentially no intermolecular correlations; only the HH, HR, and RR (i.e. the substituent–substituent) distributions display a well-defined maximum at a distance roughly equal to the sum of the van der Waals radii of the species involved. This *first coordination shell* appears particularly pronounced in the RR case. The high noise level in this case is due to the small number of methyl substituents in the model system. Skeletal carbons of different chains are kept apart by the substituents surrounding them, so that the principal feature of the carbon–carbon intermolecular distribution is a broad hump in the vicinity of 6.3 Å. The C-R intermolecular distribution displays two humps around 3.9 Å (first coordination shell) and 5 Å. Broad humps of intermolecular origin at distances beyond 5 Å can also be discerned in the H-C and H-R distributions. Note also here that the pair distribution functions have been calculated from a static ensemble of structures. In reality, thermal motion will tend to broaden the peaks and to reduce their height. Also, in reality, in atactic PP there are no R groups, but only H and C atoms, whose spatial arrangement is characterized by the three element pair distribution functions $g_{\text{HH}}^{\text{el}}(r)$, $g_{\text{HC}}^{\text{el}}(r)$, and $g_{\text{CC}}^{\text{el}}(r)$. To obtain these, each methyl group can be resolved into a carbon, at the end of the carbon-methyl bond, and three hydrogens, symmetrically bonded to this carbon by bonds of length l_{H} at an angle of 110° with the methyl stem. In this methyl *unfolding* procedure, all methyls were taken in a staggered conformation with respect to the backbone. The element pair distribution functions were computed in exactly the same way as the group pair distributions presented earlier, and are displayed as bold lines in Figure 3.8. Also shown in the figure are the purely intermolecular contributions (thin lines). The element-pair distribution functions can be used to predict X-ray and neutron scattering patterns, and thus provide a further test of the model structures against experiment.

In Section 3.5 we describe how one can utilize the minimum energy structures to predict the elastic constants of glassy atactic PP by simulating mechanical deformations of various types and degrees.

3.5 Estimation of elastic constants through quasistatic deformation

3.5.1 Thermodynamics and statistical mechanics of deformation

Our objective in this section is to perform a thermodynamic and statistical mechanical analysis of an elastic material subjected to mechanical deformation. We will consider a general (not necessarily small) deformation imposed on an atomistic model of the material subject to 3D periodic boundary conditions and show how the macroscopic *deformation gradient* tensor and *Green strain* tensor, which play a central role in continuum thermoelasticity, are related to changes in the edge vectors of

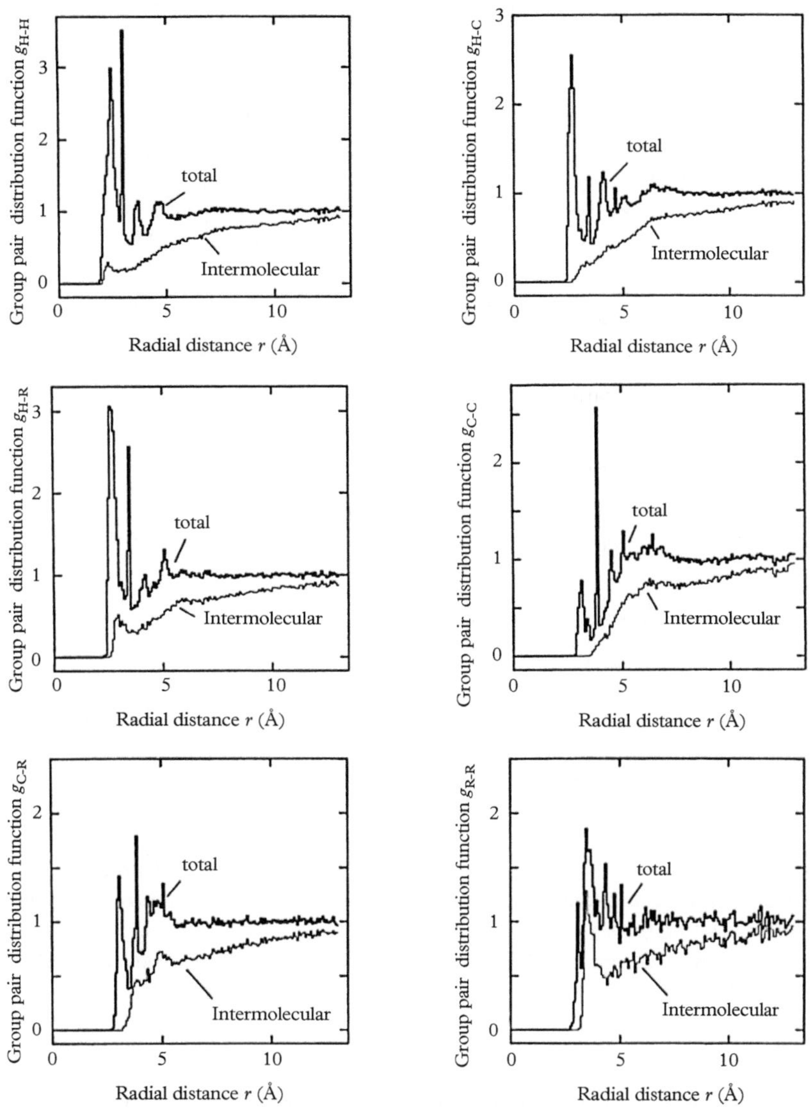

Figure 3.7 Group pair distribution functions for the six species pairs (from groups CH_3, skeletal C, and main-chain H) in glassy atactic PP at −40°C. Intermolecular contributions are shown as thin lines. Results were computed as averages over fifteen energy minimized structures.

Reprinted (adapted) with permission from Theodorou, D.N., and Suter, U.W. *Macromolecules* **18**, 1467 (1985). Copyright 1985 American Chemical Society.

the periodic box. We will derive an expression for the Helmholtz energy of the deformed material from equilibrium statistical mechanics, apply the continuum mechanical definition of stress on that expression, and recover the virial theorem. We will write down a fundamental equation for the Helmholtz energy as a function of temperature and deformation. We will consider how the general

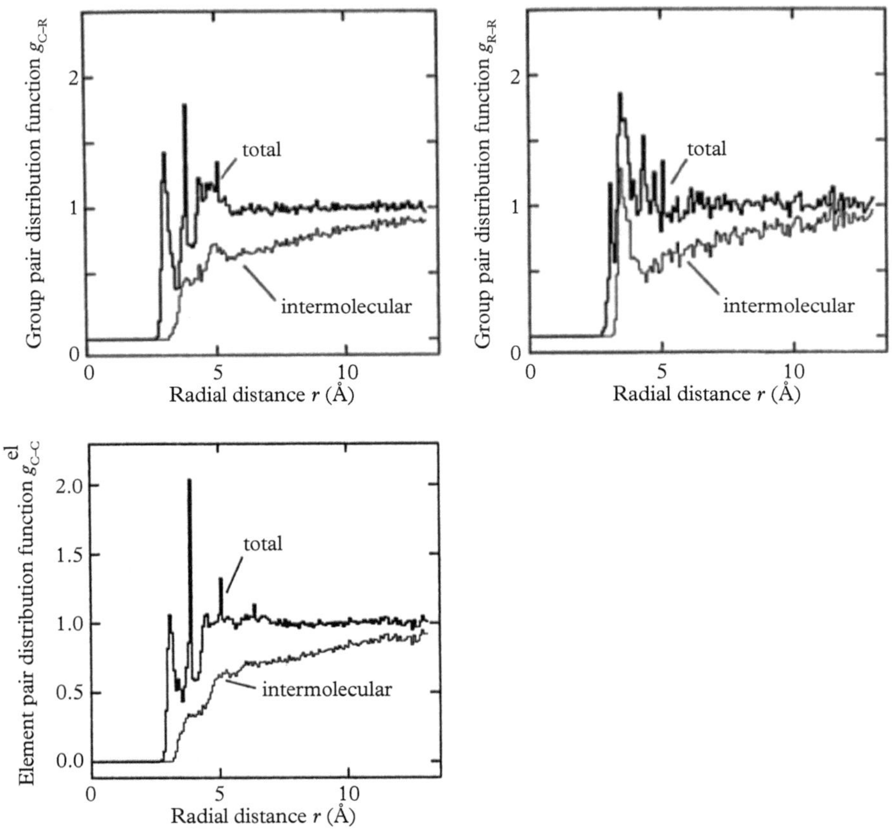

Figure 3.8 Element pair distribution functions for the three element pairs (from atoms C and H) in glassy atactic PP at −40°C. Computed as averages over fifteen energy minimized structures. Intermolecular contributions are shown as thin lines.

Reprinted (adapted) with permission from Theodorou, D.N., and Suter, U.W. *Macromolecules* **18**, 1467 (1985). Copyright 1985 American Chemical Society.

formulation is simplified in the case of small deformations in the linear elastic regime and define the isothermal elastic coefficients and the *Grüneisen* tensor. Finally, we will show how these quantities can be estimated from molecular-level information for a system confined to the vicinity of an energy minimum in its configuration space, such as a crystalline solid or a glass.

3.5.1.1 *The simulation box as a material element: connection with continuum thermoelasticity*

At the molecular level we postulate that the total potential energy function of an atomistic model of a material consisting of n interaction sites numbered from 0 to $n-1$ can be expressed as a function of interatomic distance vectors of the general form

$$\begin{aligned}&\mathcal{V}(\boldsymbol{r}_1-\boldsymbol{r}_0,\boldsymbol{r}_2-\boldsymbol{r}_0,\ldots,\boldsymbol{r}_{n-1}-\boldsymbol{r}_0,\boldsymbol{r}_2-\boldsymbol{r}_1,\ldots,\boldsymbol{r}_{n-1}-\boldsymbol{r}_1,\ldots,\boldsymbol{r}_{n-1}-\boldsymbol{r}_{n-2})\\&\quad=\mathcal{V}(\boldsymbol{r}_{01},\boldsymbol{r}_{02},\ldots,\boldsymbol{r}_{0\,n-1},\boldsymbol{r}_{12},\ldots,\boldsymbol{r}_{1n-1},\ldots,\boldsymbol{r}_{n-2n-1}).\end{aligned}\tag{3.5.1}$$

A fully flexible model is postulated here, as in Section 3.2. The presence of holonomic constraints can be dealt with as discussed in Section 3.1. In this section we will use the shorthand notation

$$\boldsymbol{r}_{ij}=\boldsymbol{r}_j-\boldsymbol{r}_i,\quad 0\leq i\leq n-1,\quad 0\leq j\leq n-1.\tag{3.5.2}$$

Note that the indices of all arguments $\boldsymbol{r}_{ij}$ of $\mathcal{V}$ are taken to satisfy the constraint $i<j$.

In a system subject to periodic boundary conditions, all $\boldsymbol{r}_{ij}$ are to be understood in a minimum image sense, i.e. as vectors from i to the minimum image of j with respect to i, where i and j lie in the primary simulation box but the minimum image of j with respect to i does not necessarily lie in the primary simulation box. Clearly, the function $\mathcal{V}$ defined by Eq. (3.5.1) is invariant to rigid translations which change all atomic positions by a constant vector.

The total force on site i can be determined as

$$\boldsymbol{F}_i=-\frac{\partial\mathcal{V}}{\partial\boldsymbol{r}_i}=\sum_{k>i}\frac{\partial\mathcal{V}}{\partial\boldsymbol{r}_{ik}}-\sum_{l<i}\frac{\partial\mathcal{V}}{\partial\boldsymbol{r}_{li}},\tag{3.5.3}$$

where site indices k and l take on all integer values from 0 to $n-1$. The symbolism $\frac{\partial f}{\partial\boldsymbol{r}}\equiv\nabla_{\boldsymbol{r}}f$ denotes the gradient of scalar function f with respect to the vectorial variable $\boldsymbol{r}$. On the right-hand side of Eq. (3.5.3), in taking the gradient with respect to a specific distance vector $\boldsymbol{r}_{ij}$, it is understood that all other distance vectors remain unchanged.

We *define* the force on atom i due to atom j as

$$\boldsymbol{F}_{ij}=\begin{cases}-\dfrac{\partial\mathcal{V}}{\partial\boldsymbol{r}_{ji}}, & j<i\\[2ex] \dfrac{\partial\mathcal{V}}{\partial\boldsymbol{r}_{ij}}, & j>i\end{cases}\tag{3.5.4}$$

Equation (3.5.4) is equivalent to Eq. (3.2.2). It just makes a bit clearer what happens for $j>i$. It is important to realize that the force $\boldsymbol{F}_{ij}$ is not necessarily central. In polymer systems, $\boldsymbol{F}_{ij}$ will be central in case i and j interact exclusively through a central nonbonded force, as happens if they belong to different chains or are sufficiently far apart along the same chain such that they are not involved in the same bonded contribution to the energy. $\boldsymbol{F}_{ij}$ between nearby atoms on the same image of the same parent chain, however, may be strongly noncentral because of the common involvement of i and j in bending, torsional, or improper torsional terms of the potential energy function.

With the definition, Eq. (3.5.4), Eq. (3.5.3) gives

$$F_i=\sum_{j\neq i}\boldsymbol{F}_{ij}.\tag{3.5.5}$$

That is, pairwise additivity of forces holds, even though the potential energy function is not pairwise additive and the forces are not necessarily central. Furthermore, the definition, Eq. (3.5.4), ensures that the *action–reaction principle* (Newton's third law) is satisfied:

$$\boldsymbol{F}_{ij} = -\boldsymbol{F}_{ji}. \tag{3.5.6}$$

Envision a change in the configuration of the system, which brings about small changes $\mathrm{d}\boldsymbol{r}_i$ in the position vectors of the atoms. Then, the potential energy will change (see Problem 3.6) by

$$\mathrm{d}\mathcal{V} = \sum_i \sum_{j>i} \boldsymbol{F}_{ij} \cdot \mathrm{d}\boldsymbol{r}_{ij}. \tag{3.5.7}$$

Let $\boldsymbol{h} = [\boldsymbol{h}_1\ \boldsymbol{h}_2\ \boldsymbol{h}_3]$ be the cell matrix formed from the three edge vectors of the simulation box. In a periodic box, these would be the *periodic continuation vectors*. An elastic material upon deformation goes through a sequence of thermodynamic equilibrium states characterized by different values of $\boldsymbol{h}$. In a classical treatment, the Helmholtz energy of the system containing n interaction sites at temperature T and edge vectors $\boldsymbol{h}$ is (compare Eq. (3.1.3))

$$A(n, \boldsymbol{h}, T) = -k_{\mathrm{B}} T \ln \left\{ \frac{1}{\prod_i \Lambda_i^3} \int \exp\left[-\frac{1}{k_{\mathrm{B}} T} \mathcal{V}(\boldsymbol{r}_{01}, \boldsymbol{r}_{02}, \ldots, \boldsymbol{r}_{0n-1}, \boldsymbol{r}_{12}, \ldots, \boldsymbol{r}_{1n-1}, \ldots, \boldsymbol{r}_{n-2n-1}; \boldsymbol{h}) \right] \mathrm{d}^3 r_0 \ldots \mathrm{d}^3 r_{n-1} \right\} \tag{3.5.8}$$

with $\Lambda_i = \dfrac{h}{\sqrt{2\pi m_i k_{\mathrm{B}} T}}$ being the thermal wavelength of site i. Factors of $\dfrac{1}{N_{\mathrm{c}}!}$ have not been included in the expression, assuming that all atoms are distinguishable, as they would be in a solid.

The configurational integral in Eq. (3.5.8) can be transformed by introducing scaled coordinates $\boldsymbol{s}_i$ in place of the Cartesian coordinates $\boldsymbol{r}_i$, such that

$$\boldsymbol{r}_i = \boldsymbol{r}_{\mathrm{ref}} + s_{i,1}\boldsymbol{h}_1 + s_{i,2}\boldsymbol{h}_2 + s_{i,3}\boldsymbol{h}_3 = \boldsymbol{r}_{\mathrm{ref}} + \boldsymbol{h} \cdot \boldsymbol{s}_i, \quad 0 \le s_{i,\alpha} \le 1 \quad \alpha = 1, 2, 3. \tag{3.5.9}$$

In Eq. (3.5.9), $\boldsymbol{r}_{\mathrm{ref}}$ is the position vector of the origin of the simulation box in a fixed (laboratory) frame of reference. The position vector of any atom in the box relative to the origin of the box is expressed as a linear combination of the box edge vectors. From Eq. (3.5.9) it follows that the Jacobian of transformation from laboratory to scaled coordinates for a given atom is

$$\left| \frac{\partial(r_{i,1}, r_{i,2}, r_{i,3})}{\partial(s_{i,1}, s_{i,2}, s_{i,3})} \right| = \det(\boldsymbol{h}). \tag{3.5.10}$$

Equation (3.5.8) for the Helmholtz energy becomes

$$A(n,\boldsymbol{h},T) = -k_{\mathrm{B}}T\ln\left\{\frac{[\det(\boldsymbol{h})]^n}{\prod\limits_i \Lambda_i^3}\int \exp\left[-\frac{1}{k_{\mathrm{B}}T}\mathcal{V}(\boldsymbol{s}_{01},\boldsymbol{s}_{02},\ldots,\boldsymbol{s}_{0n-1}, \boldsymbol{s}_{12},\ldots,\boldsymbol{s}_{1n-1},\ldots,\boldsymbol{s}_{n-2n-1};\boldsymbol{h})\right]\mathrm{d}^3s_0\ldots\mathrm{d}^3s_{n-1}\right\}. \tag{3.5.11}$$

In Eq. (3.5.11) the domain of integration is 0 to 1 for each $s_{i,\alpha}$. Now, consider a reversible isothermal change in thermodynamic state wherein one of the elements of $\boldsymbol{h}$, to be indicated by $\boldsymbol{h}_{\alpha\beta}$ (the αth element of edge vector $\boldsymbol{h}_\beta$) is changed, all other elements of $\boldsymbol{h}$ being held constant. From Eq. (3.5.11) we obtain that

$$\frac{\partial A}{\partial h_{\alpha\beta}} = -nk_{\mathrm{B}}T\frac{\partial \ln[\det(\boldsymbol{h})]}{\partial h_{\alpha\beta}} + \frac{\displaystyle\int \frac{\partial\mathcal{V}(\boldsymbol{s}_{01},\ldots,\boldsymbol{s}_{n-2n-1};\boldsymbol{h})}{\partial h_{\alpha\beta}}\exp\left[-\frac{1}{k_{\mathrm{B}}T}\mathcal{V}(\boldsymbol{s}_{01},\ldots,\boldsymbol{s}_{n-2n-1};\boldsymbol{h})\right]\mathrm{d}^3s_0\ldots\mathrm{d}^3s_{n-1}}{\displaystyle\int \exp\left[-\frac{1}{k_{\mathrm{B}}T}\mathcal{V}(\boldsymbol{s}_{01},\ldots,\boldsymbol{s}_{n-2n-1};\boldsymbol{h})\right]\mathrm{d}^3s_0\ldots\mathrm{d}^3s_{n-1}}. \tag{3.5.12}$$

The first term on the right-hand side of Eq. (3.5.12) can be shown to be equal to $-nk_{\mathrm{B}}Th^{-1}_{\beta\alpha}$ (see Problem 3.7) with $\boldsymbol{h}^{-1}$ being the inverse matrix of $\boldsymbol{h}$, while the second term is equal to $\left\langle \frac{\partial\mathcal{V}(\boldsymbol{s}_{01},\ldots,\boldsymbol{s}_{n-2n-1};\boldsymbol{h})}{\partial h_{\alpha\beta}}\right\rangle$ where the angular brackets denote averaging in the canonical ensemble, i.e. constant $[n,\boldsymbol{h},T]$. The quantity being averaged is the derivative of the potential energy with respect to $h_{\alpha\beta}$ at constant scaled coordinates of all atoms. In other words, an infinitesimal *affine transformation* of the contents of the box is considered in forming the quantity to be averaged. Furthermore, one can make use of Eq. (3.5.7) to obtain that

$$\left\langle \frac{\partial\mathcal{V}(\boldsymbol{s}_{01},\ldots,\boldsymbol{s}_{n-2n-1};\boldsymbol{h})}{\partial h_{\alpha\beta}}\right\rangle = \left\langle \sum_i\sum_{j>i}\boldsymbol{F}_{ij}\cdot\frac{\partial \boldsymbol{r}_{ij}}{\partial h_{\alpha\beta}}\right\rangle. \tag{3.5.13}$$

Now, from Eq. (3.5.9), $\boldsymbol{r}_{ij} = \boldsymbol{h}\cdot(\boldsymbol{s}_j - \boldsymbol{s}_i)$ and so $\frac{\partial \boldsymbol{r}_{ij}}{\partial h_{\alpha\beta}}$ is a vector whose αth element is $s_{j,\beta} - s_{i,\beta}$ and its other two elements are 0. Thus, Eq. (3.5.13) can be rewritten as

$$\left\langle \frac{\partial\mathcal{V}(\boldsymbol{s}_{01},\ldots,\boldsymbol{s}_{n-2n-1};\boldsymbol{h})}{\partial h_{\alpha\beta}}\right\rangle = \left\langle \sum_i\sum_{j>i}F_{ij,\alpha}(s_{j,\beta} - s_{i,\beta})\right\rangle. \tag{3.5.14}$$

Furthermore, $(\boldsymbol{s}_j - \boldsymbol{s}_i) = \boldsymbol{h}^{-1} \cdot \boldsymbol{r}_{ij}$, so $(s_{j,\beta} - s_{i,\beta}) = \sum_{\gamma=1}^{3} h_{\beta\gamma}^{-1} \; r_{ij,\gamma}$ and thus Eq. (3.5.14) gives

$$\left\langle \frac{\partial \mathcal{V}(\boldsymbol{s}_{01}, \ldots, \boldsymbol{s}_{n-2n-1}; \boldsymbol{h})}{\partial h_{\alpha\beta}} \right\rangle = \left\langle \sum_i \sum_{j>i} F_{ij,\alpha} \left(\sum_{\gamma=1}^{3} h_{\beta\gamma}^{-1} r_{ij,\gamma} \right) \right\rangle = \sum_{\gamma=1}^{3} h_{\beta\gamma}^{-1} \left\langle \sum_i \sum_{j>i} F_{ij,\alpha} \, r_{ij,\gamma} \right\rangle. \tag{3.5.15}$$

Combining the above, we get the desired expression

$$\frac{\partial A}{\partial h_{\alpha\beta}} = -nk_{\mathrm{B}} T h_{\beta\alpha}^{-1} + \sum_{\gamma=1}^{3} h_{\beta\gamma}^{-1} \left\langle \sum_i \sum_{j>i} F_{ij,\alpha} \; r_{ij,\gamma} \right\rangle. \tag{3.5.16}$$

Continuum mechanical descriptions of deformation make use of the macroscopic *deformation gradient tensor*,[20] $\mathcal{F}$. We should thus try to relate $\mathcal{F}$ to the matrix $\boldsymbol{h}$. An arbitrary deformation takes our system from origin $\boldsymbol{r}_{\mathrm{ref}}^{(0)}$ and edge vectors $\boldsymbol{h}_1^{(0)}, \boldsymbol{h}_2^{(0)}, \boldsymbol{h}_3^{(0)}$ to origin $\boldsymbol{r}_{\mathrm{ref}}$ and edge vectors $\boldsymbol{h}_1, \boldsymbol{h}_2, \boldsymbol{h}_3$. The superscript (0) is used to denote the reference undeformed state. Note that the points $\boldsymbol{r}_{\mathrm{ref}}$, $\boldsymbol{r}_{\mathrm{ref}} + \boldsymbol{h}_1, \boldsymbol{r}_{\mathrm{ref}} + \boldsymbol{h}_2, \boldsymbol{r}_{\mathrm{ref}} + \boldsymbol{h}_3$ are material points which can be considered infinitesimally far apart by macroscopic standards. By definition, the deformation gradient tensor $\mathcal{F}$ must satisfy the following relation involving the edge vectors in the undeformed and deformed states

$$\boldsymbol{h}_i = \mathcal{F} \cdot \boldsymbol{h}_i^{(0)}, \text{or } h_{i,\alpha} = \sum_{\beta=1}^{3} \mathcal{F}_{\alpha\beta} \, h_{i,\beta}^{(0)}, \; 1 \leq i \leq 3, 1 \leq \alpha \leq 3, \tag{3.5.17}$$

which means that also for the entire matrix of edge vectors

$$\boldsymbol{h} = \mathcal{F} \cdot \boldsymbol{h}^{(0)}, \text{or } h_{\alpha\gamma} = \sum_{\beta=1}^{3} \mathcal{F}_{\alpha\beta} \, h_{\beta\gamma}^{(0)}. \tag{3.5.18}$$

From Eq. (3.5.18),

$$\mathcal{F} = \boldsymbol{h} \cdot (\boldsymbol{h}^{(0)})^{-1}, \text{or } \mathcal{F}_{\alpha\beta} = \sum_{\gamma=1}^{3} h_{\alpha\gamma} \, (h^{(0)})_{\gamma\beta}^{-1}. \tag{3.5.19}$$

Then, we can show that the derivative of the Helmholtz free energy with respect to an element of the deformation gradient tensor is given by the expression (see Problem 3.8)

$$\frac{\partial A}{\partial \mathcal{F}_{\alpha\beta}} = -nk_{\mathrm{B}} T \mathcal{F}_{\beta\alpha}^{-1} + \sum_{\zeta=1}^{3} \mathcal{F}_{\beta\zeta}^{-1} \left\langle \sum_i \sum_{j>i} F_{ij,\alpha} \, r_{ij,\zeta} \right\rangle. \tag{3.5.20}$$

Using Eq. (3.5.20) for $\beta = \gamma$ and post-multiplying both sides of the equation with $\mathcal{F}^{\mathrm{T}}_{\gamma\beta}$,

$$\frac{\partial A}{\partial \mathcal{F}_{\alpha\gamma}}\mathcal{F}^{\mathrm{T}}_{\gamma\beta} = \frac{\partial A}{\partial \mathcal{F}_{\alpha\gamma}}\mathcal{F}_{\beta\gamma} = -nk_{\mathrm{B}}T\mathcal{F}_{\beta\gamma}F^{-1}_{\gamma\alpha} + \sum_{\zeta=1}^{3}\mathcal{F}_{\beta\gamma}F^{-1}_{\gamma\zeta}\left\langle\sum_{i}\sum_{j>i}F_{ij,\alpha}\,r_{ij,\zeta}\right\rangle.$$

Thus,

$$\sum_{\gamma=1}^{3}\frac{\partial A}{\partial \mathcal{F}_{\alpha\gamma}}\mathcal{F}^{\mathrm{T}}_{\gamma\beta} = -nk_{\mathrm{B}}T\sum_{\gamma=1}^{3}\mathcal{F}_{\beta\gamma}F^{-1}_{\gamma\alpha} + \sum_{\zeta=1}^{3}\sum_{\gamma=1}^{3}\mathcal{F}_{\beta\gamma}F^{-1}_{\gamma\zeta}\left\langle\sum_{i}\sum_{j>i}F_{ij,\alpha}\,r_{ij,\zeta}\right\rangle$$

and, since $\boldsymbol{\mathcal{F}}\cdot\boldsymbol{\mathcal{F}}^{-1}$ is the unit matrix,

$$\sum_{\gamma=1}^{3}\frac{\partial A}{\partial \mathcal{F}_{\alpha\gamma}}\mathcal{F}^{\mathrm{T}}_{\gamma\beta} = -nk_{\mathrm{B}}T\delta_{\alpha\beta} + \sum_{\zeta=1}^{3}\delta_{\beta\zeta}\left\langle\sum_{i}\sum_{j>i}F_{ij,\alpha}\,r_{ij,\zeta}\right\rangle.$$

or

$$\sum_{\gamma=1}^{3}\frac{\partial A}{\partial \mathcal{F}_{\alpha\gamma}}\mathcal{F}^{\mathrm{T}}_{\gamma\beta} = -nk_{\mathrm{B}}T\delta_{\alpha\beta} + \left\langle\sum_{i}\sum_{j>i}F_{ij,\alpha}\,r_{ij,\beta}\right\rangle. \tag{3.5.21}$$

It is reminded that the ensemble average appearing in Eq. (3.5.21) is a canonical average over all atomistic configurations of the deformed system at constant $[n, \boldsymbol{h}, T]$.

Equation (3.5.21) defines the elements of the tensor $\frac{\partial A}{\partial \boldsymbol{\mathcal{F}}}\cdot\boldsymbol{\mathcal{F}}^{\mathrm{T}}$. It is remarkable that the deformation gradient tensor has fallen out of the expression on the right-hand side. The calculation of this tensor involves a canonical average only in the current deformed state. The trace of the tensor $\frac{\partial A}{\partial \boldsymbol{\mathcal{F}}}\cdot\boldsymbol{\mathcal{F}}^{\mathrm{T}}$ defined by Eq. (3.5.21) is

$$\mathrm{tr}\left(\frac{\partial A}{\partial \boldsymbol{\mathcal{F}}}\cdot\boldsymbol{\mathcal{F}}^{\mathrm{T}}\right) = \sum_{\alpha=1}^{3}\left[-nk_{\mathrm{B}}T\delta_{\alpha\alpha} + \left\langle\sum_{i}\sum_{j>i}F_{ij,\alpha}\,r_{ij,\alpha}\right\rangle\right] = -3nk_{\mathrm{B}}T + \left\langle\sum_{i}\sum_{j>i}\left(\sum_{\alpha=1}^{3}F_{ij,\alpha}\,r_{ij,\alpha}\right)\right\rangle.$$

or

$$\mathrm{tr}\left(\frac{\partial A}{\partial \boldsymbol{\mathcal{F}}}\cdot\boldsymbol{\mathcal{F}}^{\mathrm{T}}\right) = -3nk_{\mathrm{B}}T + \left\langle\sum_{i}\sum_{j>i}\boldsymbol{F}_{ij}\cdot\boldsymbol{r}_{ij}\right\rangle. \tag{3.5.22}$$

In continuum thermoelasticity, the stress tensor is defined as[21,22]

$$\boldsymbol{\tau} = \frac{1}{V}\frac{\partial A}{\partial \boldsymbol{\mathcal{F}}}\cdot\boldsymbol{\mathcal{F}}^{\mathrm{T}} = \frac{1}{V^{(0)}}\frac{1}{\det(\boldsymbol{\mathcal{F}})}\frac{\partial A}{\partial \boldsymbol{\mathcal{F}}}\cdot\boldsymbol{\mathcal{F}}^{\mathrm{T}}. \tag{3.5.23}$$

where V is the volume of the system, whose Helmholtz energy is A, *in its current* (deformed) state, and $V^{(0)}$ is the volume of the same model system in its reference undeformed state. By virtue of Eq. (3.5.23),

$$\tau_{\alpha\beta} = -\frac{n}{V}k_{\mathrm{B}}T\delta_{\alpha\beta} + \frac{1}{V}\left\langle\sum_i\sum_{j>i}F_{ij,\alpha}\, r_{ij,\beta}\right\rangle = -\frac{n}{V}k_{\mathrm{B}}T\delta_{\alpha\beta} + \frac{1}{2V}\left\langle\sum_i\sum_{j\neq i}F_{ij,\alpha}\, r_{ij,\beta}\right\rangle. \quad (3.5.24)$$

Equation (3.5.24) is equivalent to the atomic virial expression for the stress, Eq. (3.2.26). We have thus proven the atomic virial expression for the stress $\boldsymbol{\tau}$ starting from a consideration of the work done by interatomic forces in the course of an arbitrary isothermal deformation and invoking the definition of stress in continuum thermoelasticity in terms of the derivative of the Helmholtz energy with respect to the deformation gradient tensor.

The symmetry of the stress tensor, as calculated by Eq. (3.5.24), is explored in Problem 3.9.

3.5.1.2 Quantum statistical mechanical treatment for the Helmholtz energy and stress at an inherent structure: quasiharmonic approximation

The development in Section 3.5.1.1 has assumed validity of classical statistical mechanics. In practice, solids are encountered at low temperatures. Atomic motion in a solid consists mainly of high-frequency vibrations around an inherent structure. For the frequencies ν of these vibrations, $h\nu$ may well exceed $k_{\mathrm{B}}T$, and thus a quantum mechanical treatment may be necessary (compare Eq. (3.1.50) and associated discussion).

A reasonable approach to use for a quantum statistical mechanics treatment of a solid trapped in the vicinity of an inherent structure is the *quasiharmonic approximation*. As detailed in Ref. 16, the Helmholtz energy can be written as

$$A = \mathcal{V}^{\mathrm{inh}} + A^{\mathrm{vib}}, \quad (3.5.25)$$

where $\mathcal{V}^{\mathrm{inh}}$ is the potential energy of the inherent structure, which constitutes a local minimum of the potential energy with respect to atomic coordinates at given n and $\boldsymbol{h}$, and A^{vib} is the Helmholtz energy of the $3n-3$ vibrational modes, considered as harmonic oscillators with angular frequencies $\omega_i = 2\pi\nu_i$ determined by diagonalization of the Hessian matrix of second derivatives of the potential energy with respect to atomic coordinates at the inherent structure (compare Eq. (3.1.27))

$$A^{\mathrm{vib}} = \frac{1}{2}\hbar\sum_{l=1}^{3n-3}\omega_l + k_{\mathrm{B}}T\sum_{l=1}^{3n-3}\ln\left[1-\exp\left(-\frac{\hbar\omega_l}{k_{\mathrm{B}}T}\right)\right]. \quad (3.5.26)$$

In forming the stress tensor $\boldsymbol{\tau}$ by applying Eq. (3.5.23) on the Helmholtz energy of Eq. (3.5.26), $\mathcal{V}^{\mathrm{inh}}$ can be treated exactly as the $\mathcal{V}$ of each configuration was treated in the classical approach. One obtains the contribution to stress

$$\sigma_{\alpha\beta}^{\mathrm{inh}} = \frac{1}{V}\sum_i\sum_{j>i}F_{ij,\alpha}^{\mathrm{inh}}\, r_{ij,\beta}^{\mathrm{inh}} = \frac{1}{2V}\sum_i\sum_{j\neq i}F_{ij,\alpha}^{\mathrm{inh}}\, r_{ij,\beta}^{\mathrm{inh}}. \quad (3.5.27)$$

As regards the vibrational part, this will be a sum over all modes l of terms of the form

$$\begin{aligned}
\tau^{\text{vib}}_{\alpha\beta,l} &= \frac{1}{V}\sum_{\gamma=1}^{3}\frac{\partial A^{\text{vib}}_{l}}{\partial \mathcal{F}_{\alpha\gamma}}\mathcal{F}_{\beta\gamma} = \frac{1}{V}\sum_{\gamma=1}^{3}\mathcal{F}_{\beta\gamma}\frac{\partial}{\partial \mathcal{F}_{\alpha\gamma}}\left\{\frac{1}{2}\hbar\omega_l + k_{\text{B}}T\ln\left[1-\exp\left(-\frac{\hbar\omega_l}{k_{\text{B}}T}\right)\right]\right\} \\
&= \frac{1}{V}\sum_{\gamma=1}^{3}\mathcal{F}_{\beta\gamma}\left\{\frac{1}{2}\hbar + k_{\text{B}}T\frac{\hbar}{k_{\text{B}}T}\frac{\exp\left(-\frac{\hbar\omega_l}{k_{\text{B}}T}\right)}{1-\exp\left(-\frac{\hbar\omega_l}{k_{\text{B}}T}\right)}\right\}\frac{\partial\omega_l}{\partial \mathcal{F}_{\alpha\gamma}} \\
&= \frac{\hbar}{V}\sum_{\gamma=1}^{3}\left\{\frac{1}{2} + \frac{\exp\left(-\frac{\hbar\omega_l}{k_{\text{B}}T}\right)}{1-\exp\left(-\frac{\hbar\omega_l}{k_{\text{B}}T}\right)}\right\}\mathcal{F}_{\beta\gamma}\frac{\partial\omega_l}{\partial \mathcal{F}_{\alpha\gamma}} \\
&= \frac{\hbar\omega_l}{V}\left\{\frac{1}{2} + \frac{\exp\left(-\frac{\hbar\omega_l}{k_{\text{B}}T}\right)}{1-\exp\left(-\frac{\hbar\omega_l}{k_{\text{B}}T}\right)}\right\}\sum_{\gamma=1}^{3}\mathcal{F}_{\beta\gamma}\frac{\partial\ln\omega_l}{\partial \mathcal{F}_{\alpha\gamma}}.
\end{aligned} \tag{3.5.28}$$

From Eqs. (3.5.26) and (3.5.27), the stress tensor according to the quantum mechanical treatment in the quasiharmonic approximation becomes

$$\tau_{\alpha\beta} = \frac{1}{V}\sum_{l=1}^{3N-3}\hbar\omega_l\left\{\frac{1}{2} + \frac{1}{\exp\left(\frac{\hbar\omega_l}{k_{\text{B}}T}\right)-1}\right\}\left(\sum_{\gamma=1}^{3}\frac{\partial\ln\omega_l}{\partial \mathcal{F}_{\alpha\gamma}}\mathcal{F}_{\beta\gamma}\right) + \frac{1}{V}\sum_{i}\sum_{j>i}F^{\text{inh}}_{ij,\alpha}\,r^{\text{inh}}_{ij,\beta}. \tag{3.5.29}$$

Its calculation obviously hinges upon the ability to quantify how the vibrational frequencies at the inherent structure change with deformation.

3.5.1.3 Strain tensor and fundamental equation in the Helmholtz energy representation

In thermoelasticity, it is customary to express the Helmholtz energy as a function of temperature and an appropriately defined strain tensor. We define the *Green strain tensor* as

$$\boldsymbol{\mathcal{E}} = \frac{1}{2}(\boldsymbol{\mathcal{F}}^{\text{T}}\cdot\boldsymbol{\mathcal{F}} - \boldsymbol{I}). \tag{3.5.30}$$

The Green strain tensor is symmetric, by construction. It depends only on the stretch, and not on the rotational component of $\boldsymbol{\mathcal{F}}$. Only the stretch component is expected to affect the Helmholtz energy. $\boldsymbol{I}$ is the unit tensor. For a differential deformation, the Green strain tensor changes by

$$\text{d}\boldsymbol{\mathcal{E}} = \frac{1}{2}(\text{d}\boldsymbol{\mathcal{F}}^{\text{T}}\cdot\boldsymbol{\mathcal{F}} + \boldsymbol{\mathcal{F}}^{\text{T}}\cdot\text{d}\boldsymbol{\mathcal{F}}). \tag{3.5.31}$$

In thermoelasticity it is postulated that the intensive Helmholtz energy $\frac{A}{n}$ is a function of T and $\mathcal{F}$. If we consider an isothermal process on our closed atomistic model system that is representative of a macroscopic material element, the change in Helmholtz energy will be

$$\mathrm{d}A = \frac{\partial A}{\partial \mathcal{F}^{\mathrm{T}}} : \mathrm{d}\mathcal{F}^{\mathrm{T}} = \mathcal{F}^{-1} \cdot \mathcal{F} \cdot \frac{\partial A}{\partial \mathcal{F}^{\mathrm{T}}} : \mathrm{d}\mathcal{F}^{\mathrm{T}} = \mathcal{F}^{-1} \cdot \left(\frac{\partial A}{\partial \mathcal{F}} \cdot \mathcal{F}^{\mathrm{T}} \right)^{\mathrm{T}} : \mathrm{d}\mathcal{F}^{\mathrm{T}}. \tag{3.5.32}$$

Combining Eqs. (3.5.23) and (3.5.32) and using that the tensor $\boldsymbol{\tau}$ is symmetric and so is the tensor $\mathcal{F}^{-1} \cdot \boldsymbol{\tau} \cdot (\mathcal{F}^{\mathrm{T}})^{-1}$, it can be shown (see Appendix) that

$$\mathrm{d}A = V\left(\mathcal{F}^{-1} \cdot \boldsymbol{\tau} \cdot \left(\mathcal{F}^{\mathrm{T}}\right)^{-1}\right) : \mathrm{d}\mathcal{E}, \tag{3.5.33}$$

from which we can conclude that

$$\frac{1}{V}\frac{\partial A}{\partial \mathcal{E}} = \mathcal{F}^{-1} \cdot \boldsymbol{\tau} \cdot (\mathcal{F}^{\mathrm{T}})^{-1}. \tag{3.5.34}$$

On the left-hand side of Eq. (3.5.34) appears the matrix of partial derivatives of the Helmholtz energy with respect to the elements of the Green strain tensor. Each such partial derivative is taken at constant temperature (isothermally), keeping all other elements of the Green strain tensor constant.

Equation (3.5.34) can give an expression for the stress tensor in terms of the derivative of the Helmholtz energy with respect to the Green strain tensor:

$$\boldsymbol{\tau} = \frac{1}{V}\mathcal{F} \cdot \frac{\partial A}{\partial \mathcal{E}} \cdot \mathcal{F}^{\mathrm{T}}. \tag{3.5.35}$$

Note that $\frac{1}{V^{(0)}}\frac{\partial A}{\partial \mathcal{E}} = \frac{V}{V^{(0)}}\mathcal{F}^{-1} \cdot \boldsymbol{\tau} \cdot (\mathcal{F}^{\mathrm{T}})^{-1} = \mathcal{S}$ is known as the second Piola–Kirchhoff stress tensor; $\frac{V}{V^{(0)}}\boldsymbol{\tau} \cdot (\mathcal{F}^{\mathrm{T}})^{-1} = \mathcal{P}$ is known as the first Piola–Kirchhoff stress tensor; and $\frac{V}{V^{(0)}}\mathcal{F}^{-1} \cdot \boldsymbol{\tau} = \mathcal{N}$ is known as the nominal stress tensor.[21]

The fundamental equation of thermodynamics for our material, in the Helmholtz energy representation, becomes

$$\mathrm{d}A = -S\mathrm{d}T + V\left[\mathcal{F}^{-1} \cdot \boldsymbol{\tau} \cdot \left(\mathcal{F}^{\mathrm{T}}\right)^{-1}\right] : \mathrm{d}\mathcal{E}. \tag{3.5.36}$$

In the notation adopted by Parrinello–Rahman[23] and Ray,[24] the second Piola–Kirchhoff stress tensor is called the thermodynamic tension tensor $\boldsymbol{t}$:

$$\boldsymbol{t} = \frac{V}{V^{(0)}}\left[\mathcal{F}^{-1} \cdot \boldsymbol{\tau} \cdot \left(\mathcal{F}^{\mathrm{T}}\right)^{-1}\right] = \frac{V}{V^{(0)}}\boldsymbol{h}^{(0)} \cdot \boldsymbol{h}^{-1} \cdot \boldsymbol{\tau} \cdot (\boldsymbol{h}^{\mathrm{T}})^{-1} \cdot (\boldsymbol{h}^{(0)})^{\mathrm{T}}. \tag{3.5.37}$$

In this notation, the fundamental equation in the Helmholtz energy representation, Eq. (3.5.36), becomes

$$\mathrm{d}A = -S\mathrm{d}T + V^{(0)}\boldsymbol{t} : \mathrm{d}\mathcal{E}. \tag{3.5.38}$$

On the other hand, from its definition, Eq. (3.5.30), the differential of the Green strain tensor becomes

$$\mathrm{d}\boldsymbol{\mathcal{E}} = \frac{1}{2}\mathrm{d}(\boldsymbol{\mathcal{F}}^{\mathrm{T}} \cdot \boldsymbol{\mathcal{F}}) = \frac{1}{2}\mathrm{d}\left[\left((\boldsymbol{h}^{0})^{\mathrm{T}}\right)^{-1} \cdot \boldsymbol{h}^{\mathrm{T}} \cdot \boldsymbol{h} \cdot (\boldsymbol{h}^{(0)})^{-1}\right] = \frac{1}{2}\left((\boldsymbol{h}^{0})^{\mathrm{T}}\right)^{-1} \cdot \mathrm{d}\boldsymbol{G} \cdot (\boldsymbol{h}^{(0)})^{-1}, \quad (3.5.39)$$

where we have defined the so-called metric tensor through[23,24]

$$\boldsymbol{G} = \boldsymbol{h}^{\mathrm{T}} \cdot \boldsymbol{h}. \quad (3.5.40)$$

Combining Eqs. (3.5.38)–(3.5.40), we obtain that

$$\begin{aligned} \mathrm{d}A &= -S\mathrm{d}T + V^{(0)}\frac{1}{2}\boldsymbol{t} : \left[\left((\boldsymbol{h}^{(0)})^{\mathrm{T}}\right)^{-1} \cdot \mathrm{d}\boldsymbol{G} \cdot \left(\boldsymbol{h}^{(0)}\right)^{-1}\right] \\ &= -S\mathrm{d}T + V^{(0)}\frac{1}{2}\left[\left(\boldsymbol{h}^{(0)}\right)^{-1} \cdot \boldsymbol{t} \cdot \left((\boldsymbol{h}^{(0)})^{\mathrm{T}}\right)^{-1}\right] : \mathrm{d}\boldsymbol{G}. \end{aligned} \quad (3.5.41)$$

The corresponding fundamental equation in the energy representation is

$$\mathrm{d}U = T\mathrm{d}S + V^{(0)}\frac{1}{2}\left[(\boldsymbol{h}^{(0)})^{-1} \cdot \boldsymbol{t} \cdot \left((\boldsymbol{h}^{(0)})^{\mathrm{T}}\right)^{-1}\right] : \mathrm{d}\boldsymbol{G}. \quad (3.5.42)$$

Ray[24] defines the tensor $\tilde{\boldsymbol{M}}$ through

$$\tilde{M}_{\alpha\beta} = \left.\frac{\partial U}{\partial G_{\alpha\beta}}\right|_{S, G_{[\alpha\beta]}}, \quad (3.5.43)$$

where the symbolism $G_{[\alpha\beta]}$ indicates that all elements of $\boldsymbol{G}$ but $G_{\alpha\beta}$ are kept constant. In view of Eqs. (3.5.42) and (3.5.37),

$$\tilde{\boldsymbol{M}} = \frac{1}{2}V^{(0)}\left[(\boldsymbol{h}^{(0)})^{-1} \cdot \boldsymbol{t} \cdot \left((\boldsymbol{h}^{(0)})^{\mathrm{T}}\right)^{-1}\right] = \frac{1}{2}V\left(\boldsymbol{h}^{-1} \cdot \boldsymbol{\tau} \cdot (\boldsymbol{h}^{\mathrm{T}})^{-1}\right). \quad (3.5.44)$$

The thermodynamic derivative of Eq. (3.5.43) can be written as an ensemble average in the micro-nanonical ensemble, where the number of molecules, the model system boundaries, and the total energy are kept constant:

$$\tilde{M}_{\alpha\beta} = \left\langle \left.\frac{\partial \mathcal{H}}{\partial G_{\alpha\beta}}\right|_{G_{[\alpha\beta]}} \right\rangle_{nhE}, \quad (3.5.45)$$

with $\mathcal{H}$ being the system Hamiltonian (total energy function)

$$\mathcal{H}(\boldsymbol{r},\boldsymbol{p}) = \sum_{i=1}^{n} \frac{\boldsymbol{p}_i^2}{2m_i} + \mathcal{V}(\boldsymbol{r}_1, \boldsymbol{r}_2, \ldots, \boldsymbol{r}_n) = \sum_{i=1}^{n} \frac{1}{2m_i} \boldsymbol{\pi}_i^{\mathrm{T}} \cdot \boldsymbol{G}^{-1} \cdot \boldsymbol{\pi}_i + \mathcal{V}(\boldsymbol{h}\cdot\boldsymbol{s}_1, \boldsymbol{h}\cdot\boldsymbol{s}_2, \ldots, \boldsymbol{h}\cdot\boldsymbol{s}_n). \tag{3.5.46}$$

In differentiating $\mathcal{H}$ within the brackets, all scaled coordinates $\boldsymbol{s}_i = \boldsymbol{h}^{-1}\cdot(\boldsymbol{r}_i - \boldsymbol{r}_{\mathrm{ref}})$ and momenta $\boldsymbol{\pi}_i = \boldsymbol{h}^{\mathrm{T}}\cdot\boldsymbol{p}_i$ with respect to the dimensions of the model system are kept constant.

3.5.1.4 *Thermodynamics of small deformations: isothermal elastic coefficients and their energy and entropy components*

The formulation presented in Sections 3.5.1.1 and 3.5.1.3 is generally applicable to materials exhibiting elasticity over large deformations, e.g. rubbery materials formed from crosslinked polymer networks which can be strained to several times their original length and recover practically their original dimensions upon removal of the externally imposed stress. Stress–strain curves for such materials may be highly nonlinear.

Polymer crystals and glassy polymers, on the other hand, have a limited elastic range. Polymer glasses generally behave elastically for strains below 1%. The stress–strain relation in most of this elastic range is linear, to a good approximation, hence well described by its slope.

In the special case of small deformations (on the order of 10^{-3}) the general thermoelastic formulation presented earlier admits considerable simplification. Matrices $\boldsymbol{h}$ in the deformed state and $\boldsymbol{h}^{(0)}$ in the reference undeformed state are not substantially different. As a consequence of Eq. (3.5.19), then, the deformation gradient tensor $\mathcal{F}$ is close to a unit tensor. Setting $\delta\boldsymbol{h} = \boldsymbol{h} - \boldsymbol{h}^{(0)}$, we have $\mathcal{F} = \boldsymbol{I} + \delta\boldsymbol{h}\cdot(\boldsymbol{h}^{(0)})^{-1}$, so, from Eq. (3.5.30), the Green strain tensor becomes

$$\mathcal{E} = \frac{1}{2}\left[\delta\boldsymbol{h}\cdot\left(\boldsymbol{h}^{(0)}\right)^{-1} + \left(\left(\boldsymbol{h}^{(0)}\right)^{-1}\right)^{\mathrm{T}}\cdot(\delta\boldsymbol{h})^{\mathrm{T}} + \left(\left(\boldsymbol{h}^{(0)}\right)^{-1}\right)^{\mathrm{T}}\cdot(\delta\boldsymbol{h})^{\mathrm{T}}\cdot\delta\boldsymbol{h}\cdot\left(\boldsymbol{h}^{(0)}\right)^{-1}\right] \simeq \varepsilon, \tag{3.5.47}$$

where

$$\varepsilon = \frac{1}{2}\left[\delta\boldsymbol{h}\cdot\left(\boldsymbol{h}^{(0)}\right)^{-1} + \left(\left(\boldsymbol{h}^{(0)}\right)^{-1}\right)^{\mathrm{T}}\cdot(\delta\boldsymbol{h})^{\mathrm{T}}\right]. \tag{3.5.48}$$

In identifying $\mathcal{E}$ with ε, the Cauchy strain tensor, all terms higher than first order in $\delta\boldsymbol{h}$ have been neglected, given that small deformations are considered.

For a reference box with the shape of a rectangular parallelepiped, $\boldsymbol{h}^{(0)}$ is diagonal and so is $(\boldsymbol{h}^{(0)})^{-1}$. Equation (3.5.48) gives

$$\varepsilon_{\alpha\beta} = \frac{1}{2}\left(\frac{\delta h_{\alpha\beta}}{h_{\beta\beta}^{(0)}} + \frac{\delta h_{\beta\alpha}}{h_{\alpha\alpha}^{(0)}}\right). \tag{3.5.49}$$

The fundamental equation, Eq. (3.5.36), for small deformations around the reference state ($\mathcal{F} = \boldsymbol{I}$) reduces to[23]

$$\mathrm{d}A = -S\mathrm{d}T + V^{(0)}\boldsymbol{\tau} : \mathrm{d}\varepsilon. \tag{3.5.50}$$

In the linear elastic regime, then, we can set

$$\tau_{\alpha\beta} = \frac{1}{V^{(0)}} \frac{\partial A}{\partial \varepsilon_{\alpha\beta}} \bigg|_{T,\varepsilon_{[\alpha\beta]}}, \tag{3.5.51}$$

where $V^{(0)}$ is the volume in the reference undeformed state and the symbolism $\varepsilon_{[\alpha\beta]}$ indicates that all elements of the strain tensor $\boldsymbol{\varepsilon}$ but $\varepsilon_{\alpha\beta}$ are kept constant.

We define the fourth-order tensor of isothermal elastic coefficients[5,23] $\boldsymbol{C}$ as

$$C_{\alpha\beta\zeta\eta} = \frac{\partial \tau_{\alpha\beta}}{\partial \varepsilon_{\zeta\eta}} \bigg|_{T,\varepsilon_{[\zeta\eta]}} = \frac{1}{V^{(0)}} \frac{\partial^2 A}{\partial \varepsilon_{\alpha\beta} \partial \varepsilon_{\zeta\eta}} \bigg|_{T,\varepsilon_{[\alpha\beta,\zeta\eta]}}. \tag{3.5.52}$$

The tensors $\boldsymbol{\tau}$ and $\boldsymbol{\varepsilon}$ are symmetric. As a consequence, for a concise description of the thermodynamics of elasticity, each one of these tensors can be condensed into a vector of six independent elements, adopting the so-called Voigt single-index notation:[5]

$$\begin{gathered} \varepsilon_1 = \varepsilon_{11}, \varepsilon_2 = \varepsilon_{22}, \varepsilon_3 = \varepsilon_{33}, \varepsilon_4 = 2\varepsilon_{23}, \varepsilon_5 = 2\varepsilon_{31}, \varepsilon_6 = 2\varepsilon_{12} \\ \tau_1 = \tau_{11}, \tau_2 = \tau_{22}, \tau_3 = \tau_{33}, \tau_4 = \tau_{23}, \tau_5 = \tau_{31}, \tau_6 = \tau_{12}. \end{gathered} \tag{3.5.53}$$

Indices 1, 2, 3 correspond to normal stresses and elongational strains, while indices 4, 5, 6 to shear stresses and shear strains. In Voigt notation, the fundamental equation for the Helmholtz energy, Eq. (3.5.50), becomes

$$\mathrm{d}A = -S\mathrm{d}T + V^{(0)} \sum_{I=1}^{6} \tau_I \mathrm{d}\varepsilon_I. \tag{3.5.54}$$

Furthermore, the fourth-order tensor of isothermal elastic coefficients is condensed into a 6×6 matrix $\boldsymbol{C}$ with elements $C_{I\mathcal{J}}$, where the correspondence between the single indices (I, $\mathcal{J}$) and the double indices (α, β) and (ζ, η) can be seen in Eqs. (3.5.53).

For an isotropic material, such as an amorphous glassy polymer, the 6×6 matrix $\boldsymbol{C}$ is of the form

$$\boldsymbol{C} = \begin{bmatrix} 2\mu+\lambda & \lambda & \lambda & 0 & 0 & 0 \\ \lambda & 2\mu+\lambda & \lambda & 0 & 0 & 0 \\ \lambda & \lambda & 2\mu+\lambda & 0 & 0 & 0 \\ 0 & 0 & 0 & \mu & 0 & 0 \\ 0 & 0 & 0 & 0 & \mu & 0 \\ 0 & 0 & 0 & 0 & 0 & \mu \end{bmatrix}. \tag{3.5.55}$$

All elements can be expressed in terms of only two independent coefficients, λ and and μ, the Lamé constants. The tensile (Young's) modulus E, the shear modulus G, the bulk modulus B (inverse of the isothermal compressibility κ_T), and the Poisson ratio ν are related to λ and μ by

$$E = \mu\frac{3\lambda+2\mu}{\lambda+\mu}, \quad G = \mu, \quad B = \frac{1}{\kappa_T} = \lambda + \frac{2}{3}\mu, \quad \nu = \frac{\lambda}{2(\lambda+\mu)}. \tag{3.5.56}$$

The strain dependence of the entropy, at constant temperature, is given by the Grüneisen tensor $\boldsymbol{\gamma}$ with elements

$$\gamma_{\alpha\beta} = \frac{1}{\rho_0 \, c_\varepsilon \, V^{(0)}} \frac{\partial S}{\partial \varepsilon_{\alpha\beta}}\bigg|_{T,\varepsilon_{[\alpha\beta]}}, \tag{3.5.57}$$

where ρ_0 is the mass density of the reference undeformed state and c_ε the heat capacity per unit mass of material at constant strain. From a Maxwell relation on the fundamental equation, Eq. (3.5.54), we obtain that

$$\gamma_{\alpha\beta} = -\frac{1}{\rho_0 c_\varepsilon} \frac{\partial \tau_{\alpha\beta}}{\partial T}\bigg|_{\varepsilon}. \tag{3.5.58}$$

For an isotropic solid, $\gamma_{ij} = \gamma\delta_{ij}$ with

$$\gamma = \frac{1}{\rho_0 c_\varepsilon} \frac{\partial S}{\partial V}\bigg|_T = \frac{1}{\rho_0 c_\varepsilon} \frac{\partial P}{\partial T}\bigg|_V = \frac{1}{\rho_0 c_\varepsilon} \frac{\alpha_P}{\kappa_T}, \tag{3.5.59}$$

where α_P is the volumetric thermal expansion coefficient and γ the Grüneisen parameter. The quantity $\frac{1}{3}\text{tr}(\boldsymbol{\tau} + \rho_0 c_\varepsilon T\boldsymbol{\gamma}) = -P + \frac{\alpha_P T}{\kappa_T} = \frac{\partial U}{\partial V}\Big|_T$ is sometimes termed the *internal pressure*. By analogy we will call the tensor $\boldsymbol{\sigma} = \boldsymbol{\tau} + \rho_0 c_\varepsilon T\boldsymbol{\gamma}$ the *internal stress tensor*.

From Eqs. (3.5.57), (3.5.58), and (3.5.52) we can obtain the second derivatives of the entropy with respect to the strain tensor:

$$\begin{aligned}
\frac{\partial^2 S}{\partial \varepsilon_{\alpha\beta} \partial \varepsilon_{\zeta\eta}}\bigg|_{T,\varepsilon_{[\alpha\beta,\zeta\eta]}} &= \frac{\partial(\rho_0 c_\varepsilon V^{(0)} \gamma_{\alpha\beta})}{\partial \varepsilon_{\zeta\eta}}\bigg|_{T,\varepsilon_{[\zeta\eta]}} \\
&= -V^{(0)} \left(\frac{\partial}{\partial \varepsilon_{\zeta\eta}} \frac{\partial \tau_{\alpha\beta}}{\partial T}\bigg|_\varepsilon \right)_{T,\varepsilon_{[\zeta\eta]}} = -V^{(0)} \left(\frac{\partial}{\partial T} \frac{\partial \tau_{\alpha\beta}}{\partial \varepsilon_{\zeta\eta}}\bigg|_{T,\varepsilon_{[\zeta\eta]}} \right)_\varepsilon = -V^{(0)} \frac{\partial C_{\alpha\beta\zeta\eta}}{\partial T}\bigg|_\varepsilon.
\end{aligned} \tag{3.5.60}$$

Consider now an elastic solid subjected to an arbitrary isothermal small deformation. Expanding the internal energy U into a Taylor series around the undeformed state to second order and using the definitions introduced earlier, we obtain

$$\begin{aligned}
U &= A + TS \\
&= U^{(0)} + V^{(0)} \sum_\alpha \sum_\beta [\tau_{\alpha\beta} + \rho_0 c_\varepsilon T \gamma_{\alpha\beta}] \varepsilon_{\alpha\beta} \\
&\quad + \frac{1}{2} V^{(0)} \sum_\alpha \sum_\beta \sum_\zeta \sum_\eta \left[C_{\alpha\beta\zeta\eta} - T \frac{\partial C_{\alpha\beta\zeta\eta}}{\partial T}\bigg|_\varepsilon \right] \varepsilon_{\alpha\beta} \varepsilon_{\zeta\eta}.
\end{aligned} \tag{3.5.61}$$

The first terms in the brackets are due to the strain derivatives of the Helmholtz energy; the second terms originate in the corresponding derivatives of the entropy.

By the definition of the isothermal elastic coefficients, Eq. (3.5.52), one obtains that

$$\begin{aligned} V_0 C_{\alpha\beta\zeta\eta} &= \left.\frac{\partial^2 A}{\partial\varepsilon_{\alpha\beta}\partial\varepsilon_{\zeta\eta}}\right|_{T,\boldsymbol{\varepsilon}_{[\alpha\beta,\zeta\eta]}} \\ &= \left.\frac{\partial^2 U}{\partial\varepsilon_{\alpha\beta}\partial\varepsilon_{\zeta\eta}}\right|_{T,\boldsymbol{\varepsilon}_{[\alpha\beta,\zeta\eta]}} - T\left.\frac{\partial^2 S}{\partial\varepsilon_{\alpha\beta}\partial\varepsilon_{\zeta\eta}}\right|_{T,\boldsymbol{\varepsilon}_{[\alpha\beta,\zeta\eta]}} \\ &= \left.\frac{\partial^2 U}{\partial\varepsilon_{\alpha\beta}\partial\varepsilon_{\zeta\eta}}\right|_{T,\boldsymbol{\varepsilon}_{[\alpha\beta,\zeta\eta]}} + TV^{(0)}\left.\frac{\partial C_{\alpha\beta\zeta\eta}}{\partial T}\right|_{\boldsymbol{\varepsilon}}. \end{aligned} \tag{3.5.62}$$

Each isothermal elastic coefficient has been separated into an internal energy contribution and an entropic contribution, and the entropic contribution has been re-expressed in terms of the temperature dependence of the isothermal elastic coefficient by virtue of Eq. (3.5.60).

It is useful to assess the relative importance of energetic and entropic contributions to the isothermal elastic coefficients. We see that entropic effects will be relatively unimportant if

$$\left|\frac{1}{C_{\alpha\beta\zeta\eta}}\left(T\left.\frac{\partial C_{\alpha\beta\zeta\eta}}{\partial T}\right|_{\boldsymbol{\varepsilon}}\right)\right| = \left|\left.\frac{\partial \ln C_{\alpha\beta\zeta\eta}}{\partial \ln T}\right|_{\boldsymbol{\varepsilon}}\right| << 1. \tag{3.5.63}$$

Since temperature derivatives of the elastic coefficients at constant strain are not usually available experimentally, we restrict our attention to an isotropic solid under the condition that the reference (undeformed) state is characterized by an isotropic stress distribution. Then, the previous criterion can be transformed into (see Appendix in Ref. [6])

$$\left|\frac{1}{C_{\alpha\beta\zeta\eta}}\left[T\left.\frac{\partial C_{\alpha\beta\zeta\eta}}{\partial T}\right|_{P} + \frac{\alpha_P T}{\kappa_T}\left.\frac{\partial C_{\alpha\beta\zeta\eta}}{\partial P}\right|_{T}\right]\right| << 1. \tag{3.5.64}$$

For uniform hydrostatic compression, for example, the relevant elastic constant is the bulk modulus

$$9B = C_{11} + C_{22} + C_{33} + 2C_{23} + 2C_{31} + 2C_{12}, \tag{3.5.65}$$

implying that the relevant criterion to consider in order to assess the importance of entropic effects in the elasticity of amorphous polymers is

$$\left|\frac{1}{B}\left[T\left.\frac{\partial B}{\partial T}\right|_{P} + \alpha_P T B\left.\frac{\partial B}{\partial P}\right|_{T}\right]\right| << 1 \tag{3.5.66}$$

or

$$\left|\frac{T}{\kappa_T}\left.\frac{\partial \kappa_T}{\partial T}\right|_{P} + \frac{\alpha_P T}{\kappa_T^2}\left.\frac{\partial \kappa_T}{\partial P}\right|_{T}\right| << 1 \tag{3.5.67}$$

with $\kappa_T = 1/B$ being the isothermal compressibility. For pure shear, on the other hand, e.g. perpendicular to z and in the direction of x, the relevant elastic constant is

$$G = \mu = C_{3131} = C_{55}, \tag{3.5.68}$$

implying that the relevant criterion to consider in order to assess the importance of entropic effects in the elasticity of amorphous polymers is

$$\left| \frac{T}{G}\frac{\partial G}{\partial T}\bigg|_P + \frac{\alpha_P T}{\kappa_T}\frac{1}{G}\frac{\partial G}{\partial P}\bigg|_T \right| << 1. \tag{3.5.69}$$

For typical polymeric glasses (e.g. PMMA) at room temperature (e.g. at $T = 20°C$), experimental data yield values of 0.071 and 0.14 for the left-hand sides of Eqs. (3.5.67) and (3.5.69), respectively.[6]

Thus, for an amorphous polymer below its glass transition temperature, far from frequency regions in which subglass relaxation transitions are observed, the energy contribution to the isothermal coefficients is expected to be much more important than the entropic contribution. This is, of course, in stark contrast to what happens in rubbery polymers, well above the glass transition temperature, where the entropic contribution to the elastic response is dominant.[5]

3.5.2 Estimation of elastic constants from static model glassy microstates

3.5.2.1 Energy approach

As discussed in Section 3.5.1, experimental data support that entropic contributions to the isothermal elastic coefficients of a polymer glass are very small in comparison to energy contributions; i.e. Eq. (3.5.63) holds to a good approximation. If we use Eq. (3.5.63) within Eq. (3.5.61), we see that we can obtain good estimates of the isothermal elastic coefficients $C_{\alpha\beta\zeta\eta}$ from the second derivatives of the internal energy with respect to the components of strain. Theodorou and Suter[6] used this as the basis of their *energy approach* for predicting elastic constants of an amorphous glass through molecular mechanics.

If, as discussed in Section 3.4, we consider the configuration of a glass as locally trapped in the vicinity of a local minimum of the energy (inherent structure), the energy U will be a sum of the inherent structure energy $\mathcal{V}^{\text{inh}}$ and the energy of the normal modes of vibration around the inherent structure (compare Eqs. (3.5.25), (3.5.26)). Considering small volume changes of glassy atactic polypropylene and utilizing experimental estimates of the set of parameters $\{\alpha_P, \kappa_T, \rho_0, c_\varepsilon, \gamma\}$ as well as of the Einstein and Debye temperatures for this polymer, they concluded[6] that the derivatives $\frac{\partial U}{\partial V}\Big|_T$ and $\frac{\partial^2 U}{\partial V^2}\Big|_T$ can be estimated through the corresponding derivatives $\frac{\partial \mathcal{V}^{\text{inh}}}{\partial V}\Big|_T$ and $\frac{\partial^2 \mathcal{V}^{\text{inh}}}{\partial V^2}\Big|_T$, respectively, with an error of a few per cent. Thus, in analysing deformations of each inherent structure, they chose to identify U with $\mathcal{V}^{\text{inh}}$. In other words, they rewrote Eq. (3.5.61) for the energy as

$$\mathcal{V}^{\text{inh}} = \mathcal{V}^{\text{inh},(0)} + V^{(0)} \sum_\alpha \sum_\beta \sigma^{(0)}_{\alpha\beta}\, \varepsilon_{\alpha\beta} + \frac{1}{2} V^{(0)} \sum_\alpha \sum_\beta \sum_\zeta \sum_\eta C_{\alpha\beta\zeta\eta}\, \varepsilon_{\alpha\beta}\, \varepsilon_{\zeta\eta} \tag{3.5.70}$$

with $\mathcal{V}^{\text{inh}}$ being the inherent structure energy at strain ε, $\mathcal{V}^{\text{inh},(0)}$ and $V^{(0)}$ being the inherent structure energy and volume in the undeformed state, and $\boldsymbol{\sigma} = \boldsymbol{\tau} + \rho_0\, c_\varepsilon T \boldsymbol{\gamma}$ being the internal stress tensor.

The deformation computer experiments proceeded as follows:[6]

1. Start with an undeformed structure in detailed mechanical equilibrium. The continuation geometry in the undeformed state was cubic; i.e. the continuation vectors were mutually perpendicular and equal in length, with $\boldsymbol{h}^{(0)} = h\,\mathrm{diag}(1,1,1)$. The starting structure was a member of the collection of minimum energy structures described earlier and satisfied the condition

$$\mathcal{V}(\boldsymbol{r}_0^{(0)}, \psi^{(0)}, \phi^{(0)}; \boldsymbol{h}^{(0)}) = \min = \mathcal{V}^{\text{inh},(0)}, \tag{3.5.71}$$

where the superscript (0) denotes undeformed conditions.

2. Choose the type of deformation to be imposed. Choose a small degree of deformation ε such that $|\varepsilon| << 1$. (Most calculations were conducted with $\varepsilon = \pm 0.001$.)
3. Modify the location of the cube origin $\boldsymbol{O}$ and the continuation vectors $\boldsymbol{h}_i$ so that they correspond to the chosen deformation. For the simple deformations studied in Ref. 6, initial system quantities were transformed in such a way as to leave the centre of the simulation box unchanged during the deformation, as follows.
Uniform hydrostatic compression

$$\text{cube origin}: \boldsymbol{O} = \boldsymbol{O}^{(0)} + \frac{1-(1-\varepsilon)^{1/3}}{2}\left[\boldsymbol{h}_1^{(0)} + \boldsymbol{h}_2^{(0)} + \boldsymbol{h}_3^{(0)}\right] \tag{3.5.72a}$$

$$\text{continuation vectors}: \begin{array}{l} \boldsymbol{h}_1 = (1-\varepsilon)^{1/3}\boldsymbol{h}_1^{(0)} \\ \boldsymbol{h}_2 = (1-\varepsilon)^{1/3}\boldsymbol{h}_2^{(0)} \\ \boldsymbol{h}_3 = (1-\varepsilon)^{1/3}\boldsymbol{h}_3^{(0)} \end{array}. \tag{3.5.72b}$$

Affine transformation of atomic positions would correspond to

$$\boldsymbol{r}_i^{\text{affine}} = \boldsymbol{O}^{(0)} + \frac{1-(1-\varepsilon)^{1/3}}{2}\left[\boldsymbol{h}_1^{(0)} + \boldsymbol{h}_2^{(0)} + \boldsymbol{h}_3^{(0)}\right] + (1-\varepsilon)^{1/3}\left(\boldsymbol{r}_i^{(0)} - \boldsymbol{O}^{(0)}\right). \tag{3.5.72c}$$

Pure shear (perpendicular to z, along x)

$$\text{cube origin}: \boldsymbol{O} = \boldsymbol{O}^{0} - \frac{1}{2}\varepsilon\boldsymbol{h}_1^{(0)} \tag{3.5.73a}$$

$$\text{continuation vectors}: \begin{array}{l} \boldsymbol{h}_1 = \boldsymbol{h}_1^{(0)} \\ \boldsymbol{h}_2 = \boldsymbol{h}_2^{(0)} \\ \boldsymbol{h}_3 = \boldsymbol{h}_3^{(0)} + \varepsilon\,\boldsymbol{h}_1^{(0)} \end{array}. \tag{3.5.73b}$$

Affine transformation of atomic positions would correspond to

$$\boldsymbol{r}_i^{\text{affine}} = \boldsymbol{r}_i^{(0)} + \varepsilon\left\{\frac{\left(\boldsymbol{r}_i^{(0)} - \boldsymbol{O}^{(0)}\right)\cdot\left(\boldsymbol{h}_1^{(0)} \times \boldsymbol{h}_2^{(0)}\right)}{\boldsymbol{h}_3^{(0)}\cdot\left(\boldsymbol{h}_1^{(0)} \times \boldsymbol{h}_2^{(0)}\right)} - \frac{1}{2}\right\}\boldsymbol{h}_1^{(0)}. \tag{3.5.73c}$$

Pure uniaxial tension (along x)

$$\text{cube origin}: \boldsymbol{0} = \boldsymbol{0}^{(0)} - \frac{1}{2}\varepsilon \boldsymbol{h}_1^{(0)} \tag{3.5.74a}$$

$$\text{continuation vectors}: \begin{array}{l} \boldsymbol{h}_1 = (1+\varepsilon)\boldsymbol{h}_1^{(0)} \\ \boldsymbol{h}_2 = \boldsymbol{h}_2^{(0)} \\ \boldsymbol{h}_3 = \boldsymbol{h}_3^{(0)} \end{array} . \tag{3.5.74b}$$

Affine transformation of atomic positions would correspond to

$$\boldsymbol{r}_i^{\text{affine}} = \boldsymbol{r}_i^{(0)} + \varepsilon \left\{ \frac{\left(\boldsymbol{r}_i^{(0)} - \boldsymbol{0}^{(0)}\right) \cdot \left(\boldsymbol{h}_2^{(0)} \times \boldsymbol{h}_3^{(0)}\right)}{\boldsymbol{h}_1^{(0)} \cdot \left(\boldsymbol{h}_2^{(0)} \times \boldsymbol{h}_3^{(0)}\right)} - \frac{1}{2} \right\} \boldsymbol{h}_1^{(0)}. \tag{3.5.74c}$$

Other shear and tension deformations are obtained by cyclic permutation.

4. Reminimize the total potential energy with respect to the microscopic degrees of freedom, keeping the continuation vectors equal to their deformed values:

$$\mathcal{V}(\boldsymbol{r}_0, \psi, \phi; \boldsymbol{h}) = \min = \mathcal{V}^{\text{inh}}. \tag{3.5.75}$$

   The resulting structure will again satisfy the conditions of detailed mechanical equilibrium, but under the continuation geometry dictated by the mode and degree of deformation chosen in step 2; it is a deformed model structure.

5. Repeat the procedure with different initial (minimum energy) structures and obtain a set of deformed model structures. To get estimates of the mechanical properties from each structure, analyse the results for the potential energy function in the three types of deformation experiments considered in step 3 according to the following expressions, which are special cases of Eq. (3.5.70):

   Uniform hydrostatic compression

$$\begin{aligned} \mathcal{V}^{\text{inh}} &= \mathcal{V}^{\text{inh},(0)} - \frac{1}{3} V^{(0)} \operatorname{tr}(\boldsymbol{\sigma}^{(0)})\varepsilon \\ &\quad + \frac{1}{2} V^{(0)} (C_{1111} + C_{1122} + C_{1133} + C_{2211} + C_{2222} \\ &\qquad + C_{2233} + C_{3311} + C_{3322} + C_{3333}) \frac{\varepsilon^2}{9} \\ &= \mathcal{V}^{\text{inh},(0)} - V^{(0)} \left(-P + \frac{\alpha_P T}{\kappa_T}\right)^{(0)} \varepsilon + \frac{1}{2} V^{(0)} B \varepsilon^2. \end{aligned} \tag{3.5.76a}$$

Thus,

$$\frac{\partial \mathcal{V}^{\text{inh}}}{\partial \varepsilon} \equiv -V^{(0)} \left(-P + \frac{\alpha_P T}{\kappa_T}\right)^{(0)}, \; \frac{\partial^2 \mathcal{V}^{\text{inh}}}{\partial \varepsilon^2} \equiv V_0 B. \tag{3.5.76b}$$

Pure shear (perpendicular to z, along x)

$$\begin{aligned}\mathcal{V}^{\text{inh}} &= \mathcal{V}^{\text{inh},(0)} + V^{(0)}\left(\sigma_{13}^{(0)} + \sigma_{31}^{(0)}\right)\frac{\varepsilon}{2} \\ &\quad + \frac{1}{2}V^{(0)}(C_{1313} + C_{1331} + C_{3113} + C_{3131})\frac{\varepsilon^2}{4} \\ &= \mathcal{V}^{\text{inh},(0)} + V^{(0)}\,\sigma_5^{(0)}\,\varepsilon + \frac{1}{2}V_0\,C_{55}\,\varepsilon^2. \end{aligned} \quad (3.5.77a)$$

Thus,

$$\frac{\partial \mathcal{V}^{\text{inh}}}{\partial \varepsilon} \equiv V^{(0)}\sigma_5^{(0)} = V^{(0)}(\tau_5 + \rho_0 c_\varepsilon T \gamma_5)^{(0)}, \; \frac{\partial^2 \mathcal{V}^{\text{inh}}}{\partial \varepsilon^2} \equiv V^{(0)} C_{55}. \quad (3.5.77b)$$

Pure uniaxial tension (along x)

$$\begin{aligned}\mathcal{V}^{\text{inh}} &= \mathcal{V}^{\text{inh},(0)} + V^{(0)}\sigma_{11}^{(0)}\varepsilon + \frac{1}{2}V_0 C_{1111}\varepsilon^2 \\ &= \mathcal{V}^{\text{inh},(0)} + V^{(0)}\sigma_{11}^{(0)}\varepsilon + \frac{1}{2}V_0 C_{11}\varepsilon^2. \end{aligned} \quad (3.5.78a)$$

Thus,

$$\frac{\partial \mathcal{V}^{\text{inh}}}{\partial \varepsilon} \equiv V^{(0)}\,\sigma_1^{(0)} = V^{(0)}(\tau_1 + \rho_0 c_\varepsilon T \gamma_1)^{(0)}, \; \frac{\partial^2 \mathcal{V}^{\text{inh}}}{\partial \varepsilon^2} \equiv V^{(0)}\,C_{11}. \quad (3.5.78b)$$

According to the above expressions, the first derivatives of the potential energy lead in the case of compression to the internal pressure $\left(-P + \frac{\alpha_P T}{\kappa_T}\right)^{(0)}$, and in the case of shear and tension to the internal stress tensor $\boldsymbol{\sigma}^{(0)}$ (all first derivative information pertains to the undeformed state only). From the second derivatives, on the other hand, we see that we obtain the bulk modulus B in the case of compression, and the diagonal elements of the 6×6 matrix of isothermal elastic coefficients $\boldsymbol{C}$ in the case of shear and tension. A test of internal consistency is provided by the fact that the internal pressure $-P + \frac{\alpha_P T}{\kappa_T}$ from (3.5.76b) must be identical with $\frac{1}{3}\,\text{tr}(\boldsymbol{\sigma}^{(0)})$ obtained from (3.5.78b) and the two equivalent forms for tension along y and z.

3.5.2.2 Force approach

An alternative strategy, termed the *force approach*, was developed by Theodorou and Suter for the calculation of isothermal elastic coefficients from static model microstates subjected to small deformations.[6] This entailed the direct calculation of the internal stress $\boldsymbol{\sigma}$ in the reference undeformed and in each deformed state and subsequent estimation of $C_{\alpha\beta\zeta\eta}$ as $\frac{\partial \sigma_{\alpha\beta}}{\partial \varepsilon_{\zeta\eta}}$ by finite differences (compare Eq. (3.5.52)). The use of the internal stress tensor $\boldsymbol{\sigma}$ in place of the full stress tensor $\boldsymbol{\tau}$ in this calculation is a consequence of neglecting entropic contributions to the elastic constants (compare Eq. (3.5.62) and subsequent discussion). The internal stress tensor was estimated as $\boldsymbol{\sigma}^{\text{inh}}$ from interatomic forces and distances at the inherent structure corresponding to each mode and degree of deformation. As discussed in Section 3.5.1.2, this amounts to a quasiharmonic approximation with neglect of vibrational contributions to the stress. To obtain $\boldsymbol{\sigma}^{\text{inh}}$ from the undeformed and from each deformed inherent structure, Eq. (3.5.27) was essentially used. Theodorou and Suter developed a special calculation scheme for this equation for the constant bond length and bond angle model they invoked.[6] This scheme considers nonbonded forces as well as torques generated by the torsional potentials. From the slopes of the elements of $\boldsymbol{\sigma}^{\text{inh}}$ with respect to ε in tension and shear, the full 6×6 matrix $\boldsymbol{C}$ was calculated. A single tension experiment provides one of the first three columns of the $\boldsymbol{C}$ matrix; a single shear experiment provides one of the last three columns. In fact, by performing a pair of computer experiments in each mode of deformation ($\varepsilon = \pm 0.001$), one can have two estimates for each element of the $\boldsymbol{C}$ matrix. These estimates should agree very well with each other, since changes upon deformation must be symmetric around the undeformed state (harmonic behaviour).

3.5.2.3 *Example application: elastic constants of glassy atactic polypropylene*

As an example of how molecular mechanics can be used to estimate elastic constants, we present some results from the early work of Theodorou and Suter[6] on glassy atactic polypropylene at −40°C. Eight of the fifteen static model microstates generated to model the polymer, as discussed in Section 3.4, were subjected to the model deformations outlined in Section 3.5.2.1.

Figure 3.9 displays the potential energy $\mathcal{V}^{\text{inh}}$ of one static model microstate as a function of the fractional increase in volume $-\varepsilon$ in the course of uniform hydrostatic compression–tension simulations with $\varepsilon = \pm 0.001$, $\varepsilon = \pm 0.002$, $\varepsilon = \pm 0.004$, and $\varepsilon = \pm 0.008$. All points lie on a smooth curve, reflecting the fact that, for small deformations, the system remains in the same basin in configuration space, which is continuously distorted with increasing deformation. These small deformations were found to be perfectly reversible, confirming elastic response. As expected from Eqs. (3.5.76), $\mathcal{V}^{\text{inh}}$ as a function of $-\varepsilon$ exhibits a positive slope and positive curvature. The curvature gives an estimate of the bulk modulus B via Eq. (3.5.76b), which is averaged with the corresponding estimates obtained from all other static model microstructures to arrive at a final prediction for B. As seen from Figure 3.9, the undeformed static model microstate along with the deformed static model microstates corresponding to $\varepsilon = \pm 0.001$ suffice for obtaining an estimate of the second derivative by central differences. This is why simulations were confined to these ε values for the purpose of calculating elastic constants via the energy approach.

As discussed in Section 3.5.2.1, with the deformations conducted here the energy approach yields the diagonal elements of the 6×6 matrix of isothermal elastic coefficients, as well as the bulk modulus. The values of these quantities, as obtained from the eight static model microstates of glassy atactic polypropylene studied in Ref. 6 (averages ± standard deviations), were[6]

$$\boldsymbol{C} = \begin{bmatrix} 4777 \pm 497 & & & & & \\ & 5871 \pm 774 & & & & \\ & & 3507 \pm 495 & & & \\ & & & 896 \pm 439 & & \\ & & & & 1230 \pm 136 & \\ & & & & & 1230 \pm 326 \end{bmatrix} \text{MPa} \tag{3.5.79}$$

$$B = 3270 \pm 415\text{MPa}. \tag{3.5.80}$$

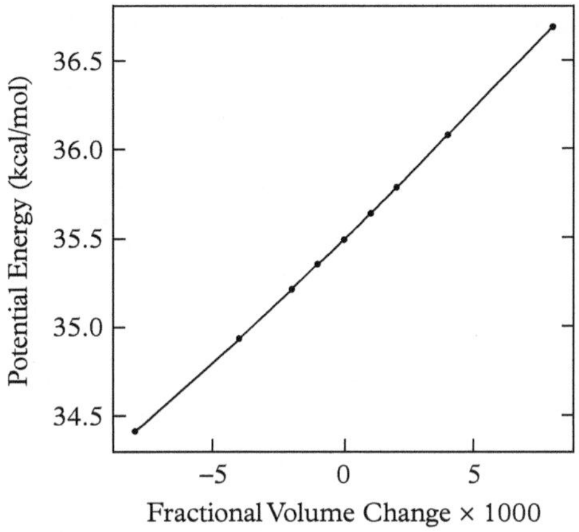

Figure 3.9 Total potential energy versus fractional increase in volume for one static model microstate of glassy atactic polypropylene subjected to hydrostatic compression–tension computer experiments.
Reprinted (adapted) with permission from Theodorou, D.N., and Suter, U.W. *Macromolecules* **19**, 139 (1986). Copyright 1986 American Chemical Society.

The force approach, on the other hand, yielded the following estimate of the entire $\boldsymbol{C}$ matrix:

$$\boldsymbol{C} = \begin{bmatrix} 4773 \pm 498 & 3244 \pm 542 & 2047 \pm 355 & 93 \pm 242 & -243 \pm 274 & 51 \pm 231 \\ 3244 \pm 542 & 5871 \pm 772 & 2943 \pm 557 & 777 \pm 707 & -162 \pm 167 & 502 \pm 384 \\ 2047 \pm 355 & 2943 \pm 557 & 3492 \pm 497 & 695 \pm 497 & -367 \pm 191 & -31 \pm 158 \\ 93 \pm 242 & 777 \pm 707 & 695 \pm 497 & 858 \pm 472 & 8 \pm 64 & -696 \pm 477 \\ -243 \pm 274 & -162 \pm 167 & -367 \pm 191 & 8 \pm 64 & 1230 \pm 135 & 176 \pm 179 \\ 51 \pm 243 & 502 \pm 384 & -31 \pm 158 & -696 \pm 477 & 176 \pm 179 & 1221 \pm 333 \end{bmatrix} \text{MPa.} \tag{3.5.81}$$

Those elements of $\boldsymbol{C}$ which were computed by both the energy and force approaches exhibit excellent consistency. For a macroscopic system, $\boldsymbol{C}$ should be of the form of Eq. (3.5.55). The small size of the static model microstates employed in Ref. 6 (1.815 nm on edge) induces considerable fluctuations in different directions within a given microstate, and also between different microstates. Nevertheless, the basic structure of Eq. (3.5.55) can be discerned in Eq. (3.5.81).

To obtain estimates of the elastic constants by the energy and by the force approach, results were averaged over all structures and all orientations.[6] The best estimates of the Lamé constants λ and μ that describe the results, along with the values of the set $\{E, G, B, \nu\}$ obtained from them, are listed in Table 3.3, along with the corresponding experimental values. Experimental values are thought to be accurate within 15%. The energy approach predicts all elastic constants with a relative error of less than 15%. Estimates obtained from the force approach are in even better agreement with experiment.

Table 3.3 *Comparison of Experimental and Predicted Values of the Elastic Constants of Glassy Atactic Polypropylene at −40° C, as Predicted by Molecular Mechanics*[6]

Property	Experimental value	Simulation			
		Energy approach		Force approach	
		Value	Deviation from expt (%)	Value	Deviation from expt (%)
Lamé constant λ (MPa)	2690	2520	−6.3	2700	+0.4
Lamé constant μ (MPa)	970	1110	+14.4	1020	+5.2
Shear modulus G (MPa)	970	1110	+14.4	1020	+5.2
Young's modulus E (MPa)	2650	2990	+12.8	2790	+5.3
Bulk modulus B (MPa)	3340	3250	−2.6	3390	+1.5
Poisson ratio ν	0.37	0.35	−5.4	0.36	−2.7

3.5.3 Accounting for vibrational contributions: quasiharmonic approximation

The early work of Theodorou and Suter[6,15] employed local minima of the potential energy generated at prescribed density to predict the structure of a polymer glass and its elastic constants. These minima were completely devoid of thermal motion. It is certainly possible, with some additional computational effort, to take into account the effects of vibrational motion within the basin of the potential energy hypersurface surrounding the minimum. This can be done by invoking the quasiharmonic approximation (QHA) outlined in Section 3.5.1.2. An advantage of such an approach is that the density is not required as input. The system confined in its basin can be allowed to change its dimensions subject to an externally imposed stress tensor, which may simply be an isotropic tensor of −1 bar in the absence of external load. The elastic response to the imposition of external stresses can likewise be studied by the QHA.

A study of glassy atactic polystyrene (aPS) based on the QHA was undertaken by Lempesis et al.[16] The model system was a box formed by a single parent chain with chemical constitution CH_3-$[CH(C_6H_5)CH_2]_{79}$-$CH(C_6H_5)CH_3$ (80mer aPS) generated as a Bernoullian sequence of meso- and racemic dyads. A fully flexible united-atom model was used.[16] Two-hundred independent, uncorrelated inherent structures were generated by chain growth and energy minimization at an initial density of 0.97 g/cm^3. To study the system confined in the vicinity of one of these inherent structures, QHA was invoked in an isothermal isostress ensemble. An in-basin Gibbs energy was defined as

$$G = \mathcal{V}^{\text{inh}} + A^{\text{vib}} - \frac{1}{3}\text{tr}(\boldsymbol{\tau})\ V^{(0)} - V^{(0)}\,\boldsymbol{\tau} : \boldsymbol{\varepsilon}. \qquad (3.5.82)$$

As explained in Sections 3.5.1.2 and 3.5.1.3, $\mathcal{V}^{\text{inh}}$, A^{vib}, and the small-deformation strain tensor $\boldsymbol{\varepsilon}$ are functions of the spatial extent of the model system, i.e. of the cell matrix $\boldsymbol{h}$. A^{vib} is also a function of temperature. It is computed from the frequencies of the vibrational modes around the inherent structure, determined by diagonalization of the Hessian matrix of second derivatives of the potential energy with respect to mass-weighted atomic coordinates (compare Eqs. (3.1.27) to (3.1.29)). At given stress $\boldsymbol{\tau}$, the system will assume that $\boldsymbol{h}$ which minimizes G. The undeformed state, characterized by volume $V^{(0)}$, is taken as that corresponding to a stress tensor $\boldsymbol{\tau}^{(0)} = \text{diag}(-1\text{bar}, \ -1\text{bar}, \ -1\text{bar})$.

Figure 3.10 displays a specific volume versus temperature curve at $\boldsymbol{\tau}^{(0)}$, obtained by applying the QHA on five of the 200 aPS inherent structures. QHA results are based on both the quantum mechanical vibrational partition function, leading to Eq. (3.5.26) for A^{vib}, and its classical approximation obtained by assuming $\hbar\omega_l << k_B T$ (compare Eq. (3.1.49) and the following discussion). Results from these two implementations of the QHA do not differ substantially. Temperatures in the considered range are high in relation to the Debye temperature of glassy aPS, so the classical partition function constitutes a reasonable approximation to the quantum mechanical one. Volumes predicted by the QHA are in good agreement with corresponding volumes from molecular dynamics (MD) simulation (see Chapter 4), which incorporate anharmonicities in the vibrational motion and allow for transitions from a basin into neighbouring basins of the potential energy hypersurface. This indicates that the assumptions underlying the QHA are reasonable. Furthermore, results from all simulations are in good agreement with experimental specific volumes. In the melt range, this reflects that the force field has been developed so as to provide a reasonable representation of real aPS. In the glassy range, the agreement of simulation with experiment is partly fortuitous. The cooling rate employed in MD (0.5 K/ns) is many orders of magnitude higher than typical experimental cooling rates and should cause the glass temperature T_g to appear higher in simulation than in experiment. On the other hand, the molar mass of the simulated aPS is small relative to that studied experimentally, causing a decrease in T_g. As a consequence, the glass temperature estimated from simulation, $T_g = 380\text{K}$, appears only slightly higher than the experimental $T_g = 373\text{K}$.

Each one of the 200 uncorrelated inherent structures of aPS was subjected to stress-controlled uniaxial tension experiments, keeping the normal stress at -1 bar in the y and z directions and applying tensile or compressive stress in the x direction up to ± 30 MPa. At each $\boldsymbol{\tau}$ studied, the Gibbs energy of Eq. (3.5.82) was minimized with respect to the diagonal components of $\boldsymbol{h}$ and the strain tensor was determined via Eq. (3.5.49). The relationship between τ_{xx} and ε_{xx} turned out to be linear, to an excellent approximation. From its slope, averaged over the 200 structures, a Young's modulus of $E = 3.55$ GPa was estimated at 300 K. The Young's modulus reported for aPS at 300 K is in the range 3.2–3.7 GPa.[6] The uniaxial tension/compression computer experiments were repeated at 200, 250, and 350 K. Based on the results obtained, shown in Figure 3.11, the predicted temperature dependence of the Young's modulus was $\frac{dE}{dT} = -6\text{MPa K}^{-1}$, which compares favourably to the experimentally reported $\left.\frac{dE}{dT}\right|_{\text{exp}} = -4.48\text{MPa K}^{-1}$.

Another observable from the QHA-based uniaxial tension/compression computer experiments conducted by Lempesis et al.[16] was the Poisson ratio ν of the polymer glass, computed directly from the changes in lateral dimensions accompanying the deformation. At 300 K, the Poisson ratio

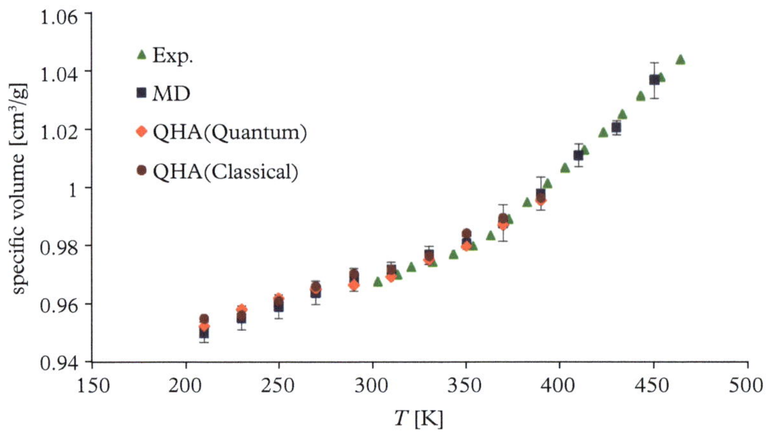

Figure 3.10 Volumetric behaviour of atactic polystyrene (aPS) at 1 bar as calculated by (a) MD computational cooling experiments in the nPT statistical ensemble; (b) QHA with quantum mechanical vibrational partition function; and (c) QHA with classical vibrational partition function. Results represent averages over calculations initiated at five uncorrelated inherent structures. Experimental data measured from high molar mass aPS are also shown.

Reprinted (adapted) from Lempesis, N., Vogiatzis, G.G., Boulougouris, G.C., van Breemen, L.C.A., Hütter, M., and Theodorou, D.N. *Molec. Phys.* **111**, 3430 (2013) by permission of Taylor & Francis Ltd, http://www.tandfonline.com

was estimated as $\nu = 0.35$, which coincides with the experimental value of $\nu_{\text{exp}} = 0.35$. The temperature dependence of the Poisson ratio was predicted to be $\frac{\mathrm{d}\nu}{\mathrm{d}T} = 0.00022\ \mathrm{K}^{-1}$, compared to $\left.\frac{\mathrm{d}\nu}{\mathrm{d}T}\right|_{\text{exp}} = 0.00015\ \mathrm{K}^{-1}$. A possible reason for this deviation is that QHA simulations do not allow the glassy polymer to depart from its original basin, while such departures do take place, especially at higher temperatures, close to T_g.

3.5.4 Mechanism of response to elastic deformation in a glassy polymer

To characterize local structure in amorphous polymers (but also in other solid materials) several methods and concepts can be used such as pair distribution functions (see Section 3.4), coordination numbers, symmetry coefficients,[24] Voronoi polyhedra,[25,26] and free volume[27–29] (see Chapter 6). For deformed solids, a more elegant way of characterizing local structure is by atomic-level stresses.[25,30–32]

In view of the atomic virial expression for stress, Eq. (3.5.24), one can define an atomic-level stress tensor for atom i via

$$\tau_{\alpha\beta,i} = -\frac{1}{V_i} k_{\mathrm{B}} T \delta_{\alpha\beta} + \frac{1}{2V_i} \left\langle \sum_{j \neq i} F_{ij,\alpha}\, r_{ij,\beta} \right\rangle, \tag{3.5.83}$$

where V_i is a characteristic volume assigned to atom i. The reader is reminded that $\boldsymbol{r}_{ij} = \boldsymbol{r}_j - \boldsymbol{r}_i$ (Eq. (3.5.2)), taken in the minimum image sense in a periodic model system, while $\boldsymbol{F}_{ij}$, the force

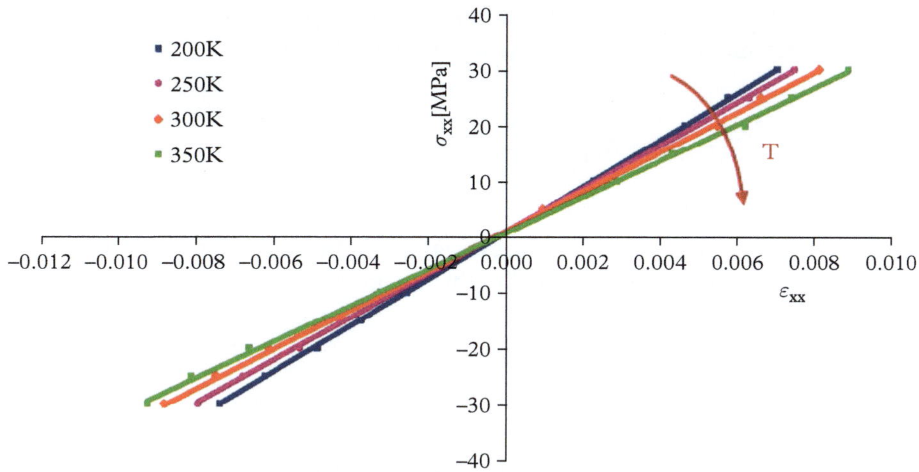

Figure 3.11 Stress–strain curves from stress-controlled uniaxial deformation QHA-based computer experiments at a pressure of 1 bar, at four different temperatures below the glass temperature of atactic polystyrene (aPS). The points represent averages over 200 uncorrelated inherent structures subjected to deformation. For each temperature, a linear fit through the simulation points is shown. The arrow indicates the direction of increasing temperature.

Reprinted (adapted) from Lempesis, N., Vogiatzis, G.G., Boulougouris, G.C., van Breemen, L.C.A., Hütter, M., and Theodorou, D.N. *Molec. Phys.* **111**, 3430 (2013) by permission of Taylor & Francis Ltd, http://www.tandfonline.com

on i due to j, is not in general central and contains contributions from stretching, bending, torsion, improper torsion, and nonbonded interactions. A general definition for $\boldsymbol{F}_{ij}$ in terms of the gradient of the total potential energy function $\mathcal{V}$ with respect to the interatomic distance vectors is given in Eq. (3.5.4).

The configurational contribution to Eq. (3.5.83) is an ensemble-averaged instantaneous *internal* atomic stress of the form

$$\sigma_{\alpha\beta,i} = \frac{1}{2V_i} \sum_{j \neq i} F_{ij,\alpha}\, r_{ij,\beta}. \tag{3.5.84}$$

If one works with static model microstates (inherent structures), Eq. (3.5.84) is invoked with all forces and distances evaluated at the inherent structure (compare Eq. (3.5.27)). Depending on the form of $\mathcal{V}$, special forms may be derived for $\boldsymbol{\sigma}$. For example, Theodorou and Suter[32] derive an expression for $\boldsymbol{\sigma}$ applicable to their static model microstates in detailed mechanical equilibrium with the constant bond length and bond angle model they invoke; this expression involves nonbonded interactions and torques due to the intrinsic torsional potentials.

With the definitions, Eq. (3.5.83) and Eq. (3.5.84), the total stress can be recovered by superposition of the atomic-level stresses:

$$\tau_{\alpha\beta} = \frac{1}{V} \sum_i V_i \, \tau_{\alpha\beta,i}, \quad \sigma_{\alpha\beta} = \frac{1}{V} \sum_i V_i \, \sigma_{\alpha\beta,i}. \tag{3.5.85}$$

Likewise, one can define a local stress tensor for any subregion Ω of the simulated system by superposition of the atomic-level stresses contained in the subregion

$$\tau_{\alpha\beta,\Omega} = \frac{\sum_{i\in\Omega} V_i \tau_{\alpha\beta,i}}{\sum_{i\in\Omega} V_i}, \quad \sigma_{\alpha\beta,\Omega} = \frac{\sum_{i\in\Omega} V_i \, \sigma_{\alpha\beta,i}}{\sum_{i\in\Omega} V_i}. \tag{3.5.86}$$

The precise way in which atomic volumes V_i are assigned for the definition of the atomic-level stress tensor may differ from study to study. Usually, this assignment conforms to the volume balance condition

$$\sum_i V_i = V. \tag{3.5.87}$$

For example, Theodorou and Suter[32] assigned atomic volumes in proportion to the van der Waals volumes of atoms:

$$V_i = V \frac{r^3_{\mathrm{vdW},i}}{\sum_i r^3_{\mathrm{vdW},i}}. \tag{3.5.88}$$

By this definition, all atoms belonging to the same species have the same V_i. A more recent practice, that is perhaps more representative, is to equate V_i with the Voronoi volume of atom i, i.e. the volume of the locus of points in 3D space which lie closer to the centre of i than to the centre of any other atom in the system. This volume can be determined readily by Voronoi tessellation of the model system.[33]

Egami and Vitek define two invariants for the atomic-level stress tensor of static model microstates.[31] One is the *atomic-level hydrostatic pressure*,

$$p_i = \frac{1}{3}\mathrm{tr}(\boldsymbol{\sigma}_i), \tag{3.5.89}$$

and a second is the *atomic von Mises shear stress*,

$$\tau_i^2 = \frac{1}{2}\mathrm{tr}[(\boldsymbol{\sigma}_i - p_i\boldsymbol{I})^2]. \tag{3.5.90}$$

The term *pressure* is actually a misnomer for p_i. It is really an atomic-level hydrostatic tension or negative pressure. p_i is sensitive to the local density. A high, positive value of p_i is associated with a high coordination number around atom i and with a lower than average atomic density. A low, negative value of p_i is associated with a low coordination number around atom i and with a higher than average local atomic density. The quantity τ_i reflects the degree of asymmetry of the local environment around atom i.

Atomic-level stresses for bonded systems have been employed by Theodorou and Suter[32] to probe local structure and explore the microscopic mechanisms of small-strain elastic deformation

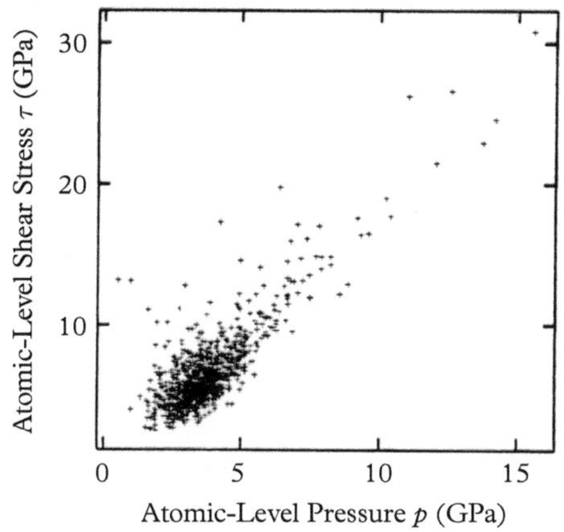

Figure 3.12 Correlation between atomic-level pressure and atomic-level shear stress (carbon atoms, eight undeformed structures) in glassy atactic PP at −40°C.

Reprinted (adapted) with permission from Theodorou, D.N., and Suter, U.W. *Macromolecules* **19**, 379 (1986). Copyright 1986 American Chemical Society.

in their collection of undeformed and deformed static model microstates of atactic polypropylene, discussed in Sections 3.4 and 3.5.2. In general, atomic-level stresses are much larger in magnitude than the total stress. While the average diagonal element of the internal stress tensor, $\frac{1}{3}\text{tr}(\boldsymbol{\sigma})$, over the collection of static model microstates was 82.2 MPa, a typical atomic stress was at least thirty times larger, the tails of the p_i and τ_i distributions extending to 20 GPa. In other words, there is a strong compensation effect in the summation of atomic-level stresses to the overall stress, Eq. (3.5.83). Skeletal carbon atoms (C) were found to be under tension ($p_i > 0$) with the p_i distribution skewed to the right, whereas substituents (hydrogens H and methyl groups CH_3) were under compression ($p_i < 0$), with the p_i distributions skewed to the left. The reason for this lies in the bonded forces experienced by carbon atoms. Figure 3.12 shows the correlation of the p_i and τ_i values experienced by carbon atoms in glassy atactic polypropylene, with each point corresponding to a single carbon atom in the undeformed state. A strong positive correlation is evident from the figure, since large values of p_i are accompanied by large values of τ_i and vice versa. Similar correlations have been computed for the other two species (H and CH_3).

It is interesting to analyse the changes upon deformation:

$$\begin{aligned}\Delta p_i &= p_i^{\text{deformed}} - p_i^{\text{undeformed}} \\ \Delta \tau_i &= \tau_i^{\text{deformed}} - \tau_i^{\text{undeformed}}.\end{aligned} \qquad (3.5.91)$$

These changes in a typical static model microstructure of glassy atactic polypropylene subjected to shear are shown in Figure 3.13. The figure reveals a pronounced positive correlation between Δp_i and $\Delta \tau_i$, implying the occurrence of large changes in local density upon elastic deformation under retention of the shape of the atomic stress ellipsoids. A remarkable feature of the pattern displayed

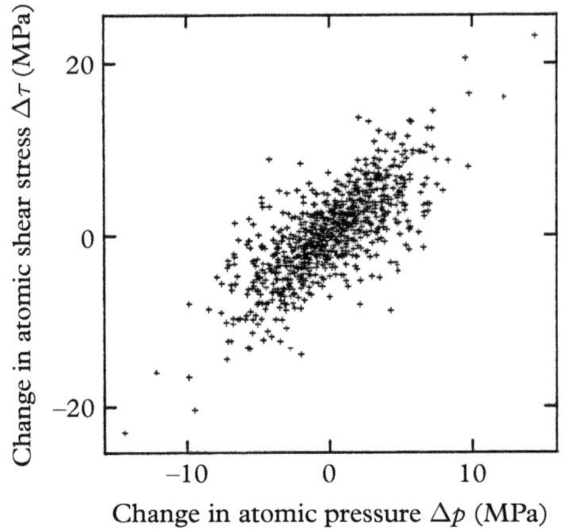

Figure 3.13 Changes in the atomic-level hydrostatic pressure p, plotted against changes in the atomic-level shear stress τ, for carbon atoms in a typical model structure of atactic PP at −40°C, subjected to shear at $\varepsilon = \pm 0.001$. The plotted points are from six shear deformations on a single static microstate.

Reprinted (adapted) with permission from Theodorou, D.N., and Suter, U.W. *Macromolecules* **19**, 379 (1986). Copyright 1986 American Chemical Society.

in Figure 3.13 is its almost perfect central symmetry. This indicates that, within the range of deformations considered in the computational experiments, for a *normal* structure opposite macroscopic deformations induce exactly opposite microscopic changes in local atomic environment. That is, the displacement vectors of all atoms are strictly antiparallel, and of very similar magnitude, for such pairs of deformations, although deviations from affine behaviour are strong. Some of the static model microstructures studied in Ref. 32 were *abnormal* in that they yielded negative shear moduli in some shearing modes and a nonpositive definite $\boldsymbol{C}$ matrix. For such structures, the correlation plots of the type of Figure 3.13 displayed changes in atomic-level stresses that were much higher (by at least a factor of 5) that in a *normal* structure and were not symmetric, especially near their fringes. Local mechanical stability is clearly related to the magnitude and symmetry of the changes in atomic-level stresses upon deformation.

Additional information on the atomic-level mechanism of deformation is provided by the inhomogeneous displacement vectors of atoms, defined as

$$\Delta \boldsymbol{r}_i = \boldsymbol{r}_i^{\text{deformed}} - \boldsymbol{r}_i^{\text{affine}}, \tag{3.5.92}$$

where $\boldsymbol{r}_i^{\text{deformed}}$ is the position atom i has in the deformed state and $\boldsymbol{r}_i^{\text{affine}}$ is the position it would have if the entire contents of the model box were subjected to a hypothetical affine (homogeneous) deformation characterized by the same macroscopic strain tensor and leaving the centre of mass of the box invariant. For the calculation of affinely transformed positions, see Eqs. (3.5.72c), (3.5.73c), (3.5.74c).

The distribution of inhomogeneous displacement lengths $|\Delta \boldsymbol{r}_i|$ for all hydrogen atoms upon shear, with shear strain ± 0.001, is shown in Figure 3.14. For comparison, the average affine displacement length for the same deformation is 4.54×10^{-3}Å. We see that $|\Delta \boldsymbol{r}_i|$ values are entirely

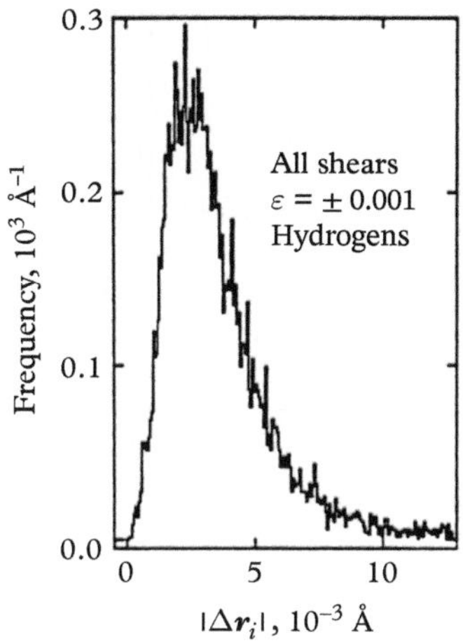

Figure 3.14 Distribution of inhomogeneous displacement lengths for all hydrogens upon shear with shear strain $\varepsilon = \pm 0.001$ (eight static model microstates of glassy atactic polypropylene at $-40°$C, six shear experiments per microstate).

comparable to the expected mean affine displacement; i.e. there is a strong nonaffine component to deformation, imposed by the stiff bond lengths and bond angles along polymer chains.

Figure 3.15 displays the directional correlation between nonaffine displacement vectors of different hydrogen atoms as a function of distance, in the undeformed state, of these atoms.[32] For a definition of the directional correlation function $S(r)$ between two vectors see Eq. (3.4.5). Figure 3.15 provides direct evidence on the range of cooperativity of atomic motions upon small-strain deformation. Both the total $S(r)$ and its strictly intermolecular part are shown. We see strong directional correlation (tendency for parallelity) in the nonaffine displacements of neighbouring atoms. This correlation dies out quasiexponentially with increasing distance with a range of approximately 7 Å. The intermolecular $S(r)$ is a measure of how deformation hops from chain to chain in the bulk. It, too, displays some correlation at short distances, which is considerably weaker than the total. By studying $S(r)$ as a function of the number of skeletal bonds intervening between the moving atoms, it was established that elastic deformation in the polymer glass is accompanied by coordinated displacement of chain segments approximately ten bonds long.

3.6 Quasistatic modelling of plastic deformation

In Sections 3.4 and 3.5 we showed how the static model microstates and atomic-level stresses can be used to study the response of a polymeric glass to small elastic deformations. The same approach has been invoked to investigate the kinematics of plastic deformation of a glassy polymer at large strains

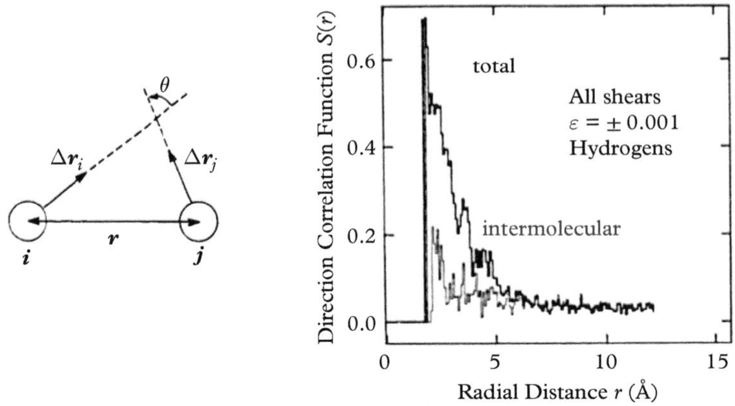

Figure 3.15 Directional correlation of inhomogeneous atom displacements as a function of interatomic distance in the bulk, for hydrogen atoms in all shear deformations studied on glassy atactic polypropylene at −40°C (eight static model microstates, six shear experiments per microstate, $\varepsilon = \pm 0.001$).

Reprinted (adapted) with permission from Theodorou, D.N., and Suter, U.W. *Macromolecules* **19**, 379 (1986). Copyright 1986 American Chemical Society.

(for example, 20% strain) and analyse the micro-structural changes accompanying the stress–strain behaviour. It has been followed by Mott et al.[34] and Argon et al.[35] in the case of glassy atactic polypropylene and revealed several important aspects which we will discuss in this section. Atomistic and mesoscale approaches to plastic deformation in glassy polymers have also been presented by Clarke and Brown[36] and Shenogin and Ozisik.[37] In a subsequent work, Lee et al.[38] reported direct measurement of molecular mobility in actively deformed polymer glasses.

3.6.1 Stress–strain response of a single static microstate

As discussed in Section 3.5.2.1, deformation of model polymer glasses can be introduced by imposing small strain steps on the continuation vectors of the periodic box, which changes the position of the parent chain(s) with respect to the positions of the image chains. The perturbation of the relative molecular positions shifts the system out of the undeformed local energy minimum. Reminimizing the energy of the system causes the polymer segments to seek a new orientation and conformation. Repeatedly straining the box and minimizing the energy can be used to simulate large deformations, in small incremental steps. For a given deformation simulation, a series of related configurations is therefore obtained, all in detailed mechanical equilibrium, the configuration being a function of the dimensions of the periodic box. In this picture, the static model employed has again no provision for the kinetic time-dependent response to deformation. Segmental motion is activated only by the deformation-induced energy increase when the periodic box is deformed. In addition, time does not enter into the description of the system; thus, the corresponding strain rate is not defined. Based on such a consideration, Mott et al.[34] simulated plastic behaviour in two types of deformation, both of them volume-preserving: axial extension and pure shear. For axial extension, the increment of strain was an extension in the x, y, or z direction with concomitant contractions in the transverse directions, such that the volume was held constant. If the y direction was extended, the incremental strain step $(\mathrm{d}\boldsymbol{\varepsilon})^{\mathrm{SYS}}$ on the periodic box was (in tensor form)

$$(\mathrm{d}\boldsymbol{\varepsilon})^{\mathrm{SYS}} = \begin{bmatrix} -1 & 0 & 0 \\ 0 & 2 & 0 \\ 0 & 0 & -1 \end{bmatrix} \times 10^{-3}. \tag{3.6.1}$$

For pure shear, the increment of strain $(d\boldsymbol{\varepsilon})^{\mathrm{SYS}}$ was imposed on the x–y, x–z, or y–z plane and was designed to keep the periodic box orthogonal. If the distortion was chosen to be in the x–y plane, the imposed strain increment was (in tensor form)

$$(\mathrm{d}\boldsymbol{\varepsilon})^{\mathrm{SYS}} = \begin{bmatrix} 1 & 0 & 0 \\ 0 & -1 & 0 \\ 0 & 0 & 0 \end{bmatrix} \times 10^{-3}, \tag{3.6.2}$$

where the z direction displacements remain unchanged.

Atomic-level strain increments were also calculated by Delaunay tessellation of the set of atomic positions before and after the imposition of each system strain increment.[34] Of an atomic-level strain increment $(\mathrm{d}\boldsymbol{\varepsilon})$, two scalar invariants are of special interest: the local dilatation increment $(\mathrm{d}\varepsilon)$ and the work equivalent strain increment $(\mathrm{d}\varepsilon)_{\mathrm{eq}}$, defined according to

$$(\mathrm{d}\varepsilon) = \mathrm{tr}[(\mathrm{d}\boldsymbol{\varepsilon})] = (\mathrm{d}\varepsilon)_{11} + (\mathrm{d}\varepsilon)_{22} + (\mathrm{d}\varepsilon)_{33} \tag{3.6.3}$$

and

$$\begin{aligned}(\mathrm{d}\varepsilon)_{\mathrm{eq}} &= \left\{\frac{2}{3}\mathrm{tr}\left[(\mathrm{d}\boldsymbol{\varepsilon}) - \frac{1}{3}(\mathrm{d}\varepsilon)\boldsymbol{I}\right]^2\right\}^{1/2} \\ &= \left\{\frac{2}{9}\left[((\mathrm{d}\varepsilon)_{22} - (\mathrm{d}\varepsilon)_{11})^2 + ((\mathrm{d}\varepsilon)_{33} - (\mathrm{d}\varepsilon)_{22})^2 + ((\mathrm{d}\varepsilon)_{33} - (\mathrm{d}\varepsilon)_{11})^2\right]\right. \\ &\quad \left. + \frac{4}{3}((\mathrm{d}\varepsilon)_{12}^2 + (\mathrm{d}\varepsilon)_{23}^2 + (\mathrm{d}\varepsilon)_{13}^2)\right\}^{1/2},\end{aligned} \tag{3.6.4}$$

respectively, where $\boldsymbol{I}$ is the identity matrix. Another form of the second invariant of the strain increment, the work equivalent shear distortion, $(\mathrm{d}\gamma)$, was introduced by

$$(\mathrm{d}\gamma) = \sqrt{3}(\mathrm{d}\varepsilon)_{\mathrm{eq}}. \tag{3.6.5}$$

The strain increment tensor of the entire simulation cell was defined as the volume-weighted average of all the atomic-level strain increment tensors $(\mathrm{d}\boldsymbol{\varepsilon}_i)$ of atoms i:

$$(\mathrm{d}\boldsymbol{\varepsilon})^{\mathrm{SYS}} = \frac{1}{V}\sum_i V_i(\mathrm{d}\boldsymbol{\varepsilon}_i) \tag{3.6.6}$$

with atomic volumes V_i being assigned via Voronoi tessellation. $\varepsilon^{\mathrm{SYS}}$ equals the strain imposed on the system.

The system stress tensor $\boldsymbol{\sigma}^{\mathrm{SYS}}$, on the other hand, was calculated as the weighted volume average of the atomic-level stress tensors (see Eq. (3.5.85))

$$\boldsymbol{\sigma}^{\mathrm{SYS}} = \frac{1}{V}\sum_i V_i \boldsymbol{\sigma}_i \tag{3.6.7}$$

Two invariants of the atomic-level stress tensor were considered: the mean normal stress $\boldsymbol{\sigma}$ defined by Eq. (3.6.8), and the von Mises equivalent tensile stress defined by Eq. (3.6.9). Note the similarity of Eqs. (3.6.8) and (3.6.9) with (3.5.89) and (3.5.90), respectively:

$$\sigma = \frac{1}{3}\mathrm{tr}[\boldsymbol{\sigma}] = \frac{1}{3}(\sigma_{11} + \sigma_{22} + \sigma_{33}) = -p \tag{3.6.8}$$

$$\sigma_{\mathrm{eq}} = \tau\sqrt{3} = \left\{\frac{3}{2}\mathrm{tr}[\boldsymbol{\sigma} - \sigma\boldsymbol{I}]^2\right\}^{1/2} =$$

$$= \left\{\frac{1}{2}[(\sigma_{22} - \sigma_{11})^2 + (\sigma_{33} - \sigma_{22})^2 + (\sigma_{33} - \sigma_{11})^2] + 3(\sigma_{12}^2 + \sigma_{23}^2 + \sigma_{13}^2)\right\}^{1/2}. \tag{3.6.9}$$

Figure 3.16 shows a representative total stress–total integrated strain plot for a single structure subjected to uniaxial extension along the y-direction. The (total) strain is the integral of all the equivalent strain increments, i.e. the integral of Eq. (3.6.3). The elastic loading line is the slope for an elastically responding system, as given by

$$(\mathrm{d}\sigma_{\alpha\beta}) = 2\mu(\mathrm{d}\varepsilon_{\alpha\beta}) + \lambda(\mathrm{d}\varepsilon)\delta_{\alpha\beta}, \tag{3.6.10}$$

with the elastic constants μ and λ determined from small-scale elastic deformations (see Table 3.3). Several prominent features that are the direct result of the very small size of the system can be seen, giving Figure 3.16 a significantly different appearance in comparison to typical stress–strain curves seen in macroscopic experiments:

a) The unstrained system does not start out at zero stress, implying that the initial structure is under substantial stress, which is not surprising since the atomic-level stresses are typically in the range 2–3 GPa and are summed over only a few hundreds of atoms. The nonzero stress in the initial system is an atomic-level stress *noise*. In real glassy materials, any microscopic region would be under a stress resulting from the misfit of the region in the surrounding elastic material in which it is embedded, as a consequence of the disorder of the glassy structure.

b) No elastic-to-plastic transition can be seen in the figure. At first inspection, this implies that the usual continuum concepts of elasticity, yield, and plasticity are macroscopic notions that do not apply to small amorphous systems, which act more like discrete structures rather than as continua.

c) At the beginning of the deformations, the figure shows a long and nearly constant stress response. The initial undeformed material apparently has little resistance to increments of

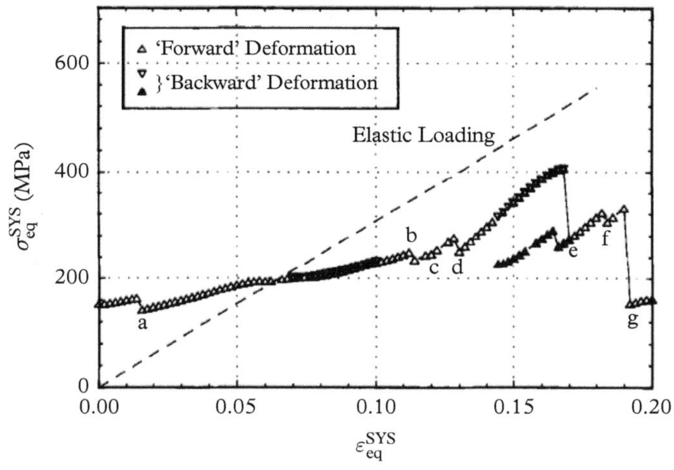

Figure 3.16 System stress–strain behaviour for a single representative structure (glassy atactic PP at −40°C). The structure was deformed in axial extension, lengthening the *y*-direction. The elastic loading line is obtained from Eq. (3.6.10), using the elastic constants from Table 3.3.

strain, up to about 10%. During this range, the material appears to become mechanically polarized by the imposed deformation, and soft parts are *squeezed out.*

d) Following this early mechanical polarization, the response of the stress takes a jagged sawtooth course which shows a smooth increase in the stress with increasing applied strain, nearly parallel to the expected loading line, followed by sharp downward jumps, whereupon the smooth increase in stress resumes. Several such sharp drops in the stress–strain curve can be seen in the figure, labelled a–g. It is also found that, as the system strain accumulates, the size of the stress drops increases.

Each sharp change in the sawtooth stress–strain curve corresponds to a transition into a new basin on the energy hypersurface, as it is distorted by deformation. This is why, after such a sharp change, the curve does not retrace itself upon reversal of the deformation.

Figure 3.17 gives the corresponding von Mises shear stress (τ^{SYS})–work-equivalent shear strain (γ^{SYS}) behaviour of the same atomistic system deformed by pure shear. In this case, the quasiperiodic two-stage sawtooth character of the work equivalent shear stress response prevails without much of an anelastic mechanical polarization type of response as seen in Figure 3.16. There are again stretches of reversible loading in the shear stress response, building up elastic energy, followed by irreversible unloading events, when the system rearranges. The simulation again shows the reversible stress responses to be roughly parallel to the shear modulus. Note that some of the stress jumps are upwards, where the plastic rearrangement increases the shear stress felt in the system.

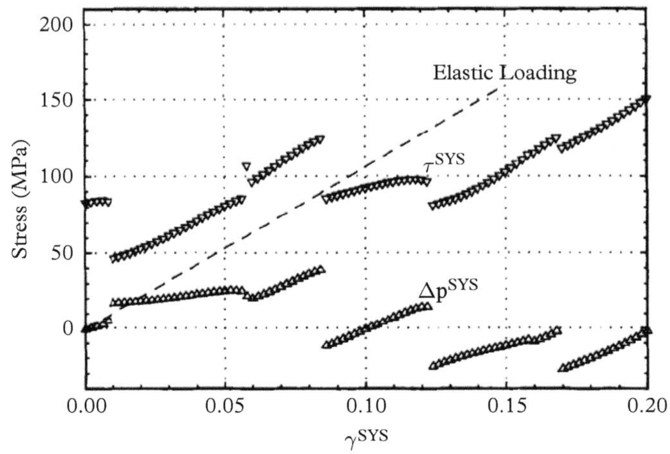

Figure 3.17 Stress–strain curve characteristic of a single structure (glassy atactic PP at −40°C) subjected to pure shear.

Reprinted (adapted) from Mott, P.H., Argon, A.S., and Suter, U.W. *Philos. Mag.* A **67**, 931 (1993) by permission of Taylor & Francis Ltd, https://www.tandfonline.com

All the results shown in Figures 3.16–3.17 have been obtained by Mott et al.[34] with a simulation box that had initially the shape of a cube with edge length equal to 1.815 nm. Additional simulations with two larger systems (one with a 2.615-nm edge length and another with a 3.396-nm edge length) showed that, although the stress of the initial structure was closer to the initial stress seen in the average over the entire collection of static microstates (especially for the larger box), the stress jumps were of the same magnitude as the stress jumps in the small system. This means that the plastic strain (per unit volume) dissipated during these elementary processes is independent of the system size and of the molecular weight, at least for the range of atomistic sizes covered in the simulations. The same was seen for the pressure increase: it was about 25 MPa and comparable for all system sizes deformed in pure shear.

3.6.2 Stress–strain behaviour averaged over a collection of microstates

The results for the stress tensors $\boldsymbol{\sigma}^{SYS}$ and stress invariants σ_{eq}^{SYS} and Δp^{SYS} averaged over the entire collection of static model microstates in the two types of deformation considered by Mott et al.[34] (axial extension and pure shear) are shown in Figure 3.18a. Three static microstates were considered, and each of them was subjected to three computer *experiments* in each of the two considered types of deformation, by exchanging the axes. Thus, there were nine independent simulations at each strain level for each type of deformation. To make the comparison easier with the stress–strain response under axial extension, the pure shear results for the von Mises shear stress τ^{SYS} and work equivalent shear strain γ^{SYS} were converted to tensile representations σ_{eq}^{SYS} and ϵ_{eq}^{SYS} using Eqs. (3.6.5) and (3.6.9).

Figure 3.18a shows the von Mises equivalent tensile stress σ_{eq}^{SYS} developing in the simulated system in response to deformation. The collection of microstates behaves similarly in both shear and axial extension. The initial stress for the collection is much closer to 0 than for the individual microstates (compare Figure 3.17), because the sample size is larger by a factor of 9, reducing statistical fluctuations. At the early stages of deformation, the tensile stress follows the elastic loading line reasonably well. Deviation from the initial linear behaviour defines the yield stress, which occurs in the range 5–7% strain, corresponding to roughly 140 MPa von Mises tensile stress.

The behaviour of pressure P^{SYS} for the collection of static microstates is shown in Figure 3.18b. In both types of deformation there is substantial increase in pressure, although the pressure rise in the axial extension experiments is significantly larger than in the shear experiments. The accumulation of the pressure continues to around 12% strain and then remains relatively constant; the steady-state pressures are about 25 and 100 MPa for the shear and axial extension experiments, respectively. If these systems were embedded in an elastic background, the build-up of pressure would be relieved somewhat by the elastic compliance of the surrounding material; this would amount to expansions of the system by 0.7 and 2.9%, respectively. The build-up of pressure in these constant-volume simulations must be attributed to an elastic dilatancy that is trapped by the segmental repacking that occurs during plastic events. When complex segmental exchanges are to occur, extra room must be made available to permit molecular segments to move by each other. The cause of the different behaviour in axial extension and pure shear is not clear.

While dilatation accompanying plastic rearrangement is a frequent occurrence in glassy polymers, it is by no means universal. In a similar study to that of Mott et al.[34] performed on polycarbonate, Hutnik et al. found the opposite response, that is negative dilatation, in the early stages of plastic deformation.[39]

3.6.3 The plastic relaxation events

From the evidence presented in Sections 3.6.1 and 3.6.2, it is clear that the stress drops result from release of built-up elastic strain energy, converting this energy to plastic rearrangement of the structure. To develop a better understanding of the distribution of the intensities of plastic rearrangements, Mott et al.[34] were guided by the simulation results which seem to suggest that the plastic strain developed is equal to the elastic strain unloaded during the event, so the net accumulated strain during a plastic event is 0. Hence

$$\begin{aligned}\Delta\gamma &= \Delta\gamma^E + \Delta\gamma^P = 0\\ \Delta\varepsilon &= \Delta\varepsilon^E + \Delta\varepsilon^P = 0,\end{aligned} \tag{3.6.11}$$

where the superscripts E and P indicate the elastic and plastic components, respectively, of the deformation. We now define the transformation shear strains and their accompanying transformation dilatations $\Delta\varepsilon^T$ as

$$\begin{aligned}\Delta\gamma^P &= \Delta\gamma^T\\ \Delta\varepsilon^P &= \Delta\varepsilon^T,\end{aligned} \tag{3.6.12}$$

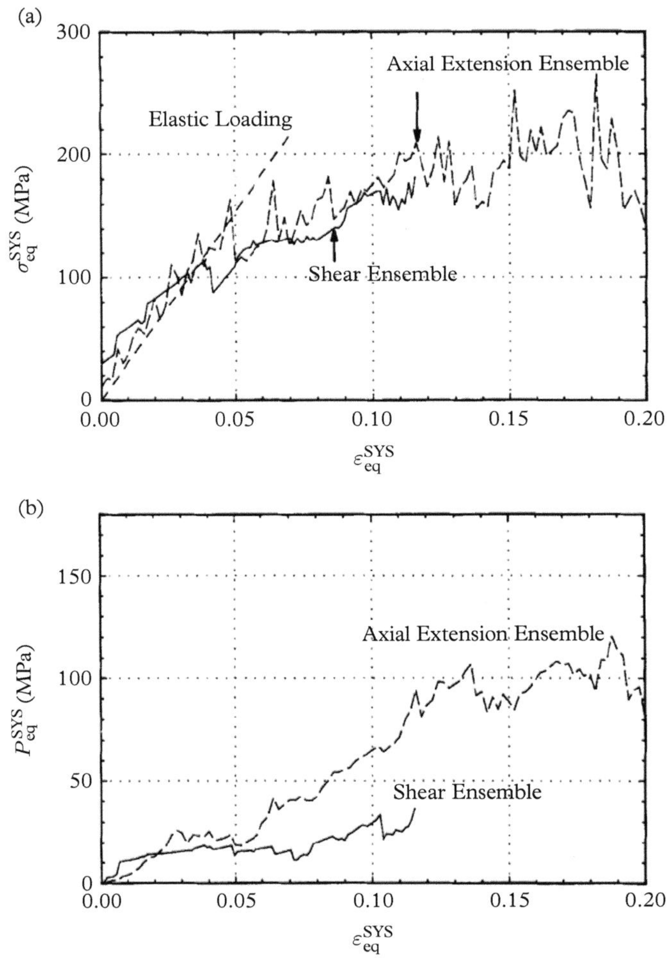

Figure 3.18 (a) Axial extension and pure shear ensemble-averaged von Mises tensile stress against system work equivalent strain. For a more convenient comparison, the von Mises shear stress τ for the collection of sheared static model microstates was converted to the von Mises tensile stress σ_{eq} by Eq. (3.6.9), while the work equivalent shear strain γ was converted to the work equivalent tensile strain ε_{eq} by Eq. (3.6.5).
(b) Ensemble-averaged change in the system pressure as a function of system work equivalent strain. All results refer to glassy atactic PP at −40°C.

where

$$\Delta\gamma^E = \frac{\Delta\tau}{\mu}$$
$$\Delta\varepsilon^P = \frac{\Delta\sigma}{B} = -\frac{\Delta p}{B}. \qquad (3.6.13)$$

These imply that

$$\Delta\gamma^T = -\frac{\Delta\tau}{\mu}$$
$$\Delta\varepsilon^T = \frac{\Delta p}{B}. \tag{3.6.14}$$

In the above expressions, τ is the von Misses shear (or tensile) stress, equal to $\sigma_{eq}/\sqrt{(3)}$. According to Eqs. (3.6.14) the plastic transformation shear strain is defined to be positive when it is associated with a drop in the von Mises shear (or tensile) stress and negative when the stress rises. Similarly, the transformation dilatations are treated as positive when the pressure rises, and negative when the pressure drops.

The sudden stress changes in pure shear can be evaluated on the basis of the above equations and the resulting distributions of the magnitude of both the transformation shear strain $\Delta\gamma^T$ and transformation dilatation $\Delta\varepsilon^T = \frac{\Delta p}{B}$, plotted in Figures 3.19a and 3.19b, respectively. Examination of the associated transformation dilatations in Figure 3.19b shows that they are symmetrically distributed around 0, indicating that the plastic events do not continually dilate the structure.

3.6.4 Comparison of simulation results for plastic deformation with experiments

The simulation results for the plastic deformation of glassy polymers with different systems indicate that the entire system modelled is affected by the relaxation and that there is no unaffected exterior region. From this, one concludes that the size of a coherently relaxing plastic unit is still larger than the largest system simulated by Mott et al.[34] (a cube with an edge length on the order of a few nanometers). Nevertheless, one can get a reliable estimate of the size of these coherent plastic relaxation units by comparing the simulation results with experimental measurements. Strain rate change experiments during plastic deformation under constant pressure, in particular, allow one to determine the so-called activation volume defined as

$$\Delta\nu^* = k_B T\left(\frac{\partial \ln \dot{\gamma}^P}{\partial\tau}\right)_\sigma = \Delta\gamma^T \Omega_c. \tag{3.6.15}$$

This gives the product of the transformation shear strain ($\Delta\gamma^T$) and the volume Ω_c in which it occurs. Argon and Bessonov[40] have made such measurements for a series of glassy polymers ranging from the very flexible chain materials of polystyrene and poly(methyl-methacrylate) to some very stiff chain polyimides, but not on atactic polypropylene. If it is assumed then that the transformation shear strain determined for PP is typical of all glassy polymers, one can get an estimate of the size of the coherent plastic relaxation cluster for all polymers. Such an idea is supported by additional simulations carried out on polycarbonate which indicated that the magnitude of this shear strain is practically the same as that in PP, even though the polycarbonate molecule is stiffer than PP. Then, on the basis of Eq. (3.6.15), estimates of coherent relaxation volumes Ω_c are indeed very large, in the range of $(10\ \text{nm})^3$ or even larger.

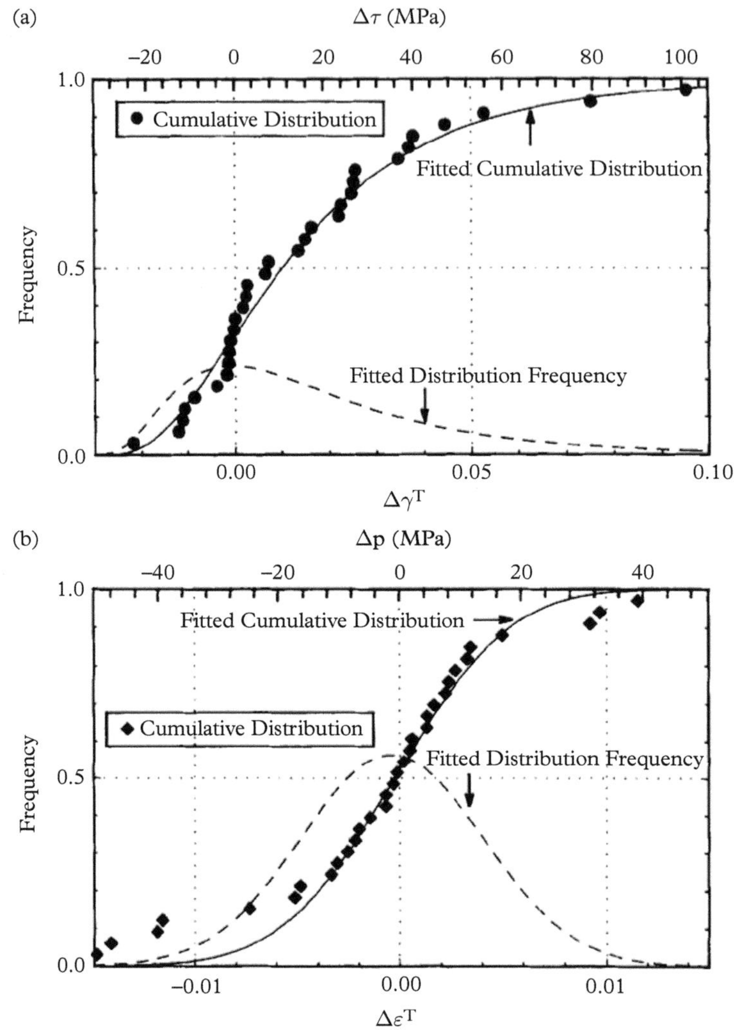

Figure 3.19 (a) Distribution of transformation shear strains in the pure shear experiments (the cumulative distribution of the data taken from the individual stress–strain curves). A displaced log-normal cumulative distribution has been fitted to the data points and the distribution frequency has been found by taking the derivative of the fitted log-normal functions. (b) Distribution of transformation dilatations (or pressure changes) in the shear experiments fitted to a Gaussian distribution function. All results refer to glassy atactic PP at −40°C.

Reprinted (adapted) from Mott, P.H., Argon, A.S., and Suter, U.W. *Philos. Mag.* A **67**, 931 (1993) by permission of Taylor & Francis Ltd, https://www.tandfonline.com

Clearly, the segmental degrees of freedom in clusters of these magnitudes are very many; moreover, the kinematical makeup of the deformations in every cluster must be different. The fact that the stress jumps are very sharp and well-defined indicates, however, that most of these large numbers of degrees of freedom must be linked, leaving only those that collectively result in shear along a well-defined activation path of the system over the saddle point.

Mott et al.[34] and Argon[41] have indicated that the jerky features of the stress–strain curves shown in Figures 3.16 and 3.17 can be fully explained as a series of mechanically activated shear transformations as they would manifest themselves at 0 K, without any of the effects of thermal smoothing, where the repeated sawtooth nature of the response is a direct result of terminating stable shear at the inflection point of a roughly periodic fundamental shear resistance curve. In this sense, the jerky plastic response of the simulation is characteristic of athermal behaviour at 0 K, even though the actual elastic properties of the material have been those at −40°C by virtue of the structural energy minimization process, targeted to achieve a material density appropriate for this temperature. If thermal motion had been specifically included (without, of course, fixing the material density at −40°C), then the plastic response would have been smoother without much of the sawtooth behaviour, as thermal penetration of activation energy barriers would become possible at stress levels well below those corresponding to the 0 K inflection point.

Appendix

We will prove the fundamental expression, Eq. (3.5.33), in the main text. From Eqs. (3.5.23) and (3.5.32) we have

$$\begin{aligned}
\mathrm{d}A &= V\mathcal{F}^{-1}\cdot\boldsymbol{\tau}^{\mathrm{T}} : \mathrm{d}\mathcal{F}^{\mathrm{T}} \\
&= V(\mathcal{F}^{-1}\cdot\boldsymbol{\tau}) : \mathrm{d}\mathcal{F}^{\mathrm{T}} \\
&= \frac{1}{2}[V(\mathcal{F}^{-1}\cdot\boldsymbol{\tau}) : \mathrm{d}\mathcal{F}^{\mathrm{T}} + V(\mathcal{F}^{-1}\cdot\boldsymbol{\tau}^{\mathrm{T}}) : \mathrm{d}\mathcal{F}^{\mathrm{T}}] \\
&= \frac{V}{2}\left[(\mathcal{F}^{-1}\cdot\boldsymbol{\tau}) : \mathrm{d}\mathcal{F}^{\mathrm{T}} + \left(\boldsymbol{\tau}\cdot(\mathcal{F}^{-1})^{\mathrm{T}}\right)^{\mathrm{T}} : \mathrm{d}\mathcal{F}^{\mathrm{T}}\right],
\end{aligned}$$

where we have invoked the symmetry of $\boldsymbol{\tau}$. The latter expression can be rewritten as

$$\begin{aligned}
\mathrm{d}A &= \frac{V}{2}\left[(\mathcal{F}^{-1}\cdot\boldsymbol{\tau}) : \mathrm{d}\mathcal{F}^{\mathrm{T}} + \mathrm{d}\mathcal{F} : \left(\boldsymbol{\tau}\cdot(\mathcal{F}^{\mathrm{T}})^{-1}\right)\right] \\
&= \frac{V}{2}\left[\left\{\left(\mathcal{F}^{-1}\cdot\boldsymbol{\tau}\cdot(\mathcal{F}^{\mathrm{T}})^{-1}\right)\cdot\mathcal{F}^{\mathrm{T}}\right\} : \mathrm{d}\mathcal{F}^{\mathrm{T}} + \mathrm{d}\mathcal{F} : \left\{\mathcal{F}\cdot\left(\mathcal{F}^{-1}\cdot\boldsymbol{\tau}\cdot(\mathcal{F}^{\mathrm{T}})^{-1}\right)\right\}\right],
\end{aligned} \qquad \text{(A.1)}$$

where the last step involved post- and premultiplication by the unit tensor, written as the product of the deformation gradient tensor times its inverse, or of the inverse transpose deformation gradient tensor times the transpose deformation gradient tensor.

If the stress tensor is symmetric, so is the tensor $\mathcal{F}^{-1} \cdot \tau \cdot (\mathcal{F}^{\mathrm{T}})^{-1}$. So, Eq. (A.1) can be rewritten as

$$
\begin{aligned}
\mathrm{d}A &= \frac{V}{2}\left[\left\{\left(\mathcal{F}^{-1} \cdot \tau \cdot (\mathcal{F}^{\mathrm{T}})^{-1}\right)^{\mathrm{T}} \cdot \mathcal{F}^{\mathrm{T}}\right\} : \mathrm{d}\mathcal{F}^{\mathrm{T}} + \mathrm{d}\mathcal{F} : \left\{\mathcal{F} \cdot \left(\mathcal{F}^{-1} \cdot \tau \cdot (\mathcal{F}^{\mathrm{T}})^{-1}\right)\right\}\right] \\
&= \frac{V}{2}\left[\left(\mathcal{F}^{-1} \cdot \tau \cdot (\mathcal{F}^{\mathrm{T}})^{-1}\right) : (\mathcal{F}^{\mathrm{T}} \cdot \mathrm{d}\mathcal{F}) + (\mathrm{d}\mathcal{F}^{\mathrm{T}} \cdot \mathcal{F}) : \left(\mathcal{F}^{-1} \cdot \tau \cdot (\mathcal{F}^{\mathrm{T}})^{-1}\right)^{\mathrm{T}}\right] \\
&= \frac{V}{2}\left[\left(\mathcal{F}^{-1} \cdot \tau \cdot (\mathcal{F}^{\mathrm{T}})^{-1}\right) : (\mathcal{F}^{\mathrm{T}} \cdot \mathrm{d}\mathcal{F}) + \left(\mathcal{F}^{-1} \cdot \tau \cdot (\mathcal{F}^{\mathrm{T}})^{-1}\right) : (\mathrm{d}\mathcal{F}^{\mathrm{T}} \cdot \mathcal{F})\right] \\
&= \frac{V}{2}\left(\mathcal{F}^{-1} \cdot \tau \cdot (\mathcal{F}^{\mathrm{T}})^{-1}\right) : \left[(\mathcal{F}^{\mathrm{T}} \cdot \mathrm{d}\mathcal{F}) + (\mathrm{d}\mathcal{F}^{\mathrm{T}} \cdot \mathcal{F})\right].
\end{aligned}
\tag{A.2}
$$

Combining Eq. (A.2) with Eq. (3.5.30) in the main text leads to the desired expression (Eq. (3.5.32) in the main text).

Problems

Problem 3.1: Prove Eq. (3.1.7) for the Jacobian of transformation from Cartesian to generalized coordinates in a system of linear chains.

Problem 3.2: Calculate the Jacobian of transformation from Cartesian to generalized coordinates for a branch point along a chain (see Figure 3.1).

Problem 3.3: Evaluate the matrix $(\boldsymbol{a}^{-1})_{\mathrm{hh}}$ for the case of a linear chain of $N+1$ segments of equal mass m, all bond lengths of which are constrained at length l_0.

Problem 3.4: Derive a recursive formula for the calculation of the following $N \times N$ tridiagonal matrix:

$$
\boldsymbol{T} = \begin{bmatrix}
a_1 & b_1 & 0 & 0 & \dots & & & 0 \\
b_1 & a_2 & b_2 & 0 & \dots & & & 0 \\
0 & b_2 & a_3 & b_3 & \dots & & & 0 \\
\cdot & & & & \dots & & & 0 \\
\cdot & & & & \dots & & & \\
\cdot & & & & & & & \\
0 & & & & \dots & b_{N-2} & a_{N-1} & b_{N-1} \\
0 & & & & \dots & 0 & b_{N-1} & a_N
\end{bmatrix}.
$$

Problem 3.5: Derive the Fixman potential for a united-atom model of a polypropylene chain, in which bond lengths are kept constant (see Figure 3.2).

Problem 3.6: Prove Eq. (3.5.7) for the change in potential energy in terms of interatomic forces and the changes in interatomic distance vectors.

Problem 3.7: Prove that $\dfrac{\partial \ln[\det(\boldsymbol{h})]}{\partial h_{\alpha\beta}} = h^{-1}_{\beta\alpha}$.

Problem 3.8: Prove Eq. (3.5.20) for the derivatives of the Helmholtz energy with respect to the elements of the deformation gradient tensor.

Problem 3.9: Investigate whether the stress tensor defined by Eq. (3.5.24) is symmetric.

References

1. Modell, M., and Reid, R.C. *Thermodynamics and Its Applications.* Prentice Hall, Englewood Cliffs, NJ (1983).
2. Stillinger, F.H., and Weber, T.A. *Phys. Rev. A* **28**, 2408 (1983).
3. Fixman, M. *Proc. Natl. Acad. Sci. USA* **71**, 3050 (1974).
4. Gō, N., and Scheraga, H.A. *Macromolecules* **9**, 535 (1976).
5. Weiner, J.H. *Statistical Mechanics of Elasticity.* John Wiley & Sons, New York (1983).
6. Theodorou, D.N., and Suter, U.W. *Macromolecules* **19**, 139 (1986).
7. Greenfield, M.L., and Theodorou, D.N. *Macromolecules* **31**, 7068 (1998).
8. Zervopoulou, E., Mavrantzas, V.G., and Theodorou, D.N. *J. Chem. Phys.* **115**, 2860 (2001).
9. Peristeras, L.D., Economou, I.G., and Theodorou, D.N., *Macromolecules* **38**, 386 (2005).
10. McQuarrie, D.A. *Statistical Mechanics.* Harper and Row, New York, 1976
11. Greenfield, M.L. *Molecular Modeling of Dilute Penetrant Gas Diffusion in a Glassy Polymer Using Multidimensional Transition-State Theory.* Ph.D. Thesis, University of California, Berkeley, 1996.
12. Antoniadis, S.J., Samara, C.T., and Theodorou, D.N. *Macromolecules* **31**, 7944 (1998).
13. Theodorou, D.N., Boone, T.D., Dodd, L.R., and Mansfield, K.F. *Macromolecul. Chem., Theory Simul.* **2**, 191 (1993).
14. Zhou, M. *Proc. Roy. Soc. Lond. A* **459**, 2347–2392 (2003).
15. Theodorou, D.N., and Suter, U.W. *Macromolecules* **18**, 1467 (1985).
16. Lempesis, N., Vogiatzis, G.G., Boulougouris, G.C., van Breemen, L.C.A., Hütter, M., and Theodorou, D.N. *Molec. Phys.* **111**, 3430 (2013).
17. Mathioudakis, I.G., Vogiatzis, G.G., and Theodorou, D.N. *Soft Matter* **12**, 7585 (2016).
18. Hillstrom, K. *Nonlinear Optimization Routines in AMDLIB,* Technical Memorandum No. 297. Argonne National Laboratory, Applied Mathematics Division, 1976.
19. Theodorou, D.N., and Suter, U.W., *J. Chem. Phys.* **82**, 955 (1985).
20. Tanner, R.I. *Engineering Rheology.* Oxford University Press, New York (1986).
21. Wikiuniversity notes on continuum mechanics-thermoelasticity, https://en.wikiversity.org/wiki/Continuum_mechanics/Thermoelasticity, accessed 29 Jan 2017.
22. Lustig, S. R., Shay, R. M., and Caruthers, J. M. *J. Rheol.* **40**, 69 (1996).
23. Parrinello M., and Rahman, A. *J. Chem. Phys.* **76**, 2662 (1982).
24. Ray, J.R. *Comp. Phys. Rep.* **8**, 109 (1988).
25. Egami, T., Maeda, K., and Vitek, V. *Philos. Mag.* **41**, 883 (1980).
26. Maeda, K., and Takeuchi, S. *Philos. Mag. A* **44**, 643 (1981).

27. Weber, T.A., and Stillinger, F.H. *J. Chem. Phys.* **74**, 4020 (1981).
28. Cohen, M.H., and Grest, G.S. *Phys. Rev. B* **20**, 1077 (1979).
29. Grest, G.S., and Cohen, M.H. *Adv. Chem. Phys.* **48**, 455 (1981).
30. Srolovitz, D., Vitek, V., and Egami, T. *Acta Metall.* **31**, 335 (1983).
31. Egami, T., and Vitek, V. *J. Non-Cryst. Solids* **61–62**, 499 (1984).
32. Theodorou, D.N., and Suter, U.W. *Macromolecules* **19**, 379 (1986).
33. Vogiatzis, G.G., and Theodorou, D.N. *Macromolecules* **47**, 387 (2014).
34. Mott, P.H., Argon, A.S., and Suter, U.W. *Philos. Mag. A* **67**, 931 (1993).
35. Argon, A.S., Bulatov, V.V., Mott, P.H., and Suter, U.W. *J. Rheol.* **39**, 377 (1995).
36. Clarke, J.H.R., and Brown, D. *Molecular Simulation* **3**, 27 (1989).
37. Shenogin, S., and Ozisik, R. *J. Polymer Science Part B: Polymer Physics* **43**, 994 (2005).
38. Lee, H.-N., Paeng, K., Swallen, S.F., and Ediger, M.D. *Science* **323**, 231 (2009).
39. Hutnik, M., Argon, A.S., and Suter, U.W. *Macromolecules* **26**, 1097 (1993).
40. Argon, A.S., and Bessonov, M.I. *Philos. Mag.* **35**, 917 (1977).
41. Argon, A.S. In Mughrabi, H. (ed.), *Materials Science and Technology*, vol. 6: *Plastic Deformation and Fracture of Materials*, p. 461. VCH, Weinheim, Germany (1993).

4

Molecular Dynamics

4.1 Integrating the equations of motion

4.1.1 Introduction

We return to the problem we started discussing in Chapter 1, namely the dynamics of a system consisting of a number of polymer molecules, each of which comprises a certain number of atoms, so that the total number of atoms is n in volume $\mathcal{V}$. In Chapter 1 we showed that a point $\boldsymbol{\Gamma}$ in the phase space of the system corresponds to a microstate which is defined by the $6n$-dimensional state vector $\boldsymbol{\Gamma} \equiv (\boldsymbol{q}, \boldsymbol{p}) \equiv (\boldsymbol{q}_1, \boldsymbol{q}_2, \ldots, \boldsymbol{q}_n; \boldsymbol{p}_1, \boldsymbol{p}_2, \ldots, \boldsymbol{p}_n)$ of the generalized coordinates and generalized momenta. In Chapter 1 we also discussed the equations of motion for such a system in terms of the Lagrangian $\mathcal{L}$ or the Hamiltonian $\mathcal{H}$. Of particular importance are Hamilton's equations of motion, Eqs. (1.1.33) and (1.1.34), which in Cartesian coordinates $\boldsymbol{q}_i = \boldsymbol{r}_i$ read[1,2]

$$\dot{\boldsymbol{r}}_i \, (\equiv \boldsymbol{\nu}_i) = \frac{\boldsymbol{p}_i}{\boldsymbol{m}_i} \tag{4.1.1a}$$

$$\dot{\boldsymbol{p}}_i = -\nabla_{\boldsymbol{r}_i} \mathcal{V} \left(\equiv -\frac{\partial \mathcal{V}}{\partial \boldsymbol{r}_i} \right) = \boldsymbol{F}_i, \tag{4.1.1b}$$

where $\mathcal{V}$ is the total potential energy function of the system, m_i the mass of atom i, and $\boldsymbol{F}_i$ the total force on i. Combining Eq. (4.1.1), we obtain that

$$m_i \ddot{\boldsymbol{r}}_i = \boldsymbol{F}_i. \tag{4.1.2}$$

Solving the equations of motion then involves the integration of the $3n$ second-order differential Eq. (4.1.2), also known as Newton's equations.

In Chapter 1 we point out that if the kinetic energy $\mathcal{K}$ and the potential energy $\mathcal{V}$ do not depend explicitly on time t, then $\dot{\mathcal{H}} = \mathrm{d}\mathcal{H}/\mathrm{d}t$ is 0; i.e. the Hamiltonian is a constant of the motion. In actual calculations this conservation law is satisfied if there exist no explicitly time- or velocity-dependent forces acting on the system. A second important property is that Hamilton's equations of motion are reversible in time. This means that if we change the signs of all the velocities, we will cause the molecules to retrace their trajectories backwards. The computer-generated trajectories should also possess this property.

Multiscale Modelling of Polymers. Doros N. Theodorou and Vlasis G. Mavrantzas, Oxford University Press.
 DOI: 10.1093/oso/9780199226191.003.0004

There are many methods for solving ordinary differential equations of the form of Eq. (4.1.1) or (4.1.2).[3–5] A well-chosen algorithm for the numerical solution of these equations should (a) not require an expensively large number of force evaluations per integration time step, (b) satisfy the energy conservation law, (c) permit the use of a long time step, (d) be fast, and (e) require little memory. It is also highly desirable that the algorithm be time reversible and conserve volume in phase space (i.e. be symplectic). In Section 4.1.2 we describe the most popular family of algorithms used to solve the classical equations of motion today, the Verlet algorithms.

4.1.2 Verlet algorithms

The Verlet algorithms[6] are simple, accurate and, as we will see in the following, time reversible, which explains why they are very broadly used in MD simulations. The initial form of the Verlet equations is obtained from a Taylor expansion of the $3n$-dimensional vector of positions at times $t + \mathrm{d}t$ and $t - \mathrm{d}t$:

$$\boldsymbol{r}(t+\mathrm{d}t) = \boldsymbol{r}(t) + \mathrm{d}t\,\boldsymbol{\nu}(t) + \frac{(\mathrm{d}t)^2}{2}\ddot{\boldsymbol{r}}(t) + \frac{(\mathrm{d}t)^3}{6}\dddot{\boldsymbol{r}}_i(t) + O\left((\mathrm{d}t)^4\right) \tag{4.1.3a}$$

$$\boldsymbol{r}(t-\mathrm{d}t) = \boldsymbol{r}(t) - \mathrm{d}t\,\boldsymbol{\nu}(t) + \frac{(\mathrm{d}t)^2}{2}\ddot{\boldsymbol{r}}(t) - \frac{(\mathrm{d}t)^3}{6}\dddot{\boldsymbol{r}}(t) + O\left((\mathrm{d}t)^4\right). \tag{4.1.3b}$$

Summing the two equations gives

$$\boldsymbol{r}(t+\mathrm{d}t) = 2\boldsymbol{r}(t) - \boldsymbol{r}(t-\mathrm{d}t) + (\mathrm{d}t)^2\,\ddot{\boldsymbol{r}}(t) + O\left((\mathrm{d}t)^4\right) \tag{4.1.4}$$

with $\ddot{\boldsymbol{r}}(t)$ calculated from the forces at the current positions. This is *the coordinate version* of the Verlet algorithm. Remarkably, the error in calculated positions at $t + \mathrm{d}t$ is on the order of $(\mathrm{d}t)^4$, because of the cancellation of third-order terms upon summation.

Despite its good numerical characteristics, in the coordinate version of the Verlet algorithm the velocities are not among the variables used in integrating the equations of motion. Thus, two modifications are in wide use. The first is the *leap-frog algorithm*, where positions and velocities are calculated at integer time steps and velocities at half-steps as follows:

$$\boldsymbol{\nu}\left(t+\frac{\mathrm{d}t}{2}\right) = \boldsymbol{\nu}\left(t-\frac{\mathrm{d}t}{2}\right) + \mathrm{d}t\,\ddot{\boldsymbol{r}}(t) \tag{4.1.5a}$$

$$\boldsymbol{r}(t+\mathrm{d}t) = \boldsymbol{r}(t) + (\mathrm{d}t)\,\boldsymbol{\nu}\left(t+\frac{\mathrm{d}t}{2}\right). \tag{4.1.5b}$$

An estimate of the velocity at time t (needed in the calculation of the Hamiltonian $\mathcal{H}$) is obtained as the average of its value at times $t + \frac{\mathrm{d}t}{2}$ and $t - \frac{\mathrm{d}t}{2}$:

$$\boldsymbol{\nu}(t) = \frac{1}{2}\left[\boldsymbol{\nu}\left(t+\frac{\mathrm{d}t}{2}\right) + \boldsymbol{\nu}\left(t-\frac{\mathrm{d}t}{2}\right)\right]. \tag{4.1.5c}$$

The second modification is the *velocity Verlet algorithm*, where positions and velocities are calculated at the integer time steps as follows:

$$\boldsymbol{r}(t+\mathrm{d}t) = \boldsymbol{r}(t) + \mathrm{d}t\,\boldsymbol{\nu}(t) + \frac{(\mathrm{d}t)^2}{2}\ddot{\boldsymbol{r}}(t) \tag{4.1.6a}$$

$$\boldsymbol{\nu}(t+\mathrm{d}t) = \boldsymbol{\nu}(t) + \frac{\mathrm{d}t}{2}\left[\ddot{\boldsymbol{r}}(t) + \ddot{\boldsymbol{r}}(t+\mathrm{d}t)\right]. \tag{4.1.6b}$$

4.1.3 Constant temperature and constant pressure (stress) methods

The methods described previously address the solution of Newton's equations of motion in the microcanonical (nVE) ensemble, in which the Hamiltonian H is a constant, equal to the total energy E of the system. From a statistical mechanics point of view, the energy conservation condition $\mathcal{H}(\boldsymbol{p}, \boldsymbol{q}) = E$ defines a $(3n-1)$-dimensional hypersurface in phase space on which the dynamical trajectory evolves, which we will call the *constant energy surface*. A system evolving according to Hamilton's equations of motion will remain on this surface. Then, an average over all possible states generated by an MD simulation at fixed n, V, and E will equal the average over the microcanonical ensemble. In statistical mechanics, the assumption that a system, given an infinite amount of time, will cover the entire constant energy hypersurface is known as the *ergodic hypothesis*. The assumption that, at equilibrium, the system will assume a uniform distribution over the constant energy hypersurface is known as the *principle of equal a priori probabilities*.

In practice, there is usually the need to perform MD simulations under specified conditions of temperature and/or pressure. In the literature there exists a variety of methodologies for performing MD simulations under isochoric and/or isothermal conditions. Most of these constitute a reformulation of Lagrangian or Hamilton's equations of motion to include the constraints of constant T and/or P by extending the phase space beyond $2\,d\,n$ dimensions, with d (usually equal to 3) being the spatial dimension.[7–17] Among these methodologies, the most widely used is that of Nosé and Hoover;[8–11] a different method, the Berendsen thermostat–barostat,[12] whose equations are not reversible in time, also enjoys extensive application. Extensions to cells of variable shape subjected to a specified stress tensor have also been formulated;[13–15] here, the MD cell shape and size can change according to dynamical equations as prescribed by the postulated extended Lagrangian of the system. The use of explicit reversible integrators based on the Liouville operator formalism and operator factorization techniques[16] and theoretical advances in the field of non-Hamiltonian molecular dynamics[17] have made it possible to generate canonical and isothermal–isobaric ensembles in a more systematic way. We discuss algorithms generating various statistical ensembles in Sections 4.1.3.1–4.1.3.4.

4.1.3.1 The Nosé–Hoover nVT algorithm with a single thermostat

The Nosé–Hoover equations are[8–10]

$$\dot{\boldsymbol{r}}_i = \frac{\boldsymbol{p}_i}{m_i}$$

$$\dot{\boldsymbol{p}}_i = \boldsymbol{F}_i - \frac{p_\eta}{Q}\boldsymbol{p}_i$$

$$\dot{\eta} = \frac{p_\eta}{Q}$$

$$\dot{p}_\eta = \sum_{i=1}^{n} \frac{\boldsymbol{p}_i^2}{m_i} - Lk_B T, \tag{4.1.7}$$

where the forces $\boldsymbol{F}_i$ are derived from an n-particle potential $\mathcal{V}(\mathbf{r}_1, \ldots, \mathbf{r}_n, V)$ through $\boldsymbol{F}_i = -\nabla_{\mathbf{r}_i}\mathcal{V}$ and L is a numerical constant. These equations contain the additional variable η, and its conjugate momentum p_η. The additional variable is regarded as a *thermostat* and controls fluctuations in the total kinetic energy of the system around the average kinetic energy value $\frac{L}{2}k_B T$. The parameter Q is the *effective mass* of the thermostat η, determining the time scale on which η evolves. It is equal to $Q = Lk_B T\tau^2$, where τ is a time scale relevant to the physical system. According to the non-Hamiltonian formulation of equations of motion by Tuckerman et al.,[17] if the total energy

$$\mathcal{H}' = \mathcal{H}(\boldsymbol{r}, \boldsymbol{p}) + \frac{p_\eta^2}{2Q} + Lk_B T\eta \tag{4.1.8}$$

is the *only* conserved quantity and $L = dn$, where d denotes the number of space dimensions, then the Nosé–Hoover equations generate a canonical distribution in the space of the physical variables. On the other hand, if there are no external forces so that $\sum_{i=1}^{n} \boldsymbol{F}_i = 0$, Nosé[9c] has shown that there are d additional conservation laws satisfied by Eqs. (4.1.7). Any difficulties resulting from the presence of multiple conservation conditions in the Nosé–Hoover system can be eliminated by using the Nosé–Hoover chain method.[11b,15a]

4.1.3.2 *Hoover's isothermal–isobaric (nPT) ensemble with uniform dilation*

The ensemble is generated by allowing the volume of the system to fluctuate such that the average internal pressure $\langle P_{\text{int}} \rangle$ is equal to an externally applied pressure P_{ext}. The volume is thus introduced as a dynamical variable in the equations of motion. In Hoover's original algorithm,[10] volume fluctuations are assumed to be purely isotropic (this can be used to simulate liquids), in which the box varies uniformly in all directions. For the Hoover algorithm, the equations of motion are

$$\dot{\boldsymbol{r}}_i = \frac{\boldsymbol{p}_i}{m_i} + \frac{p_\varepsilon}{W}\boldsymbol{r}_i$$

$$\dot{\boldsymbol{p}}_i = \boldsymbol{F}_i - \frac{p_\eta}{Q}\boldsymbol{p}_i - \frac{p_\varepsilon}{W}\boldsymbol{p}_i$$

$$\dot{\eta} = \frac{p_\eta}{Q}$$

$$\dot{p}_\eta = \sum_{i=1}^{n} \frac{\boldsymbol{p}_i^2}{m_i} + \frac{p_\varepsilon^2}{W} - Lk_B T$$

$$\dot{V} = \frac{dVp_\varepsilon}{W}$$

$$\dot{p}_\varepsilon = dV(P_{\text{int}} - P_{\text{ext}}) - \frac{p_\eta}{Q}p_\varepsilon. \tag{4.1.9}$$

The volume V is indeed incorporated into the equations of motion, and the momentum p_ε is not related to the variable $\frac{\mathrm{d}V}{\mathrm{d}t}$ but to the variable $\frac{\mathrm{d}\varepsilon}{\mathrm{d}t}$, where $\varepsilon = \frac{1}{d}\ln\left(\frac{V}{V_0}\right)$ with d being the space dimensionality and V_0 an arbitrary reference volume usually set equal to the initial system volume $V(0)$. The variable p_ε acts as a barostat which drives the system to the steady state $\langle P_{\text{int}}\rangle = P_{\text{ext}}$, where

$$P_{\text{int}} = \frac{1}{dV}\left[\sum_{i=1}^{n}\frac{\boldsymbol{p}_i^2}{m_i} + \sum_{i=1}^{n}\boldsymbol{r}_i\cdot\boldsymbol{F}_i - dV\frac{\partial \mathcal{V}}{\partial V}\right] \tag{4.1.10}$$

is the *instantaneous internal pressure* in the system, also discussed in Chapter 3. For simplicity, in the above equations, both the particles and the barostat are coupled to a single thermostat. If the forces do not sum up to 0, i.e. when $\sum_{i=1}^{n}\boldsymbol{F}_i \neq 0$, the only conservation law valid is that of the total energy

$$\mathcal{H}' = \mathcal{H}(\boldsymbol{r},\boldsymbol{p}) + \frac{p_\varepsilon^2}{2W} + \frac{p_\eta^2}{2Q} + Lk_{\text{B}}T\eta + P_{\text{ext}}V \tag{4.1.11}$$

with $L = dn + 1$ now. Unfortunately, this Hoover algorithm (as Hoover himself recognized) cannot generate the correct nPT ensemble distribution because when $\sum_{i=1}^{n}\boldsymbol{F}_i \neq 0$, the volume distribution is incorrect by a factor of $1/V$. This problem remains even when a chain of thermostats is employed, but in this case the centre-of-mass momentum distribution is correct, indicating an $(n-1)PT$ ensemble. When $\sum_{i=1}^{n}\boldsymbol{F}_i = 0$, additional momentum conservation equations arise, and the centre-of-mass position $\boldsymbol{R}$ is a driven variable. In this case, serious pathologies result if a single Nosé–Hoover thermostat is used: the volume distribution is $\sim\frac{1}{V^{1+d}}$ and the centre-of-mass momentum distribution is $\sim\frac{1}{P^{d(n-1)+2}}$, independently of the potential. All these deficiencies of the original Hoover nPT algorithm are addressed by the *Martyna–Tobias–Klein algorithm*[15b] for which the equations of motion read

$$\dot{\boldsymbol{r}}_i = \frac{\boldsymbol{p}_i}{m_i} + \frac{p_\varepsilon}{W}\boldsymbol{r}_i$$

$$\dot{\boldsymbol{p}}_i = \boldsymbol{F}_i - \frac{p_\eta}{Q}\boldsymbol{p}_i - \left(1 + \frac{1}{n}\right)\frac{p_\varepsilon}{W}\boldsymbol{p}_i$$

$$\dot{\eta} = \frac{p_\eta}{Q}$$

$$\dot{p}_\eta = \sum_{i=1}^{n} \frac{\boldsymbol{p}_i^2}{m_i} + \frac{p_\varepsilon^2}{W} - Lk_{\mathrm{B}}T$$

$$\dot{V} = \frac{dVp_\varepsilon}{W}$$

$$\dot{p}_\varepsilon = dV(P_{\mathrm{int}} - P_{\mathrm{ext}}) - \frac{p_\eta}{Q}p_\varepsilon + \frac{1}{n}\sum_{i=1}^{n}\frac{\boldsymbol{p}_i^2}{m_i} \tag{4.1.12}$$

for a single thermostat, with the instantaneous internal pressure P_{int} given by Eq. (4.1.10). The Martyna–Tobias–Klein algorithm generates a correct nPT ensemble, even for the case $\sum_{i=1}^{n} \boldsymbol{F}_i = 0$.

4.1.3.3 *Fully flexible cells*

In the simulation of phase changes in solids (e.g. crystalline polymers) we should allow for changes in the course of the MD simulation in the dimensions and angles of the unit cell. An algorithm through which one can achieve this was proposed for the first time by Parrinello and Rahman,[13] who extended Andersen's method[7a] to allow for changes not only in the size, but also in the shape of the simulation box. It is the fully flexible cell model, in which all cell lengths and angles are allowed to vary independently.[15b] The method starts by introducing the cell matrix $\boldsymbol{h}$, whose d columns contain the components of each of the d vectors describing the box parallelepiped (see Section 3.5.1). The volume is given then by $V = \det[\boldsymbol{h}]$. According to the work by Martyna et al.,[15b] the following equations of motion produce the desired ensemble:

$$\dot{\boldsymbol{r}}_i = \frac{\boldsymbol{p}_i}{m_i} + \frac{\boldsymbol{p}_\varepsilon}{W}\cdot\boldsymbol{r}_i$$

$$\dot{\boldsymbol{p}}_i = \boldsymbol{F}_i - \frac{p_\eta}{Q}\boldsymbol{p}_i - \frac{\boldsymbol{p}_\varepsilon}{W}\cdot\boldsymbol{p}_i - \frac{1}{dn}\frac{\mathrm{tr}[\boldsymbol{p}_\varepsilon]}{W}\boldsymbol{p}_i$$

$$\dot{\eta} = \frac{p_\eta}{Q}$$

$$\dot{p}_\eta = \sum_{i=1}^{n}\frac{\boldsymbol{p}_i^2}{m_i} + \frac{\mathrm{tr}\left[\boldsymbol{p}_\varepsilon^{\mathrm{T}}\cdot\boldsymbol{p}_\varepsilon\right]}{W} - Lk_{\mathrm{B}}T$$

$$\dot{\boldsymbol{h}} = \frac{\boldsymbol{p}_\varepsilon}{W}\cdot\boldsymbol{h}$$

$$\dot{\boldsymbol{p}}_\varepsilon = V(\boldsymbol{P}_{\mathrm{int}} - P_{\mathrm{ext}}\boldsymbol{I}) - \frac{p_\eta}{Q}\boldsymbol{p}_\varepsilon + \frac{1}{dn}\left[\sum_{i=1}^{n}\frac{p_i^2}{m_i}\right]\boldsymbol{I}, \tag{4.1.13}$$

where now $\boldsymbol{p}_\varepsilon$ is a 3×3 matrix and $L = dn + d^2$. Also, $\boldsymbol{I}$ is the identity matrix, $\mathrm{tr}\left[\boldsymbol{p}_\varepsilon^{\mathrm{T}}\cdot\boldsymbol{p}_\varepsilon\right]$ is the sum of the squares of all the elements of the matrix $\boldsymbol{p}_\varepsilon$, and

$$P_{\mathrm{int},\alpha\beta} = \frac{1}{V}\left[\sum_{i=1}^{n}\frac{p_{i,\alpha}p_{i,\beta}}{m_i} + \sum_{i=1}^{n}F_{i,\alpha}r_{i,\beta} - \sum_{\gamma}\frac{\partial\mathcal{V}(\boldsymbol{r},\boldsymbol{h})}{\partial h_{\alpha\gamma}}h_{\beta\gamma}\right] \tag{4.1.14}$$

defines the instantaneous internal pressure tensor (compare with a similar definition for the stress tensor in Chapter 3 but with a minus sign). These equations of motion have the conserved quantity

$$\mathcal{H}' = \sum_{i=1}^{n} \frac{p_i^2}{2m_i} + \mathcal{V}(\boldsymbol{r}, \mathbf{h}) + \frac{p_\eta^2}{2Q} + \frac{\mathrm{tr}\left[\boldsymbol{p}_\varepsilon^{\mathrm{T}} \cdot \boldsymbol{p}_\varepsilon\right]}{2W} + Lk_{\mathrm{B}} T\eta + P_{\mathrm{ext}} \det\left[\mathbf{h}\right]. \tag{4.1.15}$$

As discussed by Martyna et al.,[15b] two modifications of the above equations of motion for fully flexible cells include the separation of the isotropic fluctuations from the rest of changes in the simulation cell, and the elimination of cell rotations to facilitate data analysis. For more details on these two algorithms, the interested reader is referred to the original papers.[14–17]

4.1.3.4 The Berendsen thermostat–barostat

In practical applications of the molecular dynamics method, use is often made of the simple method proposed by Berendsen et al.[12] for performing isothermal and/or isobaric simulations without the need to use an extended Lagrangian, by coupling the system to a temperature and/or pressure bath. This is achieved by forcing the system to obey

$$\frac{\mathrm{d}T}{\mathrm{d}t} = -\frac{T - T_{\mathrm{ext}}}{\tau_T} \tag{4.1.16}$$

and

$$\frac{\mathrm{d}P}{\mathrm{d}t} = -\frac{P - P_{\mathrm{ext}}}{\tau_p}, \tag{4.1.17}$$

where T_{ext} and P_{ext} are the desired temperature and pressure values and τ_T and τ_p time constants characterizing the frequency of the system coupling to temperature and pressure baths. T and P are the instantaneous values of temperature and pressure, calculated from the momenta and configuration of the system. The solution of these equations forces velocities and positions to be scaled at every time step by the factors χ_T and χ_P, respectively, with values

$$\chi_T = \left(1 + \frac{\mathrm{d}t}{\tau_T}\left(\frac{T_{\mathrm{ext}}}{T} - 1\right)\right)^{1/2} \tag{4.1.18}$$

$$\chi_P = \left[1 - \kappa_T \frac{\mathrm{d}t}{\tau_P}\left(P_{\mathrm{ext}} - P\right)\right]^{1/3}, \tag{4.1.19}$$

where κ_T denotes the isothermal compressibility of the system and $\mathrm{d}t$ is the integration time step. The method is much simpler and easier to program than those proposed or derived by Nosé and Hoover. However, the phase-space probability density it defines does not conform to a specific statistical ensemble (e.g. nVT, nPT); consequently, there exists no Hamiltonian that should be conserved during the MD simulation.

4.1.4 Dealing with holonomic constraints

As mentioned in Chapter 1, the lengths of bonds and angles between successive bonds are usually restricted to fairly narrow ranges with fluctuations occurring more or less symmetrically about the mean values of these parameters. It is not a bad approximation, therefore, to treat the polymer system as if its bond lengths and/or bond angles were fixed. Such constraints, which can be written as equations involving only the coordinates of interaction sites, are called *holonomic*.[2] Integrating the dynamical equations in the presence of holonomic constraints requires special treatment. Here we will focus mainly on cases where system dynamics should satisfy the constraint of constant bond lengths. Constant bond angles can be easily incorporated by fixing additional lengths along the chain. Fixed bond angles are seldom adopted in practice in MD simulations, as the cooperativity between flexible bond angles and torsion angles is thought to play a significant role in the dynamics. We assume that the system is subject to l holonomic constraints of the form

$$\sigma_k(\boldsymbol{r}) \equiv (\boldsymbol{r}_i - \boldsymbol{r}_j)^2 - d_{ij}^2 = 0, \quad k = 1, 2, \ldots, l, \tag{4.1.20}$$

meaning that atoms i and j are held at a fixed distance d_{ij}. By defining, as in Eq. (3.5.2), $\boldsymbol{r}_{ij}(t) \equiv \boldsymbol{r}_j(t) - \boldsymbol{r}_i(t)$, the previous equation is equivalently written as

$$\sigma_k(\mathbf{r}) \equiv r_{ij}^2 - d_{ij}^2 = 0, \quad k = 1, 2, \ldots, l. \tag{4.1.21}$$

To proceed, we consider the following extended Lagrangian (compare Eq. (1.1.30)) for the constrained system

$$\mathcal{L}^c = \mathcal{L} - \sum_{k=1}^{l} \lambda_k \sigma_k(\boldsymbol{r}), \tag{4.1.22}$$

where k enumerates the constraints and λ_k is the Lagrange multiplier associated with constraint k. Based on this, a set of constrained equations of motion which typically have the form

$$m_i \ddot{\boldsymbol{r}}_i = \boldsymbol{F}_i - \sum_{k=1}^{l} \lambda_k \frac{\partial \sigma_k}{\partial \boldsymbol{r}_i} = \boldsymbol{F}_i + \boldsymbol{g}_i \tag{4.1.23}$$

is obtained. The second term on the right-hand side of Eq. (4.1.23) is the force on atom i due to all constraints (i.e. the constraint force on i). Also, the gradients $\frac{\partial \sigma_k}{\partial \boldsymbol{r}_i}$ and $\frac{\partial \sigma_k}{\partial \boldsymbol{r}_j}$ appearing in the equation are simply equal to $-2\boldsymbol{r}_{ij}$ and $+2\boldsymbol{r}_{ij}$, respectively. The $3n$ equations of motion and the l constraint equations constitute a system of $3n + l$ equations in $3n + l$ unknowns which should be solved subject to an initial phase-space point $\{\boldsymbol{r}_0 = \boldsymbol{r}(t=0), \boldsymbol{p}_0 = \boldsymbol{p}(t=0)\}$ of the system. The key question is how to calculate the set of Lagrange multipliers λ_k. There are several methods in the literature addressing this problem to which we refer interested readers, among which we mention the Edberg–Evans–Morriss method,[18] the SHAKE method[19] and its implementation for velocity Verlet types of integration algorithms by Palmer,[20] and the RATTLE method.[21]

4.1.5 Liouville operator and rRESPA multiple time-step algorithms

The Liouville operator formalism of classical mechanics offers a powerful method for constructing time reversible integration methods for MD simulations that possess the correct metric. For a Hamiltonian system the dynamical equations can be cast in the general form

$$\dot{\boldsymbol{\Gamma}} = iL\boldsymbol{\Gamma}. \tag{4.1.24}$$

where $\boldsymbol{\Gamma} = \{\boldsymbol{q}, \boldsymbol{p}\}$ is the phase-space vector with $\boldsymbol{q}$ and $\boldsymbol{p}$ being the $3n$-long vectors of generalized coordinates (here identified with the Cartesian coordinates of the atoms, $\boldsymbol{r}$) and generalized momenta used to describe the system[1,2] and iL is the Liouville operator, given by

$$iL = \{\ldots, \mathcal{H}\} \equiv \sum_{i=1}^{n} \frac{\partial \mathcal{H}}{\partial \boldsymbol{p}_i} \cdot \frac{\partial}{\partial \boldsymbol{q}_i} - \frac{\partial \mathcal{H}}{\partial \boldsymbol{q}_i} \cdot \frac{\partial}{\partial \boldsymbol{p}_i} = \sum_{i=1}^{n} \frac{\boldsymbol{p}_i}{m_i} \cdot \frac{\partial}{\partial \boldsymbol{q}_i} + \boldsymbol{F}_i \cdot \frac{\partial}{\partial \boldsymbol{p}_i}. \tag{4.1.25}$$

Equation (4.1.24) has the formal solution

$$\boldsymbol{\Gamma}(t) = e^{iLt}\boldsymbol{\Gamma}(0). \tag{4.1.26}$$

This is the starting point for the derivation of numerical integration procedures based on the classical propagator, the unitary operator e^{iLt}. We suppose that the Liouville operator iL can be written as the sum of two parts,

$$iL = iL_1 + iL_2, \tag{4.1.27}$$

such that the action of the classical propagator on $\boldsymbol{\Gamma}(0)$ for each part can be evaluated analytically. Using the Trotter theorem in the form[16,17]

$$\exp(iLt) = \exp[(iL_1 + iL_2)t] = \lim_{P\to\infty} \left[\exp\left(\frac{iL_2 t}{2P}\right) \exp\left(\frac{iL_1 t}{P}\right) \exp\left[\frac{iL_2 t}{2P}\right]\right]^P, \tag{4.1.28}$$

and considering a finite P such that $\Delta t = \frac{t}{P}$, we can make the approximations that

$$\begin{aligned} \exp(iL\Delta t) &\approx \exp\left(\frac{iL_2\Delta t}{2}\right) \exp(iL_1\Delta t) \exp\left(\frac{iL_2\Delta t}{2}\right) + O\left((\Delta t)^3\right) \\ \exp(iPL\Delta t) &\approx \prod_{k=1}^{P} \exp\left(\frac{iL_2\Delta t}{2}\right) \exp(iL_1\Delta t) \exp\left(\frac{iL_2\Delta t}{2}\right) + O\left((\Delta t)^2\right), \end{aligned} \tag{4.1.29}$$

which yield a numerical integration procedure that is accurate to second order in the time step at long times. Let us consider, in particular, the choice

$$iL_1 = \sum_{i=1}^{n} \frac{\boldsymbol{p}_i}{m_i} \cdot \frac{\partial}{\partial \boldsymbol{r}_i}$$

$$iL_2 = \sum_{i=1}^{n} \boldsymbol{F}_i \cdot \frac{\partial}{\partial \boldsymbol{p}_i} \tag{4.1.30}$$

and further assume the forces $\boldsymbol{F}_i$ to depend only on the positions. Then, using that $e^{c\frac{\partial}{\partial q}} f(q) = f(q+c)$, the operator $\exp(iL_1\Delta t)$ is a translation operator on the positions: $\boldsymbol{r}_i \rightarrow \boldsymbol{r}_i + \Delta t \left(\frac{\boldsymbol{p}_i}{m_i}\right)$. Similarly, $\exp\left(\frac{iL_2\Delta t}{2}\right)$ is a translation operator on the momenta: $\boldsymbol{p}_i \rightarrow \boldsymbol{p}_i + \frac{\Delta t}{2}\boldsymbol{F}_i(\boldsymbol{r})$. Combining these two factors allows the action of the operator in Eq. (4.1.26) on the full set of positions and momenta to be evaluated analytically, yielding

$$\boldsymbol{r}_i(\Delta t) = \boldsymbol{r}_i(0) + (\Delta t)\,\boldsymbol{\nu}_i(0) + \frac{(\Delta t)^2}{2m_i}\boldsymbol{F}_i(0)$$

$$\boldsymbol{\nu}_i(\Delta t) = \boldsymbol{\nu}_i(0) + \frac{\Delta t}{2m_i}\left[\boldsymbol{F}_i(0) + \boldsymbol{F}_i(\Delta t)\right], \tag{4.1.31}$$

which we can recognize as the velocity Verlet integrator (compare Eqs. (4.1.6)). According to the operator formalism and based on the subdivision of the Liouville operator as dictated by Eqs. (4.1.27) and (4.1.30), the evolution equations corresponding to the velocity Verlet algorithm can be viewed as the result of the following set of three sequential update steps: (i) a velocity translation by an amount $\frac{\Delta t}{2m_i}\boldsymbol{F}_i(0)$ using the forces calculated with the positions at the previous time step, (ii) a position translation by an amount $(\Delta t)\left(\mathrm{v}_i(0) + \frac{\Delta t}{2m_i}\boldsymbol{F}_i(0)\right)$, i.e. by using the velocities obtained at the end of step (i), and (iii) a velocity translation by an amount $\frac{\Delta t}{2m_i}\boldsymbol{F}_i(\Delta t)$, i.e. by using the forces calculated at the positions obtained at the end of step (ii). Similarly, with the choice

$$iL_1 = \sum_{i=1}^{n} \boldsymbol{F}_i \cdot \frac{\partial}{\partial \boldsymbol{p}_i}$$

$$iL_2 = \sum_{i=1}^{n} \frac{\boldsymbol{p}_i}{m_i} \cdot \frac{\partial}{\partial \boldsymbol{r}_i} \tag{4.1.32}$$

one derives the following set of evolution equations:

$$\boldsymbol{r}_i\left(\frac{\Delta t}{2}\right) = \boldsymbol{r}_i(0) + \frac{\Delta t}{2}\boldsymbol{\nu}_i(0)$$

$$\boldsymbol{v}_i(\Delta t) = \boldsymbol{v}_i(0) + \frac{\Delta t}{m_i}\boldsymbol{F}_i\left(\frac{\Delta t}{2}\right)$$
$$\boldsymbol{r}_i(\Delta t) = \boldsymbol{r}_i\left(\frac{\Delta t}{2}\right) + \frac{\Delta t}{2}\boldsymbol{v}_i(\Delta t), \tag{4.1.33}$$

which is equivalent to the Verlet leapfrog scheme, Eq. (4.1.5).

The procedure described in the previous paragraph is known as the *direct translation technique* and can be extended to treat systems with multiple time scale motion. The method[16,17] is based on the introduction of a reference system and a subdivision of the force on each atom i into a contribution from the reference system and a deviation

$$\boldsymbol{F}_i = \boldsymbol{F}_i^{\mathrm{ref}} + \boldsymbol{F}_i^{\mathrm{dev}}. \tag{4.1.34}$$

If the reference force is chosen such that $\boldsymbol{F}_i^{\mathrm{ref}}$ is a good approximation of the true force, then propagation of the system using the reference force with an occasional correction due to the deviation force $\boldsymbol{F}_i^{\mathrm{dev}}$ would yield an efficient and accurate scheme. In polymeric systems, $\boldsymbol{F}_i^{\mathrm{ref}}$ is usually taken to be the sum of *fast-changing* local intramolecular forces which are usually computationally inexpensive to evaluate. Then an efficient multiple time-step numerical algorithm can be developed by introducing a time step $\delta t = \frac{\Delta t}{P}$ appropriate for the reference system, a Liouville operator iL^{ref} for the reference system, and a deviation iL^{dev} for the Liouville operator according to

$$iL = iL^{\mathrm{ref}} + iL^{\mathrm{dev}}, \tag{4.1.35}$$

where

$$iL^{\mathrm{ref}} = \sum_{i=1}^{n}\left[\frac{\boldsymbol{p}_i}{m_i}\cdot\frac{\partial}{\partial \boldsymbol{r}_i}\right] + \sum_{i=1}^{n}\left[\boldsymbol{F}_i^{\mathrm{ref}}\cdot\frac{\partial}{\partial \boldsymbol{p}_i}\right] = iL_1^{\mathrm{ref}} + iL_2^{\mathrm{ref}}$$
$$iL^{\mathrm{dev}} = \sum_{i=1}^{n}\left[\boldsymbol{F}_i^{\mathrm{dev}}\cdot\frac{\partial}{\partial \boldsymbol{p}_i}\right]. \tag{4.1.36}$$

For a given time step Δt, $iL^{\mathrm{ref}}\left(= iL_1^{\mathrm{ref}} + iL_2^{\mathrm{ref}}\right)$ will be used to propagate the system by P steps with a time step δt while the correction iL^{dev} will be applied before and after the reference system propagation, namely, as

$$\exp(iL\Delta t)$$
$$\approx \exp\left(iL^{\mathrm{dev}}\frac{\Delta t}{2}\right)\left\{\exp\left(iL_2^{\mathrm{ref}}\frac{\delta t}{2}\right)\exp\left(iL_1^{\mathrm{ref}}\delta t\right)\exp\left(iL_2^{\mathrm{ref}}\frac{\delta t}{2}\right)\right\}^P \exp\left(iL^{\mathrm{dev}}\frac{\Delta t}{2}\right) + O\left((\Delta t)^3\right), \tag{4.1.37}$$

to ensure that the overall numerical scheme is time reversible. In Eq. (4.1.37), the operator in brackets corresponds to P steps of reference system propagation (using e.g. the algorithm of Eqs. (4.1.31) or the algorithm of Eqs. (4.1.33)) with a time step of δt. The numerical integration procedure is called the r-RESPA method, which stands for reversible reference system propagation algorithm. The advantage of this method is that, if there is a wide separation in time scales, choosing the reference system force equal to the *fast* changing force will be a good approximation to the true force. This will allow P to be large. If, in addition, the fast forces are computationally inexpensive to evaluate, then there will be a large gain in CPU time, as the expensive slow forces will only need to be updated once every large time step Δt.

In applying r-RESPA to polymer simulations in the NVE ensemble, we make use of the above factorization together with an integration of each part of the Liouville operator with a different time step. The forces $\boldsymbol{F}$ are decomposed into fast changing (short-range) forces $\boldsymbol{F}^{\mathrm{f}}$, and slowly changing (long-range) forces $\boldsymbol{F}^{\mathrm{s}}$, according to $\boldsymbol{F}(\boldsymbol{r}) = \boldsymbol{F}^{\mathrm{f}}(\boldsymbol{r}) + \boldsymbol{F}^{\mathrm{s}}(\boldsymbol{r})$. Then, following (e.g.) Eqs. (4.1.30) and (4.1.35), the total evolution operator is broken up as

$$iL_1^{\mathrm{ref}} = \sum_{i=1}^{n}\left[\dot{\boldsymbol{r}}_i \cdot \frac{\partial}{\partial \boldsymbol{r}_i}\right],\; iL_2^{\mathrm{ref}} = \sum_{i=1}^{n}\left[\boldsymbol{F}_i^{\mathrm{f}}(\boldsymbol{r}) \cdot \frac{\partial}{\partial \boldsymbol{p}_i}\right],\; iL^{\mathrm{dev}} = \sum_{i=1}^{n}\left[\boldsymbol{F}_i^{\mathrm{s}}(\boldsymbol{r}) \cdot \frac{\partial}{\partial \boldsymbol{p}_i}\right], \tag{4.1.38}$$

and integration proceeds according to Eq. (4.1.37). A multiple time-step method is thus designed: the slow modes (slow forces, iL^{dev}) are integrated with a longer time step, Δt, and the fast modes (fast forces and velocities iL_2^{ref}, iL_1^{ref}) with a smaller time step, δt ($\delta t = \frac{\Delta t}{P}$). The force calculated P times (fast force) per Δt is the reference force.

In a similar way one can write the r-RESPA in other ensembles (such as the nVT and nPT ones) by dividing the Liouville operator into a part that is purely Hamiltonian in the system variables $(\boldsymbol{r}, \boldsymbol{p})$ and a part that contains all the non-Hamiltonian contributions[17b]

$$iL = iL^{\mathrm{H}} + iL^{\mathrm{nH}}, \tag{4.1.39}$$

where $iL^{\mathrm{H}} = iL^{\mathrm{ref}} + iL^{\mathrm{dev}} = iL_1^{\mathrm{ref}} + iL_2^{\mathrm{ref}} + iL^{\mathrm{dev}}$ (with iL_1^{ref}, iL_2^{ref}, iL^{dev} as defined above) is the Hamiltonian operator which is subdivided according to Eq. (4.1.37) and iL^{nH} is the non-Hamiltonain operator following (e.g.) Eq. (4.1.8), Eq. (4.1.11), or Eq. (4.1.15). There are, in particular, two interesting modifications of the standard RESPA method, depending on how one expands the extended operator $\exp\left[iL^{\mathrm{nH}}t\right]$. The first is useful when the evolution prescribed by the operator $\exp\left[iL^{\mathrm{nH}}t\right]$ is slow compared to the time scale associated with the reference force and reads

$$\begin{aligned}\exp(iL\Delta t) = {} & \exp\left(iL^{\mathrm{nH}}\frac{\Delta t}{2}\right)\exp\left(iL^{\mathrm{dev}}\frac{\Delta t}{2}\right)\\ & \left\{\exp\left(iL_2^{\mathrm{ref}}\frac{\delta t}{2}\right)\exp\left(iL_1^{\mathrm{ref}}\delta t\right)\exp\left(iL_2^{\mathrm{ref}}\frac{\delta t}{2}\right)\right\}^{P}\\ & \exp\left(iL^{\mathrm{dev}}\frac{\Delta t}{2}\right)\exp\left(iL^{\mathrm{nH}}\frac{\Delta t}{2}\right) + O\left((\Delta t)^3\right)\end{aligned} \tag{4.1.40}$$

and is named the eXtended system Outside-Reference System Propagator Algorithm or XO-RESPA.[16] In general, XO-RESPA can be applied to systems characterized by fast vibrations, as the time scale associated with the extended system variable is usually chosen to be quite slow compared with these motions. The second modification is of relevance when the motion prescribed by the operator $\exp\left[iL^{\text{nH}}t\right]$ occurs on the same time scale as that generated by the *fast* forces; then, a useful RESPA algorithm must include the application of this operator for the small time step δt. In this case, the evolution operator takes the form

$$\begin{aligned}\exp\left(iL\Delta t\right) = &\exp\left(iL^{\text{nH}}\frac{\delta t}{2}\right)\exp\left(iL^{\text{dev}}\frac{\Delta t}{2}\right)\exp\left(-iL^{\text{nH}}\frac{\delta t}{2}\right)\\ &\left\{\exp\left(iL^{\text{nH}}\frac{\delta t}{2}\right)\exp\left(iL_2^{\text{ref}}\frac{\delta t}{2}\right)\exp\left(iL_1^{\text{ref}}\delta t\right)\exp\left(iL_2^{\text{ref}}\frac{\delta t}{2}\right)\exp\left(iL^{\text{nH}}\frac{\delta t}{2}\right)\right\}^P\\ &\exp\left(-iL^{\text{nH}}\frac{\delta t}{2}\right)\exp\left(iL^{\text{dev}}\frac{\Delta t}{2}\right)\exp\left(iL^{\text{nH}}\frac{\delta t}{2}\right) + \mathcal{O}\left(\left(\Delta t\right)^3\right)\end{aligned} \tag{4.1.41}$$

and corresponds to the eXtended system Inside-REference System Propagator Algorithm or XI-RESPA.[16]

4.1.6 Nonequilibrium methods

To calculate the nonlinear rheological properties of a polymeric fluid, one must explicitly simulate the system in the presence of a flow field.[22] The first molecular dynamics algorithm for the simulation of a system under flow was proposed by Hoover et al.[23] and is known as the DOLLS algorithm. If $\nabla\boldsymbol{\nu}$ is the applied velocity gradient tensor, the DOLLS algorithm starts with the following expression for the Hamiltonian,

$$\mathcal{H}\left(\boldsymbol{p},\boldsymbol{q}\right) = \sum_{i=1}^{n}\frac{p_i^2}{2m_i} + \mathcal{V}\left(\boldsymbol{q}\right) + \sum_{i=1}^{n}\boldsymbol{q}_i\boldsymbol{p}_i : \left(\nabla\boldsymbol{\nu}\right)^{\text{T}}, \tag{4.1.42}$$

and generates the following set of canonical equations of motion,

$$\begin{aligned}\dot{\boldsymbol{q}}_i &= \frac{\partial\mathcal{H}}{\partial\boldsymbol{p}_i} = \frac{\boldsymbol{p}_i}{m_i} + \boldsymbol{q}_i\cdot\nabla\boldsymbol{\nu}\\ \dot{\boldsymbol{p}}_i &= -\frac{\partial\mathcal{H}}{\partial\boldsymbol{q}_i} = \boldsymbol{F}_i - \nabla\boldsymbol{\nu}\cdot\boldsymbol{p}_i,\end{aligned} \tag{4.1.43}$$

where $\boldsymbol{F}_i = -\dfrac{\partial\mathcal{V}}{\partial\boldsymbol{q}_i}$. In the above equations, $\boldsymbol{q}_i$ and $\boldsymbol{p}_i$ are the vectors of generalized momentum and generalized position of particle i with mass m_i.

Although DOLLS is derived from a Hamiltonian and generates canonical equations of motion, it results in incorrect predictions at high shear rates. The second and most widely used algorithm for NEMD simulations was proposed by Evans and Morriss,[24] and it is the SLLOD algorithm. For this algorithm, which is not generated by a Hamiltonian, the microscopic equations of motion read

$$\dot{\boldsymbol{q}}_i = \frac{\boldsymbol{p}_i}{m_i} + \boldsymbol{q}_i \cdot \nabla \boldsymbol{\nu}$$
$$\dot{\boldsymbol{p}}_i = \boldsymbol{F}_i - \boldsymbol{p}_i \cdot \nabla \boldsymbol{\nu}. \tag{4.1.44}$$

For flows such that $(\nabla \boldsymbol{\nu}) = (\nabla \boldsymbol{\nu})^{\mathrm{T}}$, these equations are identical to those proposed by the DOLLS algorithm. As reported by Tuckerman et al.,[25,26] a drawback of the SLLOD algorithm is that it does not satisfy Newton's equations of motion, namely, $m_i \ddot{\boldsymbol{q}}_i = \boldsymbol{F}_i$. Indeed, from Eqs. (4.1.44), it can be shown that

$$m_i \ddot{\boldsymbol{q}}_i = \boldsymbol{F}_i + m_i \boldsymbol{q}_i \cdot \nabla \boldsymbol{\nu} \cdot \nabla \boldsymbol{\nu}, \tag{4.1.45}$$

which is inconsistent with Newton's law except for the case of the simple shear flow for which the last term on the right-hand-side of Eq. (4.1.45) disappears. As a result, the application of SLLOD to NEMD simulations of polymers has been practically limited to systems subjected to shear, together with the use of the Lees–Edwards noncyclic boundary conditions.[27] By combining the SLLOD algorithm (even if it does not satisfy Newton's law in extensional flows) with Kraynik and Reinelt's idea of temporal and spatial periodicity of lattice vectors in planar extensional flow,[28] NEMD simulations of short polymers subjected to planar extensional flow have also been reported.[29,30]

Tuckerman et al.[25a] proposed a generalization of the SLLOD algorithm, G-SLLOD, that correctly generates a distribution function which is canonical in the laboratory velocities and positions for a general strain rate. The G-SLLOD equations of motion with a Nosé–Hoover thermostat have the form

$$\dot{\boldsymbol{q}}_i = \frac{\boldsymbol{p}_i}{m_i} + \boldsymbol{q}_i \cdot \nabla \boldsymbol{\nu}$$
$$\dot{\boldsymbol{p}}_i = \boldsymbol{F}_i - \boldsymbol{p}_i \cdot \nabla \boldsymbol{\nu} - m_i \boldsymbol{q}_i \cdot \nabla \boldsymbol{\nu} \cdot \nabla \boldsymbol{\nu} - \frac{p_\eta}{Q} \boldsymbol{p}_i$$
$$\dot{\eta} = \frac{p_\eta}{Q}$$
$$\dot{p}_\eta = \sum_{i=1}^{n} \frac{\boldsymbol{p}_i^2}{m_i} - dn\, k_{\mathrm{B}} T, \tag{4.1.46}$$

where η is the thermostat variable, p_η its momentum, and Q its effective mass, and where d denotes the space dimensionality.

In a later study, Edwards and collaborators[31] proposed the p-SLLOD algorithm (p for proper) in which they distinguished between the laboratory velocity $\boldsymbol{\nu}_i$ of atom i (the velocity with respect to a fixed frame of reference) and the peculiar velocity $\boldsymbol{\nu}'_i$ (the velocity relative to the externally imposed flow velocity $\boldsymbol{\nu}$); the corresponding momenta are denoted as $\boldsymbol{p}_i$ and $\boldsymbol{p'}_i$. The corresponding sets $(\boldsymbol{p}, \boldsymbol{q})$ and $(\boldsymbol{p'}, \boldsymbol{q'})$ of generalized coordinates are related through

$$\begin{aligned} \boldsymbol{p'}_i &= \boldsymbol{p}_i + m_i \boldsymbol{q}_i \cdot \nabla \boldsymbol{\nu} \\ \boldsymbol{q'}_i &= \boldsymbol{q}_i. \end{aligned} \tag{4.1.47}$$

The Hamiltonian of the system is written as[31]

$$\mathcal{H}'(\boldsymbol{p'}, \boldsymbol{q'}) = \sum_{i=1}^{n} \frac{1}{2m_i} \boldsymbol{p'}^2_i + \mathcal{V}(\boldsymbol{q'}), \tag{4.1.48}$$

from which the following canonical equations of motion are derived:

$$\begin{aligned} \dot{\boldsymbol{q}}'_i &= \frac{\partial \mathcal{H}'}{\partial \boldsymbol{p}'_i} = \frac{\boldsymbol{p}'_i}{m_i} \\ \dot{\boldsymbol{p}}'_i &= -\frac{\partial \mathcal{H}'}{\partial \boldsymbol{q}'_i} = \boldsymbol{F}_i \end{aligned} \tag{4.1.49}$$

or, equivalently, in the $(\boldsymbol{p}, \boldsymbol{q})$ space

$$\begin{aligned} \dot{\boldsymbol{q}}_i &= \frac{\boldsymbol{p}_i}{m_i} + \boldsymbol{q}_i \cdot \nabla \boldsymbol{\nu} \\ \dot{\boldsymbol{p}}_i &= \boldsymbol{F}_i - \boldsymbol{p}_i \cdot \nabla \boldsymbol{\nu} - m_i \boldsymbol{q}_i \cdot \nabla \boldsymbol{\nu} \cdot \nabla \boldsymbol{\nu}. \end{aligned} \tag{4.1.50}$$

These equations satisfy Newton's law, namely that $m_i \ddot{\boldsymbol{q}}_i = \boldsymbol{F}_i$; also, for flows such that $\nabla \boldsymbol{\nu} \cdot \nabla \boldsymbol{\nu} = 0$, the p-SLLOD algorithm reduces to the SLLOD algorithm. With a Nosé–Hoover thermostat, the p-SLLOD equations of motion take a form identical to the G-SLLOD algorithm, Eq. (4.1.46).

4.2 Use of fluctuation relations in molecular simulations

Equilibrium molecular dynamics and Monte Carlo (Chapter 5) simulations make it possible to calculate not only averages, but also fluctuations in the instantaneous properties of a molecular system. These fluctuations and, more generally, the entire probability distribution of instantaneous properties (i.e. of functions of the representative point in phase space or in configuration space) hold valuable information about the volumetric, thermal, mechanical, electrical, and magnetic behaviour of the system.

Fluctuation relations are particularly useful in calculating susceptibilities under conditions of thermodynamic equilibrium. By susceptibility we mean the thermodynamic derivative of a property used as an independent variable in a fundamental equation, with respect to its conjugate property in the fundamental equation. The heat capacity at constant volume, $C_V = \left.\frac{\partial U}{\partial T}\right|_{n,V} = \left.\frac{\partial \langle E \rangle}{\partial T}\right|_{n,V}$, the isothermal compressibility, $\kappa_T = -\left.\frac{1}{V}\frac{\partial V}{\partial P}\right|_{n,T} = -\left.\frac{1}{\langle V \rangle}\frac{\partial \langle V \rangle}{\partial P}\right|_{n,T}$, and the derivative of the molecular density with respect to chemical potential, $\left.\frac{\partial \rho}{\partial \mu}\right|_T = \frac{1}{V}\left.\frac{\partial \langle n \rangle}{\partial \mu}\right|_{T,V}$, are all examples of susceptibilities. Often we may be interested in the thermodynamic derivative of a property used as an independent variable in a fundamental equation with respect to the conjugate property of another independent variable in the same fundamental equation. An isothermal elastic coefficient in the elastic regime $C_{\alpha\beta\zeta\eta} = \left.\frac{\partial \tau_{\alpha\beta}}{\partial \varepsilon_{\zeta\eta}}\right|_{T,\varepsilon_{[\zeta\eta]}}$, the elements of the electric susceptibility tensor, $\chi_{e,\alpha\beta} = \frac{1}{\varepsilon_0}\left.\frac{\partial \mathcal{P}_\alpha}{\partial \mathcal{E}_\beta}\right|_{n,T,\mathcal{E}_{[\beta]}}$, with $\boldsymbol{\mathcal{E}}$ being the electric field, $\boldsymbol{P}$ the polarization density, and ε_0 the electric permittivity of free space, and the elements of the magnetic susceptibility tensor $\chi_{v,\alpha\beta} = \left.\frac{\partial M_\alpha}{\partial H_\beta}\right|_{T,H_{[\beta]}}$ with $\boldsymbol{H}$ being the magnetic field strength and $\boldsymbol{M}$ the magnetization, are all examples of such derivatives. Fluctuation relations enable the calculation of these derivatives without actually imposing the considered change in the field, i.e. merely by analysing the spontaneous fluctuations of the system at equilibrium.

To illustrate how fluctuation relations are derived from equilibrium statistical mechanics, consider a homogeneous system being described in terms of the independent variables $(\boldsymbol{x}, \boldsymbol{\omega})$. The number of variables needed to provide a correct thermodynamic description is dictated by the postulates of thermodynamics. (For example, it is equal to the number of components plus two for a homogeneous fluid.) Let the variables $\boldsymbol{x}$ be f in number. A full thermodynamic description is provided by an appropriate thermodynamic potential $Y^{(0)}(\boldsymbol{x}, \boldsymbol{\omega})$. We call the functional relationship $Y^{(0)} = Y^{(0)}(\boldsymbol{x}, \boldsymbol{\omega})$ a fundamental equation of thermodynamics in the $Y^{(0)}$ representation. If we know the functional dependence $Y^{(0)} = Y^{(0)}(\boldsymbol{x}, \boldsymbol{\omega})$, we can derive any thermodynamic property of the system. Let

$$\xi_i = \left.\frac{\partial Y^{(0)}}{\partial x_i}\right|_{\omega, x_{[i]}}, \quad i = 1, 2, \ldots, f \tag{4.2.1}$$

be the partial derivative of the thermodynamic potential with respect to x_i. ξ_i is the thermodynamic variable conjugate to x_i with respect to the fundamental equation in the $Y^{(0)}$ representation.

An equivalent thermodynamic description is provided by the Legendre transform of $Y^{(0)}$ with respect to the variables $\boldsymbol{x}$:

$$Y^{(f)} = Y^{(0)} - \sum_{i=1}^{f} \xi_i x_i = Y^{(0)} - \boldsymbol{\xi} \cdot \mathbf{x}. \tag{4.2.2}$$

The Legendre transform $Y^{(f)}(\boldsymbol{\xi}, \boldsymbol{\omega})$ is a function of the independent variables $\boldsymbol{\xi}$ and $\boldsymbol{\omega}$. From the definition, Eq. (4.2.2), and by considering the differentials of $Y^{(0)}$ and $Y^{(f)}$, one can readily show that

$$\left.\frac{\partial Y^{(f)}}{\partial \xi_i}\right|_{\omega, \xi_{[i]}} = -x_i, \quad i = 1, 2, \ldots, f. \tag{4.2.3}$$

In statistical mechanics, each one of the fundamental equations $Y^{(0)} = Y^{(0)}(\boldsymbol{x}, \boldsymbol{\omega})$ and $Y^{(f)} = Y^{(f)}(\boldsymbol{\xi}, \boldsymbol{\omega})$ corresponds to a different equilibrium ensemble. Let $Q^{(0)}(\boldsymbol{x}, \boldsymbol{\omega})$ be the partition function of the system under fixed macroscopic constraints $\boldsymbol{x}$ and $\boldsymbol{\omega}$, i.e. in the $[\boldsymbol{x}, \boldsymbol{\omega}]$ ensemble. Then,

$$Y^{(0)}(\boldsymbol{x}, \boldsymbol{\omega}) = -\frac{1}{\beta} \ln Q^{(0)}(\boldsymbol{x}, \boldsymbol{\omega}). \tag{4.2.4}$$

In the $[\boldsymbol{\xi}, \boldsymbol{\omega}]$ ensemble, the system is governed by a partition function $Q^{(f)}(\boldsymbol{\xi}, \boldsymbol{\omega})$. The relationship between this partition function and the thermodynamic potential $Y^{(f)}$ is again

$$Y^{(f)}(\boldsymbol{\xi}, \boldsymbol{\omega}) = -\frac{1}{\beta} \ln Q^{(f)}(\boldsymbol{\xi}, \boldsymbol{\omega}). \tag{4.2.5}$$

The partition function $Q^{(f)}(\boldsymbol{\xi}, \boldsymbol{\omega})$ is obtained from the partition function $Q^{(0)}(\boldsymbol{x}, \boldsymbol{\omega})$ through the mathematical operation

$$Q^{(f)}(\boldsymbol{\xi}, \boldsymbol{\omega}) = \text{const.} \int Q^{(0)}(\boldsymbol{x}, \boldsymbol{\omega}) \exp(\beta \boldsymbol{\xi} \cdot \boldsymbol{x}) \, \mathrm{d}^f x, \tag{4.2.6}$$

where, unless otherwise indicated, $\beta = 1/(k_B T)$.

Indeed, in the thermodynamic limit where the phase-space distribution defined by $Q^{(f)}(\boldsymbol{\xi}, \boldsymbol{\omega})$ becomes sharply peaked around the average values $\langle \boldsymbol{x} \rangle_{\xi\omega}$, Eq. (4.2.6) gives

$$Q^{(f)}(\boldsymbol{\xi}, \boldsymbol{\omega}) = \text{const}'.Q^{(0)}\left(\langle \boldsymbol{x} \rangle_{\xi\omega}, \boldsymbol{\omega}\right) \exp\left(\beta \boldsymbol{\xi} \cdot \langle \boldsymbol{x} \rangle_{\xi\omega}\right)$$

or

$$-\frac{1}{\beta} \ln Q^{(f)}(\boldsymbol{\xi}, \boldsymbol{\omega}) = -\frac{1}{\beta} \ln Q^{(0)}\left(\langle \boldsymbol{x} \rangle_{\xi\omega}, \boldsymbol{\omega}\right) - \boldsymbol{\xi} \cdot \langle \boldsymbol{x} \rangle_{\xi\omega} - \frac{1}{\beta} \ln(\text{const}'.)$$

or

$$Y^{(f)}(\boldsymbol{\xi}, \boldsymbol{\omega}) = Y^{(0)}\left(\langle \boldsymbol{x} \rangle_{\xi\omega}, \boldsymbol{\omega}\right) - \boldsymbol{\xi} \cdot \langle \boldsymbol{x} \rangle_{\xi\omega} - \frac{1}{\beta} \ln(\text{const.}'), \tag{4.2.7}$$

which is Eq. (4.2.2) up to an additive constant which does not depend on $\boldsymbol{\xi}$ or $\boldsymbol{\omega}$. Changing between fundamental representations in thermodynamics, therefore, corresponds to changing between ensembles in statistical mechanics, and Eq. (4.2.6) is the analogue of performing a Legendre transformation.

Equation (4.2.6) defines a probability density for $\boldsymbol{x}$ in the $[\boldsymbol{\xi}, \boldsymbol{\omega}]$ ensemble:

$$\rho^{\xi\omega}(\boldsymbol{x}) = \frac{Q^{(0)}(\boldsymbol{x}, \boldsymbol{\omega}) \exp(\beta \boldsymbol{\xi} \cdot \boldsymbol{x})}{\int Q^{(0)}(\boldsymbol{x}, \boldsymbol{\omega}) \exp(\beta \boldsymbol{\xi} \cdot \boldsymbol{x}) \, \mathrm{d}^f x}. \tag{4.2.8}$$

Some concrete examples are presented below.

From the micronanonical (nVE) *to the canonical* (nVT) *ensemble:* Here $f=1$, $\boldsymbol{x}=E$, $\boldsymbol{\omega}=(V,n)$, $\beta=1/k_{\mathrm{B}}$, $Y^{(0)}=-S(E,V,n)$ the negative entropy, $\boldsymbol{\xi}=-1/T$, $Q^{(0)}=W(E,V,n)$ the volume of the E-hypersurface in phase space (number of microstates), $Y^{(1)}=\frac{A}{T}\left(-\frac{1}{T},V,n\right)=\frac{1}{T}A(n,V,T)$ the Helmholtz energy divided by temperature, $Q^{(1)}=\text{const.}\int W(E,V,n)\exp\left(-\frac{E}{k_{\mathrm{B}}T}\right)\mathrm{d}E=Q(n,V,T)$ the canonical partition function,

and $\rho^{\xi\omega}(\boldsymbol{x})=\rho^{nVT}(E)=\dfrac{W(E,V,n)\exp\left[-\frac{E}{k_{\mathrm{B}}T}\right]}{\int W(E,V,n)\exp\left[-\frac{E}{k_{\mathrm{B}}T}\right]\mathrm{d}E}$.

From the canonical (nVT) *to the isothermal-isobaric* (nPT) *ensemble:* Here $f=1$, $\boldsymbol{x}=V$, $\boldsymbol{\omega}=(T,n)$, $Y^{(0)}=A(V,T,n)$, $Q^{(0)}=Q(V,T,n)$ the canonical partition function, $\boldsymbol{\xi}=-P$, $Y^{(1)}=G(-P,T,n)$ $=G(n,P,T)$ the Gibbs energy, $Q^{(1)}=\text{const.}\int Q(V,T,n)\exp(-PV)\,\mathrm{d}V=Q(n,P,T)$ the isothermal–isobaric partition function, and $\rho^{\xi\omega}(\boldsymbol{x})=\rho^{nPT}(V)=\dfrac{Q(V,T,n)\exp(-PV)}{\int Q(V,T,n)\exp(-PV)\,\mathrm{d}V}$.

From the canonical (nVT) to the grand canonical (μVT) ensemble in a pure component system: Here $f=1$, $\boldsymbol{x}=n$, $\boldsymbol{\omega}=(V,T)$, $Y^{(0)}=A(n,V,T)$, $Q^{(0)}=Q(n,V,T)$ the canonical partition function, $\boldsymbol{\xi}=\mu$, $Y^{(1)}=\Omega(\mu,V,T)$ the grand potential, $Q^{(1)}=\sum_{n=1}^{\infty}Q(n,V,T)\exp(\mu n)=\Xi(\mu,V,T)$ the grand canonical partition function, and $\rho^{\xi\omega}(\boldsymbol{x})=\rho^{\mu VT}(n)=\dfrac{Q(n,V,T)\exp(\mu n)}{\sum_{n=1}^{\infty}Q(n,V,T)\exp(\mu n)}$ a discrete probability distribution.

From the canonical (n_1n_2VT) *to the grand canonical* $(\mu_1\mu_2VT)$ *ensemble in a binary system:* Here $f=2$, $\boldsymbol{x}=(n_1,n_2)$, $\boldsymbol{\omega}=(V,T)$, $Y^{(0)}=A(n_1,n_2,V,T)$, $Q^{(0)}=Q(n_1,n_2,V,T)$ the canonical partition function, $\boldsymbol{\xi}=(\mu_1,\mu_2)$, $Y^{(2)}=\Omega(\mu_1,\mu_1,V,T)$ the grand potential, $Q^{(2)}=\sum_{n_1=1}^{\infty}\sum_{n_2=1}^{\infty}Q(n_1,n_2,V,T)\exp(\mu_1n_1+\mu_2n_2)=\Xi(\mu_1,\mu_2,V,T)$ the grand canonical partition function, and $\rho^{\xi\omega}(\boldsymbol{x})=\rho^{\mu_1\mu_2VT}(n_1,n_2)=\dfrac{Q(n_1,n_2,V,T)\exp(\mu_1n_1+\mu_2n_2)}{\sum_{n_1=1}^{\infty}\sum_{n_2=1}^{\infty}Q(n_1,n_2,V,T)\exp(\mu_1n_1+\mu_2n_2)}$ a discrete probability distribution.

From the isothermal–isostrain $(n\varepsilon T)$ *to the isothermal–isostress* $(n\boldsymbol{\tau}T)$ *ensemble for small deformations* (compare Section 3.5.3 and Eq. (3.5.78)): Here $f=9$, $\boldsymbol{x}=\boldsymbol{\varepsilon}$ the strain tensor, $\boldsymbol{\omega}=(T,n)$, $Y^{(0)}=A(n,\varepsilon,T)$, $Q^{(0)}=Q(n,\varepsilon,T)$ the canonical partition function, $\boldsymbol{\xi}=\boldsymbol{\tau}$ the stress tensor, $Y^{(1)}=G(n,\boldsymbol{\tau},T)+\frac{1}{3}\mathrm{tr}(\boldsymbol{\tau})V^{(0)}$ with G being the Gibbs energy and $V^{(0)}$ the volume in the reference undeformed state, $Q^{(f)}=\text{const.}\int Q(n,\varepsilon,T)\exp\left(\beta V^{(0)}\boldsymbol{\tau}:\varepsilon\right)\mathrm{d}\varepsilon$, and

$$\rho^{\xi\omega}(\boldsymbol{x})=\rho^{n\tau T}(\varepsilon)=\frac{Q(n,\varepsilon,T)\exp[\beta\boldsymbol{\tau}:\varepsilon]}{\int Q(n,\varepsilon,T)\exp[\beta\boldsymbol{\tau}:\varepsilon]\,d\varepsilon}.$$

Consider the thermodynamic derivative $\left.\frac{\partial\langle x_i\rangle_{\xi,\omega}}{\partial\xi_i}\right|_{\xi_{[i]},\omega}$, with $1 \le i \le f$. By Eq. (4.2.8),

$$\langle x_i\rangle_{\xi,\omega} = \int x_i \rho^{\xi\omega}(\boldsymbol{x})\mathrm{d}^f x = \frac{\int x_i Q^{(0)}(\boldsymbol{x},\boldsymbol{\omega})\exp(\beta\boldsymbol{\xi}\cdot\boldsymbol{x})\,\mathrm{d}^f x}{\int Q^{(0)}(\boldsymbol{x},\boldsymbol{\omega})\exp(\beta\boldsymbol{\xi}\cdot\boldsymbol{x})\,\mathrm{d}^f x} \tag{4.2.9}$$

Hence,

$$\begin{aligned}
\left.\frac{\partial\langle x_i\rangle_{\xi,\omega}}{\partial\xi_i}\right|_{\xi_{[i]},\omega} &= \frac{\left[\beta\int x_i^2 Q^{(0)}(\boldsymbol{x},\boldsymbol{\omega})\exp(\beta\boldsymbol{\xi}\cdot\boldsymbol{x})\,\mathrm{d}^f x\right]\left[\int Q^{(0)}(\boldsymbol{x},\boldsymbol{\omega})\exp(\beta\boldsymbol{\xi}\cdot\boldsymbol{x})\,\mathrm{d}^f x\right] - \beta\left[\int x_i Q^{(0)}(\boldsymbol{x},\boldsymbol{\omega})\exp(\beta\boldsymbol{\xi}\cdot\boldsymbol{x})\,\mathrm{d}^f x\right]^2}{\left[\int Q^{(0)}(\boldsymbol{x},\boldsymbol{\omega})\exp(\beta\boldsymbol{\xi}\cdot\boldsymbol{x})\,\mathrm{d}^f x\right]^2} \\
&= \beta\left\{\frac{\int x_i^2 Q^{(0)}(\boldsymbol{x},\boldsymbol{\omega})\exp(\beta\boldsymbol{\xi}\cdot\boldsymbol{x})\,\mathrm{d}^f x}{\int Q^{(0)}(\boldsymbol{x},\boldsymbol{\omega})\exp(\beta\boldsymbol{\xi}\cdot\boldsymbol{x})\,\mathrm{d}^f x} - \left(\frac{\left[\int x_i Q^{(0)}(\boldsymbol{x},\boldsymbol{\omega})\exp(\beta\boldsymbol{\xi}\cdot\boldsymbol{x})\,\mathrm{d}^f x\right]}{\left[\int Q^{(0)}(\boldsymbol{x},\boldsymbol{\omega})\exp(\beta\boldsymbol{\xi}\cdot\boldsymbol{x})\,\mathrm{d}^f x\right]}\right)^2\right\} \\
&= \beta\left[\langle x_i^2\rangle_{\xi\omega} - \langle x_i\rangle^2_{\xi\omega}\right] = \beta\left\langle\left(x_i - \langle x_i\rangle_{\xi\omega}\right)^2\right\rangle_{\xi\omega} = \beta\left\langle(\delta x_i)^2\right\rangle_{\xi\omega}.
\end{aligned} \tag{4.2.10}$$

Thus, the susceptibility $\left.\frac{\partial\langle x_i\rangle_{\xi,\omega}}{\partial\xi_i}\right|_{\xi_{[i]},\omega}$ is expressed in terms of the variance of x_i in the course of its equilibrium fluctuations in the $[\boldsymbol{\xi},\boldsymbol{\omega}]$ ensemble.

Let us also consider the thermodynamic derivative $\left.\frac{\partial\langle x_i\rangle_{\xi,\omega}}{\partial\xi_j}\right|_{\xi_{[j]},\omega}$, with $1 \le i \le f, 1 \le j \le f, i \ne j$. Using Eq. (4.2.9),

$$\begin{aligned}
\left.\frac{\partial\langle x_i\rangle_{\xi,\omega}}{\partial\xi_j}\right|_{\xi_{[j]},\omega} &= \frac{\left[\beta\int x_i x_j Q^{(0)}(\boldsymbol{x},\boldsymbol{\omega})\exp(\beta\boldsymbol{\xi}\cdot\boldsymbol{x})\,\mathrm{d}^f x\right]\left[\int Q^{(0)}(\boldsymbol{x},\boldsymbol{\omega})\exp(\beta\boldsymbol{\xi}\cdot\boldsymbol{x})\,\mathrm{d}^f x\right]}{\left[\int Q^{(0)}(\boldsymbol{x},\boldsymbol{\omega})\exp(\beta\boldsymbol{\xi}\cdot\boldsymbol{x})\,\mathrm{d}^f x\right]^2} \\
&\quad - \frac{\left[\int x_i Q^{(0)}(\boldsymbol{x},\boldsymbol{\omega})\exp(\beta\boldsymbol{\xi}\cdot\boldsymbol{x})\,\mathrm{d}^f x\right]\left[\beta\int x_j Q^{(0)}(\boldsymbol{x},\boldsymbol{\omega})\exp(\beta\boldsymbol{\xi}\cdot\boldsymbol{x})\,\mathrm{d}^f x\right]}{\left[\int Q^{(0)}(\boldsymbol{x},\boldsymbol{\omega})\exp(\beta\boldsymbol{\xi}\cdot\boldsymbol{x})\,\mathrm{d}^f x\right]^2} \\
&= \beta\left(\frac{\int x_i x_j Q^{(0)}(\boldsymbol{x},\boldsymbol{\omega})\exp(\beta\boldsymbol{\xi}\cdot\boldsymbol{x})\,\mathrm{d}^f x}{\int Q^{(0)}(\boldsymbol{x},\boldsymbol{\omega})\exp(\beta\boldsymbol{\xi}\cdot\boldsymbol{x})\,\mathrm{d}^f x}\right. \\
&\quad \left. - \frac{\int x_i Q^{(0)}(\boldsymbol{x},\boldsymbol{\omega})\exp(\beta\boldsymbol{\xi}\cdot\boldsymbol{x})\,\mathrm{d}^f x}{\int Q^{(0)}(\boldsymbol{x},\boldsymbol{\omega})\exp(\beta\boldsymbol{\xi}\cdot\boldsymbol{x})\,\mathrm{d}^f x}\frac{\int x_j Q^{(0)}(\boldsymbol{x},\boldsymbol{\omega})\exp(\beta\boldsymbol{\xi}\cdot\boldsymbol{x})\,\mathrm{d}^f x}{\int Q^{(0)}(\boldsymbol{x},\boldsymbol{\omega})\exp(\beta\boldsymbol{\xi}\cdot\boldsymbol{x})\,\mathrm{d}^f x}\right) \\
&= \beta\left(\langle x_i x_j\rangle_{\xi\omega} - \langle x_i\rangle_{\xi\omega}\langle x_j\rangle_{\xi\omega}\right) = \beta\left\langle\left(x_i - \langle x_i\rangle_{\xi\omega}\right)\left(x_j - \langle x_j\rangle_{\xi\omega}\right)\right\rangle_{\xi\omega} \\
&= \beta\langle(\delta x_i)(\delta x_j)\rangle_{\xi\omega}.
\end{aligned} \tag{4.2.11}$$

The thermodynamic derivative $\left.\dfrac{\partial \langle x_i \rangle_{\xi,\omega}}{\partial \xi_j}\right|_{\xi_{[j]},\omega}$ is thus seen to equal the covariance of x_i and x_j in the course of their equilibrium fluctuations in the $[\boldsymbol{\xi}, \omega]$ ensemble. In fact, Eq. (4.2.10) can be considered as a special case of Eq. (4.2.11) for $i = j$.

Using Eq. (4.2.10) on the transition from the microcanonical to the canonical ensemble outlined above, we obtain that

$$\left.\frac{\partial \langle E \rangle_{nVT}}{\partial \left(-\frac{1}{T}\right)}\right|_{n,V} = \frac{1}{k_\mathrm{B}} \left\langle (\delta E)^2 \right\rangle_{nVT}, \text{ or } T^2 \left.\frac{\partial \langle E \rangle_{nVT}}{\partial T}\right|_{n,V} = \frac{1}{k_\mathrm{B}} \left\langle (\delta E)^2 \right\rangle_{nVT}$$

or

$$\left.\frac{\partial \langle E \rangle_{nVT}}{\partial T}\right|_{n,V} = \frac{1}{k_\mathrm{B} T^2} \left\langle (\delta E)^2 \right\rangle_{nVT}, \text{ or } \left.\frac{\partial U}{\partial T}\right|_{n,V} = \frac{1}{k_\mathrm{B} T^2} \left\langle (\delta E)^2 \right\rangle_{nVT}$$

or

$$C_V = \frac{1}{k_\mathrm{B} T^2} \left\langle (\delta E)^2 \right\rangle_{nVT}. \tag{4.2.12}$$

Since $E = \mathcal{V} + \mathcal{K}$ and fluctuations of $\mathcal{V}$ and $\mathcal{K}$ are uncorrelated, Eq. (4.2.12) can be written equivalently as

$$C_V = \frac{1}{k_\mathrm{B} T^2} \left[\left\langle (\delta \mathcal{V})^2 \right\rangle_{nVT} + \left\langle (\delta \mathcal{K})^2 \right\rangle_{nVT} \right] = \frac{1}{k_\mathrm{B} T^2} \left\langle (\delta \mathcal{V})^2 \right\rangle_{nVT} + C_V^{\mathrm{ig}}. \tag{4.2.13}$$

where C_V^{ig} is the heat capacity the system would have if it were an ideal gas, devoid of any interactions, at the same n, V, and T. For a classical system, by the equipartition theorem, $C_V^{\mathrm{ig}} = \frac{3}{2} n k_\mathrm{B}$ and therefore

$$C_V - \frac{3}{2} n k_\mathrm{B} = \frac{1}{k_\mathrm{B} T^2} \left\langle (\delta \mathcal{V})^2 \right\rangle_{nVT}. \tag{4.2.14}$$

Equation (4.2.12) expresses the heat capacity of a system at constant volume in terms of equilibrium fluctuations in its total energy in the canonical ensemble. Its special case, Eq. (4.2.14), expresses the heat capacity at constant volume in terms of the equilibrium fluctuations of the system's total potential energy in the canonical ensemble.

Using Eq. (4.2.10) for the transition from canonical to isothermal–isobaric ensemble in a single-component system, outlined earlier, we obtain

$$\left.\frac{\partial\langle V\rangle_{nPT}}{\partial(-P)}\right|_{n,T} = \frac{1}{k_{\mathrm{B}}T}\left\langle(\delta V)^2\right\rangle_{nPT}$$

or

$$-\frac{1}{\langle V\rangle_{nPT}}\left.\frac{\partial\langle V\rangle_{nPT}}{\partial P}\right|_{n,T} = \frac{1}{k_{\mathrm{B}}T}\frac{\left\langle(\delta V)^2\right\rangle_{nPT}}{\langle V\rangle_{nPT}}$$

or

$$\kappa_T = \frac{1}{k_{\mathrm{B}}T}\frac{\left\langle(\delta V)^2\right\rangle_{nPT}}{\langle V\rangle_{nPT}}. \qquad (4.2.15)$$

Equation (4.2.15) allows one to estimate the isothermal compressibility κ_T from the fluctuations of volume in the isothermal–isobaric ensemble.

Using Eq. (4.2.10) for the transition from canonical to grand canonical ensemble outlined earlier, we obtain that

$$\left.\frac{\partial\langle n\rangle_{\mu VT}}{\partial\mu}\right|_{V,T} = \frac{1}{k_{\mathrm{B}}T}\left\langle(\delta n)^2\right\rangle_{\mu VT}. \qquad (4.2.16)$$

If we define the molecular density as $\rho = \frac{\langle n\rangle_{\mu VT}}{V}$, Eq. (4.2.16) can also be written as

$$\left.\frac{\partial\rho}{\partial\mu}\right|_{T} = \frac{1}{k_{\mathrm{B}}TV}\left\langle(\delta n)^2\right\rangle_{\mu VT}. \qquad (4.2.17)$$

Equations (4.2.16) and (4.2.17) are useful in a variety of situations. Their left-hand sides can be expressed in terms of readily measurable properties.

In a pure homogeneous substance, in the thermodynamic limit, the chemical potential per molecule is also the Gibbs energy per molecule,

$$\mu = \frac{G}{\langle n\rangle_{\mu VT}} = \frac{1}{N_{\mathrm{Avo}}}G_{\mathrm{m}}, \qquad (4.2.18)$$

with G_{m} being the molar Gibbs energy and N_{Avo} being Avogadro's number. Then,

$$\mathrm{d}\mu = \frac{1}{N_{\mathrm{Avo}}}\mathrm{d}G_{\mathrm{m}} = \frac{1}{N_{\mathrm{Avo}}}\left(-S_{\mathrm{m}}\mathrm{d}T + V_{\mathrm{m}}\mathrm{d}P\right)$$

and therefore

$$\left.\frac{\partial \mu}{\partial \rho}\right|_T = \frac{1}{N_{\mathrm{Avo}}} V_{\mathrm{m}} \left.\frac{\partial P}{\partial \rho}\right|_T = \frac{1}{\rho}\left.\frac{\partial P}{\partial \rho}\right|_T = \frac{1}{(n/V)}\left.\frac{\partial P}{\partial (n/V)}\right|_{n,T} = \frac{V}{n}\left(-\frac{V^2}{n}\left.\frac{\partial P}{\partial V}\right|_{n,T}\right) = \left(\frac{V}{n}\right)^2 \frac{1}{-\frac{1}{V}\left.\frac{\partial V}{\partial P}\right|_{n,T}}$$

or

$$\left.\frac{\partial \mu}{\partial \rho}\right|_T = \frac{1}{\rho^2 \kappa_T}, \tag{4.2.19}$$

where κ_T is the isothermal compressibility. From Eqs. (4.2.17) and (4.2.19) we obtain that

$$\rho^2 \kappa_T = \frac{1}{k_{\mathrm{B}} T V}\left\langle (\delta n)^2 \right\rangle_{\mu VT}$$

or

$$\rho \kappa_T = \frac{1}{k_{\mathrm{B}} T} \frac{\left\langle (\delta n)^2 \right\rangle_{\mu VT}}{\langle n \rangle_{\mu VT}}$$

or

$$\kappa_T = \frac{V}{k_{\mathrm{B}} T} \frac{\left\langle (\delta n)^2 \right\rangle_{\mu VT}}{\langle n \rangle^2_{\mu VT}}, \tag{4.2.20}$$

from which it is clear that the isothermal compressibility is obtainable from the fluctuations in number of particles or density in an open system, governed by the grand canonical ensemble. Equation (4.2.20) forms the basis for measuring isothermal compressibility by scattering radiation from the system, in the limit of small wave vectors (large length scales). For a fluid approaching its critical point, density fluctuations diverge and so does the isothermal compressibility.

Equation (4.2.17) is also useful in the analysis of equilibrium sorption isotherms of fluids in microporous solids. One can often neglect the dimensional or structural changes caused in the solid by the sorbed fluid, in which case the solid phase can be modelled as a rigid collection of atoms occupying a constant volume and exerting a potential energy field on the sorbed molecules. The solid phase can exchange fluid molecules with the surrounding bulk fluid phase which is at constant chemical potential and the grand canonical ensemble is applicable. The sorption isotherm is a functional relationship between the concentration or molecular density ρ of sorbed molecules in the solid and the chemical potential μ or, equivalently, the fugacity $f = k_{\mathrm{B}} T \frac{Z^{\mathrm{sorb}}_{\mathrm{intra}}}{\prod_{i=1}^{n_{\mathrm{at,sorb}}} \Lambda_i^3} \exp\left(\frac{\mu}{k_{\mathrm{B}} T}\right)$ of the sorbate. In the latter expression for the fugacity, Λ_i is the thermal wavelength of each of the $n_{\mathrm{at,sorb}}$ atoms constituting a sorbate molecule and $Z^{\mathrm{sorb}}_{\mathrm{intra}}$ is the intramolecular configurational integral of an isolated sorbate molecule over all but its three translational degrees of freedom. Λ_i and $Z^{\mathrm{sorb}}_{\mathrm{intra}}$ depend only on temperature and molecular characteristics of the sorbate. The slope of the isotherm at a particular fugacity is $\left.\dfrac{\partial \rho}{\partial f}\right|_T$. This slope is often expressed in logarithmic coordinates, $\left.\dfrac{\partial \ln \rho}{\partial \ln f}\right|_T = \dfrac{k_B T}{\rho} \left.\dfrac{\partial \rho}{\partial \mu}\right|_T$. Considering Eq. (4.2.17),

$$\left.\frac{\partial \ln \rho}{\partial \ln f}\right|_T = \frac{k_B T}{\rho} \frac{1}{k_B T V} \left\langle (\delta n)^2 \right\rangle_{\mu VT} = \frac{\left\langle (\delta n)^2 \right\rangle_{\mu VT}}{\langle n \rangle_{\mu VT}}. \tag{4.2.21}$$

Thus, the factor $\left.\dfrac{\partial \ln \rho}{\partial \ln f}\right|_T$ along an equilibrium sorption isotherm informs us about density fluctuations of the sorbed fluid inside the sorbent. At very low fugacities, different sorbate molecules hardly see each other; density fluctuations are uncorrelated and the right-hand side of Eq. (4.2.21) approaches 1. One can understand this limiting value if one applies Eq. (4.2.20) to the case of an ideal gas. Thus, in the limit of very low fugacities the sorption isotherm is linear (Henry's law region) and the sorption equilibrium is dominated by sorbent–sorbate interactions. At higher fugacities, however, sorbate density fluctuations are usually constrained by the confining environment of the sorbent and the sorption isotherm becomes concave. A steeply rising step in the isotherm may be indicative of a phase transition in the sorbed fluid that causes density fluctuations to diverge.

By using Eqs. (4.2.10) and (4.2.11) for the transition from canonical to grand canonical ensemble in a two-component system, one derives equations (see e.g. Problem 4.3) relating the partial derivatives of molecular densities with respect to chemical potentials in an open binary system to the fluctuations in the number of molecules of each species that are found within a fixed macroscopic volume V. These equations can serve as starting points for determining the derivatives of chemical potentials with respect to mole fractions under constant temperature and pressure, hence the Gibbs energy of mixing of the binary system. This is done in the context of the theory of Kirkwood and Buff, an exact thermodynamic theory of multicomponent systems.[32,33] Kirkwood–Buff theory has been used to extract mixing thermodynamics from molecular simulations.[34]

Using Eq. (4.2.10) for the transition from isothermal–isostrain to isothermal–isostress ensemble discussed earlier, we obtain that

$$\left.\frac{\partial \langle \varepsilon_{\alpha\beta} \rangle_{n\tau T}}{\partial \tau_{\zeta\eta}}\right|_{n,\tau_{[\zeta\eta]},T} = V^{(0)} \frac{\langle (\delta \varepsilon_{\alpha\beta})(\delta \varepsilon_{\zeta\eta}) \rangle_{n\tau T}}{k_B T} = V^{(0)} \frac{\langle \varepsilon_{\alpha\beta} \varepsilon_{\zeta\eta} \rangle_{n\tau T} - \langle \varepsilon_{\alpha\beta} \rangle_{n\tau T} \langle \varepsilon_{\zeta\eta} \rangle_{n\tau T}}{k_B T}. \tag{4.2.22}$$

By definition, the left-hand side of Eq. (4.2.22) is the $\alpha\beta\zeta\eta$-element of the fourth-order isothermal compliance tensor $\boldsymbol{S}$. The isothermal compliance tensor is the inverse of the isothermal fourth-order tensor of isothermal elastic coefficients $\boldsymbol{C}$, defined in Eq. (3.5.52):

$$\boldsymbol{S} = \boldsymbol{C}^{-1}. \tag{4.2.23}$$

Equation (4.2.22) can therefore be rewritten as

$$S_{\alpha\beta\zeta\eta} = V^{(0)} \frac{\langle \varepsilon_{\alpha\beta} \varepsilon_{\zeta\eta} \rangle_{n\tau T} - \langle \varepsilon_{\alpha\beta} \rangle_{n\tau T} \langle \varepsilon_{\zeta\eta} \rangle_{n\tau T}}{k_B T}. \tag{4.2.24}$$

Equation (4.2.24) is strategically important, for it allows calculating the compliance tensor $\boldsymbol{S}$, hence the tensor of isothermal elastic coefficients $\boldsymbol{C}$ through Eq. (4.2.23), from spontaneous fluctuations of the system dimensions under isothermal isostress conditions. If the simulation is carried out in the undeformed reference state, $\langle \varepsilon_{\alpha\beta} \rangle_{n\tau T}$ and $\langle \varepsilon_{\zeta\eta} \rangle_{n\tau T}$ are 0 and all that matters is the covariance in Eq. (4.2.24) is the different elements of the strain tensor. One then obtains that

$$C_{\alpha\beta\zeta\eta} = \frac{k_{\mathrm{B}} T}{\langle V \rangle_{n\tau T}} \langle \varepsilon_{\alpha\beta} \varepsilon_{\zeta\eta} \rangle_{n\tau T}^{-1}. \tag{4.2.25}$$

Equation (4.2.25) is generally known as the Parrinello–Rahman fluctuation formula.[35]

Gusev, Zehnder, and Suter[36] found Eq. (4.2.25) to be slowly converging in actual variable-shape Monte Carlo simulations of a first nearest neighbour Lennard-Jones fcc crystal. They proposed an alternative equation for the elements $C_{\alpha\beta\zeta\eta}$ of the fourth-order tensor of isothermal elastic coefficients, which involves fluctuations in both strain and instantaneous stress. At low temperatures, where the difference between the instantaneous $\boldsymbol{h}$ and the average $\langle \boldsymbol{h} \rangle_{n\tau T}$ cell matrices can be neglected (compare Section 3.5.1), their equation reads

$$C_{\alpha\beta\zeta\eta} = \sum_{\gamma=1}^{3} \sum_{\xi=1}^{3} \langle \varepsilon_{\alpha\beta} \sigma_{\gamma\xi} \rangle_{n\tau T} \langle \varepsilon_{\gamma\xi} \varepsilon_{\zeta\eta} \rangle_{n\tau T}^{-1}. \tag{4.2.26}$$

In Eq. (4.2.26), $\boldsymbol{\sigma}$ is the instantaneous stress tensor given by Eq. (3.2.24). The estimate of elastic constants given by Eq. (4.2.26) converges more quickly than that from Eq. (4.2.25) because instantaneous stresses and strains are strongly correlated.[3]

Several other fluctuation relations for the volumetric, thermal, and mechanical properties of atomic systems in various ensembles can be found in the original articles.[37–46]

4.3 Probing segmental dynamics

4.3.1 Time-correlation functions and their Fourier transforms

Let $\mathcal{A}\left(\boldsymbol{p}(t), \boldsymbol{q}(t)\right)$ be some function of the phase-space coordinates; we usually write $\mathcal{A}\left(\boldsymbol{p}(t), \boldsymbol{q}(t)\right) = \mathcal{A}(t)$. If then $\rho\left(\boldsymbol{p}, \boldsymbol{q}\right)$ is the equilibrium phase-space distribution function, we define the classical time-correlation function of $\mathcal{A}(t)$ as

$$C_{\mathcal{A}}(t) = \langle \mathcal{A}(0)\mathcal{A}(t) \rangle = \int \ldots \int \mathcal{A}(0)\mathcal{A}(t)\rho\left(\boldsymbol{p}, \boldsymbol{q}\right) \mathrm{d}\boldsymbol{p}\mathrm{d}\boldsymbol{q}, \tag{4.3.1}$$

where $\mathrm{d}\boldsymbol{p}\mathrm{d}\boldsymbol{q}$ stands for $\mathrm{d}\boldsymbol{p}_1(0)\mathrm{d}\boldsymbol{p}_2(0)\ldots\mathrm{d}\boldsymbol{p}_n(0)\mathrm{d}\boldsymbol{q}_1(0)\mathrm{d}\boldsymbol{q}_2(0)\ldots\mathrm{d}\boldsymbol{q}_n(0)$ and n is the total number of interaction sites (atoms) in the system. The integration takes place over all initial phase-space points of the system, assumed to be distributed according to the requirements of equilibrium. We can also define time-correlation functions for two different phase-space functions, $\mathcal{A}\left(\boldsymbol{p}(t), \boldsymbol{q}(t)\right)$ and $\mathcal{B}\left(\boldsymbol{p}(t), \boldsymbol{q}(t)\right)$:

$$C_{\mathcal{A}\mathcal{B}}(t) = \langle \mathcal{A}(0)\mathcal{B}(t) \rangle = \int \ldots \int \mathcal{A}(0)\mathcal{B}(t)\rho\left(\boldsymbol{p}, \boldsymbol{q}\right) \mathrm{d}\boldsymbol{p}\mathrm{d}\boldsymbol{q}. \tag{4.3.2}$$

When $A = B$, the correlation function is an autocorrelation function. According to the ergodic hypothesis, the equilibrium ensemble averages in Eqs. (4.3.1) and (4.3.2) can be replaced by time averages along a very long equilibrium trajectory evolving according to the macroscopic constraints of the considered ensemble.

Time-correlation functions have played an important role in statistical mechanics for two main reasons: first, thanks to the work initiated by Green[47] and Kubo,[48] one can show that the phenomenological coefficients describing many transport processes can be written as integrals over such time-correlation functions for specific choices of the variables $\mathcal{A}$ and $\mathcal{B}$. For example, the coefficient of self-diffusion D can be expressed in terms of the velocity time-correlation function by

$$D = \frac{1}{3} \int_0^\infty \langle \boldsymbol{\nu}(0) \cdot \boldsymbol{\nu}(t) \rangle \, \mathrm{d}t. \tag{4.3.3}$$

Similarly, the shear stress relaxation modulus can be written in terms of the time-correlation function of the off-diagonal elements of the instantaneous stress tensor by

$$G(t) = \frac{V}{k_{\mathrm{B}} T} \langle \sigma_{\alpha\beta}(0) \sigma_{\alpha\beta}(t) \rangle, \quad \alpha \neq \beta, \tag{4.3.4}$$

from which the zero shear rate viscosity can also be obtained as

$$\eta_0 = \int_0^\infty G(t) \mathrm{d}t. \tag{4.3.5}$$

Second, the Fourier transforms of certain time-correlation functions describe the way the system responds to a time-dependent perturbation. Through this, we analyse the response in terms of a time-dependent susceptibility $\sigma^*(\omega)$ in the frequency domain ω given by

$$\sigma^*(\omega) = \beta \int_0^\infty \left\langle \dot{\mathcal{A}}(0) \mathcal{A}(t) \right\rangle \exp\left(-i\omega t\right) \mathrm{d}t, \tag{4.3.6}$$

where $\beta = \frac{1}{k_{\mathrm{B}} T}$, the dot above $\mathcal{A}$ signifies a time derivative, and the superscript * is used to emphasize that σ is a complex quantity.

4.3.2 Calculating the observables of dielectric spectroscopy, quasielastic neutron scattering, and ^{13}C-NMR spectroscopy, and comparison with experimental measurements

Segmental dynamics refers to very local motions along a polymer chain associated with the mobility of small units making up the chain. These could be atoms (e.g. hydrogens), or groups of atoms (e.g. a methyl unit), or pendant or side units within a monomer (e.g. a pendant phenyl ring), or individual monomers (e.g. the methylene or ethylene unit), or even groups of monomers or side groups. In polypropylene (PP), for example, segmental dynamics refers to the dynamics of individual hydrogen atoms but also to the dynamics of bond angle bending vibrations, of torsional oscillations, and, at longer times, of correlated conformational transitions of the skeletal bonds, and to torsional librations (within an energy well) and transitions (between different energy wells) of the side methyl units. In general, the more complex the chemical structure or the molecular architecture of the polymer chain, the richer its segmental dynamics.

Segmental dynamics is important because it is connected with a wealth of phenomena such as the glass transition and the secondary relaxations responsible for shaping the mechanical properties of glassy polymers. It can also be responsible for the different barrier properties of polymers with practically the same chemical structure but with a slightly different molecular architecture. For example, the different permeability properties of two poly-isomers, poly(ethylene terephthalate) (PET) and poly(ethylene isophthalate) (PEI), are believed to be connected with the different mobilities of the phenyl rings. Although the structural and conformational properties of the two polyesters are practically identical, the phenyl rings in PET exhibit a significantly higher mobility than in PEI and this is related to the way they are connected to their adjacent ester groups in the two cases. Thus, PEI is less permeable to small molecules (such as oxygen) than PEI by a factor of 2–2.5 in the amorphous state.[49]

Experimentally, segmental dynamics is probed with a number of techniques, the most widely employed ones being dielectric spectroscopy (DS), quasielastic neutron scattering (QENS), and nuclear magnetic resonance (NMR). Their theoretical underpinnings are discussed in Sections 4.3.2.1–4.3.2.4.

4.3.2.1 Dielectric spectroscopy

Dielectric spectroscopy probes the relaxation of a set of dipoles in a time-dependent external electric field. It is a method useful to apply in cases where the polymer chain is characterized by a dipole moment vector either parallel (as in *cis*-1,4 polyisoprene, PI) or perpendicular (as in polyvinyl chloride, polystyrene, and *cis*-1,4 polybutadiene, PB) to the chain skeleton, or in a flexible side group (as in poly(methyl methacrylate)). Let us consider a nonconducting isotropic material sample and an externally applied electric field $\boldsymbol{E}(t)$. Since the system is nonconducting, the electric field induces a net dipole moment in the system. If $\boldsymbol{M}_{\mathrm{e}}(t)$ is the total dipole moment (electric polarization density $\mathcal{P} = \langle \boldsymbol{M}_{\mathrm{e}} \rangle / V$), then the Hamiltonian of the perturbed system is[1]

$$\mathcal{H} = \mathcal{H}_0 - \boldsymbol{M}_{\mathrm{e}}(t) \cdot \boldsymbol{E}(t), \tag{4.3.7}$$

where $\mathcal{H}_0$ is the Hamiltonian of the system in the absence of the electric field. In the frequency domain, the response of the system is analysed in terms of the susceptibility $\sigma^*(\omega)$; here this is the electric susceptibility $\chi_e^*(\omega)$ which is related to the dielectric constant or relative permittivity through

$$\varepsilon_r^*(\omega) = 1 + \chi_e^*(\omega). \tag{4.3.8}$$

According to Eq. (4.3.6),

$$\chi_e^*(\omega) = \int_0^\infty \Phi(t) \exp(-i\omega t)\, \mathrm{d}t, \tag{4.3.9}$$

where the function $\Phi(t)$ is given now in terms of the total dipole moment $\boldsymbol{M}_e(t)$ induced in the system as[1,50,51]

$$\Phi(t) = \frac{\beta}{\varepsilon_0 V}\left\langle \dot{\boldsymbol{M}}_e(0) \cdot \boldsymbol{M}_e(t) \right\rangle, \tag{4.3.10}$$

with brackets $\langle \ldots \rangle$ representing an ensemble average in the system of volume V. The constant ε_0 is the electric permittivity of free space, $\varepsilon_0 = 8.85419 \times 10^{-12}\mathrm{C}/(\mathrm{Vm})$. If we neglect cross terms, the total dipole moment vector $\boldsymbol{M}_e(t)$ can be written as the sum of single-chain dipole moments $\boldsymbol{\mu}(t)$; each $\boldsymbol{\mu}(t)$ is the sum of monomer dipole moment vectors along the chain at time t. Then, we find that

$$\Phi(t) = \frac{\beta}{\varepsilon_0 V} N_{\mathrm{ch}} \langle \dot{\boldsymbol{\mu}}(0) \cdot \boldsymbol{\mu}(t) \rangle = -\frac{\beta N_{\mathrm{ch}}}{\varepsilon_0 V}\frac{\mathrm{d}}{\mathrm{d}t}\langle \boldsymbol{\mu}(0) \cdot \boldsymbol{\mu}(t) \rangle = -\frac{\beta N_{\mathrm{ch}}}{\varepsilon_0 V}\frac{\mathrm{d}\phi(t)}{\mathrm{d}t}, \tag{4.3.11}$$

where

$$\phi(t) = \langle \boldsymbol{\mu}(0) \cdot \boldsymbol{\mu}(t) \rangle \tag{4.3.12}$$

is the single-chain dipole moment autocorrelation function (DACF), a measure of the time correlation of $\boldsymbol{\mu}(t)$ with $\boldsymbol{\mu}(0)$. Initially, this is equal to $\langle \boldsymbol{\mu}^2(0) \rangle$, but as the chains reconform and diffuse due to thermal motion, it decays to 0.

The Debye theory of dielectric relaxation is obtained[1] by simply assuming that $\langle \boldsymbol{\mu}(0) \cdot \boldsymbol{\mu}(t) \rangle$ decays exponentially,

$$\langle \boldsymbol{\mu}(0) \cdot \boldsymbol{\mu}(t) \rangle = \langle \boldsymbol{\mu}^2(0) \rangle \exp\left(-\frac{t}{\tau}\right), \tag{4.3.13}$$

where τ is a time constant called the relaxation time. This results in the following Debye equation for the function $\varepsilon_r^*(\omega)$:[1]

$$\varepsilon_r^*(\omega) = \left[1 + \frac{\beta N_{\mathrm{ch}}}{\varepsilon_0 V}\langle \boldsymbol{\mu}^2(0) \rangle \frac{1}{1+\omega^2\tau^2}\right] + i\left[\frac{\beta N_{\mathrm{ch}}}{\varepsilon_0 V}\langle \boldsymbol{\mu}^2(0) \rangle \frac{\omega\tau}{1+\omega^2\tau^2}\right]. \tag{4.3.14}$$

The complex dielectric constant is usually written as

$$\varepsilon_r{}^*(\omega) = \varepsilon_r'(\omega) - i\varepsilon_r''(\omega), \tag{4.3.15}$$

where ε_r' and ε_r'' denote the real and imaginary parts. The complex dielectric permittivity $\varepsilon_r^*(\omega)$ is an experimental observable, and the expressions of the linear response theory above indicate that it can also be extracted from MD simulation data for the time autocorrelation function of the system dipole moment. By substituting Eq. (4.3.11) into Eqs. (4.3.8)–(4.3.9) and rearranging terms, we obtain that

$$\varepsilon_r^*(\omega) = 1 + \chi_e^*(\omega) = 1 + \frac{\beta N_{\text{ch}}}{\varepsilon_0 V} \int_0^\infty \left[-\frac{\mathrm{d}\phi(t)}{\mathrm{d}t} \right] \exp(-i\omega t)\, \mathrm{d}t, \tag{4.3.16}$$

which in actual applications is used in the form of the normalized equation [51–55]

$$\frac{\varepsilon_r^* - \varepsilon_{r,\infty}}{\Delta\varepsilon_r} = \int_0^\infty \left[-\frac{\mathrm{d}\tilde{\phi}(t)}{\mathrm{d}t} \right] \exp(-i\omega t)\, \mathrm{d}t = 1 - i\omega \int_0^\infty \tilde{\phi}(t) \exp(-i\omega t)\, \mathrm{d}t, \tag{4.3.17}$$

where $\Delta\varepsilon_r$ is given by

$$\Delta\varepsilon_r = \frac{\beta N_{\text{ch}} \langle \boldsymbol{\mu}^2(0) \rangle}{\varepsilon_0 V} \tag{4.3.18}$$

and $\tilde{\phi}(t)$ is the normalized chain dipole moment autocorrelation function

$$\tilde{\phi}(t) = \frac{\langle \boldsymbol{\mu}(0) \cdot \boldsymbol{\mu}(t) \rangle}{\langle \boldsymbol{\mu}^2(0) \rangle}. \tag{4.3.19}$$

$\Delta\varepsilon_r$ is equal to the difference $\Delta\varepsilon_r = \varepsilon_{r,0} - \varepsilon_{r,\infty}$ of the limiting low- and high-frequency values $\varepsilon_{r,0}$ and $\varepsilon_{r,\infty}$, respectively. In practice, $\Delta\varepsilon_r = \varepsilon_{r,\text{relaxed}} - \varepsilon_{r,\text{unrelaxed}}$, where $\varepsilon_{r,\text{relaxed}}$ and $\varepsilon_{r,\text{unrelaxed}}$ are the values of the relaxed and unrelaxed dielectric constants. The unrelaxed dielectric constant $\varepsilon_{r,\text{unrelaxed}}$ is the dielectric constant that includes all relaxation processes at frequencies higher than the process of interest (i.e. electronic polarization and relaxation due to vibrations and librations), whereas the relaxed dielectric constant $\varepsilon_{r,\text{relaxed}}$ is the value obtained after the relaxation process of interest (i.e. dipole orientational relaxation) is complete.

The dielectric loss curves in most polymers are asymmetric. Thus, to carry out the Fourier transform in Eq. (4.3.17) one should fit the data with an analytical function. A function that often provides an excellent description of dielectric loss spectra is the Havriliak–Negami (HN) one.[56] Alternatively, in the case of polymeric materials, and based on the time scale of the computational experiment, one should account often for dielectric relaxation due to two mechanisms: segmental dynamics and overall chain diffusion. Typically, then, one chooses to analyse the dielectric data by employing the Rouse model[57,58] (see Chapter 7) for the normal modes and a Kohlrausch–Williams

Watts (KWW) stretched exponential function[59] and its variants for the segmental modes. In the case of Rouse dynamics, the function $\tilde{\phi}(t)$ is assumed to decay in the same way as the chain end-to-end vector (Chapter 7), i.e.

$$\tilde{\phi}(t) = \sum_{p=1,3,\ldots} \frac{8}{p^2\pi^2} \exp\left(-\frac{t}{\tau_p}\right), \tag{4.3.20}$$

where $\tau_p, p = 1, 2, 3, \ldots$ denotes the relaxation spectrum, in which case the result for the dielectric loss $\varepsilon_r''(\omega)$ is

$$\varepsilon_r''(\omega) = \Delta\varepsilon \left(\frac{8}{\pi^2}\right) \sum_{p=1,3,5,\ldots} \left(\frac{1}{p^2}\right) \frac{\omega\tau_p}{1 + (\omega\tau_p)^2}. \tag{4.3.21}$$

The KWW function, on the other hand, which has the form

$$\tilde{\phi}(t) = \exp\left(-\left(\frac{t}{\tau_{\mathrm{KWW}}}\right)^{\beta}\right), \tag{4.3.22}$$

does not lead to an analytical expression in the frequency domain. In this case, Eq. (4.3.17) can be approximated as[59,60]

$$\frac{\varepsilon_r^*(\omega) - \varepsilon_{r,\infty}}{\varepsilon_{r,0} - \varepsilon_{r,\infty}} = \sum_{n=1}^{\infty} \frac{(-1)^{n-1}}{(\omega\tau_{\mathrm{KWW}})^{n\beta}} \frac{\Gamma(n\beta + 1)}{\Gamma(n+1)} \left[\cos\left(\frac{n\beta\pi}{2}\right) - i\sin\left(\frac{n\beta\pi}{2}\right)\right] \tag{4.3.23}$$

for high frequencies (i.e. $-1 \leq \log(\omega\tau_{\mathrm{KWW}}) \leq +4$ when $\beta > 0.25$), and as

$$\frac{\varepsilon_r^*(\omega) - \varepsilon_{r,\infty}}{\varepsilon_{r,0} - \varepsilon_{r,\infty}} = \sum_{n=1}^{\infty} (-1)^{n-1} \frac{(\omega\tau_{\mathrm{KWW}})^{n-1}}{\Gamma(n)} \Gamma\left(\frac{n+\beta-1}{\beta}\right) \left[\cos\left(\frac{(n-1)\pi}{2}\right) + i\sin\left(\frac{(n-1)\pi}{2}\right)\right] \tag{4.3.24}$$

for low frequencies. In Eqs. (4.3.23) and (4.3.24), $\Gamma(x)$ is the gamma function.

4.3.2.2 Quasielastic neutron scattering

Direct insight into polymer dynamics in real space can be obtained by analysing the *space-time van Hove correlation functions*, which are the Fourier transforms of scattering functions that can be measured by neutron scattering techniques. For any system in the classical limit, the van Hove correlation function is defined as[1,61–63]

$$G(\mathbf{r}, t) = \frac{1}{n} \left\langle \sum_{i=1}^{n} \sum_{j=1}^{n} \delta\left[\mathbf{r} + \mathbf{r}_j(0) - \mathbf{r}_i(0)\right] \right\rangle. \tag{4.3.25}$$

$G(\mathbf{r}, t)$ is interpreted as the conditional probability density that a particle is located at position $\mathbf{r}$ at time t given that the same or another (for example, tagged) particle was at the origin at time $t = 0$.

Equivalently, $G(\boldsymbol{r},t)$ represents the conditional probability to find a particle that has experienced a displacement vector $\boldsymbol{r}$ relative to a given (tagged) particle within time t. In Eq. (4.3.25), δ is the Dirac delta function, $\boldsymbol{r}_j$ the position vector of particle j, and n the total number of particles. According to Eq. (4.3.25), $G(\boldsymbol{r},t)$ splits naturally into self and distinct parts: The self part describes the correlation of a given particle at different times and the distinct part the correlation between different particles at different times. If we denote the self part by $G_\mathrm{s}(\boldsymbol{r},t)$ and the distinct part by $G_\mathrm{d}(\boldsymbol{r},t)$, we write[1]

$$\begin{aligned} G(\boldsymbol{r},t) &= G_\mathrm{s}(\boldsymbol{r},t) + G_\mathrm{d}(\boldsymbol{r},t) \\ G_\mathrm{s}(\boldsymbol{r},t) &= \frac{1}{n}\left\langle \sum_{i=1}^{n} \delta\left[\boldsymbol{r} + \boldsymbol{r}_i(0) - \boldsymbol{r}_i(t)\right]\right\rangle \\ G_\mathrm{d}(\boldsymbol{r},t) &= \frac{1}{n}\left\langle \sum_{i=1}^{n}\sum_{\substack{j=1 \\ j\neq i}}^{n} \delta\left[\boldsymbol{r} + \boldsymbol{r}_j(0) - \boldsymbol{r}_i(0)\right]\right\rangle. \end{aligned} \tag{4.3.26}$$

$G_\mathrm{s}(\boldsymbol{r},t)$ gives the probability that within time t a given particle moves by $\boldsymbol{r}$, while $G_\mathrm{d}(\boldsymbol{r},t)$ gives the probability of finding any different particle at distance vector $\boldsymbol{r}$ from the former position of some reference particle. For time $t=0$ these reduce to

$$\begin{aligned} G_\mathrm{s}(\boldsymbol{r},0) &= \delta(\boldsymbol{r}) \\ G_\mathrm{d}(\boldsymbol{r},0) &= \frac{n}{V} g(\boldsymbol{r}), \end{aligned} \tag{4.3.27}$$

where $g(\boldsymbol{r})$ is the *pair distribution function*[64] defined in a homogeneous system by

$$g(\boldsymbol{r}) = \frac{V}{n^2}\left\langle \sum_{i=1}^{n}\sum_{\substack{j=1 \\ j\neq i}}^{n} \delta\left[\boldsymbol{r} + \boldsymbol{r}_j - \boldsymbol{r}_i\right]\right\rangle. \tag{4.3.28}$$

In Eq. (4.3.28), the n denotes the total number of particles, e.g. chain repeat units in a polymer melt. In the case of isotropic systems, $G(\boldsymbol{r},t)$ depends only on $r = |\boldsymbol{r}|$.

The van Hove space-time correlation function is closely related to the *scattering function* (or dynamic structure factor) $S(\boldsymbol{q},\omega)$ of slow or thermal neutrons from fluids. More specifically

$$S(\boldsymbol{q},\omega) = \frac{1}{2\pi}\int G(\boldsymbol{r},t)\exp\left[i\left(\boldsymbol{q}\cdot\boldsymbol{r} - \omega t\right)\right]\mathrm{d}\boldsymbol{r}\,\mathrm{d}t \tag{4.3.29}$$

and

$$G(\boldsymbol{r},t) = \frac{1}{(2\pi)^3}\int S(\boldsymbol{q},\omega)\exp\left[-i\left(\boldsymbol{q}\cdot\boldsymbol{r} - \omega t\right)\right]\mathrm{d}\boldsymbol{q}\,\mathrm{d}\omega. \tag{4.3.30}$$

It is also convenient to define the intermediate scattering function $I(\boldsymbol{q}, t)$ through

$$I(\boldsymbol{q}, t) = \frac{1}{n} \sum_{i=1}^{n} \sum_{j=1}^{n} \langle \exp\left[-i\boldsymbol{q} \cdot (\boldsymbol{r}_i(0) - \boldsymbol{r}_j(t))\right] \rangle. \tag{4.3.31}$$

$I(\boldsymbol{q}, t)$ is related to the van Hove function through

$$I(\boldsymbol{q}, t) = \int G(\boldsymbol{r}, t) \exp(i\boldsymbol{q} \cdot \boldsymbol{r})\, \mathrm{d}\boldsymbol{r} \tag{4.3.32}$$

or

$$G(\boldsymbol{r}, t) = \frac{1}{(2\pi)^3} \int I(\boldsymbol{q}, t) \exp(-i\boldsymbol{q} \cdot \boldsymbol{r})\, \mathrm{d}\boldsymbol{q}. \tag{4.3.33}$$

It is also related to the scattering function $S(\boldsymbol{q}, \omega)$ through

$$S(\boldsymbol{q}, \omega) = \frac{1}{2\pi} \int I(\boldsymbol{q}, t) \exp(-i\omega t)\, \mathrm{d}t. \tag{4.3.34}$$

Experimentally, the scattering function determines the differential *scattering cross section* per unit solid angle Ω and unit interval of outgoing energy ε of the scattered particle (the neutron) through the following equation based on the first Born approximation:[1,62–66]

$$\frac{\mathrm{d}^2\sigma}{\mathrm{d}\Omega \mathrm{d}\varepsilon} = A(q)\, S(\boldsymbol{q}, \omega)$$
$$A(q) = \frac{m^2 n}{4\pi^2 \hbar^5} \frac{k}{k_0} \left| \int \exp(i\boldsymbol{q} \cdot \boldsymbol{r})\, \mathcal{V}(\boldsymbol{r})\, \mathrm{d}\boldsymbol{r} \right|^2. \tag{4.3.35}$$

In Eq. (4.3.55), m, $\boldsymbol{k}_0$, and $\boldsymbol{k}$ are the mass and the initial and final wave vectors of the scattered neutron, respectively, and $\boldsymbol{q}$ is defined as $\boldsymbol{q} = \boldsymbol{k}_0 - \boldsymbol{k}$. Also, $\mathcal{V}(\boldsymbol{r})$ is the interaction potential between the neutron and the nuclei of the sample. The assumption is that all nuclei interact in the same way with the incident neutrons.

Neutron scattering allows one to experimentally probe both the spatial and temporal behaviour of the molecules in a fluid. On the other hand, if the incident energy is sufficiently large compared to the energy transfers (as in the case of X-ray scattering), the momentum transfer for a given scattering angle is independent of the outgoing energy, and the differential cross section per unit solid angle becomes[1]

$$\frac{d\sigma}{d\Omega} = \int \frac{d^2\sigma}{d\Omega d\varepsilon} d\varepsilon = \hbar A(q) \int S(\boldsymbol{q}, \omega)\, d\omega = \hbar A(q) \int \exp(i\boldsymbol{q}\cdot\boldsymbol{r})\, \delta(t) G(\boldsymbol{r}, t)\, d\boldsymbol{r}\, dt$$

or

$$\frac{d\sigma}{d\Omega} = \hbar A(q) \int \exp(i\boldsymbol{q}\cdot\boldsymbol{r}) [G_s(\boldsymbol{r}, 0) + G_d(\boldsymbol{r}, 0)]\, d\boldsymbol{r} = \hbar A(q) \int \exp(i\boldsymbol{q}\cdot\boldsymbol{r}) \left[\delta(\boldsymbol{r}) + \frac{n}{V} g(\boldsymbol{r})\right] d\boldsymbol{r}$$

or

$$\frac{d\sigma}{d\Omega} \propto \left[1 + \rho \int \exp(i\boldsymbol{q}\cdot\boldsymbol{r})\, g(\boldsymbol{r})\, d\boldsymbol{r}\right], \tag{4.3.36}$$

with $\rho = \frac{n}{V}$ being the density of scatterers, in agreement with the X-ray scattering formulas.

The above analysis is based on the consideration that all nuclei in the system interact in exactly the same way with the incident neutrons. Generally, however, this is not the case. The scattering usually depends on either isotopic effects or nuclear internal variables, and so it is not the same from all nuclei. In particular, we wish to extend the above results to the case of spin-dependent scattering. If particles are distinguished by an operator which depends on their spin, then we should extend the definition of the van Hove space-time correlation function to be also spin-dependent. We replace $G(\boldsymbol{r}, t)$ in this case by $\boldsymbol{\Gamma}(\boldsymbol{r}, t)$ defined by

$$\Gamma(\boldsymbol{r}, t) = \frac{1}{n} \left\langle \sum_{l=1}^{n} \sum_{j=1}^{n} \int d\boldsymbol{r}' a_l(0) \delta[\boldsymbol{r} + \boldsymbol{r}_l(0) - \boldsymbol{r}'] a_j(t) \delta[\boldsymbol{r}' - \boldsymbol{r}_j(t)] \right\rangle, \tag{4.3.37}$$

where a_l and a_j are operators that depend on the spins of the l th and j th particles, respectively. If there is no correlation between spins nor between spins and positions, $\boldsymbol{\Gamma}(\boldsymbol{r}, t)$ can be written as

$$\Gamma(\boldsymbol{r}, t) = \langle a^2 \rangle G_s(\boldsymbol{r}, t) + \langle a \rangle^2 G_d(\boldsymbol{r}, t). \tag{4.3.38}$$

By replacing the interaction between a slow neutron and the nuclei of the scattering system by the so-called Fermi pseudo-potential, namely, $\mathcal{V}(\boldsymbol{r}) = \left(\frac{2\pi a\hbar^2}{m}\right) \delta(\boldsymbol{r})$, where m is the mass of the neutron and a the scattering length of the nucleus (this may depend on the nuclear spin state), we can write

$$\frac{d^2\sigma}{d\Omega d\varepsilon} = \frac{d^2\sigma_{coh}}{d\Omega d\varepsilon} + \frac{d^2\sigma_{inc}}{d\Omega d\varepsilon}, \tag{4.3.39}$$

where

$$\frac{d^2\sigma_{coh}}{d\Omega d\varepsilon} = \frac{\langle a \rangle^2 n}{2\pi\hbar} \frac{k}{k_0} \int \exp(i\boldsymbol{q}\cdot\boldsymbol{r} - i\omega t)\, G(\boldsymbol{r}, t)\, d\boldsymbol{r}\, dt \tag{4.3.40a}$$

and

$$\frac{d^2\sigma_{\text{inc}}}{d\Omega d\varepsilon} = \frac{\left(\langle a^2\rangle - \langle a\rangle^2\right) n}{2\pi\hbar} \frac{k}{k_0} \int \exp\left(i\boldsymbol{q}\cdot\boldsymbol{r} - i\omega t\right) G_s\left(\boldsymbol{r}, t\right) d\boldsymbol{r}\, dt. \tag{4.3.40b}$$

These are the so-called *coherent* and *incoherent* scattering cross sections. Note also that, in the static approximation discussed around Eq. (4.3.36), we have that

$$\frac{d\sigma_{\text{coh}}}{d\Omega} = \langle a\rangle^2 n \left[1 + \rho \int \exp\left(i\boldsymbol{q}\cdot\boldsymbol{r}\right) g\left(\boldsymbol{r}\right) d\boldsymbol{r}\right] \tag{4.3.41a}$$

$$\frac{d\sigma_{\text{inc}}}{d\Omega} = \left(\langle a^2\rangle - \langle a\rangle^2\right) n. \tag{4.3.41b}$$

If all nuclei in the sample interact in exactly the same way with the neutron beam, then there will be interference between the scattered waves from each nucleus (just as in X-ray scattering), and this is referred to as coherent scattering. Suppose now that the sample consists of two types of scatterers and that one interacts with a neutron with an interaction $+b$, the other with $-b$, and that there are equal numbers of these species randomly distributed through the sample. Since there is no correlation of the scattering amplitudes in space, on different nuclei, the spectrum does not contain any information on the collective modes, only on the motion of the individual scatterers, or the so-called single particle motions, or self-correlation functions. This is incoherent scattering. Coherent scattering contains both self-correlations and contributions from other particles being at the space-time point $(\boldsymbol{r}, t)$.

Neutron scattering experimental techniques probe the relaxation of polymeric systems through measurements of either the scattering function $S(\boldsymbol{q}, \omega)$ or the dynamic structure factor $I(\boldsymbol{q}, t)$ of the system defined according to Eq. (4.3.31). In these equations, the magnitude of the scattering vector $\boldsymbol{q}$ is equal to $q = \frac{4\pi}{\lambda}\sin\left(\frac{\theta}{2}\right)$, where λ is the wavelength of the radiation and θ the scattering angle (angle between incident and scattered beam), and $\boldsymbol{r}_i(t)$ the position vector of scatterer i (an atom or chain segment) at time t. Clearly, since the functions $\boldsymbol{r}_i(t) - \boldsymbol{r}_j(t)$ that appear in Eq. (4.3.31) are readily available from MD simulation data, one can employ MD in a rather straightforward manner to obtain predictions of neutron scattering experiments. In fact, by changing the magnitude q of the scattering vector, one can explore the entire spectrum of length scales for a given polymer: small q values (usually $0.01 \le q \le 0.1\,\text{Å}^{-1}$) will probe the overall chain (or terminal) relaxation, while large q values (usually $1 \le q \le 5\,\text{Å}^{-1}$) will probe the dynamics at shorter length scales (segmental or local relaxation). When $i \neq j$ in Eq. (4.3.31), the results correspond to the normalized single-chain *coherent* dynamic structure factor, which, for an isotropic system, equals

$$\tilde{I}_{\text{coh}}(q, t) = \frac{\sum\limits_{i=1}^{n}\sum\limits_{\substack{j=1\\ j\neq i}}^{n}\left\langle \frac{\sin\left[q\Delta r_{ij}(t)\right]}{q\Delta r_{ij}(t)}\right\rangle}{\sum\limits_{i=1}^{n}\sum\limits_{\substack{j=1\\ j\neq i}}^{n}\left\langle \frac{\sin\left[q\Delta r_{ij}(0)\right]}{q\Delta r_{ij}(0)}\right\rangle}, \tag{4.3.42}$$

where $\Delta r_{ij}(t) = |\boldsymbol{r}_i(t) - \boldsymbol{r}_j(0)|$ denotes the magnitude of the vector $\boldsymbol{r}_i(t) - \boldsymbol{r}_j(0)$ with $i \neq j$ and the angular brackets denote an ensemble average over all time origins for atoms with a significant coherent cross section.

When $i = j$ in Eq. (4.3.31), the results correspond to the intermediate single-chain *incoherent* scattering function

$$I_{\mathrm{inc}}(\boldsymbol{q}, t) = \left\langle \sum_{i=1}^{n} \exp\left[i\boldsymbol{q} \cdot (\boldsymbol{r}_i(t) - \boldsymbol{r}_i(0))\right] \right\rangle. \tag{4.3.43}$$

This, for isotropic systems and according to Eq. (4.3.42), reduces to the expression

$$\tilde{I}_{\mathrm{inc}}(q, t) = \frac{\sum_{i=1}^{n} \left\langle \dfrac{\sin[q\Delta r_i(t)]}{q\Delta r_i(t)} \right\rangle}{\sum_{i=1}^{n} \left\langle \dfrac{\sin[q\Delta r_i(0)]}{q\Delta r_i(0)} \right\rangle}. \tag{4.3.44}$$

where $\Delta r_i(t) = |\boldsymbol{r}_i(t) - \boldsymbol{r}_i(0)|$ denotes the magnitude of the vector $\boldsymbol{r}_i(t) - \boldsymbol{r}_i(0)$ and the angular brackets denote again an ensemble average over all time origins for atoms with a significant incoherent cross section (i.e. hydrogen atoms). $\tilde{I}_{\mathrm{inc}}(q, t)$ is an extremely important observable, since it provides information about the self-motion of protons, which dominate the incoherent scattering in a polymer chain. Again, by choosing different values of the wavevector q one can study the self-motion of protons over a range of characteristic length scales.

The intermediate incoherent scattering function is often expressed through the cumulant expansion as an exponential of a sum of terms involving even powers of q,

$$I_{\mathrm{inc}}(q, t) = \sum_{i=1}^{n} \exp\left[-q^2 \gamma_{i,1}(t) + q^4 \gamma_{i,2}(t) + \ldots\right], \tag{4.3.45}$$

where

$$\gamma_{i,1}(t) = \frac{1}{6} \left\langle [\boldsymbol{r}_i(t) - \boldsymbol{r}_i(0)]^2 \right\rangle \tag{4.3.46}$$

corresponds to a Gaussian $G_s(\boldsymbol{r}_i, t)$, in which case the term of $I_{\mathrm{inc}}(q, t)$ corresponding to i is also Gaussian. The higher terms can be neglected both for small times (ballistic motion) and for long times (diffusive motion), giving

$$I_{\mathrm{inc}}(q, t) = \sum_{i=1}^{n} \exp\left[-q^2 \gamma_{i,1}(t)\right]. \tag{4.3.47}$$

Non-Gaussian effects are taken into account by extending the previous equation as

$$I_{\mathrm{inc}}(q, t) = \sum_{i=1}^{n} \exp\left[-q^2 \gamma_{i,1}(t)\right] \left\{1 + \frac{1}{2} a_{i,2}(t) \left[q^2 \gamma_{i,1}(t)\right]^2 + \ldots\right\}, \tag{4.3.48}$$

where

$$a_{i,2}(t) = \frac{\left\langle (\Delta \boldsymbol{r}_i)^4 \right\rangle}{3\left\langle (\Delta \boldsymbol{r}_i)^2 \right\rangle^2} - 1 \tag{4.3.49}$$

Equation (4.3.46) indicates that $\gamma_{i,1}(t) = Dt$ for a diffusing particle i, and so, as already mentioned, the self-part of the van Hove correlation function for a diffusing particle i is a Gaussian in the displacement

$$G_s(\boldsymbol{r}_i, t) = (4\pi Dt)^{-3/2} \exp\left[-\frac{(\Delta \boldsymbol{r}_i)^2}{4Dt} \right], \; \Delta \boldsymbol{r}_i = \boldsymbol{r}_i(t) - \boldsymbol{r}_i(0). \tag{4.3.50}$$

implying that the contribution to the incoherent intermediate scattering function from the diffusing particle i is also a Gaussian in the wavevector

$$I_{\text{inc},i}(q, t) = \exp\left(-Dq^2 t\right) \tag{4.3.51a}$$

and the corresponding contribution to the scattering function is a Lorentzian in frequency:

$$S_{\text{inc},i}(q, \omega) = \frac{1}{\pi} \frac{Dq^2}{\omega^2 + (Dq^2)^2} \tag{4.3.51b}$$

We expect this to be most valid at small frequencies since the form of $G_s(\boldsymbol{r}_i, t)$ is most valid at long times.

An apparent mean square displacement of hydrogen atoms can be obtained from the measured elastic intensity of incoherently scattered neutrons by postulating that $S_{\text{inc}}(q, \omega \approx 0) \propto \exp\left(-q^2 \langle r^2 \rangle_{\text{app}}/6\right)$, or, more precisely,

$$\frac{\langle r^2 \rangle_{\text{app}}}{6} = -\frac{\partial \ln S_{\text{inc}}(q, \omega \approx 0)}{\partial (q^2)}. \tag{4.3.52a}$$

Inspired by the picture of harmonic motion within a three-dimensional potential well, Eq. (4.3.52a) gives only an apparent mean square displacement of atoms. It provides a convenient way of expressing the elastic incoherent structure factor (EISF) $S_{\text{inc}}(q, \omega \approx 0)$ in units of squared length.

If the motion over large length scales is diffusive, we readily obtain the diffusivity from Eq. (4.3.51b)

$$D = \left. \frac{\partial (\text{HWHM})_{S_{\text{inc}}}}{\partial (q^2)} \right|_{q \to 0} \tag{4.3.52b}$$

where HWHM stands for *half-width at half-maximum* of the S_{inc} curve as a function of ω.

At short times, on the other hand,

$$\gamma_{i,1}(t) = \frac{1}{6}\langle \boldsymbol{\nu}_i \cdot \boldsymbol{\nu}_i \rangle t^2 = \frac{k_B T}{2m_i} t^2, \; t \to 0. \tag{4.3.53}$$

This displays the initial free motion that the particle undergoes before colliding with another particle. Thus, although the exact behaviour of the mean square displacement $\left\langle [\boldsymbol{r}_i(t) - \boldsymbol{r}_i(0)]^2 \right\rangle$ is difficult to calculate for all times, we see that it goes as t^2 for short times and as t for long times. From the general expression, Eq. (4.3.43), it is seen that $I_{\rm inc}(q,t)$ can be determined by a Fourier inversion with respect to frequency of the experimentally determined incoherent scattering function. Then, using the form of $I_{\rm inc}(q,t)$ given by Eq. (4.3.48), one can determine $\gamma_{i,1}(t)$ and $a_{i,2}(t)$ numerically from the data and check their time dependence, namely if $\gamma_{i,1}(t)$ goes as t^2 for short times and as Dt for long times.

We turn our attention now to the coherent and incoherent scattering for polymers. We will adopt here a notation very similar to that of Chapter 1, where a chain of N segments numbered by $i = 1, 2, \ldots, N$ is considered, whose position vectors are denoted by $\boldsymbol{R}_i$. If chain monomers are assumed to undergo displacements such that the components of $\boldsymbol{R}_i(t) - \boldsymbol{R}_j(0)$ follow Gaussian distributions, then, according to the above discussion,

$$\langle \exp [i\boldsymbol{q} \cdot (\boldsymbol{R}_i(t) - \boldsymbol{R}_j(0))] \rangle = \exp\left\{ -\frac{q^2}{6} \left\langle (\boldsymbol{R}_i(t) - \boldsymbol{R}_j(0))^2 \right\rangle \right\}, \tag{4.3.54}$$

so that

$$\begin{aligned} I(q,t) &\equiv \frac{1}{N}\sum_{i=1}^{N}\sum_{j=1}^{N} \exp\left\{ -\frac{q^2}{6} \left\langle (\boldsymbol{R}_i(t) - \boldsymbol{R}_j(0))^2 \right\rangle \right\} \\ &= \frac{1}{N}\sum_{i=1}^{N}\sum_{j=1}^{N} \exp\left\{ -\frac{q^2}{6}\phi_{ij}(t) \right\}. \end{aligned} \tag{4.3.55}$$

Further analysis requires the explicit calculation of the quantity $\phi_{ij}(t) = \left\langle (\boldsymbol{R}_i(t) - \boldsymbol{R}_j(0))^2 \right\rangle$.

For unentangled polymer melts one can resort to the Rouse model and, as we will discuss in Chapter 7, the result is[57]

$$\phi_{ij}(t) = 6Dt + |i-j|\, b_K^2 + \frac{4\langle r^2 \rangle}{\pi^2} \sum_{p=1}^{\infty} \frac{1}{p^2} \cos\left(\frac{ip\pi}{N}\right) \cos\left(\frac{jp\pi}{N}\right) \left[1 - \exp\left(-\frac{t}{\tau_p}\right)\right], \tag{4.3.56}$$

where D denotes the centre-of-mass diffusivity; $\tau_p, p = 1, 2, \ldots$ is the spectrum of relaxation times; $\langle r^2 \rangle$ is the equilibrium mean square end-to-end distance in the melt; and b_K is the length of the Kuhn segment (compare Eq. (1.3.24)). According to the Rouse model (Chapter 7), $\tau_p = \dfrac{\tau_1}{p^2}$ where τ_1 is the longest relaxation time (also known as the Rouse time τ_R) and $D = \dfrac{k_B T}{N\zeta}$ where ζ is the monomer friction constant. Equation (4.3.56) results in the following expression for the scaled $\tilde{I}(q,t) = \dfrac{I(q,t)}{I(q,0)}$:

$$\tilde{I}(q,t) = \exp\left[-q^2 Dt\right] \sum_{i=1}^{N}\sum_{j=1}^{N} \exp\left\{-\frac{2q^2\left\langle r^2\right\rangle}{3\pi^2}\sum_{p=1}^{N}\frac{1}{p^2}\cos\left(\frac{ip\pi}{N}\right)\right.$$
$$\left.\cos\left(\frac{jp\pi}{N}\right)\left[1-\exp\left(-\frac{t}{\tau_p}\right)\right]\right\}. \tag{4.3.57a}$$

For $q\left\langle r^2\right\rangle << 1$, only the overall diffusion of the chains is monitored and

$$\tilde{I}(q,t) = \frac{1}{N}\exp\left[-q^2 Dt\right]. \tag{4.3.57b}$$

A modification of the Rouse model, the semiflexible chain model (SFCM), was proposed by Harnau et al.[67] that incorporates local chain stiffness effects. According to the SFCM, the single-chain dynamic structure factor is calculated by using again Eq. (4.3.57a) but the relaxation times τ_p are taken as

$$\tau_p^b = \frac{\tau_R}{p^2 + \dfrac{\pi^2 C_\infty^2}{4N^2 \sin^4\left(\dfrac{\theta}{2}\right)}p^4} = \frac{\tau_R}{p^2 + a_b p^4}, \tag{4.3.58}$$

where the term proportional to p^4 corrects for rigidity effects, and C_∞ denotes the characteristic ratio at infinite chain length.

For the reptation model (see also Chapter 7), the resulting expression for the mean square displacement (msd) of the atomistic segments is difficult to obtain analytically. Still, however, one can come up with a closed-form expression for the single-chain dynamic structure factor by transforming from atomistic displacements to displacements of primitive chain segments:[68–70]

$$\tilde{I}(q,t) = \left\{1-\exp\left[-\left(\frac{qd_t}{6}\right)^2\right]\right\}\exp\left(\frac{t}{\tau_0}\right)\operatorname{erfc}\left(\sqrt{\frac{t}{\tau_0}}\right)$$
$$+\frac{8}{\pi^2}\exp\left[-\left(\frac{qd_t}{6}\right)^2\right]\sum_{p=\mathrm{odd}}^{\infty}\frac{1}{p^2}\exp\left(-\frac{p^2 t}{\tau_d}\right), \tag{4.3.59}$$

with τ_0 being a short time constant given as $\tau_0 = \dfrac{12\zeta}{k_B T b_K^2 q^4}$ (here ζ is the monomer friction coefficient and b_K the Kuhn statistical length), d_t the entanglement tube diameter, and τ_d the chain disengagement (or reptation) time. Equation (4.3.59) for the single-chain dynamic structure factor has been derived on the basis of negligible contour length fluctuation and constraint release phenomena, so it should be expected to be applicable only in high molecular weight melts (chain length $N > 100N_e$) where these phenomena are less important, and for wavevectors q such that

$$\frac{1}{\sqrt{\langle r^2 \rangle}} << \frac{q}{2\pi} << \frac{1}{d_t}. \tag{4.3.60}$$

We also note that computed or measured $I(q, t)$ functions usually serve to estimate correlation times for segmental motion according to

$$\tau_c(q) = \int_0^\infty I(q, t)\mathrm{d}t. \tag{4.3.61}$$

The q-dependence of the correlation time provides information about the spatial resolution of polymer chain dynamics in the sample, and this demonstrates the usefulness of neutron scattering experiments.

4.3.2.3 ^{13}C-NMR spectroscopy

In ^{13}C-NMR spectroscopy, we probe the dynamics of ^{1}H nuclei in the proximity of ^{13}C nuclei. More specifically, we determine the longitudinal or spin–lattice relaxation time T_1, the transverse or spin–spin relaxation time T_2, and the nuclear Overhauser enhancement (NOE) factor of a ^{13}C nucleus relaxing purely by a dipolar interaction with one proton (^{1}H). The analysis[71–74] proceeds by considering a particle of spin $\boldsymbol{I}$ interacting with a particle of spin $\boldsymbol{S}$ and writing down the Hamiltonian of such an interacting pair in a magnetic field along the z direction. This Hamiltonian is the sum of the Hamiltonian of the motion of the two particles, of the Zeeman energies of the two spins in the constant magnetic field H_0, and of a time-dependent spin–spin interaction term considered as a perturbation.[72] By focusing on pairs of spins of value $\frac{1}{2}$ (e.g. ^{13}C and ^{1}H) so that there is no quadrupole interaction, and by taking the two spins to be unlike (i.e. their gyromagnetic ratios γ_I and γ_S are different, $\gamma_I \neq \gamma_S$; this implies that the two Larmor or angular velocities $\omega_I = \gamma_I H_0$ and $\omega_S = \gamma_S H_0$ are also different, $\omega_I \neq \omega_S$), it is found that $\langle I_z \rangle$, the expectation value of the magnetic moment I_z which is the experimentally observable quantity, decays exponentially with a time constant $\frac{1}{T_1}$ given by[74]

$$\frac{1}{T_1} = W_0 + 2W_{1I} + W_2, \tag{4.3.62}$$

where the W_j's are transition probabilities for the energy levels of the two spin-$\frac{1}{2}$ system. For pure $^{13}\mathrm{C} - {}^1\mathrm{H}$ dipolar relaxation, these are given by[74]

$$W_0 = \frac{1}{20} K^2 \mathcal{J}_0 (\omega_S - \omega_I) \tag{4.3.63a}$$

$$W_{1I} = \frac{3}{40} K^2 \mathcal{J}_1 (\omega_I) \tag{4.3.63b}$$

$$W_2 = \frac{3}{10} K^2 \mathcal{J}_2 (\omega_S + \omega_I). \tag{4.3.63c}$$

where $\mathcal{J}_m(\omega)$, $m = 0, 1, 2$ are the so-called spectral densities, namely, the Fourier transforms of the autocorrelation functions $G_m(t)$ describing the orientational decorrelation of the internuclear C-H vector[73]

$$\mathcal{J}_m(\omega) = \int_{-\infty}^{+\infty} G_m(t) \exp(-i\omega t)\, dt$$

$$= 2 \int_{0}^{+\infty} G_m(t) \cos(\omega t)\, dt \tag{4.3.64}$$

and $K = \hbar \gamma_S \gamma_I$ with $\hbar = h/(2\pi)$, $h = 6.62607 \times 10^{-34}$J s being Planck's constant.[74] The expression $K = \hbar \gamma_S \gamma_I$ is missing a term $\mu_0/4\pi$ proportional to the vacuum magnetic permeability μ_0 $(= 1.25664 \times 10^{-6}$ N A$^{-2})$ that is typically left out in the definition of the dipole–dipole interaction term. If we also include this term, then the full expression is $K = \frac{\mu_0}{4\pi} \hbar \gamma_S \gamma_I$. The autocorrelation functions $G_m(t)$ appearing in Eq. (4.3.64) are[73]

$$G_m(t) = \left\langle \frac{Y_{2,m}(t_0)}{r^3} \frac{Y_{2,m}(t_0 + t)}{r^3} \right\rangle, \tag{4.3.65}$$

where[73]

$$Y_{2,0} = \left(\frac{5}{4}\right)^{\frac{1}{2}} (1 - 3\cos^2\theta) \tag{4.3.66a}$$

$$Y_{2,\pm 1} = \pm \left(\frac{15}{2}\right)^{\frac{1}{2}} \cos\theta \, \sin\theta \, \exp(\pm i \phi) \tag{4.3.66b}$$

$$Y_{2,\pm 2} = -\left(\frac{15}{8}\right)^{\frac{1}{2}} \sin^2\theta \, \exp(\pm 2 i \phi) \tag{4.3.66c}$$

are properly normalized second-order spherical harmonics. In the above expressions (r, θ, ϕ) denote the internuclear vector having polar coordinates with respect to the externally applied magnetic field (which is referred to some fixed direction).[73]

Substitution of Eqs. (4.3.63a)–(4.3.63c) into Eq. (4.3.62) yields the following result for T_1:[74]

$$\frac{1}{T_1} = \frac{1}{20} K^2 \left[\mathcal{J}_0(\omega_S - \omega_I) + 3\mathcal{J}_1(\omega_I) + 6\mathcal{J}_2(\omega_S + \omega_I)\right]. \tag{4.3.67}$$

Analogous expressions are obtained for the NOE and the transverse relaxation time T_2.

To proceed, it can be assumed that the particles containing the magnetic nuclei undergo Brownian motion; thus, the functions $Y_{2,0}$, $Y_{2,\pm 1}$, and $Y_{2,\pm 2}$ defined in Eqs. (4.3.66) will vary in time, and this motion is on the average random and isotropic.[71] We can also take the molecule containing the C-H pair to be rigid; i.e. we can neglect molecular vibrations because of their high frequency and small amplitude. Thus, $r =$ const. $= l_{\text{C-H}}$ (the equilibrium value of the C-H bond length), leaving only the angle coordinates θ and ϕ of the C-H group to be variable in time. Then, the functions G_m become[71]

$$G_m(t) = \frac{1}{l_{\text{H-C}}^6} \langle Y_{2,m}(t_0)\, Y_{2,m}(t_0 + t) \rangle = \frac{1}{l_{\text{H-C}}^6} F_m(t) \qquad (4.3.68)$$

where $F_m(t) = \langle Y_{2,m}(t_0)\, Y_{2,m}(t_0 + t) \rangle$. The analysis is greatly simplified by further assuming that molecular reorientation is isotropic and descriable by means of a single rotational correlation time τ_0, and that this time is practically the same for all three correlation functions:[71]

$$F_m(t) = \exp\left(-\frac{t}{\tau_0}\right), \quad m = 0, 1, 2. \qquad (4.3.69)$$

Then, the spectral densities become[74]

$$\mathcal{J}_m(\omega) = \frac{1}{l_{\text{H-C}}^6} \frac{2\tau_0}{1 + \omega^2 \tau_0^2}, \quad m = 0, 1, 2. \qquad (4.3.70)$$

Introducing Eq. (4.3.70) into Eq. (4.3.67) yields[74]

$$\frac{1}{T_1} = \frac{1}{10} \frac{K^2}{l_{\text{H-C}}^6} \left[\frac{\tau_0}{1 + (\omega_{\text{H}} - \omega_{\text{C}})^2 \tau_0^2} + \frac{3\tau_0}{1 + \omega_{\text{C}}^2 \tau_0^2} + \frac{6\tau_0}{1 + (\omega_{\text{H}} + \omega_{\text{C}})^2 \tau_0^2} \right], \qquad (4.3.71)$$

where we have now used the notation $I = {}^{13}\text{C}$ and $S = {}^{1}\text{H}$. Analogous expressions are derived for T_2 and NOE.[74] Clearly, the quantity $\frac{K^2}{l_{\text{H-C}}^6}$ must have units of inverse time squared, which can be easily verified if we use the expression for K mentioned above that includes the factor $\mu_0/4\pi$.

The above theory,[71–74] which holds for isotropic reorientation, was later generalized to account for a rotating C-H bond attached to a body undergoing isotropic reorientation. In particular, if the angle between the C-H vector and the axis of internal reorientation is fixed and equal to Δ and the correlation time for internal motion is τ_1, then Eq. (4.3.70) becomes[75,76]

$$\mathcal{J}_m(\omega) = \frac{1}{l_{\text{H-C}}^6} \left(A \frac{2\tau_0}{1 + \omega^2 \tau_0^2} + B \frac{2\tau_B}{1 + \omega^2 \tau_B^2} + C \frac{2\tau_C}{1 + \omega^2 \tau_C^2} \right), \quad m = 0, 1, 2, \qquad (4.3.72)$$

where $\frac{1}{\tau_B} = \frac{1}{\tau_0} + \frac{1}{6\tau_1}$, $\frac{1}{\tau_C} = \frac{1}{\tau_0} + \frac{2}{3\tau_1}$, $A = \frac{1}{4}(3\cos^2\Delta - 1)^2$, $B = 3\sin^2\Delta\cos^2\Delta$, and $C = \frac{3}{4}\sin^4\Delta$. The above expressions were generalized by Howarth[77] to allow for internal librations, namely for the vector C-H to move freely within a solid cone of semiangle X about its axis of rotation or rest direction. Let us assume that these can be characterized by a single correlation time τ_{lib}. Then, for $F_1(t)$, in particular, Howarth reports the following result (normalized here to start from the value of 1 at $t = 0$),[77]

$$F_1(t) = a' \exp\left(-\frac{t}{\tau_0}\right) + (1 - a') \exp\left(-\frac{t}{\tau_D}\right). \tag{4.3.73}$$

where $a' = \left[\frac{\cos X - \cos^3 X}{2(1 - \cos X)}\right]^2$ and $\frac{1}{\tau_D} = \frac{1}{\tau_0} + \frac{1}{\tau_{lib}}$. The case of no librations is recovered for $a' \to 1$ (or, equivalently, for $X \to 0$).

For polymers, the C-H bond relaxation is even more complicated to describe mathematically, since one must account for a multitude of mechanisms including both independent but also compensating correlated transitions.[78] It also depends on the type of C-H vector (in a polyisoprene (PI) monomer, e.g., one can distinguish between four different types of C-H vector). Assuming that the long time chain diffusive motion does not contribute to the C-H bond relaxation, we are left with several other mechanisms: independent and correlated torsional rearrangements (often referred to as segmental dynamics or conformational jumps), fast rearrangements of end groups or tails, swinging motion of pendant groups, bond librations, even bond angle bending. Hall and Helfand[78] studied relaxation processes in polymer molecules which proceed via conformational transitions of the chain backbone from one rotational isomeric state (RIS) to another (see Section 1.2) and obtained time-correlation functions using a two- and a three-state model. In both cases, the final expressions contained a modified Bessel function. Motivated by this finding, and if conformational jumps and bond librations can be considered to occur independently, Dejean de la Batie et al.[79] suggested to use the Howarth description for the latter and the Hall–Helfand model for the former. This is the Dejean de la Batie–Lauprêtre–Monnerie (or DLM) model,[79] which is mathematically described by the equation

$$\begin{aligned} F_m(t) &= (1 - a_{lib}) \exp\left(-\frac{t}{\tau_0}\right) \exp\left(-\frac{t}{\tau_1}\right) I_0\left(\frac{t}{\tau_1}\right) \\ &+ a_{lib} \exp\left(-\frac{t}{\tau_0}\right) \exp\left(-\frac{t}{\tau_1}\right) \exp\left(-\frac{t}{\tau_{lib}}\right) I_0\left(\frac{t}{\tau_1}\right), \quad m = 0, 1, 2, \end{aligned} \tag{4.3.74}$$

where $1 - a_{lib} = \left[\frac{\cos X - \cos^3 X}{2(1 - \cos X)}\right]^2$ and I_0 denotes the modified Bessel function of order 0. In Eq. (4.3.74), τ_0 is the correlation time associated with damping, τ_1 is the characteristic time associated with correlated jumps responsible for orientation diffusion along the chain, and τ_{lib} is the correlation time associated with the fast bond librations. Clearly, $\tau_{lib} << \tau_0$ and $\tau_{lib} << \tau_1$. The spectral density corresponding to Eq. (4.3.74) is[79]

$$\mathcal{J}_m(\omega) \propto (1 - a_{lib}) \frac{1}{(c_1 + ic_2)^{\frac{1}{2}}} + a_{lib} \frac{\tau_{lib}}{1 + \omega^2 \tau_{lib}^2}, \quad m = 0, 1, 2, \tag{4.3.75a}$$

where $c_1 = \tau_0^{-2} + 2\tau_0^{-1}\tau_1^{-1} - \omega^2$, $c_2 = -2\omega\left(\tau_0^{-1} + \tau_1^{-1}\right)$, the real part of the right-hand side is considered, and the proportionality symbol has been used to denote that this spectral density is missing a factor of $\frac{2}{l_{\mathrm{H-C}}^6}$ (compare Eqs. (4.3.67)–(4.3.70)).

A second model for C-H relaxation is based on a modified Kohlrausch–Williams–Watts (or mKWW) function:[80]

$$F_m(t) = (1 - a_{\mathrm{lib}}) \exp\left[-\left(\frac{t}{\tau_{\mathrm{KWW}}}\right)^{\beta}\right] + a_{\mathrm{lib}} \exp\left(-\frac{t}{\tau_{\mathrm{lib}}}\right), \quad m = 0, 1, 2. \tag{4.3.75b}$$

It consists of a slow KWW decay associated with segmental motion of the polymer and local conformational changes or other local dynamics at the level of monomers or pendant groups, and a fast exponential decay with amplitude a_{lib} to describe the libration of C-H vectors. Thus, τ_{KWW} and β describe the characteristic segmental relaxation time and its distribution while a_{lib} and τ_{lib} characterize the amplitude and relaxation time for librational motion. This model also provides an analytical expression for the corresponding total correlation time τ_c defined as the time integral of $F_m(t)$:

$$\tau_c = \int_0^{\infty} F_m(t)\mathrm{d}t = (1 - a_{\mathrm{lib}}) \frac{\tau_{\mathrm{KWW}}}{\beta}\Gamma\left(\frac{1}{\beta}\right) + a_{\mathrm{lib}}\tau_0, \quad m = 0, 1, 2. \tag{4.3.76}$$

The corresponding expression for the spectral density can be found by approximating the KWW function as a superposition of exponentials:[81]

$$\exp\left[-\left(\frac{t}{\tau_{\mathrm{KWW}}}\right)^{\beta}\right] = \int_0^{\infty} e^{-\frac{t}{\tau}} \rho_{\mathrm{KWW}}(\tau)\, \mathrm{d}\tau, \tag{4.3.77}$$

where the distribution function $\rho_{\mathrm{KWW}}(\tau)$ is given as[81]

$$\rho_{\mathrm{KWW}}(\tau) = -\frac{\tau_{\mathrm{KWW}}}{\pi\tau^2} \sum_{k=0}^{\infty} \frac{(-1)^k}{k!} \sin(\pi\beta k)\, \Gamma(\beta k + 1) \left(\frac{\tau}{\tau_{\mathrm{KWW}}}\right)^{\beta k + 1}. \tag{4.3.78}$$

By using a range of relaxation times τ (namely τ_i, $i = 1, 2, \ldots, m$) to transform the integral in Eq. (4.3.77) to a sum, one finds the spectral density[80]

$$\mathcal{J}_m(\omega) \propto (1 - a_{\mathrm{lib}}) \sum_{i=1}^{m} F_{\mathrm{KWW}}(\tau_i) \left[\frac{\tau_i}{1 + (\omega\tau_i)^2}\right] + a_{\mathrm{lib}} \frac{\tau_0}{1 + (\omega\tau_0)^2}, \quad m = 0, 1, 2, \tag{4.3.79}$$

where

$$F_{\mathrm{KWW}}(\tau_i) = -\frac{\tau_{\mathrm{KWW}}}{\pi\tau_i^2} \sum_{k=0}^{\infty} \frac{(-1)^k}{k!} \sin(\pi\beta k)\, \Gamma(\beta k + 1) \left(\frac{\tau_i}{\tau_{\mathrm{KWW}}}\right)^{\beta k + 1} \tag{4.3.80}$$

and the sum of the functions $F_{\mathrm{KWW}}(\tau_i)$ satisfies $\sum_{i=1}^{m} F_{\mathrm{KWW}}(\tau_i) = 1.0$. Again, the proportionality symbol in Eq. (4.3.79) denotes that the spectral densities $\mathcal{J}_m(\omega)$, $m = 0, 1, 2$ are missing a factor of $\frac{2}{l_{\mathrm{H-C}}^6}$. The resulting spin–lattice relaxation time T_1 is obtained by substituting the above expression for $\mathcal{J}_m(\omega)$ in Eq. (4.3.67) for all three functions $\mathcal{J}_0(\omega)$, $\mathcal{J}_1(\omega)$, and $\mathcal{J}_2(\omega)$, with a suitable numerical value for the parameter K, accounting also for the number of protons attached to a particular carbon atom as well as for the total number density of C-H bonds in the sample.

4.3.2.4 Comparison with experimental measurements

Doxastakis et al.[82] have performed MD simulations of PI systems of molar mass around 1300 g/mol and compared their results concerning local and chain dynamics with other data from the literature as well as with experimental measurements on similar samples from several techniques, including ^{13}C-NMR relaxation, pulsed field gradient NMR (PFG-NMR), QENS, and dielectric spectroscopy at both high and low temperatures. From our analysis in this chapter, we understand that ^{13}C relaxation probes the orientational decorrelation of C-H bonds. Thus, to predict the ^{13}C-NMR spectrum, Doxastakis et al.[82] computed the C-H bond orientational decorrelation function $P_2(t)$ defined as

$$P_2(t) = \frac{3}{2}\left\langle \cos^2\vartheta(t) \right\rangle - \frac{1}{2}, \tag{4.3.81}$$

where $\vartheta(t)$ denotes the angle of a C-H bond at time t relative to its initial position (at $t = 0$), and the brackets denote the ensemble average over all possible orientations of the bond. Figure 4.1 presents the comparison with the ^{13}C-NMR measurements by showing how experimentally measured data for the spin–lattice relaxation time T_1 for the c-C-H vector (one of the four different C-H vectors present in a PI monomer) as a function of temperature compare with simulation results after fitting the data with the mKWW function, Eq. (4.3.75b). Overall, the simulation values are somewhat lower than the experimental results, indicating slightly slower C-H reorientational dynamics. This can be partly attributed to differences in microstructure between the modelled systems (only *cis*-1,4) and the experimental samples. The comparison between experimental and simulation data was also extended to the level of the correlation time τ_c of the incoherent intermediate scattering function $I_{\mathrm{inc}}(q, t)$ at $T \approx 320$K; see Eqs. (4.3.61) here applied to $I_{\mathrm{inc}}(q, t)$ and (4.3.76).

Ahumada et al.[83] have used MD simulations to explore dynamics in atactic polypropylene in the temperature range 260–600 K, using both a fully atomistic model and a model with united-atom methyl groups. Their results were compared with quasielastic neutron scattering (QENS) measurements in the temperature range 4–460 K and momentum transfer range $1 \leq q \leq 5\,\text{Å}^{-1}$. The logarithms of the correlation times $\tau_c(q)$ from the MD simulations at a value of $q = 1\,\text{Å}^{-1}$ were plotted as a function of the inverse temperature, and a Vogel–Fulcher–Tammann (VFT) function of the form

$$\log \tau_c = \log \tau_\infty + \frac{B}{T - T_\infty} \tag{4.3.82}$$

was seen to capture the temperature dependence of τ_c very well. Representative results are shown in Figures 4.2 and 4.3.

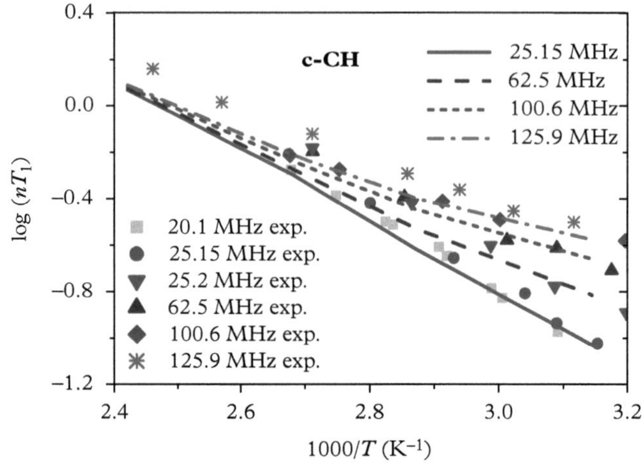

Figure 4.1 Spin–lattice relaxation times T_1 for the c-CH vector, emanating directly from sp^2 carbons in a model polyisoprene system, as measured experimentally and as predicted from simulation after fitting the data with the mKWW model. Points are data from the literature on different PI samples, while lines are predicted values from the simulations at the same frequencies.

Reprinted from Doxastakis, M., Theodorou, D.N., Fytas, G., Kremer, F., Faller, R., Müller-Plathe, F., and Hadjichristidis, N. *J. Chem. Phys.* **119**, 6883 (2003), with the permission of AIP Publishing.

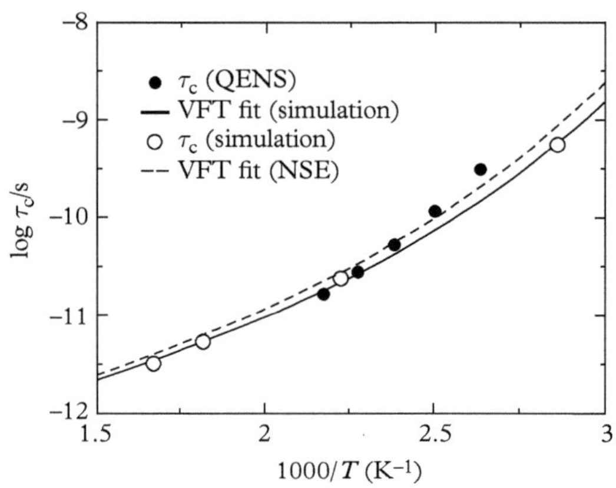

Figure 4.2 Temperature dependence of the correlation time τ_c of atactic polypropylene extracted from the time decay of the simulated $I_{inc}(q, t)$ at $q = 1$ Å^{-1} (open circles), along with its fit via a VFT function (solid line). Correlation times extracted from analysis of the QENS measurements at $q = 1$Å^{-1} are also shown for comparison (filled circles), as is a VFT fit to correlation time values obtained from coherent NSE measurements[84] at $q = 1$ Å^{-1} (broken line).

Reprinted (adapted) with permission from Ahumada, O., Theodorou, D.N., Triolo, A., Arrighi, V., Karatasos, C., and Ryckaert, J.-P. *Macromolecules* **35**, 7110 (2002). Copyright 2002 American Chemical Society.

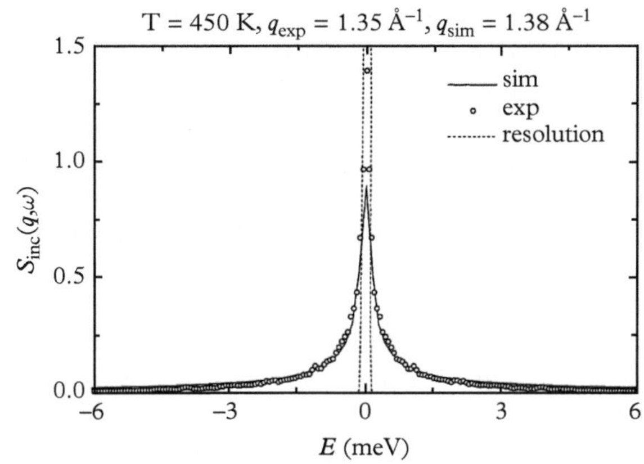

Figure 4.3 Predicted incoherent dynamic structure factor $S_{\text{inc}}(q,\omega)$ of atactic polypropylene at 450 K is compared to an experimental spectrum. The instrumental resolution function is shown as a broken line. The wavevector q is 1.35 Å^{-1} in the experiment and 1.38 Å^{-1} in the simulation.

Reprinted (adapted) with permission from Ahumada, O., Theodorou, D.N., Triolo, A., Arrighi, V., Karatasos, C., and Ryckaert, J.-P. *Macromolecules* **35**, 7110 (2002).

Also shown in Figure 4.2 is a VFT fit of the temperature dependence of τ_c values determined experimentally[84] through (coherent) neutron spin echo (NSE) measurements on deuterated aPP of mean molar mass 2570 g/mol and narrow chain length distribution, at $q = 1$ Å^{-1} in the temperature range 300–360 K.

4.4 Towards chain dynamics and rheological properties

Of particular importance among the properties of polymers is their viscoelastic rheological behaviour,[85–89] since this governs their processing in the melt state. Viscoelasticity is related to the large length scale (conformational) characteristics of polymers, at the level of the end-to-end distance, whose relaxation takes place over times which typically exceed those of segmental motion by many orders of magnitude. For example, as we will see in Chapter 7, the Rouse model predicts that the longest relaxation time of a chain in a melt grows with chain length N as $\sim N^2$, while reptation theory for chain motion in an infinite fixed network calculates the exponent to be significantly higher, equal to 3. This means that, if the molecular length of a polymer chain is doubled, then the corresponding relaxation time will increase by at least eight times. For long polymer chains, these relaxation times are on the order of milliseconds to seconds. By comparing this time with the maximum time spans that can be simulated today on readily available hardware with atomistic MD methods (which are on the order of a few hundreds of nanoseconds to a few microseconds on usually available computational platforms[90]), we immediately recognize that it is impossible to simulate truly long polymeric systems by brute-force atomistic MD.

Restricting our analysis to equilibrium conditions (nonequilibrium phenomena are discussed in Chapter 10), there exist two ways in which one can overcome the problem of terminal relaxation in simulations of long polymer melts. The first involves an intermediate step of coarse-graining to a simpler representation using e.g. iterative Boltzmann schemes and is analysed in Chapter 7. The second makes use of a hierarchical strategy where the results of equilibrium MD simulations are mapped onto the predictions of mesoscopic (preferably molecular) theories for the linear viscoelastic properties of polymers. On one hand, this helps validate and refine the theory; on the other hand, it makes it possible to extract the parameters invoked by the theory from molecular simulations. This approach is also presented in Chapter 7. Here we will discuss what other type of information one can get for the transport and rheological properties of polymer melts *directly* from molecular simulations. One such property is[91–93] the chain self-diffusion coefficient, D, in the melt and its chain length dependence. Figures 4.4 and 4.5 present results from MD simulations[91–93] for the chain length dependence of the self-diffusion coefficient, D, of three different polymers: linear PE (at $T = 450$ K and $P = 1$ atm), H-shaped PE (at $T = 450$ K and $P = 1$ atm), and *cis*-1,4-PB (at $T = 413$ K and $P = 1$ atm). In all cases, a gradual, very smooth change with chain length at an intermediate value of N is observed, marking the passage from a Rouse- to a reptation-like behaviour. Moreover, for linear PE, for chain lengths N below C_{140}, chain end effects are found to combine with Rouse dynamics to result in D values that exhibit a power-law dependence on N of the form $D \propto N^{-b}$ with $b = 1.70(\pm 0.05)$. For chain lengths N above C_{170}, reptation dynamics dominates chain mobility, resulting in D values that exhibit a stronger dependence on N, described again by a power law of the form $D \propto N^{-b}$ but now with $b = 2.20(\pm 0.10)$. These values are practically the same for all three different polymers simulated. The value of 2.20 predicted for the exponent b is in agreement with experimental data[94,95] for a number of entangled polymer melts.

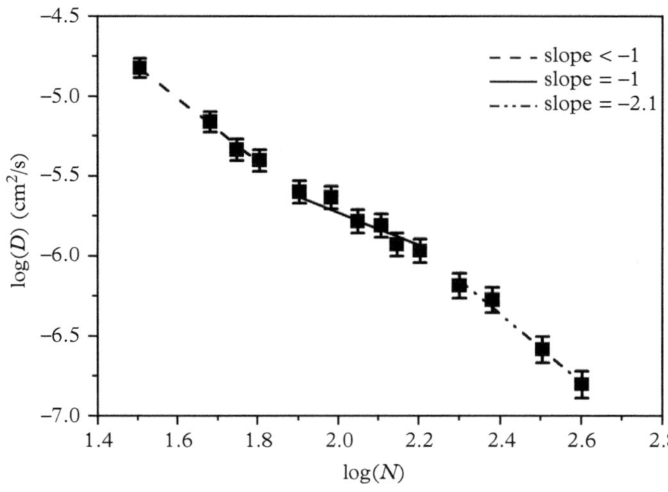

Figure 4.4 Log–log plot of the dependence of the self-diffusion coefficient D on chain length N for various cis-1,4-PB systems as obtained from MD simulations at $T = 413\,K$ and $P = 1$ atm.

Reprinted (adapted) with permission from Tsolou, G., Mavrantzas, V.G., and Theodorou, D.N. *Macromolecules* **38**, 1478 (2005). Copyright 2005 American Chemical Society.

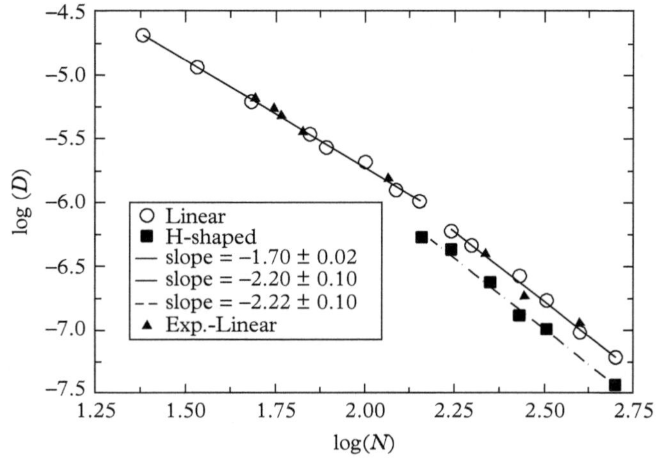

Figure 4.5 Chain centre-of-mass self-diffusion coefficient D (in cm^2/s) versus chain length N in a log–log plot for a number of linear and H-shaped PE melts as obtained from nPT MD simulations at $T = 450\,K$ and $P = 1$ atm.

Another property that can be directly obtained from MD simulations is the zero shear rate viscosity of the polymer, formally calculated as the time integral of the shear stress modulus of relaxation $G(t)$, Eqs. (4.3.4) and (4.3.5). This route was used quite early by Mondello and Grest[96] to calculate the zero shear rate viscosity of n-alkanes and demonstrate the equivalence of the Green–Kubo and Einstein approaches in both the atomic and molecular representations. In principle, the instantaneous stress tensor $\boldsymbol{\sigma}(t)$ that is needed in Eq. (4.3.4) can be obtained by an equilibrium MD simulation with the model system under study, using the virial theorem. Unfortunately, because of the strong fluctuations in the values of the off-diagonal components of $\boldsymbol{\sigma}(t)$, extremely long configurational averaging is required in order to calculate $G(t)$ by equilibrium MD with the required accuracy. In the literature, calculations of $G(t)$ for polymers based on Eqs. (4.3.4) and (4.3.5) for atomistic systems have also been reported by Harmandaris et al.[97] for a C_{24} and a C_{78} PE system, but this was possible only for times shorter than 10 ps; for longer times, the statistical noise in the calculated $\langle \sigma_{\alpha\beta}(t)\sigma_{\alpha\beta}(0)\rangle$ was too large to make it possible to obtain meaningful results.

An alternative way to obtain $G(t)$ is to monitor the stress relaxation of prestrained polymer configurations through a series of MD simulations in an extended statistical ensemble where, in addition to temperature and number of atoms, also certain components of the stress tensor and/or the simulation cell are kept constant, exactly as described in Section 4.1. The method has its origin in the principles of polymer dynamics, where $G(t)$ is interpreted as the function that describes how the shear stress relaxes after a sudden shearing displacement for a generalized linear viscolastic fluid,

$$G(t - t_0) = \frac{\tau_{xy}(t)}{\gamma_0}, \tag{4.4.1}$$

where γ_0 is the imposed sudden shear strain and t_0 denotes the time instant at which the sudden deformation is applied. Although, strictly speaking, Eq. (4.4.1) has not been employed in MD simulations, variants have appeared in the literature. We will discuss some of them in Chapter 10 where a method will be presented to obtain prestrained configurations of polymer melts by means of a Monte Carlo algorithm under conditions that mimic an applied flow.

By directly simulating a polymer melt in the presence of a flow field with one of the NEMD methods discussed in Section 4.1.6, one can also obtain the full viscosity function $\eta = \eta(\dot{\gamma})$ as well as the rest of the material functions.[98–100] Baig et al.,[98] for example, have carried out NEMD simulations with the p-SLLOD algorithm of a number of short alkane melts in shear (using the Lees–Edwards boundary conditions) and planar extension (using the Kraynik–Reinelt boundary conditions), and compared their results against the predictions of a number of well-known viscoelastic models for linear polymers, some of which will be reviewed in Chapter 10. NEMD simulations have also been employed by Baig and Mavrantzas[100] to simulate an unentangled H-shaped polymer melt in steady shear and planar elongational flow over a range of deformation rates, in full atomistic detail. The simulated system was an H-shaped PE melt containing seventy-eight carbon atoms per chain along its main backbone and twenty-five carbon atoms per chain along each one of its four dangling arms, thus denoted as H_78_25 PE. The chain length of the system was deliberately selected to lie in the unentangled regime in order to separate out reptation phenomena from the dynamics and, instead, place the emphasis on atomistic interactions and their interplay with local chain conformation and architecture. Results for planar extension showed that the primary steady-state elongational viscosity η_1 exhibits a tension-thickening behaviour as the strain rate $\dot{\varepsilon}$ is increased, which can be analysed in terms of (a) perturbations in the instantaneous intrinsic chain shape (Figure 4.6) and (b) differences in the stress distribution along the chain contour.

The authors also presented results (Figure 4.7) for the flow birefringence induced by the two different flows (planar shear and planar elongation) in the simulated H_78_25 PE melt. To this end, the anisotropy $\Delta n = n_{xx} - n_{yy}$ in the refractive index tensor $\boldsymbol{n}$ was plotted as a function of the anisotropy $\Delta\tau = \tau_{xx} - \tau_{yy}$ in the stress tensor $\boldsymbol{\tau}$, and results were compared for the two flows. Figure 4.7 shows that, to an excellent approximation, for small enough stresses (less than about 10 MPa), Δn varies linearly with $\Delta\tau$, a result known as the *stress optical rule* (*SOR*). By fitting the simulation data with straight lines, one can obtain the stress optical coefficient C; the result is $C = (1.3 \pm 0.1) \times 10^{-9}\ \mathrm{Pa}^{-1}$. According to experimental measurements using elongational flow opto-rheometry,[101] for high molar mass and high density, linear PE melts, C comes out to be independent of the molar mass and of the applied elongational flow rate, equal to $C_{\mathrm{exp}} = 2.20 \times 10^{-9}\ \mathrm{Pa}^{-1}$.

4.5 Molecular dynamics at surfaces and interfaces

To carry out MD simulations at surfaces and interfaces,[102–109] one should include in the Hamiltonian or the Lagrangian extra terms accounting for the additional interactions with the substrate. In the microcanonical ensemble,[110] the modified Lagrangian has the form

$$\mathcal{L} = \frac{1}{2}\sum_{i=1}^{n} m_i \dot{\boldsymbol{r}}_i^2 - \sum_i \mathcal{V}_{\mathrm{substrate}}(\boldsymbol{r}_i) - \sum_{i<j} \mathcal{V}_{ij}^{\mathrm{NB}} - \sum_{i<j} \mathcal{V}_{ij}^{\mathrm{tails}} - \sum_{i<j} \mathcal{V}_{ij}^{l} - \sum_{i<j<k} \mathcal{V}_{ijk}^{\theta} - \sum_{i<j<k<l} \mathcal{V}_{ijkl}^{\phi}. \tag{4.5.1}$$

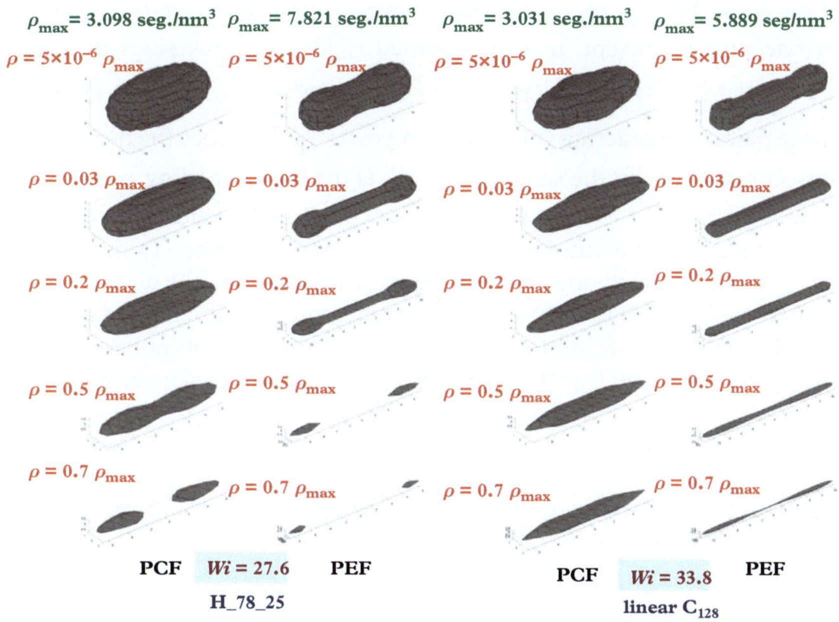

Figure 4.6 Intrinsic molecular shapes of the H_78_25 and linear C_{128} PE melts under shear or planar Couette flow (PCF) and planar extensional flow (PEF). *Wi* is the Weissenberg number.

Reprinted from Baig, C., and Mavrantzas, V.G. *J. Chem. Phys.* **132**, 014904 (2010), with the permission of AIP Publishing.

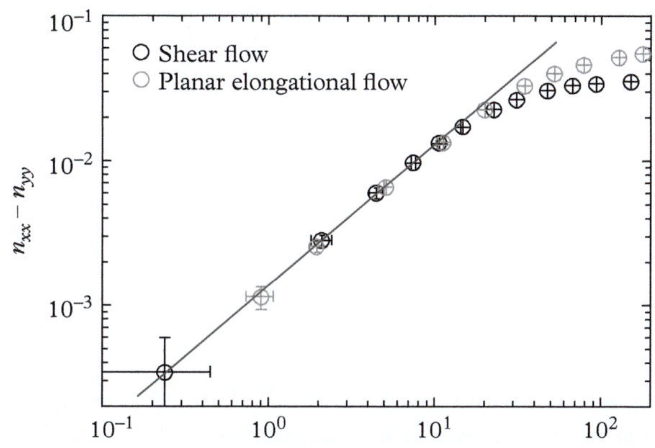

Figure 4.7 Anisotropy in the refractive index tensor $\boldsymbol{n}$ versus anisotropy in the stress tensor $\boldsymbol{\tau}$ in a H_78_25 PE melt simulated under shear (or planar Couette flow (PCF)) and under planar extensional flow (PEF). The slope in the plot indicates the linear relationship between the two anisotropies (the *stress-optical rule*) at small stresses.

Reprinted from Baig, C., and Mavrantzas, V.G. *J. Chem. Phys.* **132**, 014904 (2010), with the permission of AIP Publishing.

In Eq. (4.5.1), the term $\mathcal{V}_{\text{substrate}}(\boldsymbol{r}_i)$ is the potential energy field exerted on atom i of the polymer by all substrate phases present, usually assumed rigid. $\mathcal{V}_{ij}^{\text{NB}}$ describes the pairwise additive Lennard-Jones potential $\left(\mathcal{V}^{\text{LJ}}\right)$ for the calculation of nonbonded interatomic interactions between all minimum image pairs of interacting sites in the polymer; $\mathcal{V}_{ij}^{l}$ describes the bond-stretching potential between consecutive atoms in the same chain; $\mathcal{V}_{ijk}^{\theta}$ is the bond-bending potential between three bonded atoms; and $\mathcal{V}_{ijkl}^{\phi}$ is the torsional potential associated with skeletal bonds. Usually, $\mathcal{V}_{ij}^{\text{NB}}$ is truncated and smoothed to 0 at a cutoff distance R_{c} by a quintic spline. To do this, the long-range attractions between atom pairs beyond the truncation distance should be accounted for in the simulations by the integrated potential functions $\mathcal{V}_{ij}^{\text{tails}}$, taking into account the actual density profile of the film.[110,111] With expression (4.5.1) for the Lagrangian, equations of motion are then derived for the n atoms in the system by using Eq. (1.1.32). The resulting equations make it possible to simulate the system in the microcanonical (NVE) ensemble. To carry out a simulation in the nVT statistical ensemble, one should make use of the extended ensemble approach by Nosé–Hoover discussed in Section 4.1.

If the nonbonded Lennard-Jones potential is truncated at some cutoff distance R_{c} and no smoothing is used, i.e. if

$$\mathcal{V}_{ij}^{\text{NB}} = \mathcal{V}^{\text{LJ}}(r_{ij})\,\Pi\left(r_{ij} - R_{\text{c}}\right) = \begin{cases} 4\varepsilon\left[\left(\dfrac{\sigma}{r_{ij}}\right)^{12} - \left(\dfrac{\sigma}{r_{ij}}\right)^{6}\right], & r \leq R_{\text{c}} \\ 0, & r > R_{\text{c}} \end{cases}, \tag{4.5.2}$$

where

$$\Pi\left(r_{ij} - R_{\text{c}}\right) = \begin{cases} 1, & r_{ij} \leq R_{\text{c}} \\ 0, & r_{ij} > R_{\text{c}} \end{cases} \tag{4.5.3}$$

denotes the pulse function, then the corresponding force

$$F^{\text{NB}}(r_{ij}) = -\left.\frac{d\mathcal{V}^{\text{NB}}(r)}{dr}\right|_{r=r_{ij}} = \begin{cases} -\left.\dfrac{d\mathcal{V}^{\text{LJ}}(r)}{dr}\right|_{r=r_{ij}} + \mathcal{V}^{\text{LJ}}(R_{\text{c}})\,\delta\left(r_{ij} - R_{\text{c}}\right), & r_{ij} \leq R_{\text{c}} \\ 0, & r_{ij} > R_{\text{c}} \end{cases} \tag{4.5.4}$$

will exhibit a discontinuity at $r = R_c$ due to the term $\mathcal{V}^{\text{LJ}}(R_{\text{c}})\,\delta\left(r - R_{\text{c}}\right)$ in the previous equation. This is quite difficult to deal with, particularly in simulations of interfacial systems which are often characterized by relatively abrupt changes in the local mass density. In actual MD simulations of interfacial systems, the δ-term can be handled by following the approximation proposed by Trokhymchuk and Alejandre,[112]

$$\delta\left(r_{ij} - R_{\text{c}}\right) \approx \lim_{\Delta r \to 0} \frac{\Pi\left(r_{ij} - R_{\text{c}}\right) - \Pi\left(r_{ij} - R_{\text{c}} - \Delta r\right)}{\Delta r}, \tag{4.5.5}$$

where Δr is a fixed parameter chosen sufficiently small ($\Delta r \approx 0.01\sigma$) so that its influence on the simulation results is infinitesimal.

To compute the surface tension in an MD simulation, one can simulate a planar liquid film that is sufficiently thick for bulk conditions to prevail in its middle. Sgouros et al.[113] have undertaken such a calculation for molten PE surfaces and molten polyethylene/graphite interfaces. They focused on monodisperse linear C_{16}, C_{100}, C_{260}, and C_{520} melts at temperatures of 400 to 550 K in two interfacial geometries: (a) a free film, exposed to its own vapour on both sides; and (b) a capped film, sandwiched between undeformable semiinfinite graphite phases. The film thickness L_z (width L_x, L_y) was at least four (two) times the melt root mean square radius of gyration, $\langle R_g^2 \rangle^{1/2}$, in the aperiodic z (periodic x, y) direction, such that the pressure tensor in the middle region was isotropic with diagonal elements equal to the pressure of a bulk melt phase in equilibrium with the film. All model systems (thoroughly pre-equilibrated via connectivity-altering MC moves that we will discuss in Chapter 5) were simulated in the canonical ensemble by properly accounting for tail contributions to the nonbonded interactions beyond the cutoff distance in the anisotropic environment of the film.

An instantaneous atomic-level stress tensor $\boldsymbol{\tau}_i^{\text{inst}}$ was computed for each interaction site i in the polymer film via the equation (compare Eq. (3.5.85))

$$\tau_{i_{\alpha\beta}}^{\text{inst}} = -\frac{1}{V_i}\left[\begin{array}{l} m u_{i_\alpha} u_{i_\beta} + \frac{1}{2}\sum\limits_{\text{pairs}} \left(r_{i_\alpha} F_{\text{pair},i_\beta} + r_{j_\alpha} F_{\text{pair},j_\beta}\right) + \frac{1}{2}\sum\limits_{\text{bonds}} \left(r_{i_\alpha} F_{\text{bond},i_\beta} + r_{j_\alpha} F_{\text{bond},j_\beta}\right) \\ + \frac{1}{3}\sum\limits_{\text{angles}} \left(r_{i_\alpha} F_{\text{angle},i_\beta} + r_{j_\alpha} F_{\text{angle},j_\beta} + r_{k_\alpha} F_{\text{angle},k_\beta}\right) \\ + \frac{1}{4}\sum\limits_{\text{dihedrals}} \left(r_{i_\alpha} F_{\text{dihedral},i_\beta} + r_{j_\alpha} F_{\text{dihedral},j_\beta} + r_{k_\alpha} F_{\text{dihedral},k_\beta} + r_{l_\alpha} F_{\text{dihedral},l_\beta}\right) \\ + \boldsymbol{k}\text{-space}\left(r_{i_\alpha}, F_{i_\beta}\right) + \sum\limits_{\text{wall}} \left(d_{i_\alpha} F_{\text{wall},i_\beta}\right) \end{array}\right], \tag{4.5.6}$$

where the indices i, j, k, l denote interaction sites and the indices α and β denote spatial directions. Equation (4.5.6) contains contributions from pairwise nonbonded interactions, bond stretching, bond angle bending, and dihedral torsional potentials, a $\boldsymbol{k}$-space contribution from long-range nonbonded forces computed by Ewald summation, as well as contributions due to the interactions of the polymer with the graphite walls, if present. $F_{\text{dihedral},i_\beta}$ is the β-component of the force on atom i due to the torsional potential associated with the quartet $\{i, j, k, l\}$ considered in the summation. F_{pair,i_β}, F_{bond,i_β}, and F_{angle,i_β} have similar interpretations. F_{wall,i_β} is the β-component of the force on site i of the polymer film due to the considered graphite wall, while d_{i_α} is the α-component of the position vector of atom i relative to the wall. Separation vectors and forces were computed using the minimum image convention.

Based on the atomic-level stresses, one can define local instantaneous stress tensors $\boldsymbol{\tau}_s^{\text{inst}}$ within slabs s of the polymer normal to the z-direction, of thickness d_s (compare Eq. (3.5.82)),

$$\tau_{s_{\alpha\beta}}^{\text{inst}} = \frac{1}{V_s}\sum_{i=1}^{n_s} V_i \tau_{i_{\alpha\beta}}^{\text{inst}} \tag{4.5.7}$$

with n_s being the total number of sites whose centres lie in slab s and $V_s = d_s L_x L_y$ the slab volume.

The surface tension γ_{fv} at a free polymer surface is computed from the free film simulation as

$$\gamma_{\text{fv}} = \frac{1}{2}\int_{L_{z,\min}}^{L_{z,\max}} \left[\tau_{\text{T}}(z) - \tau_{\text{N}}(z)\right] dz \approx \frac{1}{2}\left\langle \sum_{s=1}^{N_{\text{slabs}}} \left[\frac{\tau_{s_{xx}}^{\text{inst}} + \tau_{s_{yy}}^{\text{inst}}}{2} - \tau_{s_{zz}}^{\text{inst}}\right] d_s \right\rangle \tag{4.5.8}$$

with N_{slabs} being the total number of slabs in which the model system is divided. The lower ($L_{z,\min}$) and upper ($L_{z,\max}$) limits of integration are in the interior of the bulk vapour (practically vacuum) phases flanking the free polymer film on either side.

The adhesion tension at a polymer–solid interface, defined as the difference between the surface free energy of the bare solid γ_{sv} and the interfacial tension γ_{sf} between the solid and the polymer, is computed from the capped film simulation as

$$\gamma_{\text{sv}} - \gamma_{\text{sf}} = -\frac{1}{2}\int_{L_{z,\min}}^{L_{z,\max}} \left[\tau_{\text{T}}(z) - \tau_{\text{N}}(z)\right] dz \approx -\frac{1}{2}\left\langle \sum_{s=1}^{N_{\text{slabs}}} \left[\frac{\tau_{s_{xx}}^{\text{inst}} + \tau_{s_{yy}}^{\text{inst}}}{2} - \tau_{s_{zz}}^{\text{inst}}\right] d_s \right\rangle. \tag{4.5.9}$$

$L_{z,\min}$ and $L_{z,\max}$ mark the extreme edges of the polymer in the capped film geometry.

The factors of 1/2 in Eqs. (4.5.8) and (4.5.9) are due to the presence of two macroscopically symmetric interfaces in each of the geometries considered.

From the quantities appearing on the left-hand sides of Eqs. (4.5.8) and (4.5.9), one can define four important thermodynamic quantities: (a) the work of cohesion (W_{C}), (b) the work of adhesion (W_{A}), (c) the work of spreading (W_{S}), and (d) the work of immersion (W_{I}):

$$W_{\text{C}} = 2\gamma_{\text{fv}} \tag{4.5.10}$$

$$W_{\text{A}} = \gamma_{\text{sv}} + \gamma_{\text{fv}} - \gamma_{\text{sf}} = \gamma_{\text{fv}}\left(\cos\theta_{\text{sf}} + 1\right) \tag{4.5.11}$$

$$W_{\text{S}} = \gamma_{\text{sv}} - \gamma_{\text{fv}} - \gamma_{\text{sf}} = \gamma_{\text{fv}}\left(\cos\theta_{\text{sf}} - 1\right) \tag{4.5.12}$$

$$W_{\text{I}} = \gamma_{\text{sv}} - \gamma_{\text{sf}} = \gamma_{\text{fv}}\cos\theta_{\text{sf}}, \tag{4.5.13}$$

where θ_{sf} is the equilibrium contact angle between solid and polymer, if it exists.

Figure 4.8a compares the temperature dependence of the surface tension γ_{fv} of a C_{100} melt, as predicted by molecular simulation, to an experimental correlation for long-chain PE in the same temperature range. The slope $\partial\gamma_{\text{fv}}/\partial T$ is seen to be in very good agreement between simulation and experiment;[114] the actual values of γ_{fv} are somewhat lower in the simulation, because of the difference in molar mass. The temperature dependence of the interfacial tension γ_{sf} (dotted line) between the C_{100} melt and the basal plane of graphite is shown in Figure 4.8b. Simulation estimates of γ_{sf} are captured well by the Girifalco and Good approximation[115]

$$\gamma_{\text{sf}}^{\text{G}} = \gamma_{\text{sv}} + \gamma_{\text{fv}} - 2\Phi\left(\gamma_{\text{sv}}\gamma_{\text{fv}}\right)^{1/2} \tag{4.5.14}$$

(where Φ is an interfacial interaction parameter that is usually close to unity, $\Phi = 1$) shown in Figure 4.8b as a broken line.

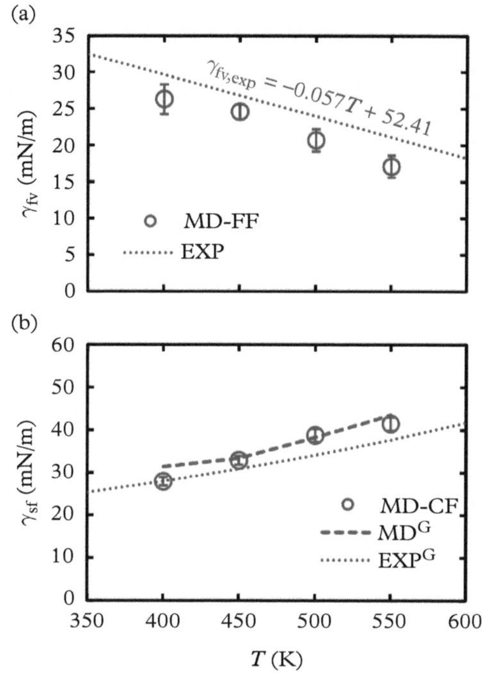

Figure 4.8 Temperature dependence of (a) the surface tension γ_{fv} of a C_{100} linear PE melt and (b) the interfacial tension γ_{sf} between the C_{100} linear PE melt and the basal plane of graphite. The dotted line in (a) presents the experimental dependence of γ_{fv} on temperature T for high molar mass linear PE. The dashed (dotted) line in (b) represents the predictions of the Girifalco and Good relation, Eq. (4.5.14), using the MD (experimental[114]) values of the surface tension and the experimental surface free energy of graphite. FF and CF stand for "free film" and "confined film", respectively.

Reprinted (adapted) with permission from Sgouros, A.P., Vogiatzis, G.G., Kritikos, G., Boziki, A., Nikolakopoulou, A., Liveris, D., and Theodorou, D.N. *Macromolecules*, **50** 8827 (2017). Copyright 2017 American Chemical Society.

Figure 4.9 displays the molar mass and temperature dependence of the work of cohesion, the work of adhesion, the work of spreading, and the work of immersion for the C_{100}-basal plane of graphite system, defined in Eqs. (4.5.10)–(4.5.13).

Another system that has been extensively studied by simulations is a polyethylene (PE) melt supported by a crystalline graphitic phase on its one side and left to be free on the other side. Simulations in this case have been carried out at $T = 450\,K$ with PE chains modelled as sequences of united atoms. The simulation box or unit cell, at the two sides of which the graphite and vacuum interfaces are located, was orthorhombic. Along the x- and y-directions of the coordinate system, periodic boundary conditions were implemented; i.e. the system behaved as infinite in these directions. Through this, a thin film of liquid of effectively infinite extent along x and y and finite width along the third (z) direction was formed. As in Ref. 113, to calculate the potential energy of interaction between polymer atoms and the semi-infinite graphite substrate, the method designed by Steele[102,103] was implemented,[110,111] capable of incorporating the exact crystallographic structure

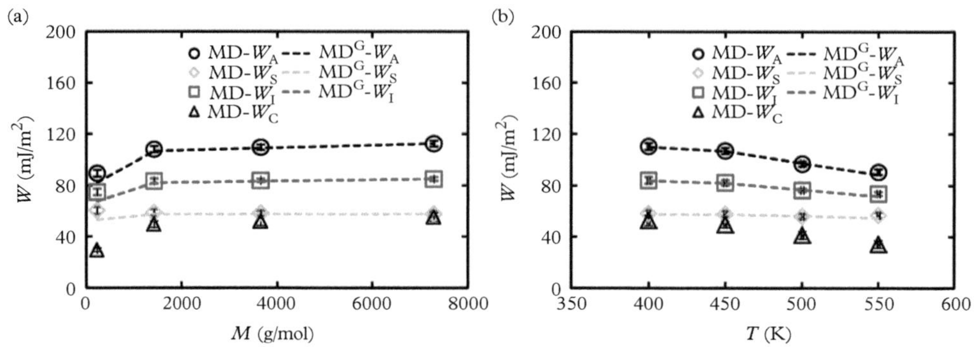

Figure 4.9 Work of adhesion (circles), work of spreading (squares), work of immersion (diamonds) and work of cohesion (triangles) versus (a) molar mass and (b) temperature. The dashed lines show the predictions based on the Girifalco and Good relation, Eq. (4.5.14), using the MD data from free film simulations.
Reprinted (adapted) with permission from Sgouros, A.P., Vogiatzis, G.G., Kritikos, G., Boziki, A., Nikolakopoulou, A., Liveris, D., and Theodorou, D.N. *Macromolecules*, **50** 8827 (2017). Copyright 2017 American Chemical Society.

of graphite. According to this, the graphite phase is represented as a set of LJ carbon atoms centred at their crystallographically correct positions, while the solid structure is assumed to remain unperturbed by the adsorption of polymer. Taking advantage of the symmetry in the crystalline substrate, the potential energy of interaction between a polymer atom and the semi-infinite graphite phase at normal distance z is written as

$$\mathcal{V}_{\text{graphite}}(z) = \mathcal{V}_0(z) + \mathcal{V}_n(z), \tag{4.5.15}$$

where $\mathcal{V}_0$ and $\mathcal{V}_n$ are given through

$$\begin{aligned} \mathcal{V}_0(\boldsymbol{z}) &= \frac{4\pi\sigma_{\text{ps}}^2\varepsilon_{\text{ps}}}{\alpha_{\text{s}}} \sum_{\alpha=0}^{\infty} \left[\frac{2}{5} \left(\frac{\sigma_{\text{ps}}}{z+\alpha d} \right)^{10} - \left(\frac{\sigma_{\text{ps}}}{z+\alpha d} \right)^{4} \right] \\ \mathcal{V}_n(\boldsymbol{z}) &= \frac{2\pi\sigma_{\text{ps}}^6\varepsilon_{\text{ps}}}{\alpha_{\text{s}}} \sum_{n>0}^{5} \left[\frac{\sigma_{\text{ps}}^6}{30} \left(\frac{g_n}{2z} \right)^5 K_5(g_n) - 2\left(\frac{g_n}{2z} \right)^2 K_2(g_n) \right] f_n(s_1, s_2). \end{aligned} \tag{4.5.16}$$

In Eqs. (4.5.16), σ_{ps} and ε_{ps} are the values of the collision diameter and well depth, respectively, of the LJ potential describing the pair interaction between a polymer-united atom and a graphite atom (calculated using the Lorentz–Berthelot rule), α_{s} is the area of the surface unit cell, d the distance between successive graphene sheets in the graphite, g_n are numerical coefficients, K_2 and K_5 are modified Bessel functions of the second kind, and s_1 and s_2 are parameters defining the x and y coordinates of the projection of the polymer atom on the first graphite basal layer ($z = 0$) relative to the surface unit cell; they range from 0 to 1. Values for the coefficients g_n and the functions $f_n(s_1, s_2)$ can be found in the original work of Steele.[102,103] The 10–4 sum is performed explicitly[110] over the first sixty-three underlying layers of graphite atoms; for the rest of the layers, the interaction is accounted for by smearing graphite carbons in all three directions, resulting in a 9–3 potential for $\mathcal{V}_0(z)$ of the form

$$\mathcal{V}_0(z) = \frac{4\pi\sigma_{ps}^2\varepsilon_{ps}}{\alpha_s}\sum_{\alpha=0}^{63}\left[\frac{2}{5}\left(\frac{\sigma_{ps}}{z+\alpha d}\right)^{10} - \left(\frac{\sigma_{ps}}{z+\alpha d}\right)^4\right] + \frac{4\pi\sigma_{ps}^3\varepsilon_{ps}}{3\alpha_s d}\left[\frac{2}{15}\left(\frac{\sigma_{ps}}{z+64d}\right)^9 - \left(\frac{\sigma_{ps}}{z+64d}\right)^3\right] \quad (4.5.17)$$

Simulations conducted with monodisperse PE samples confined between a semi-infinite graphite phase on the one side and vacuum on the other have made it possible to study how chains organize in the two interfacial regions. The simulations were performed at temperature $T = 450\,\text{K}$ with a united atom model for PE and samples of molecular length up to C_{400} were studied. Figure 4.10 displays the mass density distribution across the film for the C_{78} system, clearly revealing a characteristic oscillatory behaviour near the graphitic plane wherein at least three peaks can be discerned at distances equal to 4, 8.5, and 13 Å from the graphite plane, while a fourth peak at 18 Å is also discerned but with a considerably smaller height. In the middle region of the film the density attains a constant value (equal to the density of the free or unconstrained C_{78} PE melt at the same density). At the polymer–vacuum interface, on the other hand, the melt density gradually drops to 0, exhibiting the characteristic sigmoidal shape typical of a melt free surface.

From the same simulations one can also extract information about segmental mobility as a function of the width or thickness d of the chosen interfacial region that segments belonged to at the beginning of the diffusive process. One way to accomplish this is to map simulation results onto the solution of an unsteady 1D diffusion equation describing concentration in a semi-infinite

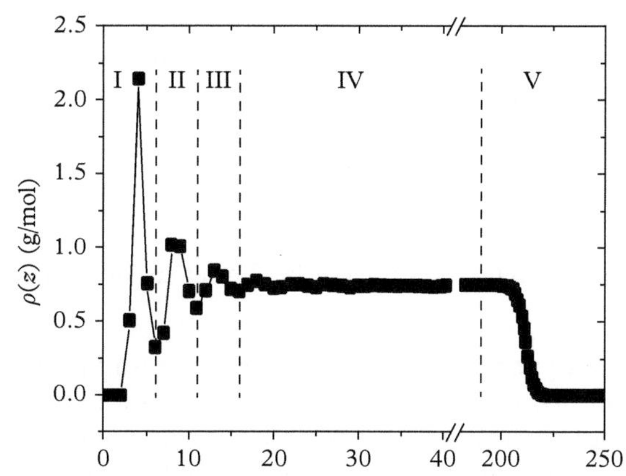

Figure 4.10 Density profile along the z (normal) direction of a C_{78} PE melt adsorbed on graphite on one side and exposed to vacuum on the other ($T = 450$ K).

Reprinted (adapted) with permission from Harmandaris, V.A., Daoulas, K., and Mavrantzas, V.G. *Macromolecules* **38**, 5796 (2005). Copyright 2005 American Chemical Society.

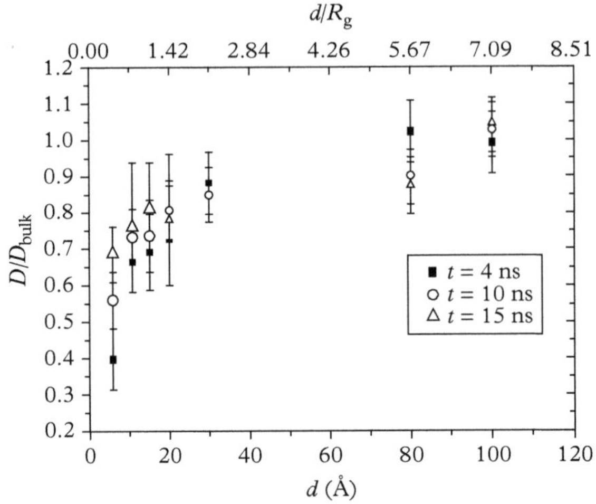

Figure 4.11 Dependence of the apparent diffusion coefficient D (scaled with its value in the bulk D_{bulk}) on the thickness d of the layer considered, after times $t = 4$, 10, and 15 ns from the time origin, for a C_{78} PE melt film adsorbed on graphite. Layer thicknesses are given as multiples of the unperturbed radius of gyration R_g at the top of the figure.

Reprinted (adapted) with permission from Harmandaris, V.A., Daoulas, K., and Mavrantzas, V.G. *Macromolecules* **38**, 5796 (2005). Copyright 2005 American Chemical Society.

medium.[111b] By plotting the ratio $\dfrac{D(t)}{D(t \to \infty)}$ as a function of d, one can then find the exact value of d beyond which the transport properties of the adsorbed film approach those of the bulk, unconstrained melt. This behaviour is shown in Figure 4.11 for times t equal to 4, 10, and 15 ns. It is concluded that d must be at least 90 Å, i.e. 6–7R_g, for the diffusive properties in an adsorbed C_{78} PE film to be equal to those in a bulk C_{78} melt.

Problems

Problem 4.1: Derive the coordinate version of the Verlet algorithm starting from the velocity Verlet algorithm.

Problem 4.2: Show that the dynamical integration scheme resulting from the Trotter factorization of the Liouville operator according to Eq. (4.1.28) with iL_1 and iL_2 defined by Eq. (4.1.30) reproduces the velocity Verlet algorithm, Eqs. (4.1.6). Similarly, show that when iL_1 and iL_2 are defined by Eq. (4.1.32), the same scheme reproduces the Verlet leapfrog algorithm, Eqs. (4.1.5).

Problem 4.3: Use Eqs. (4.2.10) and (4.2.11) to find how the partial derivatives of molecular densities with respect to chemical potentials in an open binary system are related to fluctuations in the number of molecules of each species within a fixed macroscopic volume V.

References

1. McQuarrie, D.A. *Statistical Mechanics*. Harper & Row, New York (1973).
2. Goldstein, H. *Classical Mechanics*. Addison-Wesley, MA (1980).
3. Gear, C.W. *Numerical Initial Value Problems in Ordinary Differential Equations*. Prentice-Hall, Englewood Cliffs, NJ (1971).
4. Allen, M.P., and Tildesley, D.J. *Computer Simulation of Liquids*. Oxford University Press, Oxford (1987).
5. Frenkel, D., and Smit, B. *Understanding Molecular Simulations: From Algorithms to Applications*. Academic Press, New York (1996).
6. Verlet, L. *Phys. Rev.* **159**, 98 (1967); Verlet, L. *Phys. Rev.* **165**, 201 (1968).
7. (a) Andersen, H.C. *J. Chem. Phys.* **72**, 2384 (1980); (a) Andersen, H.C. *J. Chem. Phys.* **76**, 637 (1982).
8. (a) Andersen, H.C. *Phys. Rev. Lett.* **50**, 1207 (1983); (a) Andersen, H.C. *J. Chem. Phys.* **78**, 6928 (1983).
9. (a) Nosé, S. *Mol. Phys.* **52**, 255 (1984); (b) Nosé, S. *J. Chem. Phys.* **81**, 511 (1984); (c) Nosé, S. *Mol. Phys.* **57**, 187 (1986).
10. Hoover, W.G. *Phys. Rev. A* **31**, 1695 (1985); Hoover, W.G. *Phys. Rev. A* **34**, 2499 (1986).
11. (a) Cho, K., Joannopoulos, J.D., and Kleinmann, L. *Phys. Rev. E* **47**, 3145 (1993); (b) Martyna, G.J. *Phys. Rev. E* **50**, 3234 (1995).
12. Berendsen, H.J.C., Postma, J.P.M., van Gunsteren, W.F., DiNola, A., and Haak, J.R. *J. Chem. Phys.* **81**, 3684 (1984).
13. (a) Parrinello, M., and Rahman, A. *Phys. Rev. Lett.* **45**, 1196 (1980); (b) Parrinello, M., and Rahman, A. *J. Appl. Phys.* **52**, 7182 (1981).
14. Melchionna, S., Ciccotti, G., and Holian, B.L. *Mol. Phys.* **78**, 533 (1993).
15. (a) Martyna, G.J., Klein, M.L., and Tuckerman, M. *J. Chem. Phys.* **97**, 2635 (1992); (b) Martyna, G.J., Tobias, D.J., and Klein, M.L. *J. Chem. Phys.* **101**, 4177 (1994).
16. (a) Tuckerman, M., Berne, B.J., and Martyna, G.J. *J. Chem. Phys.* **97**, 1990 (1992); (b) Martyna, G.J., Tuckerman, M.E., Tobias, D.J., and Klein, M.L. *Mol. Phys.* **87**, 1117 (1996).
17. (a) Tuckerman, M.E., Mundy, C.J., and Martyna, G.J. *Europhys. Lett.* **45**, 149 (1999); (b) Tuckerman, M.E., and Martyna, G.J. *J. Phys. Chem. B* **104**, 159 (2000); (c) Tuckerman, M.E., Liu, Y., Ciccotti, G., and Martyna, G.J. *J. Chem. Phys.* **115**, 1678 (2001).
18. Edberg, R., Evans, D.J., and Morriss, G.P. *J. Chem. Phys.* **84**, 6933 (1986).
19. Ryckaert, J.-P., Ciccoti, G., and Berendsen, H.J.C. *J. Comp. Phys.* **23**, 327 (1977).
20. Palmer, B.J. *J. Comp. Phys.* **104**, 470 (1993).
21. Andersen, H.C. *J. Comp. Phys.* **52**, 24 (1983).
22. Evans, D.J., and Morriss, G.P. *Statistical Mechanics of non-equilibrium liquids*. Academic, New York (1990).
23. Hoover, W.G., Evans, D.J., Hickman, R.B., Ladd, A.J.C., Ashurst, W.T., and Moran, B. *Phys. Rev. A* **22**, 1690 (1980).
24. Evans, D.J. and Morriss, G. P. *Phys. Rev. A* **30**, 1528 (1984).
25. (a) Tuckerman, M.E., Mundy, C. J., Balasubramanian, S., and Klein, M.L. *J. Chem. Phys.* **106**, 5615 (1997); Tuckerman, M.E., Mundy, C. J., Balasubramanian, S., and Klein, M.L. *J. Chem. Phys.* **108**, 4353 (1998); (b) Evans, D.J., Searles, D.J., Hoover, Wm.G., Hoover, C.G., Holian, B.L., and Posch, H.A. *J. Chem. Phys.* **108**, 4351 (1998).
26. (a) Tuckerman, M.E., Mundy, C.J., and Klein, M.L. *Phys. Rev. Lett.* **78**, 2042 (1997); Tuckerman, M.E., Mundy, C.J., and Klein, M.L. *Phys. Rev. Lett.* **80**, 4105 (1998); (b) Hoover, Wm.G., Evans, D.J., Posch, H.A., Holian, B.L., and Morriss, G.P. *Phys. Rev. Lett.* **80**, 4103 (1998); (c) Reimann, P. *Phys. Rev. Lett.* **80**, 4104 (1998).

27. Lees, W., and Edwards, S.F. *J. Phys. C* **5**, 1921 (1972).
28. Kraynik, A.M., and Reinelt, D.A. *Int. J. Multiphase Flow* **18**, 1045 (1992).
29. (a) Todd, B.D., and Daivis, P.J. *Phys. Rev. Lett.* **81**, 1118 (1998); (b) Todd, B.D., and Daivis, P.J. *Comp. Phys. Comm.* **117**, 191 (1999); (c) Todd, B.D., and Daivis, P.J. *J. Chem. Phys.* **112**, 40 (2000).
30. Baranyai, A., and Cummings, P.T. *J. Chem. Phys.* **110**, 42 (1999).
31. (a) Baig, C., Edwards, B.J., Keffer, D.J., and Cohran, H.D. *J. Chem. Phys.* **122**, 114103 (2005); Edwards, B.J., Baig, C., and Keffer, D.J. *J. Chem. Phys.* **123**, 114106 (2005); Edwards, B.J., Baig, C., and Keffer, D.J. *J. Chem. Phys.* **124**, 194104 (2006).
32. Kirkwood, B., and Buff, F. *J. Chem. Phys.* **19**, 774 (1950).
33. Ben Naim, A. J. *Phys. Chem. B* **111**, 2896 (2007).
34. Galata, A., Anogiannakis, S., and Theodorou, D.N. *Fluid Phase Equilibria* **470**, 25 (2018).
35. Parrinello, M., and Rahman, A. J. *Chem. Phys.* **76**, 2662 (1982).
36. Gusev, A.A., Zehnder, M.M., and Suter, U.W. *Phys. Rev. B* **54**, 1 (1996).
37. Lebowitz, J.L., Percus, J.K., and Verlet, L. *Phys. Rev.* **153**, 250 (1967).
38. Cheung, P.S.Y. *Mol. Phys.* **33**, 519 (1977).
39. (a) Ray, J.R., and Graben, H.W. *Mol. Phys.* **43**, 1923 (1981); (b) Ray, J.R. *Am. J. Phys.* **50**, 1035 (1982).
40. Rowlinson, J.S. *Liquid and Liquid Mixtures*, Butterworths (1959).
41. (a) Çağin, T., and Ray, J.R. *Phys. Rev. A* **37**, 247 (1988); (b) Çağin, T., and Ray, J.R. *Phys. Rev. B* **37**, 699 (1988); Çağin, T., and Ray, J.R. *Phys. Rev. B* **38**, 7940 (1988).
42. Adams, D.J. *Mol. Phys.* **29**, 307 (1975).
43. Ray, J.R., and Graben, H.W. *Phys. Rev. A* **34**, 2517 (1986).
44. Ray, J.R., and Rahman, A. J. *Chem. Phys.* **80**, 4423 (1984).
45. Ray, J.R., Moody, M.C., and Rahman, A. *Phys. Rev. B* **32**, 733 (1985); Ray, J.R., Moody, M.C., and Rahman, A. *Phys. Rev. B* **33**, 895 (1986).
46. (a) Ray, J.R. *J. Appl. Phys.* **53**, 6441 (1982); (b) Ray, J.R. *Comp. Phys. Rep.* **8**, 109 (1988); (c) Fay, P.J., and Ray, J.R. *Phys. Rev. A* **46**, 4645 (1992).
47. Green, M.S. *J. Chem. Phys.* **22**, 398 (1954).
48. (a) Kubo, R. *J. Phys. Soc. Japan* **12**, 570 (1957); (b) Kubo, R. In *Lectures in Theoretical Physics*, Vol. 1, Wiley, New York (1958); (c) Kubo, R. In Meixner, J. (ed.), *Statistical Mechanics of Equilibrium and Nonequilibrium*. North-Holland Park, Amsterdam (1965); (d) Kubo, R. *Rep. Prog. Phys.* **29**, 255 (1966).
49. Karayiannis, K. Ch., Mavrantzas, V.G., and Theodorou, D.N. *Macromolecules* **37**, 2978 (2004).
50. Pecora, R. *Annu. Rev. Biophys. Bioengineering* **1**, 257 (1972).
51. (a) Berne, B.J. *Accts. Chem. Res.* **6**, 318 (1973); (b) Berne, B.J., and Pecora, R. *Introduction to the Molecular Theory of Light Scattering*. Wiley, New York (1976).
52. Edwards, D.M.F., Madden, P.A., and McDonald, I.R. *Mol. Phys.* **51**, 1141 (1984).
53. Borodin, O., Douglas, R., Smith, G.D., Trouw, F., and Petrucci, S. *J. Phys. Chem. B* **107**, 6813 (2003).
54. Doxastakis, M., Kitsiou, M., Fytas, G., Theodorou, D.N., Hadjichristidis, N., Meier, G., and Frick, B. *J. Chem. Phys.* **112**, 8687 (2000).
55. Paul, W., and Smith, G.D. *Rep. Prog. Phys.* **67**, 1117 (2004).
56. Havriliak, S., and Negami, S. *Polymer* **8**, 161 (1967).
57. Rouse, P.E. *J. Chem. Phys.* **21**, 1272 (1953).
58. Doi, M., and Edwards, S.F. *The Theory of Polymer Dynamics*. Oxford University Press, Oxford (1987).
59. Williams, G., Watts, D.C., Dev, S.B., and North, A.M. *Trans. Faraday Soc.* **67**, 1323 (1971).
60. Saiz, E., Riande, E., and Diaz Calleja, R. *J. Phys. Chem.* **101**, 7324 (1997).
61. van Hove, L. *Phys. Rev.* **95**, 249 (1954).
62. Marshall, W., and Lovesey, S.W. *Theory of Thermal Neutron Scattering*. Oxford University Press, Oxford (1971).

63. Egelstaff, P.A. *Thermal Neutron Scattering*. Academic, New York (1965).
64. Hansen, J.-P., and McDonald, I.R. *Theory of Simple Liquids*. Academic Press, New York (1976).
65. Higgins, J., and Benoit, H.C. *Polymers and Neutron Scattering*. Oxford University Press, New York (1996).
66. Sköld, K., Rowe, J.M., Ostrowski, G., and Randolph, P.D. *Phys. Rev. A* **6**, 1107 (1972).
67. Harnau, L., Winkler, R.G., and Reineker, P. *J. Chem. Phys.* **102**, 7750 (1995); Harnau, L., Winkler, R.G., and Reineker, P. *J. Chem. Phys.* 104, 6355 (1996).
68. De Gennes, P.G. *J. Phys.* **42**, 735 (1981).
69. Kremer, K., and Binder, K. *J. Chem. Phys.* **81**, 6381 (1984).
70. Faller, R., and Müller-Plathe, F. *ChemPhysChem* **3**, 180 (2001).
71. Bloembergen, N., Purcell, E.M., and Pound, R.V. *Phys. Rev.* **73**, 679 (1948).
72. Solomon, I. *Phys. Rev.* **99**, 559 (1955).
73. Kuhlmann, K.F., Grant, D.M., and Harris, R.K. *J. Chem. Phys.* **52**, 3439 (1970).
74. Doddrell, D., Glushko, V., and Allerhand, A. *J. Chem. Phys.* **56**, 3683 (1972).
75. Stejskal, E.O., and Gutowsky, H.S. *J. Chem. Phys.* **28**, 388 (1958).
76. Allerhand, A., Doddrell, D., and Komoroski, R. *J. Chem. Phys.* **55**, 189 (1971).
77. Howarth, O.H. *J. Chem. Soc., Faraday Trans. 2* **75**, 863 (1979).
78. Hall, C.K., and Helfand, E. *J. Chem. Phys.* **77**, 3275 (1982).
79. Dejean de la Batie, R., Lauprêtre, F., and Monnerie, L. *Macromolecules* **21**, 2045 (1988).
80. Moe, N.E., Qiu, X.H., and Ediger, M.D. *Macromolecules* **33**, 2145 (2000).
81. Lindsey, C.P., and Patterson, G.D. *J. Chem. Phys.* **73**, 3348 (1980).
82. Doxastakis, M., Theodorou, D.N., Fytas, G., Kremer, F., Faller, R., Müller-Plathe, F., and Hadjichristidis, N. *J. Chem. Phys.* **119**, 6883 (2003).
83. Ahumada, O., Theodorou, D.N., Triolo, A., Arrighi, V., Karatasos, C., and Ryckaert, J.-P. *Macromolecules* **35**, 7110 (2002).
84. Arrighi, V., Pappas, C., Triolo, A., and Pouget, S. *Physica B* **301**, 157 (2001).
85. Middleman, S. *Fundamentals of Polymer Processing*. McGraw-Hill, New York (1977).
86. Larson R.G. *Constitutive Equations for Polymer Melts and Solutions*. Butterworth-Heinemann, London (1988).
87. Bird, R.B., Armstrong, R.C., and Hassager, O. *Dynamics of Polymeric Liquids*, Vol. 1, *Fluid Mechanics*, 2nd ed. Wiley-Interscience, New York (1987).
88. Bird, R.B., Curtiss, C.F., Armstrong, R.C., and Hassager, O. *Dynamics of Polymeric Liquids*, Vol. 2, *Kinetic Theory*, 2nd ed. Wiley-Interscience, New York (1987).
89. Ferry, J.D. *Viscoelastic Properties of Polymers*. John Wiley & Sons, New York (1980).
90. Note, however, that microsecond-long MD simulations have been possible on specially designed machines. See, for example, Lindorff-Larsen, K., Piana, S., Dror, R.O., Shaw, D.E. *Science* **334**, 517 (2011).
91. (a) Harmandaris, V.A., Mavrantzas, V.G., and Theodorou, D.N. *Macromolecules* **31**, 7934 (1998); (b) Harmandaris, V.A., Mavrantzas, V.G., Theodorou, D.N., Kröger, M., Ramírez, J., Öttinger, H.C., and Vlassopoulos, D. *Macromolecules* **36**, 1376 (2003).
92. Tsolou, G., Mavrantzas, V.G., and Theodorou, D.N. *Macromolecules* **38**, 1478 (2005).
93. Karayiannis, N.C., and Mavrantzas, V.G. *Macromolecules* **38**, 8583 (2005).
94. Pearson, D.S., Ver Strate, G., Von Meerwall, E., and Schilling, F.C. *Macromolecules* **20**, 1133 (1987); Pearson, D.S., Ver Strate, G., Von Meerwall, E., and Schilling, F.C. *Macromolecules* **27**, 711 (1994).
95. Wang, S.-Q. *J. Polym. Sci., Polym. Phys.* **41**, 1589 (2003).

96. Mondello, M., and Grest, G.S. *J. Chem. Phys.* **106**, 9327 (1997).
97. Harmandaris, V.A., Mavrantzas, V.G., and Theodorou, D.N. *Macromolecules* **33**, 8062 (2000).
98. (a) Baig, C., Edwards, B.J., Keffer, D.J., and Cohran, H.D. *J. J. Chem. Phys.* **122**, 184906 (2005); (b) Baig, C., Jiang, B., Edwards, B.J., Keffer, D.J., and Cohran, H.D. *J. Rheol.* **50**, 625 (2006); (c) Baig, C., Edwards, B.J., Keffer, D.J., Cohran, H.D., and Harmandaris, V.A. *J. Chem. Phys.* **124**, 084902 (2006).
99. Stephanou, P.S., Baig, C., and Mavrantzas, V.G. *J. Rheol.* **53**, 309 (2009).
100. Baig, C., and Mavrantzas, V.G. *J. Chem. Phys.* **132**, 014904 (2010).
101. Okamoto, M., Kojima, A., Kotaka, T., and Muenstedt, H. *Macromolecules* **31**, 5158 (1998).
102. Steele, W.A. *Surf. Sci.* **36**, 317 (1973).
103. Steele, W.A. *The Interaction of Gases with Solid Surfaces*. Pergamon Press, Oxford (1974).
104. Nicholson, D., and Parsonage, N.G. *Computer Simulation and the Statistical Mechanics of Adsorption*. Academic Press, London (1982).
105. Wu, S. *Polymer Interface and Adhesion*. Marcel Dekker, New York (1982).
106. Rowlinson, J.S., and Widom, B. *Molecular Theory of Capillarity*. Oxford University Press, Oxford (1989).
107. Kinloch, A.J. *Adhesion and Adhesives: Science and Technology*. Chapman & Hall, New York (1987).
108. Fleer, G.J., Cohen Stuart, M.A., Scheutjens, J.M.H.M., Cosgrove, T., and Vincent, B. *Polymers at Interfaces*. Chapman & Hall, London (1993).
109. Jones, R.A.L., and Richards, R.W. *Polymers at Surfaces and Interfaces*. Cambridge University Press, Cambridge, UK (1999).
110. Mansfield, K.F., and Theodorou, D.N. *Macromolecules* **24**, 6283 (1991); Mansfield, K.F., and Theodorou, D.N. *Macromolecules* **24**, 4295 (1991); Mansfield, K.F., and Theodorou, D.N. *Macromolecules* **23**, 4430 (1990); Mansfield, K.F., and Theodorou, D.N. *Macromolecules* **22**, 3143 (1989).
111. (a) Daoulas, K., Harmandaris, V.A., and Mavrantzas, V.G. *Macromolecules* **38**, 5780 (2005); (b) Harmandaris, V.A., Daoulas, K., and Mavrantzas, V.G. *Macromolecules* **38**, 5796 (2005).
112. Trokhymchuk, A., and Alejandre, J. *J. Chem. Phys.* **111**, 8510 (1999).
113. Sgouros, A.P., Vogiatzis, G.G., Kritikos, G., Boziki, A., Nikolakopoulou, A., Liveris, D., and Theodorou, D.N. *Macromolecules*, **50** 8827 (2017).
114. Wu, S. *J. Colloid Interface Sci.* **31**, 153 (1969).
115. Girifalco, L.A., and Good, R.J. *J. Phys. Chem.* **61**, 904 (1957).

5

Monte Carlo

5.1 Principles of Monte Carlo simulation

5.1.1 Introduction and basic principles

Monte Carlo (MC) is a method for simulating the properties of matter that relies on stochastic (probabilistic) computer experiments. Like molecular dynamics (MD), classical MC aims at solving exactly the statistical mechanics of a system given an expression for its potential energy as a function of its microscopic degrees of freedom. In contrast to MD, however, where the atoms are moved according to the inter- and intramolecular forces derived from the potential energy function by solving Newton's equations of motion, MC is a stochastic method: it relies on transition probabilities between different microscopic states (configurations) of the simulated system, appropriately designed such that, at steady state, the system adopts the probability distribution among states that is dictated by an equilibrium ensemble.[1–8] These transitions are traced through a scheme that involves, in general, three calculations: (a) generation of an initial configuration, (b) attempt of a move that takes the system to a randomly generated new (trial) configuration, and (c) evaluation of the attempted move according to an *acceptance criterion* that involves the properties of the old and new configurations and decision whether the move should be accepted or rejected. If the move is accepted, the new configuration becomes the current configuration of the system; if it is rejected, the old configuration is retained as the current configuration. Calculations (b) and (c) constitute what we call a MC step. They are repeated for millions to billions of times, resulting in the generation of a long *Markov chain* of states, where a given state may appear multiple times in the chain. If the attempt and acceptance criteria are designed appropriately, the Markov chain of states asymptotically samples a prescribed *equilibrium distribution*, ρ^{eq}. Macroscopic equilibrium properties are computed as averages over the equilibrated part of the chain of sampled states.

Let us call $a\,(o \to n)$ the attempt probability of an elementary move from the current (or old) state (o) into a new (or trial) state (n) within a MC step. In the original Metropolis MC algorithm,[1] the *stochastic matrix* $\boldsymbol{a}$ of attempt probabilities is symmetric in the sense that it satisfies the condition

$$a\,(o \to n) = a\,(n \to o)\,; \tag{5.1.1}$$

Multiscale Modelling of Polymers. Doros N. Theodorou and Vlasis G. Mavrantzas, Oxford University Press.
 DOI: 10.1093/oso/9780199226191.003.0005

i.e. the *underlying matrix* $\boldsymbol{a}$ of the Markov chain is symmetric. Then, the attempted move is accepted with probability

$$p_{\text{acc}}(o \to n) = \min\left[1, \frac{\rho^{\text{eq}}(n)}{\rho^{\text{eq}}(o)}\right], \tag{5.1.2}$$

where $\rho^{\text{eq}}(o)$ is the configuration-space probability density of the system being in the old state (o) and $\rho^{\text{eq}}(n)$ the probability of the system being in the new state (n). The generated Markov chain of states asymptotically samples the probability distribution ρ^{eq}. In the production phase of a MC simulation, all configuration-dependent properties fluctuate around constant average values representative of thermodynamic equilibrium, and each state is sampled with a frequency proportional to its equilibrium probability. This is called *importance sampling*.

There is considerable freedom in choosing $\boldsymbol{a}$, while respecting the requirement $a(o \to n) = a(n \to o)$. For example, instead of one (as in the original MC algorithms), several or all atoms of the system may be simultaneously moved. Moreover, one can devise totally unphysical ways for moving atoms in space that substantially depart from the system's natural trajectory. These may make it possible to move through configuration space much more efficiently than by MD. For systems of chain molecules (e.g. synthetic polymers, highly branched macromolecules, and biopolymers), this is of paramount importance.

In a Metropolis MC scheme, the transition probabilities $\pi(o \to n) = a(o \to n)\, p_{\text{acc}}(o \to n)$ from state (o) to state (n) and $\pi(n \to o) = a(n \to o)\, p_{\text{acc}}(n \to o)$ from state (n) to state (o) $(o \neq n)$ in one step satisfy the condition of microscopic reversibility, or detailed balance,

$$\rho^{\text{eq}}(o)\pi(o \to n) = \rho^{\text{eq}}(n)\pi(n \to o) \tag{5.1.3}$$

or, equivalently,

$$\rho^{\text{eq}}(o)a(o \to n)\, p_{\text{acc}}(o \to n) = \rho^{\text{eq}}(n)a(n \to o)\, p_{\text{acc}}(n \to o). \tag{5.1.4}$$

Equation (5.1.4) implies that one may relax the symmetry condition for the underlying matrix $\boldsymbol{a}$, Eq. (5.1.1), and modify the acceptance probability according to the *generalized* Metropolis criterion

$$p_{\text{acc}}(o \to n) = \min\left(1, \frac{\rho^{\text{eq}}(n)a(n \to o)}{\rho^{\text{eq}}(o)a(o \to n)}\right). \tag{5.1.5}$$

Over the years, and in order to enhance sampling efficiency, *smart* MC schemes have been developed that use a nonsymmetric matrix of attempt probabilities, i.e. one that satisfies Eq. (5.1.4) but not Eq. (5.1.1). To realize some of these in practice, new statistical ensembles had to be defined as well; we discuss many of them in this chapter.

To shed further light into what happens in a generalized Metropolis MC step, let us define the *transition probability matrix* $\boldsymbol{\pi}$ with elements

$$\pi_{no} = \pi(o \to n) = a(o \to n)\, p_{\text{acc}}(o \to n)$$
$$= \begin{cases} a(o \to n), & \text{if } \rho^{\text{eq}}(n)a(n \to o) \geq \rho^{\text{eq}}(o)a(o \to n),\ n \neq o \\ a(n \to o)\dfrac{\rho^{\text{eq}}(n)}{\rho^{\text{eq}}(o)}, & \text{if } \rho^{\text{eq}}(n)a(n \to o) < \rho^{\text{eq}}(n)a(n \to o),\ n \neq o \end{cases}$$
$$\pi_{oo} = 1 - \sum_{n \neq o} \pi_{no}. \tag{5.1.6a}$$

Note that, in this definition of the elements of $\boldsymbol{\pi}$, the index of the destination state comes first and the index of the origin state second; rows correspond to destination states and columns to origin states.

Let us also consider the row vector $\boldsymbol{\rho}_t$ containing the a priori probabilities of all states after step t of the MC:

$$\boldsymbol{\rho}_t = (\rho_t(1), \rho_t(2), \ldots, \rho_t(o), \ldots, \rho_t(n), \ldots). \tag{5.1.6b}$$

For a given state n, the a priori probability for occupancy of that state after step t is related to the probabilities of occupancy of all states o after step $t-1$ via the recursive equation

$$\rho_t(n) = \sum_o \rho_{t-1}(o)\pi_{no}, \tag{5.1.7a}$$

where n itself is included in the summation over o. Equation (5.1.7a) can be written, for all states, in matrix form as

$$\boldsymbol{\rho}_t = \boldsymbol{\pi} \cdot \boldsymbol{\rho}_{t-1}. \tag{5.1.7b}$$

As the MC progresses, $\boldsymbol{\rho}_t$ of the Markov chain of states converges to the desired equilibrium distribution, $\lim_{t \to \infty} \boldsymbol{\rho}_t = \boldsymbol{\rho}^{\text{eq}}$. Thus, the recursive Eq. (5.1.7b) becomes

$$\boldsymbol{\rho}^{\text{eq}} = \boldsymbol{\pi} \cdot \boldsymbol{\rho}^{\text{eq}}. \tag{5.1.8}$$

That is, the equilibrium distribution among states is an eigenvector of the transition probability matrix $\boldsymbol{\pi}$ with corresponding eigenvalue equal to 1. By definition (see Eq. (5.1.6a)), $\boldsymbol{\pi}$ is *stochastic*: it is a matrix of nonnegative elements such that each one of its columns adds to 1. In designing MC algorithms we make sure that they correspond to *irreducible* or *ergodic Markov chains*, i.e. that every state can be reached from another state in a finite number of moves.[2] For ergodic Markov chains, the transition matrix $\boldsymbol{\pi}$ cannot be reduced to a block diagonal form, no matter how states are reordered.

In the above discussion we have implicitly assumed a discrete set of states and interpreted the elements of $\boldsymbol{\rho}_t$ and $\boldsymbol{\rho}^{\mathrm{eq}}$ as probabilities. The discussion carries over to continuous state space, such as the configuration space we deal with in classical statistical mechanics. There, the probability associated with a point in state space is the probability density at that point times an elementary volume in state space. In the simplest description, where each point in configuration space is described through the Cartesian coordinates of all atoms, these little volumes assigned to the states are all equal. The elementary volumes cancel from both sides of equations such as (5.1.3), (5.1.4), (5.1.7), (5.1.8) or in ratios such as those appearing in Eqs. (5.1.2), (5.1.5), (5.1.6a). Thus, we can interpret $\boldsymbol{\rho}_t$ and $\boldsymbol{\rho}^{\mathrm{eq}}$ in these equations as *probability densities* in the continuous state space, rather than as probabilities.

From now on, the superscript 'eq' on the symbol $\boldsymbol{\rho}^{\mathrm{eq}}$ for the equilibrium probability density will be omitted, unless otherwise indicated.

The Metropolis MC method was originally developed to sample the canonical NVT ensemble for which the equilibrium probability density $\boldsymbol{\rho}^{NVT}$ in configuration space is proportional to the Boltzmann factor of the potential energy $\mathcal{V}$. If all steps of the MC are conducted in the set of Cartesian atomic coordinates in which $\boldsymbol{\rho}^{NVT}$ is typically defined and if the matrix of attempt probabilities $\boldsymbol{a}$ is symmetric, the acceptance probability from Eq. (5.1.5) becomes

$$p_{\mathrm{acc}}\left(o \rightarrow n\right) = \min\left(1, \exp\left[-\beta\left(\mathcal{V}^{(n)} - \mathcal{V}^{(o)}\right)\right]\right), \tag{5.1.9}$$

where $\mathcal{V}^{(n)} - \mathcal{V}^{(o)}$ denotes the difference in total potential energies between new and old configurations of the system and $\beta = \dfrac{1}{k_{\mathrm{B}}T}$ is the inverse thermal energy (k_{B} being Boltzmann's constant and T the temperature). Equation (5.1.9) guarantees that energetically more *favourable* states are accepted preferentially. MC can be readily extended to the isothermal–isobaric NPT and grand canonical μVT ensembles.

In general, one may choose to conduct an elementary move in a set of coordinates that differs from the configuration-space coordinates in which the equilibrium probability density $\boldsymbol{\rho}$ was defined. Then, in going from the configuration-space probability density to the probability of a state point, a Jacobian $\mathcal{J}$ of transformation is required. $\mathcal{J}$ is the ratio of the elementary volume around a point in the set of configuration-space coordinates in which ρ is defined divided by the elementary volume around the corresponding point in the set of coordinates used in the move. Thus, a more general expression than Eq. (5.1.5) for the acceptance probability is

$$p_{\mathrm{acc}}\left(o \rightarrow n\right) = \min\left(1, \frac{\rho(n)\,\mathcal{J}(n)\,a\left(n \rightarrow o\right)}{\rho(o)\,\mathcal{J}(o)\,a\left(o \rightarrow n\right)}\right). \tag{5.1.10}$$

5.1.2 Statistical ensembles for dense phases of long macromolecules

5.1.2.1 The semi-grand canonical ensemble

In many applications of the MC method, one has to simulate polydisperse systems, or one applies moves (such as the chain connectivity altering ones that we will discuss in Section 5.2) that cause changes in the molecular length of the polymer chains employed in the move. It is then necessary to invent a new ensemble particularly suitable for polydisperse systems.[9–11] To deal with polydispersity effects, simulations are conveniently carried out in a *semi-grand canonical ensemble*, in which a spectrum $\boldsymbol{\mu}^*$ of relative chemical potentials is invoked to control the chain length distribution. Such a semi-grand ensemble is denoted as $N_c nPT\boldsymbol{\mu}^*$, since the following variables are kept constant: the total number of chains N_c, the total number of mers n, the pressure P, the temperature T, and the spectrum of relative chemical potentials $\boldsymbol{\mu}^*$ of all chain species in the system except two (see analysis of Eqs. (5.1.11) to (5.1.24)), which are taken as reference species. In the semi-grand ensemble, although the molecular weight distribution can change from realization to realization of the system, the *average* distribution is controlled by specifying its conjugate variable, the spectrum of relative chemical potentials $\boldsymbol{\mu}^*$.

To develop the $N_c nPT\boldsymbol{\mu}^*$ ensemble, we start by writing down the expression for the differential Helmholtz free energy of a mixture of m components:

$$\mathrm{d}A = -S\mathrm{d}T - P\mathrm{d}V + \sum_{k=1}^{m} \mu_k \mathrm{d}N_k. \tag{5.1.11}$$

This is linked to statistical mechanics by the relation

$$A(V, T, N_1, N_2, \ldots, N_m) = -k_\mathrm{B} T \ln Q(V, T, N_1, N_2, \ldots, N_m), \tag{5.1.12}$$

where $Q(V, T, N_1, N_2, \ldots, N_m)$ is the canonical partition function for an m-component system defined as

$$Q(V, T, N_1, N_2, \ldots, N_m) = \frac{1}{N_1!N_2!\ldots, N_m!}\frac{1}{\Lambda^{3n}} \mathrm{Z}(V, T, N_1, N_2, \ldots, N_m) \tag{5.1.13}$$

with

$$\mathrm{Z}(V, T, N_1, N_2, \ldots, N_m) = \int \exp\left[-\beta \mathcal{V}(\boldsymbol{r}_1, \boldsymbol{r}_2, \ldots, \boldsymbol{r}_n)\right] \mathrm{d}^{3n} r. \tag{5.1.14}$$

In the above, N_i stands for the number of i-mer chains in the system, $\mathrm{Z}(V, T, N_1, N_2, \ldots, N_m)$ is the classical configurational integral, $\Lambda = \left(\frac{h^2}{2\pi\, m_\mathrm{m}\, k_\mathrm{B} T}\right)^{1/2}$ is the mer thermal wavelength, m_m is the mass of a mer, $n = \sum_{k=1}^{m} kN_k$ is the total number of mers, and $N_c = \sum_{k=1}^{m} N_k$ is the total number of chains.

Selecting an arbitrary pair of components of lengths (i, j) as reference species, N_i and N_j can then be eliminated in favour of the total number of monomers n and total number of chains N_c:

$$\mathrm{d}A = -S\mathrm{d}T - P\mathrm{d}V + \sum_{\substack{k=1 \\ k\neq i,\, j}}^{m} \mu_k^* \mathrm{d}N_k + \mu^{(n)}\mathrm{d}n + \mu^{(N_c)}\mathrm{d}N_c, \tag{5.1.15}$$

where

$$\mu_k{}^* = \mu_k - \left(\frac{k-i}{j-i}\right)\mu_j - \left(\frac{k-i}{i-j}\right)\mu_i \tag{5.1.16}$$

$$\mu^{(n)} = \frac{\mu_i - \mu_j}{i-j} \tag{5.1.17}$$

$$\mu^{(N_c)} = \frac{i\mu_j - j\mu_i}{i-j}. \tag{5.1.18}$$

Legendre-transforming[12] the above equation with respect to $\{N_k, k \neq i,j\}$ and volume V, the natural thermodynamic potential for the variables of interest $N_c nPT\boldsymbol{\mu}^*$ is obtained,[11]

$$Y = A + PV - \sum_{\substack{k=1\\k\neq i,j}}^{m} \mu_k^* N_k, \tag{5.1.19}$$

and, in differential form,

$$\mathrm{d}Y = -S\mathrm{d}T + V\mathrm{d}P - \sum_{\substack{k=1\\k\neq i,j}}^{m} N_k \mathrm{d}\mu_k{}^* + \mu^{(n)}\mathrm{d}n + \mu^{(N_c)}\mathrm{d}N_c. \tag{5.1.20}$$

Y is related to a partition function $\tilde{Q}$ as $Y = -k_B T \ln \tilde{Q}$ with

$$\tilde{Q}(T, P, \boldsymbol{\mu}^*, n, N_c) = \sum_{\{N_1,\ldots,N_m\}} \exp\left[\beta \sum_{\substack{k=1\\k\neq i,j}}^{m} \mu_k^* N_k\right] \frac{1}{V_0} \int_0^\infty \exp(-\beta PV)\, Q(V, T, N_1, N_2, \ldots, N_m)\, \mathrm{d}V, \tag{5.1.21}$$

where $Q(V, T, N_1, N_2, \ldots, N_m)$ is given by Eq. (5.1.13). In Eq. (5.1.21), V_0 is a reference volume, and the summation over species populations is limited to those distributions satisfying the constraints

$$\sum_{k=1}^{m} N_k = N_c \tag{5.1.22}$$

and

$$\sum_{k=1}^{m} kN_k = n. \tag{5.1.23}$$

In simulating the $N_c nPT\boldsymbol{\mu}^*$ ensemble, monomers are treated as distinguishable entities, while exchanging the identity of two chains of the same length does not generate a new configuration.

From Eq. (5.1.21), then, the probability density in connectivity and configuration space that is sampled in $N_c nPT\boldsymbol{\mu}^*$ simulations is

$$\rho^{N_c nPT\mu_k^*}(N_1, N_2, \ldots, N_m, V; \boldsymbol{r}_1, \ldots, \boldsymbol{r}_n; \text{connectivity})$$

$$= \text{const.} \times \exp\left\{\beta \sum_{\substack{k=1\\k\neq i,j}}^{m} \mu_k^* N_k - \beta PV - \beta\mathcal{V}(N_1, N_2, \ldots, N_m, V; \boldsymbol{r}_1, \ldots, \boldsymbol{r}_n; \text{connectivity})\right\}. \tag{5.1.24}$$

5.1.2.2 The $f_1 N_2 PT$ ensemble

This ensemble[13] is suitable for the prediction of solubilities of small molecules in polymers. It has also been invoked within an iterative scheme involving the simultaneous simulation of two coexisting phases to capture high-pressure equilibria of a light solvent and a heavy hydrocarbon while avoiding transfers of the heavy hydrocarbon between the phases. The latter iterative scheme has been termed SPECS[14] (simulation of phase equilibria of chain systems). Here we consider the formulation of the $f_1 N_2 PT$ ensemble in a single-phase binary mixture consisting of a small molar-mass component (component 1) and a polymer (component 2). The formulation can be readily generalized for many small molar-mass components.[13] The ensemble describes the binary mixture under the constraints of fixed fugacity of the small molar-mass component f_1, fixed number of polymer molecules N_2, temperature T, and pressure P. To predict the sorption isotherm of a gaseous component 1 in the polymer, one conducts a series of simulations at constant N_2 and T, systematically varying P and f_1. Note that P and f_1 are interdependent. It is usually reasonable to assume that the polymer is completely involatile, in which case f_1, the fugacity of component 1 in the considered polymer phase (equal to the fugacity of component 1 in the gaseous phase with which the polymer phase is in equilibrium) can be calculated as the fugacity of pure gaseous 1 under T and P from a reliable equation of state.

The chemical potential μ_1 of 1 is related to its fugacity f_1 by $\mu_1 = k_B T \ln\left(\dfrac{f_1 \prod_{i=1}^{s_1} \Lambda_i^3}{k_B T Z_1^{\text{intra}}}\right)$, where Λ_i is the thermal wavelength of atom i in a molecule of type 1, the total number of atoms (or mers) in the molecule being s_1, and Z_1^{intra} is the configurational integral of a molecule of type 1 over all but its three translational degrees of freedom (see Eq. (5.3.5)). Under constant temperature, fixing f_1 is equivalent to fixing μ_1. The $f_1 N_2 PT$ ensemble can be thought of as a hybrid between an isothermal–isobaric ensemble and a grand canonical ensemble for component 1. Its probability density function is given by

$$\rho^{f_1 N_2 TP}(N_1, V; \boldsymbol{r}_1, \boldsymbol{r}_2)$$

$$= \text{const.} \frac{1}{N_1!}\left(\frac{f_1 \prod_{i=1}^{s_1} \Lambda_i^3}{k_B T Z_1^{\text{intra}}}\right)^{N_1} \exp(-\beta PV) \exp\left[-\beta\mathcal{V}(N_1, N_2, V; \boldsymbol{r}_1, \boldsymbol{r}_2)\right], \tag{5.1.25}$$

where $\boldsymbol{r}_1$ and $\boldsymbol{r}_2$ are sets of atomic position vectors defining the configuration of the molecules of species 1 and 2, respectively, V is the volume of the phase, and $\mathcal{V}$ is the total potential energy.

5.1.2.3 The $f_1 N_c nPT\boldsymbol{\mu}^*$ and $f_1' N_c n_0 PT\boldsymbol{\mu}^*$ statistical ensembles for the simulation of oligomer solubility in polymers

These are extensions of the $f_1 N_2 PT$ ensemble that are particularly appropriate for the study of the solubility in a polymer melt of oligomers of the same chemical constitution as the polymer chains. They combine ideas from the $f_1 N_2 PT$ ensemble and the semigrand ensemble for long-chain polymer melts discussed in Section 5.1.2.1. Component 1 is the oligomeric species whose solubility in the polymer we want to calculate (e.g. *n*-decane in linear polyethylene), N_c the number of chains constituting the polymer, and $\boldsymbol{\mu}^*$ the set of relative chemical potentials fixing the molar mass distribution in the polymer, whose configurations are vigorously sampled via connectivity-altering MC moves (see Section 5.2). In general, the rate of equilibration of composition in the $f_1 N_c nPT\boldsymbol{\mu}^*$ ensemble is subject to the drawback of the classical simulation methods for phase equilibria, namely, the difficulty of inserting and deleting long, flexible molecules in dense polymer phases. Even when insertions are performed using the configurational bias technique (see Section 5.2), the performance of the algorithm ultimately degrades as the length of component 1 increases. One can overcome this difficulty by designing the more complex $f_1' N_c n_0 PT\boldsymbol{\mu}^*$ ensemble.[15] In this, two new moves are introduced:[15] *scission* and *fusion*, each of them being the inverse of the other. Scission (see Section 5.2) involves creation of an oligomer from the end of a polymer molecule with the simultaneous removal of a trimer; fusion, on the other hand, leads to formation of a new and longer polymer chain by joining a polymer and an oligomer end-to-end through construction of a new trimer. In this new ensemble, f_1' is a modified fugacity of the oligomer which is essentially a measure of the difference in chemical potentials between an oligomer molecule and an equivalent amount of polymer. It helps control interconversion of polymers and oligomers through scission and fusion moves and its precise definition is given in Ref. 15.

5.1.2.4 Reaction ensemble Monte Carlo (RxMC)

The $N_c nPT\boldsymbol{\mu}^*$ and $f_1' N_c n_0 PT\boldsymbol{\mu}^*$ ensembles for dense phases of long-chain polymers can be thought of as special cases of the reaction ensemble, which makes it possible to simulate chemically reactive systems at equilibrium.[16–18] RxMC is an important general ensemble, which uses MC sampling to simulate complete forward and reverse chemical reaction events, yielding the equilibrium composition of a reacting mixture. To mimic chemical reactions, RxMC removes reactant molecules from the simulation cell, while product molecules are simultaneously inserted into the cell. The acceptance criterion for such a move must incorporate the free-energy change associated with the attempted chemical conversion in an ideal gas state, plus the change in energy of physical interactions associated with the removal of reactants and the insertion of products. Inputs to a RxMC simulation are (a) the set of reactions to be simulated; (b) internal molecular partition functions for the reactants and products in the ideal gas state, including vibrational, rotational, and electronic contributions; and (c) a set of intermolecular potentials describing interactions among reactants, products, and any environment in which the reaction takes place, such as the walls of a porous medium. The internal molecular partition functions for isolated molecules are directly related to standard chemical potentials available in compilations of thermochemical data, which are combined to obtain reaction equilibrium constants. RxMC does not require reactive potentials that mimic bond breaking or forming. It is not limited by the rates of the reactions occurring and, using the chemical interconversion moves mentioned earlier, it can reach chemical equilibrium very quickly. Combination with the Gibbs ensemble MC has made it possible to study simultaneous phase and reaction equilibria. The method can be used to assess the effects of intermolecular interactions, or of confinement in microporous media, on the equilibrium composition of reacting mixtures. An interesting extension is transition-state RxMC,[19] in which transition-state species (activated complexes) are introduced as additional chemical species assumed to be at equilibrium with reactants and products, and shifts in chemical reaction kinetics due to environmental effects (e.g. confinement in carbon micropores) are assessed.

5.1.2.5 *Parallel tempering*

In simulations of complex molecules or at low enough temperatures, the acceptance rates of practically all MC moves drops dramatically, thereby necessitating too long MC runs in order to reach the state of thermodynamic equilibrium. In these cases, one way to enhance system equilibration and avoid problems associated with incomplete sampling of configuration space (leading to inaccurate calculation of physical properties) is to resort to the technique of parallel tempering (PT).[20–22] Although PT was introduced in the context of spin glass simulations,[23] its capabilities were mostly appreciated in efforts to study the conformational properties of complex biological molecules.[24,25] Similar techniques include (a) *replica exchange MD* methods to simulate protein folding and overcome the multiple-minima problem,[26] (b) *multiple-histogram reweighting* approaches to calculate thermodynamic quantities in a wide temperature range,[26] (c) *multidimensional PT* in the context of an expanded grand canonical ensemble to simulate polymer solutions and blends,[27] and (d) *parallel excluded-volume tempering* to simulate a set of systems in parallel,[28] each of which has a slightly different repulsive core potential such that a thermodynamic path from a fully excluded volume system to an ideal gas of random walks is generated.

In PT, not a single system but an extended ensemble of M systems labelled as $i = 1, 2, \ldots, M$ is considered. Each such system is viewed as a copy of the original system, equilibrated, however, at a different temperature T_i. In general, $T_1 < T_2 < \ldots < T_M$ and T_1 is normally the temperature of the system of interest. The partition function of this extended ensemble is given by[28,29]

$$Q = \prod_{i=1}^{M} Q_i, \tag{5.1.26}$$

where Q_i denotes the individual partition function of the ith system in its relevant statistical ensemble, e.g. the $N_c nPT\boldsymbol{\mu}^*$ ensemble. Since the replicas do not interact energetically, the partition function of this larger ensemble is given by

$$Q = \prod_{i=1}^{M} \frac{q_i}{N! h^{3N}} \int \exp\left[-\beta_i \mathcal{V}\left(V, N; \boldsymbol{r}_i\right)\right] \mathrm{d}\boldsymbol{r}_i, \tag{5.1.27}$$

where $q_i = \prod_{k=1}^{N} (2\pi\, m_k\, k_B T_i)^{\frac{3}{2}}$ comes from integrating out the momenta, m_k is the mass of atom k, $\boldsymbol{r}_i$ is the set of the position vectors of all N atoms in system i, $\beta_i = \dfrac{1}{k_B T_i}$ is the reciprocal temperature, and $\mathcal{V}$ is the potential energy.

The strategy of simultaneously equilibrating not *a single* but *a number* of systems at different temperatures accelerates the equilibration of the lowest temperature systems by accepting configurations from the higher temperature ones, for which the acceptance ratio of the MC moves is higher. Thus, in the PT method, apart from the typical MC moves, one attempts to swap configurations between systems being equilibrated at different temperatures. In practice, swaps are normally attempted between systems with adjacent temperatures, $j = i + 1$.

If the probability of attempting a swap move is equal for all conditions, exchanges between ensembles i and j in the canonical NVT ensemble are accepted with probability[23]

$$p_{\text{acc}} = \min\left\{1, \exp\left[+(\beta_i - \beta_j)\left(\mathcal{V}\left(V, N; \boldsymbol{r}_i\right) - \mathcal{V}\left(V, N; \boldsymbol{r}_j\right)\right)\right]\right\}. \tag{5.1.28}$$

If the simulation is carried out in the NPT ensemble, all systems are subject to the same pressure P, but each system is at a different temperature T_i. The partition function for system i is

$$\begin{aligned} Q_i &= \frac{1}{N!\Lambda_i^{3N}} \frac{1}{V_0} \int \exp\left[-\beta_i \mathcal{V}\left(V_i, N; \boldsymbol{r}_i\right) - \beta_i P V_i\right] \mathrm{d}\boldsymbol{r}_i \,\mathrm{d}V_i = \\ &= \frac{1}{N!\Lambda_i^{3N}} \frac{1}{V_0} \int \exp\left[-\beta_i \mathcal{V}\left(V_i, N; \mathrm{s}_i\right) + N \ln V_i - \beta_i P V_i\right] \mathrm{d}\boldsymbol{s}_i \,\mathrm{d}V_i \end{aligned} \tag{5.1.29}$$

with $i = 1, 2, \ldots, M$ and $\boldsymbol{s}_i$ being the scaled atomic coordinates with respect to volume V_i.

In a PT simulation of systems sampling the NPT ensemble, two different types of swap move can be envisioned: (a) exchange of both *configurations* and *volumes* between the pair of systems (i, j), or swap type 1; and (b) exchange only of *scaled* configurations between the pair of systems (i, j), or swap type 2. The acceptance criterion for swap type 1 (exchange of both *configurations* and *volumes*) depends on the ratio

$$\begin{aligned} \frac{\rho(n)}{\rho(o)} &= \frac{\exp\left[-\beta_i \mathcal{V}(N, V_j; \boldsymbol{s}_j) + N \ln V_j - \beta_i P V_j - \beta_j \mathcal{V}(N, V_i; \boldsymbol{s}_i) + N \ln V_i - \beta_j P V_i\right]}{\exp\left[-\beta_i \mathcal{V}(N, V_i; \boldsymbol{s}_i) + N \ln V_i - \beta_i P V_i - \beta_j \mathcal{V}(N, V_j; \boldsymbol{s}_j) + N \ln V_j - \beta_j P V_j\right]} \\ &= \exp\left[-(\beta_i - \beta_j)\,\mathcal{V}(N, V_j; \boldsymbol{s}_j) + (\beta_i - \beta_j)\,\mathcal{V}(N, V_i; \boldsymbol{s}_i) - (\beta_i - \beta_j)\, P V_j + (\beta_i - \beta_j)\, P V_i\right] \\ &= \exp\left\{(\beta_i - \beta_j)\left[\mathcal{V}(N, V_i; \boldsymbol{s}_i) - \mathcal{V}(N, V_j; \boldsymbol{s}_j) + P V_i - P V_j\right]\right\} \\ &= \exp\left\{(\beta_i - \beta_j)\left[(\mathcal{V}(N, V_i; \boldsymbol{s}_i) + P V_i) - (\mathcal{V}(N, V_j; \boldsymbol{s}_j) + P V_j)\right]\right\} \\ &= \exp\left\{\Delta\beta\Delta\left(\mathcal{V} + P V\right)\right\}. \end{aligned} \tag{5.1.30}$$

The acceptance criterion for swap type 2 (exchange only of *scaled* configurations) depends on the ratio

$$\begin{aligned}\frac{\rho(n)}{\rho(o)} &= \frac{\exp\left[-\beta_i \mathcal{V}(\mathbf{s}_j, V_i) + N\ln V_i - \beta_i P V_i - \beta_j \mathcal{V}(\mathbf{s}_i, V_j) + N\ln V_j - \beta_j P V_j\right]}{\exp\left[-\beta_i \mathcal{V}(\mathbf{s}_i, V_i) + N\ln V_i - \beta_i P V_i - \beta_j \mathcal{V}(\mathbf{s}_j, V_j) + N\ln V_j - \beta_j P V_j\right]} \\ &= \exp\left[-\beta_i \mathcal{V}(\mathbf{s}_j, V_i) - \beta_j \mathcal{V}(\mathbf{s}_i, V_j) + \beta_i \mathcal{V}(\mathbf{s}_i, V_i) + \beta_j \mathcal{V}(\mathbf{s}_j, V_j)\right] \\ &= \exp\left\{-\beta_i\left[\mathcal{V}(\mathbf{s}_j, V_i) - \mathcal{V}(\mathbf{s}_i, V_i)\right] - \beta_j\left[\mathcal{V}(\mathbf{s}_i, V_j) - \mathcal{V}(\mathbf{s}_j, V_j)\right]\right\}. \end{aligned} \tag{5.1.31}$$

Swap type 2 involves extensive recalculation of energies needed to *repack* each configuration into the volume of the other. Swap type 1 is computationally more straightforward to implement.

5.1.3 Limiting molar mass distributions in simulations of bulk systems with connectivity-altering MC moves

Simulations in the previously developed ensembles yield a molar mass distribution that depends on the spectrum of chemical potentials $\boldsymbol{\mu}^*$ imposed. We discuss here expressions for $\boldsymbol{\mu}^*$ that generate the most widely known molar mass distributions for polymers.

The relationship between the spectrum of chemical potentials and the chain length distribution is greatly simplified in cases wherein the configurational integral in Eq. (5.1.14) depends on the monomer density and number-averaged molecular weight, but not on the details of the distribution, i.e. if

$$Z(V, T, N_1, N_2, \ldots, N_m) = Z(V, T, N_c, n). \tag{5.1.32}$$

When this condition is satisfied, the molar mass distribution is related to the chemical potential spectrum $\boldsymbol{\mu}^*$ solely by combinatoric considerations and does not depend on the energetics of the system.[11] Condition (5.1.32) would hold exactly in systems of noninteracting unperturbed chains (no intermolecular, no nonbonded intramolecular interactions), where Z can be reduced to a product of Z's of individual chains, each depending exponentially on chain length. This condition may also hold in systems with realistic potentials, provided short-chain species are absent, under which conditions the properties of the melt should depend only on the monomer density and the density of chain ends. We will assume here that the configurational integral $Z(V, T, N_1, N_2, \ldots, N_m)$ does not vary with the specifics of the chain length distribution within the bounds imposed upon its fluctuations.

When N_c and n are large, a maximum term approximation on the partition function $\tilde{Q}$ defined by Eq. (5.1.21) for the $N_c nPT\boldsymbol{\mu}^*$ ensemble subject to the constraints (5.1.22) and (5.1.23) may be invoked to map the distribution of chemical potentials onto the corresponding distribution of chain lengths. Using Stirling's approximation for the logarithms of factorial terms, this yields $(m-2)$ equations of the form

$$\ln N_k = \beta\mu_k^* + \xi + \psi k, \tag{5.1.33}$$

where ξ and ψ are Lagrange multipliers corresponding to the two constraints (5.1.22) and (5.1.23). These equations simplify to

$$N_k = Cy^k \exp\left(\beta\mu_k^*\right). \tag{5.1.34}$$

The reference species are governed by the relations

$$N_i = Cy^i \tag{5.1.35}$$

$$N_j = Cy^j, \tag{5.1.36}$$

which are analogous to (5.1.34) with $\mu_k^* = 0$. The m equations above are substituted into the constraints (5.1.22) and (5.1.23) to solve for the constants C and y, thereby yielding the entire distribution.

5.1.3.1 *The most probable distribution*

This distribution is commonly observed in the products of condensation polymerization. The number fraction of an oligomeric species is an exponentially decaying function of chain length.

Under the condition that Eq. (5.1.32) is satisfied, this distribution results if the reduced chemical potential spectrum is set to 0 for all chain lengths. A truncated form of the most probable distribution results if a minimum chain length is imposed, i.e. if the chemical potentials are specified as

$$\mu_k^* = \begin{cases} -\infty & \text{for } k < k_{\min}; \\ 0 & \text{for } k \geq k_{\min}, k \neq i, k \neq j \text{ with } k_{\min} \leq i \leq j \end{cases}, \tag{5.1.37}$$

which has the solution

$$N_k = \begin{cases} 0, & \text{for } 1 \leq k < k_{\min} \\ cy^k, & \text{for } k_{\min} \leq k \end{cases}. \tag{5.1.38}$$

In conjunction with the two constraints (5.1.22) and (5.1.23), a truncated exponential distribution results:

$$\frac{N_k}{N_c} = \begin{cases} \dfrac{1}{X - k_{\min} + 1}\left[\dfrac{X - k_{\min}}{X - k_{\min} + 1}\right]^{k - k_{\min}} & \text{if } k \geq k_{\min} \\ 0 & \text{otherwise} \end{cases}, \tag{5.1.39}$$

where $X = \dfrac{n}{N}$ is the number-average degree of polymerization.

5.1.3.2 *The uniform distribution*

It is straightforward to generate a uniform distribution of chain lengths of width $(2l + 1)$ centred at the number-average degree of polymerization, X. The chemical potential specification for this case is

$$\mu_{\kappa}{}^{*} = \begin{cases} -\infty & \text{for } k < X - l \text{ or } k > X + l \\ 0 & \text{for } X - l \leq k \leq X + l,\ k \neq i, j \text{ with } X - l \leq i < j \leq X + l \end{cases}. \tag{5.1.40}$$

Again, the solution for the N_k is of the form

$$N_k = \begin{cases} cy^k & \text{for } X - l \leq k \leq X + l \\ 0 & \text{for } k < X - l \text{ or } k > X + l \end{cases}. \tag{5.1.41}$$

The two constraints (5.1.22) and (5.1.23) are satisfied by the solution $y = 1$, and the resulting distribution is

$$\frac{N_k}{N_c} = \begin{cases} \frac{1}{2l+1} & \text{for } X - l \leq k \leq X + l \\ 0 & \text{for } k < X - l \text{ or } k > X + l \end{cases}. \tag{5.1.42}$$

5.1.3.3 The Gaussian distribution

A Gaussian distribution may also be used to control the deviation from monodispersity. With the condition (5.1.32) assumed here, this distribution results if the reduced chemical potential is quadratic in chain length:

$$\mu_k^* = \begin{cases} \psi\left[l^2 - (X - k)^2\right] & \text{for } X - l \leq k \leq X + l \\ -\infty & \text{for } k < \mathrm{X} - l \text{ or } k > X + l \end{cases}. \tag{5.1.43}$$

The chain length distribution obtained by substitution into the set of $(2l + 1)$ equations, Eqs. (5.1.34)–(5.1.36), is

$$N_k = c' y^k \exp\left[-\beta\psi(X - k)^2\right] \quad \text{for } X - l \leq k \leq X + l. \tag{5.1.44}$$

The constraints (5.1.22) and (5.1.23) are satisfied by $y = 1$, and the resulting distribution is a truncated Gaussian:

$$\frac{N_k}{N_c} = \begin{cases} \dfrac{\exp\left[-\beta\psi(k - X)^2\right]}{\sum\limits_{k'=-l}^{l} \exp\left[-\beta\psi k'^2\right]} & \text{for } X - l \leq k \leq X + l \\ 0 & \text{for } k < X - l \text{ or } k > X + l \end{cases}. \tag{5.1.45}$$

For $\beta\psi l^2 \gg 1, \beta\psi \ll 1$, the distribution is well approximated by

$$\frac{N_k}{N_c} = \left(\frac{\beta\psi}{\pi}\right)^{1/2} \exp\left[-\beta\psi(k - X)^2\right]. \tag{5.1.46}$$

5.2 Monte Carlo moves for continuous-space simulations of polymers

MC moves employed today in simulations of soft matter systems[30–41] may be classified, with some degree of arbitrariness, into two categories: (a) simple moves, and (b) advanced moves.[6,7] In the category of simple moves we usually include moves that operate on a single chain without changing its connectivity and their straightforward extensions, such as reptation (Rep), generalized reptation (X-Rep), atom-identity exchange (ID-Exch), end-mer rotation (Rot), libration or flip, configurational bias (CB), concerted rotation (CONROT), pivot, and parallel rotation (PARROT). In the category of advanced moves, on the other hand, we include moves that effect changes in the connectivity of at least one of the chains present in the simulation cell. Typical examples include end-bridging (EB), double-bridging (DB), and their variants to handle systems with branches: End-bridging for short chain branched (SCB) polymers (SCB-EB), configurational bias for branched polymers (Br-CB), double concerted rotation (d-CONROT), H-shaped branch rebridging (H-BR), etc. We discuss these moves in this section.

5.2.1 Simple Monte Carlo moves

5.2.1.1 *Reptation (Rep)*

It is one of the earliest moves developed for the simulation of dense polymer systems.[30] It is a *slithering-snake* type of move that deletes a segment from one end of a randomly selected chain and appends it to the other end, typically with the same bond angle and a randomly selected torsion angle; through this, the chain *slides* along its contour by one segment (Figure 5.1a).

5.2.1.2 *X-reptation (X-Rep)*

This move[11,31] is a generalization of the reptation move in the sense that one cuts an end atom (or atomistic unit) from a randomly selected chain and attaches it to the end of a *different* chain, usually with a fixed bond length and/or bond angle (Figure 5.1b). That is, two chains are involved in the move. As a result, in addition to displacing chains in space, X-Rep induces polydispersity in the system since the molecular length of one chain decreases and of the other increases.

5.2.1.3 *End-mer rotation*

This is a very simple move involving the random selection of an end chain segment and its rotation about the axis joining it with the previous segment on the same chain by a small randomly chosen dihedral angle (Figure 5.1c). It modifies only one dihedral angle in the chain.

5.2.1.4 *Flip (or libration)*

The flip is a simple move involving the random selection of an internal chain segment and its rotation about the axis joining its two neighbouring segments on the same chain by a small random angle (Figure 5.2a).[31] It leaves bond lengths unchanged but modifies up to two bond angles and up to four dihedral angles.

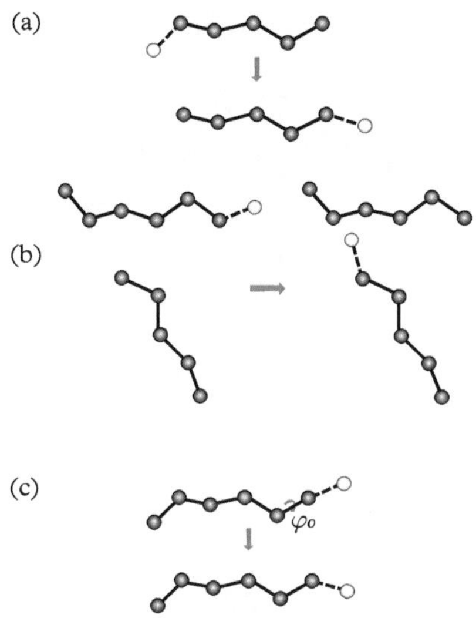

Figure 5.1 Schematic representation of simple moves for chain molecules: (a) reptation, (b) X-reptation, and (c) end-mer rotation.

5.2.1.5 *Dimer flip (DF)*

The DF move (Figure 5.2b) is an extension of the simple flip move.[32,33] Its development was motivated by the need for a larger number of fluctuating degrees of freedom during a single move. In the move, the position of atom 3 is altered by changing torsional angle ω_3 that defines the atom position on the circle C_3, which constitutes the intersection of a sphere S_1 centred at $\boldsymbol{r}_1$ with radius $|\boldsymbol{l}_2 + \boldsymbol{l}_3|$ and sphere S_2 centred at $\boldsymbol{r}_4$, of radius l_4. Here $\boldsymbol{l}_i$ is used to denote the bond vector of skeletal bond i, of length $|\boldsymbol{l}_i| = l_i$. Note that, during the DF construction, the distance l_{13} between atoms 1 and 3 and bond length l_4 are kept equal to their old values. Once the position of atom 3 has been fixed, the position of atom 2 is changed by applying a simple flip move (i.e. by changing angle ω_2 that defines the position of atom 2 on the circle C_2, which is the locus of this atom for given bond lengths l_2 and l_3).

5.2.1.6 *n*-mer flip

The n-mer flip move is a generalization of the dimer flip move performed by applying the following operations for atoms i, $i = 2, \ldots, n+1$: given the positions of atoms 1 ($\boldsymbol{r}_1$) and $i+1$ ($\boldsymbol{r}_{i+1}$) and the distances $|\boldsymbol{r}_i - \boldsymbol{r}_1|$ and $|\boldsymbol{r}_{i+1} - \boldsymbol{r}_i|$, atom i is displaced on the periphery of the circle C_i, which is the locus of positions for atom i given the constraints imposed by the molecular geometry. This displacement is brought about by changing the rotational degree of freedom ω_i.

5.2.1.7 *Configurational bias (CB)*

CB was developed originally for lattice MC simulations of polymers[34] and reinvented for MC calculations of phase equilibria.[35,36] It has been formulated in the context of both the

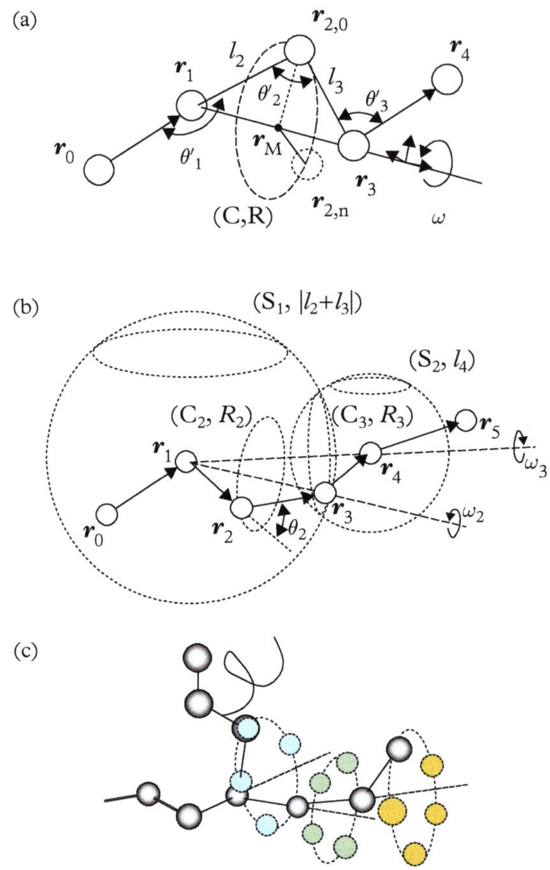

Figure 5.2 Schematic representation of simple moves for chain molecules: (a) single flip ($\theta_i{}' = \pi - \theta_i$). (b) Dimer flip; the atoms displaced are 2 and 3. (c) Configurational bias.

Gibbs ensemble MC method[37] and an expanded grand canonical ensemble;[38] the latter formulation alleviates somewhat problems associated with the insertion/deletion or exchange of large chain molecules in dense systems. In a CB move (Figure 5.2c), a randomly selected chain is cut at a random point to form two subchains. One of these subchains is removed from the system and subsequently regrown segment by segment in a biased fashion which avoids overlaps with the surrounding segments in the same or in neighbouring chains and/or weights solutions based on their intramolecular energy. The bias introduced by this selective construction is removed from the system with proper design of the acceptance criterion (see Section 5.2.5).

5.2.1.8 *Concerted rotation (CONROT)*

The CONROT move[39] was introduced in order to effect significant local rearrangements in the interior of the chains (and not only at the ends). The move (Figure 5.3a) is initiated by excising a randomly selected internal trimer of skeletal atoms in the polymer. In the original formulation[39] (unidirectional CONROT), one atom neighbouring the trimer is displaced by changing its *driver* torsion angle by an amount randomly selected within prespecified bounds $(-\delta\varphi_{\max}, +\delta\varphi_{\max})$, and the trimer is then reconstructed. In the more general formulation[11,31] (also termed intramolecular rebridging move or double-driven CONROT), both atoms neighbouring the excised trimer are displaced by imposing randomly selected changes to both torsion angles neighbouring the trimer on its two sides, thus removing any directionality from the move; the trimer is then reconstructed. In the symmetric version of CONROT, which we will focus on in the following, a total of five skeletal segments are ultimately relocated and eight torsion angles are altered. The move can be considered as a special case of the more general EB move discussed in Section 5.2.2.1, since its implementation essentially reduces to the same mathematical problem (the trimer bridging problem). In the double-driven CONROT move (Figure 5.3a), first a trimer $\{k, l, m\}$ is excised from a random position along the chain backbone. Next, the backbone atoms j and n neighbouring the trimer are displaced to positions j' and n', respectively, by imposing independent random changes in the corresponding *driver* torsion angles. Finally, atoms j' and n' are *rebridged* with specified bond lengths and bond angles; trimer $\{k', l', m'\}$ now has a new position and orientation.

5.2.1.9 *Extended configurational bias*

Extended configurational bias MC moves enable the equilibration of rather long chain length systems without effecting alterations in chain connectivity. These moves were first introduced[40] as extensions of the original continuum CB method. Today, they have developed to what are known as internal configurational bias (ICB) and self-adapting fixed end point configurational bias (SAFE-CB) methods which, by combining the philosophies of CONROT and CB methods, enable sampling of arbitrarily long internal chain sections.[41,42] Generalized CB algorithms have proven very useful in simulations of crosslinked-network structures, but most importantly in designing simulation strategies based on density-of-states sampling methods.[43–45] A typical design of an extended configurational bias move is shown in Figure 5.3b.

5.2.1.10 *Pivot*

The move,[46] which is very effective for rearranging single-chain configurations but has low probability of success in dense phases, is schematically illustrated in Figure 5.3c.

5.2.1.11 *Parallel rotation*

The move[47] is schematically illustrated in Figure 5.4a.

5.2.1.12 *Atom identity exchange*

A move that can accelerate the simulation of the adsorption of small alkane-like molecules with a head group on substrates is atom identity exchange (Figure 5.4b):[48] this consists of swapping the types between the head- and the terminal groups in a randomly selected chain while simultaneously inverting the direction of the chain backbone. In the example of Figure 5.4b the

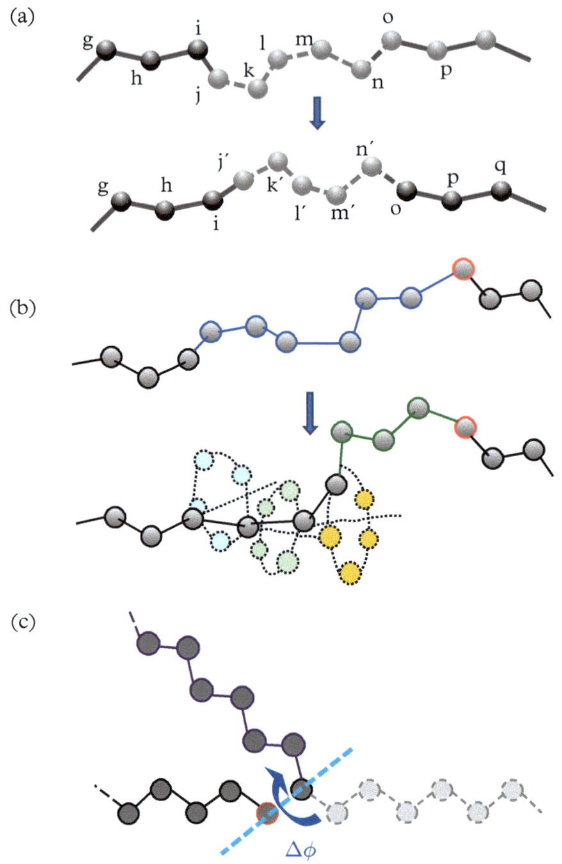

Figure 5.3 Schematic representation of simple moves for chain molecules. (a) Concerted rotation (CONROT). (b) Extended configurational bias. In this example, a hexamer sequence, shown in blue, is excised from the original configuration of a chain. The three first mers of the hexamer are regrown by configurational bias, in the presence of an attractive potential that draws them towards the target mer shown in red. The last regrown mer is finally bridged to the target mer by construction of a new trimer, shown in green. (c) Pivot.

move is implemented through an *elongation* of the CH_2-CH_3 bond and a *reduction* of the S-CH_2 bond without changing the relevant bending and torsional angles.

5.2.1.13 Volume fluctuation

This move allows for sampling fluctuations in the density of the system by allowing for changes in the size of the simulation box; for the simplest case of a cubic cell, it is schematically illustrated in Figure 5.4c. Volume fluctuations are initiated by randomly expanding or contracting all edge lengths of the periodic box by a random amount. Then, usually, the starts of all chains are affinely displaced

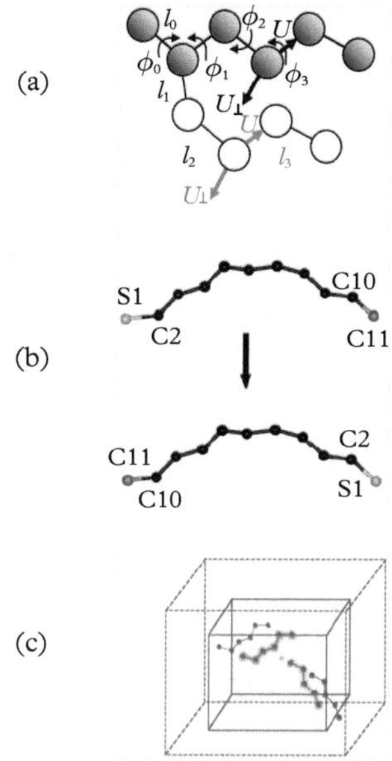

Figure 5.4 Schematic representation of simple moves for chain molecules. (a) Parallel rotation (PARROT); the move is a concerted change in the torsion angles $\{\phi_1, \phi_2, \phi_3\}$ driven by a change in ϕ_0. The vectors $\boldsymbol{U}$ and $\boldsymbol{U}_\perp$ remain unchanged after the PARROT move. (b) Identity exchange. (c) Volume fluctuation; the scaled coordinates of all chain starts and the internal configuration of all chains remain unchanged.

following the isotropic deformation of the periodic box, with all Euler angles, bond lengths, bond angles, and torsion angles left unchanged.

5.2.2 Advanced Monte Carlo moves for linear polymers

With the MC moves described in Section 5.2.1, only melts made up of chains up to approximately seventy units long can be equilibrated. To simulate longer-chain systems, moves capable of inducing drastic reconfiguration of large internal sections along the chain are needed. Such moves were first introduced in the 1980s in the form of *chain breaking* or *pseudokinetic* MC algorithms for lattice models of polymers.[49] These algorithms alter the connectivity among polymer segments in the lattice at the expense of introducing some polydispersity (i.e. a distribution of chain lengths) in the polymer.

The first such method to be developed was EB,[11] a very drastic move whose implementation is strongly dependent on the presence of chain ends. MC moves independent of the presence of chain ends were also proposed. They appeared initially as *intrachain* and *interchain concerted rotation* moves developed as generalizations of the CONROT move for simulating the morphology and elasticity of semicrystalline polymer interphases consisting of chain segments whose terminus is fixed to the crystal surface.[50,51] Through a novel design,[52–54] these moves developed later into the so-called *double bridging (DB)* and *intramolecular double rebridging (IDR)* moves. They are generalizations of the EB move, since they involve the construction of two trimer bridges (instead of one). We discuss all these moves in the following.

5.2.2.1 *End-bridging*

EB is initiated (Figure 5.5) by selecting at random an end-mer p of a randomly chosen chain in the system (numbered as chain 1 here) and one of its potential neighbours j from the list of $N_{\text{bridge}}(p)$ bridgeable candidates, at random; we assume that this atom belongs to the chain numbered as chain 2 in the system. The move proceeds by (a) having trimer $\{k, l, m\}$ adjacent to atom j either on the right or on the left along chain 2 being cut off, and (b) connecting atoms p and j with a new trimer $\{k', l', m'\}$. To this end, a geometric bridging procedure needs to be invoked,[31] capable of determining all geometrically admissible candidate configurations for the trimer bridge $\{k', l', m'\}$. A prospective solution is chosen from the reduced set of feasible solutions, either at random or by a weighting scheme designed to enhance acceptance of the move. Typically, all geometric solutions are subjected to two screening procedures: one based on hard-sphere overlaps of the bridge atoms with other atoms in the system, and one based on the change in torsional energy. These two screening procedures are effective in efficiently culling the collection of geometrically valid solutions to a smaller, energetically viable set.

Clearly, the geometric problem of trimer bridging is at the heart of the EB move (and, more generally, of all variable connectivity MC moves for chain molecules developed on the same idea). This problem is mathematically formulated as follows: *given two dimers in space, connect them with a trimer such that the resulting heptamer has prescribed bond lengths and angles*. Geometrically, the general problem of constructing a bridging trimer is understood by considering Figure 5.6. Given the two dimers $\{\boldsymbol{r}_0, \boldsymbol{r}_1\}$ and $\{\boldsymbol{r}_5, \boldsymbol{r}_6\}$ in space, the goal is to join them by a trimer $\{\boldsymbol{r}_2, \boldsymbol{r}_3, \boldsymbol{r}_4\}$, such that the bond lengths $\{l_2, l_3, l_4, l_5\}$ and bond angle supplements $\{\theta_1, \theta_2, \theta_3, \theta_4, \theta_5\}$ have specified values. The problem has been addressed by two methods, purely analytically[55] and semi-analytically.[31] Its solution typically involves an implicit transformation from the Cartesian coordinates of atoms to a set of *constraint* variables. As a result, Jacobians that relate differential volume elements in the two coordinate systems enter the acceptance criterion (see Eq. (5.1.10) and Section 5.2.5).

The computation of nonbonded energies of the initial (o) and final (n) states forms the next stage in the implementation of the EB move. The final step is the evaluation of the acceptance criterion which must take into account the probability of selection of a given bridge site, the probability of selecting one of the multiple screened solutions at that site, the Jacobians of transformation from constraint to Cartesian coordinates in the initial and final states, and the composite energies of the initial and final states; see detailed expressions in Section 5.2.5.

5.2.2.2 *Intramolecular end-bridging (IEB)*

IEB constitutes the intramolecular version of EB. As shown in Figure 5.7, the move is initiated by randomly selecting end mer i, which *attacks* an internal mer j in the *same* chain. Then a trimer next to mer j is excised and a new trimer bridge is constructed to maintain chain connectivity. In general, in the IEB move, the maximum number of allowed bridging combinations is one half those allowed in the EB move.

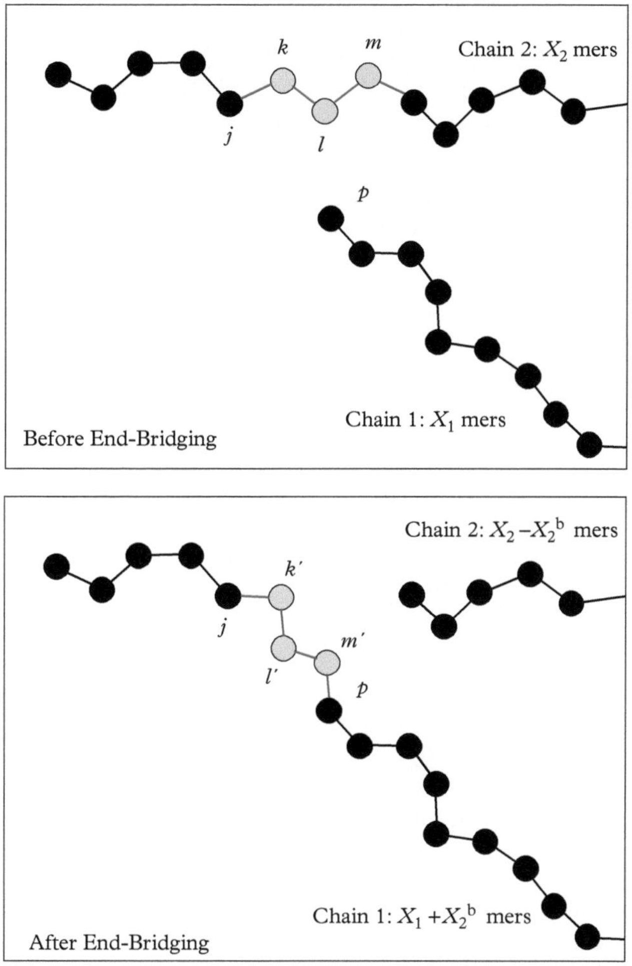

Figure 5.5 Schematic of the intermolecular end-bridging (EB) problem. Dark atoms are stationary, and lightly shaded atoms are mobile. Pendant groups are omitted for clarity. Top: local configuration of the two chains prior to the EB move. The attacking chain is denoted as 1 and the victim chain as 2. Bottom: trimer $\{k, l, m\}$ of the victim chain is excised and end p of the attacking chain is bridged to internal mer j of the victim chain. Trimer $\{k', l', m'\}$ is the bridging trimer.

Reprinted (adapted) with permission from Mavrantzas, V.G., Boone, T.D., Zervopoulou, E., and Theodorou, D.N. *Macromolecules* **32**, 5072 (1999). Copyright 1999 American Chemical Society.

5.2.2.3 Double bridging (DB)

The move (Figure 5.8) is initiated by randomly selecting a predator atom i in chain *ich* and a bridgeable intermolecular neighbour j in chain *jch*, from the list of all possible $N_{\mathrm{DB},(i)}$ bridgeable candidates. From all possible DB combinations between i and j, one is picked randomly. Based on the selected combination, atoms i_2 and j_2 are then identified, together with the two trimers to be

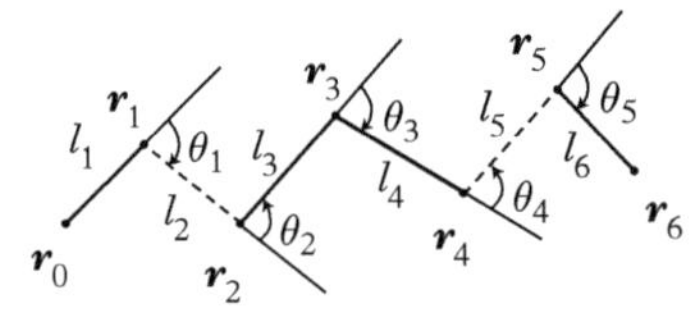

Figure 5.6 Schematic of the trimer bridging construction. Given are the dimers $\{\boldsymbol{r}_0, \boldsymbol{r}_1\}$ and $\{\boldsymbol{r}_5, \boldsymbol{r}_6\}$ in 3D space. Sought is the construction of a trimer $\{\boldsymbol{r}_2, \boldsymbol{r}_3, \boldsymbol{r}_4\}$, such that the bond lengths $\{l_2, l_3, l_4, l_5\}$ and the bond angle supplements $\{\theta_1, \theta_2, \theta_3, \theta_4, \theta_5\}$ have prescribed values.

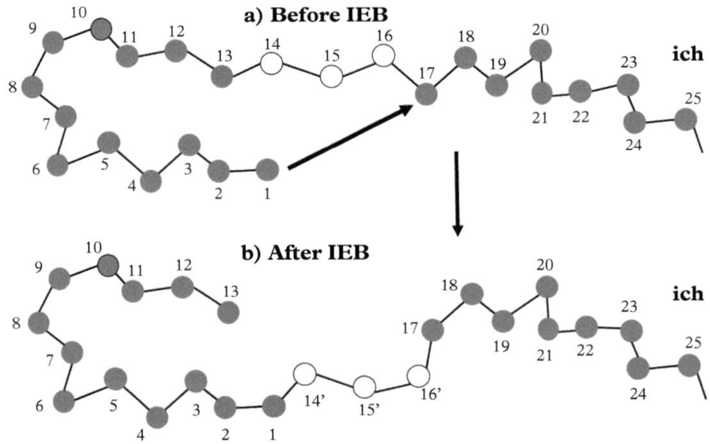

Figure 5.7 Schematic representation of the intramolecular end-bridging (IEB) move.

excised from the chains. The geometric bridging procedure is then invoked twice to determine all geometrically admissible candidate configurations for the two new trimer bridges $\{j'_a, j'_b, j'_c\}$ and $\{i'_a, i'_b, i'_c\}$, for the primary and secondary bridging from i to j and from j_2 to i_2, respectively. In the case of monodisperse systems, given atoms i and j, the allowed DB combination is uniquely specified.

5.2.2.4 *Intramolecular double rebridging (IDR)*

The main difference between DB and IDR is that the latter applies to a single chain; apart from this, the two moves are completely analogous. Thus, the implementation of the IDR algorithm is conceptually the same as that of the DB move, except that it involves only a single chain. The move is shown schematically in Figure 5.9.

5.2.3 Advanced Monte Carlo moves for branched polymers

All moves discussed in Sections 5.2.1 and 5.2.2 are designed for simulations with linear chain molecules. In practice, one often encounters the problem of simulating polymers with a nonlinear

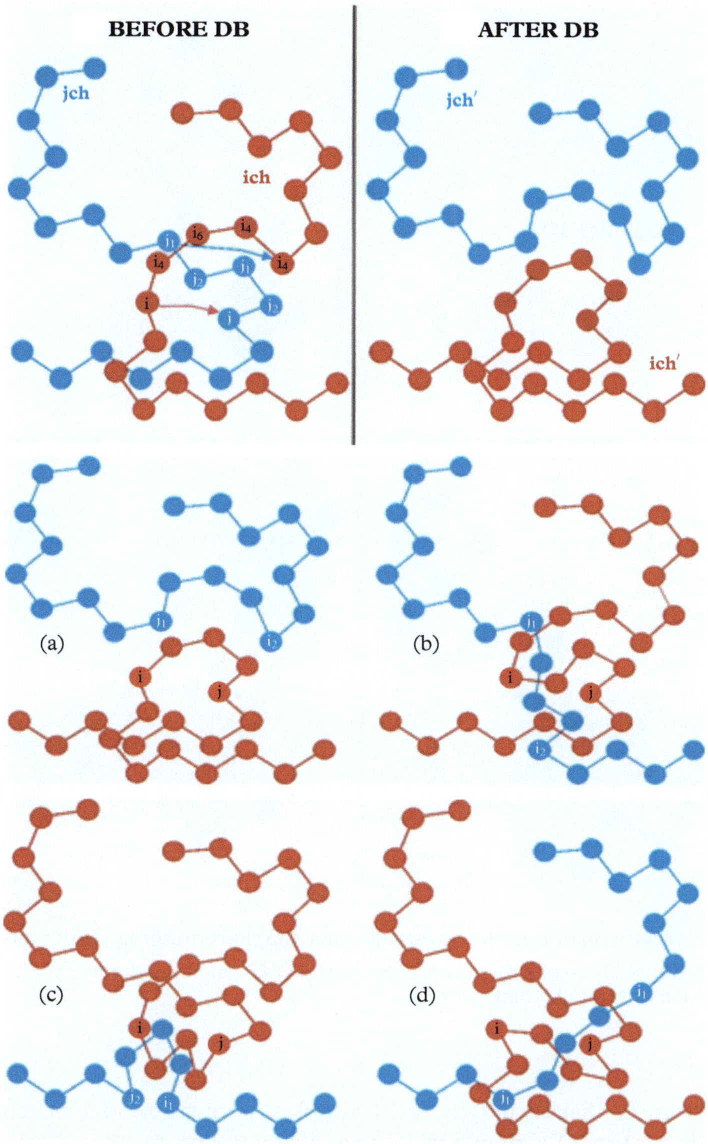

Figure 5.8 Schematic representation of the double-bridging (DB) move and of the four possible configurations that can result from that.

Reprinted from Karayiannis, N.C., Giannousaki, A.E., Mavrantzas, V.G., and Theodorou, D.N. *J. Chem. Phys.* **117**, 5465 (2002), with the permission of AIP Publishing.

chain architecture, for example branched polymer molecules. For the moves to be applicable to these systems, nontrivial modifications are required. In Figure 5.10 we provide schematics of common nonlinear architectures encountered today in polymer research and technology. For example, let us

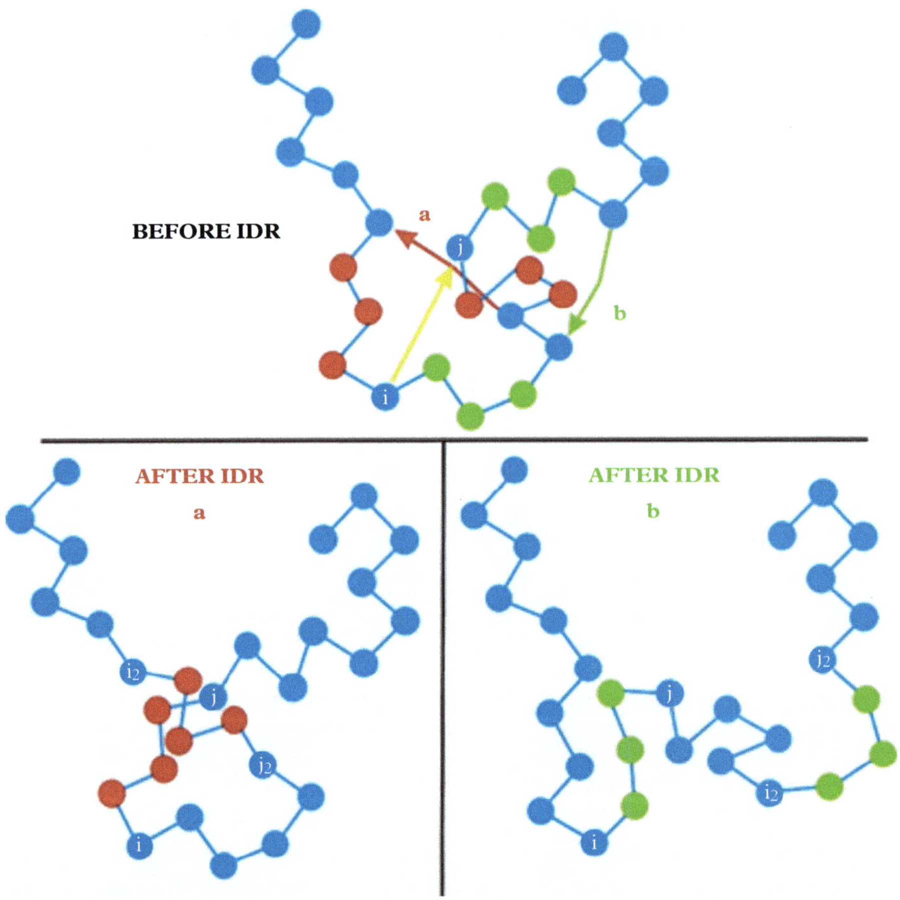

Figure 5.9 Schematic representation of the intramolecular double-rebridging (IDR) move.
Reprinted from Karayiannis, N.C., Giannousaki, A.E., Mavrantzas, V.G., and Theodorou, D.N. *J. Chem. Phys.* **117**, 5465 (2002), with the permission of AIP Publishing.

consider the case of an H-shaped polymer, the simplest perhaps model structure for a branched polymer molecule. It consists of a main backbone which connects two branch points (or junctions), from each one of which two dangling arms (i.e. branches) emanate. To be able to simulate an H-shaped polymer melt, we should properly redesign the moves in order to allow for bridgings between (a) the main backbones of two different chains, (b) the branches of two different chains, and (c) the branches of the same chain. We should also introduce special moves to effect atom displacements at the junction points.

In the following we discuss MC moves that can simulate a variety of nonlinear polymer systems such as H-shaped, short-chain branched (SCB) ones, and long $A_3\mathbf{A}A_3$ multiarm or pom-pom molecules.

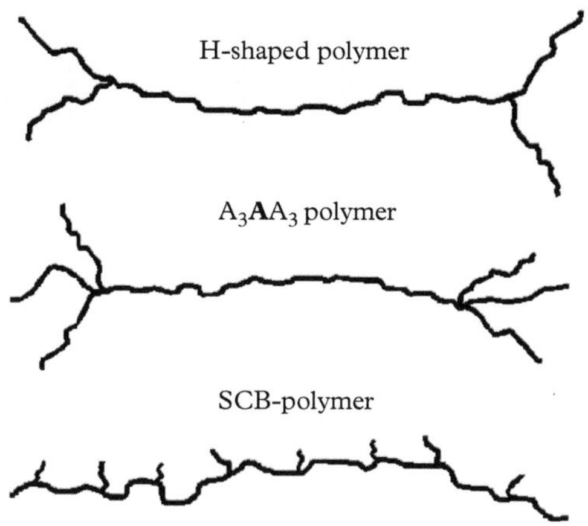

Figure 5.10 Typical branched molecular architectures.

Reprinted (adapted) with permission from Baig, C., Alexiadis, O., and Mavrantzas, V.G. *Macromolecules* **43**, 986 (2010). Copyright 2010 American Chemical Society.

5.2.3.1 ***End configurational bias (E-CB)***

This move (Figure 5.11a) aims at altering the configuration of the chain ends in a SCB polymer melt. Following the corresponding implementation for linear polymers, a chain end is randomly chosen and then N_{cut} backbone mers from the chain end are rebuilt using a configurational bias strategy. If a branch point is encountered during the regrowth procedure, the branch and backbone mers are rebuilt one by one in an alternating fashion using the CB method.

5.2.3.2 ***Branch configurational bias (Br-CB)***

Br-CB is a modified version of the classical configurational bias MC move designed especially for the simulation of SCB polymers (Figure 5.11b). Its aim is to equilibrate the configuration of the short chain branches along the main chain backbone by rebuilding (regrowing) all segments of the branch in a biased way. If N_{br} is the number of mers along the branches, then the move consists of a combination of branch point (bp) rebuild for the first mer in the branch and configurational bias regrowth for the remaining $N_{\text{br}} - 1$ mers in the branch.

5.2.3.3 ***CONROT for SCB polymers (SCB-CONROT)***

The traditional CONROT move is based on the reconstruction of a randomly chosen internal sequence of mers along a chain by solving the rebridging problem numerically or analytically. To apply the move to SCB polymer chains, lateral branches must be appropriately displaced in order to preserve the architecture of the polymer. A typical implementation of the CONROT move for SCB polymers known as SCB-CONROT is as follows (see Figure 5.11c). (i) A monomer is randomly chosen and the bridge trimer segment is defined (I). (ii) The ϕ_{L} and ϕ_{R} dihedrals are rotated, causing monomers 1 and 5 to move. These rotations break the connectivity of 1–2 and 4–5 bonds. Then, reconstruction of bridge atoms (2, 3, and 4) is accomplished (II) by solving the bridging problem. (iii) If there is a branch attached to any bridge atom, this is rebuilt from the new bridge configuration using the Br-CB move described in Section 5.2.3.2 (III, IV).

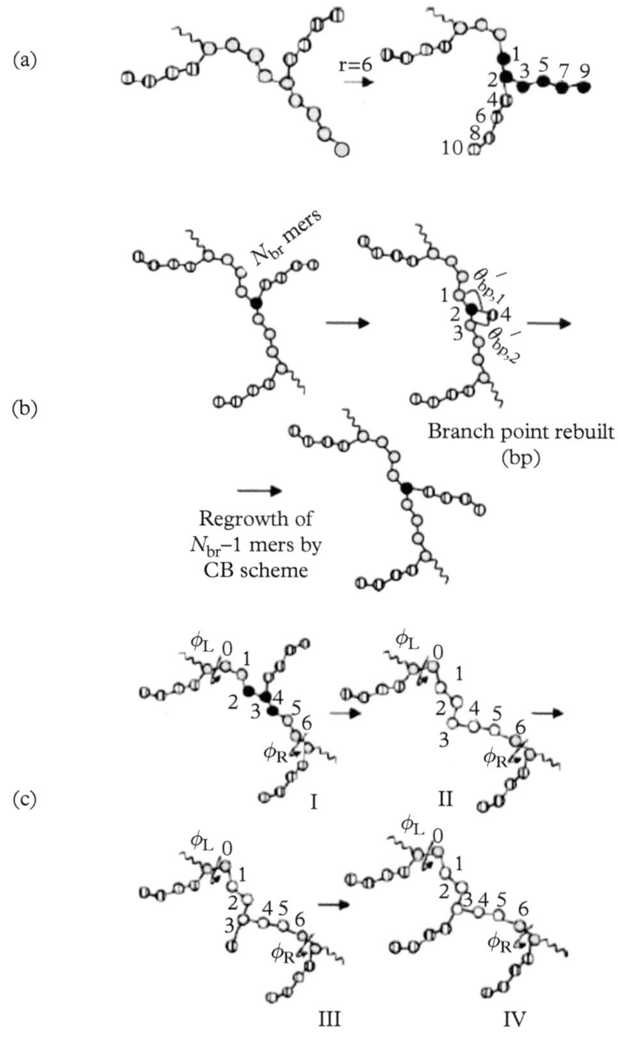

Figure 5.11 Intramolecular rearrangement moves for short-chain branched (SCB) molecules: (a) End configurational bias. (b) Branch configurational bias (Br-CB). (c) SCB-CONROT.

(a–c) Reprinted (adapted) with permission from Ramos, J., Peristeras, L.D., and Theodorou, D.N. *Macromolecules* **40**, 9640 (2007). Copyright 2007 American Chemical Society.

5.2.3.4 Double concerted rotation (d-CONROT)

d-CONROT is a generalization of the standard CONROT move designed to induce large-scale system rearrangements near branch points. It is applicable to all types of branched polymers: H-shaped, A_3AA_3, and SCB ones. Figure 5.12a illustrates its basic concepts. The move is initiated by removing two trimer sets $\{A_1, A_2, A_3\}$ and $\{B_1, B_2, B_3\}$ connected to a branch point C. The two atoms (A_4, B_4) neighbouring the removed trimers and the branch point C are then driven to new positions $\{A'_4, B'_4, C'\}$, by altering the set of torsion angles $\{\phi_1, \phi_2, \phi_3\}$ within prescribed bounds $(-\delta\phi_{\max}, +\delta\phi_{\max})$. Given the new coordinates of the three driven atoms $\{A'_4, B'_4, C'\}$, the geometric problem of trimer bridging is solved simultaneously two times: one for the new construction connecting atoms A'_4 and C' through the $\{A'_1, A'_2, A'_3\}$ trimer, and another for the new construction connecting atoms B'_4 and C' with the $\{B'_1, B'_2, B'_3\}$ trimer. The bridging process takes place in such a way that the bonds of the altered segments $\{C_1, C', A'_1, A'_2, A'_3, A'_4, A_5\}$ and $\{C_1, C', B'_1, B'_2, B'_3, B'_4, B_5\}$ preserve their equilibrium lengths, given a set of preselected values for all bond-bending angles involved in the move.

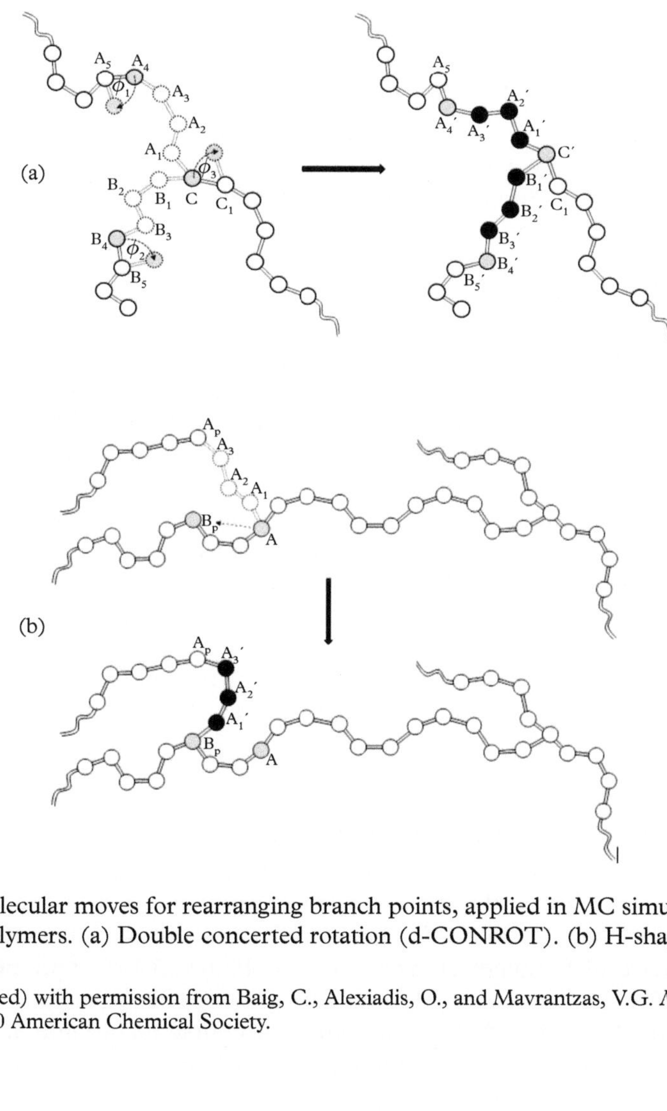

Figure 5.12 Intramolecular moves for rearranging branch points, applied in MC simulations of H-shaped, $A_3\mathbf{A}A_3$, and SCB polymers. (a) Double concerted rotation (d-CONROT). (b) H-shaped branch rebridging (H-BR).

(a, b) Reprinted (adapted) with permission from Baig, C., Alexiadis, O., and Mavrantzas, V.G. *Macromolecules* **43**, 986 (2010). Copyright 2010 American Chemical Society.

5.2.3.5 *H-shaped branch rebridging (H-BR)*

The H-BR move (Figure 5.12b)[54,56] begins by choosing one of the two branch points (e.g. atom A in Figure 5.12b) in a randomly selected chain and an atom four bonds apart from the branch point along the backbone or one of the neighbouring branches (e.g. A_p in Figure 5.12b). Then, a trimer such as $\{A'_1, A'_2, A'_3\}$ between the branch point and the selected atom is excised and rebridged to a new selected branch point (e.g. B_p). As a result, and depending on the movement direction of the branch point, the length of each branch and the backbone of an H-shaped or $A_3\mathbf{A}A_3$ chain may increase or decrease. For example, for the case considered in Figure 5.12b, the length of the branch containing atom B_p is to be shortened and the backbone length to be lengthened after acceptance of the move, while the length of the branch containing the trimer $\{A_1, A_2, A_3\}$ remains constant.

5.2.3.6 *Branch point flip (BrP-Flip) and branch point slithering (BrP-S)*

These two moves are applicable to branch points with functionality 3.[32,33] During the BrP-Flip move, shown graphically in Figure 5.13 (top), the branch point is flipped in one of three possible ways, corresponding to the three branches. Each way involves rotation about the axis defined by a pair of atoms adjacent to the branch point, keeping bond lengths fixed. In each case, the first atom of the third branch (i.e. the branch that does not contain any of the atoms in the axis pair) is reconstructed by selecting the two bond angles that define its position with respect to the branch point and the axis pair. Subsequently, a rebridge scheme is applied in order to reconnect that atom with the fifth atom of the third branch. In this way, the positions of the second through the fourth atoms of the branch are changed. This is the branch attachment scheme (Br-AS).

The BrP-S move is based on the concept that a branch point can slither along the *backbone* defined by any two of the branches attached to it (see bottom image of Figure 5.13). Slithering of the branch point is constrained by the molecular weight distribution of the branches. Once the branch point is displaced, the third branch is reconnected to the backbone using the Br-AS. The first atom of the branch is constructed by using the corresponding bond length and two bond angles. BrP-S makes it possible to sample new configurations for the degrees of freedom associated with short branches along the chain. In addition, it induces fluctuations (within a prespecified interval) to the locations of branch points (and thus to branch–branch distance), which is very significant, since most SCB polyolefins synthesized by polymer industries are characterized by a certain degree of polydispersity not only in the total chain length but also in the interbranch spacing.

5.2.3.7 *End-bridging for SCB systems (SCB-EB)*

This is a generalization of the end-bridging move for SCB polymers in order to enhance the performance of a MC algorithm at the level of global properties (end-to-end distance vectors). It is based on the traditional EB move for linear polymers but slightly reformulated to take into account the presence of branches along the chain (Figure 5.14). The move is realized by randomly selecting two chains and then allowing the end of one of them to attempt a trimer bridging with an internal backbone atom of the other, of course taking into account the restrictions imposed by the allowed values of the branch-to-branch and total (or backbone) molecular lengths.

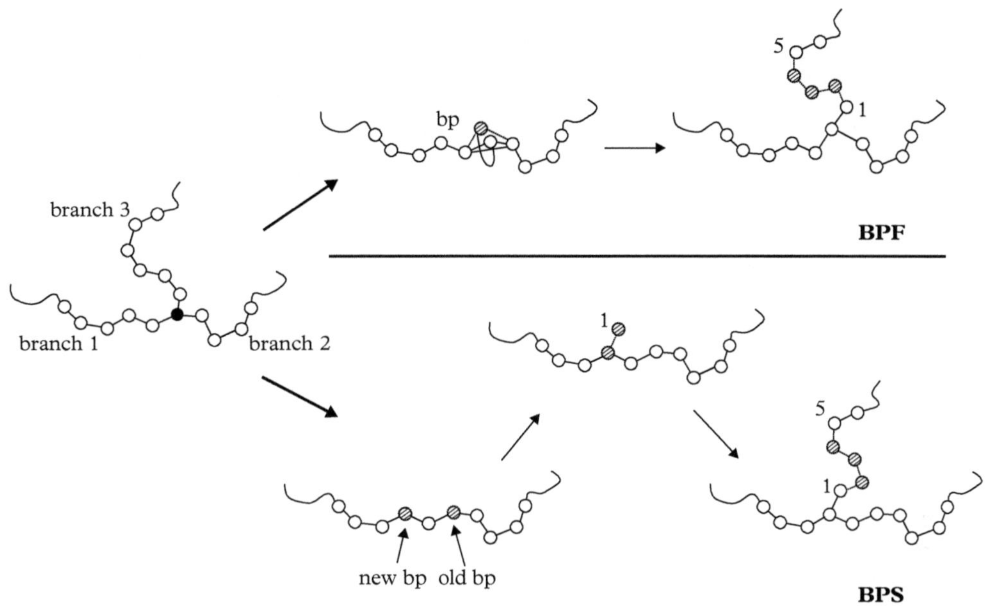

Figure 5.13 Schematic representation of the BrP-Flip (top) and BrP-S (bottom) moves.
Reprinted (adapted) with permission from Peristeras, L.D., Economou, I.G., and Theodorou, D.N. *Macromolecules* **38**, 386 (2005). Copyright 2005 American Chemical Society.

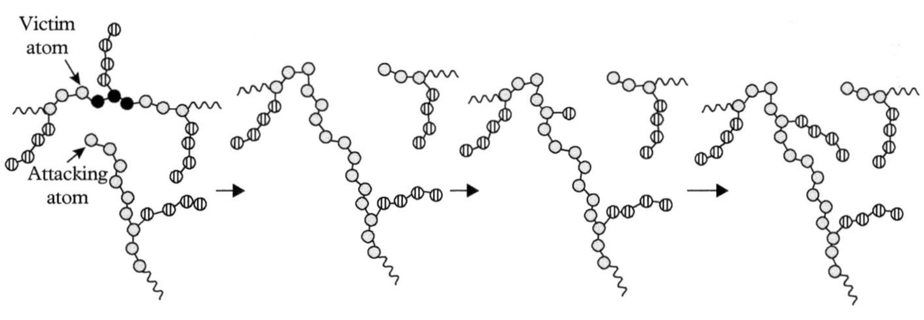

Figure 5.14 Schematic representation of short-chain branch end-bridging (SCB-EB) move.
Reprinted (adapted) with permission from Ramos, J., Peristeras, L.D., and Theodorou, D.N. *Macromolecules* **40**, 9640 (2007). Copyright 2007 American Chemical Society.

5.2.3.8 Double-bridging for branched systems (Br-DB)

This is a generalization of the double-bridging move for SCB polymers and its aim is also to enhance the performance of a MC algorithm at the level of global properties (end-to-end distance vectors). It is based on the traditional DB move for linear polymers but reformulated to take into account the

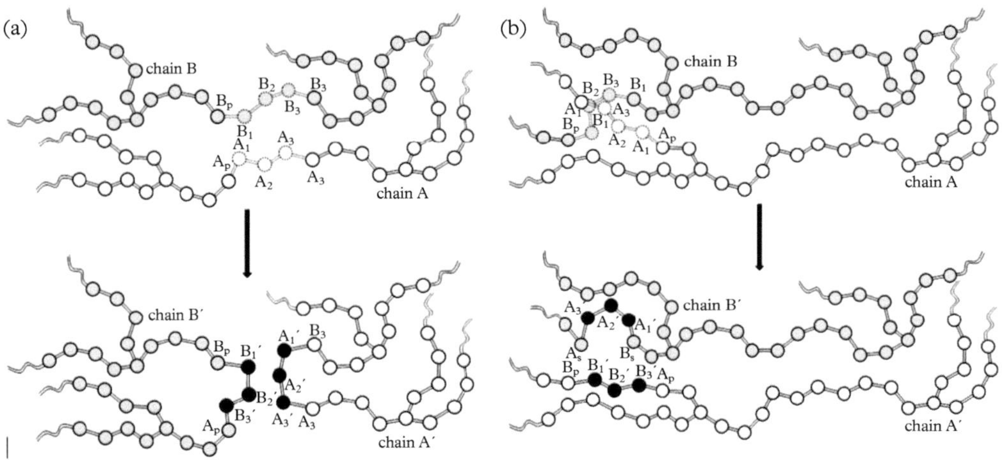

Figure 5.15 Schematic representation of the intermolecular double bridging (DB) move, applied to chain backbones (a) or to chain branches (b), in MC simulations of H-shaped and $A_3\mathbf{A}A_3$ polymers.

presence of branches along the chain. Figure 5.15 gives a schematic of intermolecular DB moves between the crossbars or between the branches of H-shaped polymers.[54]

5.2.3.9 IDR for branched systems (Br-IDR)

It is the intramolecular version of the previous move, and is shown in Figure 5.16.[54]

5.2.4 Monte Carlo moves for solubility calculations

5.2.4.1 Particle insertion

This was the first move proposed for the MC simulation of the solubility of small molecules in a polymer matrix. It involves the random insertion of a particle in the system.

5.2.4.2 Particle deletion

This is the inverse of the direct particle insertion move and involves the random deletion of a particle from the system.

5.2.4.3 Polymer–oligomer end-bridging

The polymer–oligomer end-bridging move is completely analogous to the EB move developed for polymer melts (Figure 5.5), but now the attacking chain is an oligomer molecule of length s_1, which is joined to a polymer chain via a trimer bridge. The part of the original polymer chain that is cut off, through excision of a trimer, becomes a new oligomer molecule. Clearly, this move does not alter the lengths of the polymer chain and oligomer participating in it, nor does it change the number of oligomers present in the system. It permits, however, moving the oligomer molecules around in the

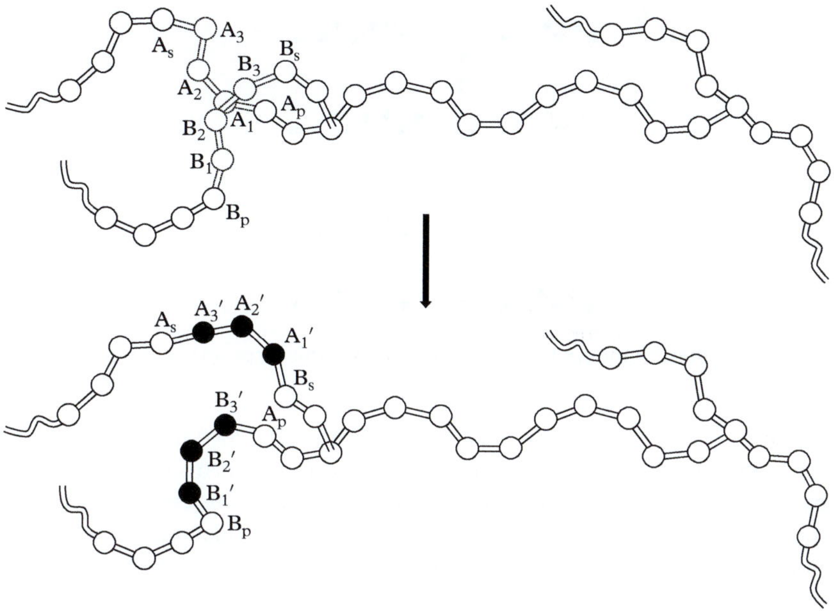

Figure 5.16 Schematic representation of the intramolecular double rebridging (IDR) MC move adopted in the simulations of H-shaped and $A_3\mathbf{A}A_3$ polymers.

Reprinted (adapted) with permission from Baig, C., Alexiadis, O., and Mavrantzas, V.G. *Macromolecules* **43**, 986 (2010). Copyright 2010 American Chemical Society.

melt, and this enhances system equilibration. It makes it possible to equilibrate polyethylene–alkane solutions with alkane length typically up to C_{10}. This happens because the oligomer content in the polymer matrix is still controlled via insertion and deletion moves only.

5.2.4.4 Fusion–scission

These two moves are sketched in Figure 5.17 and have been proposed for simulating the phase equilibrium between an oligomer and a polymer of the same chemistry. Scission involves creation of an oligomer from the end of a polymer molecule with the simultaneous removal of a trimer; fusion, on the other hand, leads to formation of a new, longer polymer chain by joining a polymer and an oligomer end-to-end. MC algorithms using these two moves should be executed in the $f_1{}'N_c n_0 PT\boldsymbol{\mu}^*$ statistical ensemble already discussed in Section 5.1.2. The geometric problem of trimer bridging involved in the two moves is identical to that involved in the traditional EB move.

5.2.5 Acceptance criteria of the various Monte Carlo moves

The starting point for the derivation of the acceptance criterion of a MC move is the microscopic detailed balance condition, Eq. (5.1.4). The correct use of this equation, however, depends on the coordinate system employed to describe the phase space of the simulated system. Here, we assume that we make use of the flexible model in the limit of infinitely stiff bond lengths discussed in detail in

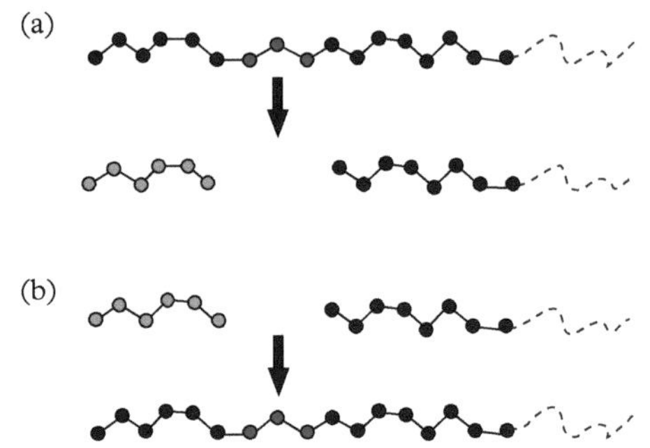

Figure 5.17 Connectivity-altering moves for oligomer–polymer mixtures: (a) scission, and (b) fusion.

Section 3.1.2. In this model, we typically move between different coordinate systems describing the phase space. For example, we usually move from a description in terms of Cartesian coordinates to a description involving generalized (chain relative) coordinates. In this case, an appropriate Jacobian must be introduced to calculate $\rho^{\text{eq}}(o)$ and $\rho^{\text{eq}}(n)$ correctly. If $\boldsymbol{q}$ is selected as the coordinate system, the a priori probability of the system being in state (o) is

$$\rho^{\text{eq}}(o)\text{d}\boldsymbol{r} = \rho(o)\mathcal{J}(o)\text{d}\boldsymbol{q}, \tag{5.2.1}$$

where $\rho(o)$ is the normalized probability density function in Cartesian space determined by the statistical ensemble in use. Elementary volumes in Cartesian space and in $\boldsymbol{q}$-space around state (o) are related by $\text{d}\boldsymbol{r} = \boldsymbol{J} \cdot \text{d}\boldsymbol{q}$.[57]

Moreover, for several attempted moves, the conditional attempt probability $a\,(o \to n)$ may include a number of bias selections to enhance subsequent acceptance of the move. If the move consists of more than one independent step, then the overall probability $a\,(o \to n)$ as well as the Jacobian are calculated as products of the corresponding quantities (bias selection weight w_s and attempt probability a_s) for each step,

$$a\,(i \to j) = \prod_{s=1}^{\text{no of steps}} w_s\,(i \to j)\,a_s\,(i \to j)$$
$$\mathcal{J}\,(i) = \prod_{s=1}^{\text{no of steps}} \mathcal{J}_s\,(i), \tag{5.2.2}$$

where with $w_s\,(i \to j)$ and $a_s\,(i \to j)$ we have denoted the bias selection weighting factor and attempt probability, respectively, for the intermediate step s.

From the above equations then, and provided all chain species involved in both states are probable, the generalized Metropolis acceptance criterion for the transition from an old state (o) to a new state (n) is obtained by imposing the condition of microscopic reversibility in the form

$$\rho(o)\mathcal{J}(o)\text{d}\boldsymbol{q}\,a\,(o \to n) = \rho(n)\mathcal{J}(n)\text{d}\boldsymbol{q}\,a\,(n \to o) \tag{5.2.3}$$

and reads

$$p_{\text{acc}}(o \to n) = \min\left(1, \frac{\rho(n) \prod_{s=1}^{\text{no of steps}} w_s(n \to o)\, a_s(n \to o)\, \mathcal{J}_s(n)}{\rho(o) \prod_{s=1}^{\text{no of steps}} w_s(o \to n)\, a_s(o \to n)\, \mathcal{J}_s(o)}\right). \tag{5.2.4}$$

For a move that does not involve a change in volume or number of molecules, Eq. (5.2.4) assumes the special form

$$p_{\text{acc}}(o \to n) = \min\left(1, \frac{\prod_{s=1}^{\text{no of steps}} w_s(n \to o)\, a_s(n \to o)\, \mathcal{J}_s(n)}{\prod_{s=1}^{\text{no of steps}} w_s(o \to n)\, a_s(o \to n)\, \mathcal{J}_s(o)} \exp\left[-\beta\left(\mathcal{V}^{(n)} - \mathcal{V}^{(o)}\right)\right]\right). \tag{5.2.5}$$

The key step then in order to get an acceptance criterion for the move under consideration is to derive the analytical expressions for the bias selection weights and attempt probabilities (i.e. the quantities $w_s(n \to o)$, $a_s(n \to o)$, $w_s(o \to n)$, and $a_s(o \to n)$).

For most of the moves introduced and discussed in Section 5.2, the acceptance probabilities are reported in Table 5.1.

5.3 Methods for the calculation of free-energy differences

Developing molecular methods for computing free-energy differences is of paramount importance in many applications: in evaluating the chemical potential or the fugacity for fluid-phase equilibria calculations of monomer–polymer, solvent–polymer, and polymer–polymer mixtures, in computing sorption and permeability properties of membranes, in separating enthalpic and entropic contributions to mechanical properties, in testing analytical theories for the above, and many others. In the following we describe how one can compute free-energy differences based on MC simulations.

5.3.1 Widom insertion

Widom's test particle insertion method is an early example of free-energy perturbation for the calculation of the chemical potential.[58,59] Without loss of generality, let us consider N_c molecules of a pure fluid, each consisting of s atoms, at temperature T and volume V. The chemical potential per molecule of the fluid is thermodynamically defined by the relation

$$\mu = \left(\frac{\partial A}{\partial N_c}\right)_{T,V}, \tag{5.3.1}$$

where A is the total Helmholtz energy of the fluid. To derive a convenient expression for computing μ from a MC simulation, we consider that an additional molecule is added to the system. Then, in the thermodynamic limit $N_c \to \infty$, $V \to \infty$, $\rho = N_c/V = \text{const.}$,

Table 5.1 *Acceptance Probabilities for Some Moves Employed in Continuous-Space Simulations of Polymers*

Move	Ratio	Reference
Reptation (Figure 5.1a)	$\exp\left[-\beta\left(\mathcal{V}^{(n)}-\mathcal{V}^{(o)}\right)\right] \equiv \exp(-\beta\Delta\mathcal{V})$	
End-mer rotation (Figure 5.1c)	$\exp(-\beta\Delta\mathcal{V})$	
Flip (Figure 5.2a)	$\frac{\mathcal{J}^{\mathrm{F}}(n)}{\mathcal{J}^{\mathrm{F}}(o)}\exp(-\beta\Delta\mathcal{V})$, where $\mathcal{J}^{\mathrm{F}} = \frac{l_2 l_3 \sin^2(\theta_2)}{l_2^2 + l_3^2 + 2 l_2 l_3 \cos(\theta_2)}, \theta_2 = \pi - \theta'_2$	32 (App. A)
Dimer flip (Figure 5.2b)	$\frac{a^{\mathrm{DF}}(n \to o)\mathcal{J}^{\mathrm{DF}}(n)}{a^{\mathrm{DF}}(o \to n)\mathcal{J}^{\mathrm{DF}}(o)}\exp(-\beta\Delta\mathcal{V})$, where $a^{\mathrm{DF}}(o \to n) = \frac{1}{N_{\mathrm{c}}}\frac{1}{n_{\mathrm{ch}}^{\mathrm{DF}}}$ $n_{\mathrm{ch}}^{\mathrm{DF}}$ = number of atoms in selected chain that can participate in the move $\mathcal{J}^{\mathrm{DF}} = \frac{l_2 l_3 l_4}{l_{14}}\sin^2(\theta_2)\left[1-\left(\frac{\boldsymbol{l}_{13}\cdot\hat{\boldsymbol{l}}_4}{l_{13}}\right)^2\right], \boldsymbol{l}_i = \boldsymbol{r}_i - \boldsymbol{r}_{i-1}, \hat{\boldsymbol{l}}_i = \boldsymbol{l}_i/l_i, l_i = \lvert\boldsymbol{l}_i\rvert, l_{ij} = \lvert\boldsymbol{r}_j - \boldsymbol{r}_i\rvert$	32 (App. B)
n-mer flip	$\frac{a^{n-\mathrm{F}}(n \to o)\mathcal{J}^{n-\mathrm{F}}(n)}{a^{n-\mathrm{F}}(o \to n)\mathcal{J}^{n-\mathrm{F}}(o)}\exp(-\beta\Delta\mathcal{V})$, where $a^{n-\mathrm{F}}(o \to n) = \frac{1}{N_{\mathrm{c}}}\frac{1}{n_{\mathrm{ch}}^{n-\mathrm{F}}} n_{\mathrm{ch}}^{n-\mathrm{F}}$ = number of atoms in selected chain that can participate in the move $\mathcal{J}^{n-\mathrm{F}} = \frac{1}{l_{1(n+2)}}\sin^2(\theta_2)\left(\prod_{j=2}^{n+2} l_j\right)\prod_{j=3}^{n+1}\left[1-\left(\frac{\boldsymbol{l}_{1j}\cdot\hat{\boldsymbol{l}}_{j+1}}{l_{1j}}\right)^2\right],$ $\boldsymbol{l}_i = \boldsymbol{r}_i - \boldsymbol{r}_{i-1}, \hat{\boldsymbol{l}}_i = \boldsymbol{l}_i/l_i, l_i = \lvert\boldsymbol{l}_i\rvert, l_{ij} = \lvert\boldsymbol{r}_j - \boldsymbol{r}_i\rvert$	
Atom identity exchange (Figure 5.4b)	$\exp(-\beta\Delta\mathcal{V})$	48

Volume fluctuation (Figure 5.4c)	$\exp\left[-\beta\Delta\mathcal{V}-\beta P\Delta V+N_c k_B T\ln\left(\frac{V^{(n)}}{V^{(o)}}\right)\right]$	
Configurational bias (Figure 5.2c)	$\frac{\left(\prod_{l=1}^{N_{\text{cut}}} w_{\text{LJ}}^{l}(o\to n)\right)}{\left(\prod_{l=1}^{N_{\text{cut}}} w_{\text{LJ}}^{l}(n\to o)\right)}\exp\left(-\beta\Delta\mathcal{V}_{\text{tor}}\right),$ where $w_{\text{LJ}}^{l}=\sum_{k=1}^{N_{\text{trial}}}\exp\left[-\beta\Delta\mathcal{V}_{l}^{\text{NB}}\left(\boldsymbol{r}_{l,k}\right)\right]$ Products in numerator and denominator of the ratio are called Rosenbluth weights.	5, 35, 36
End-bridging (Figures 5.5, 5.6)	$\frac{w_{\text{tor}}^{\text{EB}}(n\to o)\,a^{\text{EB}}(n\to o)\,\mathcal{J}^{\text{bridge}}(n)}{w_{\text{tor}}^{\text{EB}}(o\to n)\,a^{\text{EB}}(o\to n)\,\mathcal{J}^{\text{bridge}}(o)}\exp\left(-\beta\Delta\mathcal{V}\right),$ where $w_{\text{tor}}^{\text{EB}}(o\to n)=\frac{\exp\left(-\frac{\mathcal{V}_{\text{tor}}(n)}{k_B T}\right)}{\sum_{l=1}^{N_{\text{sol}}(n)}\exp\left(-\frac{\mathcal{V}_{\text{tor}}(l)}{k_B T}\right)},\ w_{\text{tor}}^{\text{EB}}(n\to o)=\frac{\exp\left(-\frac{\mathcal{V}_{\text{tor}}(o)}{k_B T}\right)}{\sum_{l=1}^{N_{\text{sol}}(o)}\exp\left(-\frac{\mathcal{V}_{\text{tor}}(l)}{k_B T}\right)}$ N_{sol}: number of solutions to the bridging problem $a^{\text{EB}}(o\to n)=\frac{1}{N_{\text{bridge}}(o)},\ a^{\text{EB}}(n\to o)=\frac{1}{N_{\text{bridge}}(n)}$ N_{bridge}: number of possible bridge sites $\mathcal{J}^{\text{bridge}}=\frac{l_2^2 l_3^2 l_4^2 l_5^2\sin\theta_1\sin\theta_2\sin\theta_3\sin\theta_4\sin\theta_5}{\left\lvert\begin{array}{l}-l_2 l_3 l_5\left[\left(\hat{\boldsymbol{l}}_1\times\hat{\boldsymbol{l}}_2\right)\cdot\hat{\boldsymbol{l}}_3\right]\left[\left(\hat{\boldsymbol{l}}_2\times\hat{\boldsymbol{l}}_5\right)\cdot\hat{\boldsymbol{l}}_6\right]\left[\left(\hat{\boldsymbol{l}}_3\times\hat{\boldsymbol{l}}_4\right)\cdot\hat{\boldsymbol{l}}_5\right]\\+l_2 l_4 l_5\left[\left(\hat{\boldsymbol{l}}_1\times\hat{\boldsymbol{l}}_2\right)\cdot\hat{\boldsymbol{l}}_5\right]\left[\left(\hat{\boldsymbol{l}}_2\times\hat{\boldsymbol{l}}_3\right)\cdot\hat{\boldsymbol{l}}_4\right]\left[\left(\hat{\boldsymbol{l}}_4\times\hat{\boldsymbol{l}}_5\right)\cdot\hat{\boldsymbol{l}}_6\right]\\+l_3 l_4\left[\left(\hat{\boldsymbol{l}}_1\times\hat{\boldsymbol{l}}_2\right)\cdot\hat{\boldsymbol{l}}_3\right]\left[l_2\left(\hat{\boldsymbol{l}}_2\times\hat{\boldsymbol{l}}_3\right)\cdot\hat{\boldsymbol{l}}_4-l_5\left(\hat{\boldsymbol{l}}_3\times\hat{\boldsymbol{l}}_4\right)\cdot\hat{\boldsymbol{l}}_5\right]\left[\left(\hat{\boldsymbol{l}}_4\times\hat{\boldsymbol{l}}_5\right)\cdot\hat{\boldsymbol{l}}_6\right]\\+l_2 l_4^2\left[\left(\hat{\boldsymbol{l}}_1\times\hat{\boldsymbol{l}}_2\right)\cdot\hat{\boldsymbol{l}}_4\right]\left[\left(\hat{\boldsymbol{l}}_2\times\hat{\boldsymbol{l}}_3\right)\cdot\hat{\boldsymbol{l}}_4\right]\left[\left(\hat{\boldsymbol{l}}_4\times\hat{\boldsymbol{l}}_5\right)\cdot\hat{\boldsymbol{l}}_6\right]\\-l_3^2 l_5\left[\left(\hat{\boldsymbol{l}}_1\times\hat{\boldsymbol{l}}_2\right)\cdot\hat{\boldsymbol{l}}_3\right]\left[\left(\hat{\boldsymbol{l}}_3\times\hat{\boldsymbol{l}}_4\right)\cdot\hat{\boldsymbol{l}}_5\right]\left[\left(\hat{\boldsymbol{l}}_3\times\hat{\boldsymbol{l}}_5\right)\cdot\hat{\boldsymbol{l}}_6\right]\end{array}\right\rvert}$ $\hat{\boldsymbol{l}}_i=\boldsymbol{l}_i/l_i$	5, 11, 31 (App. C), 53 (App.)

Continued

Table 5.1 *Continued*

Move	Ratio	Reference
Intramolecular end-bridging (Figure 5.7)	As in end-bridging above	
Concerted rotation (Figure 5.3a)	As in end-bridging above	5, 31, 39
Double bridging (Figure 5.8)	$\frac{w^{DB}_{tor,\,LJ}(n \to o)\, a^{DB}(n \to o)\, \mathcal{J}^{DB}(n)}{w^{DB}_{tor,\,LJ}(o \to n)\, a^{DB}(o \to n)\, \mathcal{J}^{DB}(o)} \exp(-\beta \Delta \mathcal{V})$, where $w^{DB}_{tor,\,LJ}(o \to n) = \frac{\exp\left(\frac{-\mathcal{V}_{tor}(n) - \mathcal{V}^{NB}(n)}{k_B T}\right)}{\sum_{m=1}^{N_{sol(o \to n)}} \exp\left(\frac{-\mathcal{V}_{tor}(m) - \mathcal{V}^{NB}(m)}{k_B T}\right)}$, $w^{DB}_{tor,\,LJ}(n \to o) = \frac{\exp\left(\frac{-\mathcal{V}_{tor}(o) - \mathcal{V}^{NB}(o)}{k_B T}\right)}{\sum_{m=1}^{N_{sol(n \to o)}} \exp\left(\frac{-\mathcal{V}_{tor}(m) - \mathcal{V}^{NB}(m)}{k_B T}\right)}$ $a^{DB}(o \to n) = \frac{1}{N_{DB,i}(o)}$, $a^{DB}(n \to o) = \frac{1}{N_{DB,j_2}(n)}$ $N_{DB,i}$: number of bridgeable candidates for the selected predator atom i $\mathcal{J}^{DB} = \left(\mathcal{J}^{bridge}\right)_i \left(\mathcal{J}^{bridge}\right)_{j_2}$	52, 53, 54
Intramolecular double rebridging (Figure 5.9)	As in double bridging above	52, 53, 54
End configurational bias (Figure 5.11a)	As in configurational bias above	33

Branch point flip and branch point slithering (Figure 5.13)	$\dfrac{w^{\mathrm{Br-AS}}(n \to o)\, w^{\mathrm{bridge}}(n \to o)\, \rho(n) \mathcal{J}^{\mathrm{Br-AS}}(n) \mathcal{J}^{\mathrm{bridge}}(n)}{w^{\mathrm{Br-AS}}(o \to n)\, w^{\mathrm{bridge}}(o \to n)\, \rho(o) \mathcal{J}^{\mathrm{Br-AS}}(o) \mathcal{J}^{\mathrm{bridge}}(o)}$ where $w^{\mathrm{Br-AS}}(i \to j) = \dfrac{1}{2} \exp\left[-\beta \sum_{k=2,3} \mathcal{V}_{\mathrm{bend}}\left(\theta^{j}_{\mathrm{bp},k}\right)\right]$ $\theta^{j}_{\mathrm{bp},k}\ (k = 2, 3)$: bond angle supplements $\theta_{\mathrm{bp},2}, \theta_{\mathrm{bp},3}$ at the branch point (Figure 3.1) in state j $w^{\mathrm{bridge}}(i \to j) = \dfrac{\exp\left[-\beta \sum_{k=1}^{6} \mathcal{V}_{\mathrm{tor}}\left(\phi^{j}_{s,k}\right)\right]}{\sum_{m=1}^{N_{\mathrm{sol}}} \exp\left[-\beta \sum_{k=1}^{6} \mathcal{V}_{\mathrm{tor}}\left(\phi^{j}_{m,k}\right)\right]} \exp\left[-\beta \sum_{k=1}^{5} \mathcal{V}_{\mathrm{bend}}\left(\theta^{j}_{\mathrm{br},k}\right)\right]$ $\theta^{j}_{\mathrm{br},k}\ (k = 1, \ldots, 5)$: bond angle k involved in the bridge leading to the j state $\phi^{j}_{m,k}\ (k = 1, \ldots, 6)$: torsion angle k of the m solution of the j state $\phi^{j}_{s,k}\ (k = 1, 2, \ldots, 6)$: torsion angle of the selected solution N_{sol}: total number of solutions to the bridging problem $\mathcal{J}^{\mathrm{Br-AS}} = \dfrac{l_q^2 \sin(\theta_{\mathrm{bp},2}) \sin(\theta_{\mathrm{bp},3})}{\left[1 - \cos^2(\theta_2) - \cos^2(\theta_{\mathrm{bp},2}) - \cos^2(\theta_{\mathrm{bp},3}) - 2\cos(\theta_2)\cos(\theta_{\mathrm{bp},2})\cos(\theta_{\mathrm{bp},3})\right]^{\frac{1}{2}}}$ $\mathcal{J}^{\mathrm{bridge}}$ as in end-bridging	32
Branch configura-tional bias (Figure 5.11b)	$\dfrac{\mathcal{J}^{\mathrm{Br-AS}}(n) w^{\mathrm{bp}}(o \to n) \prod_{l=2}^{N_{\mathrm{br}}} w^{\mathrm{br},l}(o \to n) \exp\left[-\beta \mathcal{V}^{(n)}_{\mathrm{tor}}(\phi_1, \phi_2, \ldots, \phi_l)\right]}{\mathcal{J}^{\mathrm{Br-AS}}(o) w^{\mathrm{bp}}(n \to o) \prod_{l=2}^{N_{\mathrm{br}}} w^{\mathrm{br},l}(n \to o) \exp\left[-\beta \mathcal{V}^{(o)}_{\mathrm{tor}}(\phi_1, \phi_2, \ldots, \phi_l)\right]},$ where $\mathcal{J}^{\mathrm{Br-AS}}$ as in branch point flip $w^{\mathrm{bp}} = \sum_{k=1}^{N_{\mathrm{trial}}} \exp\left[-\beta \mathcal{V}^{\mathrm{NB}}(\boldsymbol{r}_{\mathrm{bp},k})\right] \prod_{p=1,2} \exp\left[-\beta \mathcal{V}_{\mathrm{bend}}(\theta_{\mathrm{bp},p,k})\right]$ $w^{\mathrm{br},l} = \sum_{k=1}^{N_{\mathrm{trial}}} \exp\left[-\beta \mathcal{V}_{\mathrm{bend},l}(\theta_{\mathrm{br},k}) - \beta \mathcal{V}^{\mathrm{NB}}_{l}(\boldsymbol{r}_{\mathrm{br},k})\right]$	33

Continued

Table 5.1 *Continued*

Move	Ratio	Reference
SCB-CONROT (Figure 5.11c)	$\dfrac{w_{tor}^{bridge}(o \to n)\, \mathcal{J}^{bridge}(n) \exp\left[-\beta U^{NB(n)}(\boldsymbol{r}_{bridge})\right]}{w_{tor}^{bridge}(o \to n)\, \mathcal{J}^{bridge}(n) \exp\left[-\beta U^{NB(n)}(\boldsymbol{r}_{bridge})\right]}$ $\dfrac{\mathcal{J}^{bp}(n) w^{bp}(o \to n)\, w^{br}(o \to n) \exp\left[\beta \mathcal{V}_{tor}^{(n)}(\boldsymbol{\phi}_{bridge})\right]}{\mathcal{J}^{bp}(o) w^{bp}(n \to o)\, w^{br}(n \to o) \exp\left[\beta \mathcal{V}_{tor}^{(o)}(\boldsymbol{\phi}_{bridge})\right]}$, where $w_{tor}^{bridge} = \sum_{i=1}^{N_{sol}} \exp\left[\beta \mathcal{V}_{tor}(\boldsymbol{\phi}_{bridge,i})\right]$ w^{bp}, w^{br} as in branch configurational bias $\mathcal{J}^{bridge}$ as in end-bridging	33
d-CONROT, H-shaped branch rebridging (Figure 5.12)	As in double bridging	56
End-bridging for SCB polymers (Figure 5.14)	$\dfrac{w_{tor}^{bridge}(o \to n)\, N_{bridge}(o) \mathcal{J}^{bridge}(n) \exp\left[-\beta \mathcal{V}^{NB(n)}(\boldsymbol{r}_{bridge})\right]}{w_{tor}^{bridge}(n \to o)\, N_{bridge}(n) \mathcal{J}^{bridge}(o) \exp\left[-\beta \mathcal{V}^{NB(o)}(\boldsymbol{r}_{bridge})\right]}$ $\dfrac{\mathcal{J}^{Br-AS}(n) w^{bp}(o \to n)\, w^{br}(o \to n) \exp\left[\beta \mathcal{V}_{tor}^{(n)}(\phi_{bridge})\right]}{\mathcal{J}^{Br-AS}(o) w^{bp}(n \to o)\, w^{br}(n \to o) \exp\left[\beta \mathcal{V}_{tor}^{(o)}(\phi_{bridge})\right]}$, where w_{tor}^{bridge} as in SCB-CONROT N_{bridge} : number of candidate victims for the attacking end $\mathcal{J}^{bridge}$ as in end-bridging $\mathcal{J}^{Br-AS}$ as in branch point flip w^{bp}, w^{br} as in branch configurational bias	33
Double bridging for branched systems (Figure 5.15)	As in double bridging for linear chains	56

Intramolecular double rebridging for branched systems (Figure 5.16)	As in intramolecular double rebridging for linear chains	5
s_1-mer insertion in $f_1'N_c n_0 PT\boldsymbol{\mu}^*$ ensemble	$\frac{1}{N_1+1}(\sin\psi\sin\theta_1\cdots\sin\theta_{s_1-2})_{N_1+1}\frac{\beta f_1 V}{I_1^{\text{intra}}}\exp(-\beta\Delta\mathcal{V}),$ where $I_i^{\text{intra}}=8\pi^2\int \mathrm{d}\theta_1\cdots\mathrm{d}\theta_{s_i-2}\mathrm{d}\phi_2\cdots\mathrm{d}\phi_{s_i-2}\sin\theta_1\cdots\sin\theta_{s_i-2}\exp(-\beta\mathcal{V}_i^{\text{intra}})$ $f'_1=f_1\exp\left[-\beta(s_1+3)\mu_{\text{seg},i}^{\text{ex}}\right]\left(\frac{x_i}{x_{i+1}}\right)^{(s_i+3)}\left(\frac{I_{i+1}^{\text{intra}}}{I_i^{\text{intra}}}\right)^{(s_i+3)}$ N_1: number of oligomer molecules in the current configuration ψ: first Euler angle, θ_i: bond angle complements, ϕ_i: torsion angles s_1: chain length (number of mers) of oligomer molecules s_i: chain length (number of mers) of reference species i x_i: mole fraction of first reference species i x_{i+1}: mole fraction of second reference species whose chain length (number of mers) has been chosen to be equal to $s_{i+1}=s_i+1$	15
s_1-mer deletion in $f_1'N_c n_0 PT\boldsymbol{\mu}^*$ ensemble	$N_1\frac{1}{(\sin\psi\sin\theta_1\cdots\sin\theta_{s_1-2})_{N_1}}\frac{1}{\frac{\beta f_1 V}{I_1^{\text{intra}}}}\exp(-\beta\Delta\mathcal{V})$	15
Fusion of s_1-mer with s_k-mer (Figure 5.17b)	$N_1\frac{\sin\theta_{s_k-1}\cdots\sin\theta_{s_k+3}}{\sin\psi}\frac{1}{\frac{\beta f_1' V}{I_1^{\text{intra}}}}\exp\left\{\beta(\mu_{s_k+s_1+3}^*-\mu_{s_k}^*)-\beta\Delta\mathcal{V}\right\}$	15
Scission of s_k-mer to create a s_1-mer (Figure 5.17a)	$\frac{1}{N_1+1}\frac{\sin\psi_{N_1+1}}{\sin\theta_{s_k-4-s_1}\cdots\sin\theta_{s_k-s_1}}\frac{\beta f_1' V}{I_1^{\text{intra}}}\exp\left[\beta(\mu_{s_k-s_1-3}^*-\mu_{s_k}^*)-\beta\Delta\mathcal{V}\right]$	15

Note: In all cases, $p_{\text{acc}}(o\rightarrow n)=\min(1,\text{ratio})$ with ratio given for each move. Detailed derivations are provided in the respective references.

$$\mu = \frac{A(N_c + 1, V, T) - A(N_c, V, T)}{(N_c + 1) - N_c} = A(N_c + 1, V, T) - A(N_c, V, T). \quad (5.3.2)$$

But in the canonical ensemble, according to a purely classical formulation which assumes separable kinetic and potential energy contributions, the former depending on the atomic momenta and the latter depending on the atomic coordinates,

$$\begin{aligned} \exp[-\beta A(N_c, V, T)] &= Q(N_c, V, T) = \frac{1}{N_c!\left(\prod_{k=1}^{s} \Lambda_k^3\right)^{N_c}} Z(N_c, V, T) \\ Z(N_c, V, T) &= \int\int\ldots\int \exp[-\beta \mathcal{V}_{N_c}(\boldsymbol{r}_1, \boldsymbol{r}_2, \ldots, \boldsymbol{r}_{N_c})]\, \mathrm{d}\boldsymbol{r}_1\, \mathrm{d}\boldsymbol{r}_2 \ldots \mathrm{d}\boldsymbol{r}_{N_c} \end{aligned}, \quad (5.3.3)$$

where $\boldsymbol{r}_i$ $(i = 1, \ldots, N_c)$ is the $3s$-long vector of Cartesian coordinates of the atoms comprising molecule i, $\Lambda_k = \left(\frac{h^2}{2\pi m_k k_B T}\right)^{1/2}$ is the thermal wavelength of the kth atom in a molecule, the mass of that atom being denoted by m_k, and $\mathcal{V}_{N_c}(\boldsymbol{r}_1, \boldsymbol{r}_2, \ldots, \boldsymbol{r}_{N_c})$ is the total potential energy (intra- and intermolecular) of the system of N_c molecules. $Q(N_c, V, T)$ is the canonical partition function and $Z(N_c, V, T)$ is the configurational integral of the system of N_c molecules in volume V at temperature T. Using Eq. (5.3.3) within Eq. (5.3.2), we obtain that

$$\begin{aligned} \mu &= -k_B T \ln \frac{Q(N_c + 1, V, T)}{Q(N_c, V, T)} = \\ &= k_B T \ln \frac{(N_c + 1)\left(\prod_{k=1}^{s} \Lambda_k^3\right)}{V Z^{\text{intra}}} - k_B T \ln \left[\frac{1}{V Z^{\text{intra}}} \frac{Z(N_c + 1, V, T)}{Z(N_c, V, T)}\right] \end{aligned}, \quad (5.3.4)$$

where we have subtracted and added a term $-k_B T \ln Z^{\text{intra}}$, with Z^{intra} being the intramolecular configurational integral for a single molecule, say molecule $(N_c + 1)$, over all but its three translational degrees of freedom:

$$Z^{\text{intra}} = \int \exp\left[-\beta \mathcal{V}^{\text{intra}}(\boldsymbol{r}'_{N_c+1})\right] \mathrm{d}\boldsymbol{r}'_{N_c+1}. \quad (5.3.5)$$

The coordinates $\boldsymbol{r}'_{N_c+1}$ over which the integration takes place in Eq. (5.3.5) are by three fewer than the coordinates $\boldsymbol{r}_{N_c+1}$. They are obtained from $\boldsymbol{r}_{N_c+1}$ by subtracting from each atomic position vector the position vector of a reference atom in the molecule (e.g. the first atom or the molecular centre) and removing the position vector of the reference atom from the set. Clearly, $\mathcal{V}^{\text{intra}}$ can be written as a function of $\boldsymbol{r}'_{N_c+1}$, since it remains invariant to rigid translations of the molecule. On the other hand,

$$VZ^{\text{intra}} = \int \exp\left[-\beta\mathcal{V}^{\text{intra}}(\boldsymbol{r}_{N_c+1})\right] d\boldsymbol{r}_{N_c+1}, \tag{5.3.6}$$

where now the integration takes place over all internal positional degrees of freedom and over the position vector of the reference atom within the system volume V.

In the thermodynamic limit, we recognize the first term of Eq. (5.3.4) as the chemical potential of an *ideal gas of molecules* identical to those of the considered system, at the same temperature T and molecular density $\rho = N_c/V$ as the real system:

$$\mu^{\text{ig}}(\rho, T) = k_B T \ln \left[\frac{\rho \left(\prod_{k=1}^{s} \Lambda_k^3 \right)}{Z^{\text{intra}}} \right]. \tag{5.3.7}$$

We can thus define the *excess chemical potential* as

$$\mu^{\text{ex}}(\rho, T) = \mu(\rho, T) - \mu^{\text{ig}}(\rho, T). \tag{5.3.8}$$

Then, from Eqs. (5.3.4)–(5.3.8),

$$\begin{aligned}\mu^{\text{ex}} &= -k_B T \ln \left[\frac{1}{VZ^{\text{intra}}} \frac{Z(N_c+1, V, T)}{Z(N_c, V, T)} \right] \\ &= -k_B T \ln \left[\frac{\int\int\ldots\int\int \exp\left[-\beta\mathcal{V}_{N_c+1}(\boldsymbol{r}_1, \boldsymbol{r}_2, \ldots, \boldsymbol{r}_{N_c}, \boldsymbol{r}_{N_c+1})\right] d\boldsymbol{r}_1 d\boldsymbol{r}_2 \ldots d\boldsymbol{r}_{N_c} d\boldsymbol{r}_{N_c+1}}{\int\int\ldots\int \exp\left[-\beta\mathcal{V}_{N_c}(\boldsymbol{r}_1, \boldsymbol{r}_2, \ldots, \boldsymbol{r}_{N_c})\right] d\boldsymbol{r}_1 d\boldsymbol{r}_2 \ldots d\boldsymbol{r}_{N_c} \int \exp\left[-\beta\mathcal{V}^{\text{intra}}(\boldsymbol{r}_{N_c+1})\right] d\boldsymbol{r}_{N_c+1}} \right]\end{aligned} \tag{5.3.9}$$

Following Widom,[58,59] we now view the argument of the logarithm on the right-hand side of Eq. (5.3.9) as an ensemble average. Let us define the *test molecule energy* $\mathcal{V}_{\text{test}}$ by

$$\mathcal{V}_{\text{test}}(\boldsymbol{r}_{N_c+1}; \boldsymbol{r}_1, \boldsymbol{r}_2, \ldots, \boldsymbol{r}_{N_c}) = \mathcal{V}_{N_c+1}(\boldsymbol{r}_1, \boldsymbol{r}_2, \ldots, \boldsymbol{r}_{N_c}, \boldsymbol{r}_{N_c+1}) - \mathcal{V}_{N_c}(\boldsymbol{r}_1, \boldsymbol{r}_2, \ldots, \boldsymbol{r}_{N_c}) - \mathcal{V}^{\text{intra}}(\boldsymbol{r}_{N_c+1}). \tag{5.3.10}$$

Physically, $\mathcal{V}_{\text{test}}$ measures the increase in *intermolecular* energy of interaction when molecule $N_c + 1$ is inserted in configuration $\boldsymbol{r}_{N_c+1}$ within the system of N_c molecules in configuration $(\boldsymbol{r}_1, \boldsymbol{r}_2, \ldots, \boldsymbol{r}_{N_c})$. With the definition of Eq. (5.3.7), Eq. (5.3.9) becomes then

$$\begin{aligned}\mu^{\text{ex}} \\ = -k_B T \ln \left[\frac{\begin{array}{c}\int\int\ldots\int \exp\left[-\beta\mathcal{V}_{N_c}(\boldsymbol{r}_1, \boldsymbol{r}_2, \ldots, \boldsymbol{r}_{N_c})\right] d\boldsymbol{r}_1 d\boldsymbol{r}_2 \ldots d\boldsymbol{r}_{N_c} \\ \int \exp\left[-\beta\mathcal{V}^{\text{intra}}(\boldsymbol{r}_{N_c+1})\right] d\boldsymbol{r}_{N_c+1} \exp\left[-\beta\mathcal{V}_{\text{test}}(\boldsymbol{r}_{N_c+1}; \boldsymbol{r}_1, \boldsymbol{r}_2, \ldots, \boldsymbol{r}_{N_c})\right]\end{array}}{\int\int\ldots\int \exp\left[-\beta\mathcal{V}_{N_c}(\boldsymbol{r}_1, \boldsymbol{r}_2, \ldots, \boldsymbol{r}_{N_c})\right] d\boldsymbol{r}_1 d\boldsymbol{r}_2 \ldots d\boldsymbol{r}_{N_c} \int \exp\left[-\beta\mathcal{V}^{\text{intra}}(\boldsymbol{r}_{N_c+1})\right] d\boldsymbol{r}_{N_c+1}} \right]\end{aligned} \tag{5.3.11}$$

or

$$\mu^{\text{ex}} = -k_B T \ln \langle \exp [-\beta \mathcal{V}_{\text{test}} (\boldsymbol{r}_{N_c+1}; \boldsymbol{r}_1, \boldsymbol{r}_2, \ldots, \boldsymbol{r}_{N_c})] \rangle_{\text{Widom}}. \tag{5.3.12}$$

The average indicated by $\langle\rangle_{\text{Widom}}$ is taken over all possible insertions of the $(N_c + 1)$th molecule in the system of N_c molecules. In these insertions, the configurations of the N_c pre-existing molecules are distributed according to the canonical ensemble, subject to their potential energy $\mathcal{V}_{N_c}(\boldsymbol{r}_1, \boldsymbol{r}_2, \ldots, \boldsymbol{r}_{N_c})$. The inserted configurations (overall orientations, bond lengths, bond angles, and conformations) of the $(N_c + 1)$th molecule, on the other hand, are sampled according to the Boltzmann factor of its intramolecular potential energy. In other words, they are ideal gas configurations. The insertion takes place at all positions within the system of N_c molecules, which evolves without ever *feeling* the inserted molecule. For this reason, the inserted molecule is often referred to as the *ghost* molecule.

Widom's test molecule insertion method is very useful in the calculation of solubilities and fluid-phase equilibria.

5.3.2 Configurationally biased Widom insertion and excess segmental chemical potential

A serious drawback of Widom's test particle insertion method is that it becomes inefficient at high densities and low temperatures, because most random insertions lead to overlap with $\dfrac{\mathcal{V}_{\text{test}}}{k_B T} \to \infty$. This insertion problem is especially severe for bulky molecules, molecules of complex shape, and molecules that experience strong and specific interactions, such as hydrogen bonding, with their surroundings. To overcome this, one should resort to more efficient methods: for flexible molecules, a useful method is to implement insertions in a configurationally biased way, as proposed by Maginn et al.[60] for the case of long alkanes in zeolite pores. For bulky molecules, particle deletion (or inverse Widom[61]) methods are useful. For problems involving insertions of chain molecules, in particular, one can develop rigorous formulations for calculating not only the excess chain chemical potential μ^{ex} but also the excess segmental chemical potential $\mu^{\text{ex}}_{\text{seg}}$ from detailed atomistic simulations of a chain fluid. An advantage of having independently available methodologies for calculating μ^{ex} and $\mu^{\text{ex}}_{\text{seg}}$ is the capability next to test whether their values (obtained from well-converged atomistic simulations of a chain fluid in the liquid and vapour states) conform to the so-called *chain increment ansatz*, which is stated and discussed in Section 5.5.

Since the chain conformation in the liquid state is known to be close to unperturbed, liquid state chemical potentials are typically computed with respect to two reference systems: *ideal gas* and *continuous unperturbed chains*. The validity of the chain increment ansatz can then be examined for both reference states (subject to the same force field as chains in the considered phase). By definition, *ideal gas chains* retain all bonded and nonbonded intramolecular interactions along their contour but there exist no interchain interactions to be considered. *Continuous unperturbed chains*, on the other hand, are assumed to be unperturbed by nonlocal interactions; in the case of polyethylene, nonbonded interactions between segments separated by more than four bonds along the backbone are omitted in sampling continuous unperturbed chains.[62]

Let us consider[14] a chain of N_{test} segments or mers at infinite dilution within a fluid of N-mer chains at temperature T and pressure P. The N_{test}-mer chains are so far apart from each other that there are no interactions between N_{test}-mers. Let there be one N_{test}-mer chain for every N_c N-mer chain. The chemical potential (partial molecular Gibbs energy) of the test chain can be obtained from the following difference in the thermodynamic limit $N_c \to \infty$:

$$\begin{aligned}\beta\mu(T,P) &= -\ln Q_P(N_c,N,T,P;1,N_{\text{test}}) + \ln Q_P(N_c,N,T,P) \\ &= -\ln \frac{\int \exp(-\beta PV)\, Q(N_c,N,T,V;1,N_{\text{test}})\,\mathrm{d}V}{\int \exp(-\beta PV)\, Q(N_c,N,T,V)\,\mathrm{d}V} \\ &= -\ln \frac{\int \exp(-\beta PV)\, Q(N_c,N,T,V)\,\frac{Q(N_c,N,T,V;1,N_{\text{test}})}{Q(N_c,N,T,V)}\,\mathrm{d}V}{\int \exp(-\beta PV)\, Q(N_c,N,T,V)\,\mathrm{d}V} \\ &= -\ln \left\langle \frac{Q(N_c,N,T,V;1,N_{\text{test}})}{Q(N_c,N,T,V)} \right\rangle_{N_cPT}. \end{aligned} \qquad (5.3.13)$$

In the above equation:

- $Q_P(N_c,N,T,P;1,N_{\text{test}})$ stands for the isothermal–isobaric partition function of a system of N_c N-mer chains and one N_{test}-mer chain at T and P,
- $Q_P(N_c,N,T,P)$ is the isothermal–isobaric partition function for a system of N_c N-mer chains at T and P, and
- $Q(N_c,N,T,V;1,N_{\text{test}})$ and $Q(N_c,N,T,V)$ are the corresponding canonical partition functions at temperature T and volume V.

The subscript N_cPT on the right-hand side of Eq. (5.3.13) denotes that fluctuations in volume are sampled according to the isothermal–isobaric ensemble for N_c N-mer chains at T and P.

Let us symbolize by $\mathcal{V}(N_c,N,\boldsymbol{r};1,N_{\text{test}},\boldsymbol{r}_{\text{test}})$ the potential energy function of a system of N_c N-mer chains and one N_{test}-mer chain, and by $\mathcal{V}(N_c,N,\boldsymbol{r})$ the potential energy function of a system of N_c N-mer chains. The $3N_cN$-dimensional vector $\boldsymbol{r}$ comprises the Cartesian coordinates of all segments of all N-mer chains, while the $3N_{\text{test}}$-dimensional vector $\boldsymbol{r}_{\text{test}}$ consists of the coordinates of all segments of the N_{test}-mer chain, of which the first three correspond to overall translation of the chain. Let us also define the *test*-energy of the N_{test}-mer chain as the difference

$$\mathcal{V}_{\text{test}}(N_c,N,\boldsymbol{r};N_{\text{test}},\boldsymbol{r}_{\text{test}}) = \mathcal{V}(N_c,N,\boldsymbol{r};1,N_{\text{test}},\boldsymbol{r}_{\text{test}}) - \mathcal{V}(N_c,N,\boldsymbol{r}). \qquad (5.3.14)$$

Then, one can work out the ratio of canonical partition functions on the right-hand side of Eq. (5.3.13) to find (Problem 5.1):

$$\begin{aligned}
&\beta\mu^{\mathrm{ex}}(T,\rho)\\
&\equiv \beta\mu(T,\rho) - \beta\mu_{\mathrm{ref}}(T,\rho)\\
&= -\ln\left[\frac{1}{\langle V\rangle_{N_cPT}}\left\langle \frac{\int\int \exp\left[-\beta\mathcal{V}(N_c,N,\boldsymbol{r})\right]\exp\left[-\beta\mathcal{V}_{\mathrm{test}}(N_c,N,\boldsymbol{r};N_{\mathrm{test}},\boldsymbol{r}_{\mathrm{test}})\right]\mathrm{d}^{3N_cN}r\,\mathrm{d}^{3N_{\mathrm{test}}}r_{\mathrm{test}}}{\int\int \exp\left[-\beta\mathcal{V}(N_c,N,\boldsymbol{r})\right]\exp\left[-\beta\mathcal{V}_{\mathrm{ref}}(\boldsymbol{r}'_{\mathrm{test}})\right]\mathrm{d}^{3N_cN}r\,\mathrm{d}^{3N_{\mathrm{test}}-3}r'_{\mathrm{test}}}\right\rangle_{NPT}\right].\\
&= -\ln\left[\frac{1}{\langle V\rangle_{N_cPT}}\left\langle V\langle\exp\left[-\beta\mathcal{V}_{\mathrm{test}}\right]\rangle_{\mathrm{Widom}}\right\rangle_{N_cPT}\right] + \ln\left\langle\exp\left[-\beta\mathcal{V}_{\mathrm{ref}}\right]\right\rangle_{\mathrm{single\ chain}}
\end{aligned} \tag{5.3.15}$$

The prime on $\boldsymbol{r}'_{\mathrm{test}}$ should be understood in the sense of Eq. (5.3.5). In the first term on the right-hand side of Eq. (5.3.15), the inner (Widom) average refers to a N_{test}-mer chain being inserted with random position, orientation, and conformation within a system of N_c N-mer chains in volume V, whose configuration is distributed according to the canonical ensemble at temperature T. The outer average requires that fluctuations of V be sampled according to the isothermal–isobaric ensemble of N_c N-mer chains at temperature T and pressure P. The average in the second term of Eq. (5.3.15) is taken over all conformations of a single chain, distributed according to the Boltzmann factor of the energy $\mathcal{V}_{\mathrm{ref}}$.

We can rewrite Eq. (5.3.15) so as to show explicitly the role of inter- and intramolecular contributions to the two terms:

$$\begin{aligned}
&\beta\mu^{\mathrm{ex}}(T,\rho) =\\
&-\ln\left[\frac{1}{\langle V\rangle_{N_cPT}}\left\langle V\left\langle\exp\left[-\beta\mathcal{V}_{\mathrm{test}}^{\mathrm{intra}} - \beta\mathcal{V}_{\mathrm{test}}^{\mathrm{inter}}\right]\right\rangle_{\mathrm{Widom}}\right\rangle_{N_cPT}\right]\\
&+\ln\left\langle\exp\left[-\beta\mathcal{V}_{\mathrm{ref}}\right]\right\rangle_{\mathrm{single\ chain}}
\end{aligned} \tag{5.3.16}$$

Then, it is interesting to examine how Eq. (5.3.16) is written in terms of generalized coordinates. To this end, we consider a polymer model with bond lengths governed by harmonic potentials whose spring constants tend to infinity (flexible model in the limit of infinite stiffness; see Section 3.1.2). By conversion to a set of generalized coordinates (chain start positions $\{\boldsymbol{r}_0\}$, Euler angles $\{\boldsymbol{\psi}\}$, bond lengths $\{\boldsymbol{l}\}$, bond angle supplements $\{\boldsymbol{\theta}\}$, and torsion angles $\{\boldsymbol{\phi}\}$) the Boltzmann factors associated with the infinitely stiff bond length potentials give rise to δ-functions in the bond length. Thus, configurational averages reduce to averages over $\{\boldsymbol{r}_0,\boldsymbol{\psi},\boldsymbol{\theta},\boldsymbol{\phi}\}$ for each chain with the additional requirement of constant bond length, and the energies $\mathcal{V}_{\mathrm{test}}^{\mathrm{intra}}$ and $\mathcal{V}_{\mathrm{ref}}^{\mathrm{intra}}$ in Eq. (5.3.16) can be considered as devoid of any bond length contributions.[60] Then, from Eq. (5.3.16), an alternative expression for $\beta\mu^{\mathrm{ex}}$ is

$$\beta\mu^{\mathrm{ex}}(T,\rho) = \\ -\ln\left[\frac{1}{\langle V\rangle_{N_cPT}}\left\langle\int \exp\left[-\beta\mathcal{V}^{\mathrm{intra}}_{\mathrm{test}} - \beta\mathcal{V}^{\mathrm{inter}}_{\mathrm{test}}\right] \mathrm{d}^{2N_{\mathrm{test}}+1} q_{\mathrm{test}}\right\rangle_{N_cPT}\right] \\ +\ln\left\langle\int \exp\left[-\beta\mathcal{V}^{\mathrm{intra}}_{\mathrm{ref}}\right] \mathrm{d}^{2N_{\mathrm{test}}-2} q'_{\mathrm{test}}\right\rangle \tag{5.3.17}$$

where $\boldsymbol{q} \equiv \{\boldsymbol{r}_0, \boldsymbol{\psi}, \boldsymbol{\theta}, \boldsymbol{\phi}\}$, $\boldsymbol{q}' \equiv \{\boldsymbol{\psi}, \boldsymbol{\theta}, \boldsymbol{\phi}\}$, and the volume elements $\mathrm{d}^{2N_{\mathrm{test}}+1} q_{\mathrm{test}}$ and $\mathrm{d}^{2N_{\mathrm{test}}-2} q'_{\mathrm{test}}$ incorporate a Jacobian of transformation from Cartesian to internal or generalized coordinates[60] (see Section 3.1.1). The configurational integrals in Eq. (5.3.17) are identical to those computed by Maginn et al.[60] in the context of their study of low-occupancy sorption thermodynamics of alkanes in zeolites, the only difference being that, in the first term of Eq. (5.3.17), the test molecule insertion takes place not in a rigid porous framework but in a fluctuating matrix of N-mer chains, whose configurations are distributed according to the isothermal–isobaric N_cPT ensemble.

In analysing the thermodynamic properties of polymer fluids it is useful to define a *segmental chemical potential*, μ_{seg}. Under given T and P, μ_{seg} in a system of N-mer chains is defined as the change in total Gibbs energy brought about by replacing one N-mer chain in the system by one $(N+1)$-mer chain:

$$\begin{aligned}
\beta\mu_{\mathrm{seg}}(T,P) &= -\ln Q_{\mathrm{P}}(N_c-1,N,T,P;1,N+1) + \ln Q_{\mathrm{P}}(N_c-1,N,T,P;1,N) \\
&= -\ln Q_{\mathrm{P}}(N_c-1,N,T,P;1,N+1) + \ln\left[N_c Q_{\mathrm{P}}(N_c,N,T,P)\right] \\
&= -\ln\frac{Q_{\mathrm{P}}(N_c-1,N,T,P;1,N+1)}{N_c Q_{\mathrm{P}}(N_c,N,T,P)} \\
&= -\ln\frac{\int \exp(-\beta PV)\, Q(N_c-1,N,T,V;1,N+1)\,\mathrm{d}V}{N_c\int \exp(-\beta PV)\, Q(N_c,N,T,V)\,\mathrm{d}V} \\
&= -\ln\frac{\int \exp(-\beta PV)\, Q(N_c,N,T,V)\,\dfrac{Q(N_c-1,N,T,V;1,N+1)}{N_c Q(N_c,N,T,V)}\mathrm{d}V}{\int \exp(-\beta PV)\, Q(N_c,N,T,V)\,\mathrm{d}V} \\
&= -\ln\left\langle\frac{Q(N_c-1,N,T,V;1,N+1)}{N_c Q(N_c,N,T,V)}\right\rangle_{N_cPT} \\
&= -\ln\left\langle\frac{Q(N_c-1,N,T,V;1,N+1)}{Q(N_c,N,T,V;1,N)}\right\rangle_{N_cPT}.
\end{aligned} \tag{5.3.18}$$

From comparing Eqs. (5.3.13) and (5.3.18) one sees that

$$\beta\mu_{\mathrm{seg}} = \beta\mu(N_{\mathrm{test}} = N+1) - \beta\mu(N_{\mathrm{test}} = N) \tag{5.3.19}$$

with $\mu(N_{\mathrm{test}})$ defined by Eq. (5.3.13).

The ratio of canonical partition functions in Eq. (5.3.18) can be associated with a Widom process of appending an additional segment at the end of one of the chains of the original N-mer system. The appended segment experiences intramolecular interactions from its own chain and intermolecular interactions from the other chains. Note that there may be a change in intermolecular energy associated with converting what was previously an end segment of the augmented chain to an internal (penultimate) segment. The energy change associated with adding the segment to the N-mer system will be denoted as $\mathcal{V}^{\text{seg}}(N_c, N, \boldsymbol{r}; \boldsymbol{r}_{\text{seg}})$, with $\boldsymbol{r}_{\text{seg}}$ being the position vector of the added segment and $\boldsymbol{r}$ the $3N_cN$-dimensional vector describing the configuration of the N-mer system, as earlier. Then, it can be shown (Problem 5.2) that

$$\beta\mu^{\text{ex}}_{\text{seg}} = -\ln\left\langle \exp\left(-\beta\mathcal{V}^{\text{intra}}_{\text{seg}} - \beta\mathcal{V}^{\text{inter}}_{\text{seg}}\right)\right\rangle_{\text{incr.},\, N_cPT} + \ln\left\langle \exp\left(-\beta\mathcal{V}^{\text{intra}}_{\text{seg,ref}}\right)\right\rangle_{\text{incr., single chain}}, \tag{5.3.20}$$

where we have set $\mathcal{V}^{\text{intra}}_{\text{seg,ref}}(\boldsymbol{r}_{\text{seg}}; \boldsymbol{r}_{\text{test}}) = \mathcal{V}^{\text{intra}}_{\text{ref},N+1}(\boldsymbol{r}_{\text{test}}, \boldsymbol{r}_{\text{seg}}) - \mathcal{V}^{\text{intra}}_{\text{ref},N}(\boldsymbol{r}_{\text{test}})$. The first average on the right-hand side of Eq. (5.3.20) corresponds to a procedure of incrementing one chain by one segment within a system of N-mer chains governed by the probability density distribution of the isothermal–isobaric ensemble. The second average corresponds to incrementing by one segment a single N-mer reference chain whose conformations are distributed according to the Boltzmann factor of $\mathcal{V}^{\text{intra}}_{\text{ref}}$. As mentioned in conjunction with μ^{ex}, Eqs. (5.3.19) and (5.3.20) are readily recast in generalized coordinates, and bond length contributions to $\mathcal{V}^{\text{intra}}_{\text{seg}}$ and $\mathcal{V}^{\text{intra}}_{\text{seg,ref}}$ need not be included. Moreover, the integrals on the right-hand side of Eq. (5.3.20) (particularly the first one) are best computed through a configurational bias MC scheme.[60]

Equations (5.3.18) and (5.3.19) for the segmental chemical potential and Eq. (5.3.20) for $\mu^{\text{ex}}_{\text{seg}}$ are readily generalized to N-mer chains in a fluid phase of any composition (i.e. not necessarily a pure N-mer chain fluid as was used in the derivation above).

When ideal gas chains are used as a reference, the excess chemical potential computed from molecular simulations can be compared against that obtained from a macroscopic description of the thermodynamics of mixtures of N_{test}-mer and N-mer chains, such as that provided by an equation of state. From its definition,

$$\beta\mu^{\text{ex}} = \ln\phi^{\infty} + \ln\left(\frac{\beta P}{\rho}\right), \tag{5.3.21}$$

where ϕ^{∞} is the fugacity coefficient of N_{test}-mer chains at infinite dilution within an N-mer fluid at T and P, and ρ is the molecular density of the N-mer fluid under these conditions. ϕ^{∞} is calculable by standard methods[63,64] from a mixture equation of state, such as the Soave–Redlich–Kwong (SRK) or the Peng–Robinson (PR) cubic equations of state. As we will see in Section 5.5, estimates based on the SRK and PR equations for vapour–liquid equilibria of alkane mixtures in the range of molecular weights addressed in the MC simulations are very accurate and can be considered as experimental information for all practical purposes. Equation (5.3.21) can also be used to calculate μ^{ex} for a pure chain fluid, with ϕ^{∞} replaced by the fugacity coefficient ϕ, obtainable from the same equation of state for the pure fluid.

As we will also discuss in Section 5.5, an estimate of $\mu^{\text{ex}}_{\text{seg}}$ is obtainable from macroscopic thermodynamics through

$$\beta\mu^{\text{ex}}_{\text{seg}} = \beta\mu^{\text{ex}}(N+1; N) - \beta\mu^{\text{ex}}(N), \tag{5.3.22}$$

where $\mu^{ex}(N+1; N)$ is the excess chemical potential of an $(N+1)$-mer infinitely dilute in N-mer at T and P, and $\mu^{ex}(N)$ the excess chemical potential of pure N-mer, both estimated through Eq. (5.3.21).

5.3.3 Other methods for computing the chemical potential

Other methods developed to calculate the chemical potential through either insertions or deletions of test particles include

(a) the staged particle deletion[61] and direct particle deletion methods;[61,65–68]
(b) the configurational-bias method,[69] which grows the solute molecule in a bond-by-bond fashion;
(c) the extended ensemble MD method, where the coupling of the solute with the rest of the system is continuously changing, allowing the solute to escape from low-energy pockets and explore all of the accessible volume;[70]
(d) the minimum mapping (minmap) method,[71] which brings about local configurational changes in the vicinity of the inserted molecule to alleviate unfavourable (e.g. excluded volume) interactions;
(e) the free-energy perturbation, where a coupling parameter modulates the solute–matrix interactions and the solubility is obtained by thermodynamic integration (TI) over a number of simulations conducted at different values of the coupling parameter;[72] and
(f) the fast growth TI method,[73] based on Jarzynski's nonequilibrium work theorem,[74] which makes it possible to calculate the chemical potential via several independent TI runs.

5.3.4 General methods for calculating Helmholtz energy differences

As exemplified by the discussion of chemical potential calculations in Sections 5.3.1 and 5.3.2, obtaining free energies from simulations is generally not very easy. In this section we focus on calculating the Helmholtz energy difference $A_2 - A_1$ between two systems 1 and 2 characterized by the same temperature T, volume V, and number of molecules N, but governed by different potential energy functions. Methods for accomplishing this by simulation fall into two broad categories: density-of-states methods and work-based methods.[75]

In density-of-states methods, $A_2 - A_1$ is calculated as

$$A_2 - A_1 = -k_B T \ln\left(\frac{p_2}{p_1}\right), \quad (5.3.23)$$

where p_i is the probability of sampling system i in an equilibrium simulation, which is given complete freedom to explore both systems. In practice, bias techniques may be used to ensure efficient sampling of both systems.

Work-based methods, on the other hand, can be envisioned as applications of the Jarzynski equality,[74]

$$A_2 - A_1 = -k_B T \ln \left\langle \exp\left(-\frac{W_{1\to 2}}{k_B T}\right)\right\rangle, \tag{5.3.24}$$

where $W_{1\to 2}$ is the work associated with a process (not necessarily reversible) that transforms system 1 into system 2, usually by modulating a parameter in the Hamiltonian of the system, and the average is taken over an ensemble of such transformations, beginning from an equilibrated system 1. When the transformation of 1 into 2 occurs reversibly, Eq. (5.3.24) leads to the TI method.[73,76] This relies on a nonphysical path and a coupling parameter λ which varies continuously from 0 (representative of system 1) to 1 (representative of system 2), through which the two states are connected. Standard TI assumes that one obtains full equilibration along this λ-path which makes it possible to calculate the free-energy difference by integrating the derivative of the Hamiltonian (or of the potential energy of the system) with respect to λ:

$$\Delta A_{12} = A(\lambda = 1) - A(\lambda = 0) = \int_0^1 \left\langle \frac{\partial \mathcal{H}(\lambda)}{\partial \lambda}\right\rangle_\lambda d\lambda = \int_0^1 \left\langle \frac{\partial \mathcal{V}(\lambda)}{\partial \lambda}\right\rangle_\lambda d\lambda. \tag{5.3.25}$$

When, on the other hand, an instantaneous transformation of individual configurations of system 1 into the corresponding configurations of system 2 is undertaken, Eq. (5.3.24) is reduced to the *free-energy perturbation method* originally developed by Zwanzig,[77] which states that

$$A_2 - A_1 = -k_B T \ln \left\langle \exp\left(-\frac{\mathcal{V}_2 - \mathcal{V}_1}{k_B T}\right)\right\rangle_1, \tag{5.3.26}$$

where the ensemble average is taken over all configurations of system 1 distributed according to the canonical ensemble, and $\mathcal{V}_1$ and $\mathcal{V}_2$ denote the potential energy functions of systems 1 and 2, respectively, evaluated at each sampled configuration. For free-energy perturbation to work, the region of configuration space that shapes the properties of system 2 must be a subset of that of system 1. If this is not the case, staging schemes can be invoked.[75]

5.3.5 Umbrella sampling

In order to investigate free-energy barriers to rare event processes, such as bubble nucleation in a liquid, it is necessary to examine infrequently sampled, high-energy configurations. A technique that offers itself for accomplishing this is umbrella sampling MC, originally developed by Torrie and Valleau.[78] According to Eq. (5.3.26), one can measure the free-energy difference between the system of interest and a suitably chosen *reference* system, via the relationship

$$\begin{aligned}\beta(\Delta A) \equiv \beta(A - A_0) = -\ln\left(\frac{Q}{Q_0}\right) &= -\ln\left(\frac{\int \exp[-\beta(\mathcal{V} + \mathcal{V}')]\, d^N q}{\int \exp[-\beta \mathcal{V}]\, d^N q}\right)\\ &= -\ln\langle \exp(-\beta \mathcal{V}')\rangle_0,\end{aligned} \tag{5.3.27}$$

where $\mathcal{V}' = \mathcal{V} - \mathcal{V}_0$ is the difference in the total potential energy between the system of interest and the reference system, and $\langle \ldots \rangle_0$ denotes the configurational average on the N degrees of freedom $\boldsymbol{q}$ over a canonical ensemble of reference systems; as such it can in principle be estimated by a conventional MC simulation using Metropolis sampling. As pointed out above, a prerequisite for the free-energy perturbation scheme, Eq. (5.3.27), to work is that $\mathcal{V}'/(k_B T)$ be small over the region of $\boldsymbol{q}$-space that contributes to the equilibrium properties of the reference system. This becomes more and more difficult as the size of the reference system and the system of interest increases, since fluctuations away from the mean properties become more improbable as system size increases.

In umbrella sampling this problem is circumvented by biasing the reference system to sample rare fluctuations. The key idea in umbrella sampling is to assume that there exists an order parameter Φ that characterizes the evolution of a system under the application of an external field. Let us assume that we focus on states where this order parameter has a fixed mean value around which it fluctuates.[79] Then, in the isothermal–isobaric *NPT* ensemble, the probability density of finding the system in a state with a value Φ^* for the order parameter is

$$p(\Phi^*) = \frac{\int \exp\left[-\beta\left(\mathcal{V}(\boldsymbol{q}) + PV\right)\right] \delta\left(\Phi(V, q) - \Phi^*\right) \mathrm{d}V \mathrm{d}^N q}{\int \exp\left[-\beta\left(\mathcal{V}(\boldsymbol{q}) + PV\right)\right] \mathrm{d}V \mathrm{d}^N q}, \tag{5.3.28}$$

where $\mathcal{V}(q^N)$ is the total potential energy of interaction between molecules in the system and $\delta(x)$ is the Dirac delta function. The integration in Eq. (5.3.28) is carried out over all possible values of system volume V, and all possible values of molecular coordinates $\boldsymbol{q}$. The Gibbs energy is given by $G(N, P, T) = -k_B T \ln Q_P(N, P, T)$, where $Q_P(N, P, T)$ is the isothermal–isobaric partition function. Defining, by analogy, a Gibbs energy $G(N, P, T; \Phi^*)$ for the system subject to the constraint that the order parameter is kept equal to a given value Φ^* and comparing against Eq. (5.3.28), we obtain that

$$G(N, P, T; \Phi^*) = \text{const.} - k_B T \ln p(\Phi^*), \tag{5.3.29}$$

where const. is a numerical constant whose value does not depend on Φ^*. Usually we are only concerned with Gibbs energy differences:

$$G_2 - G_1 = G(N, P, T; \Phi_2^*) - G(N, P, T; \Phi_1^*) = -k_B T \ln \frac{p(\Phi_2^*)}{p(\Phi_1^*)}. \tag{5.3.30}$$

Thus, the task of calculating $G_2 - G_1$ reduces to the calculation of $p(\Phi^*)$ given in Eq. (5.3.28). We note that $p(\Phi^*)$ is an equilibrium property of the system. In practice, if we were to use conventional Metropolis MC to calculate this probability distribution, we would find that the probability of sampling high-energy configurations being governed by the Boltzmann factor would be small, resulting in poor statistics when Φ^* is far removed from the equilibrium average value of Φ. In order to sample high-energy configurations efficiently, we multiply and divide the integrands in Eq. (5.3.28) by a biasing function $\omega(\Phi)$:

$$p(\Phi^*) = \frac{\int \exp\left[-\beta\left(\mathcal{V}(\boldsymbol{q}) + PV\right)\right]\delta\left(\Phi(V,\boldsymbol{q}) - \Phi^*\right)\omega^{-1}\left(\Phi(V,\boldsymbol{q})\right)\omega\left(\Phi(V,\boldsymbol{q})\right)\mathrm{d}V\mathrm{d}^N q}{\int \exp\left[-\beta\left(\mathcal{V}(\boldsymbol{q}) + PV\right)\right]\omega^{-1}\left(\Phi(V,\boldsymbol{q})\right)\omega\left(\Phi(V,\boldsymbol{q})\right)\mathrm{d}V\mathrm{d}^N q}. \quad (5.3.31)$$

Introducing $\omega(\Phi)$ allows one to calculate equilibrium properties while sampling from a non-Boltzmann probability distribution $\omega(\Phi(V,\boldsymbol{q}))\exp\left[-\beta\left(V(\boldsymbol{q}) + PV\right)\right]$, no matter how unphysical the biased distribution might be. This is the technique of umbrella sampling. Essentially, one is simply biasing the simulation to explore infrequently sampled regions of phase space, and then correcting for the bias at the end. The correction that must be applied can be seen more clearly by expressing Eq. (5.3.31) in the manner

$$p(\Phi^*) = \frac{\left\langle \dfrac{\delta(\Phi - \Phi^*)}{\omega(\Phi)} \right\rangle_\omega}{\left\langle \dfrac{1}{\omega(\Phi)} \right\rangle_\omega}, \quad (5.3.32)$$

where the $\langle \ldots \rangle_\omega$ denotes average quantities as determined using the biased distribution. The biasing function is chosen so that sampling is performed within an order parameter window of width $\Delta\Phi$. Thus, a series of umbrella sampling runs are performed on several adjacent order parameter windows to obtain an entire free-energy curve. The size of the sampling windows need not be uniform, and the windows can overlap if desired. Within each window, the relative free-energy curve is calculated by MC simulation and then curve-fitted to a polynomial in Φ. The Gibbs energy curves from adjacent windows are then *linked* by adjusting the value of the constant in the curve fit.[79] One can think of this series of simulations as the computational equivalent of moving the system reversibly in order parameter space so that all of the important states in configuration space are well sampled.[80]

For example,[79] a relevant order parameter that characterizes a system as it changes from liquid to vapour is the reduced number density ρ which can be conveniently expressed as

$$\Phi \equiv \frac{\rho - \rho_{\text{liquid}}}{\rho_{\text{gas}} - \rho_{\text{liquid}}}, \quad (5.3.33)$$

where ρ_{liquid} and ρ_{gas} are the reduced number densities of the coexisting liquid and vapour phases at the saturation pressure $\left(\rho = \frac{N\sigma^3}{V}\right)$. On the other hand, the biasing function $\omega(\Phi)$ can be chosen to be of the following form within each sampling window:

$$\omega(\Phi) = \exp\left\{a - b(\Phi_{\max} - \Phi)^2\right\}. \quad (5.3.34)$$

The logarithm of such an expression is a parabola, which we demand to be concave down. $\Phi_{\max}$ is the location of the parabola's maximum, b adjusts the width of the sampling window, and a is a constant whose value is chosen for numerical convenience. Adjustment of b and $\Phi_{\max}$ biases the simulation to preferentially sample configurations within the desired order parameter window.

The result of umbrella sampling is the calculation of Gibbs energy curves at the given T and P conditions (or Helmholtz energy curves under given T and V conditions), for several (imposed) values of the order parameter Φ. That is, the technique allows for developing MC techniques for estimating the free energies of fluids by sampling on distributions designed for this purpose rather than based on the Boltzmann distribution.

5.3.6 Potential of mean force

As a special case of free-energy calculation, we discuss the extraction of *potentials of mean force* from atomistic simulations. This is strategically important in coarse-graining approaches for addressing long time- and length-scale properties. Generally, if we partition the configuration space of a system into a subset of (slowly evolving) degrees of freedom ($\boldsymbol{X}$) upon which we wish to focus and a complementary subset of (fast) degrees of freedom ($\boldsymbol{x}$), the potential energy function of the system will be of the form $\mathcal{V}(\boldsymbol{X}, \boldsymbol{x})$. The configurational integral of the system in the canonical ensemble will be

$$Z = \int \exp\left[-\beta \mathcal{V}(\boldsymbol{X}, \boldsymbol{x})\right] \mathrm{d}\boldsymbol{X}\, \mathrm{d}\boldsymbol{x}. \tag{5.3.35}$$

Z is a definite integral, where each position variable is integrated over the system volume V. The configuration-space probability density of the system is

$$\rho(\boldsymbol{X}, \boldsymbol{x}) = \frac{1}{Z} \exp\left[-\beta \mathcal{V}(\boldsymbol{X}, \boldsymbol{x})\right]. \tag{5.3.36}$$

We define the potential of mean force with respect to the variables $\boldsymbol{X}$, $W(\boldsymbol{X})$, as

$$W(\boldsymbol{X}) = -k_B T \ln \int \exp\left[-\beta \mathcal{V}(\boldsymbol{X}, \boldsymbol{x})\right] \mathrm{d}\boldsymbol{x} + C. \tag{5.3.37}$$

Note that, in the definition of $W(\boldsymbol{X})$, a *restricted partition function* is involved, in which the Boltzmann factor of the potential energy is integrated over all $\boldsymbol{x}$ at fixed $\boldsymbol{X}$. The fast degrees of freedom are allowed to sample all values accessible to them according to the equilibrium probability density dictated by the canonical ensemble, subject to given values of the slow degrees of freedom. The idea is that the fast degrees of freedom $\boldsymbol{x}$ quickly equilibrate at given $\boldsymbol{X}$. For dimensional consistency, the additive constant C should be of the form $k_{\mathrm{B}} T \ln C'$ with C' having dimensions of $\int \mathrm{d}\boldsymbol{x}$.

The marginal equilibrium probability density of the degrees of freedom $\boldsymbol{X}$ is

$$\begin{aligned} \rho_{\boldsymbol{X}}(\boldsymbol{X}) &= \frac{\int \exp\left[-\beta \mathcal{V}(\boldsymbol{X}, \boldsymbol{x})\right] \mathrm{d}\boldsymbol{x}}{\int \exp\left[-\beta \mathcal{V}(\boldsymbol{X}, \boldsymbol{x})\right] \mathrm{d}\boldsymbol{X}\, \mathrm{d}\boldsymbol{x}} = \frac{1}{Z} \int \exp\left[-\beta \mathcal{V}(\boldsymbol{X}, \boldsymbol{x})\right] \mathrm{d}\boldsymbol{x} = \\ &= \frac{1}{Z} \exp\left[-\beta \left(W(\boldsymbol{X}) - C\right)\right] = \frac{\exp\left[-\beta W(\boldsymbol{X})\right]}{\int \exp\left[-\beta W(\boldsymbol{X})\right] \mathrm{d}\boldsymbol{X}}. \end{aligned} \tag{5.3.38}$$

Now, if we take the gradient of $W(\boldsymbol{X})$ with respect to a position vector $\boldsymbol{X}_i$ comprising three of the configuration-space variables which are included in $\boldsymbol{X}$, we find that

$$
\begin{aligned}
-\nabla_{\boldsymbol{X}_i} W(\boldsymbol{X}) &= k_B T\beta \frac{\int [-\nabla_{\boldsymbol{X}_i}\mathcal{V}(\boldsymbol{X},\boldsymbol{x})]\exp[-\beta\mathcal{V}(\boldsymbol{X},\boldsymbol{x})]\,\mathrm{d}\boldsymbol{x}}{\int \exp[-\beta\mathcal{V}(\boldsymbol{X},\boldsymbol{x})]\,\mathrm{d}\boldsymbol{x}} \\
&= \frac{\int \boldsymbol{F}_i(\boldsymbol{X},\boldsymbol{x})\exp[-\beta\mathcal{V}(\boldsymbol{X},\boldsymbol{x})]\,\mathrm{d}\boldsymbol{x}}{\int \exp[-\beta\mathcal{V}(\boldsymbol{X},\boldsymbol{x})]\,\mathrm{d}\boldsymbol{x}} \\
&= \langle \boldsymbol{F}_i \rangle_{\boldsymbol{X}}
\end{aligned}
\tag{5.3.39}
$$

with $\langle \boldsymbol{F}_i \rangle_{\boldsymbol{X}}$ denoting the force on i averaged over all configurations at fixed $\boldsymbol{X}$ according to the equilibrium distribution of the canonical ensemble. This explains the name *potential of mean force* for $W(\boldsymbol{X})$.

The potential of mean force is a free energy, comprising both energetic and entropic contributions. To separate out its entropic contribution we can invoke the common thermodynamic relation $S = -\frac{\partial A}{\partial T}\big|_{N,V}$, obtaining that

$$
S(\boldsymbol{X}) = -\frac{\partial W(\boldsymbol{X})}{\partial T}. \tag{5.3.40}
$$

We also define a configurational internal energy $U(\boldsymbol{X})$ as a function of the subset of degrees of freedom $\boldsymbol{X}$ as

$$
U(\boldsymbol{X}) = \frac{\int \mathcal{V}(\boldsymbol{X},\boldsymbol{x})\exp[-\beta\mathcal{V}(\boldsymbol{X},\boldsymbol{x})]\,\mathrm{d}\boldsymbol{x}}{\int \exp[-\beta\mathcal{V}(\boldsymbol{X},\boldsymbol{x})]\,\mathrm{d}\boldsymbol{x}} = \langle \mathcal{V}(\boldsymbol{X},\boldsymbol{x})\rangle_{\boldsymbol{X}}. \tag{5.3.41}
$$

Then, it can be shown (Problem 5.3) that

$$
\mathcal{U}^{\mathrm{ex}} = \int U(\boldsymbol{X})\,\rho_{\boldsymbol{X}}(\boldsymbol{X})\,\mathrm{d}\boldsymbol{X}, \tag{5.3.42}
$$

i.e. the configurational internal energy of the system $\mathcal{U}^{\mathrm{ex}}$ is obtained as an average of $U(\boldsymbol{X})$ weighted by the equilibrium distribution of $\boldsymbol{X}$.

It can also be shown (Problem 5.4) that

$$
S(\boldsymbol{X}) = -k_B \frac{\int \rho(\boldsymbol{X},\boldsymbol{x})\ln\rho(\boldsymbol{X},\boldsymbol{x})\,\mathrm{d}\boldsymbol{x}}{\int \rho(\boldsymbol{X},\boldsymbol{x})\,\mathrm{d}\boldsymbol{x}} + k_B \ln\rho_{\boldsymbol{X}}(\boldsymbol{X}) - k_B\ln C' - k_B T\frac{\partial \ln C'}{\partial T}. \tag{5.3.43}
$$

Equation (5.3.43) is a generalization of the Gibbs entropy formula.

We have already seen examples of potentials of mean force. The reader will recognize the quantity $-k_B T \ln\Phi(\boldsymbol{r},N)$ with $\Phi(\boldsymbol{r},N)$ defined through Eq. (1.3.27) as the potential of mean force associated with the end-to-end vector $\boldsymbol{r}$ of a freely jointed chain, where an integration over the positions of all internal segments is considered; see also Problem 5.5. Another example of potential of mean force is the function $\hat{\mathcal{V}}(\phi_2,\phi_3,\ldots,\phi_{N-1})$ appearing in the definition of the single-chain conformational partition function, Eq. (1.2.2). This is a potential of mean force with respect to all torsional degrees of freedom of a single chain, where the vibrational (bond stretching, bond angle bending), overall translational and orientational degrees of freedom have been integrated out, as have the degrees of freedom of the environment surrounding the chain (solvent molecules, other chains).

Another example from liquid theory is the potential of mean force between two molecules at positions $\boldsymbol{r}_1$ and $\boldsymbol{r}_2$, within a simple fluid consisting of a total of N molecules in volume V at temperature T. This is worked out in Problem 5.6. For a thorough treatment of the potential of mean force in fluid mixtures, which employs *coupling parameters* λ_i to modulate pairwise intermolecular interactions, the reader is referred to the work of Kirkwood.[81]

5.3.7 Example calculations of potentials of mean force from simulation

The potential of mean force $W(\boldsymbol{X})$ defined by Eq. (5.3.37) provides a correct and consistent thermodynamic description of the degrees of freedom $\boldsymbol{X}$, having projected out $\boldsymbol{x}$. A correct *dynamical* description cast exclusively in terms of the subset $\boldsymbol{X}$ generally is much more difficult to formulate. It is provided by Zwanzig[82] and Mori's[83] projection operation formalism (see Chapter 7), and it is commonly approximated via Brownian dynamics, Langevin dynamics, dissipative particle dynamics, and related schemes. In all these schemes, the gradient of the potential of mean force must be used as the thermodynamic driving force, to ensure consistency with the full atomistic description.

Efficient MC schemes are valuable for calculating potentials of mean force, $W(\boldsymbol{X})$, with respect to selected coarse-grained degrees of freedom $\boldsymbol{X}$. An early example is provided by the work of Maginn et al. on sorption and diffusion of long linear alkanes in the zeolite silicalite-1.[60,84] As shown in Figures 5.18a and 5.18b, at the coarse-grained level, the zeolite was reduced to a set of intersecting curved lines representing the axes of straight and sinusoidal channels as computed directly from the force field exerted on a test particle inside the zeolite crystal. A sorbed alkane was coarse-grained into a linear wormlike object that is defined by the sequence of channel segments and intersections it goes through and by the positions (x_1, x_2) of its two ends along the channel segments in which they find themselves. The potential of mean force $W(x_1, x_2)$ was computed by CBMC integration. Figure 5.18c shows the typical shape of the potential of mean force. It refers to an n-octane molecule sorbed along a straight channel of silicalite in an extended conformation, in such a way as to straddle a channel intersection. A distinctive characteristic of $W(x_1, x_2)$ is a low-energy trough along which $x_1 - x_2$ is more or less constant. Moving along this trough corresponds to sliding the molecule along the straight channel. The favourable energy it experiences in the trough is due to dispersive interactions with the zeolite lattice. If one tries to extend the contour of the sorbed molecule (increasing $x_1 - x_2$), one encounters a steep repulsive wall, because one must work against intramolecular bending potentials. If one tries to compress the contour of the molecule (decreasing $x_1 - x_2$), the potential of mean force increases again, because of confinement, but not as steeply as in the case of stretching the molecule. Potentials of mean force, such as that of Figure 5.18c, were used within a Brownian dynamics/transition state theory formulation to track the intracrystalline diffusion of C_4-C_{20} molecules inside silicalite-1.[84] Predictions were confirmed a posteriori via quasielastic neutron scattering measurements.[85]

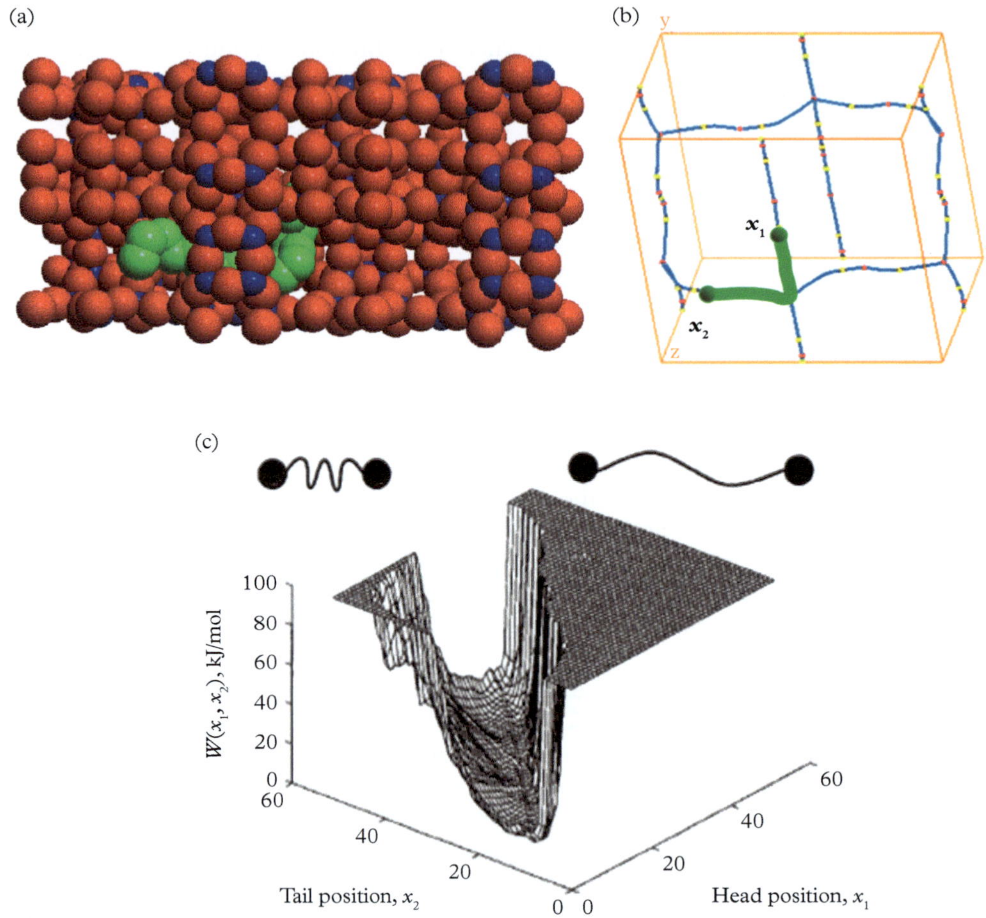

Figure 5.18 (a) Atomistic and (b) coarse-grained representation of an *n*-eicosane molecule sorbed in silicalite-1. At the coarse-grained level, the molecule is represented by the sequence of channel segments and intersections it goes through and by the positions (x_1, x_2) of its two ends along their respective channels. (c) Potential of mean force $W(x_1, x_2)$ for an *n*-octane molecule straddling an intersection, with its two ends in straight channel segments.[84] Large (small) values of $x_1 - x_2$ correspond to extended (compressed) conformations of the molecule inside the channels.

Another example of a direct calculation of a potential of mean force from atomistic simulation is exemplified by the calculation of the free energy of deformation (per chain) of a multichain polymer melt.[86] It involves a coarse-graining of a polymer chain at the level of the conformation tensor $\mathbf{c} = 3\left\langle \mathbf{rr} \right\rangle / \left\langle r^2 \right\rangle$ (dyadic product of the end-to-end vector with itself, averaged over all chains and reduced by one-third of its trace in the equilibrium undeformed state).

5.4 Example: equilibration of high-molecular weight linear, H-shaped, and short-chain branched polyethylene melts

5.4.1 Chain relaxation in the melt

A good measure of efficiency of any MC algorithm developed for the simulation of chain systems is provided by the rate at which the orientational autocorrelation function $\langle \boldsymbol{u}(0) \cdot \boldsymbol{u}(t) \rangle$ of the unit vector $\boldsymbol{u}$ directed along the chain end-to-end vector drops to 0 with CPU time. This is because $\langle \boldsymbol{u}(0) \cdot \boldsymbol{u}(t) \rangle$ describes how fast a system forgets its initial configuration. Figure 5.19 displays results for the evolution of $\langle \boldsymbol{u}(0) \cdot \boldsymbol{u}(t) \rangle$ with CPU time from simulations with a linear polyethylene (PE) melt consisting of ten C_{78} chains by four different simulation methods.[31] Starting from the bottom of Figure 5.19, the lowermost (solid) curve represents the results obtained by the end-bridging MC method using the following mix of moves: 6% reptations, 6% rotations, 6% flips, 32% CONROT, 49.5% end bridging, and 0.5% volume fluctuations; this is clearly the most effective of all. In the second curve (dotted), half of the end-bridging moves have been replaced by CB ones (keeping the type and attempt probabilities of all other moves in the mix unaltered), which seems to considerably degrade the performance of the algorithm. If, in fact, all end-bridging moves are replaced by CB (keeping always the mix of the rest of the moves unaltered), the curve obtained is the uppermost (dot-dashed) curve. CB, without any end-bridging moves, is clearly unable to equilibrate long-range conformational properties for this chain length after 20×10^3 CPU s. Also shown in the figure is the curve obtained from an NVE MD run using the same potentials and incorporating a Fixman potential to ensure that the configuration-space probability density sampled is identical to that of the MC. MD is more effective than the CB MC method; however, the figure shows it to be almost two orders of magnitude less effective than EB for this chain length.

Figure 5.20 shows the performance of the end-bridging MC method when applied to five long-chain PE melt systems (C_{78}, C_{156}, C_{200}, C_{400}, C_{500}), all characterized by uniform chain length distributions with reduced half-width $\Delta = l/X = 0.5$ (compare Eq. (5.1.42)). CPU times reported in this figure have been normalized for possible differences in the number of effective interaction sites (different n values). The figure shows quite convincingly that end-bridging MC can fully equilibrate all of these systems within modest CPU time. What is more surprising about the results shown in this figure is that the longer the mean chain length $\bar{N} \equiv X$ (compare Eq. (5.1.42)), the faster the equilibration of the system with the end-bridging method! As $\bar{N}$ increases, the $\langle \boldsymbol{u}(0) \cdot \boldsymbol{u}(t) \rangle$ curve decays faster. This is a unique feature of the end-bridging MC method. It sets it apart from all other simulation techniques currently available, whose performance in equilibrating long-length scale characteristics is always a strongly decreasing function of chain length.

From the Monte Carlo simulations one can get a wealth of results for the volumetric, structural, and conformational properties of polymers. For example, from the Fourier transform of the total pair correlation function between atomistic units one can calculate the static structure factor $S(k)$, which can be measured experimentally by X-ray diffraction. In Figure 5.21 we show the $S(k)$ curve for the C_{1000} PE system (solid line) together with experimental points (open circles) from X-ray diffraction measurements at $P = 1$ atm and $T = 430$K reported by Honnell et al.[87] The agreement between the two curves is excellent, even for the first peak in $S(k)$ (amorphous halo), which reflects

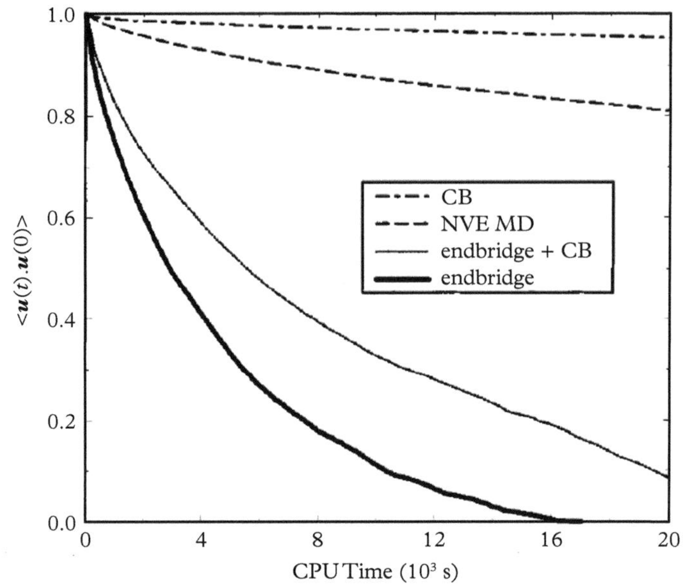

Figure 5.19 Decay of the chain end-to-end vector orientational autocorrelation function $\langle \boldsymbol{u}(0) \cdot \boldsymbol{u}(t) \rangle$ for a small melt model system consisting of ten linear C_{78} chains, as obtained with the N_cPT configurational bias MC method (dot-dashed line), the N_cVE MD method (dashed line), and the end-bridging MC method (solid line), allowing for a polydispersity index of 1.083. Also shown is the curve obtained with the end-bridging MC method where 50% of the attempted end-bridging moves have been replaced by configurational bias moves (dotted line) ($T = 450$K and $P = 1$ atm).

Reprinted (adapted) with permission from Mavrantzas, V.G., Boone, T.D., Zervopoulou, E., and Theodorou, D.N. *Macromolecules* **32**, 5072 (1999). Copyright 1999 American Chemical Society.

mainly intermolecular correlations. It is noted here that $S(k)$ is very similar for all high molar mass linear PE systems studied with connectivity-altering MC.[31]

5.4.2 Intrinsic molecular shape

According to the pioneering work of Šolc and Stockmayer,[88] the intrinsic shape of random walk chains at equilibrium in the eigenvector space of the instantaneous radius of gyration tensor is far more like a *cake of soap* than a sphere or ellipsoid. Theodorou and Suter[89] verified this by analysing a large number of conformations of unperturbed polypropylene chains with MC using a rotational isomeric state model. Mavrantzas and Theodorou,[86] on the other hand, reported that, under the application of a strong enough elongational flow field, short PE chains (e.g. C_{24}) are partially unravelled to intrinsically more elongated shapes; thus, instead of a cake of soap, the segment cloud takes an elongated, cigar-like shape. Guided by these studies, in Figure 5.22 are presented additional results revealing the intrinsic shape of the highly branched molecular architectures simulated in Ref. 56. For an explanation of symbols used to represent the various systems, see Ref. 56. Figure 5.22 shows the segment cloud (i.e. the density distribution of segments around the chain centre-of-mass)

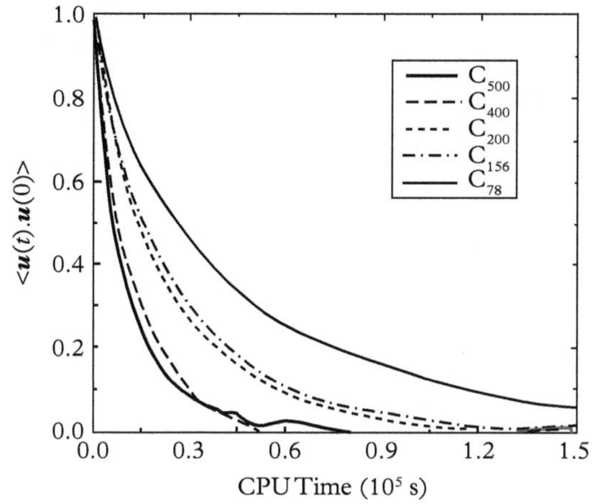

Figure 5.20 Decay of the chain end-to-end vector orientational autocorrelation function $\langle \boldsymbol{u}(0) \cdot \boldsymbol{u}(t) \rangle$ with CPU time, as obtained with the end-bridging MC algorithm, for various PE melt systems with mean chain length $\bar{N}$ ranging from C_{78} to C_{500} and $\Delta = 0.5$. CPU times have been corrected so that they correspond to the same number $n = 4000$ of mers in the periodic box for all systems, $T = 450$K and $P = 1$ atm.

Reprinted (adapted) with permission from Mavrantzas, V.G., Boone, T.D., Zervopoulou, E., and Theodorou, D.N. *Macromolecules* **32**, 5072 (1999). Copyright 1999 American Chemical Society.

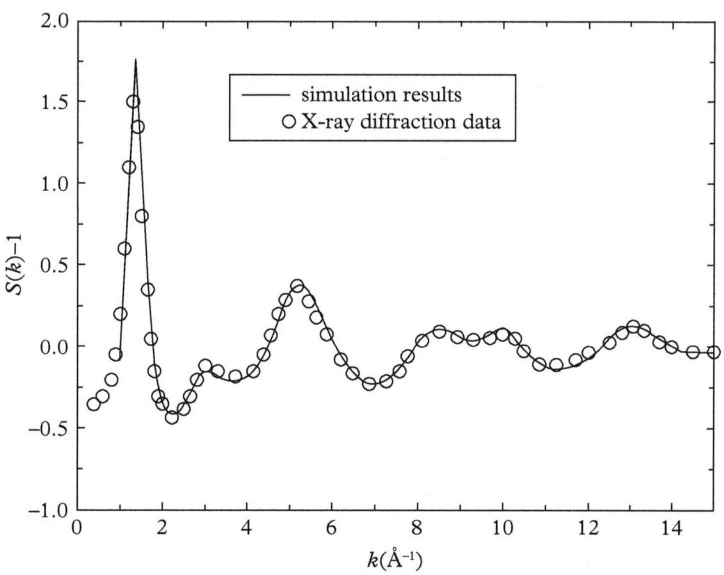

Figure 5.21 Simulated and experimental X-ray diffraction patterns of linear polyethylene at $P = 1$ atm. The simulation data are at $T = 450$K, whereas the experimental values at $T = 430$K.

Reprinted (Figure 6) with permission from Karayiannis, N.C., Mavrantzas, V.G., and Theodorou, D.N. *Phys. Rev. Lett.* **88**, 105, 503 (2002). Copyright 2002 by the American Physical Society.

Figure 5.22 Intrinsic molecular shapes in the L_1072, H_512_140, A_512_140, and SCB_48×16.5_47×6 PE melts and their comparison based on isosurface plots computed at equal values of the (number-averaged) monomer (or segment) number density in the frame of the principal axes (corresponding to the three eigenvectors) of the instantaneous radius of gyration tensor. For each system, the pictures from the bottom to the top represent the molecular shapes that include 95, 70, 50, 20, 10, and 5% of the total number of segments, respectively.

as computed in the space of eigenvectors of the instantaneous radius of gyration tensor, for various values of the monomer density. The notation for the simulated systems in Figure 5.22 is as follows: a) L_1072 denotes a linear PE melt containing on the average 1072 carbon atoms per chain; b) H_512_140 denotes an H-shaped PE melt containing on the average 512 carbon atoms along the main chain backbone and 140 carbon atoms along each one of its four arms; c) A_512_140 denotes a multi-arm (pom-pom) A3AA3 PE melt containing on the average 512 atoms along the main chain backbone and 140 carbon atoms along each one of its six arms; and d) SCB_48 × 16.5_47 × 6 denotes a short-chain-branched (SCB) PE melt containing on the average 792 carbon atoms along the main backbone and 48 short branches, each of which is 6 carbon atoms long. For all systems, the resulting shapes are consistent with those expected for random walks: they resemble a *cake of soap* instead of having a spherical or ellipsoidal shape. However, as the segment density increases (from the bottom to the top in Figure 5.22), we observe significant differences in the shapes of the core of the segment distributions among different polymer architectures. Compared, for example, to the linear analogue of the same total chain length (L_1072), the characteristic stem or neck between the two areas of high concentration exhibited at intermediate values of the segment

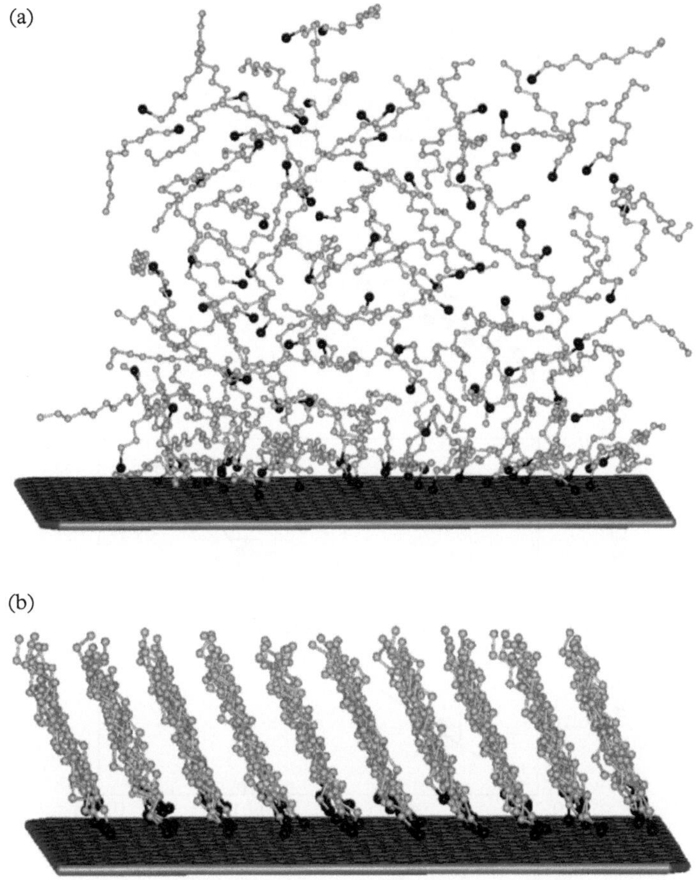

Figure 5.23 Typical atomistic view of the simulated 110-chain $CH_3(CH_2)_9$-SH/Au(111) system: (a) at the end of the energy minimization step, and (b) at the end of the thermal equilibration with the MC algorithm at $T = 300$K.

Reprinted (adapted) with permission from Alexiadis, O., Daoulas, K.Ch., and Mavrantzas, V.G. *J. Phys. Chem. B* **112**, 1198 (2008). Copyright 2008 American Chemical Society.

density is thinner in the H_512_140 PE melt, implying a higher number of segments near branch points than close to the centre of the crossbar. This is even more pronounced in the case of the A_512_140 polymer, for which the number of arms emanating from each of its branch points is 3 (compared to 2 in H-polymers).

5.4.3 Polymers at interfaces

Monte Carlo has also been used to simulate (a) systems of end-grafted chains or mixtures of end-grafted and free chains,[90–92] (b) polymers adsorbed on solid substrates,[93] and (c) highly ordered monolayers formed by the adsorption of short-chain molecules with a functional head group on a metallic substrate.[48,94] A typical example is the simulation of alkanethiol self-assembled monolayers on Au(111), Pt(111), and Ag(111) substrates by combining MC moves and parallel tempering. Figure 5.23 shows a typical system configuration obtained from these simulations. Monte Carlo

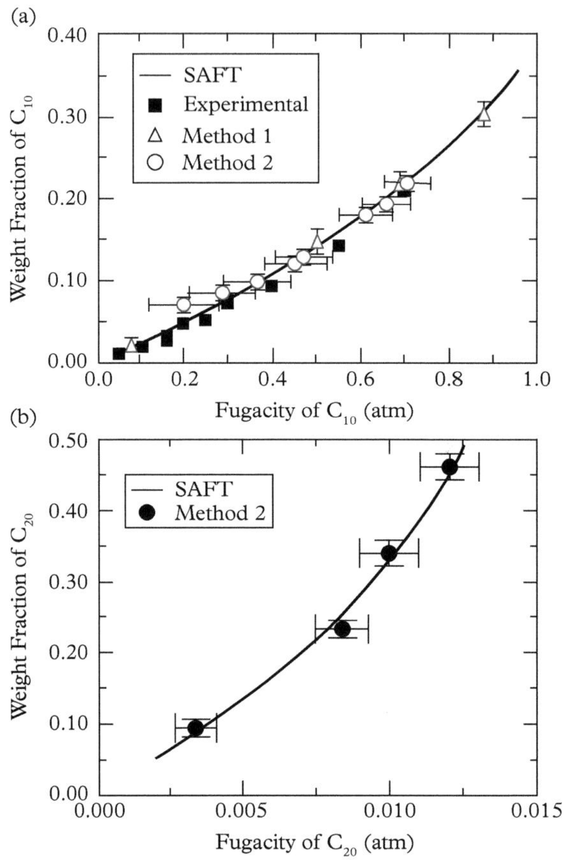

Figure 5.24 Sorption isotherm of (a) C_{10} at $T = 458K$ and (b) C_{20} at $T = 474K$ in a PE melt as predicted using the NERD force field based on either $f_1 N_c n_0 PT\boldsymbol{\mu}^*$ (Method 1) or $f_1' N_c n_0 PT\boldsymbol{\mu}^*$ (Method 2) MC simulations. In both cases, the melt polydispersity index was $I = 1.08$. The experimental data shown in (a) are from Ref. 98.

has also been used to simulate the glass transition temperature, T_g, of bulk and grafted amorphous polyethylene,[95] with remarkable accuracy with respect to experimental data.[96,97]

5.5 Example: solubilities and phase equilibria

5.5.1 Solubility of alkanes in a polyethylene melt

In Figure 5.24a are presented simulation results for the sorption isotherm of decane in PE (weight fraction as a function of fugacity). A C_{10}-C_{78} mixture was simulated in the $f_1 N_c nPT\boldsymbol{\mu}^*$ ensemble (Method 1) while for longer polymer chains, simulations in the $f_1' N_c nPT\boldsymbol{\mu}^*$ ensemble were employed (Method 2).[15] The experimental results are from the work of Bonner et al.[98] We also present comparisons of the simulation results with the predictions of the statistical associating fluid theory (SAFT) EoS.[99–102] As we can observe, both sets of simulation results and the EoS predictions are in very good agreement with the experiment for this mixture, also. The large error bars in the fugacity, calculated from f_1' at the end of the simulation with Method 2, are due to the small but finite error in the estimation of the reduced segmental chemical potential.

When the length of the alkane is increased even further, Method 1 fails completely to produce any results within reasonable simulation time. For example, it was seen that when C_{20} was attempted to be inserted in a C_{78} PE matrix, only one C_{20} molecule had been inserted in the matrix after about 3×10^5 CPU s. Obviously, the C_{20} molecules are too large to be inserted in PE, even when a configurational bias scheme is used. In remarkable contrast, it was seen that with Method 2, about 50 C_{20} molecules had been created in the system in about 1×10^5 CPU s, thus clearly demonstrating the value of Method 2. Only with the help of *scission* and *fusion* moves was it possible to achieve equilibrium for the concentration of C_{20} and longer alkane or oligomer molecules in PE. The ability of these moves to provide fast equilibration for the concentration of long oligomers in the polymer melt is impressive in this case and becomes even more so for longer penetrants. Thus, with the help of Method 2, the solubility of C_{20} in PE was also simulated. The sorption isotherm obtained, reported in Figure 5.24b, shows very convincingly that the simulation results are completely consistent with the SAFT EoS, exactly as was seen with the solubilities of the C_5 and C_{10} molecules. Unfortunately, a comparison of the simulation predictions with experimental results was not possible for this particular system, because no experimental data were found in the literature.

5.5.2 The chain increment ansatz

The chain increment ansatz,[103,104] mentioned in Section 5.3.2, can be used as a basis for calculating the chemical potentials of polymer chains in condensed phases from MC simulations using insertions of small oligomers and virtual augmentations of chains by one segment. The expression for the excess chemical potential μ^{ex} of a chain molecule in a fluid is given by Eq. (5.3.12) and more explicitly by Eq. (5.3.16). Application of these equations to polymeric systems would involve the insertion of a test polymer chain at random into the melt to determine its chemical potential. Although such insertions have been applied successfully to dense systems comprised of short lattice chains,[105–110] when one deals with longer chains the insertion of a test chain will almost always result in a situation where it overlaps with one or more of the chains in the system. Obtaining an accurate value of μ^{ex} will thus not be possible through random insertion of a test polymer chain. The chain increment ansatz could come to our help in this case.

According to the chain increment ansatz, the excess segmental chemical potential of an N-mer chain in a phase of given temperature, pressure, and composition becomes independent of N beyond a certain value of N. As a consequence, the excess chemical potential of a long N-mer chain becomes a linear function of its length N beyond a minimum length N_{small}:

$$\mu^{\text{ex}}(N) = \mu^{\text{ex}}(N_{\text{small}}) + (N - N_{\text{small}})\,\mu^{\text{ex}}_{\text{seg}}. \tag{5.5.1}$$

If Eq. (5.5.1) holds, with the definition of excess quantities based on an appropriate single-chain reference state (ideal gas chain or unperturbed single chain), it offers a way out of the computational difficulties associated with Widom insertion of long chains. $\mu^{\text{ex}}(N_{\text{small}})$ and $\mu^{\text{ex}}_{\text{seg}}$ can be computed at modest computational cost through (configurationally biased) Widom insertions of N_{small}-mer chains and virtual augmentations of N_{small}-mer chains by one segment within the phase of interest. From $\mu^{\text{ex}}\,(N_{\text{small}})$ and $\mu^{\text{ex}}_{\text{seg}}$, the excess chemical potential of long N-mer chains can then be estimated.

The chain increment ansatz has been tested for pure n-alkanes by Spyriouni et al.[104] with two realistic united-atom models, the Smit et al.[111] and the Dodd–Theodorou,[62] in both liquid and vapour phases. The simulations were carried out with MC in the N_cPT ensemble. The running average system pressure in each MC run was calculated using the molecular virial theorem.[112] In Figure 5.25, values of the excess chemical potential obtained from such calculations for $N = 6 - 16$ are compared against EoS predictions. The reference state in all cases is the ideal gas. μ^{ex} varies linearly with the carbon number of the inserted molecule for the entire range of chain lengths both from simulation and from the cubic EoS. A similar linear dependence is seen in the gas phase. This is a proof that the chain increment ansatz is valid in both phases for the system studied.

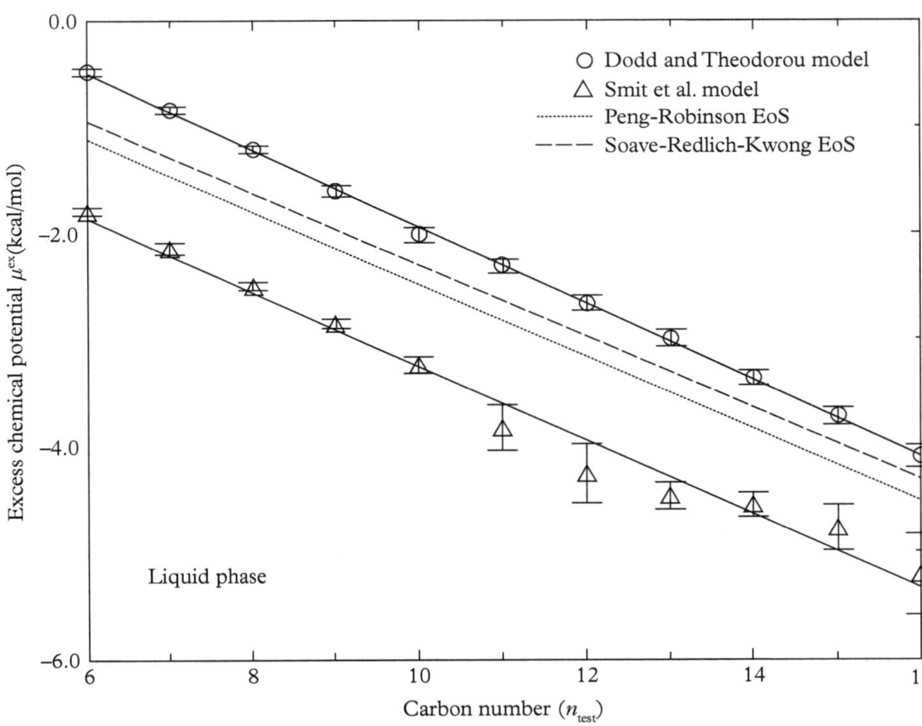

Figure 5.25 Excess chemical potential of chains of different lengths within a liquid phase of pure n-hexadecane (580 K, 50 atm), using an ideal gas chain of the same length as a reference state. EoS predictions are shown for comparison.

Reprinted (adapted) with permission from Spyriouni, T., Economou, I.G., and Theodorou, D.N. *Macromolecules* **30**, 4744 (1997). Copyright 1997 American Chemical Society.

5.5.3 The SPECS method

The simulation of phase equilibria for chain systems (SPECS) method[14] was developed for carrying out calculations of the phase equilibria of light solvent–heavy hydrocarbon mixtures, which are of great importance for the petrochemical, polymer, and oil industry. Traditional approaches for the calculation of such equilibria are based on cubic (e.g. Peng–Robinson and Soave–Redlich–Kwong[63] and higher order EoS, such as SAFT[99] and PHSC (perturbed hard sphere chain)).[113] Several adjustable parameters are required in order to obtain quantitative agreement with experimental data.

In SPECS, a low-density, solvent-rich phase I and a high-density, heavy hydrocarbon-rich phase II are simulated in the f_1N_2PT ensemble. Insertions and deletions of the solvent molecules are performed in both phases in order to attain the imposed value of the solvent fugacity. Equilibrium is achieved by equalizing the fugacities of the heavy hydrocarbon in each of the two phases, using an iterative scheme based on the Gibbs–Duhem equation. The excess chemical potential of the solvent is calculated by standard Widom insertions,[58] and its fugacity is obtained from the following expression (compare Eq. (5.3.21)):

$$\begin{aligned} \ln f_1 &= \beta\mu_1^{\mathrm{ex}} - \ln\left(\frac{\beta}{x_1\rho}\right) \\ \mu_i^{\mathrm{ex}} &\equiv \mu_i - \mu_{i,\mathrm{pure}}^{\mathrm{ig}}\left(T,\rho,x_i\right). \end{aligned} \tag{5.5.2}$$

The fugacity of the light solvent, as calculated from the low-density, solvent-rich phase I, is imposed on phase II, so that both phases are simulated under the same f_1, T, and P. The imposed value of f_1 is achieved by insertions and deletions of light solvent molecules. The condition of phase equilibrium is satisfied if the fugacities of component 2 are found equal in the two phases, since equal fugacities of component 1 are imposed and the simulation in both phases is performed under the same temperature and pressure. If, as expected, the fugacities of component 2 are not equal, both phases are simulated again in the f_1N_2PT ensemble using a revised common value of f_1, which is estimated through a Newton–Raphson scheme designed to equalize f_2 in the two phases. This scheme, based on the Gibbs–Duhem equation, has the simple form

$$\Delta\ln f_1 \equiv \left(1 - x_1^{\mathrm{I}}\right) x_2^{\mathrm{II}} \frac{\ln f_2^{\mathrm{II}} - \ln f_2^{\mathrm{I}}}{1 - x_1^{\mathrm{I}} - x_2^{\mathrm{II}}}. \tag{5.5.3}$$

The iterations proceed until $\Delta\ln f_1$ reaches a sufficiently small value.

The SPECS scheme has been used to simulate[114] several real mixtures of industrial interest, such as binary mixtures of ethane (C_2H_6) with *n*-hexadecane (*n*-$C_{16}H_{34}$) and with *n*-eicosane (*n*-$C_{20}H_{42}$), and mixtures of ethylene (C_2H_4) with *n*-$C_{20}H_{42}$ and *n*-tetracontane (*n*-$C_{40}H_{82}$) at high pressure.[14] Typical results are shown in Figure 5.26 displaying pressure-composition diagrams at constant temperature. Experimental data[115–117] and predictions from two widely used EoS, namely, SAFT and Peng–Robinson, with no adjustable binary interaction parameters are also shown for comparison.

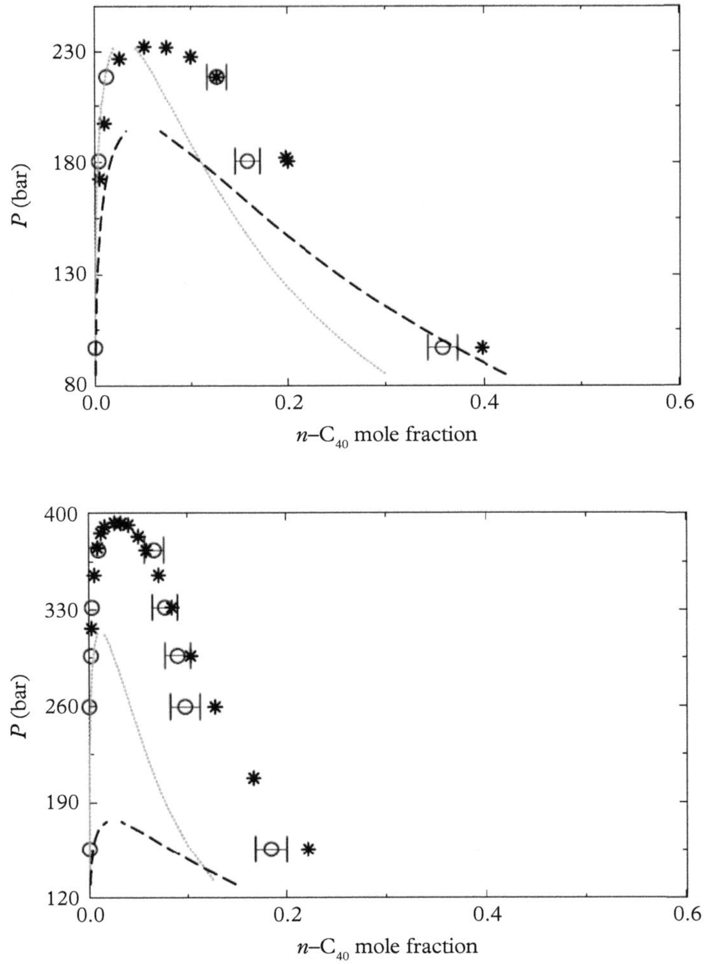

Figure 5.26 Pressure–heavy hydrocarbon mole fraction phase diagrams for binary C_2H_4-n-$C_{20}H_{42}$ (top) and C_2H_4-n-$C_{40}H_{82}$ (bottom) systems at $T = 398.15$K. Experimental data (stars),[117] SPECS (circles), SAFT (dotted lines), and Peng–Robinson EoS (dashed lines) predictions.

Mixtures with a high degree of asymmetry exhibit a lower critical solution temperature (LCST)[63] and become immiscible in the vicinity of the solvent critical point. This behaviour has been attributed to free volume effects, since the temperature increase results in higher density dissimilarities between the components. In Figure 5.27, experimental data and molecular simulation predictions are shown for C_2H_4-n-$C_{40}H_{82}$ at 296.4 bar. Simulation results are in very good agreement with the experimental data. SAFT captures this behaviour only qualitatively, while Peng–Robinson predicts complete miscibility under these conditions. Peng–Robinson was designed

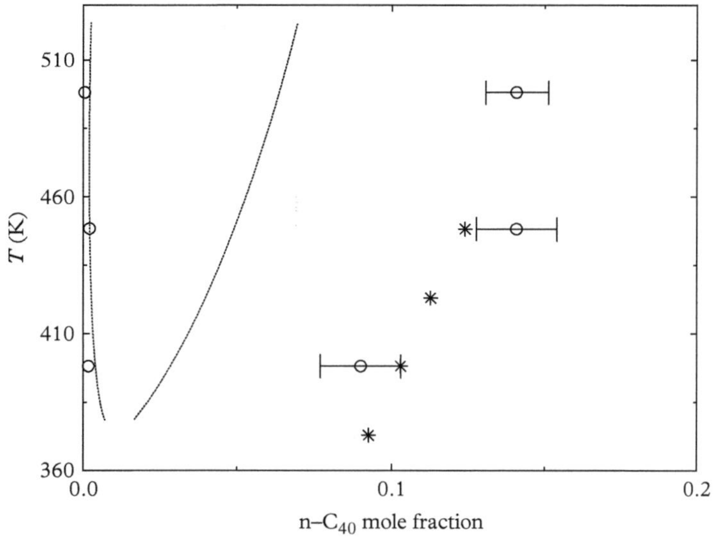

Figure 5.27 High-pressure phase equilibria of LCST-type for C_2H_4 -*n*-$C_{40}H_{82}$ mixture at $P = 296.4$bar. Experimental data (stars),[117] SPECS (circles), and SAFT (dotted lines) predictions.

Reprinted (Figure 3) with permission from Spyriouni, T., Economou, I.G., and Theodorou, D.N. *Phys. Rev. Lett.* **80**, 4466 (1998). Copyright 1998 by the American Physical Society.

for small and intermediate-size hydrocarbons, and such inaccuracy for asymmetric mixtures is expected.

Problems

Problem 5.1: Prove Eq. (5.3.15).

Problem 5.2: Prove Eq. (5.3.20).

Problem 5.3: Prove Eq. (5.3.42)

Problem 5.4: Prove Eq. (5.3.43).

Problem 5.5: Derive expressions for the potential of mean force $W(\boldsymbol{r}) - W(\mathbf{0})$ of the end-to-end vector of a freely jointed chain based on the Gaussian approximation and based on the inverse Langevin approximation. Derive also the internal energy, $U(\boldsymbol{r}) - U(\mathbf{0})$, and entropic, $T[S(\boldsymbol{r}) - S(\mathbf{0})]$, contributions to the potential of mean force.

Problem 5.6: Of particular importance in liquid theory is the potential of mean force between two molecules at positions $\boldsymbol{r}_1$ and $\boldsymbol{r}_2$, within a simple fluid consisting of a total of N molecules in volume V at temperature T. The number density of molecules is symbolized by $\rho = N/V$. The position vectors of the N molecules are denoted as $\boldsymbol{r}_1, \boldsymbol{r}_2, \ldots, \boldsymbol{r}_{N-1}, \boldsymbol{r}_N$, the potential energy function of the fluid as $\mathcal{V}(\boldsymbol{r}_1, \boldsymbol{r}_2, \ldots, \boldsymbol{r}_{N-1}, \boldsymbol{r}_N)$, and the configurational integral as $Z(N, V, T)$. Usually, the potential of mean force $W_N^{(2)}(\boldsymbol{r}_1, \boldsymbol{r}_2)$ is defined as[118]

$$W_N^{(2)}(\boldsymbol{r}_1, \boldsymbol{r}_2) = -k_B T \ln g_N^{(2)}(\boldsymbol{r}_1, \boldsymbol{r}_2), \tag{5.5.4}$$

where $g_N^{(2)}(\boldsymbol{r}_1, \boldsymbol{r}_2)$ is the pair distribution function in the fluid. Recalling the definition of the pair distribution function in the canonical ensemble,[118]

$$g_N^{(2)}(\boldsymbol{r}_1, \boldsymbol{r}_2) = \frac{N(N-1)}{\rho^2} \frac{\int \exp\left[-\beta \mathcal{V}(\boldsymbol{r}_1, \boldsymbol{r}_2, \ldots, \boldsymbol{r}_{N-1}, \boldsymbol{r}_N)\right] \mathrm{d}\boldsymbol{r}_3 \mathrm{d}\boldsymbol{r}_4 \ldots \mathrm{d}\boldsymbol{r}_{N-1} \mathrm{d}\boldsymbol{r}_N}{Z(N, V, T)}, \tag{5.5.5}$$

relate the potential of mean force between the two molecules, 1 and 2, to the mean force experienced by 1 when the positions of 1 and 2 are fixed in space and all other molecules are allowed to adopt all possible configurations accessible to them subject to the probability density of the canonical ensemble.

References

1. Metropolis, N., Rosenbluth, A.W., Rosenbluth, M.N., Teller, A.H., and Teller, E. *J. Chem. Phys.* **21**, 1087 (1953).
2. Allen, M.P., and Tildesley, D.J. *Computer Simulation of Liquids*. Clarendon Press, Oxford (1987).
3. Sadus, R.J. *Molecular Simulation of Fluids: Theory, Algorithms and Object-Orientation*. Elsevier, Amsterdam (1999).
4. Frenkel, D., and Smit, B. *Understanding Molecular Simulation: From Algorithms to Applications* (2nd ed.). Academic Press, San Diego, CA (2002).
5. Theodorou, D.N. *Variable-connectivity Monte Carlo algorithms for the atomistic simulation of long-chain polymer systems*. In P. Nielaba, M. Mareschal, G. Ciccotti (eds.), *Bridging Time Scales: Molecular Simulations for the Next Decade*. Springer Verlag, Berlin (2002).
6. Mavrantzas, V.G. *Monte Carlo simulation of chain molecules*. In Yip, S. (ed.), *The Encyclopedia of Modeling for Advanced Materials*. Marcel Dekker, New York (2004).
7. Theodorou, D.N. *Ind. Eng. Chem. Res.* **49**, 3047 (2010).
8. Mavrantzas, V.G. *Front. Phys.* **9**, 661367 (2021).
9. Briano, J.G., and Glandt, E.D. *J. Chem. Phys.* **80**, 3336 (1984).
10. Kofke, D.A., and Glandt, E.D. *Mol. Phys.* **64**, 1105 (1988).
11. Pant, P.V.K., and Theodorou, D.N. *Macromolecules* **28**, 7224 (1995).
12. McQuarrie, D.A. *Statistical Mechanics*. Harper and Row, New York (1976); Modell, M., and Reid, R.C. Thermodynamics and Its Applications (2nd ed.). Prentice Hall, Englewood Cliffs, NJ (1983).
13. Theodorou, D.N. *Molecular simulations of sorption and diffusion in amorphous polymers*. In Neogi, P. (ed.), *Diffusion in Polymers*, pp 67–142. Marcel Dekker, New York (1996).
14. Spyriouni, T., Economou, I.G., and Theodorou, D.N. *Phys. Rev. Lett.* **80**, 4466 (1998).
15. Zervopoulou, E., Mavrantzas, V.G., and Theodorou, D.N. *J. Chem. Phys.* **115**, 2860 (2001).
16. Johnson, J.K., Panagiotopoulos, A.Z., and Gubbins, K.E. *Mol. Phys.* **81**, 717 (1994).
17. Smith, W.R., and Tříska, B. *J.Chem. Phys.* **100**, 3019 (1994).
18. Turner, H.C., Brennan, J.K., Lísal, M., Smith, W.R., Johnson, J.K., and Gubbins, K.E. *Mol. Sim.* **34**, 119 (2008).

19. Turner, H.C., Brennan, J.K., Johnson, J.K., and Gubbins, K.E. *J. Chem. Phys.* **116**, 2138 (2002).
20. Newman, M.E.J., and Barkema, G.T. *Monte Carlo in Statistical Physics*. Clarendon Press, Oxford (1999).
21. Geyer, C.J. *Computing Science and Statistics*, Proceedings of the 23rd Symposium on the Interface, 156 (1991).
22. Geyer, C.J., and Thomson, E.A. *J. Am. Stat. Assoc.* **90**, 909 (1995).
23. Hukushima, K., Takayama, H., and Nemoto, K. *J. Phys. Soc. (Jap.)* **65**, 1604 (1996).
24. Hansmann, U.H.E. *Chem. Phys. Lett.* **281**, 140 (1997).
25. Deem, M.W., and Wu, M.G. *Mol. Phys.* **97**, 559 (1999).
26. Sugita, Y., and Okamoto, Y. *Chem. Phys. Lett.* **314**, 141 (1999).
27. Yan, Q., and de Pablo, J.J. *J. Chem. Phys.* **113**, 1276 (2000).
28. Bunker, A., and Dünweg, B. *Phys. Rev. E* **63**, 016701 (2000).
29. Doxastakis, M., Mavrantzas, V.G., and Theodorou, D.N. *J. Chem. Phys.* **115**, 11352 (2001).
30. Vacatello, M., Avitabile, G., Corradini, P., and Tuzi, A. *J. Chem. Phys.* **73**, 548 (1980).
31. Mavrantzas, V.G., Boone, T.D., Zervopoulou, E., and Theodorou, D.N. *Macromolecules* **32**, 5072 (1999).
32. Peristeras, L.D., Economou, I.G., and Theodorou, D.N. *Macromolecules* **38**, 386 (2005).
33. Ramos, J., Peristeras, L.D., and Theodorou, D.N. *Macromolecules* **40**, 9640 (2007).
34. Rosenbluth, M.N., and Rosenbluth, A.W. *J. Chem. Phys.* **23**, 356 (1955).
35. de Pablo, J.J., Laso, M., and Suter, U.W. *J. Chem. Phys.* **96**, 2395 (1992).
36. Siepmann, J.I., and Frenkel, D. *Mol. Phys.* **75**, 59 (1992).
37. Panagiotopoulos, A.Z. *Mol. Phys.* **9**, 1 (1992).
38. Lyubartsev, A.P., Martsinovskii, A.A., Shavkunov, S.V., and Vorontsov-Velyaminov, P.N. *J. Chem. Phys.* **96**, 1776 (1992).
39. Dodd, L.R., Boone, T.D., and Theodorou, D.N. *Mol. Phys.* **78**, 961 (1993).
40. Escobedo, F.A., and de Pablo, J.J. *J. Chem. Phys.* **102**, 2636 (1994).
41. Uhlherr, A. *Macromolecules* **33**, 1351 (2000).
42. Wick, C.D., and Siepmann, J.I. *Macromolecules* **33**, 7207 (2000).
43. Wang, F., and Landau, D.P. *Phys. Rev. Lett.* **86**, 2050 (2001).
44. Jain, T.S., and de Pablo, J.J. *J. Chem. Phys.* **116**, 7238 (2002).
45. Shell, M.S., Debenedetti, P.G., and Panagiotopoulos, A.Z. *J. Chem. Phys.* **119**, 9406 (2003).
46. Lal, M. *Mol. Phys.* **17**, 57 (1969); Madras, N., and Sokal, A.D. *J. Stat. Phys.* **50**, 109 (1988).
47. Santos, S., Suter, U.W., Müller, M., and Nievergelt, J. *J. Chem. Phys.* **114**, 9772 (2001).
48. Alexiadis, O., Daoulas, K.Ch., and Mavrantzas, V.G. *J. Phys. Chem. B* **112**, 1198 (2008).
49. Olaj, O.F., and Lantschbauer, W. *Makromol. Chem.- Rapid Commun.* **3**, 847 (1982).
50. Balijepalli, S., and Rutledge, G.C. *Macromol. Symp.* **133**, 71 (1998).
51. In 't Veld, P.J., and Rutledge, G.C. *Macromolecules* **36**, 7358 (2003).
52. Karayiannis, N.C., Mavrantzas, V.G., and Theodorou, D.N. *Phys. Rev. Lett.* **88**, 105503 (2002).
53. Karayiannis, N.C., Giannousaki, A.E., Mavrantzas, V.G., and Theodorou, D.N. *J. Chem. Phys.* **117**, 5465 (2002).
54. Karayiannis, N.C., Giannousaki, and V.G. Mavrantzas, *J. Chem. Phys.* **118**, 2451 (2003).
55. Wu, M.G., and Deem, M.W. *J. Chem. Phys.* **111**, 6625 (1999).
56. Baig, C., Alexiadis, O., and Mavrantzas, V.G. *Macromolecules* **43**, 986 (2010).
57. Gō, N., and Scheraga, H.A. *Macromolecules* **9**, 535 (1976).
58. Widom, B. *J. Chem. Phys.* **39**, 2802 (1963).
59. Widom, B. *J. Phys. Chem.* **86**, 869 (1982).
60. Maginn, E.J., Bell, A.T., and Theodorou, D.N. *J. Phys. Chem.* **99**, 2057 (1995).
61. Boulougouris, G.C., Economou, I.G., and Theodorou, D.N. *Mol. Phys.* **96**, 905 (1999).

62. Dodd, L.R., and Theodorou, D.N. *Adv. Pol. Sci.* **116**, 249 (1994).
63. Prausnitz, J.M., Lichtenthaler, R.N., and Azevedo, E.G. *Molecular Thermodynamics of Fluid-Phase Equilibria* (2nd ed.). Academic Press, New York (1986).
64. Walas, S.M. *Phase Equilibria in Chemical Engineering.* Butterworth Publishers, Stoneham, MA (1985).
65. Boulougouris, G.C. *Phase Equilibria of Water-Hydrocarbon Mixtures Using Molecular Simulations and Molecular Theory.* Ph.D. thesis, National Technical University of Athens (2001).
66. Boulougouris, G.C., Economou, I.G., and Theodorou, D.N. *J. Chem. Phys.* **115**, 8231 (2001).
67. Spyriouni, T., Boulougouris, G.C., and Theodorou, D.N. *Macromolecules* **42**, 1759 (2009).
68. De Angelis, M.G., Boulougouris, G.C., and Theodorou, D.N. *J. Phys. Chem. B* **114**, 6233 (2010).
69. Frenkel, D., Mooij, G.C.A., and Smit, B. *J. Phys.: Condens. Matter* **4**, 3053 (1992).
70. van der Vegt, N.F.A., and Briels, W.J. *J. Chem. Phys.* **109**, 7578 (1998).
71. Theodorou, D.N. *J. Chem. Phys.* **124**, 034109 (2006).
72. Swope, W.C., and Andersen, H.C. *J. Phys. Chem.* **88**, 6548 (1984).
73. Hess, B., Peter, C., Ozal, T., and van der Vegt, N.F. *Macromolecules* **41**, 2283 (2008).
74. Jarzynski, C. *Phys. Rev. Lett.* **78**, 2690 (1997).
75. Kofke, D. *Fluid Phase Equil.* **41**, 228 (2005).
76. Müller-Plathe, F. *Macromolecules* **24**, 6475 (1991).
77. Zwanzig, R.W. *J. Chem. Phys.* **22**, 1420 (1954).
78. Torrie, G.M., and Valleau, J.P. *Chem. Phys. Lett.* **28**, 578 (1974).
79. Shen, V.K., and Debenedetti, P.G. *J. Chem. Phys.* **111**, 3581 (1999).
80. Chandler, D. *Introduction to Modern Statistical Mechanics.* Oxford University Press, New York (1987).
81. Kirkwood, J.G. *J. Chem. Phys.* **3**, 300 (1935).
82. Zwanzig, R. *Phys. Rev.* **124**, 983 (1961).
83. Mori, H. *Prog. Theor. Rep.* **33**, 423 (1965).
84. Maginn, E.J., Bell, A.T., and Theodorou, D.N. *J. Phys. Chem.* **100**, 7155 (1996).
85. Jobic, H., and Theodorou, D.N. *Microporous Mesoporous Mater.* **102**, 21 (2007).
86. Mavrantzas, V.G., and Theodorou, D.N. *Macromolecules* **31**, 6310 (1998).
87. Honnell, K.J., McCoy, J.D., Curro, J.G., Schweitzer, K.S., Narten, A.H., and Habenschuss, A. *J. Chem. Phys.* **94**, 4659 (1991).
88. Šolc, K., and Stockmayer, W.H. *J. Chem. Phys.* **54**, 2756 (1971).
89. Theodorou, D.N., and Suter, U.W. *Macromolecules* **18**, 1206 (1985).
90. Daoulas, K.Ch., Terzis, A.F., and Mavrantzas, V.G. *J. Chem. Phys.* **116**, 11028 (2002).
91. Daoulas, K.Ch., Mavrantzas, V.G., and Photinos, D.J. *J. Chem. Phys.* **118**, 1521 (2003).
92. Daoulas, K.Ch., Terzis, A.F., and Mavrantzas, V.G. *Macromolecules* **36**, 6674 (2003).
93. Daoulas, K.Ch., Harmandaris, V.A., and Mavrantzas, V.G. *Macromolecules* **38**, 5780 (2005).
94. Alexiadis, O., Harmandaris, V.A., Mavrantzas, V.G., and Delle Site, L. *J. Phys. Chem. C* **111**, 6380 (2007).
95. Alexiadis, O., Mavrantzas, V.G., Khare, R., Beckers, J., and Baljon, A.R.C. *Macromolecules* **41**, 987 (2008).
96. Wunderlich, B. *J. Chem. Phys.* **37**, 1203 (1962).
97. Loufakis, K., and Wunderlich, B. *J. Phys. Chem.* **92**, 4205 (1988).
98. Bonner, D.C., Brockmeier, N.F., and Cheng, Y.I. *Ind. Eng. Chem. Process Des. Dev.* **4**, 437 (1974).
99. Chapman, W.G., Gubbins, K.E., Jackson, G., and Radosz, M. *Ind. Eng. Chem. Res.* **29**, 1709 (1990).
100. Banaszak, M., Chiew, Y.C., and Radosz, M. *Phys. Rev. E* **48**, 3760 (1993).
101. Banaszak, M. Chiew, Y.C., O'Lenick, R., and Radosz, M. *J. Chem. Phys.* **100**, 3803 (1994).
102. Huang, S.H., and Radosz, M. *Ind. Eng. Chem. Res.* **29**, 2284 (1990); Huang, S.H., and Radosz, M. *Ind. Eng. Chem. Res.* **30**, 1994 (1991).

103. Kumar, S.K., Szleifer, I., and Panagiotopoulos, A.J. *Phys. Rev. Lett.* **66**, 2935 (1991).
104. Spyriouni, T., Economou, I.G., and Theodorou, D.N. *Macromolecules* **30**, 4744 (1997).
105. Dickman, R., and Hall, C.K. *J. Chem. Phys.* **89**, 3168 (1988).
106. Honnell, K.G., and Hall, C.K. *J. Chem. Phys.* **90**, 1841 (1989).
107. Honnell, K.G., Dickman, R., and Hall, C.K. *J. Chem. Phys.* **87**, 664 (1987).
108. Sariban, A., Binder, K., and Heermann, D.W. *Phys. Rev. B* **35**, 6873 (1987).
109. Madden, W.G., Pesci, A.I., and Freed, K.F. *Macromolecules* **23**, 1181 (1990).
110. Siepmann, J.I. *Mol. Phys.* **70**, 1145 (1990).
111. Smit, B., Karaborni, S., and Siepmann, J.I. *J. Chem. Phys.* **102**, 2126 (1995).
112. Theodorou, D.N., Boone, T.D., Dodd, L.R., and Mansfield, K.F. *Makromol. Chem. Theory Simul.* **2**, 191 (1992).
113. Song, Y., Lambert, S.M., and Prausnitz, J.M. *Ind. Eng. Chem. Res.* **33**, 1047 (1994).
114. T. Spyriouni, I.G. Economou, and D.N. Theodorou, *Macromolecules* **31**, 1430 (1998).
115. De Goede, R., Peters, C.J., Van Der Kooi, H.J., and Lichtenthaler, R.N. *Phase Equil.* **50**, 305 (1989).
116. Peters, C.J., De Roo, J.L., and Lichtenthaler, R.N. *Fluid Phase Equil.* **34**, 287 (1987).
117. De Loos, T.W., Poot, W., and Lichtenthaler, R.N. *Fluid Phase Equil.* **88**, 855 (1984).
118. Hansen, J.-P., and McDonald, I.R. *Theory of Simple Liquids*. Academic Press, New York (1976).

6

Techniques for the Analysis and Simulation of Infrequent Events

6.1 Principles of infrequent event theory

6.1.1 Infrequent events, states, rate constants

Molecular dynamics (MD), discussed in Chapter 4, is an excellent simulation technique for elucidating the mechanisms of dynamical phenomena in materials and biomolecular systems and predicting their rates. The Achilles' heel of MD lies in its heavy computational requirements, as a result of which the time spans that can be simulated atomistically with commonly available resources are quite limited. Currently, MD runs undertaken on academic Beowulf clusters seldom exceed a few microseconds. The longest biomolecular simulations that have been conducted on specially developed hardware[1] covered time spans of milliseconds. Given that the properties of real-life systems may be governed by time scales of seconds to years, the need for modelling and simulation methods that can address longer length and time scales becomes imperative. While the challenge of large length scales can be addressed on parallel systems via domain decomposition, enabling the simulation of multimillion atom model systems, the challenge of long time scales appears more severe. The remainder of this book is largely devoted to strategies that can meet the challenge of long time and length scales and thus provide bridges to macroscopic descriptions. One category of strategies that have been developed for this purpose relies on the statistical mechanical theory of infrequent (rare) events.

In many cases, the dynamics of a physical system is slow because the system spends most of its time trapped within relatively narrow regions of configuration space and only infrequently jumps from one such region to another. We shall call these regions *states*. Generally, a state is a *basin* of low potential energy with respect to the (generalized) coordinates spanning configuration space, or of low free energy (potential of mean force) with respect to a set of coarse-grained variables (order parameters) providing a coarse-grained description of the system. The entire space of variables used to describe the configuration of a system is divided into states in a manner analogous to the division of a geographic map into water catchment areas. Transitions between states are infrequent events, in the sense that the mean waiting time for transition out of a state is long in comparison to the time needed for the system to establish a restricted equilibrium distribution among configurations in the state. If one represents a state in a coarse-grained sense

Multiscale Modelling of Polymers. Doros N. Theodorou and Vlasis G. Mavrantzas, Oxford University Press.
 DOI: 10.1093/oso/9780199226191.003.0006

by a point in configuration space or in order parameter space and connects all pairs of states between which a transition is possible, one obtains a *graph*, or *network of states*. Many phenomena of direct relevance to the performance of materials and biomolecular systems can be modelled as occurring through a succession of infrequent transitions in a network of states. Examples include: the diffusion of defects and impurities in metals and semiconductors;[2] the diffusion of gas molecules in amorphous polymers;[3] the diffusion of bulky hydrocarbons in nanoporous solids, such as zeolites;[4] phase transitions in molecular and atomic clusters;[5] structural relaxation in molecular and atomic glasses;[6] surface diffusion;[7] protein folding;[8] and chemical reactions.[9] Infrequent event theory has its roots in the absolute rate theory, or Transition State Theory, of Henry Eyring.[10] Its application to chemical reactions constitutes a cornerstone of chemical kinetics.[11]

If the waiting time that must elapse for the system to jump out of a state i into another state is long in comparison to the correlation time for the system to distribute itself among the configurations of state i and achieve local equilibrium in that state, then, by the time the system leaves state i, it will have lost all memory of how it entered that state. Thus, the dynamical process is reduced to a succession of independent jumps $i \to j$ between states. Once the system is in state i, the conditional probability per unit time that a jump $i \to j$ occurs under these conditions assumes a well-defined time-independent value, which is the *rate constant* $k_{i \to j}$. By focusing on the energy barrier between the states, the theory of infrequent events manages to obtain an estimate of $k_{i \to j}$ within reasonable computer time, no matter how low $k_{i \to j}$ may be.

6.1.2 Transition state theory, harmonic approximation, Kramers theory

Here we assume that states i and j are known and examine how we can estimate the transition rate constant $k_{i \to j}$, given the energy hypersurface and the atomic masses. Consider first a system whose dynamics is described in terms of a single degree of freedom x, associated with a mass m. To fix ideas, we may consider that x describes the position of a small spherical sorbate molecule along the curvilinear path taking it from one cavity into another in an amorphous glassy polymer matrix. The dynamics in x (translational motion along the path) will be governed by a potential of mean force, $U(x)$, obtained from the full potential energy hypersurface of the system by integrating (see Eq. (5.3.37) where the potential of mean force is symbolized by W) over all other degrees of freedom (e.g. vibrational modes of the surrounding polymer matrix), whose fluctuations are fast relative to (translational) progress along x. Assume[12] that $U(x)$ looks as shown in Figure 6.1, making it possible to define two states, A and B, as basins around the local minima at x_A and x_B. The states are separated by a point at $x = q$ where the potential of mean force exhibits a local maximum. We will assume that the barrier in potential of mean force at this point, $U(q) - U(x_A)$, is high relative to $k_B T$, so that the transition from A to B is an *infrequent event*. The degree of freedom x will be referred to as the *reaction coordinate* and the locus of points satisfying $x = q$ as the *dividing (hyper)surface*. In this one-dimensional example, the dividing surface reduces to a single point.

In simple transition state theory (TST), it is assumed that the degree of freedom x in the example of Figure 6.1 is equilibrated at all times with the remaining degrees of freedom of the system, which are envisioned as constituting a bath of temperature T. The probability distribution of x and of its

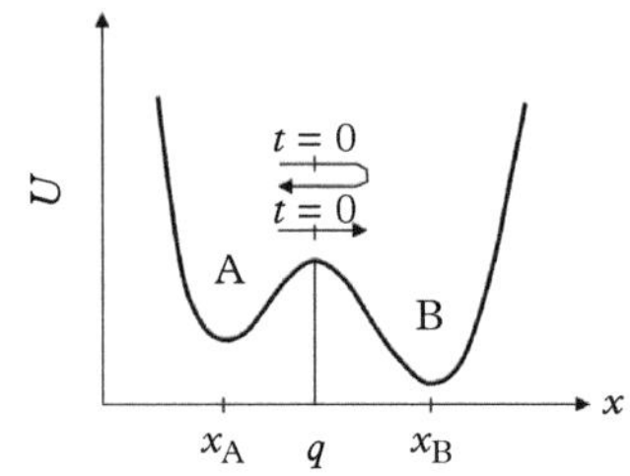

Figure 6.1 Potential of mean force $U(x)$ as a function of a single variable x assumed to evolve much more slowly than all other configurational variables of a system. The arrows show a successful and a recrossing trajectory. The time at which the system passes the barrier is defined as $t = 0$.

conjugate momentum p within state A is thus assumed to be governed by the canonical ensemble at temperature T:

$$\rho(x, p) = \frac{\exp\left[-\frac{1}{k_B T}\left(\frac{p^2}{2m} + U(x)\right)\right]}{\int_{-\infty}^{q}\int_{-\infty}^{\infty} \exp\left[-\frac{1}{k_B T}\left(\frac{p^2}{2m} + U(x)\right)\right] \mathrm{d}x\, \mathrm{d}p}. \tag{6.1.1}$$

The *TST approximation* postulates that, whenever the degree of freedom x moves from state A in the direction towards state B with momentum p that is sufficient for surmounting the potential energy barrier at q, a successful transition occurs from state A to state B. The probability of undergoing a transition from state A to state B within time $\mathrm{d}t$ with momentum between p and $p + \mathrm{d}p$ is equal[13] to $\rho(q, p)\left(\frac{p}{m}\mathrm{d}t\right)\mathrm{d}p$. The total probability of transition from A to B per unit time, i.e. the estimate of the transition rate constant according to simple TST, is then

$$k^{\mathrm{TST}} = \int_0^{\infty} \rho(q, p)\left(\frac{p}{m}\right)\mathrm{d}p = \frac{1}{m}\int_0^{\infty} \rho(q, p)\, p\, \mathrm{d}p. \tag{6.1.2}$$

Combining Eqs. (6.1.1) and (6.1.2) we obtain that

$$\begin{aligned} k^{\mathrm{TST}} &= \frac{\exp\left[-\frac{U(q)}{k_B T}\right]\int_0^{\infty} p \exp\left(-\frac{p^2}{2m\, k_B T}\right)\mathrm{d}p}{m\int_{-\infty}^{q} \exp\left[-\frac{U(x)}{k_B T}\right]\mathrm{d}x \int_{-\infty}^{\infty} \exp\left(-\frac{p^2}{2mk_B T}\right)\mathrm{d}p} \\ &= \frac{1}{2}\left(\frac{2k_B T}{\pi m}\right)^{1/2} \frac{\exp\left[-\frac{U(q)}{k_B T}\right]}{\int_{-\infty}^{q} \exp\left[-\frac{U(x)}{k_B T}\right]\mathrm{d}x}. \end{aligned} \tag{6.1.3}$$

The quantity $\left(\frac{2k_B T}{\pi m}\right)^{1/2} = \left\langle \frac{|p|}{m} \right\rangle$ is the mean magnitude of the velocity vector for motion along x at temperature T. The term $\exp\left[-\frac{U(q)}{k_B T}\right] \Big/ \int\limits_{-\infty}^{q} \exp\left[-\frac{U(x)}{k_B T}\right] dx$ is the conditional probability density in configuration space that the system finds itself on the dividing surface provided it is in state A. The prefactor 1/2 is associated with the fact that only half of the trajectories that reach the dividing surface with mean magnitude of the velocity vector equal to $\left\langle \frac{|p|}{m} \right\rangle$ will bring about a successful transition from A to B, namely those moving in the direction from A to B.

Equation (6.1.3) can also be rewritten as

$$k^{\mathrm{TST}} = \frac{1}{2}\left(\frac{2k_B T}{\pi m}\right)^{1/2} \frac{\exp\left[-\frac{U(q) - U(x_A)}{k_B T}\right]}{\int\limits_{-\infty}^{q} \exp\left[-\frac{U(x) - U(x_A)}{k_B T}\right] dx} = \nu(T) \exp\left[-\frac{E^*}{k_B T}\right], \tag{6.1.4}$$

where

$$\nu(T) = \frac{1}{2}\left(\frac{2k_B T}{\pi m}\right)^{1/2} \frac{1}{\int\limits_{-\infty}^{q} \exp\left[-\frac{U(x) - U(x_A)}{k_B T}\right] dx} \tag{6.1.5}$$

is a preexponential frequency factor and

$$E^* = U(q) - U(x_A) \tag{6.1.6}$$

is the activation (free) energy for the transition.

At low temperatures relative to the activation energy, $k_B T << E^*$, the main contribution to the integral in the denominator of Eq. (6.1.5) comes from the immediate vicinity of the minimum $x = x_A$. There, the potential of mean force can be approximated through a Taylor expansion truncated at the second-order term (*harmonic approximation*)

$$U(x) = U(x_A) + \frac{1}{2}\kappa_W (x - x_A)^2, \text{with } \kappa_W = \left.\frac{\partial^2 U}{\partial x^2}\right|_{x=x_A}. \tag{6.1.7}$$

Equation (6.1.5) gives

$$\nu(T) \approx \frac{1}{2}\left(\frac{2k_B T}{\pi m}\right)^{1/2} \frac{1}{\int\limits_{-\infty}^{+\infty} \exp\left[-\frac{1}{2k_B T}\kappa_W (x - x_A)^2\right] dx} = \frac{1}{2}\left(\frac{2k_B T}{\pi m}\right)^{1/2} \frac{1}{\left(\frac{2\pi k_B T}{\kappa_W}\right)^{1/2}}$$

$$= \frac{1}{2\pi}\left(\frac{\kappa_W}{m}\right)^{1/2} = \nu_0 \tag{6.1.8}$$

with ν_0 being the natural frequency of oscillation within the well of state A, a quantity independent of temperature. The upper limit of integration in the denominator has been replaced by $+\infty$, since the integrand is assumed to be 0 everywhere but in the immediate vicinity of x_A. Combining Eqs. (6.1.4) and (6.1.8) we obtain that

$$k_{\mathrm{HA}}^{\mathrm{TST}} = \nu_0 \exp\left[-\frac{E^*}{k_B T}\right]. \tag{6.1.9}$$

Equation (6.1.9) shows that the system attempts to overcome the barrier every time it moves in the direction from A to B in the course of its oscillatory motion within the well of state A, with frequency ν_0. Of these attempts, only a fraction $\exp\left[-E^*/(k_B T)\right]$ manage to reach the top of the barrier and effect a successful transition.

Simple TST ignores perturbations of the dynamical trajectory resulting from interactions between x and the remaining (fast) degrees of freedom of the system. As a result of such interactions, even x-trajectories from A to B that approach the separating surface with sufficient momentum to overcome the barrier may eventually be turned back and thermalize in state A (*recrossing trajectories*; see Figure 6.1). On the other hand, x-trajectories with insufficient momentum may gain momentum as a result of their interaction with the remaining degrees of freedom and ultimately overcome the barrier.

A more accurate theoretical treatment of the transition has been introduced by Kramers in a seminal paper.[14] In this treatment, the degree of freedom x is viewed as executing Brownian motion subject to the potential of mean force $U(x)$, its interactions with the remaining (fast) degrees of freedom of the system being described in terms of a frictional force $-\zeta\dot{x}$ and a random force $\boldsymbol{F}_r(t)$ satisfying the conditions $\langle \boldsymbol{F}_r(t)\rangle = 0$, $\langle \boldsymbol{F}_r(t_0)\cdot\boldsymbol{F}_r(t_0+t)\rangle = 2\zeta k_B T\delta(t)$ (fluctuation-dissipation theorem; compare Section 7.7.4). An analytical solution for the rate constant $k_{A\to B}^{\mathrm{Kramers}}$ can be obtained based on the solution of the Langevin equation for a Brownian particle subject to a parabolic potential[13] if a harmonic approximation is invoked for $U(x)$ in the vicinity of the well minimum $x = x_A$ (see Eq. (6.1.7)) and in the vicinity of the top of the barrier:

$$U(x) = U(q) - \frac{1}{2}\kappa_b (x-q)^2 \tag{6.1.10}$$

for x near q, with $\kappa_b = -\left.\dfrac{\partial^2 U}{\partial x^2}\right|_{x=q}$; clearly, κ_b is positive as defined. The resulting expression for $k_{A\to B}^{\mathrm{Kramers}}$ is[13]

$$k_{A\to B}^{\mathrm{Kramers}} = \nu_0\left(\sqrt{\xi^2+1}-\xi\right)\exp\left(-\frac{E^*}{k_B T}\right), \tag{6.1.11}$$

where, as in Eq. (6.1.8), $\nu_0 = \dfrac{1}{2\pi}\left(\dfrac{\kappa_w}{m}\right)^{1/2}$, while the dimensionless parameter

$$\xi = \frac{\zeta}{2\sqrt{\kappa_b m}} \tag{6.1.12}$$

depends on the friction coefficient ζ (strength of dynamical interaction between the transition coordinate x and the remaining degrees of freedom of the system), on the magnitude κ_b of the negative curvature of the potential of mean force at the barrier, and on the mass m associated with the transition coordinate. When $\xi << 1$ (low friction, large negative curvature of the potential at the barrier, large mass), the Kramers expression, Eq. (6.1.11), reduces to the simple TST expression in the harmonic approximation, Eq. (6.1.9).

Often, we define a *dynamical correction factor* or *transmission coefficient* f_d with reference to the transition rate constant estimate of simple TST:

$$f_{d,A \to B} = \frac{k_{A \to B}}{k_{A \to B}^{TST}}. \tag{6.1.13}$$

Then, according to Kramers theory,

$$f_{d,A \to B}^{Kramers} = \sqrt{\xi^2 + 1} - \xi < 1. \tag{6.1.14}$$

Simple TST provides an upper estimate of the transition rate constant. In reality, the transition rate constant is smaller, because of dynamical recrossing events, which are ignored by simple TST.

Of particular interest is the *high friction limit* $\xi >> 1$ (high friction, flat barrier, small mass). In this limit, $f_{d,A \to B}^{Kramers} = \sqrt{\xi^2 + 1} - \xi = \dfrac{1}{\sqrt{\xi^2 + 1} + \xi} \approx \dfrac{1}{2\xi} << 1$ and the Kramers theory estimate of the rate constant becomes

$$k_{A \to B}^{Kramers} = \frac{\nu_0}{2\xi} \exp\left(-\frac{E^*}{k_B T}\right) \text{ (high friction limit)}. \tag{6.1.15}$$

In the high friction limit, dynamical recrossing is very probable, as dynamical trajectories acquire the characteristics of a random walk on top of the barrier.[13] In this case, simple TST grossly overestimates the rate. Equation (6.1.15) is very useful in the analysis of infrequent events, such as conformational isomerizations or chemical reactions, taking place within highly viscous media, where the measured rate is often found to be inversely proportional to the viscosity.

6.1.3 Bennett–Chandler theory

A more modern formulation of infrequent event theory that is particularly well suited for estimating $k_{A \to B}$ from molecular simulations was developed through the works of Bennett[15] and Chandler.[16] Here we will briefly outline the basic elements of the Bennett–Chandler theory, borrowing from the very clear treatment of Voter and Doll.[7] Our discussion has a lot in common with Section 8.4 of Ref. 12.

We consider a canonical ensemble of systems characterized by the $U(x)$ profile of Figure 6.1 at equilibrium, at temperature T. Let N_A, N_B be the numbers of systems which find themselves in states A and B, respectively, at a particular time. We follow the dynamical trajectories of these systems. Over long times, $N_A(t)$ and $N_B(t)$ will fluctuate as a result of infrequent transitions between the two basins, but the total number of copies in the ensemble, $N_A(t) + N_B(t) = N$, which is very large, will be assumed fixed.

Under appropriate conditions of time-scale separation, which we will define, $N_A(t)$ and $N_B(t)$ evolve according to the first-order kinetic laws,

$$\begin{aligned} \dot{N}_A &= -k_{A\to B}N_A + k_{B\to A}N_B \\ \dot{N}_B &= k_{A\to B}N_A - k_{B\to A}N_B, \end{aligned} \tag{6.1.16}$$

where dots over symbols indicate time derivatives. Let $\overline{N}_A$ and $\overline{N}_B$ be the time-averaged (over very long times) values of N_A and N_B, respectively. Time-averaging Eq. (6.1.16), assuming constancy of the rate constants with time, leads to the condition of *microscopic reversibility*:

$$\overline{N}_A\, k_{A\to B} = \overline{N}_B\, k_{B\to A}. \tag{6.1.17}$$

We define the instantaneous deviations of $N_A(t)$ and $N_B(t)$ from their average values,

$$\begin{aligned} \delta N_A &= N_A - \overline{N}_A \\ \delta N_B &= N_B - \overline{N}_B. \end{aligned} \tag{6.1.18}$$

Clearly, the constancy of $N_A(t) + N_B(t)$ implies that

$$\delta N_A + \delta N_B = 0. \tag{6.1.19}$$

By virtue of Eqs. (6.1.16) and (6.1.17), δN_A satisfies the kinetic equation

$$\delta\dot{N}_A = -k_{A\to B}\delta N_A + k_{B\to A}\delta N_B. \tag{6.1.20}$$

Combining Eqs. (6.1.19) and (6.1.20),

$$\delta\dot{N}_A = -\left(k_{A\to B} + k_{B\to A}\right)\delta N_A = -k_{\text{eff}}\,\delta N_A \tag{6.1.21}$$

with $k_{\text{eff}} = k_{A\to B} + k_{B\to A}$. Multiplying Eq. (6.1.21) at time t with the deviation of the population of state A from its equilibrium value at time 0, $\delta N_A(0)$, and taking an equilibrium average over multiple time origins along the ensemble trajectories, one obtains that

$$\left\langle \delta N_A(0)\delta\dot{N}_A(t)\right\rangle = -k_{\text{eff}}\left\langle \delta N_A(0)\delta N_A(t)\right\rangle. \tag{6.1.22}$$

Note that, by virtue of the time-reversibility of the motion and the equilibrium conditions prevailing, $\left\langle \delta N_A(0)\delta\dot{N}_A(t)\right\rangle = -\left\langle \delta N_A(0)\delta\dot{N}_A(-t)\right\rangle = -\left\langle \delta N_A(t)\delta\dot{N}_A(0)\right\rangle$.

From Eqs. (6.1.19) and (6.1.22), one obtains the exact relation

$$k_{\text{eff}} \equiv k_{A\to B} + k_{B\to A} = \frac{\left\langle \delta\dot{N}_A(0)\delta N_A(t)\right\rangle}{\left\langle \delta N_A(0)\delta N_A(t)\right\rangle} = -\frac{\left\langle \delta\dot{N}_A(0)\delta N_B(t)\right\rangle}{\left\langle \delta N_A(0)\delta N_A(t)\right\rangle}. \tag{6.1.23}$$

The denominator of Eq. (6.1.23) is a slowly varying function of time, relative to the numerator; the rate of change in δN_A will fluctuate between positive and negative values as a result of transitions occurring in the various replicas of the ensemble, and therefore decorrelate from the initial value of δN_A much more rapidly than δN_A itself. It is therefore reasonable to introduce the approximation

$$\langle \delta N_A(0)\delta N_A(t)\rangle \cong \langle \delta N_A(0)\delta N_A(0)\rangle \tag{6.1.24}$$

in the denominator, which leads to

$$k_{\text{eff}} \equiv k_{A\to B} + k_{B\to A} = \frac{\left\langle \delta \dot{N}_A(0)\delta N_A(t)\right\rangle}{\langle \delta N_A(0)\delta N_A(0)\rangle} = -\frac{\left\langle \delta \dot{N}_A(0)\delta N_B(t)\right\rangle}{\langle \delta N_A(0)\delta N_A(0)\rangle}. \tag{6.1.25}$$

Without loss of generality, now, instead of an ensemble of systems, we consider a single system evolving along its equilibrium trajectory; in other words, we set $N = 1$. Invoking the Heaviside step function $H(x)$, with $H(x) = 1$ for $x > 0$, $H(x) = 0$ for $x < 0$, and $H(x) = 1/2$ for $x = 0$, we can write then

$$\begin{aligned} N_A(t) &= H(q - x(t)) \\ N_B(t) &= H(x(t) - q) \end{aligned} \tag{6.1.26}$$

and

$$\begin{aligned} \overline{N}_A &= P_A^{\text{eq}} \\ \overline{N}_B &= P_B^{\text{eq}}, \end{aligned} \tag{6.1.27}$$

where P_A^{eq} and $P_B^{\text{eq}} = 1 - P_A^{\text{eq}}$ are the fractions of times the system spends in states A and B, respectively, at equilibrium, i.e. the equilibrium probabilities of states A and B. By definition, then,

$$\begin{aligned} \delta N_A(t) &= H(q - x(t)) - P_A^{\text{eq}} \\ \delta N_B(t) &= H(x(t) - q) - P_B^{\text{eq}} \end{aligned} \tag{6.1.28}$$

and

$$\delta \dot{N}_A(0) = -\dot{x}(0)\,\delta(q - x(0)), \tag{6.1.29}$$

bearing in mind that the derivative of the Heaviside function is the Dirac delta function, $\delta(x)$. The autocorrelation function in the denominator of Eq. (6.1.25) becomes

$$\begin{aligned} \langle \delta N_A(0)\delta N_A(0)\rangle &= \left\langle [H(q - x(0)) - P_A^{\text{eq}}]^2\right\rangle \\ &= \left\langle [H(q - x(0))]^2\right\rangle - 2P_A^{\text{eq}}\langle H(q - x(0))\rangle + (P_A^{\text{eq}})^2 \\ &= \langle H(q - x(0))\rangle - 2P_A^{\text{eq}}\langle H(q - x(0))\rangle + (P_A^{\text{eq}})^2 \\ &= P_A^{\text{eq}} - 2P_A^{\text{eq}}P_A^{\text{eq}} + (P_A^{\text{eq}})^2 = P_A^{\text{eq}} - (P_A^{\text{eq}})^2 \\ &= P_A^{\text{eq}}(1 - P_A^{\text{eq}}) = P_A^{\text{eq}}P_B^{\text{eq}} \end{aligned} \tag{6.1.30}$$

On the other hand, the numerator of Eq. (6.1.25), taking into account Eqs. (6.1.28) and (6.1.29), becomes

$$\begin{aligned}\left\langle \delta\dot{N}_{\mathrm{A}}(0)\delta N_{\mathrm{B}}(t)\right\rangle &= -\langle \dot{x}(0)\,\delta(q-x(0))\,[H(x(t)-q)-P_{\mathrm{B}}^{\mathrm{eq}}]\rangle \\ &= -\langle \dot{x}(0)\,\delta(q-x(0))\,H(x(t)-q)\rangle + P_{\mathrm{B}}^{\mathrm{eq}}\,\langle \dot{x}(0)\,\delta(q-x(0))\rangle \\ &= -\langle \dot{x}(0)\,\delta(q-x(0))\,H(x(t)-q)\rangle, \end{aligned} \tag{6.1.31}$$

where we have used the fact that $\langle \dot{x}(0)\,\delta(q-x(0))\rangle$ is 0, since positive and negative values of the velocity $\dot{x}(0)$ are equally probable at equilibrium.

From Eqs. (6.1.25), (6.1.30), and (6.1.31), k_{eff} emerges as

$$k_{\mathrm{eff}} = \frac{\langle \dot{x}(0)\,\delta(q-x(0))\,H(x(t)-q)\rangle}{P_{\mathrm{A}}^{\mathrm{eq}}P_{\mathrm{B}}^{\mathrm{eq}}}. \tag{6.1.32}$$

On the other hand, the condition of detailed balance, Eq. (6.1.17), gives

$$k_{\mathrm{A}\to\mathrm{B}}P_{\mathrm{A}}^{\mathrm{eq}} = k_{\mathrm{B}\to\mathrm{A}}P_{\mathrm{B}}^{\mathrm{eq}}. \tag{6.1.33}$$

Hence

$$k_{\mathrm{eff}} = k_{\mathrm{A}\to\mathrm{B}} + k_{\mathrm{B}\to\mathrm{A}} = k_{\mathrm{A}\to\mathrm{B}}\left(1+\frac{P_{\mathrm{A}}^{\mathrm{eq}}}{P_{\mathrm{B}}^{\mathrm{eq}}}\right) = k_{\mathrm{A}\to\mathrm{B}}\frac{1}{P_{\mathrm{B}}^{\mathrm{eq}}}. \tag{6.1.34}$$

Combining Eqs. (6.1.31) and (6.1.34), we express the rate constant for transition from A to B as

$$k_{\mathrm{A}\to\mathrm{B}}(t) = \frac{\langle \dot{x}(0)\,\delta(q-x(0))\,H(x(t)-q)\rangle}{P_{\mathrm{A}}^{\mathrm{eq}}}, \tag{6.1.35}$$

where we are showing an explicit dependence of the rate constant on time.

The numerator of Eq. (6.1.35) can be interpreted as the flux of probability through the dividing surface $x = q$, modified by the step function $H(x(t)-q)$. Trajectories which cross the dividing surface in the direction from A to B and find themselves in state B after time t contribute positively to $k_{\mathrm{A}\to\mathrm{B}}(t)$. Trajectories which cross the dividing surface in the direction from A to B, but quickly recross it and find themselves in state A after time t do not contribute to $k_{\mathrm{A}\to\mathrm{B}}(t)$ (see Figure 6.1). Trajectories which cross the dividing surface from B to A, but quickly recross and find themselves in state B after time t contribute negatively to $k_{\mathrm{A}\to\mathrm{B}}(t)$. If the correlation time τ_{cor} required for the system to *thermalize*, i.e. establish a restricted equilibrium distribution within either state A or state B, is short relative to k_{eff}^{-1}, making the transitions A→B and B→A infrequent events, then recrossing events will quickly subside and $k_{\mathrm{A}\to\mathrm{B}}(t)$ will assume an asymptotic, time-independent value over a wide range of times $\tau_{\mathrm{cor}} < t < k_{\mathrm{eff}}^{-1}$.

Example 6.1

Show that by invoking the TST assumption, Eq. (6.1.35) for the rate constant $k_{A \to B}(t)$ can reduce to an expression similar to Eq. (6.1.3).

Solution: As was pointed out earlier, TST assumes that, whenever a crossing of the dividing surface occurs, the system thermalizes in the state towards which it is moving. Thus, mathematically, TST replaces the time t in the numerator of Eq. (6.1.35) with 0^+, a time just after the crossing, at which the system is guaranteed to be in its destination state. The TST approximation to the transition state is thus

$$k_{A \to B}^{TST} = \frac{\langle \dot{x}(0)\, \delta\,(q - x(0))\, H\,(x\,(0^+) - q) \rangle}{P_A^{eq}} = \frac{\langle \dot{x}(0)\, \delta\,(q - x(0))\, H\,(\dot{x}(0)) \rangle}{P_A^{eq}}. \tag{6.1.36}$$

In the canonical ensemble considered, positive and negative values of the velocity $\dot{x}(0)$ are equally probable. The positive and negative fluxes through $x(0) = q$ are equal in magnitude, and the TST expression can further be written as

$$k_{A \to B}^{TST} = \frac{1}{2} \frac{\langle |\dot{x}(0)|\, \delta\,(q - x(0)) \rangle}{P_A^{eq}}. \tag{6.1.37}$$

Furthermore, in a classical treatment where configurational and momentum degrees of freedom are separable,

$$k_{A \to B}^{TST} = \frac{1}{2} \langle |\dot{x}(0)| \rangle \frac{\langle \delta(q - x(0)) \rangle}{P_A^{eq}} = \left(\frac{k_B T}{2\pi m} \right)^{1/2} \frac{\exp\left(-\dfrac{U(q)}{k_B T} \right)}{\displaystyle\int_{\text{state A}} \exp\left(-\dfrac{U(x)}{k_B T} \right) dx}, \tag{6.1.38}$$

where we have used that, for a one-dimensional Boltzmann distribution of velocities, $\langle |\dot{x}(0)| \rangle = \left(\dfrac{2k_B T}{\pi m} \right)^{1/2}$, and the configuration-space probability density is proportional to $\exp\left(-\dfrac{U(x)}{k_B T} \right)$. Equation (6.1.38) is the same as Eq. (6.1.3). Remarkably, the TST approximation has reduced the rate constant to the product of a mean molecular speed times a conditional probability per unit length along the reaction coordinate that the system will find itself on the dividing surface, provided it is allowed to sample the origin state A according to its equilibrium distribution.

By neglecting fast recrossing events, TST typically overestimates the transition rate constant. Comparing Eq. (6.1.32) with the precursor to Eq. (6.1.36), we can define a dynamical correction factor as

$$f_d(t) = \frac{\langle \dot{x}(0)\, \delta\,(q - x(0))\, H\,(x(t) - q) \rangle}{\langle \dot{x}(0)\, \delta\,(q - x(0))\, H\,(x\,(0^+) - q) \rangle}, \tag{6.1.39}$$

so that the transition rate constant can be written as (compare Eq. (6.1.13))

$$k_{A \to B} = f_{d,A \to B} k_{A \to B}^{TST}. \tag{6.1.40}$$

Equation (6.1.40) forms a basis for computing dynamically correct transition rate constants from simulation, based on TST.

6.1.4 Infrequent event theory in many dimensions

The formulation outlined for calculating the rate constant of infrequent transitions from state A to state B can readily be extended to the case where more than one degrees of freedom are involved in effecting a transition. In this case one works with a potential of mean force U that depends on a number of degrees of freedom, constituting a vector $\boldsymbol{R}$ (compare $\boldsymbol{X}$ in Section 5.3.6). If *all* configurational degrees of freedom of the system are included in $\boldsymbol{R}$, then U is nothing else than the potential energy function $\mathcal{V}$. Multidimensional TST was first treated by Vineyard,[2] in the context of self-diffusion in crystalline solids.

For the mathematical development of multidimensional TST it is convenient to work not with the atomic coordinates themselves, but with mass-weighted atomic coordinates. These are defined as $\boldsymbol{x}_i = \boldsymbol{R}_i m_i^{1/2}$, where m_i is the mass of atom i. Thus, in place of $\boldsymbol{R}$ we will use the vector of mass-weighted atomic coordinates, $\boldsymbol{x}$, whose dimensionality will be indicated by d. The origin state A and the destination state B are basins in the d-dimensional $\boldsymbol{x}$-space constructed around neighbouring local minima of $U(\boldsymbol{x})$. Let us denote these minima by $\boldsymbol{x}_A$ and $\boldsymbol{x}_B$. The dividing surface is a $(d-1)$-dimensional hypersurface in $\boldsymbol{x}$-space, which we can represent mathematically by an equation of the form $C(\boldsymbol{x}) = 0$. Although there is some latitude in the definition of the dividing surface, it is advantageous to define it according to the following requirements:

1. The dividing surface passes through the first-order saddle point, $\boldsymbol{x}^\dagger$, lying between $\boldsymbol{x}_A$ and $\boldsymbol{x}_B$, which corresponds to the considered transition. By definition, the gradient $\nabla_{\boldsymbol{x}} U$ satisfies

$$\nabla_{\boldsymbol{x}} U\big|_{\boldsymbol{x}_A} = \nabla_{\boldsymbol{x}} U\big|_{\boldsymbol{x}_B} = \nabla_{\boldsymbol{x}} U\big|_{\boldsymbol{x}^\dagger} = \boldsymbol{0}, \tag{6.1.41}$$

while the Hessian matrix of second derivatives $\dfrac{\partial^2 U}{\partial \boldsymbol{x} \partial \boldsymbol{x}^T}$ is positive definite at $\boldsymbol{x}_A$ and $\boldsymbol{x}_B$ and has exactly one negative eigenvalue at $\boldsymbol{x}^\dagger$. For the dividing surface, as stated above, we require that

$$C\left(\boldsymbol{x}^\dagger\right) = 0. \tag{6.1.42}$$

2. At $\boldsymbol{x}^\dagger$, the dividing surface is normal to the eigenvector $\boldsymbol{n}^\dagger$ corresponding to the negative eigenvalue of the Hessian matrix, $\left.\dfrac{\partial^2 U}{\partial \boldsymbol{x} \partial \boldsymbol{x}^T}\right|_{\boldsymbol{x}^\dagger}$,

$$\left.\frac{\nabla_{\boldsymbol{x}} C(\boldsymbol{x})}{\left|\nabla_{\boldsymbol{x}} C(\boldsymbol{x})\right|}\right|_{\boldsymbol{x}=\boldsymbol{x}^\dagger} \equiv \boldsymbol{n}\left(\boldsymbol{x}^\dagger\right) = \boldsymbol{n}^\dagger, \tag{6.1.43}$$

where we have used the symbolism $\boldsymbol{n}(\boldsymbol{x})$ to denote the unit vector normal to the dividing surface at $\boldsymbol{x}$, pointing away from state A.

3. At all points other than $\boldsymbol{x}^{\dagger}$, the dividing surface is tangent to the gradient vector of $U(\boldsymbol{x})$:

$$\nabla_{\boldsymbol{x}} C(\boldsymbol{x}) \cdot \nabla_{\boldsymbol{x}} U = 0, \quad \boldsymbol{x} \neq \boldsymbol{x}^{\dagger}. \tag{6.1.44}$$

The saddle point $\boldsymbol{x}^{\dagger}$ is often referred to as the *transition state*. The reaction path is a one-dimensional line in the d-dimensional $\boldsymbol{x}$-space, which connects $\boldsymbol{x}_{\mathrm{A}}$ and $\boldsymbol{x}_{\mathrm{B}}$ and passes through $\boldsymbol{x}^{\dagger}$. It points along $\boldsymbol{n}^{\dagger}$ at $\boldsymbol{x}^{\dagger}$ (and is therefore normal to the dividing surface at this point), and along the gradient vector $\nabla_{\boldsymbol{x}} U$ everywhere else (steepest descent direction).

According to the TST approximation, whenever the system finds itself on the dividing surface with net momentum from state A to state B, a successful transition from A to B will occur. Integrating the probability flux over the entire dividing surface one obtains that

$$k^{\mathrm{TST}}_{\mathrm{A}\to\mathrm{B}} P^{\mathrm{eq}}_{\mathrm{A}} = \int \mathrm{d}^d x \int_{\boldsymbol{n}(\boldsymbol{x})\cdot\boldsymbol{p}>0} \mathrm{d}^d p \left[\boldsymbol{n}(\boldsymbol{x}) \cdot \boldsymbol{p}\right] \delta\left[C(\boldsymbol{x})\right] \left|\nabla_{\boldsymbol{x}} C(\boldsymbol{x})\right| \rho^{NVT}(\boldsymbol{x}, \boldsymbol{p}), \tag{6.1.45}$$

where $\boldsymbol{p} = \dot{\boldsymbol{x}}$ is the generalized momentum conjugate to the mass-weighted coordinate $\boldsymbol{x}$, i.e. the d-dimensional vector collecting components $m_i^{1/2} \dot{\boldsymbol{R}}_i$ from each degree of freedom participating in the transition, and $\rho^{NVT}(\boldsymbol{x}, \boldsymbol{p})$ is the phase-space probability density. The factor $\delta\left[C(\boldsymbol{x})\right] \left|\nabla_{\boldsymbol{x}} C(\boldsymbol{x})\right|$ in Eq. (6.1.45) ensures that the integral over $\boldsymbol{x}$ is restricted to the dividing surface (the term $\left|\nabla_{\boldsymbol{x}} C(\boldsymbol{x})\right|$ ensures correct normalization). On the other hand, the equilibrium probability of residing in state A is

$$P^{\mathrm{eq}}_{\mathrm{A}} = \int_{\text{state A}} \rho^{NVT}(\boldsymbol{x}, \boldsymbol{p})\, \mathrm{d}^d x\, \mathrm{d}^d p. \tag{6.1.46}$$

Performing the momentum-space integrations in Eqs. (6.1.45) and (6.1.46) leads to

$$k^{\mathrm{TST}}_{\mathrm{A}\to\mathrm{B}} = \frac{1}{(2\beta\pi)^{1/2}} \frac{\displaystyle\int_{\text{state A}} \delta\left[C(\boldsymbol{x})\right] \left|\nabla_{\boldsymbol{x}} C(\boldsymbol{x})\right| \exp\left[-\beta U(\boldsymbol{x})\right] \mathrm{d}^d x}{\displaystyle\int_{\text{state A}} \exp\left[-\beta U(\boldsymbol{x})\right] \mathrm{d}^d x}, \tag{6.1.47}$$

where $\beta = \dfrac{1}{k_{\mathrm{B}} T}$, or, in simpler form,

$$k^{\mathrm{TST}}_{\mathrm{A}\to\mathrm{B}} = \frac{1}{(2\beta\pi)^{1/2}} \frac{\displaystyle\int_{\text{div. surf.}} \exp\left[-\beta U(\boldsymbol{x})\right] \mathrm{d}^{d-1} \boldsymbol{x}}{\displaystyle\int_{\text{state A}} \exp\left[-\beta U(\boldsymbol{x})\right] \mathrm{d}^d \boldsymbol{x}}. \tag{6.1.48}$$

Equation (6.1.48) is a multidimensional analogue of Eq. (6.1.3).

A harmonic approximation to $k_{A\to B}^{TST}$ is possible also in the case of multidimensional TST. If the pathway from A to B is highly restricted, i.e. if $U(\boldsymbol{x})$ rises steeply relative to $k_B T$ as one moves away from $\boldsymbol{x}^\dagger$ on the dividing surface, then the overwhelming contribution to the integral in the numerator of Eq. (6.1.48) will come from the vicinity of the saddle point. Under these conditions, the dividing surface may be approximated by a hyperplane normal to the reaction path at the saddle point:

$$C(\boldsymbol{x}) \cong \boldsymbol{n}^\dagger \cdot \left(\boldsymbol{x} - \boldsymbol{x}^\dagger\right) = 0. \tag{6.1.49}$$

In the $(d-1)$-dimensional integral over this hyperplane, the potential of mean force $U(\boldsymbol{x})$ may be approximated through its Taylor expansion to second order around $\boldsymbol{x}^\dagger$. Similarly, if $U(\boldsymbol{x})$ presents a deep well around the minimum $\boldsymbol{x}_A$, then the overwhelming contribution to the integral in the denominator of Eq. (6.1.48) will come from the vicinity of the minimum. In the d-dimensional integral over the origin state, $U(\boldsymbol{x})$ may be approximated through its Taylor expansion to second order around $\boldsymbol{x}_A$. If these approximations are introduced, the integrals in the numerator and denominator of Eq. (6.1.48) reduce to multidimensional Gaussian integrals and the integrations can be performed analytically, the result being

$$k_{A\to B}^{TST,HA} = \frac{1}{2\pi} \frac{\prod_{\alpha=1}^{d} \omega_{\alpha,A}}{\prod_{\alpha=2}^{d} \omega_\alpha^\dagger} \exp\left[-\frac{U(\boldsymbol{x}^\dagger) - U(\boldsymbol{x}_A)}{k_B T}\right]. \tag{6.1.50}$$

The quantities $\omega_{\alpha,A}$ $(\alpha = 1, 2, \ldots, d)$ are the normal mode angular frequencies of the system at the minimum $\boldsymbol{x}_A$. They are obtained readily as square roots of the eigenvalues of the Hessian matrix of second derivatives of the potential of mean force with respect to mass-weighted atomic coordinates at the minimum, $\left.\frac{\partial^2 U}{\partial \boldsymbol{x} \partial \boldsymbol{x}^T}\right|_{\boldsymbol{x}_A}$. The quantities $\omega_\alpha^\dagger$ $(\alpha = 2, \ldots, d)$ are the normal mode angular frequencies of the system confined to the dividing surface at the saddle point $\boldsymbol{x}^\dagger$. They are obtained as square roots of the positive eigenvalues of the Hessian matrix at the saddle point, $\left.\frac{\partial^2 U}{\partial \boldsymbol{x} \partial \boldsymbol{x}^T}\right|_{\boldsymbol{x}^\dagger}$. Observe that the number of $\omega_\alpha^\dagger$ is by one less than the number of $\omega_{\alpha,A}$, as the eigendirection of negative curvature, $\boldsymbol{n}^\dagger$, does not contribute an angular frequency. Thus, the preexponential factor in Eq. (6.1.50) has dimensions of frequency, as it should. The exponential incorporates the barrier in potential of mean force, $U(\boldsymbol{x}^\dagger) - U(\boldsymbol{x}_A)$. There is an obvious analogy between Eq. (6.1.50) and its one-dimensional counterpart, Eq. (6.1.9).

An alternative expression for the TST estimate of the rate constant can be obtained if we start from Eq. (6.1.45), perform the integration only over the component of $\boldsymbol{p}$ that is normal to the dividing surface, divide the equation we obtain by Eq. (6.1.46), and recall the form of the probability density in phase space, the definition of the partition function Q (compare Eq. (5.1.13)), and the definition of the potential of mean force, Eq. (5.3.37):

$$k_{A\to B}^{TST} = \frac{k_B T}{h} \frac{Q^\dagger}{Q_A} \tag{6.1.51}$$

or, equivalently, $k_{\mathrm{A}\to\mathrm{B}}^{\mathrm{TST}} = \frac{k_{\mathrm{B}}T}{h} \exp\left(-\frac{A^{\dagger} - A_{\mathrm{A}}}{k_{\mathrm{B}}T}\right)$. Equation (6.1.51) involves the ratio of two partition functions, rather than of two configurational integrals. The partition function in the numerator, involving $(3N-1)$ configurational and $(3N-1)$ momentum degrees of freedom, is confined to the dividing surface between states A and B. The partition function in the denominator, involving $3N$ configurational and $3N$ momentum degrees of freedom, refers to the system confined in the origin state, A. The factor $k_{\mathrm{B}}T$ in front of the ratio of partition functions arises from integration over the component of momentum normal to the dividing surface, while the factor h stems from the difference in dimensionality between the systems to which the two partition functions refer.

Molecular simulations offer a number of techniques for calculating the ratio of configurational integrals in Eq. (6.1.48) or of the partition functions in Eq. (6.1.51). Some of these are the following:

(a) Identification of the dividing surface and of state A in d-dimensional space and direct evaluation of the configurational integrals by MC integration;[17,18]

(b) Expression of the ratio of configurational integrals as a product of ratios of configurational integrals involving a series of intermediate $(d-1)$-dimensional hypersurfaces normal to the reaction path between $\boldsymbol{x}$ and $\boldsymbol{x}^{\dagger}$, and calculation of the individual ratios by the free energy perturbation method via MC or MD;

(c) Calculation of density profiles within thin slices of $\boldsymbol{x}$-space normal to the reaction path through a series of MC or MD simulations constrained to these slices, Boltzmann inversion of the profiles obtained to form configurational free energy profiles along the reaction path, and matching of the latter profiles between adjacent slices to construct a global free energy profile between $\boldsymbol{x}$ and $\boldsymbol{x}^{\dagger}$, which is a form of *umbrella sampling*;[19] and

(d) Use of the *blue moon* ensemble, which effectively drags the system from $\boldsymbol{x}$ to $\boldsymbol{x}^{\dagger}$ along the reaction coordinate and calculates the free energy profile as an integral of the work done by the force required to perform the dragging.[20]

By focusing on the dividing surface, TST manages to overcome the time-scale limitation of *brute force* MD. It does, however, require that one have a good idea of the variables $\boldsymbol{x}$ playing a significant role in the transition and of the states, dividing surface, and reaction path involved. In diffusion problems, the geometry of the medium (analysis of accessible volume) can be a very helpful guide for this.[21,22]

Transition state theory is not exact. As discussed in the one-dimensional treatment in Section 6.1.3, not all trajectories penetrating the dividing surface in the direction from A to B are successful in effecting a transition, because of recrossing events. A dynamical correction factor, f_{d}, can be computed for use in Eq. (6.1.40) through the following analogue of Eq. (6.1.39),

$$f_{\mathrm{d,A}\to\mathrm{B}} = \frac{\langle \boldsymbol{n}(\boldsymbol{x}(0)) \cdot \boldsymbol{p}(0)\delta\left[C(\boldsymbol{x}(0))\right]\left|\nabla C(\boldsymbol{x}(0))\right| h_{\mathrm{B}}(\boldsymbol{x}(t))\rangle}{\langle \boldsymbol{n}(\boldsymbol{x}(0)) \cdot \boldsymbol{p}(0)\delta\left[C(\boldsymbol{x}(0))\right]\left|\nabla C(\boldsymbol{x}(0))\right| h_{\mathrm{B}}(\boldsymbol{x}(0^{+}))\rangle}, \tag{6.1.52}$$

where $h_{\mathrm{B}}(\boldsymbol{x}) = 1$ when $\boldsymbol{x}$ finds itself within state B and 0 otherwise.

6.1.5 Infrequent event theory for multistate systems

Each state i in d-dimensional space is surrounded by a $(d-1)$-dimensional hypersurface which delimits it from the remaining states. In a manner analogous to the two-state case, we will assume that this boundary to state i is described by an equation $C_i(\boldsymbol{x}) = 0$, where C_i is a continuous, differentiable function of the mass-weighted coordinates $\boldsymbol{x}$. We assume:

$$C_i(\boldsymbol{x}) \begin{cases} < 0, \text{if } \boldsymbol{x} \text{ is in state } i \\ = 0, \text{if } \boldsymbol{x} \text{ is on the boundary of state } i \\ > 0, \text{if } \boldsymbol{x} \text{ is outside state } i \end{cases} \tag{6.1.53}$$

Then, $\boldsymbol{n}_i(\boldsymbol{x}) = \nabla C_i(\boldsymbol{x}) / |C_i(\boldsymbol{x})|$ will be the unit vector normal to the boundary surface of state i at point $\boldsymbol{x}$ pointing towards the outside of state i [compare Eq. (6.1.43)]. If we introduce the function

$$h_i(\boldsymbol{x}) = 1 - H(C_i(\boldsymbol{x})) \tag{6.1.54}$$

with $H(x)$ being the Heaviside step function, then:

$$k_{i\to j} = \frac{\langle \boldsymbol{n}_i(\boldsymbol{x}(0)) \cdot \dot{\boldsymbol{x}}(0)\, \delta(C_i(\boldsymbol{x}(0))) |\nabla C_i(\boldsymbol{x}(0))| h_j(\boldsymbol{x}(t)) \rangle}{P_i^{\text{eq}}} \tag{6.1.55}$$

which is a multistate, multidimensional generalization of Eq. (6.1.35).

In the multistate case, the TST approximation amounts to assuming that, whenever the system exits state i through its boundary surface, it will remain outside state i at all subsequent times. As in the two-state case [compare derivation of Eq. (6.1.45) to (6.1.48)], this leads to an expression for the rate constant for exiting state i based on TST:

$$k_{i\to}^{\text{TST}} = \frac{1}{(2\beta\pi)^{1/2}} \frac{\displaystyle\int_{\text{bound. surf. of state } i} \exp[-\beta U(\boldsymbol{x})] \mathrm{d}^{d-1}x}{\displaystyle\int_{\text{state } i} \exp[-\beta U(\boldsymbol{x})] \mathrm{d}^{d}x} \tag{6.1.56}$$

The actual rate constant $k_{i\to j}$ for transitions from i to j is obtained from $k_{i\to}^{\text{TST}}$ through multiplication by a dynamical correction factor $f_{\text{d},i\to j}$:

$$k_{i\to j} = k_{i\to}^{\text{TST}} f_{\text{d},i\to j} \tag{6.1.57}$$

which, as in Eq. (6.1.39), can be obtained as

$$f_{\text{d},i\to j} = \frac{\langle \boldsymbol{n}_i(\boldsymbol{x}(0)) \cdot \dot{\boldsymbol{x}}(0) \delta[C_i(\boldsymbol{x}(0))] |\nabla C(\boldsymbol{x}(0))| h_j(\boldsymbol{x}(t)) \rangle}{\langle \boldsymbol{n}_i(\boldsymbol{x}(0)) \cdot \dot{\boldsymbol{x}}(0) \delta[C_i(\boldsymbol{x}(0))] |\nabla C(\boldsymbol{x}(0))| [1 - h_i(\boldsymbol{x}(0^+))] \rangle} \tag{6.1.58}$$

and simplified to

$$f_{\text{d},i\to j} = \frac{\langle \boldsymbol{n}_i(\boldsymbol{x}(0)) \cdot \dot{\boldsymbol{x}}(0) \delta[C_i(\boldsymbol{x}(0))] |\nabla C(\boldsymbol{x}(0))| h_j(\boldsymbol{x}(t)) \rangle}{\frac{1}{2}\langle |\boldsymbol{n}_i(\boldsymbol{x}(0)) \cdot \dot{\boldsymbol{x}}(0)| \delta[C_i(\boldsymbol{x}(0))] |\nabla C(\boldsymbol{x}(0))| \rangle} \tag{6.1.59}$$

6.1.6 Methods for calculating rate constants from molecular simulations

If transitions are subject to relatively low barriers (say, up to 7 $k_B T$), such that rate constants $k_{i \to j}$ are relatively high (say, up to ns^{-1}), then rate constants can be estimated by direct MD simulation. All one needs is a technique to map every configuration recorded in the course of an MD trajectory onto a state. Often, when states are defined as regions around local minima in configuration space, this mapping is accomplished by direct energy minimization leading to the closest energy minimum or *inherent structure*.[23] A reduced trajectory of states visited is thus accumulated in parallel with the MD trajectory, and switches between states can readily be identified along this reduced trajectory. Rate constants can be computed by statistical analysis of the reduced trajectory, capitalizing on the exponential distribution of waiting times that characterizes Poisson processes. A simple method that can be used for this purpose is *hazard plot analysis*.[24,25]

When energy barriers between states are high in relation to $k_B T$ and rate constants are correspondingly low, transitions between states cannot be sampled adequately by straightforward MD. One way to get around this problem is to resort to temperature-accelerated dynamics (TAD) simulations.[26] MD simulation at a higher temperature is used to access transition pathways. Waiting times obtained at the higher temperature are extrapolated down to the temperature of interest using the Arrhenius dependence of rate constants on temperature. The method has been used to great advantage in surface diffusion problems.[26] Tsalikis et al.[27] have combined microcanonical simulations at various energy levels with the histogram reweighting method to obtain rate constants in the spirit of TAD for transitions between basins in configuration space in the course of structural relaxation of a glassy binary Lennard-Jones mixture.

In recent years, a variety of advanced methods have been proposed for calculating free energy profiles along a coarse-grained variable or reaction coordinate. One such method is flux-tempered metadynamics,[28] based on the metadynamics method.[29] Metadynamics entails MD simulation in which a repulsive Gaussian potential in a few selected coarse-grained variables is periodically added to the potential energy function of the system, to encourage its escape from the vicinity of local free-energy minima with respect to these coarse-grained variables. If uniform sampling of the space of coarse-grained variables is achieved, the free energy can be estimated from the sum of added Gaussian potentials.

A general, essentially exact, but computationally intensive method for computing transition rate constants between two known states in configuration space is *transition path sampling*.[30] The method samples dynamical trajectories (generated and manipulated using importance sampling techniques) connecting the two states. Detailed information on the method can be found in a number of excellent reviews.[30,31]

6.1.7 Evolution though infrequent transitions in a network of states; kinetic Monte Carlo; analytical solution of the master equation

Forward and reverse transition rate constants between states i and j must satisfy the condition of microscopic reversibility or detailed balance, Eqs. (6.1.17) and (6.1.33), which we rewrite here as

$$P_i^{\mathrm{eq}} k_{i \to j} = P_j^{\mathrm{eq}} k_{j \to i} \equiv k_{ij}, \tag{6.1.60}$$

where P_i^{eq} denote the equilibrium probabilities of occupancy of the states and are normalized.

Let $P_i(t)$ be the probability of finding the system in state i at time t. We assume that the system starts off from an initial probability distribution $P_i(0)$ which may or may not correspond to equilibrium. The quantities $P_i(t)$ will evolve according to the *master equation*

$$\frac{\mathrm{d}P_i}{\mathrm{d}t} = -\sum_{j \neq i} k_{i \to j} P_i + \sum_{j \neq i} k_{j \to i} P_j. \tag{6.1.61}$$

In Eq. (6.1.61), the probability of occupancy of state i changes as a result of efflux of probability to other states j and as a result of influx of probability from other states. By Eq. (6.1.60), the equilibrium probabilities $\{P_i^{\mathrm{eq}}\}$ constitute a stationary solution to the master equation.

A convenient stochastic method for the numerical solution of Eqs. (6.1.61) under given initial conditions is kinetic Monte Carlo simulation (kMC) whose earliest application dates back to Beeler's 1966 simulation of radiation damage annealing,[32] although the name *kinetic Monte Carlo* was not widely adopted before 1990.[33] kMC rests on the fact that the sequence of independent interstate transitions undergone by the system constitutes a Poisson process.[34] Let us consider, as in Section 6.1.5, a system evolving through infrequent transitions in a network of n states (Figure 6.2), where the rate constants $k_{i \to j}$ for all possible state-to-state transitions (and their inverses) are known. Also known are the initial probabilities $P_i(0), i = 1, 2, \ldots, n$ for the system to reside in each state i at time $t = 0$. Then, kMC generates a set of stochastic trajectories for the system at times $t > 0$ as follows:

(i) Initially ($t = 0$), a large number $N\,(>> n)$ of independent *walkers* are deployed among the n states of the system allowing for multiple occupancy of sites, in such a way that they conform to the initial probability distribution $P_i(0)$. The walkers are allowed to jump between states without interacting with each other. Let $N_i(t)$ be the number of walkers that find themselves in state i at time t; initially, $N_i(t)/N \approx P_i(0)$.

(ii) For each state i that is occupied at the current time t, the expected fluxes $R_{i \to j}(t) = N_i(t) k_{i \to j}$ are calculated to all states j to which this is connected. Also, the overall flux $R(t) = \sum_i \sum_j R_{i \to j}(t)$ and the probabilities $q_{i \to j}(t) = R_{i \to j}(t)/R(t)$ are calculated.

(iii) A uniformly distributed pseudorandom number[35] $\xi \in [0, 1)$ is generated and the time for occurrence of the next elementary jump event in the network of states is chosen as $\Delta t = -\ln(1 - \xi)/R(t)$. Also the type of the next elementary jump event is chosen by picking one of the possible transitions $i \to j$ according to the probabilities $q_{i \to j}(t)$.

(iv) Of the $N_i(t)$ walkers present in state i, one is picked with probability $1/N_i(t)$ and moved to state j.

(v) The simulation time is advanced by Δt and the array is updated, keeping track of the current positions of all walkers. The occupancy numbers $N_i(t + \Delta t) = N_i(t) - 1$ and $N_j(t + \Delta t) = N_j(t) + 1$ are also updated.

(vi) Algorithm returns to step (ii) to implement the next transition.

The outcome from performing this stochastic simulation over a large number of steps is a set of trajectories for all N walkers. Each trajectory consists of a long sequence of transitions (jumps)

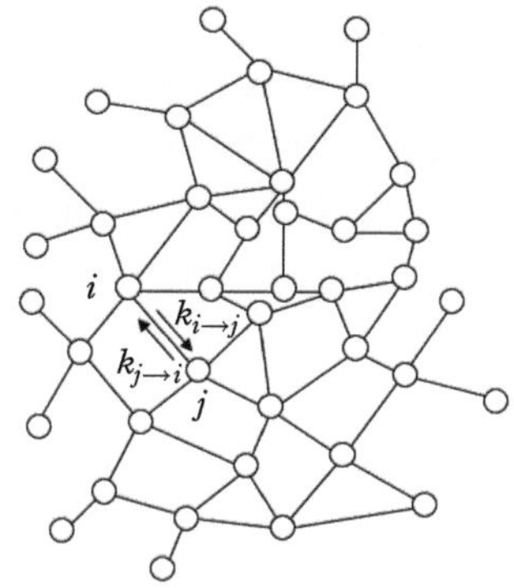

Figure 6.2 Schematic representation of a network of states adopted by a system. The system evolves through infrequent transitions in this network. States and transitions appear as nodes and links in the graph representing the network. The forward and reverse transitions between two particular states i and j, with associated transition rate constants $k_{i\to j}$ and $k_{j\to i}$, are indicated.

between states of the network. Time-dependent system properties are then estimated as ensemble averages over the trajectories at a specific time. For example, if states correspond to sites in a 3D network where a molecule can reside, one can calculate the mean square displacement (msd) of the molecule along each of the three coordinate directions as a function of time by averaging over the trajectories, and hence obtain the self-diffusivity tensor via the Einstein relation (see Section 6.2).

When all rate constants $k_{i\to j}$ are small, the kMC simulation will take large strides Δt on the time axis. When the rate constants $k_{i\to j}$ are very broadly distributed, kMC may become inefficient because time steps Δt must be short enough to track the fastest processes occurring in the system. Thus, one is faced with the same long-time problem as in MD. In such cases of extreme dynamical heterogeneity, and given an initial condition for $\{P_i(0)\}$, it may be better to resort to direct solution of the master equation, Eq. (6.1.61), for the time-dependent state occupancy probabilities $\{P_i(t)\}$, a problem for which analytical solutions have been developed,[9,36,37] through (e.g.) spectral decomposition. Indeed, one can collect the state probabilities $P_i(t)$ to form an n-dimensional vector $\boldsymbol{P}(t)$. Then, the master equation, Eq. (6.1.61), can be written in compact form as

$$\frac{\mathrm{d}\boldsymbol{P}(t)}{\mathrm{d}t} = \boldsymbol{K}\cdot\boldsymbol{P}(t) \tag{6.1.62}$$

with initial condition $\boldsymbol{P}(0)$. The $n \times n$ rate constant matrix $\boldsymbol{K}$ is defined by $K_{ij} = k_{j\to i}$ for $j \neq i$ and $K_{ii} = -\sum_{j\neq i} k_{i\to j}$. It is sometimes called stochastic, in the sense that the elements in each of its columns add to zero. We further transform the state probability vector $\boldsymbol{P}(t)$ into a reduced vector $\tilde{\boldsymbol{P}}(t)$ with elements $\tilde{P}_i(t) = P_i(t)\Big/\sqrt{P_i^{\mathrm{eq}}}$. This satisfies the reduced master equation $\frac{\mathrm{d}\tilde{\boldsymbol{P}}(t)}{\mathrm{d}t} = \tilde{\boldsymbol{K}}\cdot\tilde{\boldsymbol{P}}(t)$ with $\tilde{K}_{ij} = K_{ij}\sqrt{P_j^{\mathrm{eq}}}\Big/\sqrt{P_i^{\mathrm{eq}}}$. The matrix $\tilde{\boldsymbol{K}}$ is symmetric by virtue of detailed balance, Eq. (6.1.60), and has the same eigenvalues as $\boldsymbol{K}$. The eigenvalues are real, since $\tilde{\boldsymbol{K}}$ is symmetric. Of these eigenvalues, one (corresponding to the establishment of the equilibrium distribution among states) is 0, and the remaining eigenvalues are negative. This is because $\tilde{\boldsymbol{K}}$ is a negative semidefinite matrix (see Problem 6.1). We denote these eigenvalues by $\lambda_0 = 0 \geq \lambda_1 \geq \ldots \geq \lambda_{n-1}$. We symbolize by $\tilde{\boldsymbol{u}}_m = (\tilde{u}_{1,m}, \tilde{u}_{2,m}, \ldots, \tilde{u}_{i,m}, \ldots, \tilde{u}_{n,m})$ the eigenvector of $\tilde{\boldsymbol{K}}$ corresponding to eigenvalue $\lambda_m (0 \leq m \leq n-1)$. Note that $\tilde{u}_{i,0} = \tilde{P}_i^{\text{eq}} = \sqrt{P_i^{\text{eq}}}$, corresponding to the equilibrium distribution among states. The Euclidean norm of $\tilde{\boldsymbol{u}}_0$ is unity by the normalization of P_i^{eq}. The solution of the master equation can then be written as

$$\tilde{\boldsymbol{P}}(t) = \sum_{m=0}^{n-1} \left[\tilde{\boldsymbol{u}}_m \cdot \tilde{\boldsymbol{P}}(0) \right] e^{\lambda_m t} \, \tilde{\boldsymbol{u}}_m = \tilde{\boldsymbol{P}}^{\text{eq}} + \sum_{m=1}^{n-1} \left[\tilde{\boldsymbol{u}}_m \cdot \tilde{\boldsymbol{P}}(0) \right] e^{\lambda_m t} \, \tilde{\boldsymbol{u}}_m, \tag{6.1.63}$$

where the normalization condition $\sum_{j=1}^{n} P_j(0) = 1$ has been used in separating out the equilibrium contribution ($\lambda_0 = 0$). The eigenvectors $\tilde{\boldsymbol{u}}_m$ form an orthonormal basis set, $\tilde{\boldsymbol{u}}_m \cdot \tilde{\boldsymbol{u}}_l = \delta_{ml}, 0 \leq m, l \leq n-1$, with δ_{ml} being the Kronecker delta. They also satisfy $\sum_{m=0}^{n-1} \tilde{u}_{i,m} \tilde{u}_{j,m} = \delta_{ij}$. Once $\tilde{\boldsymbol{P}}(t)$ has been determined, the state probabilities $\boldsymbol{P}(t)$ can be calculated via $P_i(t) = \tilde{P}_i(t) \sqrt{P_i^{\text{eq}}}$.

Equation (6.1.63) has an interesting geometric interpretation, which is discussed at length by Boulougouris et al.[37] in the context of a general formulation for analytical solution of the master equation and calculation of time-dependent averages and autocorrelation functions which was dubbed *EROPHILE*, for *e*igenvalue *r*epresentation of *o*bservables and *p*robabilities in a *hi*gh-dimensiona*l* *E*uclidean space. Moreover, for studying diffusion in spatially periodic systems, in which the set of states (sites) is obtainable by replication of a *unit cell* of states in one, two, or three dimensions, Kolokathis and Theodorou[38] have developed a method for calculating the eigenvalues and eigenvectors of the matrix $\tilde{\boldsymbol{K}}$ of the whole system by diagonalizing matrices of dimension corresponding to a single unit cell, known as master equation solution by recursive reduction of dimensionality (MESoRReD) (see also Section 6.3.4).

6.2 Diffusion of gases in glassy polymers: Gusev–Suter approach

Molecular simulations offer attractive opportunities for exploring the molecular mechanisms of diffusion in polymers and developing the predictive capability needed for the molecular engineering design of materials with controlled permeability.[39–45] The self-diffusivity, equal to the transport diffusivity D in the low-concentration regime encountered in gas separation membrane and barrier applications, is typically extracted from the msd $\langle \Delta r_{\text{pen}}^2 \rangle$ of the penetrant at long times through the Einstein relation (see Section 7.7.4)

$$D = \frac{1}{6} \lim_{t \to \infty} \frac{\mathrm{d} \langle \Delta r_{\text{pen}}^2 \rangle}{\mathrm{d}t}. \tag{6.2.1}$$

A remarkable finding from MD simulations of gas diffusion in polymer melts and rubbery polymers is that, at short times, an extended *anomalous regime* is observed, wherein the msd of the penetrant scales roughly as $\langle \Delta r_{\text{pen}}^2 \rangle \propto t^n$ with $n < 1$ (typically, $n \cong 0.5$).[44] Anomalous diffusion is due to long-lived structural correlations in the polymer, which cause the diffusant to encounter a locally heterogeneous environment. From a practical viewpoint, it increases considerably the computational cost of MD simulations required for the prediction of D.

As temperature is reduced towards and below T_g, penetrant diffusion in an amorphous polymer matrix becomes too slow to be predictable by MD simulations. At these low temperatures, the penetrant spends most of its time *trapped* within clusters of accessible volume in the matrix and only *infrequently* jumps from cluster to cluster. A brute-force MD simulation exhausts itself in tracking the relatively uninteresting *rattling* motions of the penetrant within a cluster but is much too short to accumulate sufficient statistics on the jumps that govern diffusion. This was shown characteristically in the MD work of Takeuchi,[45] which gave a graphical representation of a jump process in terms of the evolution of the contours of potential energy experienced by the penetrant. It should be emphasized that this jump process is not confined to diffusion in polymers below T_g. As shown by MD work, it occurs in low-temperature melts and rubbery polymers as well, whenever the temperature is sufficiently low for the distribution of accessible volume clusters to remain relatively unchanged over the time scale required for the penetrant to move by the mean distance between clusters.

The above observations suggest a description of diffusion at low temperatures as a sequence of successive, infrequent jump events, the rate constant for each jump being estimable from TST. Focusing on the diffusivity prediction of a spherical gas molecule at infinite dilution within a glassy polymer matrix, this yields to a *hierarchical* modelling strategy, consisting of the following steps:

Step 1. Identify the *states* where the penetrant + polymer system spends most of its time, and the *reaction paths* along which transitions between neighbouring states take place (compare Sections 6.1.4 and 6.1.5) through geometric/energetic analysis of a detailed molecular model of the system.

Step 2. Calculate the rate constant for each reaction path.

Step 3. Model diffusion as a Poisson process consisting of elementary jumps between the sites determined in Step 1 with rate constants determined in Step 2 and estimate the diffusivity by tracking the msd of the penetrant as a function of time, according to Eq. (6.2.1).

In Section 6.1.7 we have discussed Step 3 in the above hierarchy, whereby $\langle \Delta r_{\text{pen}}^2 \rangle$ can be determined once all states and interstate rate constants are known by kMC or analytical solution of the master equation describing evolution of the system state. In the following we outline[46] two approaches for determining states, reaction paths, and rate constants from the atomistically represented polymer configurations. These are the 3D TST approach of Gusev and Suter[3,18,47] discussed in the remainder of this Section and the more elaborate multidimensional TST approach of Greenfield and Theodorou[48–51] discussed in Section 6.3.

Given an atomistic model for a low-temperature amorphous polymer phase, the simplest calculation one can perform to address the diffusion of a spherical penetrant is to assume that the polymer matrix is rigid and map out the potential energy of penetrant–polymer interactions $\mathcal{V}^{\text{pen}}$ as a function of the penetrant position $\boldsymbol{r}_{\text{pen}}$. To incorporate the response of the polymer matrix to the presence of the penetrant, Gusev and Suter[3,18,47] proposed that, over the residence times of the polymer + penetrant system in a sorption *state*, the polymer atoms execute harmonic vibrations about their equilibrium positions in the penetrant-free matrix. These motions (small-amplitude vibrations of bond lengths and bond angles, librations of torsion angles) are termed *elastic motion* and are distinguished from *structural relaxation* involving e.g. torsional transitions in the main chains or side groups. The polymer contribution $\mathcal{V}_\mathrm{p}$ to the potential energy is thus modelled by the simplified expression

$$\mathcal{V}_\mathrm{P}(\boldsymbol{r}_\mathrm{P}) = \mathcal{V}_\mathrm{P}(\boldsymbol{r}_{\mathrm{P},0}) + k_\mathrm{B}T\sum_k \frac{(\boldsymbol{r}_{\mathrm{P},k} - \boldsymbol{r}_{\mathrm{P},k,0})^2}{2\langle\Delta^2\rangle}, \tag{6.2.2}$$

where k enumerates atoms in the polymer, $\boldsymbol{r}_{\mathrm{P},k}$ stands for the instantaneous 3D position vector of polymer atom k, and the subscript 0 denotes the minimum energy configuration used to represent the penetrant-free polymer matrix (see Chapter 3 on molecular mechanics). The symbol $\boldsymbol{r}_\mathrm{P}$ is used to denote collectively the set of $\boldsymbol{r}_{\mathrm{P},k}$ for all polymer atoms k. $\langle\Delta^2\rangle$ is the variance in atomic position along each coordinate direction in the course of the elastic motion of the polymer.

Under the assumption of Eq. (6.2.2), and by considering that the penetrant + polymer potential energy $\mathcal{V}^{\,\mathrm{pen}}$ is modelled as the sum of central pairwise additive potential contributions:

$$\mathcal{V}^{\,\mathrm{pen}}(\boldsymbol{r}_\mathrm{P}, \boldsymbol{r}_\mathrm{pen}) = \sum_k \mathcal{V}_k^{\,\mathrm{pen}}(|\boldsymbol{r}_{\mathrm{P},k} - \boldsymbol{r}_\mathrm{pen}|). \tag{6.2.3}$$

The equilibrium probability density $\rho(\boldsymbol{r}_\mathrm{pen})$ of penetrant positions in the polymer matrix is governed by a 3D free-energy field, or potential of mean force, $U(\boldsymbol{r}_\mathrm{pen})$, made up of additive contributions from all polymer atoms. Then, in the canonical ensemble,

$$\begin{aligned}\rho(\boldsymbol{r}_\mathrm{pen}) &= \frac{\displaystyle\int \mathrm{d}\boldsymbol{r}_\mathrm{P} \exp\left[-\beta\mathcal{V}(\boldsymbol{r}_\mathrm{P}, \boldsymbol{r}_\mathrm{pen})\right]}{\displaystyle\int \mathrm{d}\boldsymbol{r}_\mathrm{P} \int \mathrm{d}\boldsymbol{r}_\mathrm{pen} \exp\left[-\beta\mathcal{V}(\boldsymbol{r}_\mathrm{P}, \boldsymbol{r}_\mathrm{pen})\right]} = \\ &= \mathrm{const.} \times \prod_k \left\{\int \mathrm{d}\boldsymbol{r}_{\mathrm{P},k} \exp\left[-\frac{(\boldsymbol{r}_{\mathrm{P},k} - \boldsymbol{r}_{\mathrm{P},k,0})^2}{2\langle\Delta^2\rangle} - \beta\mathcal{V}_k^{\,\mathrm{pen}}(|\boldsymbol{r}_{\mathrm{P},k} - \boldsymbol{r}_\mathrm{pen}|)\right]\right\}.\end{aligned} \tag{6.2.4}$$

Performing the integrations over the different $\boldsymbol{r}_{\mathrm{P},k}$ independently leads to

$$\rho^\mathrm{eq}(\boldsymbol{r}_\mathrm{pen}) = \mathrm{const.} \times \exp\left[-\beta U(\boldsymbol{r}_\mathrm{pen})\right] = \mathrm{const.} \times \exp\left[-\beta\sum_k U_k(|\boldsymbol{r}_\mathrm{pen} - \boldsymbol{r}_{\mathrm{P},k,0}|)\right] \tag{6.2.5}$$

with

$$\begin{aligned}U_k(|\boldsymbol{r}_\mathrm{pen} - \boldsymbol{r}_{\mathrm{P},k,0}|) &= -\frac{1}{\beta}\ln\left\{\int \mathrm{d}\boldsymbol{r}_{\mathrm{P},k} \exp\left[-\frac{(\boldsymbol{r}_{\mathrm{P},k} - \boldsymbol{r}_{\mathrm{P},k,0})^2}{2\langle\Delta^2\rangle} - \beta\mathcal{V}_k^{\,\mathrm{pen}}(|\boldsymbol{r}_{\mathrm{P},k} - \boldsymbol{r}_\mathrm{pen}|)\right]\right\} \\ &= -\frac{1}{\beta}\ln\left\{\int \mathrm{d}(\boldsymbol{r}_{\mathrm{P},k} - \boldsymbol{r}_{\mathrm{P},k,0}) \exp\left[-\frac{(\boldsymbol{r}_{\mathrm{P},k} - \boldsymbol{r}_{\mathrm{P},k,0})^2}{2\langle\Delta^2\rangle}\right.\right. \\ &\quad \left.\left. - \beta\mathcal{V}_k^{\,\mathrm{pen}}(|(\boldsymbol{r}_\mathrm{pen} - \boldsymbol{r}_{\mathrm{P},k,0}) - (\boldsymbol{r}_{\mathrm{P},k} - \boldsymbol{r}_{\mathrm{P},k,0})|)\right]\right\}.\end{aligned} \tag{6.2.6}$$

For given $\langle \Delta^2 \rangle$, the pair potential of mean force U_k is readily computed by numerical integration for each atomic species k in the polymer. The total potential of mean force $U(\boldsymbol{r}_{\text{pen}})$ is thus readily obtainable[18] given a minimum energy configuration $\boldsymbol{r}_{\text{P},0}$. With this model, introduction of the penetrant molecule in the matrix causes small, *elastic* displacements in the mean positions of polymer atoms relative to the penetrant-free state $\boldsymbol{r}_{\text{P},0}$. These displacements are largest for the atoms nearest the penetrant.

The equilibrium probability P_i^{eq} of a state i is computed by the straightforward numerical integration of $\rho^{\text{eq}}(\boldsymbol{r}_{\text{pen}})$,

$$P_i^{\text{eq}} = \int_{\text{state } i} \rho^{\text{eq}}(\boldsymbol{r}_{\text{pen}})\, d\boldsymbol{r}_{\text{pen}} = \text{const.} \int_{\text{state } i} \exp\left[-\beta U(\mathbf{r}_{\text{pen}})\right] d\boldsymbol{r}_{\text{pen}}, \tag{6.2.7}$$

where const. is a numerical constant which may be evaluated such that the sum of all P_i^{eq} (in the primary amorphous cell of the simulation) equals 1. The rate constant $k_{i\to j}^{\text{TST}}$ for transitions from state i to state j according to the TST assumption (i.e. setting the dynamical correction factor or transmission coefficient to 1) is obtained from Eq. (6.1.48) with $A = i$, $B = j$, $d = 3$, and $\boldsymbol{x} = m_{\text{pen}}^{1/2}\, \boldsymbol{r}_{\text{pen}}$ with m_{pen} being the mass of the penetrant. The 3D configurational integral over the origin state appearing in the denominator of Eq. (6.1.48) and the 2D configurational integral over the dividing surface appearing in the numerator of the same equation can be computed readily from the pretabulated field $U(\boldsymbol{r}_{\text{pen}})$ by numerical or Monte Carlo integration. The values of P_i^{eq} and $k_{i\to j}^{\text{TST}}$ so determined can then be used within a kMC simulation procedure, as described in Section 6.1.7, to track the diffusive progress of the penetrant and calculate the diffusivity.

The Gusev–Suter elastically moving matrix approach is computationally efficient, although the results for D depend sensitively on the value of $\langle \Delta^2 \rangle$ used to describe *elastic* fluctuations of the matrix. To avoid arbitrariness in the choice of this parameter, Gusev and Suter proposed a way of estimating it from short (*c.*5 ps in duration) atomistic MD simulations of the pure polymer matrix. The approach has been applied to He, H_2, Ar, Kr, Xe, O_2, and N_2 in matrices such as (glassy) polycarbonate and PVC, (rubbery) poly(isobutylene) and even poly(dimethylsiloxane) around room temperature.

6.3 Multidimensional transition-state theory approach to diffusion in glassy polymers

As the penetrant size increases relative to the size of accessible volume cavities in a glassy polymer, the Gusev–Suter picture of penetrant hops in a matrix of atoms executing fast, uncorrelated harmonic vibrations around their static equilibrium positions in the penetrant-free state becomes less satisfactory. The polymer degrees of freedom have to change a lot to accommodate the sorbed penetrant. More importantly, infrequent large-amplitude segmental motions of the polymer become important in opening pathways among the chains that can enable the penetrant to hop from cavity to cavity. *Relaxational* motions of the matrix are intimately involved in the infrequent events governing the translational progress of the penetrant and dynamical characteristics of the polymer (e.g. atomic masses) start playing a role. To resolve these problems, a new approach can be followed[48–51] in which polymer degrees of freedom are incorporated explicitly in the reaction coordinate of the infrequent events whereby diffusion takes place. The approach is described below and entails the following steps:

Step 1. Geometric analysis of accessible volume;
Step 2. Calculation of paths and rate constants for elementary diffusive jumps;
Step 3. Lumping of micro-states to macro-states and formulation of the master equation; and
Step 4. Calculation of the diffusivity by kinetic Monte Carlo.

6.3.1 Geometric analysis of accessible volume, cluster formation, cluster volume-shape analysis, cluster connectivity

Given a collection of points in space, Delaunay tessellation is first invoked to partition them into irregular tetrahedra such that the circumsphere of each tetrahedron is devoid of points.[48–59] Then, the *unoccupied volume* V_i^{unoc} of the ith Delaunay tetrahedron is defined as the part of the total volume V_i^{T} of this tetrahedron not occupied by atoms of the system at its four vertices. Obviously, $\sum_{i=1}^{N^{\text{T}}} V_i^{\text{T}} = V$, where N^{T} denotes the total number of tetrahedra defined for the given system configuration and V the total volume of the system (in the given configuration). Similarly, the sum $\sum_{i=1}^{N^{\text{T}}} V_i^{\text{unoc}}$ defines the total unoccupied volume V^{unoc} in the system, $\sum_{i=1}^{N^{\text{T}}} V_i^{\text{unoc}} = V^{\text{unoc}}$. Consistent with such a definition, the *unoccupied volume fraction* φ^{unoc} in the system is defined as the ratio of the sum of the *unoccupied volume* over all tetrahedra divided by the total volume of the system. Accordingly, given the radius r_{p} of a penetrant, one defines the *accessible* (by the penetrant) *volume* as the volume which remains unoccupied by the atoms (hard spheres) augmented by the radius of the penetrant. To obtain this, one first determines the accessible volume V_i^{acc} of the ith tetrahedron to a given penetrant by subtracting from the volume of the tetrahedron, V_i^{T}, the volume occupied by the spheres residing on its vertices but with their augmented radii $r_{\text{vdW}} + r_{\text{p}}$.[48,55] Clearly, for $r_{\text{p}} = 0$, $V^{\text{unoc}} = V^{\text{acc}}$. In addition, the sum $\sum_{i=1}^{N^{\text{T}}} V_i^{\text{acc}}$ defines the total accessible volume V^{acc} in the system, $\sum_{i=1}^{N^{\text{T}}} V_i^{\text{acc}} = V^{\text{acc}}$, while the corresponding *accessible volume fraction* φ^{acc} is $\varphi^{\text{acc}} = \dfrac{\sum_{i=1}^{N^{\text{T}}} V_i^{\text{acc}}}{\sum_{i=1}^{N^{\text{T}}} V_i^{\text{T}}} = \dfrac{V^{\text{acc}}}{V}$.

To examine whether the penetrant can move between two neighbouring tetrahedra through their common surface, and thus define the clusters of accessible volume to this penetrant, one can work as follows. Consider the surface of the triangle (ABC) shared by the two tetrahedra ($ABCD$) and ($ABCE$) (with D and E denoting the two noncommon vertices), with $V_i^{\text{acc}} > 0$ in both of them; assume that the two tetrahedra have in total M neighbouring vertices. One can use Monte Carlo to uniformly sample points inside the triangle (ABC). If these points do not belong to the interior of any sphere (with centre at points A, B, C, D, E or at the position of one of the M neighbours), the surface is considered as *accessible* and the procedure stops immediately. Otherwise, after many failed trials, the surface is considered as *inaccessible*. On the other hand, if for a given tetrahedron with $V^{acc} > 0$, a true neighbour tetrahedron with $V^{acc} = 0$ exists, then their common surface is characterized as *inaccessible* or *isolated.*

After one has identified all *accessible surfaces* between truly neighbouring tetrahedra, one can look for larger areas or domains in the given configuration available for the penetrant to reside or move through; these will be the so-called *accessible volume clusters*. Such clusters may be composed of up to thousands of tetrahedra, and their boundaries are *inaccessible surfaces*. Starting from an *accessible tetrahedron*, one constructs a graph whose nodes correspond to tetrahedra, and there are at most four edges (corresponding to the *accessible surfaces*) leading to other nodes. One then identifies the entire set of nodes without forming a closed loop, i.e. without returning to a node already visited.

To quantify the size and shape of clusters accessible to a particular penetrant, several measures can be used[60] such as the dimensionless asphericity b, the dimensionless acylindricity c, and the relative shape anisotropy κ^2, all of them computed from the eigenvalues of the radius of gyration tensor $\boldsymbol{S}$ of the cluster.

Clusters of accessible volume provide a geometric approximation to the sorption states. While a hard sphere penetrant in a rigid polymer matrix would be confined to the sorption states, a soft-core penetrant infrequently accesses pathways between states as it diffuses through the polymer. In the final step of the geometric analysis, therefore, the connectivity of these states (for the given penetrant size) is defined, and from this one obtains the connectivity among clusters.[48,59] Examples of clusters accessible to several penetrants from a geometric analysis with the above methodology in glassy atactic polypropylene[48] are shown in Figure 6.3.

6.3.2 Calculation of paths and rate constants for elementary diffusive jumps

As in Section 6.1.4, we introduce[2] mass-weighted Cartesian coordinates for the penetrant and for the polymer atoms, namely $\boldsymbol{x}_{\text{pen}} = m_{\text{pen}}^{1/2}\boldsymbol{r}_{\text{pen}}$ and $\boldsymbol{x}_{\text{p},k} = m_{\text{p},k}^{1/2}\boldsymbol{r}_{\text{p},k}$. We will use the symbol $\boldsymbol{x} \equiv (\boldsymbol{x}_{\text{pen}}, \boldsymbol{x}_{\text{p},k})$ to denote the mass-weighted coordinates of all atoms in the system. Let n be the number of elements in $\boldsymbol{x}$ and let also the potential energy $\mathcal{V}$ of the system be expressed as a function of $\boldsymbol{x}$. The penetrant + polymer system is envisioned as evolving though transitions $i \to j$ from the vicinity of one energy minimum $\boldsymbol{x}_i$ to the vicinity of another energy minimum $\boldsymbol{x}_j$ along a reaction path (line in $\boldsymbol{x}$-space) which passes through a saddle point $\boldsymbol{x}_{ij}$ of the function $\mathcal{V}(\boldsymbol{x})$. According to the TST approximation, whenever the system finds itself on the dividing surface S_{ij} with net momentum directed from i to j, a successful transition between i and j will occur. Setting the dynamical correction factor or transmission coefficient to 1, the corresponding rate constant $k_{i\to j}^{\text{TST}}$ is obtained from the ratio of the partition function of the system confined to the two-dimensional dividing surface S_{ij} to the partition function of the system in site i according to Eq. (6.1.51).

The search for reaction paths starts with the identification of all necks between accessible volume clusters within penetrant-free minimum-energy configurations of the polymer. These necks are located by geometric analysis of each configuration, using probe radii which range from that of the penetrant of interest down to a low value, smaller than that corresponding to percolation. Using a neck point as an initial guess for the penetrant position and keeping the polymer configuration fixed, a 3D search is undertaken to determine the closest saddle point of $\mathcal{V}$ in the position space

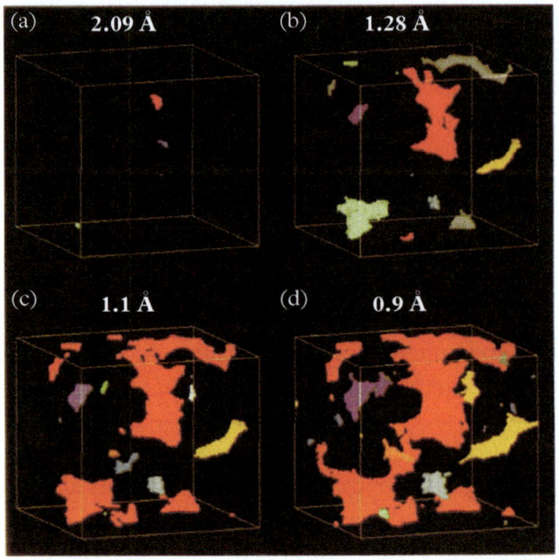

Figure 6.3 Clusters within glassy polypropylene accessible to (a) methane $\left(r_{\mathrm{p}} = 2.09\,\text{Å}\right)$, (b) helium $\left(r_{\mathrm{p}} = 1.28\,\text{Å}\right)$, (c) a hypothetical penetrant of van der Waals radius $r_{\mathrm{p}} = 1.1$ Å, and (d) a hypothetical penetrant of van der Waals radius $r_{\mathrm{p}} = 0.9$ Å. Different colours indicate different clusters; wrapping over the box edge is due to periodic boundary conditions. The box edge is approximately 23 Å. As the penetrant radius decreases, accessible clusters grow and fuse together. Percolation is observed for the red cluster accessible to the penetrant of size $r_{\mathrm{p}} = 0.9$ Å.

of the penetrant, $\boldsymbol{x}_{\mathrm{pen}}$. The set of the system degrees of freedom is then augmented to encompass generalized coordinates associated with polymer atoms within a sphere centred at the penetrant; a new saddle point of $\mathcal{V}$ is found in this augmented space, using the configuration determined from the 3D saddle point search as an initial guess. This process of augmentation of the set of generalized coordinates and of the determination of a saddle point in the augmented set is repeated, until further inclusion of coordinates leaves the saddle point energy unchanged.

At a saddle point, the system satisfies the following equations in the space of generalized coordinates,

$$\nabla_{\boldsymbol{q}} \mathcal{V} = \mathbf{0} \tag{6.3.1}$$

and

$$\boldsymbol{H}_{\boldsymbol{qq}} \cdot \mathrm{d}\boldsymbol{q} = \lambda \boldsymbol{a} \cdot \mathrm{d}\boldsymbol{q}, \tag{6.3.2}$$

with all eigenvalues λ positive except one. In the above, $\boldsymbol{H}_{\boldsymbol{qq}}$ is the Hessian matrix of second derivatives of the potential energy $\mathcal{V}$ with respect to the vector of generalized coordinates $\boldsymbol{q}$, $\mathrm{d}\boldsymbol{q}$ denotes an eigenvector in the space of generalized coordinates, and $\boldsymbol{a}$ is the covariant metric tensor in the space of generalized coordinates defined in Section 3.1.

Once a saddle point in multidimensional configuration space has been identified, then one can set out to compute the entire reaction path. This can be started off on either side of the saddle point in the direction of the eigenvector $\boldsymbol{n}_{ij} = \frac{\mathrm{d}\boldsymbol{x}}{|\mathrm{d}\boldsymbol{x}|}$ corresponding to the negative eigenvalue of the Hessian $\boldsymbol{H}$. The corresponding change in generalized coordinates is $\mathrm{d}\boldsymbol{q}$ and satisfies Eq. (6.3.2) with negative λ at the saddle point. Once the first small step away from the saddle point has been taken, the entire reaction path down to the state i on one side, or to the state j on the other, can be computed through a steepest descent construction in $\boldsymbol{x}$-space, following Fukui's intrinsic reaction coordinate (IRC).[61] The IRC is constructed by taking small steps $\mathrm{d}\boldsymbol{q}$ in the generalized coordinates according to the equation

$$\boldsymbol{a} \cdot \mathrm{d}\boldsymbol{q} = \nabla_{\boldsymbol{q}} \mathcal{V} \, \mathrm{d}\tau, \tag{6.3.3}$$

with $\mathrm{d}\tau$ being a quantity with dimensions of squared time, which measures how far we are along our way to the minimum.

In a fully flexible model representation it is convenient to identify $\boldsymbol{q}$ with $\boldsymbol{x}$ in Eqs. (6.3.1)–(6.3.3). The metric tensor $\boldsymbol{a}$ then becomes the unit tensor and the whole calculation is conducted in mass-weighted Cartesian coordinates.

Having determined a reaction path leading from a state i to a state j through a saddle point ij in multidimensional configuration space, one can proceed to compute the rate constant $k_{i\to j}^{\mathrm{TST}}$ using one of the methods discussed in Section 6.1. On the basis of the simpler harmonic approximation approach, in particular, the major contribution to the partition function of the system confined to the dividing surface (the numerator of Eq. (6.1.51)) comes from the immediate vicinity of the saddle point $\boldsymbol{x}_{ij}$. In this region, $\mathcal{V}(\boldsymbol{x})$ can be approximated by a Taylor expansion to second order in the $(n-1)$-dimensional subspace obtained by projecting the eigen direction $\boldsymbol{n}_{ij}$ out of $\boldsymbol{x}$. Similarly, the major contribution to the partition function of i in the denominator of Eq. (6.1.51) is assumed to come from the immediate vicinity of the state minimum $\boldsymbol{x}_i$ where $\mathcal{V}(\boldsymbol{x})$ can be approximated by a Taylor expansion to second order in $\boldsymbol{x}$-space. Then, the integral in the numerator of Eq. (6.1.48) becomes Gaussian and thus the partition function in the numerator of Eq. (6.1.51) reduces to the product of partition functions of $(n-1)$ independent harmonic oscillators with frequencies,

$$\nu_a^{(ij)} = \frac{1}{2\pi}\sqrt{\lambda_a^{(ij)}}, \tag{6.3.4}$$

where $\lambda_a^{(ij)}$ are the $(n-1)$ positive eigenvalues of the Hessian matrix $\boldsymbol{H}$ at the saddle point $\boldsymbol{x}_{ij}$ (compare Eq. (6.1.8)), and likewise, the partition function in the denominator of Eq. (6.1.51) reduces to a product of partition functions of n harmonic oscillators with frequencies $\nu_a^{(i)}$ related to the eigenvalues $\lambda_a^{(i)}$ of the Hessian $\boldsymbol{H}$ at the minimum $\boldsymbol{x}_i$. Then, using the quantum mechanical vibrational partition function for each harmonic oscillator, with the zero-point energy taken as the reference for energy levels,

$$Q_{\mathrm{vib},a} = [1 - \exp(\beta h \nu_a)]^{-1}, \tag{6.3.5}$$

we obtain that

$$k_{i\to j}^{\mathrm{TST}} = \frac{k_{\mathrm{B}}T}{h} \frac{\prod_{a=1}^{n}\left[1-\exp\left(-\beta h\nu_a^{(i)}\right)\right]}{\prod_{a=2}^{n}\left[1-\exp\left(-\beta h\nu_a^{(ij)}\right)\right]} \exp\left[-\frac{\mathcal{V}\left(\boldsymbol{x}_{ij}\right)-\mathcal{V}\left(\boldsymbol{x}_{i}\right)}{k_{\mathrm{B}}T}\right]. \tag{6.3.6}$$

The eigenvalues $\lambda_a^{(ij)}$ or $\lambda_a^{(i)}$, which determine the eigen frequencies appearing in Eq. (6.3.4), fall into two categories: those which remain finite as the force constants of the (possible) stiff generalized degrees of freedom approach infinity and are determined through Eq. (6.3.4); and those which approach infinity along with the force constants. For the latter category, the factors appearing in the pre-exponential of Eq. (6.3.6) reduce to unity. Ultimately, then, the rate constant assumes the form

$$k_{i\to j}^{\mathrm{TST}} = \frac{k_{\mathrm{B}}T}{h} \frac{\prod_{a=1}^{f}\left[1-\exp\left(-\beta h\nu_a^{(i)}\right)\right]}{\prod_{a=2}^{f}\left[1-\exp\left(-\beta h\nu_a^{(ij)}\right)\right]} \exp\left[-\frac{\mathcal{V}\left(\boldsymbol{x}_{ij}\right)-\mathcal{V}\left(\boldsymbol{x}_{i}\right)}{k_{\mathrm{B}}T}\right], \tag{6.3.7}$$

where $f < n$ enumerates the generalized degrees of freedom which are flexible along any part of the reaction path.

The multidimensional TST formulation invoked above is cast in terms of the mass-weighted Cartesian coordinates $\boldsymbol{x}$, or, equivalently, the generalized coordinates $\boldsymbol{q}$ describing the *full* configuration of the polymer + penetrant system. If molecular models are invoked in which some degrees of freedom (e.g. bond lengths) are envisioned as subject to infinitely stiff harmomic potentials, the reaction path must be conducted in the *flexible* subset of the generalized coordinates, considering that (a) the mass-weighted Cartesian coordinates of individual atoms are no longer independent; and (b) polymer degrees of freedom away from the penetrant are considered unchanged in the reaction path calculations. Following the notation introduced in Chapter 3, let us use the symbol $\boldsymbol{q}$ to denote the full set of generalized coordinates (these are n in number, as many as the mass-weighted Cartesian coordinates $\boldsymbol{x}$), the symbol $\boldsymbol{q}_{\mathrm{f}}$ to denote the subset of independent *flexible* generalized coordinates in which we wish to conduct our calculation, and the symbol $\boldsymbol{q}_{\mathrm{h}}$ to denote the remaining *stiff* or *hard* generalized coordinates. The gradient vector and Hessian matrix of $\mathcal{V}$ with respect to $\boldsymbol{x}$ are denoted by $\nabla_{\boldsymbol{x}}\mathcal{V}$ and $\boldsymbol{H}$, respectively. Those with respect to $\boldsymbol{q}$ are denoted by $\nabla_{\boldsymbol{q}}\mathcal{V}$ and $\boldsymbol{H}_{\boldsymbol{qq}}$, respectively (see Eqs. (6.3.1)–(6.3.3)). The gradient vectors $\nabla_{\boldsymbol{q}}\mathcal{V}$ and $\nabla_{\boldsymbol{x}}\mathcal{V}$ are related through Eqs. (3.1.30) and (3.1.31) of Chapter 3.

Since we have partitioned the vector $\{\boldsymbol{q}\}$ of generalized coordinates into *flexible* and *hard* subsets, $\{\boldsymbol{q}\} = \{\boldsymbol{q}_{\mathrm{f}}, \boldsymbol{q}_{\mathrm{h}}\}$, then, following Chapter 3, we write $\mathrm{d}\boldsymbol{q} = \begin{pmatrix}\mathrm{d}\boldsymbol{q}_{\mathrm{f}}\\ \mathrm{d}\boldsymbol{q}_{\mathrm{h}}\end{pmatrix}$ and $\nabla_{\boldsymbol{q}}\mathcal{V} = \begin{Bmatrix}\nabla_{\boldsymbol{q}_{\mathrm{f}}}\mathcal{V}\\ \nabla_{\boldsymbol{q}_{\mathrm{h}}}\mathcal{V}\end{Bmatrix}$. We can also make use of Eq. (3.1.20) of Chapter 3 for the partitioning of the covariant metric tensor $\boldsymbol{a}$ into flexible–flexible, flexible–hard, hard–flexible, and hard–hard parts, to express the eigenvalue problem at a minimum or saddle point, Eq. (6.3.2), in the form

$$\begin{pmatrix} \boldsymbol{H}_{\boldsymbol{qq},\mathrm{ff}} & \boldsymbol{H}_{\boldsymbol{qq},\mathrm{fh}} \\ \boldsymbol{H}_{\boldsymbol{qq},\mathrm{hf}} & \boldsymbol{H}_{\boldsymbol{qq},\mathrm{hh}} \end{pmatrix} \cdot \begin{pmatrix} \mathrm{d}\boldsymbol{q}_\mathrm{f} \\ \mathrm{d}\boldsymbol{q}_\mathrm{h} \end{pmatrix} = \lambda \boldsymbol{a} \cdot \begin{pmatrix} \mathrm{d}\boldsymbol{q}_\mathrm{f} \\ \mathrm{d}\boldsymbol{q}_\mathrm{h} \end{pmatrix}, \tag{6.3.8}$$

which is again the differential analogue of Eq. (3.1.37) of Chapter 3. The IRC equation can also be partitioned into flexible–flexible, flexible–hard, hard–flexible, and hard–hard blocks, as follows:

$$\begin{pmatrix} \boldsymbol{a}_\mathrm{ff} & \boldsymbol{a}_\mathrm{fh} \\ \boldsymbol{a}_\mathrm{hf} & \boldsymbol{a}_\mathrm{hh} \end{pmatrix} \cdot \begin{pmatrix} \mathrm{d}\boldsymbol{q}_\mathrm{f} \\ \mathrm{d}\boldsymbol{q}_\mathrm{h} \end{pmatrix} = \begin{pmatrix} \nabla_{\boldsymbol{q}_\mathrm{f}}\mathcal{V} \\ \nabla_{\boldsymbol{q}_\mathrm{h}}\mathcal{V} \end{pmatrix} \cdot \mathrm{d}\tau. \tag{6.3.9}$$

Now let us check the limit of *infinite* force constants for the hard degrees of freedom. The only way in which the terms of Eq. (6.3.8) can remain finite in the presence of infinite force constants is that the set of $\{\boldsymbol{q}_\mathrm{h}\}$ remain confined to their equilibrium positions, i.e. $\mathrm{d}\boldsymbol{q}_\mathrm{h} = 0$. Under this condition, Eqs. (6.3.1) to (6.3.3) reduce to their analogues with $\boldsymbol{q}_\mathrm{f}$ in place of $\boldsymbol{q}$, $\boldsymbol{a}_\mathrm{ff}$ in place of $\boldsymbol{a}$, $\nabla_{\boldsymbol{q}_\mathrm{f}}\mathcal{V}$ in place of $\nabla_{\boldsymbol{q}}\mathcal{V}$, and $\boldsymbol{H}_{\boldsymbol{qq},\mathrm{ff}}$ in place of $\boldsymbol{H}_{\boldsymbol{qq}}$.

6.3.3 From states to macrostates: formulation of the master equation

Typical distributions of transition rate constants as calculated with the analysis just described[48] indicate that interstate rate constants follow a broad distribution with many transitions connecting states communicating over energy barriers that are low relative to $k_\mathrm{B}T$. These states can be lumped together into *macrostates*, since exchange between the states within a macrostate is fast while transitions between different macrostates are rate-limiting for diffusion. To expedite computations, it makes sense to also coarse grain the master equation describing diffusion (Section 6.1.7) from a state to a macrostate level. To this end, let us consider that a typical macrostate I encompasses several states i. Summing the master equation, Eq. (6.1.61), over all i in I we obtain that

$$\frac{\mathrm{d}P_I(t)}{\mathrm{d}t} = -\sum_j \left(\sum_{i\in I} k_{i\to j} \frac{P_i}{P_I} \right) P_I + \sum_j \left(\sum_{i\in I} k_{j\to i} P_j \right), \tag{6.3.10}$$

where we have set

$$P_I = \sum_{i\in I} P_i. \tag{6.3.11}$$

In Eq. (6.3.10), the destination states j can also be grouped according to the macrostates $\mathcal{J}$ to which they belong. This leads to

$$\frac{\mathrm{d}P_I(t)}{\mathrm{d}t} = -\sum_{\mathcal{J}} \left(\sum_{i\in I} \sum_{j\in \mathcal{J}} k_{i\to j} \frac{P_i}{P_I} \right) P_I + \sum_{\mathcal{J}} \left(\sum_{j\in \mathcal{J}} \sum_{i\in I} k_{j\to i} \frac{P_j}{P_{\mathcal{J}}} \right) P_{\mathcal{J}}. \tag{6.3.12}$$

Furthermore, due to the fast exchange between states within a macrostate, one can substitute the ratios P_i/P_I and $P_j/P_{\mathcal{J}}$ by their equilibrium values $P_i^{\text{eq}}/P_I^{\text{eq}}$ and $P_j^{\text{eq}}\big/P_{\mathcal{J}}^{\text{eq}}$, respectively, obtaining

$$\frac{\mathrm{d}P_I(t)}{\mathrm{d}t} = -\sum_{\mathcal{J}} k_{I\to\mathcal{J}} P_I + \sum_{\mathcal{J}} k_{\mathcal{J}\to I} P_{\mathcal{J}}. \tag{6.3.13}$$

Equation (6.3.13) is a master equation at the level of macrostates, and is entirely analogous to Eq. (6.1.61) at the level of states with

$$k_{I\to\mathcal{J}} = \sum_{i\in I}\sum_{j\in\mathcal{J}} k_{i\to j}\frac{P_i^{\text{eq}}}{P_I^{\text{eq}}}. \tag{6.3.14}$$

Rate constants $k_{i\to j}$ for all relevant state-to-state transitions needed in Eq. (6.3.14) are already known through the procedures described in Section 6.1.6. The additional equilibrium ratios $P_i^{\text{eq}}/P_I^{\text{eq}}$ for all termini i of inter-macrostate paths within a macrostate I which are also required in Eq. (6.3.12) can be estimated through short MD simulations of the polymer–penetrant system confined in macrostate I, as ratios of time spent in the state i to time spent in the entire macrostate. Then, given a network representative of the macrostates and of the connections between them, the master equation at the level of macrostates can be solved by kMC simulation as already described in Section 6.1.7.

6.3.4 Alternative approaches to getting the diffusivity from the master equation. MESoRReD, KOBRA

Kinetic Monte Carlo simulations are usually very inexpensive in comparison to the calculations needed for extracting the rate constants. Thus, they constitute the most common method employed to solve the master equation and compute (e.g.) the elements of the self-diffusivity tensor by tracking the msd of the walkers as a function of time, via the Einstein relation. kMC calculations become problematic only when jumps of widely disparate rate constants are present in the description. In this case, alternative methods which circumvent the need to conduct kMC simulations and which allow for a safe estimation of the diffusivity directly from the geometry and connectivity of sites and the intersite rate constants have been proposed. One such method is the analytical approach known as master equation solution by recursive reduction of dimensionality (MESoRReD),[62] mentioned briefly in Section 6.1.7.

In MeSoRReD, one constructs a rate constant matrix $\boldsymbol{K}_1$ with dimension $m \times m$, where $m \times m$ is the number of states in one unit cell, taking into account the periodic boundary conditions across the faces of the unit cell. The diagonal elements of $\boldsymbol{K}_1$ equal the negative sum of the rest of the elements in a column. For example, if transitions occur between states j and i, the elements $\boldsymbol{K}_1(i,j)$ and $\boldsymbol{K}_1(j,i)$ must be defined and $\boldsymbol{K}_1(i,i) = -\sum_{j\neq i}\boldsymbol{K}_1(j,i)$. Also, transport in each coordinate direction is analysed separately. For a specific direction, considered as extending horizontally from left to right, a $\boldsymbol{K}_{R1}$ matrix is constructed. $\boldsymbol{K}_{R1}$ contains the rate constants of transition from a unit cell to the cell neighbouring it on the right. All its remaining elements are 0. Similarly, a $\boldsymbol{K}_{L1}$ matrix is constructed containing only the rate constants of transitions between a unit cell and its neighbour to the left. Knowing $\boldsymbol{K}_1$, $\boldsymbol{K}_{R1}$, and $\boldsymbol{K}_{L1}$, one constructs the matrix $\boldsymbol{A}_\nu = \boldsymbol{K}_1 - \left[1 - \cos\left(\frac{\pi}{2^{\nu-1}}\right)\right](\boldsymbol{K}_{R1} + \boldsymbol{K}_{L1}) + i\sin\left(\frac{\pi}{2^{\nu-1}}\right)(\boldsymbol{K}_{R1} - \boldsymbol{K}_{L1})$, with i being the imaginary unit. The matrix $\boldsymbol{A}_\nu$ is characteristic of a sequence of 2^ν unit cells arranged along the considered direction with periodic conditions at the ends. Then, the eigenvalue of $\boldsymbol{A}_\nu$ with the smallest absolute magnitude, $\min(|\lambda_{\boldsymbol{A}_\nu}|)$, yields the self-diffusivity in the considered direction. Using L_d to denote the length of the unit cell in the considered direction, the self-diffusivity D in this direction is obtained as[62]

$$D = \lim_{\nu \to +\infty} \left\{ \min(|\lambda_{\boldsymbol{A}_\nu}|) \left(\frac{2^{\nu-1} L_d}{\pi} \right)^2 \right\}. \tag{6.3.15}$$

The main disadvantage of the MeSoRReD method is that it is applicable only for orthogonal crystal systems at infinite dilution, in sharp contrast to the kMC method which can be used for all crystal systems at different loadings. However, Braun and Sholl[63] have proposed an analytical scheme to calculate the exact diffusion tensor at infinite dilution for any crystal system, albeit through rather complicated matrix calculations. Recently, the complex Braun–Sholl equations have been rewritten[64] in a more compact way and incorporated in a ready-to-use, graphical user interface program (KOBRA), requiring as input data only the values for the unit cell lengths and angles, and the rate constants for the transitions of the penetrant.

6.4 Multidimensional transition-state theory approach to structural relaxation in glasses

Glassy materials play an important role in our lives and have therefore constituted an object of extensive research, at both basic and applied levels. They are considered to be nonequilibrium materials because their properties depend on their formation history. Furthermore, their properties change very slowly with time in the course of *physical ageing*, whose characteristic times exceed common macroscopic observation times below the glass temperature T_g. The study of glassy materials by means of molecular simulation[65–67] faces serious challenges, because one needs to bridge time scales spanning some twenty orders of magnitude, from the period of fast atomic vibrations (10^{-14} s) up to the longest time for structural, volume, and enthalpy relaxation (on the order of years 20 K below T_g). State-of-the-art theories of the supercooled liquid state include mode coupling theory[68] and theories for enumerating stationary points[69] on the multidimensional energy hypersurface of the system.[70] *Fragile* glass-forming liquids, whose viscosity exhibits a strongly non-Arrhenius dependence on temperature, are characterized by very rugged potential energy landscapes. This is seen characteristically in the *disconnectivity graphs*.[71]

Tracking structural relaxation in a glass at temperatures below T_g over times (milliseconds to years) relevant to the applications of glasses as structural, optical, packaging, and membrane materials is clearly impossible by direct MD simulation; reverting to an infrequent event theory-based approach seems much more appropriate. On the other hand, the rugged potential energy landscape of glass-forming systems gives rise to a very broad distribution of characteristic times for elementary transitions and a complex connectivity graph among basins. kMC simulation would have to track the fastest of these transitions, and this would limit its ability to sample long-time evolution. An approach based on analytical solution of the master equation (compare Section 6.1.7), equivalent to averaging over all dynamical trajectories originating from a given initial distribution among basins, would seem more promising. However, it is impossible to build a complete map of all basins and transitions between them in the rugged potential energy landscape of a glassy system even of modest size. A way out of this difficulty is provided by the fact that, when one studies structural relaxation, one typically starts from an initial distribution among states that is highly localized (e.g. from a single basin in the potential energy landscape, where the system was trapped via the glass formation history that was followed to obtain it). The region of configuration space where the system resides is thus initially very confined and expands gradually as transitions between basins take place.

6.4.1 Determining the states on the fly; DIMW approach and results for glassy atactic polystyrene

Based on the idea outlined in the last paragraph of the previous Section, one can develop[65,72] a computational approach for tracking the temporal evolution of the distribution among basins (or *states*) around local minima of the potential energy hypersurface of a glass via infrequent transitions, starting off from a highly localized initial distribution. This approach is called *dynamic integration of a Markovian web*, or DIMW. DIMW distinguishes states that it samples into two categories: *explored* and *boundary* states. An *explored* state is a state for which an exhaustive calculation of as many as possible transition pathways leading out of it to neighbouring states has been undertaken and rate constants associated with these transitions have been computed. In the application to isothermal–isochoric structural relaxation of a polymer glass,[65,72] this calculation proceeds by computing as many as possible saddle points of the potential energy in $(3N-3)$-dimensional configuration space around the state under investigation[72] and subsequently constructing a transition path through each of these saddle points to a neighbouring state via Fukui's IRC approach.[61] *Boundary* states, on the other hand, are states connected to explored states, which, however, have not been explored themselves.

If E and B symbolize the current set of explored and boundary states, respectively, the DIMW algorithm makes use of the conditional probability $Q_i(t; B)$ that the system resides in explored state i at time t, provided it has never visited any of the boundary states in B. This probability obeys the master equation

$$\frac{\partial Q_i(t;B)}{\partial t} = \sum_{m \in E} Q_i(t;B)\, k_{m\to i} - Q_i(t;B) \sum_{j \in E \cup B} k_{i\to j}, \quad i \in E. \qquad (6.4.1)$$

According to Eq. (6.4.1), $Q_i(t; B)$ increases with time as a result of influx from other explored states but decreases with time as a result of efflux to other explored *and boundary* states.

The rate constants appearing in Eq. (6.4.1) can be grouped together to form a rate constant matrix $\boldsymbol{K}$ analogous to that of Eq. (6.1.62), the only difference being that here the diagonal elements K_{ii} contain contributions of the type $-k_{i\to j}$, $i \in E, j \in B$ leading out of explored states to boundary states. The matrix $\boldsymbol{K}$ is not stochastic in the sense of page 327. However, through an extension of the derivation presented in connection with Problem 6.1, one can show that this matrix is negative definite (all its eigenvalues are negative). In fact, if $P_i^{\mathrm{eq};E}$ is the equilibrium occupancy probability of state i for the current set of explored states, through the transformation $\tilde{K}_{ij} = K_{ij}\sqrt{P_j^{\mathrm{eq};E}} \Big/ \sqrt{P_i^{\mathrm{eq};E}}$, this $\boldsymbol{K}$ matrix can be converted into a symmetric, nonstochastic matrix $\tilde{\boldsymbol{K}}$ that is mathematically similar to (has the same eigenvalues as) $\boldsymbol{K}$. This makes it possible to express the conditional probabilities $Q_i\,(t;B)$ analytically as sums of decaying exponential functions of time involving the eigenvalues and eigenvectors of $\tilde{\boldsymbol{K}}$, as in Eq. (6.1.63):

$$Q_i\,(t;B) = \tilde{Q}_i\,(t;B)\sqrt{P_i^{\mathrm{eq};E}}, \quad \tilde{\boldsymbol{Q}}\,(t;B) = \sum_{m=0}^{n-1}\left[\tilde{\boldsymbol{u}}_m \cdot \tilde{\boldsymbol{Q}}\,(0;B)\right] \mathrm{e}^{\lambda_m t}\,\tilde{\boldsymbol{u}}_m. \tag{6.4.2}$$

Knowing $Q_i\,(t;B)$, one can compute the rate of reaching a specific boundary state $j \in B$ from the current set of explored states at time t, the total efflux of probability (probability per unit time) from the current set of explored states to the current boundary states, and the distribution of the corresponding escape times.[72]

An application of the DIMW method using as a model system glassy polystyrene of degree of polymerization equal to 80, pre-equilibrated at 300 K and 1 bar, has provided the probability flux to boundary states as a function of time, $F(t;B)$. For a sufficiently large set of explored states, this settles to an asymptotic shape with well-defined maxima at two characteristic times. These times are shown for three sub-T_g temperatures in Figure 6.4. Compared against experimental spectroscopic measurements,[73,74] they are seen to be representative of the γ and δ relaxation transitions in polystyrene. At higher temperatures, direct MD simulation in the melt[75] can capture the time scales of the main (α) relaxation, which displays strongly non-Arrhenius (Willams–Landel–Ferry (WLF) or Vogel–Fulcher–Tammann (VFT)) temperature dependence. With modest computational resources, DIMW can access time scales on the order of milliseconds, thus nicely complementing the information that can be obtained from MD by providing access to dynamical processes deep in the glassy state. Nevertheless, the method suffers from heavy linear algebra operations, i.e. the diagonalization of large matrices (which scales with the third power of the number of states), and has high storage requirements (which is not the case for either MD or MC). Additional developments are needed in the future to alleviate some of these shortcomings.

6.4.2 Lumping

A difficulty with DIMW-type approaches is that the number of states to be tracked becomes prohibitively large at long times. A way out of this problem is to group, or *lump*, states communicating via transitions that are fast in relation to the observation time into single clusters of states. If performed judiciously, this lumping does not result in loss of essential information. At long observation times, the system distributes itself among fast-communicating states according to the requirements of a restricted equilibrium, so clusters of such states behave as single *meta-states*, for all practical purposes. From the mathematical point of view, lumping has been examined in the context of networks of chemical reactions in the classic work of Wei and Kuo[76] and in several subsequent works. As shown there, lumping calls for the determination of an $\hat{n} \times n$ transformation matrix $\boldsymbol{M}$, where

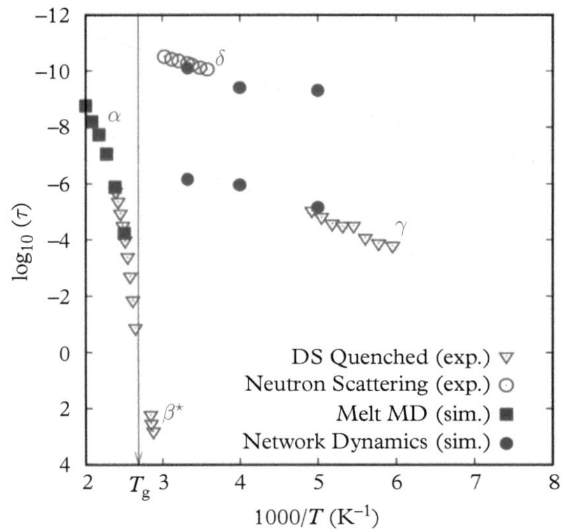

Figure 6.4 Logarithm of segmental relaxation time as a function of inverse temperature for 80-mer glassy atactic polystyrene. The filled circle symbols correspond to the maxima in $F(t; B)$ obtained through the DIMW approach at 300, 250, and 200 K. They are seen to be close to experimental estimates of time scales of the subglass γ and δ relaxation processes measured by dielectric spectroscopy[75] and neutron scattering,[76] respectively. The square points are correlation times obtained by integration of orientational autocorrelation functions of pendant bond vectors from MD simulations in the melt state at 1 bar and various temperatures, using the same force field. They are seen to be in good agreement with α relaxation measurements from dielectric spectroscopy.[75]

Reproduced (Figure 7) from [Vogiatzis, G.G., van Breemen, L.C.A., Theodorou, D.N., and Hütter, M. *Mol. Syst. Des. Eng.* **8**, 1013 (2023)], with permission from the Institution of Chemical Engineers (IChemE) and the Royal Society of Chemistry.

n is the number of original states and $\hat{n} < n$ the number of lumped states (or clusters of states). The transformation from the probability distribution among the original states to that among the lumped states at any time t takes the form

$$\hat{\boldsymbol{P}}(t) = \boldsymbol{M} \cdot \boldsymbol{P}(t). \tag{6.4.3}$$

The lumping matrix $\boldsymbol{M}$ has the following properties:

i. Its elements are either '0' or '1'.
ii. Each of its columns contains exactly one '1'. The physical meaning behind this is that every state of the original description (assigned to a column of $\boldsymbol{M}$) belongs to one cluster only (assigned to a row of $\boldsymbol{M}$).
iii. The position of '1' in every column of $\boldsymbol{M}$ (i.e. state in the original description) describes to which cluster (row of $\boldsymbol{M}$) the state of the initial system is being lumped.

Once $\boldsymbol{M}$ is known, the $\hat{n} \times n$ rate constant matrix $\hat{\boldsymbol{K}}$ to be used at the lumped level is calculated as

$$\hat{\boldsymbol{K}} = \boldsymbol{M} \cdot \boldsymbol{K} \cdot \boldsymbol{A} \cdot \boldsymbol{M}^{\mathrm{T}} \cdot \hat{\boldsymbol{A}}^{-1} \tag{6.4.4}$$

where $\boldsymbol{A}$ is an $n \times n$ diagonal matrix whose diagonal elements equal the elements of the equilibrium probability vector $\boldsymbol{P}(\infty)$ corresponding to the original rate constant matrix $\boldsymbol{K}$, the superscripts 'T' and '−1' indicate matrix transpose and matrix inverse, respectively, and $\hat{\boldsymbol{A}}$ is defined as

$$\hat{\boldsymbol{A}} = \boldsymbol{M} \cdot \boldsymbol{A} \cdot \boldsymbol{M}^{\mathrm{T}} \tag{6.4.5}$$

Lempesis et al.[77] proposed a methodology for determining the number of lumped states $\hat{n}$ and the lumping matrix $\boldsymbol{M}$ in such a way that the long-time dynamics of the original description is reproduced. The strategy is to minimize an objective function of the form

$$z\left(\hat{n}, \boldsymbol{M}\right) = z_1 E + z_2 W + z_3 \hat{n} \tag{6.4.6}$$

with z_1, z_2, and z_3 being predefined as real positive constants, E is the Frobenius norm of the $\hat{n} \times n$ error matrix $\boldsymbol{E}\left(\equiv \boldsymbol{M} \cdot \boldsymbol{K} - \hat{\boldsymbol{K}} \cdot \boldsymbol{M}\right)$

$$E = \|\boldsymbol{E}\|_F = \sqrt{\sum_{i=1}^{n} \sum_{j=1}^{\hat{n}} |E_{ij}|^2} \tag{6.4.7}$$

(for exact lumping, the lumping error E would be 0), and W is the Frobenius norm of the lumped matrix $\hat{\boldsymbol{K}}$.

The minimization of the objective function defined in Eq. (6.4.6) can be performed stochastically, using Monte Carlo moves which change the dimensionality $\hat{n}$ and the form of the lumping matrix $\boldsymbol{M}$, while respecting the constraints on the form of that matrix (stated above). To avoid trapping in local minima of the objective function, a Wang–Landau scheme can be invoked to determine the density of $\boldsymbol{M}$-matrix *states* in the space of variables $(E, W, \hat{n})$ and pick that $\boldsymbol{M}$, close to the origin of $(E, W, \hat{n})$ space, which minimizes the objective function.

The lumping strategy has been applied[77] to a mixture of Lennard-Jones particles of two different sizes at a temperature of 37 K, just below T_g. Histograms of the negative inverse eigenvalues $t_i = -1/\lambda_i$ of the rate constant matrices $\boldsymbol{K}$ (original description) and $\hat{\boldsymbol{K}}$ (lumped description) were seen to be similar. Furthermore, the eight longest t_i values were seen to agree quantitatively between the original and lumped system, testifying to the success of the lumping method in reproducing the long-time dynamics of the original system.

Problem

Problem 6.1: Show that the reduced rate constant matrix $\tilde{\boldsymbol{K}}$ defined in Section 6.1.7 is negative semidefinite.

References

1. Lindorff-Larsen, K., Piana, S., Dror, R.O., and Shaw, D.E. *Science* **334**, 517 (2011).
2. Vineyard, G.H. *J. Phys. Chem. Solids* **3**, 121 (1957).
3. Gusev, A.A., Müller-Plathe, F., van Gunsteren, W.F., and Suter, U.W. *Advan. Polym. Sci.* **116**, 207 (1994).
4. Theodorou, D.N., Snurr, R.Q., and Bell, A.T. In Alberti, G., and Bein, T. (eds.), *Comprehensive Supramolecular Chemistry*, pp. 507–548. Elsevier Science, Oxford (1994).
5. Wales, D.J., Doye, J.P.K., Miller, M.A., Mortenson, P.N., and Walsh, T.R. *Advan. Chem. Phys.* **115**, 1 (2000).
6. Stillinger, F.H. *Science* **267**, 1935 (1995).
7. Voter, A.F., and Doll, J.D. *J. Chem. Phys.* **82**, 80 (1985).
8. Wolynes, P.G. *Phil. Trans. Roy. Soc. A* **363**, 453 (2005).
9. Wei, J., and Prater, C.D. *Advan. Catal.* **13**, 204 (1962).
10. Glasstone, S., Laidler, K.J., and Eyring, H. *The Theory of Rate Processes*. McGraw-Hill, New York (1941).
11. Moore, J.W., and Pearson, R.G. *Kinetics and Mechanism* (3rd ed.). J. Wiley and Sons, New York (1981).
12. Kärger, J.; Ruthven, D.M., and Theodorou, D.N. *Diffusion in Nanoporous Materials*. Wiley-VCH, Berlin (2012).
13. Weiner, J.H. *Statistical Mechanics of Elasticity*. J. Wiley and Sons, New York (1983).
14. Kramers, H. *Physica* **7**, 284 (1940).
15. Bennett, C.H. In Nowick, A.S., and Burton, J.J. (eds.), *Diffusion in Solids: Recent Developments*, pp. 73–113. Academic Press, New York (1975).
16. Chandler, D. *J. Chem. Phys.*, **68**, 2959 (1978).
17. June, R.L., Bell, A.T., and Theodorou, D.N. *J. Phys. Chem.*, **95**, 8866 (1991).
18. Gusev, A.A., and Suter, U.W. *J. Chem. Phys.* **99**, 2228 (1993).
19. Kolokathis, P., Kali, G.; Jobic, H., and Theodorou, D.N. *J. Phys. Chem. C*, **120**, 21410 (2016).
20. Forester, T.R., and Smith, W. *J. Chem. Soc., Faraday Trans.* **93**, 3249 (1997).
21. Theodorou, D.N. In Neogi, P. (ed.), *Diffusion in Polymers*, pp. 67–142. Dekker, New York (1995).
22. Theodorou, D.N. In Yampolskii, Yu., Pinnau, I., and Freeman, B.D. (eds.), *Materials Science of Membranes for Gas and Vapor Separation*, pp. 47–92. John Wiley and Sons, Hoboken, NJ (2006).
23. Stillinger, F.H., and Weber, T.A. *Phys. Rev. A* **25**, 978 (1982).
24. Helfand, E. *J. Chem. Phys.* **69**, 1010 (1978).
25. Theodorou, D.N. In Grotendorst, J., Sutmann, G., Gompper, G., and Marx, D. (eds.), *Hierarchical Methods for Dynamics in Complex Molecular Systems*, IAS Series Volume 10, pp. 347–389. Schriften des Forschungzentrums Jülich, Jülich (2012).
26. Sørensen, M. R., and Voter, A.F. *J. Chem. Phys.* **112**, 9599 (2000).
27. Tsalikis, D.G., Lempesis, N., Boulougouris, G.C., and Theodorou, D.N. *J. Phys. Chem. B* **114**, 7844 (2010).
28. Singh, S., Chiu, C.C., and de Pablo, J. *J. Stat. Phys.* **144**, 1 (2011).
29. Laio, A., and Parrinello, M. *Proc. Natl. Acad. Sci. USA* **99**, 1256 (2002).
30. Bolhuis, P.G., Chandler, D., Dellago, C., and Geissler, P.L. *Annu. Rev. Phys. Chem.* **53**, 291 (2002).
31. Dellago, C., Bolhuis, P.G., and Geissler, P.L. *Adv. Chem. Phys.* **123**, 1 (2002).
32. Beeler, J.R. *J. Appl. Phys.* **37**, 3000 (1966).
33. Voter, A.F. In Sickafus, K.E., Kotomin, E.A., and Uberuaga, B.P. (eds.), *Radiation Effects in Solids*, NATO Science Series II: Mathematics, Physics, and Chemistry, Vol. 235, pp. 1–23. Springer, Dordrecht, The Netherlands (2007).
34. Feller, W. *An Iintroduction to Probability Theory and Its Applications*, Volume 2. John Wiley and Sons, New York (1971).

35. Knuth, D.E. *The Art of Computer Programming*, Volume 2, Chapter 3. Addison-Wesley, Reading, MA (1997).
36. Buchete, N.V., and Hummer, G. *J. Phys. Chem. B*, **112**, 6057 (2008).
37. Boulougouris, G., and Theodorou, D.N. *J. Chem. Phys.* **130**, 044905 (2009).
38. Kolokathis, P.D., and Theodorou, D.N. *J. Chem. Phys.* **137**, 034112 (2012).
39. Duda, J.L., and Zielinski, J.M. In Neogi, P. (ed.), *Diffusion in Polymers*, Chapter 3. Marcel Dekker, New York (1996).
40. Takeuchi, H., and Okazaki, K. *J. Chem. Phys.* **92**, 5643 (1990).
41. Trohalaki, S., Kloczowski, A., and Mark, J.E. In Roe, R.-J. (ed.), *Computer Simulation of Polymers*, Chapter 17, p. 220. Prentice Hall, Englewood Cliffs, NJ (1991).
42. Sok, R.M., Berendsen, H.J.C., and van Gunsteren, W.F. *J. Chem. Phys.* **97**, 4699 (1992).
43. Pant, P.V.K., and Boyd, R.H. *Macromolecules* **25**, 494 (1992).
44. Chassapis, C., Petrou, J.K., Petropoulos, J.H., and Theodorou, D.N. *Macromolecules* **29**, 3615 (1996).
45. Takeuchi, H. *J. Chem. Phys.* **93**, 2062 (1990).
46. The discussion in Sections 6.2 and 6.3 borrows heavily from Theodorou, D.N. in Berne, B.J., Ciccotti, G., and Coker, D.F. (eds.), *Classical and Quantum Dynamics in Condensed Phase Simulations*, pp. 211–249. World Scientific, Singapore (1998).
47. Gusev, A.A., Arizzi, S., Suter, U.W., and Moll, D.J. *J. Chem. Phys.* **99**, 2221 (1993).
48. (a) Greenfield, M.L., and Theodorou, D.N. *Macromolecules* **26**, 5461 (1993); (b) Greenfield, M.L., and Theodorou, D.N. *Macromolecules* 31, 7068 (1998).
49. Greenfield, M.L. *Molecular Modeling of Dilute Penetrant Gas Diffusion in a Glassy Polymer Using Multidimensional Transition-State Theory.* PhD Thesis, University of California, Berkeley (1996).
50. Greenfield, M.L., and Theodorou, D.N. *Molecular Simulation* **19**, 329 (1997).
51. Greenfield, M.L., and Theodorou, D.N. *Macromolecules* **34**, 8541 (2001).
52. Anikeenko, A.V., Alinchenko, M.G., Voloshin, V.P., Medvedev, N.N., Gavrilova, M.L., and Jedlovszky, P. In Laganà, A., Gavrilova, M.L., Kumar, V., Mun, Y., Tan, C.J.K., and Gervasi, O. (eds.), *Computational Science and Its Applications—ICCSA 2004, Pt 3*, Vol. 3045, pp. 217–226 (2004).
53. Tanemura, M., Ogawa, T., and Ogita, N., *J. Comput. Phys.* **51**, 191 (1983).
54. Dodd, L.R., and Theodorou, D.N. *Mol. Phys.* **72**, 1313 (1991).
55. Arizzi, S., Mott, P.H., and Suter, U.W. *J. Polym. Sci. B Polym. Phys.* **30**, 415 (1992).
56. In 't Veld, P.J., Stone, M.T., Truskett, T.M., and Sanchez, I.C. *J. Phys. Chem. B* **104**, 12028 (2000).
57. Voyiatzis, E., Böhm, M.C., and Müller-Plathe, F. *Comput. Phys. Commun.* **196**, 247 (2015).
58. Voyiatzis, E., Müller-Plathe, F., and Böhm, M.C. *Polymer* **101**, 107 (2016).
59. Mermigkis, P.G., and Mavrantzas, V.G. *Macromolecules* **53**, 9563 (2020).
60. Theodorou, D.N., and Suter, U.W. *Macromolecules* **18**, 1206 (1985).
61. Fukui, K. *Acc. Chem. Res.* **14**, 363 (1981).
62. Kolokathis, P.D., Pantatosaki, E., Gatsiou, C.-A., Jobic, H., Papadopoulos, G.K., and Theodorou, D.N. *Mol. Sim.* **40**, 80 (2014).
63. Braun, O.M., and Sholl, C.A. *Phys. Rev. B* **58**, 14870 (1998).
64. Kolokathis, P.D., and Braun, O.M. *J. Comput. Chem.* **40**, 2053 (2019). The KoBra code can be accessed through: https://sourceforge.net/projects/kobra/.
65. Boulougouris, G.C., and Theodorou, D.N. *J. Chem. Phys.* **125**, 084903 (2007).
66. Tsalikis, D.G., Lempesis, N., Boulougouris, G.C., and Theodorou, D.N. *J. Phys. Chem. B* **112**, 10619 (2008).
67. Tsalikis, D.G., Lempesis, N., Boulougouris, G., and Theodorou, D.N. *J. Chem. Theory Comp.* **6**, 1307 (2010).

68. Götze, W. In Hansen, J.P., Levesque, D., and Zinn-Justin, J. (eds.), *Les Houches Session LI, 1989 Liquids, Freezing and Glass Transition*, pp. 287–503. Elsevier Science Publishers B.V., Amsterdam (1991).
69. Shell, S.M., Debenedetti, P.G., and Panagiotopoulos, A.Z. *Fluid Phase Equilib.* **241**, 147 (2006).
70. Sciortino, F. *J. Stat. Mech.* P050515 (2005).
71. de Souza, V.K., and Wales, D.J. *J. Chem. Phys.* **130**, 194508 (2009).
72. Vogiatzis, G.G., van Breemen, L.C.A., Theodorou, D.N., and Hütter, M. *Mol. Syst. Des. Eng.* **8**, 1013 (2023).
73. Grigoriadi, K., Putzeys, T., Wubbenhorst, M., van Breemen, L.C.A., Anderson, P.D., and Hütter, M. *J.Polym.Sci. B Polym. Phys.* **57**, 1394 (2019).
74. Arrese-Igor, S., Arbe, A., Frick, B., and Colmenero, J. *Macromolecules* **44**, 3161 (2011).
75. Vogiatzis, G.G., and Theodorou, D.N. *Macromolecules* **47**, 387 (2014).
76. Wei, J., and Kuo, J.C.W. *Ind. Eng. Chem. Fundam.* **8**, 114 (1969).
77. Lempesis, N., Tsalikis, D.G., Boulougouris, G.C., and Theodorou, D.N. *J. Chem. Phys.* **135**, 204507 (2011).

7

Coarse-Graining

7.1 The coarse-graining problem

Coarse-graining is a modelling strategy for overcoming the problem of large length and time scales characterizing structure, conformation, and dynamics in polymers by mapping atoms or repeat units to larger groups depending on the property(ies) of interest. It is accompanied by a significant reduction in the number of degrees of freedom. Depending on the purpose, different mapping procedures may be used: for example, a coarse-grained model for studying diffusive behaviour may be different from a model for studying the structure of a complex fluid. It is not surprising, therefore, that today a large number of coarse-grained models have been proposed for polymers, including mappings to chains consisting of beads or superatoms,[1–14] to uncrossable chains made by large blobs,[15,16] to ensembles of primitive paths[17–25] and primitive path networks,[26–29] to sets of pair correlation functions,[30] to dumbbells described by a conformation tensor,[31–34] and even to the level of single particles representing entire chains.[35–40]

In general, we can distinguish coarse-graining into *bottom up*, in which a more detailed atomistic description is used as a starting point; and *top down*, which aims at reproducing experimentally observed properties, without reference to a more detailed model. Furthermore, we can broadly distinguish coarse-graining efforts into (a) *thermodynamic*, which aims at capturing the potential of mean force of the system as a function of the variables invoked in the coarse-grained representation (coarse-grained degrees of freedom), and therefore its equilibrium structure, equation of state, and phase behaviour; and (b) *dynamical*, which, in addition, strives to capture the dynamical behaviour of the system, as quantified by time-correlation functions, transport properties, nonequilibrium structure, and the response to perturbations that take it far from equilibrium. A theoretical foundation for bottom-up dynamical coarse-graining is given by the projection operation formalism of Mori and Zwanzig,[41,42] which, however, cannot be applied in practice without simplifying approximations. Thermodynamic coarse-graining is a prerequisite for dynamical coarse-graining; it is more manageable and has been addressed more extensively in the literature, although by no means fully solved.

In thermodynamic coarse-graining, the potential of mean force, which generally depends on the entire set of coarse-grained variables (see Section 5.3.6), is often approximated as a sum of intramolecular and intermolecular terms involving pairs, triplets, etc. of coarse-grained particles, in a manner analogous to the way in which the potential energy is expressed as a function of atomic coordinates in atomistic models (see Chapters 2 and 3). It is important to realize that this decomposition of the potential of mean force is only an approximation, which is generally far less

Multiscale Modelling of Polymers. Doros N. Theodorou and Vlasis G. Mavrantzas, Oxford University Press.
 DOI: 10.1093/oso/9780199226191.003.0007

satisfactory than corresponding decompositions of the atomistic potential energy; also, that the terms constituting the potential of mean force in the coarse-grained representation may exhibit strong dependence on conditions such as temperature, pressure, and composition and are therefore far less transferable than analogous terms in atomistic force fields. In general, terms of the coarse-grained potential of mean force (*effective potentials*) are much softer than atomistic potentials. While atoms experience harsh excluded volume interactions, blobs defined by grouping these atoms can interpenetrate. This softness makes the dynamical equations of motion less stiff. In addition to the reduction in the total number of degrees of freedom, it is an important factor of the reduction in computational cost afforded by coarse-graining.

In bottom-up approaches, intramolecular potentials of mean force associated with the coarse-grained degrees of freedom are usually obtained by sampling unperturbed atomistic chains (or chain fragments) in continuous space subject to the atomistic force fields. When lattice models are invoked at the coarse-grained level, local intramolecular interactions can be described through a potential that reproduces classical rotational isomeric state (RIS)-based effective bond length and bond angle distributions (e.g. in the bond fluctuation model[1]) or via an extended RIS formalism in a second nearest-neighbour diamond lattice.[43] For nonbonded effective potentials, force matching, as proposed by Voth, Noid, and collaborators,[6,7,13,14] is a broadly used method, whose theoretical basis is briefly presented in Section 7.2. It entails minimizing the ensemble averaged sum of squared deviations between the atomistically calculated forces on each coarse-grained bead (sum of forces on all atoms of which the bead consists) and the forces calculated on the basis of the coarse-grained potential. A versatile structure-based method often employed to derive effective potentials of mean force is iterative Boltzmann inversion (IBI), which is discussed in detail in Section 7.3. Another method, which does not involve fitting to experimental data or to atomistic simulation results, is that of pretabulated potentials.[8] This method is discussed in Section 7.4. Other bottom-up schemes for thermodynamic coarse-graining include the following:

(a) Inverse Monte Carlo, as introduced by Lyubartsev and Laaksonen.[44,45] This structure-based method optimizes the effective pair potential, so as to reproduce the pair distribution functions between coarse-grained centres as computed from an atomistic simulation. A Newton–Raphson scheme is used to optimize the potential, represented by a set of tabulated values as a step-wise function of distance.

(b) Relative entropy minimization methods, as proposed by Shell and collaborators.[46,47] Here the coarse-grained effective potential is defined by a condition designed to provide minimum entropy change between the atomistically represented and the coarse-grained system and thereby minimize the information loss associated with the coarse-graining process.

(c) Methods accounting for the geometry of the underlying atomic structure (excluded volume) of the interacting moieties by incorporating a shape effect (asphericity) in the intermolecular energy expression.[48] The anisotropy of the moieties can also be accounted for by a biaxial Gay–Berne potential fitted to ab initio binding energies.[49]

(d) Methods where the nonbonded coarse-grained potentials are parametrized to fit structural properties *on the fly*.[50,51] Here, a piecewise potential is adjusted to reproduce a radial distribution function via optimization of a merit function. This method requires a realistic initial estimate for the adjustable parameters and an a priori knowledge of physical properties of the actual system.

(e) Approaches where the potential of mean force is determined from the radial distribution function of chain segments[52] as an extension to the McCoy–Curro scheme[53] with one adjustable parameter, or of the entire polymer molecules, based on integral equation techniques.[35–38] Here, caution must be exercised with the use of effective pair potentials describing many-body interactions, as they may be thermodynamically inconsistent and, at best, perform well for the particular physical properties for which they were developed.[38] More recently, an analytical coarse-grained model for polymer melts based on integral equation theory has been proposed by Guenza and collaborators, which can provide a consistent representation of chains at variable levels of chemical detail.[54]

(f) A systematic coarse-graining procedure for polymers of various architectures and block copolymers in solution has been proposed by Hansen and collaborators.[55] Here each block is represented by a series of blobs, each consisting of a large number of monomers. Blobs do not overlap on average, even though entire polymer coils interpenetrate (semi-dilute regime). Ultra-soft effective interactions between blobs are determined by a rigorous inversion procedure in the low concentration limit.

In general, there exist a number of issues associated with structural coarse-graining:[56,57] The first concerns the features of the atomistic model that are the most important to preserve in mapping from atoms to particles. The second refers to the range of conditions over which reasonable estimates of the structural, thermodynamic, and conformational properties can be expected, given that the generation of the potentials of mean force was performed at a particular temperature, pressure, and composition. The third refers to the possibility of using the coarse-grained potentials to describe the system over scales of hundreds of nanometers.[58]

Mapping atomistic units to coarse-grained particles and computing their bonded and nonbonded effective potentials on the basis of equilibrium properties is one aspect of the coarse-graining problem. Developing dynamical equations describing the time evolution of the coarse-grained units at equilibrium (but mostly beyond equilibrium; e.g. for systems subjected to a flow field) is the second aspect, which we referred to as *dynamical* coarse-graining in the above discussion. This second aspect is the biggest challenge in the overall coarse-graining effort. Issues to be addressed here include the following: what is the dynamics of the system at the coarse-grained level, as derived from the atomistic level by just matching a few structural functions (of the type of a radial distribution or a probability density function)? What choice of the coarse-grained variables ensures a good representation of this dynamics with optimal savings in computational expense? What approximations are allowed in dealing with the random forces and memory kernels arising in the coarse-grained dynamical description and how should the friction factors appearing in these approximations be estimated from atomic-level information? To what degree can one capture the dynamical processes taking place in the real system at various length scales under equilibrium and nonequilibrium (flow) conditions? These questions are outstanding research issues today. We briefly discuss the projection operator formalism, which constitutes the safest starting point for dynamical coarse-graining, in Section 7.7. For systems subjected to flow, dynamical coarse-graining should conform to strict principles from fundamental sciences such as nonequilibrium statistical mechanics and nonequilibrium statistical thermodynamics.[59–61] Two modern formalisms of nonequilibrium statistical thermodynamics, in particular, have been proven useful in developing complete sets of governing equations for complex systems: these are the generalized bracket[59] and the GENERIC[60] formalisms. Both emphasize the importance of careful selection of the vector of state variables in describing a system under conditions beyond equilibrium, and then provide a strict framework for deriving evolution equations by including reversible and dissipative contributions to the dynamics. They have helped greatly in providing admissible governing equations describing polymer dynamics at the level of the conformation tensor, at the level of an orientation tensor and a scalar, and at the level of distribution functions. We discuss them in Chapter 10.

Only through guidance by the principles of nonequilibrium thermodynamics can one develop a thermodynamically admissible set of governing equations. Nevertheless, a number of more heuristic approaches have been proposed independently for the description of dynamics at the coarse-grained level, under flow conditions. The most important ones are the following:

(a) The slip-link and slip-spring models for studying dynamics of long-chain polymers based on entanglement network pictures;
(b) The dissipative particle dynamics (DPD) method, invoking a set of dynamical equations similar to those of Brownian dynamics but preserving total momentum through the use of pairwise additive frictional and random forces;
(c) Methods involving Langevin-type equations describing the dynamics of chains made up of blobs and directly accounting for entanglements to prevent chain crossing; and
(d) Methods based on Langevin- or Smoluchowski-type equations describing the dynamics of entangled polymers, where the entire polymer chain is considered as a particle. The dynamical equations generate displacements in the position vector of each polymer chain (considered as a particle), but also in the number of entanglements per chain.

We present some of these methods in Chapters 8 and 9 of this book.

7.2 Introduction to the thermodynamics of coarse-graining—the force matching method

We consider a system of N_{ch} chains, each consisting of N_{a} atoms per chain. Let us denote by $\{\boldsymbol{r}^n\}$ the positions of the $n\ (= N_{\mathrm{ch}} \times N_{\mathrm{a}})$ atoms describing the system at the atomistic level. At the mesoscopic or coarse-grained level, we assume that each collection of λ atoms along the chain is taken as a new type of particle.[62] At this level, the system consists of N_{ch} chains, each containing $N_b\ (= \frac{N_{\mathrm{a}}}{\lambda})$ particles, and is described by the set $\{\boldsymbol{R}^N\}$ of the positions of the $N\ (= N_{\mathrm{ch}} \times N_{\mathrm{b}})$ particles (superatoms or beads), where $N = n/\lambda$. A key issue of the (structural or thermodynamic) coarse-graining problem is to define the effective potentials governing particle–particle interactions, so that the coarse-grained model can describe the atomistic model as closely as possible. A useful approach for achieving this is based on the potential of mean force.

A coarse-grained configuration stands for a large number of atomistic configurations of the system. Given a set of values $\{\boldsymbol{R}^N\}$ for the coarse-grained variables, a particular atomistic configuration compatible with this set can be specified by setting the values for an additional set of variables $\left\{\tilde{\boldsymbol{r}}^{n-N}\right\}$. If, for example, $\{\boldsymbol{R}^N\}$ encompasses the $9 \times N_{\mathrm{ch}}$ Cartesian coordinates of every other aromatic carbon in each one of the N_{ch} molecules representing liquid benzene, $\left\{\tilde{\boldsymbol{r}}^{n-N}\right\}$ would consist of $27 \times N_{\mathrm{ch}}$ generalized coordinates needed to specify the positions of the remaining carbons and all hydrogen atoms in the system. The set of variables $\left\{\boldsymbol{R}^N, \tilde{\boldsymbol{r}}^{n-N}\right\}$ provides the same information content as $\{\boldsymbol{r}^n\}$. We will use the symbol $\mathcal{J}$ to denote the Jacobian of transformation $\left|\partial\boldsymbol{r}^n/\partial\left(\boldsymbol{R}^N, \tilde{\boldsymbol{r}}^{n-N}\right)\right|$ from $\left\{\boldsymbol{R}^N, \tilde{\boldsymbol{r}}^{n-N}\right\}$ to $\{\boldsymbol{r}^n\}$. The *potential of mean force* (compare Section 5.3.6) with respect to the coarse-grained variables $\{\boldsymbol{R}^N\}$ is defined, up to an additive constant, as

$$u_{\mathrm{mf}}(\boldsymbol{R}^N) = -k_{\mathrm{B}}T\ln\int e^{-\beta\mathcal{V}(\boldsymbol{r}^n)}\mathcal{J}\,\mathrm{d}\tilde{\boldsymbol{r}}^{n-N} \tag{7.2.1}$$

with k_{B} being the Boltzmann constant and $\beta = 1/k_{\mathrm{B}}T$. The function $u_{\mathrm{mf}}\left(\boldsymbol{R}^N\right)$ is a free energy with respect to the coarse-grained variables $\{\boldsymbol{R}^N\}$ obtained by integrating over all values of the degrees of freedom $\left\{\tilde{\boldsymbol{r}}^{n-N}\right\}$ for each set of $\{\boldsymbol{R}^N\}$ values. By definition, the configurational integral Z of the system can be recovered by treating $u_{\mathrm{mf}}\left(\boldsymbol{R}^N\right)$ as an effective potential energy of interactions at the coarse-grained level:

$$Z \equiv \int e^{-\beta\mathcal{V}(\boldsymbol{r}^n)}\mathrm{d}\boldsymbol{r}^n = \int e^{-\beta u_{\mathrm{mf}}\left(\boldsymbol{R}^N\right)}\mathrm{d}\boldsymbol{R}^N. \tag{7.2.2}$$

Usually, to arrive at a manageable form for $u_{\mathrm{mf}}(\boldsymbol{R}^N)$, it is *assumed* that it consists of additive contributions, each depending on a small subset of $\{\boldsymbol{R}^N\}$. In polymer systems, by analogy to atomistic force fields, $u_{\mathrm{mf}}(\boldsymbol{R}^N)$ is split into bonded interactions (from bond stretching, bond bending, and dihedrals formed among superatoms) and nonbonded interactions (referring to all intermolecular and all intramolecular interactions between superatoms separated by more than three effective bonds). Then, by making the assumption that the total nonbonded contribution to the potential of mean force is pairwise additive, we can write

$$\begin{aligned}
u_{\mathrm{mf}}\left(\boldsymbol{R}^N\right) &\approx U_\lambda^{\mathrm{CG}}\left(\boldsymbol{R}^N\right)\\
&= \sum_{k=1}^{N_{\mathrm{ch}}}\left(\sum_{l=2}^{N_{\mathrm{b}}} u_\lambda^{\mathrm{CG,str}}\left(\boldsymbol{R}_{(k-1)N_{\mathrm{b}}+l-1}, \boldsymbol{R}_{(k-1)N_{\mathrm{b}}+l}\right)\right.\\
&\quad + \sum_{l=2}^{N_{\mathrm{b}}-1} u_\lambda^{\mathrm{CG,bend}}\left(\boldsymbol{R}_{(k-1)N_{\mathrm{b}}+l-1}, \boldsymbol{R}_{(k-1)N_{\mathrm{b}}+l}, \boldsymbol{R}_{(k-1)N_{\mathrm{b}}+l+1}\right)\\
&\quad \left.+ \sum_{l=2}^{N_{\mathrm{b}}-2} u_\lambda^{\mathrm{CG,tor}}\left(\boldsymbol{R}_{(k-1)N_{\mathrm{b}}+l-1}, \boldsymbol{R}_{(k-1)N_{\mathrm{b}}+l}, \boldsymbol{R}_{(k-1)N_{\mathrm{b}}+l+1}, \boldsymbol{R}_{(k-1)N_{\mathrm{b}}+l+2}\right)\right)\\
&\quad + \sum_{i=1}^{N-1}\sum_{j=i+1}^{N} u_\lambda^{\mathrm{CG,nb}}\left(\boldsymbol{R}_i, \boldsymbol{R}_j\right),
\end{aligned} \tag{7.2.3}$$

where $u_\lambda^{\mathrm{CG,str}}$, $u_\lambda^{\mathrm{CG,bend}}$, and $u_\lambda^{\mathrm{CG,tor}}$ denote the *effective* potentials for bond stretching, bond bending, and dihedrals, respectively, and $u_\lambda^{\mathrm{CG,nb}}$ denotes the effective *pair* potentials for the nonbonded interactions.

Noid et al.[13] have presented a general statistical mechanical formulation for thermodynamic coarse-graining starting from atomically detailed (or other high-resolution) models. Let us consider a high-resolution or atomistic model for a given molecular system, involving *atoms*, and a low-resolution (coarse-grained, CG) model for the same system, involving *sites*. We restrict this discussion to models that have no rigid intramolecular constraints.

The instantaneous dynamical state of the high-resolution model is specified by the values of the sets $\{\boldsymbol{r}^n\} = \{\boldsymbol{r}_1, \boldsymbol{r}_2, \ldots, \boldsymbol{r}_n\}$ and $\{\boldsymbol{p}^n\} = \{\boldsymbol{p}_1, \boldsymbol{p}_2, \ldots, \boldsymbol{p}_n\}$ of Cartesian coordinates and momenta of the n atoms of the system, where we consider that $\boldsymbol{p}_i = m_i \dot{\boldsymbol{r}}_i$. The atomistic Hamiltonian is

$$h(\boldsymbol{r}^n, \boldsymbol{p}^n) = \sum_{i=1}^{n} \frac{1}{2m_i} \boldsymbol{p}_i^2 + u(\boldsymbol{r}^n). \tag{7.2.4}$$

The equilibrium probability density of dynamical states in the canonical ensemble is

$$\rho_{\boldsymbol{rp}}(\boldsymbol{r}^n, \boldsymbol{p}^n) = \rho_r(\boldsymbol{r}^n)\rho_p(\boldsymbol{p}^n) \tag{7.2.5}$$

where

$$\rho_{\boldsymbol{r}}(\boldsymbol{r}^n) \sim \exp[-\beta u(\boldsymbol{r}^n)] \tag{7.2.6a}$$

and

$$\rho_{\boldsymbol{p}}(\boldsymbol{p}^n) \sim \exp\left[-\beta \sum_{i=1}^{n} \frac{\boldsymbol{p}_i^2}{2m_i}\right]. \tag{7.2.6b}$$

Similarly, the instantaneous dynamical state of the CG model is specified by the sets $\{\boldsymbol{R}^N\} = \{\boldsymbol{R}_1, \boldsymbol{R}_2, \ldots, \boldsymbol{R}_N\}$ and $\{\boldsymbol{P}^N\} = \{\boldsymbol{P}_1, \boldsymbol{P}_2, \ldots, \boldsymbol{P}_N\}$ of Cartesian coordinates and momenta of the N sites of the system, where we consider that $\boldsymbol{P}_i = M_i \dot{\boldsymbol{R}}_i$. The CG Hamiltonian is

$$H(\boldsymbol{R}^N, \boldsymbol{P}^N) = \sum_{i=1}^{N} \frac{1}{2M_i} \boldsymbol{P}_i^2 + U(\boldsymbol{R}^N). \tag{7.2.7}$$

The equilibrium probability density of dynamical states in the canonical ensemble for the CG model is assumed to be of the form

$$P_{\boldsymbol{RP}}(\boldsymbol{R}^N, \boldsymbol{P}^N) = P_{\boldsymbol{R}}(\boldsymbol{R}^N) P_{\boldsymbol{P}}(\boldsymbol{P}^N) \tag{7.2.8}$$

where

$$P_{\boldsymbol{R}}(\boldsymbol{R}^N) \sim \exp[-\beta U(\boldsymbol{R}^N)] \tag{7.2.9a}$$

and

$$P_{\boldsymbol{P}}\left(\boldsymbol{P}^N\right) \sim \exp\left[-\beta \sum_{i=1}^{N} \frac{\boldsymbol{P}_i^2}{2M_i}\right]. \tag{7.2.9b}$$

To connect the two equilibrium probability densities, we assume that the CG model has been chosen such that the positions and momenta of its sites are specified in terms of atomic positions and momenta by two *linear* mapping operators $\boldsymbol{M_R}$ and $\boldsymbol{M_P}$ through the two relations

$$\boldsymbol{R}^N = \boldsymbol{M}_{\boldsymbol{R}}^N(\boldsymbol{r}^n) \Leftrightarrow \boldsymbol{R}_I = \sum_{i=1}^{n} C_{Ii}\boldsymbol{r}_i, \quad I = 1, 2, \ldots, N \tag{7.2.10a}$$

and

$$\boldsymbol{P}^N = \boldsymbol{M}_{\boldsymbol{P}}^N(\boldsymbol{p}^n) \Leftrightarrow \boldsymbol{P}_I = M_I \sum_{i=1}^{n} C_{Ii}\boldsymbol{p}_i/m_i, \quad I = 1, 2, \ldots, N. \tag{7.2.10b}$$

This is the case, for example, when a CG site corresponds to the centre of mass of a specific set of atoms along the chain. The CG model is said to be *consistent* with the atomistic model in phase space (i.e. in the space of both positions and momenta) if the equilibrium joint probability density of CG coordinates and momenta as given by Eq. (7.2.8) is equal to that implied by the atomistic probability density as given by Eq. (7.2.5) using the mapping operators $\boldsymbol{M}_{\boldsymbol{R}}^N$ and $\boldsymbol{M}_{\boldsymbol{P}}^N$. The conditions for consistency have been derived by Noid et al.[13] and read as

$$\exp\left[-\beta U\left(\boldsymbol{R}^N\right)\right] \sim \int \exp\left[-\beta u(\boldsymbol{r}^n)\right] \delta\left(\boldsymbol{M}_{\boldsymbol{R}}^N(\boldsymbol{r}^n) - \boldsymbol{R}^N\right) \mathrm{d}\boldsymbol{r}^n \tag{7.2.11a}$$

and

$$\exp\left[-\beta \sum_{i=1}^{N} \frac{\boldsymbol{P}_i^2}{2M_i}\right] \sim \int \exp\left[-\beta \sum_{i=1}^{n} \frac{\boldsymbol{p}_i^2}{2m_i}\right] \delta\left(\boldsymbol{M}_{\boldsymbol{P}}^N(\boldsymbol{p}^n) - \boldsymbol{P}^N\right) \mathrm{d}\boldsymbol{p}^n. \tag{7.2.11b}$$

Equation (7.2.11a) guarantees consistency between the two models in the space of positions, and Eq. (7.2.11b) guarantees consistency in the space of momenta. Equation (7.2.11a) implies that the effective potential $U\left(\boldsymbol{R}^N\right)$ for a consistent CG model is a many-body potential of mean force, which is completely determined (except for an undetermined additive constant) by the atomistic potential $u\{\boldsymbol{r}^n\}$ and the mapping operator $\boldsymbol{M}_{\boldsymbol{R}}^N$. One can write that

$$U\{\boldsymbol{R}^N\} = -k_{\mathrm{B}}T \ln Z\left(\boldsymbol{R}^N\right) + \text{const.}$$

with

$$Z\left(\boldsymbol{R}^N\right) = \int \exp\left[-\beta u(\boldsymbol{r}^n)\right] \delta\left(\boldsymbol{M}_{\boldsymbol{R}}^N(\boldsymbol{r}^n) - \boldsymbol{R}^N\right) \mathrm{d}\boldsymbol{r}^n \tag{7.2.12}$$

Forces in the CG model are given by the corresponding gradients of $U\{\boldsymbol{R}^N\}$:

$$\boldsymbol{F}_I\left(\boldsymbol{R}^N\right) = -\frac{\partial U\left(\boldsymbol{R}^N\right)}{\partial \boldsymbol{R}_I}$$

$$= \frac{k_\mathrm{B}T}{Z\left(\boldsymbol{R}^N\right)} \int \exp\left[-\beta u(\boldsymbol{r}^n)\right] \prod_{\mathcal{J}\neq I} \delta\left(\sum_j C_{\mathcal{J}j}\boldsymbol{r}_j - \boldsymbol{R}_\mathcal{J}\right) \frac{\partial}{\partial \boldsymbol{R}_I} \delta\left(\sum_i C_{Ii}\boldsymbol{r}_i - \boldsymbol{R}_I\right) \mathrm{d}\boldsymbol{r}^n \quad (7.2.13)$$

At this point, it is convenient to introduce the definitions of involved and specific atoms for a CG site I: (a) an atom i in the atomistic model is said to be *involved* in a CG site I if and only if that atom provides a nonzero contribution to the sum in Eq. (7.2.10a) for I; and (b) an atom i in the atomistic model is said to be *specific* to site I if and only if that atom i is involved in site I and is not involved in the definition of any other site. Mathematically, the corresponding sets $\{\boldsymbol{I}\} = \{I_1, I_2, \ldots, I_N\}$ and $\{\boldsymbol{S}\} = \{S_1, S_2, .., S_N\}$ of involved and specific atoms for the sites of the CG model, respectively, are defined as

$$I_I = \{i | C_{Ii} \neq 0\}, \quad \forall I = 1, 2, \ldots, N \quad (7.2.14)$$

and

$$S_I = \{i | C_{Ii} \neq 0 \text{ and } C_{\mathcal{J}i} = 0 \;\; \forall \mathcal{J} \neq I\}, \quad \forall I = 1, 2, \ldots, N. \quad (7.2.15)$$

Noid et al.[13] introduce a set of constant coefficients $\{d_{Ii}\}$ for each site I of the CG model as

$$d_{Ii} \neq 0 \quad \text{if and only if } i \in S_I \quad (7.2.16a)$$

and

$$\sum_{j\in S_I} d_{Ij} = 1 \quad \forall I = 1, 2, \ldots, N. \quad (7.2.16b)$$

With the set $\{d_{Ii}\}$, the integrations involved in Eq. (7.2.13) can be carried out (see Problem 7.1) and the final result is

$$\boldsymbol{F}_I\left(\boldsymbol{R}^N\right) = \langle \mathcal{F}_I(\boldsymbol{r}^n) \rangle_{\boldsymbol{R}^N}, \quad (7.2.17)$$

where

$$\mathcal{F}_I(\boldsymbol{r}^n) = \sum_{j\in S_I} \frac{d_{Ij}}{C_{Ij}} \boldsymbol{f}_j(\boldsymbol{r}^n) = -\sum_{j\in S_I} \frac{d_{Ij}}{C_{Ij}} \frac{\partial u(\boldsymbol{r}^n)}{\partial \boldsymbol{r}_j}, \quad (7.2.18)$$

while the brackets, for any continuous function $\mathcal{A}(\boldsymbol{r}^n)$ of the atomistic coordinates, denote an average of the form

$$\langle \mathcal{A}(\boldsymbol{r}^n) \rangle_{\boldsymbol{R}^N} \equiv \frac{\int \mathcal{A}(\boldsymbol{r}^n) \exp\left[-\beta u(\boldsymbol{r}^n)\right] \delta\left(\boldsymbol{M}_{\boldsymbol{R}}(\boldsymbol{r}^n) - \boldsymbol{R}^N\right) \mathrm{d}\boldsymbol{r}^n}{\int \exp\left[-\beta u(\boldsymbol{r}^n)\right] \delta\left(\boldsymbol{M}_{\boldsymbol{R}}(\boldsymbol{r}^n) - \boldsymbol{R}^N\right) \mathrm{d}\boldsymbol{r}^n}. \quad (7.2.19)$$

Equation (7.2.17) defines the force field in the CG model in terms of the forces at the atomistic level. It ensures consistency between the two models in configuration space.

For consistency in momentum space, Noid et al.[13] have found that, if no atom is involved in the definition of more than one site, then a sufficient condition is that the masses of the CG particles satisfy (see also Problem 7.2)

$$M_I = \left(\sum_{i \in I_I} \frac{C_{Ii}^2}{m_i} \right)^{-1}. \tag{7.2.20}$$

Equations (7.2.17) and (7.2.20) are the important results of the statistical mechanical analysis of Noid et al.[13]. Equation (7.2.17) expresses the force on CG site I in terms of an equilibrium average in the space of atomistic coordinates of the quantity $\mathcal{F}_I(\boldsymbol{r}^n)$, which is a linear combination of the atomistic forces acting on all atoms specific to CG site I. It has been derived on the basis of only one assumption, namely, that every CG site has at least one atom that is specific to it. It provides the basis for a variational principle and a practical algorithm for calculating the many-body potential of mean force from simulations of the atomistic system.

The above analysis was cast by Noid et al.[13,14] in the form of a variational principle for the computation of the many-body potential of mean force $U(\boldsymbol{R}^N)$ and the corresponding force field at the CG level. We consider that, for each site I, the CG force field is a set of real continuous functions, $\boldsymbol{G}_I(\boldsymbol{R}^N), I = 1, 2, \ldots, N$. Its members will be the *true* CG forces $\boldsymbol{F}_I$ given by Eqs. (7.2.17) and (7.2.18) discussed above as well as approximations to them derived, for example, from an expression for the many-body potential of mean force such as Eq. (7.2.3), involving a number of parameters. We define the functional

$$\chi^2[\boldsymbol{G}] = \frac{1}{3N} \left\langle \sum_{I=1}^{N} |\mathcal{F}_I(\boldsymbol{r}^n) - \boldsymbol{G}_I(\boldsymbol{M}_R(\boldsymbol{r}^n))|^2 \right\rangle, \tag{7.2.21}$$

where the brackets denote an equilibrium canonical average over the coordinates of the atomistic model. The variational principle is that

$$\chi^2[\boldsymbol{G}] \geq \chi^2[\boldsymbol{F}] \tag{7.2.22}$$

with the equality holding when $\boldsymbol{G}_I = \boldsymbol{F}_I, \forall I = 1, 2, \ldots, N$. Variational principle (7.2.22) allows one to determine the best approximation to the many-body potential of mean force $U(\boldsymbol{R}^N)$ through a numerical technique that boils down to a linear least squares problem for the parameter set appearing in it. Noid et al.[13,14] note that (a) for systems governed by central pair potentials, the multiscale coarse-graining (MS-CG) method based on minimizing $\chi^2[\boldsymbol{G}]$ (*force matching*) is related to the Yvon–Born–Green equation and determines the CG pair interaction from knowledge of two- and three-body correlation functions, while (b) for more complex molecular systems with bonded and nonbonded interactions, not only two- and three-body correlations between nonbonded sites, but also correlations between different bonded interactions, as well as correlations between bonded and nonbonded interactions, are considered in the CG force field.

The MS-CG method has been applied by Noid et al. in computations of the potential of mean force for a one-site CG model of methanol and a molecular CG model of the $EMIM^+/NO_3^-$ ionic liquid.[14] Comparison of the methanol radial distribution function as computed from simulations with the atomistic and with the CG model with one CG site per molecule revealed excellent agreement. The same agreement was noted for the distribution of methanol molecular velocities. These results indicate that the many-body potential of mean force for the one-site CG model of methanol is well represented by a sum of pairwise additive interactions. In the case of the ionic liquid, each cation was represented by four distinct sites and each nitrate anion by one site. Here, significant deviations in conformation were noted between the CG model and the atomistic one. This was attributed to the fact that no intramolecular nonbonded interactions were included in the CG model.

7.3 Iterative Boltzmann inversion

A versatile, structure-based coarse-graining method widely used in polymer simulations to derive the effective potentials is iterative Boltzmann inversion (IBI).[4] IBI is an optimization method which requires knowledge of the distributions of the relevant CG degrees of freedom (these are called *target distributions* and are usually extracted from atomistic simulations of shorter-chain homologues). IBI aims at matching these distributions with those predicted from the CG model. Initial estimates for the CG potentials are obtained by Boltzmann inversion of the target distribution functions. Effective potentials are then refined through an iterative scheme designed to bring the distribution functions computed from the CG model within a prespecified tolerance from the corresponding target distributions.

Let us assume that no effective torsional potentials are necessary at the CG level and that the multibody potential of mean force with respect to superatom positions can be satisfactorily represented as a sum of effective bond stretching, effective bond bending, and pairwise nonbonded effective interactions (Eq. (7.2.3) without the torsional term). If a stands for a CG bond length l or a bond angle θ, then the calculation of the corresponding effective bonded $u_{\text{mf}}^{\text{b}}(a)$ and effective nonbonded $u_{\text{mf}}^{\text{nb}}(r)$ potentials through IBI proceeds according to the iterative scheme

$$\left(u_{\text{mf}}^{\text{b}}(a)\right)_{i+1} = \left(u_{\text{mf}}^{\text{b}}(a)\right)_i + k_{\text{B}}T \ln\left(\frac{p_i(a)}{p^*(a)}\right) \tag{7.3.1}$$

and

$$\left(u_{\text{mf}}^{\text{nb}}(r)\right)_{i+1} = \left(u_{\text{mf}}^{\text{nb}}(r)\right)_i + k_{\text{B}}T \ln\left(\frac{g_i(r)}{g^*(r)}\right), \tag{7.3.2}$$

respectively, where the index i refers to the iteration number, the superscript * denotes target distributions, and $p(a)$, $g(r)$ are the bonded probability densities and pair distribution functions computed from the CG simulation. A good initial guess for the iterative procedure is provided by the effective potentials of mean force corresponding to the target distribution functions $p(l)$, $p(\theta)$, and $g(r)$, respectively, through

$$\left(u_{\text{mf}}^{\text{b}}(l)\right)_0 = -k_{\text{B}}T \ln\left(\frac{p^*(l)}{l^2}\right)$$
$$\left(u_{\text{mf}}^{\text{b}}(\theta)\right)_0 = -k_{\text{B}}T \ln\left(\frac{p^*(\theta)}{\sin\theta}\right)$$
$$\left(u_{\text{mf}}^{\text{nb}}(r)\right)_0 = -k_{\text{B}}T \ln\left(g^*(r)\right), \tag{7.3.3}$$

where $u_{\text{mf}}^{\text{b}}(l)$, $u_{\text{mf}}^{\text{b}}(\theta)$, and $u_{\text{mf}}^{\text{nb}}(r)$ denote the effective stretching, bending, and nonbonded contributions, respectively, and the terms l^2 and $\sin\theta$ on the right-hand sides are the corresponding Jacobians of the transformation from Cartesian to generalized coordinates.

When employed in actual simulations with the CG model, the corresponding optimized effective potentials yield a pressure generally different from that used in the atomistic simulations performed to derive the target distributions. A pressure correction should thus be introduced in the simulations with the CG polymer model. Usually this is achieved by adding a weak linear potential term to the attractive long-range part of the optimized potential and post-optimizing the structure until also the pressures are the same between the two representations. Such a correction, also known as the *ramp correction*, is of the form

$$\Delta u_{\text{mf}}^{\text{nb}}(r) = \Delta u_{\text{mf},0}^{\text{nb}}\left(1 - \frac{r}{r_{\text{c}}}\right), \tag{7.3.4}$$

where r_{c} denotes the cut-off distance of the non-bonded (e.g., Lennard-Jones) potential. The pressure correction affects mainly the long-range part of the attractive potential, whereas structure is dominated by short-range interactions. It is also noted that, in general, the effective or CG potential is temperature dependent; however, it has been observed that temperature changes at about the same density do not drastically affect its parameterization.

As mentioned in Section 7.1, dynamics at the level of superatoms is an intriguing subject: the effective CG models based on the definition of superatoms derived with IBI methods are softer than the atomistic ones; hence the effective friction between the beads in CG MD simulations is significantly smaller than that between the true polymer structural units. Coarse-grained simulations, therefore, should be employed with caution when one is interested in studying polymer dynamics, because the intrinsic time scales of the system are altered when one passes from the atomistic to the CG level. An empirical method for deriving the scaling factor that connects the intrinsic time scales between two different polymer models *at equilibrium* has been presented by Harmandaris and Kremer,[58] who proposed matching the mean-square displacements of the relevant units in the two models (atomistic: AA, coarse-grained: CG), both in amplitude and in slope, by defining a time-scaling parameter for the friction factor ζ,

$$S(M) = \frac{\zeta^{\text{AA}}}{\zeta^{\text{CG}}}, \tag{7.3.5}$$

where M is the molecular weight of the polymer. The approach was used by Harmandaris and Kremer[58] to predict the equilibrium dynamical properties of high-molecular-weight PS based on

CG simulations scaled to real time. It remains, however, questionable[57,61] whether dynamics at the level of superatoms can in general be captured with just the use of a time-rescaling factor (see also Section 7.7 and Chapter 10).

7.4 Pretabulation of interaction energies between inflexible moieties

Polymers used in industrial applications comprising units of fused aromatic rings, between which free torsion may be substantially hindered, have been analysed by Walsh,[49] who denoted them as *ribbon-like*. Walsh explored the possibility of parameterizing a CG potential capable of describing *ribbon-like*, torsionally stiff polymers with the Gay–Berne potential. Given the ellipsoidal shape of sites it invokes, such a model is a good candidate if one seeks to not only represent the excluded volume (i.e. the repulsive part of the potential), but also to capture the anisotropic nature of the attractive interactions as well as intrachain torsional interactions. A major shortcoming of course is that the use of the Gay–Berne potential results in an increase in the symmetry of the molecule from D_{6h} to $D_{\infty h}$.

A more fundamental treatment of the coarse-graining problem for stiff moieties has been proposed by Zacharopoulos et al.[8] Let us denote by $\boldsymbol{x}_i^{\nu}$ the vibrational degrees of freedom of moiety i (which, later, will be assumed to be frozen). For i being a benzene molecule these are thirty in number, corresponding to twelve bond lengths and eighteen bond angles fixing its internal configuration. Let us also denote by $\boldsymbol{q}_i$ the generalized degrees of freedom fixing the position and orientation of moiety i. For the benzene molecule (see Figure 7.1), these are six per molecule and can be chosen as $\boldsymbol{q}_i = \left(\boldsymbol{s}_{\mathrm{cm},i}, \tilde{\theta}_i, \phi_i, \omega_i\right)$, where $\boldsymbol{s}_{\mathrm{cm},i}$ are the scaled (divided by $V^{1/3}$) coordinates of the centre of mass, $\tilde{\theta}_i$ and ϕ_i the polar and azimuthal angles specifying the orientation of a characteristic vector of the molecule (e.g. the normal vector $\boldsymbol{u}_i$ to the benzene plane with respect to the laboratory frame of reference), and ω_i is an angle of rotation around that vector. Let us also write $\boldsymbol{q}_i = (\boldsymbol{q}_i', \omega_i)$ with $\boldsymbol{q}_i' = \left(\boldsymbol{s}_{\mathrm{cm},i}, \tilde{\theta}_i, \phi_i\right)$. In the method of pretabulated interaction energies, we coarse-grain with respect to $\boldsymbol{x}_i^{\nu}$ or with respect to $\boldsymbol{x}_i^{\nu}$ and ω_i. For a pair of molecules, then, the corresponding scheme is six-dimensional (6D) or four-dimensional (4D), respectively.

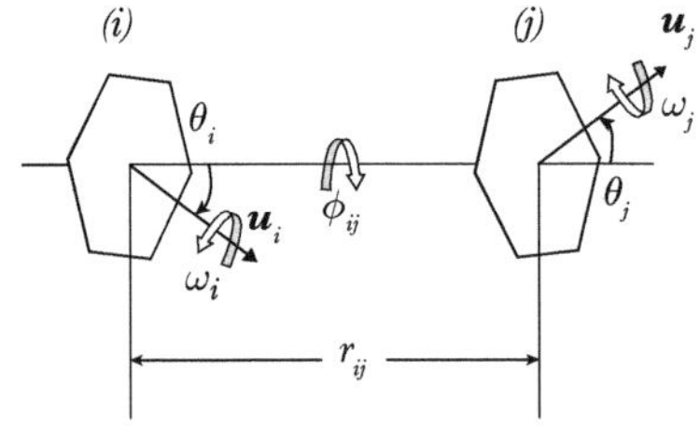

Figure 7.1 Generalized degrees of freedom specifying the configuration of two benzene molecules, considered as inflexible moieties.

Reprinted from Zacharopoulos, N., Vergadou, N., and Theodorou, D.N. *J. Chem. Phys.* **122**, 244111 (2005), with the permission of AIP Publishing.

7.4.1 Six-dimensional coarse graining scheme

We consider a fully flexible atomistic description of the fluid in the limit of *infinite stiffness*. Fluctuations in the vibrational degrees of freedom $\boldsymbol{x}_i^\nu$ around a reference geometry $\boldsymbol{x}_{i,0}^\nu$ (in the case of benzene this is a planar geometry with perfect hexagonal symmetry and appropriately chosen average values for the bond lengths) then become negligible. We are also working in the space of generalized coordinates

$$\boldsymbol{R}^N = \left(\boldsymbol{q}_1, \boldsymbol{q}_2, \ldots, \boldsymbol{q}_{N_{\text{ch}}}\right) \tag{7.4.1a}$$

and

$$\tilde{\boldsymbol{r}}^{n-N} = \left(\boldsymbol{x}_1^\nu, \boldsymbol{x}_2^\nu, \ldots, \boldsymbol{x}_{N_{\text{ch}}}^\nu\right), \tag{7.4.1b}$$

where $n = 36N_{\text{ch}}$ (since each benzene molecule contains 12 atoms, 6 carbons and 6 hydrogens) and $N = 6N_{\text{ch}}$ in the example of benzene, where N_{ch} denotes the number of benzene molecules in volume V. Using $(\boldsymbol{q}_i, \boldsymbol{x}_i^\nu)$ instead of atomic coordinates in describing the configuration of the system introduces the Jacobian $\mathcal{J}\left(\boldsymbol{q}_1', \boldsymbol{q}_2', \ldots, \boldsymbol{q}_{N_{\text{ch}}}'\right) = \left(\prod_{i=1}^{N_{\text{ch}}} \sin \tilde{\theta}_i\right) V^{N_{\text{ch}}}$ in the configurational partition function. In the original, fully flexible representation, the intermolecular potential-energy function is of the form $\mathcal{V}(\boldsymbol{r}^n) = \sum_{i=1}^{N_{\text{ch}}-1} \sum_{j=i+1}^{N_{\text{ch}}} u_{ij}\left(\boldsymbol{q}_i, \boldsymbol{x}_i^\nu, \boldsymbol{q}_j, \boldsymbol{x}_j^\nu\right)$. Upon introducing the infinite stiffness limit approximation, the effective potential energy becomes

$$U_{\text{mf}(6)}\left(\boldsymbol{R}^N; V\right) = \sum_{i=1}^{N_{\text{ch}}-1} \sum_{j=i+1}^{N_{\text{ch}}} u_{ij}\left(\boldsymbol{q}_i, \boldsymbol{x}_{i,0}^\nu, \boldsymbol{q}_j, \boldsymbol{x}_{j,0}^\nu\right) = \sum_{i=1}^{N_{\text{ch}}-1} \sum_{j=i+1}^{N_{\text{ch}}} u_{ij}^0\left(\boldsymbol{q}_i, \boldsymbol{q}_j\right) = \sum_{i=1}^{N_{\text{ch}}-1} \sum_{j=i+1}^{N_{\text{ch}}} u_{ij}^0\left(\boldsymbol{q}_{ij}\right). \tag{7.4.2}$$

The molecule–molecule pair interaction contributions u_{ij}^0 do not depend on the entire set of variables $\left(\boldsymbol{q}_i, \boldsymbol{q}_j\right)$, but rather on the *relative* configuration of i and j. To describe the latter, we use a new set of coordinates, $\boldsymbol{q}_{ij} = \left(r_{ij}, \theta_i, \theta_j, \phi_{ij}, \omega_i, \omega_j\right) = \left(\boldsymbol{q}_{ij}', \omega_i, \omega_j\right)$, where θ_i and θ_j are the angles formed by $\boldsymbol{u}_i$ and $\boldsymbol{u}_i$ with $\boldsymbol{r}_{ij}$, and ϕ_{ij} is the dihedral angle formed by $\boldsymbol{u}_i$, $\boldsymbol{r}_{ij}$, and $\boldsymbol{u}_j$ (see also Figure 7.1). With the infinite stiffness approximation introduced above, the configurational integral becomes

$$Z_6 = \int \exp\left[-\beta \sum_{i=1}^{N_{ch}-1} \sum_{j=i+1}^{N_{\text{ch}}} u_{ij}^0\left(\boldsymbol{q}_{ij}', \omega_i, \omega_j\right)\right] \mathcal{J}\left(\boldsymbol{q}_1', \boldsymbol{q}_2', \ldots, \boldsymbol{q}_N'\right) \mathrm{d}\boldsymbol{q}_1' \mathrm{d}\boldsymbol{q}_2' \ldots \mathrm{d}\boldsymbol{q}_N' \mathrm{d}\omega_1 \mathrm{d}\omega_2 \ldots \mathrm{d}\omega_N \tag{7.4.3}$$

and forms the basis for Monte Carlo (MC) simulation with a 6D pretabulation scheme: The molecular pair potential u_{ij}^0 is pretabulated as a function of the six degrees of freedom $\left(\boldsymbol{q}_{ij}', \omega_i, \omega_j\right)$ and the MC sampling is based on the use of the configurational integral Z_6. This significantly accelerates the simulations, but the method is subject to two types of errors: the first is associated with introducing the infinitely stiff model in place of the original fully flexible model, and the second with discretization errors entering the pretabulation/interpolation in $\boldsymbol{q}_{ij}$ space invoked in calculating molecular pair interactions.

7.4.2 Four-dimensional coarse-graining scheme

By introducing a preaveraging approximation, the dimensionality of the CG model can be reduced to 4. We achieve this by replacing $u_{ij}^0\left(\boldsymbol{q}'_{ij}, \omega_i, \omega_j\right)$ in the configurational integral of Eq. (7.4.3) by a preaveraged quantity $u_{\mathrm{mf},ij}^0\left(\boldsymbol{q}'_{ij}\right)$ with respect to rotation angles ω_i and ω_j, defined as

$$u_{\mathrm{mf},ij}^0\left(\boldsymbol{q}'_{ij}\right) = -\frac{1}{\beta}\ln\left(\frac{\int \exp\left[-\beta u_{ij}^0\left(\boldsymbol{q}'_{ij}, \omega_i, \omega_j\right)\right] \mathrm{d}\omega_i \mathrm{d}\omega_j}{\int \mathrm{d}\omega_i \mathrm{d}\omega_j}\right). \tag{7.4.4}$$

In the general coarse-graining notation introduced earlier, the 4D scheme corresponds to

$$\boldsymbol{R}^N = \left(\boldsymbol{q}'_1, \boldsymbol{q}'_2, \ldots, \boldsymbol{q}'_{N_{\mathrm{ch}}}\right) \tag{7.4.5a}$$

and

$$\tilde{\boldsymbol{r}}^{n-N} = \left(\boldsymbol{x}_1^{\nu}, \omega_1, \boldsymbol{x}_2^{\nu}, \omega_2, \ldots, \boldsymbol{x}_{N_{\mathrm{ch}}}^{\nu}, \omega_{N_{\mathrm{ch}}}\right) \tag{7.4.5b}$$

with $n = 36N_{\mathrm{ch}}$ and $N = 5N_{\mathrm{ch}}$. Use of $u_{\mathrm{mf},ij}^0\left(\boldsymbol{q}'_{ij}\right)$ in place of $u_{ij}^0\left(\boldsymbol{q}'_{ij}, \omega_i, \omega_j\right)$ amounts to ignoring couplings higher than pairwise among ω_i, which are present in the original energy function. Changing ω_i and ω_j will generally affect not only u_{ij}^0 but also u_{ik}^0 and u_{kj}^0 with $k \neq i, j$. Here we ignore this and assume that

$$\frac{\int \exp\left[-\beta \sum_{i=1}^{N_{\mathrm{ch}}-1} \sum_{j=i+1}^{N_{\mathrm{ch}}} u_{ij}^0\left(\boldsymbol{q}'_{ij}, \omega_i, \omega_j\right)\right] \mathrm{d}\omega_1 \mathrm{d}\omega_2 \ldots \mathrm{d}\omega_N}{\int \mathrm{d}\omega_1 \mathrm{d}\omega_2 \ldots \mathrm{d}\omega_N} = \prod_{i=1}^{N_{\mathrm{ch}}-1} \prod_{j=i+1}^{N_{\mathrm{ch}}} \frac{\int \exp\left[-\beta u_{ij}^0\left(\boldsymbol{q}'_{ij}, \omega_i, \omega_j\right)\right] \mathrm{d}\omega_i \mathrm{d}\omega_j}{\int \mathrm{d}\omega_i \mathrm{d}\omega_j}. \tag{7.4.6}$$

Consequently, in place of the potential energy $\mathcal{V}\left(\boldsymbol{r}^n\right)$ of the original, fully flexible model, we use

$$U_{\mathrm{mf}(4)}\left(\boldsymbol{R}^N; V\right) = \sum_{i=1}^{N_{\mathrm{ch}}-1} \sum_{j=i+1}^{N_{\mathrm{ch}}} u_{\mathrm{mf},ij}^0\left(\boldsymbol{q}_{ij}'\right), \tag{7.4.7}$$

implying the following form for the configurational integral:

$$Z_4 = \int \exp\left[-\beta \sum_{i=1}^{N_{\mathrm{ch}}-1} \sum_{j=i+1}^{N_{\mathrm{ch}}} u_{\mathrm{mf},ij}^0\left(\boldsymbol{q}'_{ij}\right)\right] \mathcal{J}\left(\boldsymbol{q}'_1, \boldsymbol{q}'_2, \ldots, \boldsymbol{q}'_N\right) \mathrm{d}\boldsymbol{q}'_1\, \mathrm{d}\boldsymbol{q}'_2 \ldots \mathrm{d}\boldsymbol{q}'_N. \tag{7.4.8}$$

Equation (7.4.8) forms the basis for simulations with the 4D pretabulation scheme. The potential of mean force $u_{\mathrm{mf},ij}^0$ is calculated from the atomistic force field via Eq. (7.4.4) and stored as a function of the four degrees of freedom $\boldsymbol{q}'_{ij}$. A simulation based on the 4D scheme, in comparison to an atomistic simulation based on the original, fully flexible potential, is subject to the two error types discussed in conjunction with the 6D scheme. However, it is also subjected to a third type of error due to the preaveraging approximation, Eq. (7.4.4).

Overall, the reduced dimensionality of the 4D scheme in comparison to the 6D one makes it possible to construct a finer table of potentials of mean force under given memory limitations. Note also that, by definition, $\lim_{r_{ij}\to\infty} u^0_{\mathrm{mf},ij} = 0$; therefore, $U_{\mathrm{mf}(4)}$ vanishes in the limit of an infinitely dilute gas. Application of the 4D pretabulation scheme on liquid benzene has yielded excellent results for the pressure, pair distribution function of molecular centres, and orientational correlation function between molecules.[8]

7.5 Reverse mapping to the atomistic representation

Having achieved equilibration at the CG level, one often wishes to restore the details in the representation, obtaining atomistic or united-atom representations for use in the prediction of properties that depend on local structure and dynamics. This procedure of going back to a detailed model representation is referred to as *reverse mapping* or *fine graining.* Clearly, to each CG configuration $\{\boldsymbol{R}^N\}$ correspond a large number of detailed configurations $\{\boldsymbol{r}^n\}$. At equilibrium, these comprise an ensemble subject to constraints imposed by the values of the degrees of freedom $\{\boldsymbol{R}^N\}$ in the CG configuration. Usually, by *reverse mapping* we mean the generation of only one among the detailed configurations belonging to this ensemble; this entails choosing appropriate values for the remaining (detailed, fast) degrees of freedom $\left\{\tilde{\boldsymbol{r}}^{\,n-N}\right\}$ that were projected out of the CG configuration. The single detailed configuration $\{\boldsymbol{r}^n\}$ thus obtained must be of high probability, so as to be representative of the system under study. Subsequent unconstrained simulation in the detailed representation starting from $\{\boldsymbol{r}^n\}$ should produce a set of detailed configurations that overlaps significantly with the constrained ensemble corresponding to the CG configuration $\{\boldsymbol{R}^N\}$ that was subjected to reverse mapping.

There are several ways in which one can try to go about addressing the reverse mapping problem. Here we mention two of them. The first is basically geometric.[63,64] The second involves structure optimization at the level of (united) atoms, which is achieved through local Monte Carlo moves and energy minimization to relax fine structural details. It was proposed for pure molten polystyrene[11] but was also implemented for polystyrene-containing fullerene[65] and silica[66] nanoparticles.

7.6 Topological analysis and reduction of atomistic configurations to entanglement networks

Dynamics in high-molar mass polymeric liquids is determined by topological constraints arising from chain connectivity and chain uncrossability. As discussed by Doi and Edwards[17,25] and by de Gennes,[67] the effects of topological constraints on the viscoelastic properties of the polymer can be captured in a mean-field approach by the tube model, according to which mutual entanglements with surrounding chains restrict the number of allowed conformations of a reference chain compared to that in free space. The allowed chain conformations are effectively confined within a tube-like region, whose axis is defined by the shortest path connecting the two ends of the chain having the same topology as the chain itself relative to the constraints; such a path is called the *primitive path* (PP). The motion of the chain is confined to a tube-like region around the PP, from which the PP gradually escapes by translational self-diffusion along its curvilinear axis. This *reptation* of the chain along its contour is the predominant mode of motion in melts of very high molar mass ($M \gg M_e$, with M_e being the molar mass between entanglements). As M is lowered towards M_e, fluctuations in the contour length of the primitive path and constraint release due to the motion of surrounding chains modify the basic picture of reptation.

The problem of defining PPs geometrically can be stated as follows:[24] *given a set of uncrossable curved lines in space, reduce (shrink) continuously their contour lengths while keeping their ends fixed, until the reduction process is blocked by topological constraints which organize into a network of quantifiable topology*. The network is defined by a set of rectilinear segments (entanglement strands) coming together at nodal points (entanglements). From each entanglement point four strands emanate, belonging to two chains. Starting from an ensemble of well-equilibrated and statistically independent atomistic configurations, it is possible to obtain a multitude of topologically distinct computational samples for a given polymer. Such an ensemble can greatly facilitate statistical approaches to the topology of polymers, where large topological-type fluctuations of the network formed by the diverse tortuous conformations adopted by chains are possible. In the literature, a number of approaches have been proposed for coarse-graining chains to the level of PP networks such that chain crossing is avoided and chain topology is preserved. They can be distinguished into stochastic, geometric, and energy-minimization approaches.

The energy-minimization approach is due to Everaers et al.,[19] who analysed the topological state of polymeric liquids through a methodology which makes it possible to identify primitive paths simultaneously for all polymers in the system by keeping chain ends fixed in space and disabling intrachain excluded-volume interactions while retaining the interchain ones. Energy minimization by slowly cooling the system towards zero kelvin ($T = 0$ K) causes the bond springs to reduce their length towards 0, which pulls chains taut. The result (given that the method converges to a well-defined, unique ground state that can serve as a topological invariant) is a mesh of primitive paths consisting of straight segments of strongly varying length and more or less sharp turns at the entanglement points (the points where two primitive paths come together), automatically accounting for multichain entanglements of arbitrary order in the system. If the intrachain excluded-volume interaction is disabled up to a certain chemical distance along the chain, then one can further distinguish between local self-knots and entanglements between different sections of the same chain. If different parts of the same chain are free to intersect, such self-entanglements are not accounted for.

Topological relaxation can also be achieved through an algorithm that uses only geometric operations and returns a shortest path and the related number of entanglements for a given configuration of a polymeric system in 2D and 3D space, proposed by Kröger.[22] In such an approach, called the Z algorithm, primitive paths are treated as *infinitely* thin, tensionless lines (rather than multi-bead chains) connecting the two ends of the chain, which are considered as fixed in space. They are obtained under the constraints of chain connectivity and uncrossability through a set of special geometric operations designed to minimize the contour length of the multiple disconnected path (i.e. the contour length summed over all individual PPs) simultaneously for all chains in the simulation cell. The result is the construction of the shortest multiple disconnected path (SP) for a 2D system (polymeric chain within spherical obstacles) and an optimal SP for a 3D system (a collection of polymeric chains in Euclidean space). The number of entanglements is obtained from the shortest path as either the number of interior kinks, or from the average length of a line segment.

Chains can be reduced to shortest paths also through a direct application of a stochastic method implementing random string-aligning moves and treating monomers as united-atom hard spheres (beads) of a certain diameter.[24] The associated algorithm is called CReTA (contour reduction topological analysis), and is capable of reducing the atomistic configuration of a model polymer sample to a network of corresponding PPs. A string is defined as a set of m (≥ 1) consecutive chain atoms; m is randomly chosen between $1 + (i - 1)\,\Delta j$ and $i\,\Delta j$, with i initially set to 1 and Δj equal to a small value. In such a method, a string is randomly chosen in each move and its atoms are displaced to corresponding equidistant points on the straight line segment joining the atoms on either side of the string. Chain ends are fixed throughout the process. To avoid chain crossing and preserve system topology, moves leading to overlaps between any string atom with any other atom belonging to a different chain or to a bead displacement larger than its diameter are rejected. Intrachain bead overlaps are permitted. The string-aligning moves are applied in parallel on all chains. Accepted moves result in simultaneous reduction of the chain contours and progressive shrinkage of unentangled loops to straight strands composed of fused spheres. Through the random aligning moves, contour curvature over unentangled portions of the chains is reduced, while at the same time respecting the relative topological constraints (TCs) through hard core interactions.

Topological relaxation with the CReTA method results in PPs consisting of short pairwise-blocked chain sections (kinks) connected by straight strands of fused spheres. Kinks represent entanglement points with pairwise associations to entanglement points on other chains. The whole structure resembles a network, with nodal points (vertices) at the entanglements, and edges the joining straight entanglement strands (Figure 7.2).

Application of the method to thoroughly equilibrated molecular structures of several polymeric systems provides a number of topological measures of the underlying network such as[24] (a) the distribution $p(Z)$ of the number of entanglements Z per chain, (b) the distribution $p(N_e)$ of the entanglement spacing N_e, (c) the average interentanglement spacing $\langle N_e \rangle$ (average number of skeletal atoms between successive entanglement points), (d) the distribution $p(L)$ of the contour length, and (e) the distribution of the step length between entanglements and the Kuhn length a_{pp} of the primitive path (tube diameter) which is calculated from $\langle r^2 \rangle$ and L (Section 7.9). In particular, a_{pp} and the entanglement molar mass M_e agree with experimental estimates based on the plateau modulus and correlate with melt density and chain packing length as proposed by Fetters.[68,69]

Figure 7.3 shows the variation of the normalized distribution of the reduced number of monomers (united-atoms or beads) in an entanglement strand based on the topological properties of the resulting PP networks in a number of simulated polyethylene and polybutadiene systems. The reduced number n is obtained by dividing the actual number of monomers N_{ES} in an entanglement strand by its mean value $\bar{N}_{ES}$ for the considered polymer. Data from several randomly coiled polymer melts studied are found to superimpose on each other, falling on a master curve of the form[24] $P(n) = \dfrac{bc}{c-b}\left(e^{-bn} - e^{-cn}\right)$, where b and c are numerical constants, suggesting a universal character of $P(n)$ for linear polymers. Figure 7.4 presents results for the normalized distribution $p\left(Z_{ES} - 1\right)$ of the number of topological constraints per chain. The results are consistent with the general prediction of renewal theory that the limiting distribution of Z_{ES} should be asymptotically normal with mean $\bar{Z}_{ES} - 1$ and variance $\sigma^2_{Z_{ES}}$ given by the correspondences[24] $\bar{Z}_{ES} \to \dfrac{N}{\bar{N}_{ES}}$ and $\sigma^2_{Z_{ES}} \to \sigma^2_{N_{ES}} \dfrac{N}{\bar{N}^3_{ES}}$. The solid line in Figure 7.4 showing a normalized Gaussian with mean $\bar{Z}_{ES} - 1$

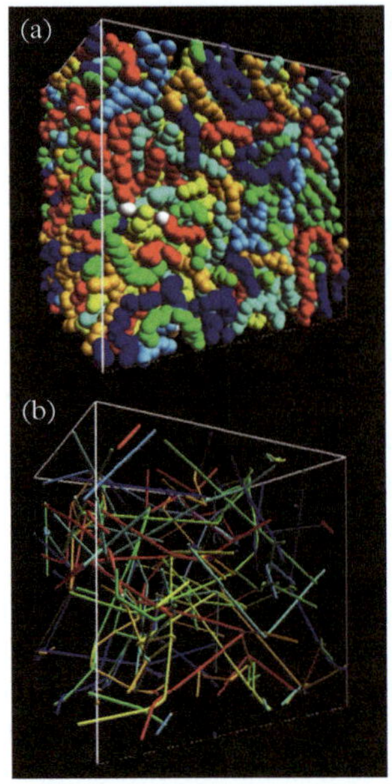

Figure 7.2 (a) Representative atomistic PE configuration and (b) the corresponding reduced network of PPs obtained from CReTA. Chain ends, represented by white spheres in the atomistic picture, remain fixed during the reduction process.

Reprinted (adapted) with permission from Tzoumanekas, C., and Theodorou, D.N. *Macromolecules* **39**, 4592 (2006). Copyright 2006 American Chemical Society.

and variance $\sigma^2_{Z_{ES}}$ provides indeed a satisfactory description of the simulation data, albeit with some deviations, since the asymptotic limit corresponds to much longer chains than those presented in the figure.

Methods for reducing atomistic configurations of polymer systems to entanglement networks are the first step towards realizing slip-link and slip-spring models for studying the dynamics and viscoelasticity of long-chain polymers; these are discussed in Chapter 9.

7.7 Elements of projection operator theory and Brownian dynamics

Up to now we have been concerned mainly with *thermodynamic* coarse-graining, aimed at capturing the structure and thermodynamic properties of atomistic model systems at equilibrium. We pointed out, however, the importance of *dynamical* coarse-graining, which, additionally, aims at reproducing the long-time dynamics of the atomistic system. In this section we introduce elements of the *theory of coarse-graining*, which forms a basis for deriving CG equations of motion from the microscopic equations of motion through a systematic procedure. The importance of dynamical coarse-graining cannot be overstated. Understanding its theoretical foundations and how they can lead, through well-controlled approximations, to widely practiced mesoscopic simulation methods

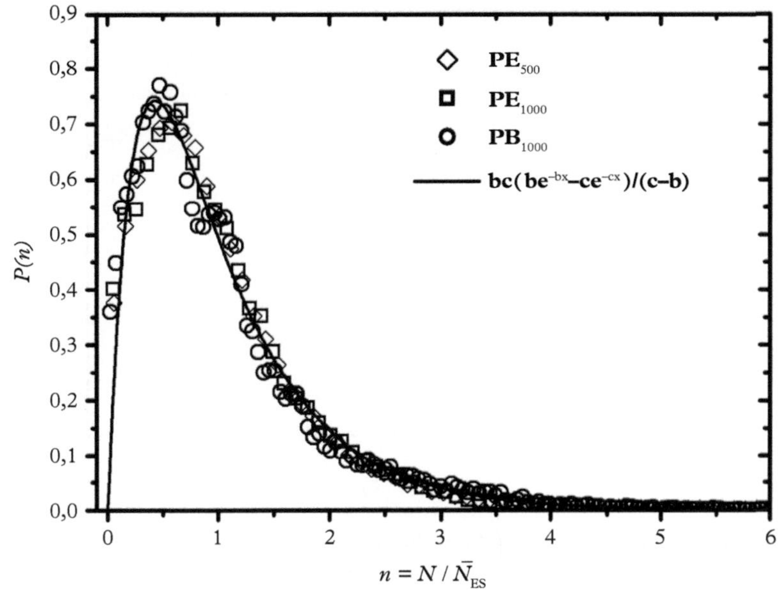

Figure 7.3 Normalized distribution of the reduced distance along the chain contour between entanglements for various polymers.

Reprinted (adapted) with permission from Tzoumanekas, C., and Theodorou, D.N. *Macromolecules* **39**, 4592 (2006). Copyright 2006 American Chemical Society.

such as Langevin dynamics, Brownian dynamics, dynamic density functional theory, dissipative Brownian dynamics, and multiparticle collision dynamics is very much an active research field today.

7.7.1 Hamiltonian dynamics of the atomistic system

Consider a classical system of n particles being described atomistically in terms of the $3n-$ dimensional vectors of generalized coordinates $\boldsymbol{q}$ and generalized momenta $\boldsymbol{p}$. We describe the microscopic state of the system by specifying the $6n$-dimensional vector $(\boldsymbol{q}, \boldsymbol{p})$ of atomic coordinates and momenta, i.e. a representative point in phase space. Consider an arbitrary collection of points in phase space at time $t = 0$. These constitute an ensemble. As time goes by, each point in phase space moves, describing a trajectory $(\boldsymbol{q}(t), \boldsymbol{p}(t))$. For an isolated system evolving solely under the influence of interactions among its particles, this trajectory (line in $6n$-dimensional phase space) can be obtained by integrating Hamilton's equations of motion, Eqs. (1.1.33) and (1.1.34), which we rewrite here as

$$\frac{\mathrm{d}}{\mathrm{d}t}\begin{bmatrix} \boldsymbol{q} \\ \boldsymbol{p} \end{bmatrix} = \begin{bmatrix} \dfrac{\partial}{\partial \boldsymbol{p}} \\ -\dfrac{\partial}{\partial \boldsymbol{q}} \end{bmatrix} \mathcal{H}, \tag{7.7.1}$$

where $\mathcal{H}(\boldsymbol{q}, \boldsymbol{p})$ is the Hamiltonian of the system, while gradients with respect to the $3n$-dimensional vectors $\boldsymbol{q}$, $\boldsymbol{p}$, and with respect to their 3D components $\boldsymbol{q}_i, \boldsymbol{p}_i$ corresponding to individual particles

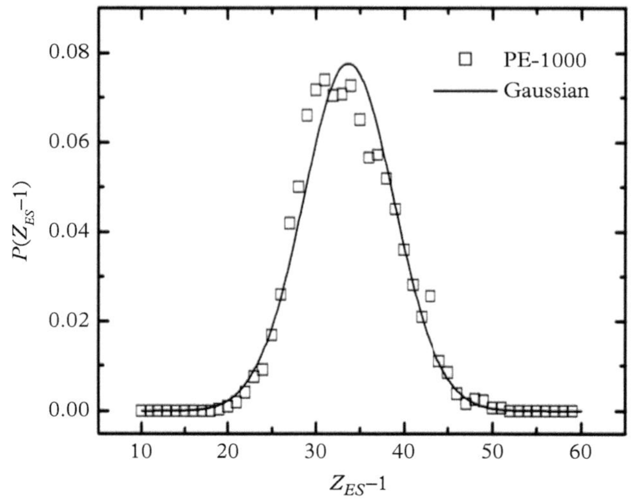

Figure 7.4 Normalized distribution $p\,(Z_{ES}-1)$ of the number of topological constraints per chain. Symbols correspond to PE-1000 data. The solid line is a Gaussian with the same variance and mean as the PE-1000 data.

Reprinted (adapted) with permission from Tzoumanekas, C., and Theodorou, D.N. *Macromolecules* **39**, 4592 (2006). Copyright 2006 American Chemical Society.

are symbolized as partial derivatives, for simplicity. As a consequence, the probability density of the ensemble in phase space, $\rho(\boldsymbol{q},\boldsymbol{p},t)$, changes with time. Given that representative points in the ensemble can be neither created nor destroyed, and using Hamilton's equations of motion, one can readily prove that the time-dependent probability density of the ensemble obeys the Liouville equation[70]

$$\frac{\partial \rho}{\partial t}+\sum_{i=1}^{n}\left(\frac{\partial \mathcal{H}}{\partial \boldsymbol{p}_i}\cdot\frac{\partial}{\partial \boldsymbol{q}_i}-\frac{\partial \mathcal{H}}{\partial \boldsymbol{q}_i}\cdot\frac{\partial}{\partial \boldsymbol{p}_i}\right)\rho=0 \tag{7.7.2a}$$

or

$$\frac{\partial \rho}{\partial t}+\mathrm{i}L\rho=0 \tag{7.7.2b}$$

with iL being the classical Liouville operator that was defined in Section 4.1.5:

$$\mathrm{i}L=\sum_{i=1}^{n}\left(\frac{\partial \mathcal{H}}{\partial \boldsymbol{p}_i}\cdot\frac{\partial}{\partial \boldsymbol{q}_i}-\frac{\partial \mathcal{H}}{\partial \boldsymbol{q}_i}\cdot\frac{\partial}{\partial \boldsymbol{p}_i}\right). \tag{7.7.2c}$$

Equations (7.7.1) and (7.7.2) provide a full, time-reversible, deterministic description of the dynamics of the system at the atomistic level. The partial differential Eq. (7.7.2a) describes the evolution of an ensemble of representative points in phase space. For an equilibrium ensemble, the left-hand side of Eq. (7.7.2a) becomes 0 and the probability density $\rho(\boldsymbol{q}, \boldsymbol{p}, t)$ converges to a stationary value $\rho^{\mathrm{eq}}(\boldsymbol{q}, \boldsymbol{p})$. For an isolated system this equilibrium distribution is expected to be uniform on the constant energy hypersurface $\mathcal{H}(\boldsymbol{q}, \boldsymbol{p}) = E$ (principle of equal a priori probabilities), the total energy E being a constant of the motion.

Let $\mathcal{A}(\boldsymbol{q}, \boldsymbol{p})$ be any function of the phase-space point of the system. As the trajectory evolves in time, $\mathcal{A}$ will change. Setting $\mathcal{A}(t) = \mathcal{A}(\boldsymbol{q}(t), \boldsymbol{p}(t))$, we have

$$\frac{\mathrm{d}\mathcal{A}}{\mathrm{d}t} = \sum_{i=1}^{n} \left(\frac{\partial \mathcal{A}}{\partial \boldsymbol{q}_i} \cdot \frac{\mathrm{d}\boldsymbol{q}_i}{\mathrm{d}t} + \frac{\partial \mathcal{A}}{\partial \boldsymbol{p}_i} \cdot \frac{\mathrm{d}\boldsymbol{p}_i}{\mathrm{d}t} \right) = \sum_{i=1}^{n} \left(\frac{\partial \mathcal{H}}{\partial \boldsymbol{p}_i} \cdot \frac{\partial}{\partial \boldsymbol{q}_i} - \frac{\partial \mathcal{H}}{\partial \boldsymbol{q}_i} \cdot \frac{\partial}{\partial \boldsymbol{p}_i} \right) \mathcal{A} = \mathrm{i}L\mathcal{A} \qquad (7.7.3a)$$

or

$$\dot{\mathcal{A}} \equiv \frac{\mathrm{d}\mathcal{A}}{\mathrm{d}t} = \mathrm{i}L\mathcal{A}, \qquad (7.7.3b)$$

where the time derivatives of generalized coordinates and momenta have been re-expressed in terms of the gradients of the Hamiltonian at the considered phase-space point using Hamilton's Eqs. (7.7.1). Note that the phase-space point itself satisfies Eq. (7.7.1) (compare Section 4.1.5).

If we know the value of the observable $\mathcal{A}(0)$ at $t = 0$, its value at any time t is given by the formal solution to Eq. (7.7.3):

$$\mathcal{A}(t) = \exp(\mathrm{i}Lt)\mathcal{A}(0). \qquad (7.7.4)$$

The exponential of an operator can really be understood as a series involving repeated application of the operator,

$$\exp(\mathrm{i}Lt) = 1 + (\mathrm{i}Lt) + \frac{1}{2}(\mathrm{i}Lt)(\mathrm{i}Lt) + \cdots. \qquad (7.7.5)$$

Equation (7.7.4) explains the name *propagator operator* which is often attributed to $\exp(\mathrm{i}Lt)$.

It can be shown (Problem 7.3) that the operator L appearing in the classical Liouville equation is Hermitian,

$$L^{\dagger} = L, \qquad (7.7.6)$$

where, as in Eq. (2.4.26), the superscript † stands for *complex conjugate transpose*. Equation (7.7.6) is a shorthand for

$$\int \phi_k^{*}(\boldsymbol{q}, \boldsymbol{p})\, L\phi_l(\boldsymbol{q}, \boldsymbol{p})\, \mathrm{d}\boldsymbol{q}\mathrm{d}\boldsymbol{p} = \int \phi_l(\boldsymbol{q}, \boldsymbol{p})(L\phi_k(\boldsymbol{q}, \boldsymbol{p}))^{*} \mathrm{d}\boldsymbol{q}\mathrm{d}\boldsymbol{p}, \qquad (7.7.7)$$

where the superscript * denotes *complex conjugate* and $\phi_k(\boldsymbol{q}, \boldsymbol{p})$, $\phi_l(\boldsymbol{q}, \boldsymbol{p})$ are any two complex functions of phase space satisfying $\phi_k(\boldsymbol{q}, \boldsymbol{p}) \to 0$, $\phi_k(\boldsymbol{q}, \boldsymbol{p}) \to 0$ as $q_{i\alpha} \to \pm\infty$, and $p_{i\alpha} \to \pm\infty$, $\alpha = 1, 2, 3$; that is, the phase space is bounded. One can further show (Problem 7.4) that the propagator operator $e^{\mathrm{i}Lt}$ is unitary, implying time-reversal symmetry in the equations of motion. The unitarity of $e^{\mathrm{i}Lt}$ (a generalization of orthogonality) also means that it preserves volume in phase space, which is a well-known property of Hamiltonian dynamics.

7.7.2 Projection onto the slow variables

The task of developing equations of motion cast in terms of a small subset of (slowly evolving) observables, which nevertheless provide a faithful description of the system dynamics over long time scales is undertaken by the *Mori–Zwanzig formalism*. It was originally developed by Nakajima,[71] Zwanzig,[72,73] and Mori,[42] with important contributions by Kawasaki,[74] Español,[75] and many others. Here we will try to present essential elements of this formalism without going into tremendous mathematical detail. We will largely follow the pedagogical treatment of te Vrugt and Wittkowski.[76] These authors distinguish between two forms of the projection operator formalism: methods where the projection operator is time-independent, and methods where the projection operator is time-dependent. Here we discuss only the first category, which is appropriate for systems at or close to thermodynamic equilibrium, characterized by a time-independent normalized phase-space probability density $\rho^{\mathrm{eq}}(\boldsymbol{q},\boldsymbol{p})$.

We think of all observables we could use to describe the n-particle system as forming a Hilbert space, i.e. a vector space equipped with a scalar product. For two observables $\mathcal{A}$ and $\mathcal{B}$ we define the scalar product as the correlation function

$$\langle \mathcal{A},\mathcal{B}\rangle = \int \mathcal{A}(\boldsymbol{q},\boldsymbol{p})\,\mathcal{B}(\boldsymbol{q},\boldsymbol{p})\,\rho^{\mathrm{eq}}(\boldsymbol{q},\boldsymbol{p})\,\mathrm{d}\boldsymbol{q}\mathrm{d}\boldsymbol{p}. \tag{7.7.8}$$

A simple example of a Hilbert space is $\mathbb{R}^n$. Given a vector $\boldsymbol{x}$ in that space, we may be interested in the part of $\boldsymbol{x}$ which points in the direction of a (not necessarily normalized) reference vector $\boldsymbol{a}$. That part is given by the projection of $\boldsymbol{x}$ on $\boldsymbol{a}$, defined in terms of scalar products as

$$P\boldsymbol{x} = \frac{\boldsymbol{x}\cdot\boldsymbol{a}}{\boldsymbol{a}\cdot\boldsymbol{a}}\boldsymbol{a}. \tag{7.7.9}$$

Again in $\mathbb{R}^n$, we may define a set of (not necessarily orthogonal or normalized) reference vectors $\{\boldsymbol{a}_1,\boldsymbol{a}_2,\ldots,\boldsymbol{a}_N\}$. Given an arbitrary vector $\boldsymbol{x}$, we can pose the question of representing $\boldsymbol{x}$ as a linear combination of the reference vectors in the best possible way. This calls for determining coefficients $\{c_1,c_2,\ldots,c_N\}$ such that the square deviation

$$(\boldsymbol{x} - c_1\boldsymbol{a}_1 - c_2\boldsymbol{a}_2 - \ldots - c_N\boldsymbol{a}_N)^2 = \min. \tag{7.7.10}$$

Setting the derivatives of the left-hand side with respect to the coefficients c_i, $i = 1(1)N$, equal to 0, we readily obtain

$$\begin{bmatrix} \boldsymbol{a}_1\cdot\boldsymbol{a}_1 & \boldsymbol{a}_1\cdot\boldsymbol{a}_2 & \cdots & \boldsymbol{a}_1\cdot\boldsymbol{a}_N \\ \boldsymbol{a}_2\cdot\boldsymbol{a}_1 & \boldsymbol{a}_2\cdot\boldsymbol{a}_2 & \cdots & \boldsymbol{a}_2\cdot\boldsymbol{a}_N \\ \vdots & \vdots & \ddots & \vdots \\ \boldsymbol{a}_N\cdot\boldsymbol{a}_1 & \boldsymbol{a}_N\cdot\boldsymbol{a}_2 & \cdots & \boldsymbol{a}_N\cdot\boldsymbol{a}_N \end{bmatrix} \begin{bmatrix} c_1 \\ c_2 \\ \vdots \\ c_N \end{bmatrix} = \begin{bmatrix} \boldsymbol{x}\cdot\boldsymbol{a}_1 \\ \boldsymbol{x}\cdot\boldsymbol{a}_2 \\ \vdots \\ \boldsymbol{x}\cdot\boldsymbol{a}_N \end{bmatrix}, \text{ or } c_j = \sum_{i=1}^{N} (\boldsymbol{a}\cdot\boldsymbol{a})^{-1}_{ji}\,(\boldsymbol{x}\cdot\boldsymbol{a}_i) \tag{7.7.11}$$

with $(\boldsymbol{a}\cdot\boldsymbol{a})$ being the $N\times N$ matrix of scalar products of the reference vectors appearing on the left hand side of Eq. (7.7.11) and $(\boldsymbol{a}\cdot\boldsymbol{a})^{-1}$ being the inverse of that matrix. We call the optimal representation $c_1\boldsymbol{a}_1 + c_2\boldsymbol{a}_2 + \cdots + c_N\boldsymbol{a}_N$ a projection of $\boldsymbol{x}$ on the set of reference vectors $\{\boldsymbol{a}_1,\boldsymbol{a}_2,\ldots,\boldsymbol{a}_N\}$

and symbolize this projection by the operator P:

$$P\boldsymbol{x} = \sum_{j=1}^{N}\left(\sum_{i=1}^{N}(\boldsymbol{a}\cdot\boldsymbol{a})_{ji}^{-1}(\boldsymbol{x}\cdot\boldsymbol{a}_i)\right)\boldsymbol{a}_j. \tag{7.7.12}$$

Returning to our atomistic system, and guided by Eq. (7.7.12), we define the projection of an arbitrary observable X on the set of observables $\{\mathcal{A}_1,\mathcal{A}_2,\ldots,\mathcal{A}_N\}$ as

$$P\mathcal{X} = \sum_{j=1}^{N}\left(\sum_{i=1}^{N}\langle\mathcal{A},\mathcal{A}\rangle_{ji}^{-1}\langle\mathcal{X},\mathcal{A}_i\rangle\right)\mathcal{A}_j, \tag{7.7.13}$$

where $\langle\mathcal{A},\mathcal{A}\rangle$ symbolizes the matrix that has as elements the scalar products (equilibrium correlations) $\langle\mathcal{A}_i,\mathcal{A}_j\rangle$ defined as in Eq. (7.7.8), $\langle\mathcal{A},\mathcal{A}\rangle^{-1}$ symbolizes the matrix inverse of $\langle\mathcal{A},\mathcal{A}\rangle$, and $\langle\mathcal{A},\mathcal{A}\rangle_{ji}^{-1}$ symbolizes the ji-element of $\langle\mathcal{A},\mathcal{A}\rangle^{-1}$. If the variables $\mathcal{A}_1,\mathcal{A}_2,\ldots,\mathcal{A}_N$ are uncorrelated to each other, then $\langle\mathcal{A},\mathcal{A}\rangle_{ji}^{-1} = \langle\mathcal{A}_j,\mathcal{A}_j\rangle^{-1}\delta_{ij}$.

We are primarily interested in cases where $\{\mathcal{A}_1,\mathcal{A}_2,\ldots,\mathcal{A}_N\}$ is a set of slowly evolving observables. PX then constitutes a projection of the arbitrary observable $\mathcal{X}$ on these slowly evolving observables. We also define the projection operator Q, orthogonal to P, by

$$Q = 1 - P. \tag{7.7.14}$$

If we have managed to include all slow dynamics of the system in the N observables $\{\mathcal{A}_1,\mathcal{A}_2,\ldots,\mathcal{A}_N\}$, then $Q\mathcal{X} = \mathcal{X} - P\mathcal{X}$ is likely to contain the fast-varying part of $\mathcal{X}$ which fluctuates rapidly around $P\mathcal{X}$, its ensemble average quickly decaying to 0.

One can show that the projection operator P defined by Eq. (7.7.13) is idempotent (Problem 7.5), i.e. that $PP\mathcal{X} = P\mathcal{X}$. As a consequence of the idempotency of P, we also see that $P^n\mathcal{X} = P\mathcal{X}$ for any observable $\mathcal{X}$ and any $n \in \mathbb{N}$.

7.7.3 Dyson identity, Mori-Zwanzig equation, and Markovian approximation

With iL being the Liouville operator, Eq. (7.7.2c), P being the projection operator onto a set of (slowly evolving) observables $\{\mathcal{A}_1,\mathcal{A}_2,\ldots,\mathcal{A}_N\}$ defined based on the scalar product in Hilbert space via Eq. (7.7.8), and Q being its orthogonal operator defined through Eq. (7.7.14), one can show the *Dyson identity* (Problem 7.6):

$$e^{\mathrm{i}Lt} = e^{\mathrm{i}QLt} + \int_0^t \mathrm{d}s\, e^{\mathrm{i}L(t-s)} P\mathrm{i}L\, e^{\mathrm{i}QLs}. \tag{7.7.15}$$

Let us apply this identity to the function $Q\,\mathrm{i}L\mathcal{A}_k(0) = Q\dot{\mathcal{A}}_k(0)$ $(k = 1,\ldots,N)$. This is the part of the rate of change with time of observable $\mathcal{A}_k$ that is orthogonal to $\{\mathcal{A}_1,\mathcal{A}_2,\ldots,\mathcal{A}_N\}$. The left-hand side of Eq. (7.7.15) operating on $Q\,\mathrm{i}L\mathcal{A}_k(0) = Q\dot{\mathcal{A}}_k(0)$ gives

$$\begin{aligned}
e^{\mathrm{i}Lt}Q\,\mathrm{i}L\mathcal{A}_k(0) &= e^{\mathrm{i}Lt}\,\mathrm{i}L\mathcal{A}_k(0) - e^{\mathrm{i}Lt}P\mathrm{i}L\mathcal{A}_k(0) = e^{\mathrm{i}Lt}\dot{\mathcal{A}}_k(0) - e^{\mathrm{i}Lt}P\mathrm{i}L\mathcal{A}_k(0)\\
&= \dot{\mathcal{A}}_k(t) - e^{\mathrm{i}Lt}\sum_{j=1}^{N}\left(\sum_{i=1}^{N}\langle\mathcal{A},\mathcal{A}\rangle_{ji}^{-1}\langle\mathrm{i}L\mathcal{A}_k,\mathcal{A}_i\rangle\right)\mathcal{A}_j(0)\\
&= \dot{\mathcal{A}}_k(t) - \sum_{j=1}^{N}\left(\sum_{i=1}^{N}\langle\mathcal{A},\mathcal{A}\rangle_{ji}^{-1}\left\langle\dot{\mathcal{A}}_k,\mathcal{A}_i\right\rangle\right)\mathcal{A}_j(t) = \dot{\mathcal{A}}_k(t) - \sum_{j=1}^{N}\mathrm{i}\,\Omega_{kj}\mathcal{A}_j(t),
\end{aligned} \tag{7.7.16}$$

where we have used Eqs. (7.7.3), (7.7.4), and (7.7.13) and introduced the *frequency matrix* $\mathrm{i}\Omega$ with elements

$$\mathrm{i}\Omega_{kj} = \sum_{i=1}^{N}\langle\mathcal{A},\mathcal{A}\rangle_{ji}^{-1}\left\langle\dot{\mathcal{A}}_k,\mathcal{A}_i\right\rangle. \tag{7.7.17}$$

Note that the scalar products appearing in the definition of $\mathrm{i}\Omega$ are not functions of time, being equilibrium ensemble averages of observables taken at the same time.

The first term on the right-hand side of Eq. (7.7.15), operating on $Q\,\mathrm{i}L\mathcal{A}_k(0) = Q\dot{\mathcal{A}}_k(0)$, gives

$$e^{\mathrm{i}QLt}Q\,\mathrm{i}L\mathcal{A}_k(0) = R_k(t). \tag{7.7.18}$$

$R_k(t)$ is the part of $\dot{\mathcal{A}}_k$, the rate of change of $\mathcal{A}_k$, that is initially orthogonal to observables $\{\mathcal{A}_1,\mathcal{A}_2,\ldots,\mathcal{A}_N\}$ and evolves in such a way as to remain orthogonal to the initial values of these observables at all subsequent times. As discussed above, we expect it to be a rapidly fluctuating function of time. We will call it the *random force* $R_k(t)$, although it does not necessarily have dimensions of force; its dimensions are the same as those of $\dot{\mathcal{A}}_k$.

The integral on the right-hand side of Eq. (7.7.15), operating on $Q\,\mathrm{i}L\mathcal{A}_k(0) = Q\dot{\mathcal{A}}_k(0)$, gives

$$\begin{aligned}
\int_0^t \mathrm{d}s e^{\mathrm{i}L(t-s)}P\mathrm{i}L e^{\mathrm{i}QLs}Q\,\mathrm{i}L\mathcal{A}_k(0) &= \int_0^t \mathrm{d}s e^{\mathrm{i}L(t-s)}P\mathrm{i}L R_k(s)\\
&= \int_0^t \mathrm{d}s e^{\mathrm{i}L(t-s)}\sum_{j=1}^{N}\left(\sum_{i=1}^{N}\langle\mathcal{A},\mathcal{A}\rangle_{ji}^{-1}\langle\mathrm{i}LR_k(s),\mathcal{A}_i(0)\rangle\right)\mathcal{A}_j(0)\\
&= \int_0^t \mathrm{d}s\sum_{j=1}^{N}\left(\sum_{i=1}^{N}\langle\mathcal{A},\mathcal{A}\rangle_{ji}^{-1}\langle\mathrm{i}LR_k(s),\mathcal{A}_i(0)\rangle\right)e^{\mathrm{i}L(t-s)}\mathcal{A}_j(0)\\
&= \int_0^t \mathrm{d}s\sum_{j=1}^{N}\left(\sum_{i=1}^{N}\langle\mathcal{A},\mathcal{A}\rangle_{ji}^{-1}\langle\mathrm{i}LR_k(s),\mathcal{A}_i(0)\rangle\right)\mathcal{A}_j(t-s)\\
&= \int_0^t \mathrm{d}s\sum_{j=1}^{N}M_{kj}(s)\mathcal{A}_j(t-s),
\end{aligned} \tag{7.7.19}$$

where we have used Eqs. (7.7.18), (7.7.13), and (7.7.4) and introduced the *memory matrix* $\boldsymbol{M}$ with elements

$$M_{kj}(s) = \sum_{i=1}^{N} \langle \mathcal{A},\mathcal{A}\rangle_{ji}^{-1} \langle \mathrm{i}LR_k(s),\mathcal{A}_i(0)\rangle . \qquad (7.7.20)$$

By virtue of the Hermitian property of L, Eq. (7.7.6), the memory matrix element of Eq. (7.7.20) can be written in a more useful form involving time-correlation functions of the random forces:

$$\begin{aligned} M_{kj}(s) &= -\sum_{i=1}^{N} \langle \mathcal{A},\mathcal{A}\rangle_{ji}^{-1} \langle R_k(s),\mathrm{i}L\mathcal{A}_i(0)\rangle = -\sum_{i=1}^{N} \langle \mathcal{A},\mathcal{A}\rangle_{ji}^{-1} \langle R_k(s),Q\mathrm{i}L\mathcal{A}_i(0)\rangle = \\ &= -\sum_{i=1}^{N} \langle \mathcal{A},\mathcal{A}\rangle_{ji}^{-1} \langle R_k(s),R_i(0)\rangle = -\sum_{i=1}^{N} \langle \mathcal{A},\mathcal{A}\rangle_{ji}^{-1} \langle R_i(s),R_k(0)\rangle . \end{aligned} \qquad (7.7.21)$$

Combining all the above terms, we finally obtain from Eq. (7.7.15) operating on $Q\,\mathrm{i}L\mathcal{A}_k(0) = Q\dot{\mathcal{A}}_k(0)$

$$\dot{\mathcal{A}}_k(t) = \sum_{j=1}^{N} \mathrm{i}\,\Omega_{kj}\mathcal{A}_j(t) + \int_0^t \mathrm{d}s \sum_{j=1}^{N} M_{kj}(s)\mathcal{A}_j(t-s) + R_k(t), \quad k = 1,2,\ldots,N \qquad (7.7.22\mathrm{a})$$

or, in vector form,

$$\dot{\boldsymbol{\mathcal{A}}}(t) = \mathrm{i}\boldsymbol{\Omega}\cdot\boldsymbol{\mathcal{A}}(t) + \int_0^t \mathrm{d}s\boldsymbol{M}(s)\cdot\boldsymbol{\mathcal{A}}(t-s) + \boldsymbol{R}(t). \qquad (7.7.22\mathrm{b})$$

Equation (7.7.22a) is known as the *Mori–Zwanzig equation.*

We can define the correlation functions among the slow variables by

$$C_{kl}(t) = \langle \mathcal{A}_k(t),\mathcal{A}_l(0)\rangle \equiv \langle \mathcal{A}_k(t)\mathcal{A}_l(0)\rangle \qquad (7.7.23)$$

with the brackets denoting, as usual, an equilibrium ensemble average. Postmultiplying both sides of the Mori–Zwanzig Eq. (7.7.22a) by $\mathcal{A}_l(0)$, taking an ensemble average, and using the fact that $\langle R_k(t)\mathcal{A}_l(0)\rangle = 0$, yields

$$\dot{C}_{kl}(t) = \sum_{j=1}^{N} \mathrm{i}\Omega_{kj}C_{kl}(t) + \int_0^t \mathrm{d}s \sum_{j=1}^{N} M_{kj}(s)C_{jl}(t-s), \quad k = 1,2,\ldots,N \qquad (7.7.24\mathrm{a})$$

or

$$\dot{\boldsymbol{C}}(t) = \mathrm{i}\boldsymbol{\Omega}\cdot\boldsymbol{C}(t) + \int_0^t \mathrm{d}s\,\boldsymbol{M}(s)\cdot\boldsymbol{C}(t-s). \qquad (7.7.24\mathrm{b})$$

That is, the correlation matrix obeys a similar equation to the Mori–Zwanzig, involving the frequency matrix and the memory matrix, but without the random force term.

The Mori–Zwanzig equation, Eq. (7.7.22), offers an exact theoretical framework for dynamical coarse-graining under conditions of equilibrium. It is conceptually valuable, but of limited practical utility in its original form. It is a time-delayed integrodifferential equation and, in general, very difficult to solve. To obtain practically more useful evolution equations from Eq. (7.7.22), one must make approximations.

One approximation we can make is that there is a clear *time-scale separation* between the *slow* variables $\{\mathcal{A}_1, \mathcal{A}_2, \ldots, \mathcal{A}_N\}$ and the remaining degrees of freedom of the atomistic system. The judicious choice of the slow variables, given the atomistic Hamiltonian of the system (how many there are, how they should best be defined) is a nontrivial and very important problem in coarse-graining. Assume that we have solved this problem and have selected $\{\mathcal{A}_1, \mathcal{A}_2, \ldots, \mathcal{A}_N\}$ such that time-correlation functions involving the rates of change of all other variables decay much faster with time than time-correlation functions involving the rates of change of the $\mathcal{A}$'s. We are interested in describing the dynamics of the system over time scales much longer than the correlation times of fast variables, which are projected out of the description, and also much shorter than the correlation times of the $\mathcal{A}$'s. On such time scales, the autocorrelation functions of the random forces $\langle R_i(s), R_k(0)\rangle$, and therefore the elements of the memory matrix, drop to 0 for times s much shorter than the correlation times of the $\mathcal{A}$'s. For such times s that are short in comparison to the rates of change of the $\mathcal{A}$'s, we can Taylor-expand the $\mathcal{A}_j(t-s)$ appearing within the memory convolution integral and keep only the first term:

$$\mathcal{A}_j(t-s) = \mathcal{A}_j(t) - \dot{\mathcal{A}}_j(t)s + \ldots \approx \mathcal{A}_j(t). \tag{7.7.25}$$

The term of Eq. (7.7.22) involving the memory integrals becomes

$$\int_0^t \mathrm{d}s \sum_{j=1}^N M_{kj}(s)\mathcal{A}_j(t-s) \approx \sum_{j=1}^N \left(\int_0^t \mathrm{d}s M_{kj}(s) \right) \mathcal{A}_j(t). \tag{7.7.26}$$

Furthermore, given that $M_{kj}(s)$ has decayed to 0 over times much shorter than t, the upper limit of integration in Eq. (7.7.24) can be replaced by $+\infty$, leading to

$$\int_0^t \mathrm{d}s \sum_{j=1}^N M_{kj}(s)\mathcal{A}_j(t-s) \approx \sum_{j=1}^N \left(\int_0^\infty \mathrm{d}s M_{kj}(s) \right) \mathcal{A}_j(t) = -\sum_{j=1}^N \xi_{kj}\mathcal{A}_j(t) \tag{7.7.27}$$

with

$$\xi_{kl} = -\int_0^\infty \mathrm{d}s\, M_{kj}(s) = \sum_{i=1}^N \int_0^\infty \mathrm{d}s\, \langle \mathcal{A}, \mathcal{A}\rangle_{ji}^{-1} \langle R_i(s), R_k(0)\rangle. \tag{7.7.28}$$

The matrix $\boldsymbol{\xi}$ defined by Eq. (7.7.28) is known as the *dissipative matrix*.

With these approximations, Eq. (7.7.22) becomes

$$\dot{\mathcal{A}}_k(t) = \sum_{j=1}^{N} \mathrm{i}\,\Omega_{kj}\mathcal{A}_j(t) - \sum_{k=1}^{N} \xi_{kj}\mathcal{A}_j(t) + R_k(t), \quad k = 1, 2, \ldots, N \tag{7.7.29a}$$

or

$$\dot{\mathcal{A}}(t) = \mathrm{i}\,\boldsymbol{\Omega}\cdot\mathcal{A}(t) - \boldsymbol{\xi}\cdot\mathcal{A}(t) + \boldsymbol{R}(t). \tag{7.7.29b}$$

Equations (7.7.29) are local in time, and therefore much easier to solve than Eq. (7.7.22), since the memory integral has been replaced by the simple dissipative term $-\sum_{k=1}^{N} \xi_{kj}\mathcal{A}_j(t)$ and the evolution of the CG variables depends only on their present state. The approximation leading to Eq. (7.7.29) from Eq. (7.7.22) is generally referred to as *Markovian approximation.*

7.7.4 Langevin equation for a free Brownian particle

We consider the very simple case of a single colloidal particle of mass m, density ρ, and diameter d suspended in a liquid solvent of small molecules of mass $m_\mathrm{b}(m_\mathrm{b} \ll m)$ at temperature T, where the mass density of the solvent is ρ_b. The particle will be considered free of any external force field; its only interactions are with the surrounding solvent. Under these conditions, it is known through the pioneering experiments of Brown[77] that the particle will execute a random flight motion within the solvent.

A full description of the dynamics would be given through Hamilton's equations, Eq. (7.7.1), in terms of the positions and momenta of both the particle and the solvent molecules. Clearly, there is a time-scale separation between the characteristic time τ_b for the motion of the solvent molecules and the characteristic time τ for motion of the Brownian particle ($\tau_\mathrm{b} \ll \tau$); these times could be quantified as integrals of the normalized velocity autocorrelation function for a solvent molecule and the free particle, respectively. We wish to develop a CG description for the motion of the particle that is appropriate for use with a time resolution Δt, where $\tau_\mathrm{b} \ll \Delta t \ll \tau$. In this description we wish to retain the degrees of freedom of the particle but project out the fast degrees of freedom of the solvent molecules. The solvent molecules should be described collectively as a *Brownian bath* of fast degrees of freedom exerting appropriate forces on the Brownian particle.

We choose the $N = 3$ components of the momentum vector $\boldsymbol{p}$ of the Brownian particle as the slow observable on which we wish to project translational motion. (The CG description could encompass the angular momentum of the particle to describe its rotational motion as well, but this is not done here, for simplicity.) In terms of the general notation we have used in the projection operator formalism, we consider

$$\mathcal{A}_k \to p_k, \quad k = 1, 2, 3. \tag{7.7.30}$$

The elements of the equilibrium fluctuation matrix become, by virtue of the equipartition theorem,

$$\langle \mathcal{A}_i, \mathcal{A}_j \rangle \to \langle p_i, p_j \rangle = \left\langle p_i^2 \right\rangle \delta_{ij} = m k_\mathrm{B} T \delta_{ij}. \tag{7.7.31}$$

The matrix $\langle \mathcal{A}, \mathcal{A} \rangle$ is diagonal, and so is its inverse, with elements

$$\langle \mathcal{A}, \mathcal{A} \rangle_{ji}^{-1} \rightarrow \frac{1}{\langle p_i^2 \rangle} \delta_{ji} = \frac{1}{m k_B T} \delta_{ji}. \tag{7.7.32}$$

The frequency matrix has elements

$$\mathrm{i}\,\Omega_{kj} = \sum_{i=1}^{N} \langle \mathcal{A}, \mathcal{A} \rangle_{ji}^{-1} \left\langle \dot{\mathcal{A}}_k, \mathcal{A}_i \right\rangle \rightarrow \sum_{i=1}^{3} \frac{1}{m k_B T} \delta_{ji} \langle \dot{p}_k, p_i \rangle = \frac{1}{m k_B T} \langle \dot{p}_k, p_j \rangle = 0 \tag{7.7.33}$$

because the total force on the particle ($\boldsymbol{F} = \dot{\boldsymbol{p}}$, determined by the atomistic configuration) is uncorrelated to its momentum, and the ensemble average of the components of the momentum is 0 at equilibrium.

The random force $\boldsymbol{R}(t)$, whose components are generally defined by Eq. (7.7.18), actually has dimensions of $\dot{\boldsymbol{p}}$, or force. As seen from the definition, assuming $\boldsymbol{p}(0) = \boldsymbol{0}$, the force $\boldsymbol{R}(0)$ at time $t = 0$ equals the total force exerted by the solvent molecules on the Brownian particle:

$$\boldsymbol{R}(0) = \boldsymbol{F}(0). \tag{7.7.34}$$

At subsequent times $t > 0$, $\boldsymbol{R}(t)$ can be physically identified with the total force that would be exerted by the solvent on the Brownian particle at time t if initially the force on it were $\boldsymbol{F}(0)$ and the whole system evolved between times 0 and t in such a way that the Brownian particle remained stationary, i.e. if the momentum of the Brownian particle were constrained at $\boldsymbol{p} = 0$. Using the subscript o to denote the force on the Brownian particle along such a trajectory,

$$\boldsymbol{R}(t) = \boldsymbol{F}_{\mathrm{o}}(t). \tag{7.7.35}$$

The elements of the memory matrix, Eq. (7.7.21), become

$$M_{kj}(s) = -\sum_{i=1}^{N} \langle \mathcal{A}, \mathcal{A} \rangle_{ji}^{-1} \langle R_i(s), R_k(0) \rangle = -\sum_{i=1}^{N} \frac{1}{m k_B T} \delta_{ji} \langle F_{\mathrm{o}i}(s) F_k(0) \rangle_{\mathrm{o}} = -\frac{1}{m k_B T} \langle F_k(0) F_{\mathrm{o}j}(s) \rangle_{\mathrm{o}}$$

and, by the independence of the motion in the three different directions and the isotropy of the system,

$$M_{kj}(s) \rightarrow -\frac{1}{3 m k_B T} \langle \boldsymbol{F}(0) \cdot \boldsymbol{F}_{\mathrm{o}}(s) \rangle_{\mathrm{o}} \delta_{kj}. \tag{7.7.36}$$

The Mori–Zwanzig equation, Eq. (7.7.22), then becomes

$$\dot{\boldsymbol{p}}(t) = -\int_0^t \mathrm{d}s \frac{1}{3 m k_B T} \langle \boldsymbol{F}(0) \cdot \boldsymbol{F}_{\mathrm{o}}(s) \rangle_{\mathrm{o}} \, \boldsymbol{p}(t - s) + \boldsymbol{F}_{\mathrm{o}}(t). \tag{7.7.37}$$

Deutsch and Oppenheim[78] derive this equation for $\left(\frac{m_\mathrm{b}}{m}\right)^{1/2} \ll 1$ and $\frac{\rho_\mathrm{b}}{\rho} \leq 1$. If, in addition, $\frac{\rho_\mathrm{b}}{\rho} \leq 1$, they show that a Markovian approximation is allowed (compare Eq. (7.7.29)), leading to

$$\dot{\boldsymbol{p}}(t) = -\left\{\int_0^\infty \mathrm{d}s \frac{1}{3mk_\mathrm{B}T} \langle \boldsymbol{F}(0) \cdot \boldsymbol{F}_\mathrm{o}(s)\rangle_\mathrm{o}\right\} \boldsymbol{p}(t) + \boldsymbol{F}_\mathrm{o}(t) = -\xi \boldsymbol{p}(t) + \boldsymbol{F}_\mathrm{o}(t). \tag{7.7.38}$$

This is the *Langevin equation* for the motion of a free Brownian particle, originally proposed by Langevin.[79] We usually see the Langevin equation written in terms of the velocity $\boldsymbol{\nu}(t)$ of the free Brownian particle,

$$m\dot{\boldsymbol{\nu}}(t) = -m\xi\,\boldsymbol{\nu}(t) + \boldsymbol{R}(t), \tag{7.7.39}$$

where the random force $\boldsymbol{R}(t)$ is a stochastic process with the properties

$$\langle \boldsymbol{R}(t) \rangle = 0 \tag{7.7.40}$$

$$\langle \boldsymbol{R}(t) \cdot \boldsymbol{\nu}(t) \rangle = 0 \tag{7.7.41}$$

$$\langle \boldsymbol{R}(t) \cdot \boldsymbol{R}(0) \rangle = 6\,m\,k_\mathrm{B}\,T\xi\,\delta(t) \tag{7.7.42a}$$

or, in component form,

$$\langle R_\alpha(t) \cdot R_\beta(0) \rangle = 2\,m\,k_\mathrm{B}\,T\xi\,\delta_{\alpha\beta}\,\delta(t). \tag{7.7.42b}$$

The Langevin equation projects the motion of the full (Brownian particle + solvent) system onto the velocity vector of the particle. The solvent is represented in a collective way, in terms of a random force and a friction force it exerts on the particle. As shown by the projection operator formalism, these forces arise upon eliminating the fast degrees of freedom and are necessary for a correct description of the dynamics at the CG level. The friction force has the usual form of a drag force derived from macroscopic hydrodynamics in the creeping flow regime, $-m\xi\,\boldsymbol{\nu}(t) = -\zeta\,\boldsymbol{\nu}(t)$, with

$$\zeta = m\xi \tag{7.7.43}$$

being the friction coefficient. In practical applications, this is usually estimated from the macroscopic viscosity η of the solvent and the diameter d of the Brownian particle as $\zeta = 3\,\pi\,\eta\,d$ (Stokes expression, no slip condition on the particle surface, usually invoked for solid particles in fluids) or $\zeta = 2\,\pi\,\eta\,d$ (Hadamard–Rybcynski expression, zero shear stress at the interface, circulation in the interior and renewal of the interface, invoked for very clean gas bubbles in liquids).

The random force $\boldsymbol{R}(t)$ is defined based on its stochastic properties, Eqs. (7.7.40)–(7.7.42). It is isotropic and uncorrelated to the velocity (hence, also the position) of the particle. Furthermore, as is obvious from Eq. (7.7.42), $\boldsymbol{R}(t)$ is viewed as losing memory of its previous value instantaneously. The autocorrelation function of the random force is proportional to a delta function of time at $t = 0$, which, when integrated over time in Eq. (7.7.38), gives merely a factor of 1/2. Equation (7.7.42) embodies the Markovian approximation in the Langevin model. From the point of view of stochastic processes, $\boldsymbol{R}(t)$ is a *white noise*; because its autocorrelation function is proportional to a delta function of time, its Fourier spectrum contains equal contributions from all frequencies.

The friction factor is directly related to the time integral of the autocorrelation function of the random force, as seen explicitly from Eq. (7.7.38), which was arrived at starting from the Mori–Zwanzig projection operator formalism. Equation (7.7.42) is a form of the *fluctuation–dissipation theorem*. It expresses the physical fact that the energy imparted by the solvent to the Brownian particle through fast, random collisions between the solvent molecules and the particle must be balanced by the energy removed from the particle through frictional forces; otherwise, the system could not be at equilibrium.

Starting from the Langevin equation, one can also show that (Problem 7.7)[80]

$$\langle \boldsymbol{\nu}(0) \cdot \boldsymbol{\nu}(t) \rangle = \frac{3k_B T}{m} \exp\left(-\xi t\right). \tag{7.7.44}$$

The exponential decay of the velocity autocorrelation function of the Brownian particle is a signature of the Markovian character of the motion and quite different from the more complex behaviour, including negative values and a long-time tail, exhibited by the velocity autocorrelation function of molecules in a liquid. The time constant of the exponential decay is seen to be $\tau = \xi^{-1}$.

With the use of the Langevin equation as a starting point and by invoking the equipartition theorem, one can also derive the following important equation for the mean-square displacement of the Brownian particle (Problem 7.8):

$$\left\langle \boldsymbol{r}^2(t) \right\rangle = \frac{6\, k_B T}{\xi\, m} \left[t - \frac{1}{\xi} + \frac{1}{\xi} \exp\left(-\xi t\right) \right]. \tag{7.7.45}$$

At short times $t \ll \frac{1}{\xi}$, Taylor-expanding the exponential we obtain ballistic motion

$$\left\langle \boldsymbol{r}^2(t) \right\rangle \simeq \frac{3\, k_B T}{m} t^2 = \left\langle \boldsymbol{\nu}^2 \right\rangle t^2.$$

At long times $t \gg \frac{1}{\xi}$ we obtain diffusive motion, $\left\langle \boldsymbol{r}^2(t) \right\rangle \simeq \frac{6\, k_B T}{\xi\, m} t = 6\, D\, t$, with the diffusivity being

$$D = \frac{k_B T}{\xi\, m} = \frac{k_B T}{\zeta}. \tag{7.7.46}$$

This is the celebrated *Einstein equation* describing the diffusivity of the particle in terms of the temperature T and friction coefficient ζ.

7.7.5 Fokker–Planck equation for a free Brownian particle, Chandrasekhar equation, Smoluchowski equation, and position Langevin equation

Consider the simple case of a free Brownian particle, whose dynamics is described by Eq. (7.7.39). Let $\rho\,(\boldsymbol{\nu}, t)$ be the probability density that characterizes an ensemble of such particles in velocity (or momentum) space at time t, all prepared under the same initial conditions (given initial velocity $\boldsymbol{\nu}(0)$). By definition, $\rho(\boldsymbol{\nu}, t)\,\mathrm{d}\boldsymbol{\nu}$ is the probability of finding the Brownian particle with velocity vector between $\boldsymbol{\nu}$ and $\boldsymbol{\nu} + \mathrm{d}\boldsymbol{\nu}$ at time t.

Let Δt be a time interval that is long enough such that the random force $\boldsymbol{R}(t)$ has undergone many fluctuations within it, yet small enough such that $\boldsymbol{\nu}(t)$ has not changed substantially ($\tau_b \ll \Delta t \ll \tau = \xi^{-1}$, see the discussion of time-scale separation and of the velocity autocorrelation function in Section 7.7.4). Let $P(\boldsymbol{\nu}; \Delta\boldsymbol{\nu})\mathrm{d}(\Delta\boldsymbol{\nu})$ be the probability density that velocity $\boldsymbol{\nu}$ experiences a change $\Delta\boldsymbol{\nu}$ within time Δt. By the Markovian character of Brownian motion, how the system will evolve at time t will depend on the state of the system at time t, but not on the prior history of the system. As a consequence, the conditional probability density $P(\boldsymbol{\nu}; \Delta\boldsymbol{\nu})$ does not carry an explicit dependence on time; it depends only on $\boldsymbol{\nu}$ and $\Delta\boldsymbol{\nu}$, and refers to the chosen time interval Δt. Moreover, $P(\boldsymbol{\nu}; \Delta\boldsymbol{\nu})$ is normalized:

$$\int P(\boldsymbol{\nu}; \Delta\boldsymbol{\nu})\,\mathrm{d}(\Delta\boldsymbol{\nu}) = 1. \tag{7.7.47}$$

With these definitions, the following recursive relation holds:

$$\rho(\boldsymbol{\nu}, t + \Delta t) = \int \rho(\boldsymbol{\nu} - \Delta\boldsymbol{\nu}, t)\,P(\boldsymbol{\nu} - \Delta\boldsymbol{\nu}; \Delta\boldsymbol{\nu})\,\mathrm{d}(\Delta\boldsymbol{\nu}). \tag{7.7.48}$$

The limits of integration are $-\infty$ and $+\infty$ for each component of $\Delta\boldsymbol{\nu}$. This is a *master equation*, entirely analogous to Eq. (5.1.7) that we discussed in connection with the Monte Carlo method, the only difference being that here we have a continuous state space. In the language of stochastic processes, Eq. (7.7.48) is a *Chapman–Kolmogorov equation.* By expanding the terms $\rho(\boldsymbol{\nu}, t + \Delta t)$, $\rho(\boldsymbol{\nu} - \Delta\boldsymbol{\nu}, t)$, and $P(\boldsymbol{\nu} - \Delta\boldsymbol{\nu}; \Delta\boldsymbol{\nu})$ in the integral Eq. (7.7.48) up to second order in $\Delta\boldsymbol{\nu}$ and up to first order in Δt, one derives the *Fokker–Planck equation* (see Problem 7.9):

$$\frac{\partial \rho}{\partial t} = \xi\,\nabla_{\nu} \cdot (\rho\boldsymbol{\nu}) + \frac{\xi\,k_B T}{m}\nabla_{\nu}^2\,\rho. \tag{7.7.49}$$

The Fokker–Planck equation is a partial differential equation governing the evolution of the probability density of the velocity vector of the free Brownian particle.

One can readily confirm that the Boltzmann distribution

$$\rho^{\mathrm{eq}}(\boldsymbol{\nu}) = \mathrm{const.}\exp\left(-\frac{m\nu^2}{2k_B T}\right) \tag{7.7.50}$$

is the steady-state solution of the Fokker–Planck equation.

The above arguments can be generalized for the case of a Brownian particle that is subject to an external force field $\mathcal{V}(\boldsymbol{r})$ generating a force $\boldsymbol{F} = -\nabla_{\boldsymbol{r}}\mathcal{V}(\boldsymbol{r})$ on the Brownian particle. In the description of the motion through a stochastic differential equation, this external force is added to the right-hand side of the Langevin equation, Eq. (7.7.39). The corresponding partial differential equation description is given in terms of the probability density $\rho(\boldsymbol{r}, \boldsymbol{\nu}, t)$ of the Brownian particle in its phase space. This obeys the *Chandrasekhar equation:*[81]

$$\frac{\partial}{\partial t}\rho(\boldsymbol{r},\boldsymbol{\nu},t)+\boldsymbol{\nu}\cdot\nabla_{\boldsymbol{r}}\rho(\boldsymbol{r},\boldsymbol{\nu},t)+\frac{1}{m}\boldsymbol{F}\cdot\nabla_{\boldsymbol{\nu}}\rho(\boldsymbol{r},\boldsymbol{\nu},t)$$
$$=\xi\,\nabla_{\boldsymbol{\nu}}\cdot[\rho\,(\boldsymbol{r},\boldsymbol{\nu},t)\,\boldsymbol{\nu}]+\frac{\xi\,k_{\mathrm{B}}T}{m}\nabla_{\boldsymbol{\nu}}^{2}\rho(\boldsymbol{r},\boldsymbol{\nu},t). \qquad (7.7.51)$$

Equation (7.7.51) provides a description of the irreversible evolution of the probability density of the Brownian particle in its phase space, in a manner analogous to how the Liouville equation, Eq. (7.7.2), provides a description of the reversible evolution of the probability density of the entire system (Brownian particle plus solvent molecules) in phase space. In the absence of an external field ($\boldsymbol{F}=\boldsymbol{0}$), integrating the Chandrasekhar equation, Eq. (7.7.51), over all space leads to the Fokker–Planck equation, Eq. (7.7.49) for the probability density in velocity space, $\rho(\boldsymbol{\nu},t)$.

In many applications we are not interested in tracking the short-time evolution of the velocity of the Brownian particle, but only its position as a function of time. We are interested, in other words, in describing the evolution, within the bath of fast degrees of freedom, of the Brownian particle generally subject to an external force field $\boldsymbol{F}=-\nabla_{\boldsymbol{r}}\mathcal{V}(\boldsymbol{r})$ using time steps Δt that are long in comparison to the relaxation time of the velocity ξ^{-1}:

$$\Delta t \gg 1/\xi. \qquad (7.7.52)$$

Over a time step Δt that satisfies Eq. (7.7.52), the velocity distribution of the Brownian particle will have decayed to the equilibrium Boltzmann distribution, Eq. (7.7.50). Under these conditions, one can derive a partial differential equation for the evolution of the probability distribution $\rho(\boldsymbol{r},t)$ of the Brownian particle in configuration space only,

$$\frac{\partial}{\partial t}\rho\,(\boldsymbol{r},t)+\frac{D}{k_{\mathrm{B}}T}\nabla_{\boldsymbol{r}}\cdot\,(\boldsymbol{F}\rho\,(\boldsymbol{r},t))=D\,\nabla_{\boldsymbol{r}}^{2}\rho\,(\boldsymbol{r},t) \qquad (7.7.53)$$

with the diffusivity given by Eq. (7.7.46). Equation (7.7.53) is usually referred to as the *Smoluchowski equation.* It is a diffusion equation in 3D space with bias due to the external force field. The steady-state solution of Eq. (7.7.53) is the Boltzmann distribution

$$\rho^{\mathrm{eq}}(\boldsymbol{r})=\mathrm{const.}\exp\left(-\frac{\mathcal{V}(\boldsymbol{r})}{k_{\mathrm{B}}T}\right). \qquad (7.7.54)$$

As one might expect, there should be a stochastic differential equation description that can be used in place of the partial differential equation description of Eq. (7.7.53). This should be entirely analogous to how the Langevin equation description, Eq. (7.7.39), can be used in place of the Fokker–Planck description, Eq. (7.7.49), or to how the Hamiltonian equations of motion, Eq. (7.7.1), can be used instead of the Liouville equation, Eq. (7.7.2). To extract this stochastic differential equation description that corresponds to the Smoluchowski equation for a single Brownian particle subject to an external field, Eq. (7.7.53), one can start from the Langevin equation for the particle, i.e. Eq. (7.7.39) with an external force term added:

$$m\dot{\boldsymbol{\nu}}(t) = -m\xi\,\boldsymbol{\nu}(t) + \boldsymbol{F}(\boldsymbol{r}) + \boldsymbol{R}(t). \tag{7.7.55}$$

We are interested in generating trajectories out of this stochastic differential equation using a long time step $\Delta t \gg 1/\xi$ that satisfies Eq. (7.7.52).

For such a long time step, the inertial term on the left-hand side of Eq. (7.7.55) is on the order of

$$|m\dot{\boldsymbol{\nu}}(t)| \sim |m\boldsymbol{\nu}/\Delta t| \ll |m\boldsymbol{\nu}\,\xi| = |-m\xi\,\boldsymbol{\nu}(t)|, \tag{7.7.56}$$

which means that the inertial term in the Langevin equation, Eq. (7.7.55), becomes negligible in comparison to the friction term and the other terms in the equation. This is why, when we study Brownian motion under the constraint of Eq. (7.7.52), we say that we study Brownian motion *in the high friction limit.*

Neglecting the inertial term in the Langevin equation, Eq. (7.7.55), we obtain the *position Langevin equation*

$$\dot{\boldsymbol{r}}(t) = \frac{1}{m\xi}\boldsymbol{F}(\boldsymbol{r}) + \boldsymbol{r}'(t), \tag{7.7.57}$$

where now $\boldsymbol{r}'(t) = \dfrac{1}{m\xi}\boldsymbol{R}(t)$ is a *random velocity process* satisfying the conditions

$$\langle r_\alpha{}'(t)\rangle = 0 \qquad (\alpha = 1,2,3) \tag{7.7.58}$$

$$\left\langle r'_\alpha(t) r'_\beta(t)\right\rangle = 2\frac{k_B T}{m\xi}\delta_{\alpha\beta}\delta(t) = 2D\,\delta_{\alpha\beta}\,\delta(t) \quad (\alpha,\beta = 1,2,3). \tag{7.7.59}$$

For a system of n interacting Brownian particles subject to an external field, the Smoluchowski equation, Eq. (4.7.53), governing their probability density in configuration space $\rho(\boldsymbol{r},t)$, with $\boldsymbol{r}$ denoting the $3n$-dimensional vector of particle positions $(\boldsymbol{r}_1, \boldsymbol{r}_2, \ldots, \boldsymbol{r}_n)$, becomes[81]

$$\frac{\partial}{\partial t}\rho(\boldsymbol{r},t) + \sum_i\sum_j \nabla_{\boldsymbol{r}_i}\cdot\frac{\boldsymbol{D}_{ij}(\boldsymbol{r})}{k_B T}\cdot\boldsymbol{F}_j(\boldsymbol{r})\,\rho(\boldsymbol{r},t) = \sum_i\sum_j \nabla_{\boldsymbol{r}_i}\cdot\boldsymbol{D}_{ij}(\boldsymbol{r})\cdot\nabla_{\boldsymbol{r}_j}\rho(\boldsymbol{r},t) \tag{7.7.60}$$

with $i,j \in \{1,2,\ldots,n\}$ and $\boldsymbol{F}_j$ denoting the total systematic force on particle j due to its (solvent-mediated) interactions with other particles and due to the external field, if present. The second-order tensors $\boldsymbol{D}_{ij}(\boldsymbol{r})$ (diffusivity tensors, diffusivity matrices) are characteristic of specific pairs of particles and generally depend on the configuration. They incorporate hydrodynamic interactions among the particles. A *hydrodynamic interaction* on particle i due to particle j arises because, as particle j moves relative to the solvent, it modifies the flow field of the solvent around particle i, thereby exerting a frictional force on particle i. For a set of solid Brownian particles of diameter d moving in a solvent of viscosity η in dilute solution ($r_{ij} \gg d$), a satisfactory approximation for $\boldsymbol{D}_{ij}(\boldsymbol{r})$ is provided by the Oseen tensor[82]

$$\boldsymbol{D}_{ij}(\boldsymbol{r}) = \begin{cases} \dfrac{k_B T}{3\pi\eta d}\boldsymbol{1}, & \text{if } i = j \\ \dfrac{k_B T}{8\pi\eta r_{ij}}\left(\boldsymbol{1} + \dfrac{\boldsymbol{r}_{ij}\boldsymbol{r}_{ij}}{r_{ij}^2}\right) & \text{if } i \neq j \end{cases}, \tag{7.7.61}$$

where η is the solvent viscosity and $\zeta = 3\pi\eta d$. The long-range nature of hydrodynamic interactions, which scale inversely with the distance between different particles, is evident from the Oseen expression.

The position Langevin equations for a system of interacting Brownian particles, integrated over a time step Δt, give

$$\boldsymbol{r}_i(t + \Delta t) = \boldsymbol{r}_i(t) + \sum_j \frac{\boldsymbol{D}_{ij}(\boldsymbol{r})}{k_B T} \cdot \boldsymbol{F}_j(\boldsymbol{r})\,\Delta t + \nabla_{\boldsymbol{r}_j} \cdot \boldsymbol{D}_{ij}(\boldsymbol{r})\,\Delta t + \Delta \boldsymbol{r}_i^{G}, \tag{7.7.62}$$

where $\Delta \boldsymbol{r}_i^G$ is a random displacement imposed on the individual particle i. The components $\Delta r_{i\alpha}^G$ of this displacement are selected from a multivariate Gaussian distribution[83] with mean

$$\langle \Delta r_{i\alpha}^G \rangle = 0 \qquad (i = 1, 2, \ldots, n; \quad \alpha = 1, 2, 3) \tag{7.7.63a}$$

and covariance matrix

$$\langle \Delta r_{i\alpha}^G \Delta r_{j\beta}^G \rangle = 2 D_{ij,\alpha\beta} \Delta t \qquad (i, j = 1, 2, \ldots, n; \quad \alpha, \beta = 1, 2, 3). \tag{7.7.63b}$$

In general, numerical algorithms for the integration of stochastic differential equations such as the Langevin equation or the position Langevin equation are based on integrating the stochastic differential equation between times t and $t + \Delta t$. Derivation of the algorithms is aided by the following theorem.

If $\phi(t)$ is any function of time and $\boldsymbol{R}(t)$ is a random force (white noise) with the statistical properties described by Eqs. (7.7.40)–(7.7.42), then the random variable

$$(\Delta\phi R_\alpha)^G \equiv \int_t^{t+\Delta t} \phi(t')\, R_\alpha(t')\, dt' \tag{7.7.64}$$

follows a Gaussian distribution with mean value

$$\left\langle (\Delta\phi R_\alpha)^G \right\rangle = 0 \tag{7.7.65a}$$

and variance

$$\left\langle \left[(\Delta\phi R_\alpha)^G \right]^2 \right\rangle = 2 m \xi k_B T \int_t^{t+\Delta t} \phi^2(t')\, dt'. \tag{7.7.65b}$$

The interested reader can easily verify that use of Eqs. (7.7.64)–(7.7.65) on the position Langevin equation, Eq. (7.7.58) with $\phi(t) = 1/(m\xi)$, leads to the difference scheme of Eq. (7.7.62) in the case of a single Brownian particle subject to an external field in the high friction limit.

In general, integration of a Langevin equation in both momentum and configuration space is referred to as a *Langevin dynamics simulation*, while integration of a position Langevin equation, describing the system in the high friction limit, in configuration space only, is referred to as a *Brownian dynamics simulation*. Algorithms for Langevin dynamics and for Brownian dynamics are discussed in Refs. 83 and 84.

From a mathematical–numerical point of view, an important issue faced with the integration of the position Langevin equation is that of the *correct calculus* to use. In the following, we have relied heavily on the material presented in Ref. 85. To simplify the analysis, let us consider the position Langevin equation, Eq. (7.7.57), in only one dimension and rewrite it in the form of a general stochastic differential equation for the x position of the Brownian particle as

$$\mathrm{d}x = \nu(x)\mathrm{d}t + g(x)\mathrm{d}W(t), \tag{7.7.66}$$

where we have defined $\nu(x) \equiv \dfrac{1}{m\xi}F(x)$, $g(x) \equiv \dfrac{1}{m\xi}$, and $\mathrm{d}W(t) \equiv R(t)\mathrm{d}t$. From this last equation we also see that $R(t) \equiv \dfrac{\mathrm{d}W(t)}{\mathrm{d}t}$; i.e. we have expressed the Markovian, Gaussian process $R(t)$ having the properties described by Eqs. (7.7.40)–(7.4.42) and known as white noise as the derivative of the Wiener process $W(t)$. The latter is defined probabilistically through

$$P(W(t)) = \frac{1}{\sqrt{2\pi Dt}} e^{-\frac{W(t)^2}{2Dt}}. \tag{7.7.67}$$

The Wiener process inherits the Markovian and Gaussian character of the random walk process and can be characterized by giving its mean value and the two-time-correlation function:

$$\langle W(t)\rangle = 0 \tag{7.7.68a}$$

$$\langle W(t')\, W(t'')\rangle = D\min(t', t''). \tag{7.7.68b}$$

Because the white noise stochastic process is introduced as the derivative of the Wiener process and the latter does not have a well-defined derivative, the precise interpretation of the white noise process depends on the way the derivative is performed. The most widely used interpretations are those of Itô[86] and Stratonovich.[87] More specifically, to integrate Eq. (7.7.66), we need to calculate terms of the form $\int g(x)\mathrm{d}W(t)$, which are stochastic integrals. A possible approach is through a Taylor expansion. By assuming for simplicity that $v(x) = 0$, then

$$\begin{aligned} x(t) - x(0) &= \int_0^t g(x(s))\,\mathrm{d}W(s) \\ &\cong \int_0^t \mathrm{d}W(s)\,[g(x(0)) + g'(x(0))\,(x(s) - x(0))]. \end{aligned} \tag{7.7.69}$$

Evaluating the integrals on the right-hand-side of Eq. (7.7.69) order by order, we find that to the lowest order $x(t) - x(0) \cong g(x(0))\, W(t)$. This means that in the next order (provided that $g'(x) \neq 0$), one should evaluate the integral $\int_0^t W(s)\, \mathrm{d}W(s)$. The standard approach to this is through discretization, in the mean square limit,

$$\int_0^t W \mathrm{d}W = \text{m.s.} \lim_{n\to\infty} \sum_{i=1}^{n} W(t_i^*)\,[W(t_i) - W(t_{i-1})], \tag{7.7.70}$$

using a suitable partition $0 = t_0 < t_1 < \ldots < t_i < \ldots < t_n = t$, and where $t_{i-1} < t_i^* < t_i$. The mean square limit is defined as

$$\text{m.s.} \lim_{n\to\infty} X_n = X \;\Leftrightarrow\; \lim_{n\to\infty} \left\langle [X_n - X]^2 \right\rangle = 0, \tag{7.7.71}$$

with the average taken over all possible realizations of the Wiener process. Let us also take

$$t_i^* = t_{i-1} + a\,(t_i - t_{i-1}),\quad 0 \le a \le 1. \tag{7.7.72}$$

In the evaluation of the mean square limit, Eq. (7.7.70), terms like $\langle W(t')\, W(t'')\rangle$ appear. Recalling that the increment $W(t'') - W(t')$ is independent of $W(t')$, it follows that $\langle W(t')\, W(t'')\rangle = \langle W(t')\, W(t')\rangle = D\,t'$. Thus,

$$\int_0^t W \mathrm{d}W = \text{m.s.} \lim_{n\to\infty} \sum_{i=1}^{n} W(t_i^*)\,[W(t_i) - W(t_{i-1})] = D \sum_{i=1}^{n} [t_{i-1} + a\,(t_i - t_{i-1}) - t_{i-1}] = D\,a\,t \tag{7.7.73}$$

Equation (7.7.73) implies that the value of the integral depends critically on the choice of a. Although in principle any value of a in the range $[0, 1]$ is possible, in the literature only two values are commonly preferred: $a = 0$ corresponding to *Itô calculus*,[86] and $a = \frac{1}{2}$ corresponding to *Stratonovich calculus*.[87] If $W(t)$ were a smooth function, clearly $\left\langle \int_0^t W \mathrm{d}W \right\rangle = \left\langle \frac{1}{2} W(t)^2 \right\rangle = \frac{1}{2} D\,t$, corresponding to $a = \frac{1}{2}$.

Even more importantly, from Eq. (7.7.66), one can derive the following Fokker–Planck equation for the probability $\rho\,(x, t)$,[88]

$$\frac{\partial \rho\,(x,t)}{\partial t} = \frac{\partial}{\partial x} \left\{ -v(x) - a g(x) g'(x) + \frac{1}{2} \frac{\partial}{\partial x} g^2(x) \right\} \rho\,(x,t), \tag{7.7.74}$$

which also depends explicitly on a. The final equilibrium distribution will also depend on a. The above has been referred to as the *Itô versus Stratonovich controversy* that appeared in the literature between the late 1970s and the early 1980s.[85] From Eq. (7.7.74), we also see that when the function $g(x)$ in Eq. (7.7.66) is constant (which is the case for the position Langevin equation), then there is no difference between the two interpretations.

7.8 The Rouse model

The Rouse model[89] is a cornerstone for describing polymer dynamics. We have briefly stated some results obtained from it concerning the chain self-diffusivity and the relaxation time spectrum in an unentangled monodisperse melt of linear polymers in Sections 1.4.3 and 4.3.2. Here we examine the Rouse model a bit more in depth, as a prime example of dynamical coarse-graining based on the theory of Brownian motion.

In deriving the Rouse model, an atomistic chain in an unentangled melt is mapped onto a CG Gaussian bead-and-spring model (Figure 7.5; see also Chapter 1). Each bead stands for a sufficiently long section of chain, such that successive interbead vectors are not directionally correlated at equilibrium; i.e. the sequence of interbead vectors conforms itself as a random flight in 3D space. Furthermore, as long as we are not considering very strong flows, we will assume that each interbead vector follows a Gaussian distribution; in other words, we will ignore the finite extensibility of interbead vectors (compare Eq. (1.3.35) and Problem 5.5). The intramolecular potential of mean force between successive beads will thus be assumed to be parabolic with a minimum at 0; i.e. each interbead vector will behave as a Gaussian spring with spring constant

$$k = \frac{3k_{\mathrm{B}}T}{b^2}, \tag{7.8.1}$$

where b^2 is the mean squared distance between successive beads. Clearly, each bead must be at least as large as a Kuhn segment. In practice, chain sections of approximately fifty atomistic skeletal bonds are lumped into single beads. The results of the theory, however, will not depend much on the precise definition of a bead.

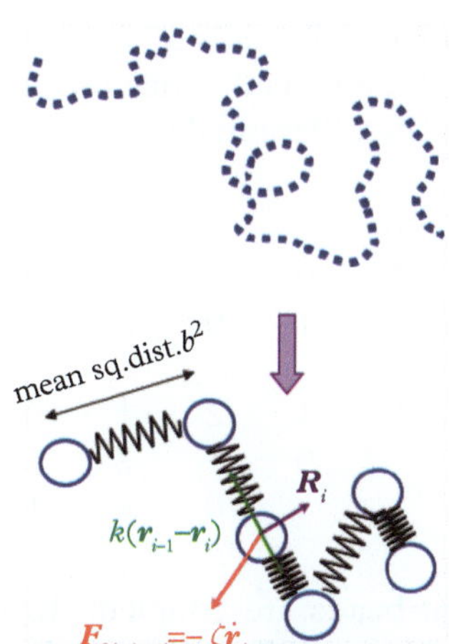

Figure 7.5 Schematic representation of the coarse-graining invoked in deriving the Rouse model, with the forces acting on bead i indicated.

Let N be the total number of beads representing the chain, their indices running from 0 to $N-1$. Each bead carries mass m. The number of springs along a chain will then be $(N-1)$ and the equilibrium mean square end-to-end distance will be related to b^2 through the random walk relation

$$\langle r^2 \rangle = (N-1)\, b^2. \tag{7.8.2}$$

Now, in the Rouse model, each bead of a chain in the (unentangled) melt is assumed to behave as a Brownian particle, experiencing random forces and frictional forces from the rest of the polymer, represented as a Brownian bath. As in Section 7.7, Eq. (7.7.43), we will assume a friction factor $\zeta = m\xi$ that is independent of configuration, to describe frictional forces between a bead and its environment. We will also ignore hydrodynamic interactions between different beads in a chain. The latter approximation is not at all good for describing the dynamics of a chain in dilute solution in a micromolecular solvent. (There, hydrodynamic interactions must be incorporated, as is done in the Zimm model of polymer dynamics.) Ignoring hydrodynamic interactions, or, in more elaborate terminology, assuming that hydrodynamic interactions are *screened*, turns out to be a very good approximation for describing the motion of the chain in a sea of chains that look exactly like itself, as happens in a melt.

As outlined in Figure 7.5, according to the assumptions of the Rouse model, an internal bead i with position vector $\boldsymbol{r}_i$ is subject to the systematic spring forces $k\,(\boldsymbol{r}_{i+1} - \boldsymbol{r}_i)$ and $k\,(\boldsymbol{r}_{i-1} - \boldsymbol{r}_i)$ from its neighbouring beads, to the Langevin force $\boldsymbol{R}_i(t)$, and to the frictional force from its surrounding bath. *In the high friction limit* (compare Eqs. (7.7.52) and (7.7.57)), the Langevin equation for the motion of bead i becomes

$$-\zeta \frac{\mathrm{d}\boldsymbol{r}_i}{\mathrm{d}t} + k\,(\boldsymbol{r}_{i+1} - \boldsymbol{r}_i) + k\,(\boldsymbol{r}_{i-1} - \boldsymbol{r}_i) + \boldsymbol{R}_i = 0 \tag{7.8.3}$$

with $\boldsymbol{R}_i(t)$ fulfilling the statistical properties of Eqs. (7.7.40) to (7.7.42) as well as Eq. (7.7.43).

Collecting the Langevin equations for all beads i, $i = 0, 1, \ldots, N-1$ and taking into account that the first bead ($i = 0$) has no preceding bead, while the last bead ($i = N-1$) has no following bead, we obtain the system of stochastic differential equations

$$\zeta \frac{\mathrm{d}}{\mathrm{d}t}\begin{bmatrix} \boldsymbol{r}_0 \\ \boldsymbol{r}_1 \\ \boldsymbol{r}_2 \\ \vdots \\ \boldsymbol{r}_{N-3} \\ \boldsymbol{r}_{N-2} \\ \boldsymbol{r}_{N-1} \end{bmatrix} = -k \begin{bmatrix} 1 & -1 & 0 & . & . & . & 0 \\ -1 & 2 & -1 & 0 & . & . & 0 \\ 0 & -1 & 2 & -1 & 0 & . & 0 \\ . & . & . & . & . & . & . \\ 0 & . & 0 & -1 & 2 & -1 & 0 \\ 0 & . & . & 0 & -1 & 2 & -1 \\ 0 & . & . & . & 0 & -1 & 1 \end{bmatrix} \begin{bmatrix} \boldsymbol{r}_0 \\ \boldsymbol{r}_1 \\ \boldsymbol{r}_2 \\ \vdots \\ \boldsymbol{r}_{N-3} \\ \boldsymbol{r}_{N-2} \\ \boldsymbol{r}_{N-1} \end{bmatrix} + \begin{bmatrix} \boldsymbol{R}_0 \\ \boldsymbol{R}_1 \\ \boldsymbol{R}_2 \\ \vdots \\ \boldsymbol{R}_{N-3} \\ \boldsymbol{R}_{N-2} \\ \boldsymbol{R}_{N-1} \end{bmatrix}. \tag{7.8.4}$$

Clearly, the motions of the different beads are coupled. To uncouple them, we diagonalize the symmetric tridiagonal matrix appearing on the right-hand side of Eq. (7.8.4). The matrix has eigenvalues[90]

$$4\sin^2\left(\frac{p\pi}{2N}\right), \quad p = 0, 1, \ldots, N-1 \tag{7.8.5}$$

and corresponding eigenvectors

$$\left(\cos\left(\frac{p\pi}{2N}\right), \cos\left(\frac{3p\pi}{2N}\right), \ldots, \cos\left(\frac{(2N-1)\,p\pi}{2N}\right)\right)\sqrt{\frac{2-\delta_{p0}}{N}} \tag{7.8.6}$$

with δ being the Kronecker delta.

Using this diagonalization, we recast the system of stochastic differential equations, Eq. (7.8.4), into the equivalent form

$$\zeta\frac{\mathrm{d}\boldsymbol{X}_p}{\mathrm{d}t} = -k_p\boldsymbol{X}_p + \tilde{\boldsymbol{R}}_p, \tag{7.8.7}$$

where

$$\boldsymbol{X}_p = \sum_{j=1}^{N}\Omega_{jp}\boldsymbol{r}_{j-1} \tag{7.8.8a}$$

with

$$\Omega_{jp} = \sqrt{\frac{2-\delta_{p0}}{N}}\cos\left(\frac{(2j-1)\,p\pi}{2N}\right), \quad j = 1, \ldots, N \tag{7.8.8b}$$

and

$$k_p = 4k\sin^2\left(\frac{p\pi}{2N}\right), \quad p = 0, 1, \ldots, N-1. \tag{7.8.9}$$

The random forces $\tilde{\boldsymbol{R}}_p$ appearing in Eq. (7.8.7) have the following statistical properties, which are derived from the statistical properties of the $\boldsymbol{R}_i$'s:

$$\begin{aligned}\left\langle\tilde{\boldsymbol{R}}_p(t)\right\rangle &= 0\\ \left\langle\tilde{R}_{p,\alpha}(t)\tilde{R}_{q,\beta}(t')\right\rangle &= 2\zeta\,k_{\mathrm{B}}T\,\delta_{pq}\,\delta_{\alpha\beta}\,\delta\,(t-t').\end{aligned} \tag{7.8.10}$$

Conversely, one can express the bead coordinates in terms of the $\boldsymbol{X}_p$'s:

$$\boldsymbol{r}_i = \sum_{p=0}^{N-1}\Omega_{i+1p}\boldsymbol{X}_p = \sum_{p=0}^{N-1}\sqrt{\frac{2-\delta_{p0}}{N}}\cos\left(\frac{(2i+1)\,p\pi}{2N}\right)\boldsymbol{X}_p, \quad i = 0, 1, \ldots, N-1. \tag{7.8.11}$$

The quantities $\boldsymbol{X}_p$, $p = 0, 1, \ldots, N-1$ in Eq. (7.8.7) are linear combinations of the bead positions, which behave as *independent* Brownian particles subject to frictional forces from their surroundings with a friction constant ζ and to random forces satisfying the fluctuation–dissipation relations, Eq. (7.8.10). The $\boldsymbol{X}_p$'s are the celebrated *Rouse modes* of the chain.

The zeroth mode ($p = 0$) describes the motion of the centre of mass of the chain. By its definition, Eq. (7.8.8),

$$\boldsymbol{X}_0 = \sqrt{\frac{1}{N}}(\boldsymbol{r}_0 + \boldsymbol{r}_1 + \cdots + \boldsymbol{r}_{N-1}) = \sqrt{N}\,\boldsymbol{r}_{\mathrm{cm}}, \tag{7.8.12}$$

where $\boldsymbol{r}_{\mathrm{cm}}$ is the position of the centre of mass. By Eq. (7.8.9), $k_0 = 0$, i.e. the spring constant for the zeroth mode is zero; in other words, the mode describes the diffusive motion of a free Brownian particle. The Langevin equation in the high friction limit, Eq. (7.8.7), for this mode becomes

$$\zeta\frac{\mathrm{d}\boldsymbol{X}_0}{\mathrm{d}t} = \tilde{\boldsymbol{R}}_0 \text{ or } (N\zeta)\frac{\mathrm{d}\boldsymbol{r}_{\mathrm{cm}}}{\mathrm{d}t} = \sqrt{N}\tilde{\boldsymbol{R}}_0, \tag{7.8.13}$$

implying that the chain centre of mass moves as a free Brownian particle subject to a friction factor ζN. By Eq. (7.7.46), the chain centre-of-mass self-diffusivity is

$$D = \frac{k_{\mathrm{B}}T}{N\zeta}, \tag{7.8.14}$$

which is Eq. (1.4.16).

For $p \geq 1$, Rouse mode p describes the motion of a section of the chain comprising roughly $1/p$ of its total contour length, i.e. N/p segments; higher p modes describe more local motion and, by Eq. (7.8.9), are subject to a stiffer spring constant k_p. In a manner entirely analogous to the derivation of Eq. (7.7.44), one can obtain the time autocorrelation function for the pth mode:

$$\begin{aligned} \langle \boldsymbol{X}_p(0)\cdot\boldsymbol{X}_p(t)\rangle &= \frac{3k_{\mathrm{B}}T}{k_p}\exp\left(-\frac{k_p}{\zeta}t\right), \quad p = 1,2,\ldots,N-1 \\ \langle \boldsymbol{X}_p(0)\cdot\boldsymbol{X}_q(t)\rangle &= 0, \quad p \neq q, \quad p,q = 1,2,\ldots,N-1. \end{aligned} \tag{7.8.15}$$

The time autocorrelation function of the pth mode decays exponentially, with a characteristic time

$$\tau_p = \frac{\zeta}{k_p} = \frac{\zeta}{4k}\frac{1}{\sin^2\left(\frac{p\pi}{2N}\right)} = \frac{\zeta b^2}{12\,k_{\mathrm{B}}T}\frac{1}{\sin^2\left(\frac{p\pi}{2N}\right)}, \quad p = 1,2,\ldots,N-1 \tag{7.8.16}$$

from an initial mean square amplitude

$$\langle \boldsymbol{X}_p^2(0)\rangle = \frac{3\,k_{\mathrm{B}}T}{k_p} = \frac{3\,k_{\mathrm{B}}T}{4\,k}\frac{1}{\sin^2\left(\dfrac{p\pi}{2N}\right)} = \frac{b^2}{4}\frac{1}{\sin^2\left(\dfrac{p\pi}{2N}\right)}, \quad p = 1,2,\ldots,N-1. \tag{7.8.17}$$

Usually we are interested in the first few modes of the chain, $1 \leq p \ll N-1$. High p modes cannot be expected to describe accurately the true dynamics of the chain, which will depend sensitively on the detailed chemical constitution and therefore not be well represented by the bead-and-spring model invoked. For these low p modes the following approximation is satisfactory:

$$\sin^2\left(\frac{p\pi}{2N}\right) \simeq \frac{p^2\pi^2}{4N^2}. \tag{7.8.18}$$

Use of Eq. (7.8.18) within (7.8.16) and (7.8.17) leads to

$$\tau_p = \frac{\zeta b^2 N^2}{3\pi^2 k_B T}\frac{1}{p^2} = \frac{\tau_1}{p^2} \quad (1 \le p \ll N-1) \text{ with } \tau_1 \equiv \tau_R = \frac{\zeta b^2 N^2}{3\pi^2 k_B T} \simeq \frac{(\zeta N)\left\langle r^2\right\rangle}{3\pi^2 k_B T} \tag{7.8.19}$$

and

$$\left\langle \boldsymbol{X}_p^2(0)\right\rangle = \frac{N^2 b^2}{p^2\pi^2}, \quad p = 1, 2, \ldots, N-1. \tag{7.8.20}$$

Equation (7.8.19) is a very important result. In the spectrum of relaxation times of a Rouse chain, the time governing the motion of mode p scales inversely with the square of p; the relaxation time drops rapidly as the motion described by the mode becomes more local. The longest relaxation time τ_R in the (unentangled) melt (Rouse time) increases in proportion to the square of the chain length.

It is remarkable that both Eq. (7.8.14) for the chain self-diffusivity and Eq. (7.8.19) for the Rouse time depend on the product $N\zeta$, i.e. on the total friction constant experienced by the chain in the free-draining limit (no hydrodynamic interactions), rather than on N and ζ individually. This means that the exact discretization used in coarse-graining the atomistically detailed macromolecule into a bead-and-spring chain does not matter for the main results of the theory. The longest relaxation time τ_R is seen to be commensurate with the time required for the centre of mass of a chain to diffuse by a length equal to its root mean squared end-to-end distance.

The time autocorrelation function of the end-to-end vector of a long Rouse chain is obtained as

$$\langle \boldsymbol{r}(0)\cdot\boldsymbol{r}(t)\rangle = Nb^2 \sum_{p=1,3,5,\ldots} \frac{8}{p^2\pi^2}\exp\left(-\frac{t}{\tau_R}p^2\right). \tag{7.8.21}$$

The mean square displacement (msd) of beads can be calculated based on Eq. (7.8.11). Given that different modes are uncorrelated, we get for bead i, $i = 0, 1, \ldots, N-1$

$$\begin{aligned}\left\langle [\boldsymbol{r}_i(t) - \boldsymbol{r}_i(0)]^2\right\rangle &= \sum_{p=0}^{N-1}\left(\frac{2-\delta_{p0}}{N}\right)\cos^2\left(\frac{(2i+1)p\pi}{2N}\right)\left\langle [\boldsymbol{X}_p(t) - \boldsymbol{X}_p(0)]^2\right\rangle = \\ &= \frac{1}{N}\left\langle [\boldsymbol{X}_0(t) - \boldsymbol{X}_0(0)]^2\right\rangle + \frac{2}{N}\sum_{p=1}^{N-1}\cos^2\left(\frac{(2i+1)p\pi}{2N}\right)\left\langle [\boldsymbol{X}_p(t) - \boldsymbol{X}_p(0)]^2\right\rangle.\end{aligned} \tag{7.8.22}$$

Substituting the mode autocorrelation functions and mode mean square amplitudes from Eqs. (7.8.15) and (7.8.17) into Eq. (7.8.22) yields

$$\left\langle [\boldsymbol{r}_i(t)-\boldsymbol{r}_i(0)]^2\right\rangle = 6Dt + \frac{b^2}{N}\sum_{p=1}^{N-1}\frac{\cos^2\left(\frac{(2i+1)\,p\pi}{2N}\right)}{\sin^2\left(\frac{p\pi}{2N}\right)}\left(1-\exp\left(-\frac{k_p}{\zeta}t\right)\right),\quad i=0,1,\ldots,N-1. \tag{7.8.23}$$

For long times $t \gg \tau_R$, the first, diffusive term $6Dt$ on the right-hand side of Eq. (7.8.23) dominates the segmental msd: the segment follows the overall diffusive motion of the chain, with self-diffusivity D.

For short times $t \ll \tau_R$, the second term in Eq. (7.8.23), descriptive of the motion of the segment relative to the chain centre of mass, dominates. In view of Eq. (7.8.18) and the spectrum of relaxation times defined by Eq. (7.8.29), it can be recast as

$$\left\langle [\boldsymbol{r}_i(t)-\boldsymbol{r}_i(0)]^2\right\rangle \simeq \frac{12k_BT}{N\zeta}\sum_{p=1}^{N-1}\tau_p\cos^2\left(\frac{(2i+1)\,p\pi}{2N}\right)\left(1-\exp\left(-\frac{t}{\tau_p}\right)\right),\quad i=0,1,\ldots,N-1. \tag{7.8.24}$$

To proceed, we replace $\cos^2\left(\frac{(2i+1)\,p\pi}{2N}\right)$ for all $i=0,1,\ldots,N-1$ by its average ½; thus,

$$\begin{aligned}
\left\langle [\boldsymbol{r}_i(t)-\boldsymbol{r}_i(0)]^2\right\rangle &\simeq 6\frac{k_BT}{N\zeta}\sum_{p=1}^{N-1}\tau_p\left(1-\exp\left(-\frac{t}{\tau_p}\right)\right) = 6\frac{k_BT}{N\zeta}\sum_{p=1}^{N-1}\int_0^t e^{-t'/\tau_p}dt' \\
&\simeq 6\frac{k_BT}{N\zeta}\sum_{p=1}^{N-1}\int_0^t e^{-\frac{t'p^2}{\tau_1}}\,dt' = 6\frac{k_BT}{N\zeta}\int_0^t dt'\left(\sum_{p=1}^{N-1}e^{-\frac{t'p^2}{\tau_1}}\right) \\
&\simeq 6\frac{k_BT}{N\zeta}\int_0^t dt'\left(\int_0^\infty e^{-\frac{t'p^2}{\tau_1}}\,dp\right) \\
&= 6\frac{k_BT}{N\zeta}\int_0^t dt'\frac{1}{2}\sqrt{\frac{\pi\tau_1}{t'}} = 3\frac{k_BT}{N\zeta}\sqrt{\pi\tau_1}\int_0^t\frac{dt'}{\sqrt{t'}} = 3\frac{k_BT}{N\zeta}\sqrt{\pi\tau_1}2\sqrt{t} = 6\frac{k_BT}{N\zeta}\sqrt{\pi\tau_1 t} \\
&= \left(\frac{12k_BTb^2}{\pi\zeta}\right)^{\frac{1}{2}}t^{\frac{1}{2}}\quad(\tau_{N-1}\ll t\ll\tau_1),\qquad i=0,1,\ldots,N-1.
\end{aligned}$$

or, alternatively,

$$\left\langle [\boldsymbol{r}_i(t) - \boldsymbol{r}_i(0)]^2 \right\rangle \simeq \frac{6k_BT}{N\zeta}\tau_1 \int_0^\infty \frac{1 - \exp\left(-\frac{t}{\tau_1}p^2\right)}{p^2} \mathrm{d}p$$

$$\simeq \frac{6k_BT}{N\zeta}\tau_1 \sqrt{\frac{\pi t}{\tau_1}}$$

$$= \left(\frac{12k_BTb^2}{\pi\zeta}\right)^{\frac{1}{2}} t^{\frac{1}{2}} \quad (\tau_{N-1} \ll t \ll \tau_1), \quad i = 0, 1, \ldots, N-1. \tag{7.8.25}$$

Thus, at short times, the msd of beads in a Rouse chain grows with time as $t^{1/2}$ and is independent of the number of segments N in the chain. A crossover from subdiffusive $\sim t^{\frac{1}{2}}$ to diffusive $\sim t$ behaviour in the msd of a bead occurs at around the Rouse time $\tau_1 = \tau_R$.

Of great interest are the linear viscoelastic properties predicted by the Rouse model. We model a melt of N_{ch} chains in volume V at temperature T, each represented as a bead-and-spring chain of N beads and mean square interbead separation b^2, each bead behaving as a Brownian particle subject to a friction factor ζ from its surroundings. All systematic forces other than the spring forces between successive beads along a chain are neglected. With these conditions, the elements of the *instantaneous* stress tensor can be obtained from the atomic virial expression, Eq. (3.2.24), as

$$\sigma_{\alpha\beta,\text{inst.}} = -\frac{1}{V}\sum_{i_{ch}=1}^{N_{ch}}\sum_{i=0}^{N-1}\frac{p_{i_{ch}i,\alpha}p_{i_{ch}i,\beta}}{m_{i_{ch}i}} - \frac{1}{V}\sum_{i_{ch}=1}^{N_{ch}}\sum_{i=1}^{N-1}\left(r_{i_{ch}i,\alpha} - r_{i_{ch}i-1,\alpha}\right)F_{i_{ch}i,i_{ch}i-1,\beta} =$$

$$= -\frac{1}{V}\sum_{i_{ch}=1}^{N_{ch}}\sum_{i=0}^{N-1}\frac{p_{i_{ch}i,\alpha}p_{i_{ch}i,\beta}}{m_{i_{ch}i}} + \frac{k}{V}\sum_{i_{ch}=1}^{N_{ch}}\sum_{i=1}^{N-1}\left(r_{i_{ch}i,\alpha} - r_{i_{ch}i-1,\alpha}\right)\left(r_{i_{ch}i,\beta} - r_{i_{ch}i-1,\beta}\right). \tag{7.8.26}$$

In the second (configurational) term on the right-hand side of Eq. (7.8.26) the contribution from each pair of bonded beads along a chain appears once; hence, there is no factor of ½ in front of the term. In the second line of the equation we have expressed the β component of the force on bead i of chain i_{ch} from bead $i-1$ of the same chain in terms of the β component of the distance vector between the two beads and the spring constant k. In the CG model invoked here, different chains behave as independent molecules. No contributions from random or frictional forces are included in the instantaneous stress tensor.

Concentrating on the ensemble averaged values of the shear components ($\alpha \neq \beta$), for which there is no kinetic contribution, these can be computed by inserting in Eq. (7.8.26) the bead positions expressed in terms of the Rouse modes or normal coordinates from Eq. (7.8.11) and noticing that $\cos(A) - \cos(B) = -2\sin\left(\frac{A+B}{2}\right)\sin\left(\frac{A-B}{2}\right)$. The *total* $\alpha\beta$ component of stress due to all chains is then

$$\sigma_{\alpha\beta} = \langle \sigma_{\alpha\beta,\text{inst.}} \rangle$$

$$= \frac{N_{\text{ch}}}{V}\frac{3k_{\text{B}}T}{b^2}\left\langle \sum_{i=1}^{N-1} (r_{i,\alpha} - r_{i-1,\alpha})(r_{i,\beta} - r_{i-1,\beta}) \right\rangle$$

$$= \frac{N_{\text{ch}}}{V}\frac{12k_{\text{B}}T}{b^2}\sum_{i=1}^{N-1}\sum_{p=0}^{N-1}\sum_{q=0}^{N-1} \frac{[(2-\delta_{p0})(2-\delta_{q0})]^{1/2}}{N} \sin\left(\frac{p\pi}{2N}\right)\sin\left(\frac{q\pi}{2N}\right)$$

$$\times \sin\left(\frac{ip\pi}{N}\right)\sin\left(\frac{iq\pi}{N}\right)\langle X_{p,\alpha}X_{q,\beta}\rangle$$

$$= \frac{N_{\text{ch}}}{V}\frac{12k_{\text{B}}T}{b^2}\sum_{i=1}^{N-1}\sum_{p=0}^{N-1} \frac{(2-\delta_{p0})}{N}\sin^2\left(\frac{p\pi}{2N}\right)\sin^2\left(\frac{ip\pi}{N}\right)\langle X_{p,\alpha}X_{p,\beta}\rangle$$

$$= \frac{N_{\text{ch}}}{V}\frac{12k_{\text{B}}T}{b^2}\frac{(N-1)}{2}\sum_{p=0}^{N-1} \frac{(2-\delta_{p0})}{N}\sin^2\left(\frac{p\pi}{2N}\right)\langle X_{p,\alpha}X_{p,\beta}\rangle$$

$$= \frac{N_{\text{ch}}}{V}\frac{12k_{\text{B}}T}{b^2}\frac{(N-1)}{2}\sum_{p=1}^{N-1} \frac{2}{N}\sin^2\left(\frac{p\pi}{2N}\right)\langle X_{p,\alpha}X_{p,\beta}\rangle$$

$$= \frac{3k_{\text{B}}TN_{\text{ch}}}{V}\sum_{p=1}^{N-1}\frac{\langle X_{p,\alpha}X_{p,\beta}\rangle}{\langle \boldsymbol{X}_p^2\rangle}. \tag{7.8.27}$$

In deriving Eq. (7.8.27), we have used that $\langle \boldsymbol{X}_p(0)\cdot\boldsymbol{X}_q(t)\rangle = 0$ for $p \neq q$ $(p, q = 1, 2, \ldots, N-1)$, that the $p = 0$ term does not contribute to the sum, and we have replaced $\sin^2\left(\frac{ip\pi}{N}\right)$ for all $i = 0, 1, \ldots, N-1$ by its average 1/2. We have also used Eqs. (7.8.9), (7.8.1) and (7.8.17).

Similarly, the ensemble average value of the correlation of the $\alpha\beta$ component of the instantaneous total stress at time 0 to the same component at time t is

$$\langle \sigma_{\alpha\beta,\text{inst.}}(t)\sigma_{\alpha\beta,\text{inst.}}(0)\rangle = N_{\text{ch}}\left(\frac{3k_{\text{B}}T}{V}\right)^2\sum_{p=1}^{N-1}\sum_{q=1}^{N-1}\frac{\langle X_{p,\alpha}(t)X_{p,\beta}(t)X_{q,\alpha}(0)X_{q,\beta}(0)\rangle}{\langle \boldsymbol{X}_p^2\rangle\langle \boldsymbol{X}_q^2\rangle}. \tag{7.8.28}$$

By using the independence of modes,[91] Eq. (7.8.28) leads to the following expression of the shear stress relaxation modulus $G(t)$ (compare Eq. (4.3.4)):

$$G(t) = \frac{V}{k_{\mathrm{B}}T}\langle\sigma_{\alpha\beta,\mathrm{inst.}}(t)\sigma_{\alpha\beta,\mathrm{inst.}}(0)\rangle$$
$$= \frac{N_{\mathrm{ch}}k_{\mathrm{B}}T}{V}\sum_{p=1}^{N-1}\left[\frac{\langle\boldsymbol{X}_p(t)\cdot\boldsymbol{X}_p(0)\rangle}{\langle\boldsymbol{X}_p^2\rangle}\right]^2$$
$$= \frac{N_{\mathrm{ch}}k_{\mathrm{B}}T}{V}\sum_{p=1}^{N-1}\exp\left(-2t/\tau_p\right). \tag{7.8.29}$$

The zero shear viscosity of the Rouse melt is obtained by Eq. (4.3.5),

$$\eta_0 = \int_0^\infty G(t)\mathrm{d}t = \frac{N_{\mathrm{ch}}k_{\mathrm{B}}T}{V}\sum_{p=1}^{N-1}\int_0^\infty \exp\left(-2t/\tau_p\right)\mathrm{d}t = \frac{N_{\mathrm{ch}}k_{\mathrm{B}}T}{2V}\sum_{p=1}^{\infty}\tau_p = \frac{N_{\mathrm{ch}}k_{\mathrm{B}}T}{2V}\tau_1\sum_{p=1}^{\infty}\frac{1}{p^2}$$
$$= \frac{N_{\mathrm{ch}}k_{\mathrm{B}}T}{2V}\tau_1\frac{\pi^2}{6} = \frac{\pi^2}{12}\frac{\rho_{\mathrm{mass}}RT}{M}\tau_1 = \frac{\rho_{\mathrm{mass}}N_{\mathrm{Avo}}}{36M}\zeta b^2N^2 = \frac{\rho_{\mathrm{mass}}N_{\mathrm{Avo}}}{36}\frac{\langle r^2\rangle}{M}\zeta N$$
$$= \frac{\rho_{\mathrm{mass}}RT}{36D}\frac{\langle r^2\rangle}{M}, \tag{7.8.30}$$

where ρ_{mass} and M are the mass density and molar mass of the melt, respectively, and N_{Avo} is Avogadro's number.

According to the Rouse model, the zero-shear viscosity of a melt scales linearly with the length of its chains. The scalings $D \sim N^{-1}$, Eq. (7.8.14); $\tau_1 \sim N^2$, Eq. (7.8.19); and $\eta_0 \sim N^1$, Eq. (7.8.30) are confirmed by experimental measurements on unentangled melts.[92]

Recently, the Rouse model has been solved for ring polymers[93] as well as for polymers adsorbed by one or both of their ends to the surface of a nanoparticle.[94]

7.9 The tube model and reptation theory

Reptation theory describes the dynamics of high-molecular-weight polymers by taking into account the topological constraints on chain diffusion arising from chain connectivity and the fact that polymers cannot cross each other, as discussed in Section 7.6. It was originally proposed[17,25,67] to address the problem of rubber elasticity, by considering that the intrinsic properties of the chain are still represented by the Rouse model consisting of N segments with bond length b and friction constant ζ; their dynamics, however, is significantly affected by the topological constraints that develop due to entanglements. As a result, each chain is effectively confined to move in a tube-like region whose axis is defined by the shortest path connecting the two ends of the chain having the same topology as the chain itself relative to the constraints. As we discussed in Section 7.6, such a path is called the primitive path (PP) and can move back and forth only along its contour. In the simplest version of the theory, the contour length L of the PP is assumed constant and is described by the expression

$$L = \frac{\langle r^2 \rangle}{a_{\text{pp}}} = \frac{Nb^2}{a_{\text{pp}}}, \tag{7.9.1}$$

where the new parameter a_{pp} introduced is called the step length of the primitive chain.

Of central importance in reptation theory is the function $\psi(s,t)$ specifying the probability that at time t the PP point s has remained in the original tube (the tube specified for the corresponding PP at time $t = 0$). Once this is known, it is a matter of simple algebraic manipulations to obtain *all* linear viscoelastic properties of the polymer melt under study. For example, one can calculate the following:[17]

- the spectrum of relaxation times, and consequently the longest relaxation or disengagement time τ_d, through

$$\int_0^L \psi(s,t)\,\mathrm{d}s = L \sum_{p:\text{odd}} \frac{8}{p^2 \pi^2} \exp\left(-\frac{p^2 t}{\tau_{\text{d}}}\right), \tag{7.9.2}$$

  where

$$\tau_{\text{d}} = \frac{(\zeta N) \langle r^2 \rangle^2}{\pi^2 a_{\text{pp}}^2 k_{\text{B}} T} \tag{7.9.3}$$

  denotes the longest relaxation or disengagement or reptation time;
- the relaxation modulus of the polymer through

$$G(t) = G_{\text{N}}^0 \frac{1}{L} \int_0^L \psi(s,t)\,\mathrm{d}s; \tag{7.9.4}$$

- the storage and loss moduli through

$$G'(\omega) = \omega \int_0^\infty G(t) \sin(\omega t)\,\mathrm{d}t \quad \text{and} \quad G''(\omega) = \omega \int_0^\infty G(t) \cos(\omega t)\,\mathrm{d}t; \tag{7.9.5}$$

- the melt zero shear viscosity through

$$\eta_0 = G_{\text{N}}^0 \frac{1}{L} \int_0^\infty \int_0^L \psi(s,t)\,\mathrm{d}s\,\mathrm{d}t = \int_0^\infty G(t)\mathrm{d}t. \tag{7.9.6}$$

For the particular case of reptation of a linear polymer in a fixed, infinite network, implying that contour length fluctuation (CLF) effects and constraint-release (CR) mechanisms are neglected, Doi and Edwards computed the following analytical expression for $\psi(s,t)$,[17]

$$\psi(s,t) = \sum_{p:\text{odd}} \frac{4}{p\pi} \sin\left(\frac{p\pi s}{L}\right) \exp\left(-\frac{p^2 t}{\tau_d}\right), \tag{7.9.7}$$

where L is the contour length of the PP. Then, the zero-shear rate viscosity and the chain centre-of-mass self-diffusion coefficient assume the forms

$$\eta_0 = \frac{\pi^2}{12} G_N^0 \tau_d \tag{7.9.8}$$

and

$$D = \frac{D_c}{3Z} = \frac{k_B T a_{pp}^2}{3N^2 \zeta b^2}, \tag{7.9.9}$$

where $Z = \dfrac{L}{a_{pp}} = \dfrac{\langle r^2 \rangle}{a_{pp}^2}$ denotes the number of entanglement strands per chain. The plateau modulus G_N^0 is usually estimated by analogy to rubber elasticity theory from the melt mass density ρ and the molar mass of an entanglement strand $M_e = \dfrac{M}{Z}$ as

$$G_N^0 = \frac{4}{5} \frac{\rho RT}{M_e} \tag{7.9.10}$$

and is independent of chain length N. Equations (7.9.3), (7.9.8) and (7.9.9) suggest that $\eta_0 \sim N^3$ and $D \sim N^{-2}$, which do not fully agree with experimental data.[95–97] This motivated refinements in the theory through consideration of phenomena present in real systems that had been neglected in the first versions, known today as CLF of the primitive chain and CR caused by the movement of surrounding chains.

Based on Eq. (7.9.7) for the function $\psi(s,t)$, reptation theory makes some other important predictions for the dynamics of entangled polymer melts: For example, it predicts the msd $\phi(n,t)$ of a Rouse segment n in the network. Although the precise calculation of $\phi(n,t)$ is a difficult task, its main characteristic features can be obtained by analysing its limiting behaviour in four regimes:[17]

- At times shorter than the characteristic time τ_e marking the onset of the effect of topological constrains on segmental motion

$$\tau_e = \frac{1}{3\pi^2} \frac{a_{pp}^4 \zeta}{k_B T b^2}, \tag{7.9.11}$$

  segments do not feel the constraints of the network; thus, $\phi(n,t)$ should be the same as that predicted by the Rouse model in free space, Eq. (7.8.23).

- At times t such that $\tau_e < t < \tau_R$ (Rouse time given by Eq. (7.8.19)), the motion of the segment perpendicular to the PP is restricted but that along the path remains free.
- At times t such that $\tau_R < t < \tau_d$ (disengagement time given in Eq. (7.9.3)) the dynamics is identical to that predicted by the PP dynamics and $\phi(n,t) \sim t^{1/2}$.
- At very long times $\tau_d < t$, the dynamics is governed by the reptation process and $\phi(n,t) \sim t^1$.

Overall, the four scalings are summarized as

$$\phi(n,t) \sim \begin{cases} t^{1/2} & t \leq \tau_e \\ t^{1/4} & \tau_e \leq t \leq \tau_R \\ t^{1/2} & \tau_R \leq t \leq \tau_d \\ t^1 & \tau_d \leq t \end{cases} \tag{7.9.12}$$

The four scalings are well supported by simulations,[98–101] starting with the pioneering work of Kremer and Grest.[102]

7.10 Mapping MD data onto the Rouse and tube models

7.10.1 The friction coefficient ζ

By mapping atomistic MD data for either the chain longest relaxation time (obtained by monitoring the decay of the time autocorrelation function for the chain end-to-end vector) or the chain centre-of-mass self-diffusion coefficient (obtained by monitoring the average msd of the chain centre of mass) onto Eqs (7.8.19) and (7.8.14) of the Rouse model, respectively, for unentangled melts or on Eqs. (7.9.3) and (7.9.9) of the reptation model for entangled melts, one can obtain estimates of the segmental friction factor as a function of chain length. The two estimates are denoted as ζ_τ and ζ_D, respectively, and early MD simulation data[98–100] showed them to be practically equal to each other; thus, in the literature only a single value of ζ per carbon atom is usually reported. Figure 7.6 shows a typical plot of the chain length dependence of ζ for linear PE at $T = 450\text{K}$ and *cis*-1,4 PB at $T = 413\text{K}$, as obtained with such mappings from MD simulations with a number of (different chain length) systems. For the mapping to reptation theory one needs first to extract the step length of the PP by analysing the topological state of the polymeric liquid in terms of PPs by reducing atomistically detailed chain configurations onto an ensemble of PPs, as discussed in Section 7.6. According to Figure 7.6 the results obtained from the mapping to the reptation theory constitute a smooth continuation of the friction factor data obtained from the MD mapping onto the Rouse model.

The results reported in Figure 7.6 for PE refer to the friction factor per carbon chain of strictly *linear* melts. For H-shaped PE systems, of interest is the value of the branch point diffusion coefficient, D_b,[103] which can be used to extract an estimate for the branch point friction coefficient, ζ_b, through an Einstein argument, namely, $\zeta_b = \dfrac{k_B T}{D_b}$. Reported MD-based ζ_b simulation results for a number of model H-shaped PE melts (at $T = 450\text{K}$ and $P = 1\text{atm}$)[104] reveal a strong dependence on the crossbar (or backbone) and arm lengths N_b and N_a; furthermore, they are all on the order

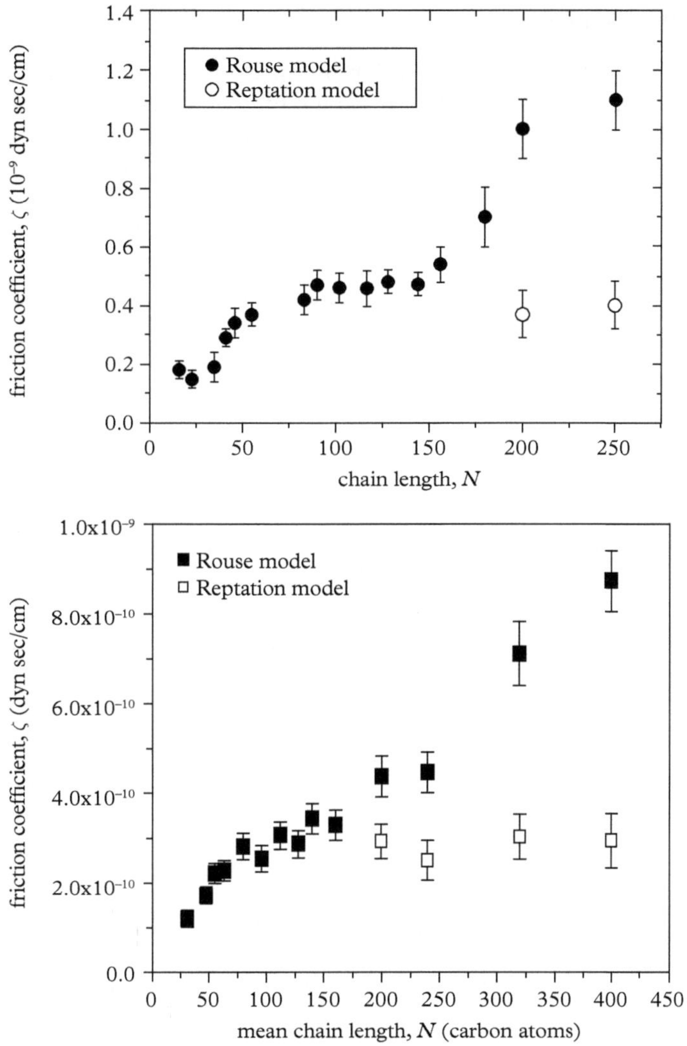

Figure 7.6 Values of the friction coefficient ζ for various chain lengths, as obtained by mapping atomistic MD data for the chain self-diffusivity D onto the Rouse (filled symbols) and the reptation model (open symbols) for (top) linear PE ($T = 450$K, $P = 1$atm) and (bottom) *cis*-1,4 PB ($T = 413$K, $P = 1$atm).
(top) Reprinted (adapted) with permission from Harmandaris, V.A., Mavrantzas, V.G., Theodorou, D.N., Kröger, M., Ramírez, J., Öttinger, H.C., and Vlassopoulos, D. *Macromolecules* **36**, 1376 (2003). Copyright 2003 American Chemical Society. (bottom) Reprinted (adapted) with permission from Tsolou, G., Mavrantzas, V.G., and Theodorou, D.N. *Macromolecules* **38**, 1478 (2005). Copyright 2005 American Chemical Society.

of 10^{-10} kg/s and higher, which is considerably larger than the value calculated for the segmental friction coefficient of linear PE melts at the same temperature. This demonstrates the large friction concentrated at the branch points of an H-shaped PE molecule as compared to the friction felt by a monomer unit along a linear chain.

7.10.2 The tube diameter

Stephanou et al.[101] have presented two different methods for determining the tube diameter $d_t \approx a_{pp}$: (a) by analysing the segmental msd versus time as obtained directly from the atomistic simulations, and (b) by monitoring the displacement of primitive chain segments orthogonal to the initial PP. Then, they implemented the two methods in simulations with a number of linear PE and PB melts and practically identical (within the statistical uncertainty) results were obtained from the two methods. A typical plot is shown in Figure 7.7. Figure 7.8 shows a reptating chain from the simulations with a PE C_{500} melt system and how its conformation and PP change with time. The primitive chain pictures reveal that the polymer chain moves mainly back and forth along its main contour, which brings more and more segments out of the original tube. Chain ends escape first and middle segments follow.

7.10.3 The function $\psi\ (s, t)$

Having obtained an estimate of the tube diameter d_t and using the mapped time trajectory of primitive chains, Stephanou et al.[101] have also described a computational methodology for obtaining the function $\psi(s, t)$ from MD simulations. Representative plots for a number of model linear PE melts consisting of C_{320}, C_{400}, and C_{500} chains (denoted as PE320, PE400, and PE500, respectively), and *cis*-1,4 PB and *trans*-1,4 PB melts consisting of C_{320} and C_{400} chains (denoted as PB-*cis*320, PB-*cis*400, PB-*trans*320, and PB-*trans*400, respectively) are shown in Figure 7.9.

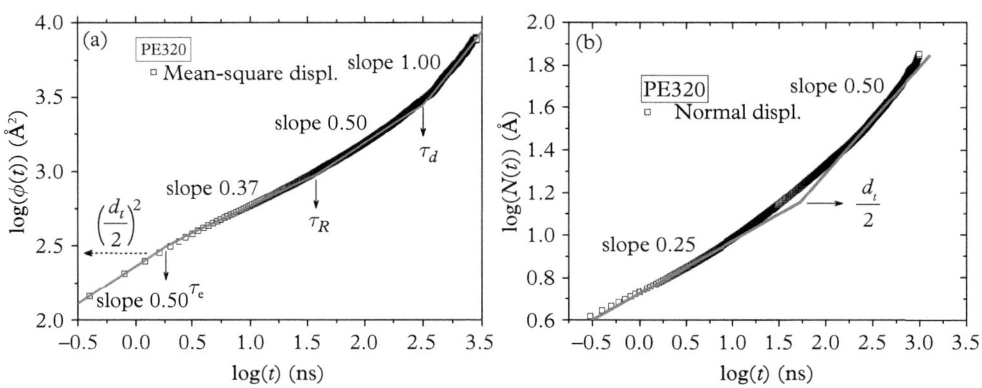

Figure 7.7 Calculation of the tube diameter d_t based on (a) the atomistic segmental mean-square displacement (msd) of the innermost chain segments versus time and (b) the displacement of the primitive chain segments orthogonal to the initial primitive path versus time, from MD simulations with a linear C_{320} PE melt at $T = 450K$ and $P = 1atm$.

Reprinted from Stephanou, P.S., Baig, C., Tsolou, G., Mavrantzas, V.G., and Kröger, M. *J. Chem. Phys.* **132**, 124904 (2010), with the permission of AIP Publishing.

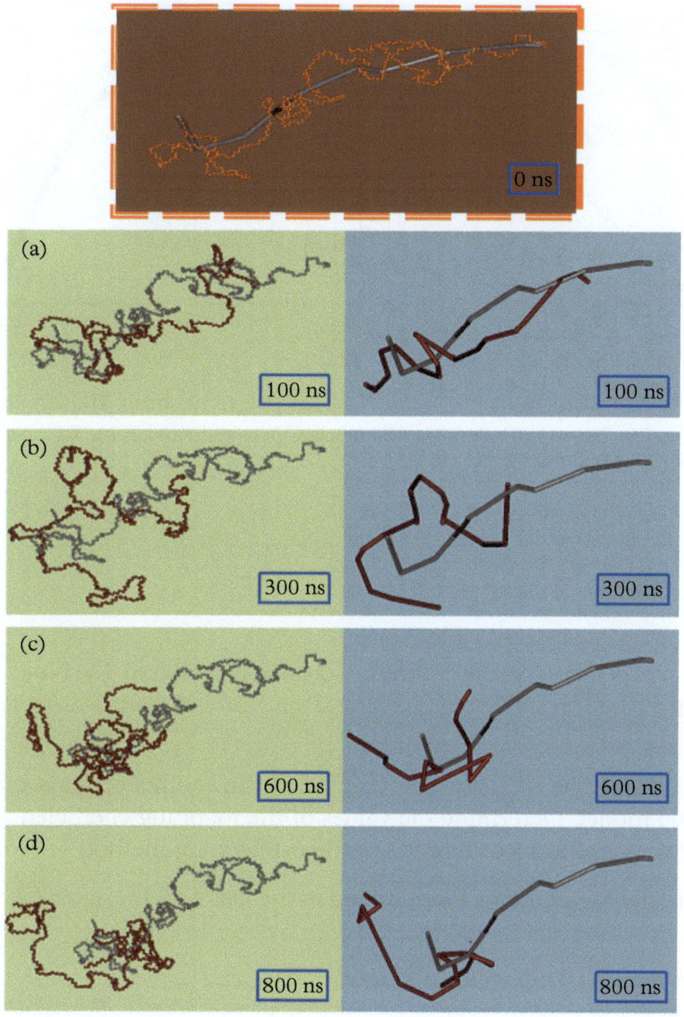

Figure 7.8 A reptating chain. Shown on the left is how the atomistic configuration of a single chain from the simulations with a C_{500} PE system changes with time. Shown on the right are the corresponding changes in its primitive path. Atomistic and primitive chain configurations at $t = 0$ are also shown.

Reprinted from Stephanou, P.S., Baig, C., Tsolou, G., Mavrantzas, V.G., and Kröger, M. *J. Chem. Phys.* **132**, 124904 (2010), with the permission of AIP Publishing.

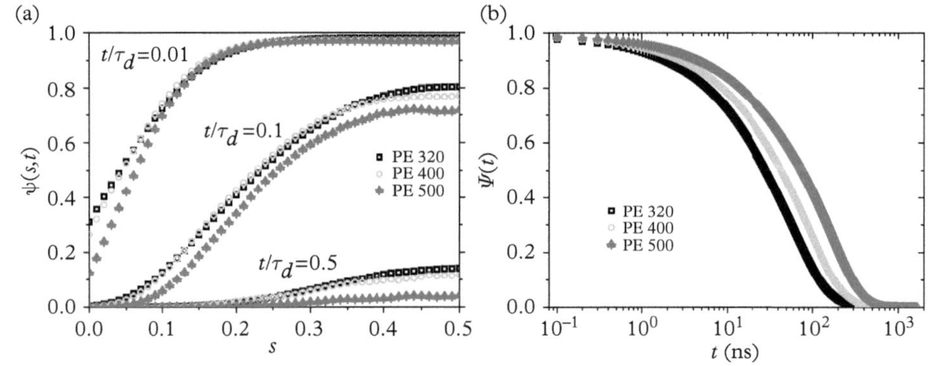

Figure 7.9 Typical plots of the functions $\psi(s,t)$ and $\Psi(t) \equiv \int_0^1 \psi(s,t)\mathrm{d}s$ from MD simulations.
Reprinted from Stephanou, P.S., Baig, C., Tsolou, G., Mavrantzas, V.G., and Kröger, M. *J. Chem. Phys.* **132**, 124904 (2010), with the permission of AIP Publishing.

Problems

Problem 7.1: For the general coarse-graining scheme considered by Noid et al.,[13] show Eq. (7.2.17) expressing the force on a CG site in terms of the atomistic forces on its constituents starting from Eq. (7.2.13).

Problem 7.2: Quite often, the CG sites are defined as the centres of mass of disjoint subsets of atoms of the atomistic model. Determine the values of the mapping coefficients C_{Ii} for use in Eqs. (7.2.10) and the site masses M_I according to the force-matching method of Noid et al.[13]

Problem 7.3: Prove that the operator L appearing in the classical Liouville equation is Hermitian, Eq. (7.7.6).

Problem 7.4: Show that the propagator operator e^{iLt} is unitary.

Problem 7.5: Prove that the projection operator P defined by Eq. (7.7.13) satisfies $PPX = PX$; thus it is idempotent.

Problem 7.6: Show the Dyson identity, Eq. (7.7.15).

Problem 7.7: Starting from the Langevin equation, obtain the velocity autocorrelation function of a Brownian particle, Eq. (7.7.44).

Problem 7.8: Derive Eq. (7.7.45) describing the translational motion of a free Brownian particle.

Problem 7.9: Starting from the *Chapman–Kolmogorov equation*, Eq. (7.7.48), and by expanding the terms $\rho(\nu, t+\Delta t)$, $\rho(\nu+\Delta\nu, t)$, $P(\nu-\Delta\nu; \Delta\nu)$ up to second order in $\Delta\nu$ and up to first order in Δt, derive the *Fokker–Planck equation*, Eq. (7.7.49).

References

1. Baschnagel, J., Binder, K, Paul, W., Laso, M., Suter, U.W., Batoulis, I., Jilge, W., and Bürger, T. *J. Chem. Phys.* **95**, 6014 (1991).
2. Tschöp, W., Kremer, K., Batoulis, J. Bürger, T., and Hahn, O. *Acta Polymerica* **49**, 61 (1998).
3. Tschöp, W., Kremer, K., Hahn, O., Batoulis, J., and Bürger, T. *Acta Polymerica* **49**, 75 (1998).
4. Müller-Plathe, F. *ChemPhysChem* **3**, 754 (2002).
5. Reith, D., Pütz, M., and Müller-Plathe, F. *J. Comp. Chem.* **24**, 1624 (2003).
6. Izvekov, S., and Voth, G.A. *J. Chem. Phys.* **123**, 134105 (2005).
7. Izvekov, S., and Voth, G. *J. Phys.Chem. B* **109**, 2469 (2005).
8. Zacharopoulos, N., Vergadou, N., and Theodorou, D.N. *J. Chem. Phys.* **122**, 244111 (2005).
9. Harmandaris, V.A., Adhikari, N.P., van der Vegt, N.F.A., and Kremer, K. *Macromolecules* **39**, 6708 (2006).
10. Harmandaris, V.A., Reith, D., van der Vegt, N.F.A., and Kremer, K. *Macromol. Chem. Phys.* **208**, 2109 (2007).
11. Spyriouni, T., Tzoumanekas, C., Theodorou, D.N., Müller-Plathe, F., and Milano, G. *Macromolecules* **40**, 3876 (2007).
12. Knotts IV, T.A., Rathore, N., Schwartz, D.C., and de Pablo, J.J. *J. Chem. Phys.* **126**, 084901 (2007).
13. Noid, W.G., Chu, J.-W., Ayton, G.S., Krishna, V., Izvekov, S., Voth, V.A., Das, A., and Andersen, H.C. *J. Chem. Phys.* **128**, 244114 (2008).
14. Noid, W.G., Liu, P., Wang,Y., Chu, J.-W., Ayton, G.S., Izvekov, S., Andersen, H.C., and Voth, G.A. *J. Chem. Phys.* **128**, 244115 (2008).
15. Akkermans, R.L.C., and Briels, W.J. *J. Chem. Phys.* **113**, 6409 (2000); Akkermans, R.L.C., and Briels, W.J. *J. Chem. Phys.* **114**, 1020 (2001).
16. Padding, J.T., and Briels, W.J. *J. Chem. Phys.* **115**, 2846 (2001); Padding, J.T., and Briels, W.J. *J. Chem. Phys.* **118**, 10276 (2003).
17. Doi, M., and Edwards, S.F. *The Theory of Polymer Dynamics*. Oxford University Press, Oxford (1986).
18. Kröger, M., Ramírez, J., and Öttinger, H.C. *Polymer* **43**, 477 (2002).
19. Everaers, R., Sukumaran, S.K., Grest, G.S., Svaneborg, C., Sivasubramanian, A., and Kremer, K. *Science* **303**, 823 (2004).
20. Sukumaran, S.K., Grest, G.S., Kremer, K., and Everaers, R. *J. Polymer Science: Part B: Polymer Physics* **43**, 917 (2005).
21. Kremer, K., Sukumaran, S.K., Everaers, R., and Grest, G.S. *Comp. Phys. Commun.* **169**, 75 (2005).
22. Kröger, M. *Comp. Phys. Commun.* **168**, 209 (2005).
23. Foteinopoulou, K., Karayiannis, N.C., Mavrantzas, V.G., and Kröger, M. *Macromolecules* **39**, 4207 (2006).
24. Tzoumanekas, C., and Theodorou, D.N. *Macromolecules* **39**, 4592 (2006).
25. Doi, M., and Edwards, S.F. *J. Chem. Soc. Faraday Trans. II*, **74**, 1789 (1978); Doi, M., and Edwards, S.F. *J. Chem. Soc. Faraday Trans. II*, **74**, 1802 (1978); Doi, M., and Edwards, S.F. *J. Chem. Soc. Faraday Trans. II*, **74**, 1818 (1978).
26. Doi, M., and Takimoto, J.-I. *Phil. Trans. R. Soc. Lond. A* **361**, 641 (2003).
27. Masubuchi, Y., Takimoto, J.-I., Koyama, K., Ianniruberto, G., Marrucci, G., and Greco, F. *J. Chem. Phys.* **115**, 4387 (2001).
28. Masubuchi, Y., Ianniruberto, G., Greco, F., and Marrucci, G. *J. Chem. Phys.* **119**, 6925 (2003).
29. Masubuchi, Y., Ianniruberto, G., Greco, F., and Marrucci, G. *Modelling Simul. Mater. Sci. Eng.* **12**, S91 (2003).

30. Gao, C.Y., Nicholson, D.M., Keffer, D.J., and Edwards, B.J. *J. Non-Newt. Fluid Mech.* **152**, 140 (2008).
31. Mavrantzas, V.G., and Theodorou, D.N. *Macromolecules* **31**, 6310 (1998).
32. Mavrantzas, V.G., and Öttinger, H.C. *Macromolecules* **35**, 960 (2002).
33. Baig, C., and Mavrantzas, V.G. *Phys. Rev. Lett.* **99**, 257801 (2007).
34. Ionescu, T.C., Edwards, B.J., Keffer, D.J., and Mavrantzas, V.G. *J. Rheology* **52**, 105 (2008); Ionescu, T.C., Edwards, B.J., Keffer, D.J., and Mavrantzas, V.G. *J. Rheology* **52**, 567 (2008).
35. Louis, A.A., Bolhuis, P.G., Hansen, J.-P., and Meijer, E.J. *Phys. Rev. Lett.* **85**, 2522 (2000).
36. Bolhuis, P.G., Louis, A.A., Hansen, J.-P. and Meijer, E.J. *J. Chem. Phys.* **114**, 4296 (2001).
37. Bolhuis, P.G., Louis, A.A., and Hansen, J.-P. *Phys. Rev. E* **64**, 021801 (2001).
38. Louis, A.A. *J. Phys.: Condens. Matter* **14**, 9187 (2002).
39. Kindt, P., and Briels, W.J. *J. Chem. Phys.* **127**, 134901 (2007).
40. van den Noort, A., den Otter, W.K., and Briels, W.J. *Europhys. Lett.* **80**, 28003 (2007).
41. Zwanzig, R. *Nonequilibrium Statistical Mechanics* (3rd ed.). Oxford University Press, New York (2001).
42. Mori, H. *Progr. Theor. Phys.* **33**, 423 (1965).
43. Doruker, P., and Mattice, W.L. *Macromolecules* **30**, 5520 (1997).
44. Lyubartsev, A.P., and Laaksonen, A. *Phys. Rev. E* **52**, 3730 (1995).
45. Lyubartsev, A.P., and Laaksonen, *Lect. Notes Phys.* **640**, 219 (2004).
46. Shell, S. *J. Chem. Phys.* **129**, 144108 (2008).
47. Chaimovitch, A., and Shell, S. *J. Chem. Phys.* **134**, 094112 (2011).
48. Hahn, O., Delle Site, L., and Kremer, K. *Macromol. Theory Simul.* **10**, 288 (2001).
49. Walsh, T.R. *Molec. Phys.* **100**, 2867 (2002).
50. Faller, R., Schmitz, H., Biermann, O., and Müller-Plathe, F. *J. Comp. Chem.* **20**, 1009 (1999).
51. Meyer, H., Biermann, O., Faller, R., Rieth, D., and Müller-Plathe, F. *J. Chem. Phys.* **113**, 6264 (2000).
52. Fukunaga, H., Takimoto, J., and Doi, M. *J. Chem. Phys.* **116**, 8183 (2002).
53. McCoy, J.D., and Curro, J.G. *Macromolecules* **31**, 9362 (1998).
54. McCarty, J., Clark, A.J., Copperman, J., and Guenza, M.G. *J. Chem. Phys.* **140**, 204813 (2014).
55. Capone, B., Coluzza, I., and Hansen, J.-P. *J. Phys.: Condens. Matt.* **23**, 194102 (2011).
56. de Pablo, J.J., and Curtin, W.A. *MRS Bulletin* **32**, 905 (2007).
57. Öttinger, H.C. *MRS Bulletin* **32**, 936 (2007).
58. Harmandaris, V.A., and Kremer, K. *Macromolecules* **42**, 791 (2009).
59. Beris, A.N., and Edwards, B.J. *Thermodynamics of Flowing Systems with Internal Microstructure.* Oxford University Press: New York (1994).
60. Öttinger, H.C. *Beyond Equilibrium Thermodynamics*. Wiley-Interscience, Hoboken, NJ (2005).
61. Ilg, P., Mavrantzas, V.H., and Öttinger, H.C. *Modeling and Simulations of Irreversible Phenomena in Polymers*. Wiley-VCH, Weinheim, Germany (2008).
62. Guerrault, X., Rousseau, B., and Farago, J. *J. Chem. Phys.* **121**, 2538 (2004).
63. Santangelo, G., Di Matteo, A., Müller-Plathe, F., and Milano, G. *J. Phys. Chem. B* **111**, 2765 (2007).
64. Peroukidis, S.D., Tsalikis, D.G., Noro, M.G., Stott, I.P., and Mavrantzas, V.G. *J. Chem. Theory Comput.* **16**, 3363 (2020).
65. Vogiatzis, G.G., and Theodorou, D.N. *Macromolecules* **47**, 387 (2014).
66. Mathioudakis, I.G., Vogiatzis, G.G., Tzoumanekas, C., and Theodorou, D.N. *Soft Matter* **12**, 7585 (2016).
67. De Gennes, P.G. *J. Chem. Phys.* **55**, 572 (1971).
68. Fetters, L.J., Lohse, D.J., Richter, D., Witten, T.A., and Zirkel, A. *Macromolecules* **27**, 4639 (1994).
69. Fetters, L.J., Lohse, D.J., and Graessley, W.W. *J. Polym. Sci., Part B: Polym. Phys.* **37**, 1023 (1999).
70. Reichl, L.E. *A Modern Course in Statistical Physics* (4th rev. ed.). Wiley-VCH, Weinheim, Germany (2016).

71. Nakajima, S. *Progr. Theor. Phys.* **20**, 948 (1958).
72. Zwanzig, R. *J. Chem. Phys.* **33**, 1338 (1960).
73. Zwanzig, R. *Phys. Rev.* **124**, 983 (1961).
74. Kawasaki, K. *J. Phys. A. Math. Nucl. Gen.* **6**, 1289 (1973).
75. Hijón, C., Español, P., Vanden-Eijnden, E., and Delgado-Buscalioni, R. *Faraday Discuss.* **144**, 301 (2010).
76. te Vrugt, M., and Wittkowski, R. *Projection operators in statistical mechanics: a pedagogical approach.* arXiv:2001.01572v1 [cond-mat.stat-mech] 29 Dec 2019.
77. Brown, R. *The Miscellaneous Botanical Works of Robert Brown*, volume I. The Ray Society, London (1866).
78. Deutsch, J.M., and Oppenheim, I. *Faraday Discuss. Chem. Soc.* **83**, 1 (1987).
79. Langevin, P. *C. R. Acad. Sci. Paris* **146**, 530 (1908).
80. Bian, X., Kim, C., and Karniadakis, G.E. *Soft Matter* **12**, 6331 (2016).
81. Chandrasekhar, S. *Rev. Mod. Phys.* **21**, 383 (1949).
82. Oseen, C.W. *Arkiv för Matematik, astronomi och fysik* **6**, 1 (1910).
83. Allen, M.P., and Tildesley, D.J. *Computer Simulation of Liquids.* Clarendon Press, Oxford (1987).
84. Leimkuhler, B., and Matthews, C. *Molecular Dynamics with Deterministic and Stochastic Numerical Methods.* Springer International Publishing, Zurich (2015).
85. Mannella, R., and McClintock, P.V.E., *Fluctuation and Noise Letters* **11**, 1240010 (2012).
86. Itô, K. *Proc. Imp. Acad.* **20**, 519 (1944).
87. Stratonovich, R.L. *SIAM J. Control* **4**, 362 (1966).
88. van Kampen, N.G. *Stochastic Processes in Physics and Chemistry.* Elsevier, Amsterdam (2007).
89. Rouse, P.E. *J. Chem. Phys.* **21**, 1272 (1953).
90. Lin, Y.-H. *Polymer Viscoelasticity: Basics, Molecular Theories, and Experiments.* World Scientific Publishing, Singapore (2003).
91. Padding, J.T. *Theory of Polymer Dynamics.* Advanced Courses in Macroscopic Physical Chemistry, Han-sur-Lesse winter school 2005. University of Cambridge (2005).
92. Graessley, W.W. In Mark, J.E, Eisenberg, A., Graessley, W.W., Mandelkern, L., and Koenig, J.L. (eds.), *Physical Properties of Polymers.* American Chemical Society, Washington, DC (1984), pp. 97–153.
93. Tsolou, G., Stratikis, N., Baig, C., Stephanou, P.S., and Mavrantzas, V.G. *Macromolecules* **43**, 10692 (2010).
94. Skountzos, E.N, Tsalikis, D.G., Stephanou, P.S., and Mavrantzas, V.G. *Macromolecules* **54**, 4470 (2021).
95. McLeish, T.C.B. *Adv. Phys.* **51**, 1379 (2002).
96. Lodge, T.P. *Phys. Rev. Lett.* **83**, 3218 (1999).
97. Tao, H., Lodge, T.P.; and von Meerwall, E.D. *Macromolecules* **33**, 1747 (2000).
98. Harmandaris, V.A., Mavrantzas, V.G., and Theodorou, D.N. *Macromolecules* **31**, 7934 (1998).
99. Harmandaris, V.A., Mavrantzas, V.G., Theodorou, D.N., Kröger, M., Ramírez, J., Öttinger, H.C., and Vlassopoulos, D. *Macromolecules* **36**, 1376 (2003).
100. Tsolou, G., Mavrantzas, V.G., and Theodorou, D.N. *Macromolecules* **38**, 1478 (2005).
101. Stephanou, P.S., Baig, C., Tsolou, G., Mavrantzas, V.G., and Kröger, M. *J. Chem. Phys.* **132**, 124904 (2010).
102. Kremer, K. and Grest, G.S. *J. Chem. Phys.* **92**, 5057 (1990).
103. McLeish, T.C.B., and Larson, R.G. *J. Rheol.* **42**, 81 (1998).
104. Karayiannis, N.C., and Mavrantzas, V.G. *Macromolecules* **38**, 8583 (2005).

8

Entanglement Network-Based Simulations of Deformation and Flow

Perhaps the greatest challenge faced by molecular modelling of polymers is that their properties are governed by very long time scales, exceeding by far the range of times that can be addressed by atomistic molecular dynamics. In Chapter 7 we described several coarse-graining strategies, whereby the range of length and time scales addressed by molecular modelling and simulation can be augmented. Most of these strategies rely upon grouping multiatom moieties into effective interaction sites or beads. The resulting effective potential (potential of mean force with respect to the bead positions) is much smoother than the detailed atomistic potential energy hypersurface of the system, interactions between coarse-grained moieties being softer than interatomic interactions. This feature, along with the drastic reduction in the number of degrees of freedom, greatly accelerates calculations. There is, however, a downside to this strategy: a very soft effective nonbonded potential allows chains to pass through each other, and therefore does not respect the uncrossability of chains. Therefore, topological constraints or *entanglements* cannot be adequately represented at the coarse-grained level. While this is not a problem in predicting equilibrium structure and thermodynamic properties, it has very serious consequences for dynamics. There is every indication that rheological properties of melts and large-deformation properties of solid polymers of high molecular weight depend very strongly on entanglements. To predict these properties from coarse-grained models, we must somehow reintroduce entanglements by hand. In doing so, we can be helped by topological analyses of well-equilibrated atomistic long-chain configurations which provide a wealth of information about the statistics of entanglement networks, as discussed in Chapter 7. Having introduced entanglements in a coarse-grained model, we must also have a way to describe how the topology of the entanglement network evolves as a consequence of thermal motion and externally imposed deformation. To this end, it is advantageous to view this evolution as a succession of elementary processes, each of which is treated as an infrequent event. The techniques for analysis and simulation of infrequent events discussed in Chapter 6 give us, at least in principle, tools for estimating rate constants for these elementary processes from detailed atomistic information.

In this chapter we outline some coarse-grained techniques for simulating the rheological properties of polymer melts and the large-strain and ultimate mechanical properties of solid polymers based on the consideration of entanglement networks. We begin with *simple* slip-link and *dual* slip-link models which do not explicitly consider the 3D geometry of the entanglement network (Section 8.1). We then proceed with *explicit* 3D models of entangled chains using slip-links

Multiscale Modelling of Polymers. Doros N. Theodorou and Vlasis G. Mavrantzas, Oxford University Press.
 DOI: 10.1093/oso/9780199226191.003.0008

(Section 8.2). Subsequently, we discuss sophisticated models using slip-springs parametrized on the basis of atomistic simulations and confront their predictions for linear and nonlinear rheological properties against experiments (Section 8.3). We then turn to entanglement network simulations of semicrystalline polymers (Section 8.4). We present an early model for large deformation stress–strain behaviour and fracture of polyethylene as a function of molecular weight and outline a more recent model for adhesion at polyamide–polypropylene interfaces strengthened by grafted polyethylene chains.

8.1 Slip-link and dual slip-link model with virtual entanglements

According to the tube model of Doi–Edwards,[1] a chain in an entangled polymeric system above T_g is confined to move within a tube-like region due to topological constraints with surrounding chains. In the original formulation of the theory, the tube was regarded as fixed in space; its shape changes in time only because the two ends of the chain evacuate the old tube and create a new tube segment. The tube, however, can also change due to the motion of neighbouring chains, which causes constraint release (as entanglements may disappear). To account for such an effect, modifications are incorporated in the model accounting for mechanisms such as *double reptation*, *dynamic tube dilution* (especially for systems of star polymers), and *convective constraint release*.[2–12] These modifications greatly improve the predictive capability of the theory. However, they cannot be accommodated easily in the same simple analytical framework; thus, descriptions of the polymer through a *slip-link* or a *dual slip-link* model have been proposed,[13,14] which can account for all the above effects cumulatively in a consistent way (Figure 8.1). The slip-links can be thought of as small rings through which the chain can slide. At equilibrium, they are separated by a distance equal to the mesh size of the underlying entanglement network, or effective tube diameter a_{pp} (compare Section 7.6). The *primitive path* is then the line joining these slip-links, while the part of the line between two consecutive slip-links corresponds to an *entanglement strand.*

In the slip-link model, no other interactions between different chains are taken into account except for the topological ones captured through the slip-links; the polymer is thus free to move between slip-links. Since this would eventually cause the chains to shrink, in the first such slip-link model proposed by Doi–Edwards[13] two Maxwell demons were assumed to act at the two ends of the chain pulling them apart with a constant tensile force equal to $\sim\frac{3k_BT}{a_{pp}}$. This tensile force actually determines the equilibrium number of chain segments between slip-links. In the dual slip-link model (see Figure 8.1), the slip-link confines a pair of chains (and not a single chain); thus, a slip-link is destroyed if any of the chains participating in it slides past the slip-link. From a dynamics point of view, the slip-link model is based on the physical picture that the drag force exerted on a reference chain by the surrounding ones in an entangled polymer is highly anisotropic, characterized also by a separation of length scales: on a small length scale, entanglements are not important, since a small movement of the chain (even in such a highly dense system as a melt) can easily be accommodated through a local rearrangement of the rest of the chains; the material would thus respond

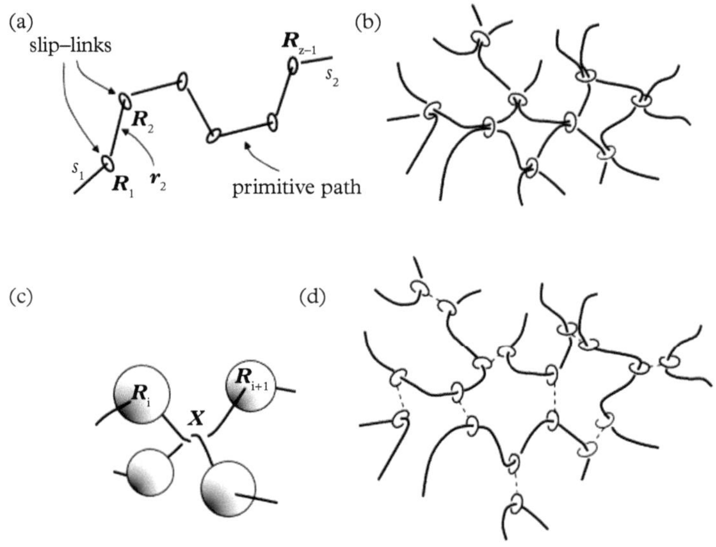

Figure 8.1 (a) Chain primitive paths and slip-links. (b) The dual slip-link model. (c) Picture invoked in TWENTANGLEMENT simulation method of Padding and Briels. (d) The slip-spring model.

as if it were a normal viscous liquid. On larger length scales, on the other hand, entanglements are very important, since they prevent the lateral motion of the chain (the motion perpendicular to its primitive path) due to the topological constraints with the rest of the chains and allow only motion along its axis (i.e. its primitive path), for which the resistance felt by the chain is significantly lower. Based on such a picture, the characteristic length a_{pp} should also be perceived as that length, above which the motion of the reference chain is felt by the rest of the chains. The number of slip-links is obviously a statistical quantity whose average at equilibrium should equal the number $Z-1$ of entanglements per chain in the real polymer. The equilibrium contour length of the primitive path will then be $L_{eq} = Z\,a_{pp}$.

The above description makes it possible to develop an algorithm that can successfully simulate the viscoelastic properties of an entangled polymer melt of linear chains starting with an initial configuration of a large number of chains bearing slip-links along their contours. Each slip-link on a given chain pairs it with a partner chain; it is identified with a slip-link selected randomly from the population of slip-links on the other chains in the system. These pairings represent binary entanglements between polymer chains. Each chain then moves in its own 3D space subjected to the topological constraints represented by the paired slip-links along its contour. At each time step of this algorithm, the following operations take place:[14]

(a) *Affine deformation due to flow.* Each slip-link position $\boldsymbol{R}_i$ is displaced affinely according to the macroscopic flow applied on the fluid.
(b) *Contour length fluctuations.* Let the lengths of the primitive path segments at the two chain ends be s_1 and s_2, respectively (see Figure 8.1a). Then the length L of each primitive path

is updated to $L+\Delta L$ by changing the lengths of the tails s_1 and s_2 by the same amount, $\Delta L/2$. ΔL is determined from the Langevin equation

$$\frac{\mathrm{d}L}{\mathrm{d}t}=\left(\frac{\mathrm{d}L}{\mathrm{d}t}\right)_{\text{affine}}-\frac{1}{\tau_{\mathrm{R}}}\left(L(t)-L_{\mathrm{eq}}\right)+g(t), \tag{8.1.1}$$

where $\left(\frac{\mathrm{d}L}{\mathrm{d}t}\right)_{\text{affine}}$ is the change due to the affine deformation, $g(t)$ a random variable accounting for contour length fluctuations, and τ_{R} the Rouse relaxation time.

(c) *Reptation.* Each primitive path is randomly displaced along itself corresponding to a diffusive motion with diffusion coefficient along the contour D_{c}. By this operation, either s_1 decreases and s_2 increases, or vice versa.

(d) *Constraint renewal.* If s_1 or s_2 becomes negative by operations (b) and (c), the last slip-link of the chain and the corresponding slip-link of its partner are removed. If, on the other hand, s_1 or s_2 becomes longer than a_{pp}, then a new slip-link is created at the end of the chain and on its chain partner, which is selected randomly.

In such a model, instantaneous stresses can be calculated according to the virial theorem, assuming that strands between entanglements behave as Gaussian *entropy springs*, through the equation (compare Eq. (7.8.26))

$$\sigma_{\alpha\beta}=\frac{1}{V}\frac{3k_{\mathrm{B}}T}{a_{\mathrm{pp}}^2}\sum_i\langle r_{i,\alpha}r_{i,\beta}\rangle, \tag{8.1.2}$$

where V is the system volume, $\boldsymbol{r}_i=\boldsymbol{R}_i-\boldsymbol{R}_{i-1}$ is the end-to-end vector of the ith interentanglement strand, $r_{i,\alpha}$ is the α-component of $\boldsymbol{r}_i$, the summation is taken over all strands in a configuration, and the angular brackets denote an ensemble average over all configurations considered in the simulation.

8.2 Dual slip-link model with entanglements

In the Doi and Takimoto formulation[14] of the dual slip-link model, the slip-links are virtual, since they do not represent entanglement junctions in real space; they rather represent effective constraints, whose statistical character is determined by other polymers. Masubuchi and collaborators,[15,17–19] on the other hand, have regarded a slip-link as an actual link in real space. In their formulation, each polymer chain is represented by a sequence of entanglement strands considered as segments (phantom entropic springs) joining consecutive entanglement points (the beads) along the chain. These Rouse-like chains are all interconnected by slip-links at the entanglement points to form a 3D primitive chain network. The system is described by the following set of dynamical state variables: the number Z of segments (terminal and interentanglement strands) in each chain, the number n of monomers in each chain segment, and the position vectors $\boldsymbol{R}$ of the slip-links or entanglement points. These state variables are postulated to obey the following set of Langevin-type governing equations.

8.2.1 Nodal points

The position vector of the nodal points obeys a 3D position Langevin equation of the form

$$\zeta \frac{\mathrm{d}\boldsymbol{R}}{\mathrm{d}t} = \boldsymbol{F}^{\text{flow}} + \boldsymbol{F}^{\text{elastic}} + \boldsymbol{F}^{\text{osmotic}} + \boldsymbol{F}^{\text{random}}. \tag{8.2.1}$$

In Eq. (8.2.1), the first term on the right-hand side represents the force due to the velocity gradient in the continuum

$$\boldsymbol{F}^{\text{flow}} = \zeta(\nabla \boldsymbol{\nu})^{\mathrm{T}} \cdot \boldsymbol{R}, \tag{8.2.2}$$

where ζ is the friction coefficient per slip-link. The second term is the sum of the elastic forces of the chain segments pulling on the node. If $\boldsymbol{r}_j$ denotes the end-to-end vector of each such segment, then

$$\boldsymbol{F}^{\text{elastic}} = \frac{3k_{\mathrm{B}}T}{b_{\mathrm{K}}^2} \sum_{j=1}^{4} \frac{\boldsymbol{r}_j}{n_j}, \tag{8.2.3}$$

where we have assumed that each such segment contains n_j Kuhn segments of average squared length b_{K}^2 and obeys Gaussian statistics. For the internal nodes, the sum contains four such terms, because entanglements (or slip-links) are binary contacts between chains. For chain ends, the sum should be replaced by a single term, and the friction coefficient ζ in Eq. (8.2.1) should be replaced by $\zeta/4$. The third term represents the *osmotic* force arising from fluctuations in the number density ρ of network nodes. In the work of Masubuchi et al.,[17–19] it is set equal to

$$\boldsymbol{F}^{\text{osmotic}} = -k_{\mathrm{B}}T\boldsymbol{\nabla} \ln \rho \tag{8.2.4}$$

or, equivalently, $\boldsymbol{F}^{\text{osmotic}} = -\boldsymbol{\nabla}\mu$, where $\mu\,(=\mu_{\text{ref}} + k_B T \ln(\rho/\rho_{\text{ref}}))$ is the node chemical potential. Finally, the fourth term is the usual Gaussian random force obeying the fluctuation–dissipation theorem (compare Eq. (7.7.42)):

$$\left\langle \boldsymbol{F}^{\text{random}} \right\rangle = \boldsymbol{0} \quad \text{and} \quad \left\langle \boldsymbol{F}^{\text{random}}(t) \cdot \boldsymbol{F}^{\text{random}}(t') \right\rangle = 6k_{\mathrm{B}}T\zeta\,\delta\,(t-t'). \tag{8.2.5}$$

Putting all force terms together in Eq. (8.2.1), the equation for node motion takes the form

$$\zeta \frac{\mathrm{d}\boldsymbol{R}}{\mathrm{d}t} = \zeta(\nabla \boldsymbol{\nu})^{\mathrm{T}} \cdot \boldsymbol{R} + \frac{3k_{\mathrm{B}}T}{b_{\mathrm{K}}^2} \sum_{i}^{4} \frac{\boldsymbol{r}_i}{n_i} - k_{\mathrm{B}}T\boldsymbol{\nabla} \ln \rho + \boldsymbol{F}^{\text{random}}. \tag{8.2.6}$$

8.2.2 Number of monomers

The rate of change of the number of monomers n in an entanglement strand m obeys a Langevin type of equation in 1D (along the curvilinear coordinate of the primitive path) reflecting monomer sliding through slip-links:

$$\zeta_m \nu_{\mathrm{sl}} = f_m + f_m^{\mathrm{osmotic}} + f_m^{\mathrm{random}}. \tag{8.2.7}$$

Here, ζ_m is a (constant) friction coefficient accounting for all monomers in the (average) chain segment between consecutive entanglements, ν_{sl} the sliding velocity along the chain, f_m the magnitude of the elastic force arising from the local tension difference of the two chain segments in and out of the slip-link, f_m^{osmotic} the force along the chain due to osmotic pressure, and f_m^{random} a 1D Gaussian random force again along the chain contour. These are given by the relationships $\zeta_m = \frac{1}{2}\zeta$, $f_m = \frac{3k_BT}{b_{\mathrm{K}}^2}\left(\frac{r_{\mathrm{out}}}{n_{\mathrm{out}}} - \frac{r_{\mathrm{in}}}{n_{\mathrm{in}}}\right)$, and $f_m^{\mathrm{osmotic}} = -k_{\mathrm{B}}T\left(\boldsymbol{u}\cdot\boldsymbol{\nabla}\ln\rho\right)$, where $\boldsymbol{u}$ is the unit vector tangent to the chain contour at the considered point, while the random force satisfies $\left\langle \boldsymbol{f}_m^{\mathrm{random}}\right\rangle = 0$ and $\left\langle \boldsymbol{f}_m^{\mathrm{random}}(t)\cdot\boldsymbol{f}_m^{\mathrm{random}}(t')\right\rangle = 2k_{\mathrm{B}}T\zeta_m\,\delta\left(t-t'\right)$. The slip velocity is connected to the corresponding rate of change of the number of monomers in each of the two chain segments in and out of the slip-link through

$$\dot{n} = d\,\nu_{\mathrm{sl}}, \tag{8.2.8}$$

where d describes the linear density of monomers in the tube. Eliminating the slip velocity in favour of the rate of change of the number of monomers between Eqs. (8.2.7) and (8.2.8) provides the following expression for the monomer motion,

$$\frac{1}{2}\zeta\frac{\dot{n}}{d} = \frac{3k_BT}{b_{\mathrm{K}}^2}\left(\frac{r_i}{n_i} - \frac{r_{i-1}}{n_{i-1}}\right) - k_{\mathrm{B}}T\left(\boldsymbol{u}\cdot\boldsymbol{\nabla}\ln\rho\right) + f_m^{\mathrm{random}}, \tag{8.2.9}$$

where $\dot{n}$ is the rate of change of the number of monomers in segment i due to monomer transport from segment $i-1$ and the local monomer density along the primitive path is estimated as $d = \frac{1}{2}\left(\frac{n_i}{r_i} + \frac{n_{i-1}}{r_{i-1}}\right)$. To proceed, one should also specify an expression for the segment chemical potential $\mu\left(= \mu_{\mathrm{ref}} + k_BT\ln\left(\rho/\rho_{\mathrm{ref}}\right)\right)$. This is assumed to be of the form

$$\mu = \begin{cases} \varepsilon\left(\dfrac{\varphi}{\langle\varphi\rangle} - 1\right)^2, & \varphi > \langle\varphi\rangle \\ 0, & \varphi \le \langle\varphi\rangle \end{cases}, \tag{8.2.10}$$

where ε is a numerical constant (chosen equal to 0.5 to avoid unphysical clustering of the nodes of the primitive path); φ denotes the local segment density and $\langle\varphi\rangle$ its average over the whole sample. In practice, μ from expression (8.2.10) is used in lieu of $k_{\mathrm{B}}T\ln\rho$ in Eqs. (8.2.4), (8.2.6), and (8.2.9). Expression (8.2.10) is phenomenological and is meant to enforce incompressibility, since chains in the slip-link model have no excluded volume. The value of the average segment density is fixed to be close to ten segments in a volume of about a_{pp}^3, a value typical of most polymer melts.

The single relevant parameter of the primitive chain network is $\langle Z \rangle$, the average value of strands per chain. In each individual chain, however, the number of entanglements is allowed to fluctuate because of chain rearrangement initiated by chain ends. Masubuchi et al.[19] have adopted the following criterion for chain rearrangement: if the number of monomers in the chain end segment becomes less than a certain minimum value, the neighbouring slip-link vanishes, implying that the value of Z for that chain (and of the companion chain in the slip-link) decreases by 1. On the other hand, if the number of monomers in the end segment exceeds a maximum value, an entanglement is created along that end segment by hooking a randomly chosen nearby segment; consequently, the value of Z for that chain and for the chain containing the hooked segment increases by 1. Due to the Gaussian assumption for chain segments (linear springs), the instantaneous stress tensor in the course of the simulation will be given by Eq. (8.1.2). To obtain rheological results from the dual slip-link model for PS melts at 453 K and compare against experimental data,[20] the number of monomers in a strand, normalized by its mean value at equilibrium n_0, was specified (arbitrarily) to be between 0.5 and 1.5, i.e. $0.5 < n/n_0 < 1.5$, in order to conserve the average value of Z initially assumed for the system. Since this window is somewhat arbitrarily defined, however, the average value of Z at the end of the system equilibration under static conditions can depart slightly from the value set at the beginning of the simulation (some integer number). Thus, noninteger values of $\langle Z \rangle$ are usually encountered with such a method.

8.2.3 The continuously unconstrained Brownian slip-link model

Schieber and collaborators[21,22] have also proposed a full-chain slip-link model with a mean-field implementation of constraint release and constant chain friction (as opposed to constant entanglement friction). The dynamical variables of such a model are the following:

- the instantaneous number $Z(t) - 1$ of entanglements (slip-links) in the chain (as above, there are $Z(t)$ strands in the chain and hence there exist $Z(t) - 1$ entanglements and $Z(t) - 2$ entangled strands; compare Figure 8.1);
- the instantaneous number $n_i(t), i = 1, 2, ..., Z(t)$ of Kuhn steps in a strand; and
- the set $\boldsymbol{R}_i(t), i = 1, 2, ..., Z(t) - 1$ of the position vectors of entanglement points.

The total number of strands $Z(t)$ fluctuates because of the creation and destruction of entanglements, and the average number of strands at equilibrium is calculated as $\langle Z \rangle_{\text{eq}} = \dfrac{N}{n_{\text{e}}}$, where N is the chain length and n_{e} is the mean length of an interentanglement strand, both measured in Kuhn segments. The length of an entangled strand is given by the magnitude of the vector $\boldsymbol{r}_i(t) = \boldsymbol{R}_i(t) - \boldsymbol{R}_{i-1}(t), i = 2, 3, ..., Z(t)$ connecting two consecutive entanglements. For the free energy A of such a model, the following expression is assumed:

$$A(T, \boldsymbol{\Omega}) = \sum_{i=2}^{Z-1} A_{\text{S}}(T, \boldsymbol{r}_i, n_i) + A_{\text{E}}(T, n_1) + A_{\text{E}}(T, n_Z), \tag{8.2.11}$$

where A_S is the contribution due to the interior entangled strands, A_E the contribution due to the end strands, and $\mathbf{\Omega}$ a shorthand for the chain conformation, $\mathbf{\Omega} = \left\{ \{\boldsymbol{r}\}_{i=2}^{Z-1}, \{n\}_{i=1}^{Z}, Z \right\}$. The mathematical expressions for the two free energies are[22]

$$\frac{A_S\left(T, \boldsymbol{r}_i, n_i\right)}{k_B T} = \frac{3\boldsymbol{r}_i^2}{2n_i b_K^2} + \frac{3}{2}\ln\left(\frac{2\pi n_i b_K^2}{3}\right) \tag{8.2.12}$$

for the entangled strands, and

$$\frac{A_E\left(T, n_i\right)}{k_B T} = -\frac{1}{2}\ln\left(n_i\right) \tag{8.2.13}$$

for the dangling ends. In Eq. (8.2.12) the reader will recognize $A_S\left(T, \boldsymbol{r}_i, n_i\right) = -k_B T \ln \Phi\left(\boldsymbol{r}_i, n_i\right)$ with $\Phi\left(\boldsymbol{r}_i, n_i\right)$ being the probability density of the end-to-end vector of a Gaussian chain of n_i steps and mean squared step size (Kuhn length) b_K^2 (see Eq. (1.1.25) and Section (5.3.6) or Problem 5.5). Equation (8.2.13) is chosen by Schieber et al. for numerical convenience and not for physical reasons.[21,22] The number of strands Z on a chain is allowed to fluctuate, subject to a constant chemical potential, while the strand lengths n_i are required to satisfy the constraint $\sum_{i=1}^{Z} n_i = N$. Based on Eqs. (8.2.11)–(8.2.13) for A, assuming $n_e >> 1$, and setting $\beta = \exp\left(-\frac{\mu}{k_B T}\right)$ with $\mu = k_B T \ln\left(1/n_e - 1/N\right)$, the chemical potential conjugate to Z, the equilibrium distributions for Z, n_i, and $\boldsymbol{r}_i$ come out to be[21,22]

$$P_{eq}\left(Z\right) = \frac{1}{\mathcal{J}\beta^Z}\frac{N^{Z-1}}{\left(Z-1\right)!} \tag{8.2.14}$$

$$p_{eq}\left(n_i\right) = \frac{1}{n_e}\exp\left(-\frac{n_i}{n_e}\right) + \exp\left(-\frac{N}{n_e}\right)\delta\left(N - n_i\right) \tag{8.2.15}$$

$$p_{eq}\left(\boldsymbol{r}_i\right) = \frac{3}{2\pi r_i n_e b_K^2}\exp\left(-\sqrt{\frac{6\boldsymbol{r}_i^2}{n_e b_K^2}}\right), \tag{8.2.16}$$

where $\mathcal{J} = \frac{1}{\beta}\exp\left(\frac{N}{\beta}\right)$. Clearly, the number of strands Z, a discrete variable, follows a Poisson distribution, while the strand length n_i, treated as a continuous variable taking values from the domain $[0, N]$, is exponentially distributed.

Inspired by the above theoretical arguments, Nair and Schieber[22] presented a consistently unconstrained Brownian slip-link (CUBS) model with constant chain frictions accounting for segment connectivity, contour length fluctuations, chain stretching, and constrained release in a mean field way by including fluctuations both as a binary interaction between chains and as a multichain interaction. Unlike previous approaches where constraint release was included by assuming the relaxation modulus to be the product of two moduli found by independent processes (reptation and contour length fluctuations on one hand, and constraint release on the other), here it is included on the level of chain dynamics. The evolution equation for the chain conformation probability density at equilibrium in the new model has three contributions:

$$\frac{\partial p}{\partial t} = \dot{p}_{\text{steps}} + \dot{p}_{\text{CR}} + \dot{p}_{\text{ends}}. \tag{8.2.17}$$

The first represents shuffling of Kuhn segments across entanglements, the second constraint represents release, and the third accounts for the creation–destruction dynamics at the dangling ends. They are explained in detail in Sections 8.2.3.1–8.2.3.3.

8.2.3.1 *Shuffling of Kuhn segments*

This arises because of imbalance in chemical potential and Brownian forces driving Kuhn steps across the entanglements. Such a process is modelled by Nair–Schieber[22] as a Fokker–Planck equation:

$$\mathrm{d}n_i = -\sum_j \hat{A}_{ij} \left(\frac{\mu_j}{k_B T} \right) \mathrm{d}t + \sqrt{2} \sum_j \hat{B}_{ij} \mathrm{d}W_j. \tag{8.2.18}$$

For constant chain friction (although, strictly speaking, the chain end effects considered here cause the chain friction not to be exactly constant), the matrix $\hat{\boldsymbol{A}}$ is defined as

$$\hat{A}_{ij} = \begin{cases} -\dfrac{1}{\tau_{i-1}} & i = j+1 \\ \dfrac{1}{\tau_i} + \dfrac{1}{\tau_{i-1}} & i = j \\ -\dfrac{1}{\tau_i} & i = j-1 \\ 0 & \text{otherwise} \end{cases}, \quad i, j = 1, 2, \ldots, Z-1, \tag{8.2.19}$$

where $\tau_i = \dfrac{\tau_K}{2n_e}(n_i + n_{i+1})$ and τ_K is a time scale associated with motion of a Kuhn step across an entanglement. Also, $\hat{\boldsymbol{B}}$ is a matrix defined in terms of the time constants above as

$$\hat{B}_{ij} = \frac{1}{\sqrt{\tau_j}} (\delta_{ij} - \delta_{i-1,j}) \tag{8.2.20}$$

(δ_{ij} being the Kronecker delta). W_j is the stochastic noise term on entanglement j (a Wiener process; ΔW_j over a time interval Δt is a Gaussian random variable with zero mean and variance Δt). From Eq. (8.2.18), the evolution equation for the number of Kuhn steps in an entanglement strand is written as

$$\begin{aligned} \mathrm{d}n_i = {} & \frac{1}{k_B T \tau_i} (\mu_{i+1} - \mu_i)\, \mathrm{d}t - \frac{1}{k_B T \tau_{i-1}} (\mu_i - \mu_{i-1})\, \mathrm{d}t \\ & + \sqrt{\frac{2}{\tau_i}} \mathrm{d}W_i - \sqrt{\frac{2}{\tau_{i-1}}} \mathrm{d}W_{i-1}, \qquad i = 1, 2, \ldots, Z-1 \end{aligned} \tag{8.2.21}$$

with μ_i denoting the segmental chemical potential of the ith strand, $\mu_i = \left(\frac{\partial A}{\partial n_i}\right)_{T,\{\boldsymbol{r}_i\},\{n_{j\neq i}\}}$; for a free energy given by Eqs. (8.2.11)–(8.2.13),

$$\mu_i = \mu_S + \mu_E, \tag{8.2.22}$$

where

$$\begin{aligned}\frac{\mu_S(n_i, \boldsymbol{r}_i)}{k_B T} &= -\frac{3r_i^2}{2n_i^2 b_K^2} + \frac{3}{2n_i}\\ \frac{\mu_E(n_i)}{k_B T} &= -\frac{1}{2n_i}.\end{aligned} \tag{8.2.23}$$

8.2.3.2 Constraint release

Following Graessley[23] and Rubinstein–Colby,[24] constraint release was modelled by Nair–Schieber[22] as a diffusive process of entanglements through the following stochastic differential equation for the position of entanglements,

$$\mathrm{d}\boldsymbol{R}_i = -(\nabla\boldsymbol{\nu})^{\mathrm{T}} \cdot \boldsymbol{R}_i\,\mathrm{d}t - \frac{n_e b_K^2}{12 k_B T \tau_i^{\mathrm{CR}}}\left(\frac{\partial A}{\partial \boldsymbol{R}_i}\right)_{T,\{n_i\},\boldsymbol{R}_{j\neq i}}\mathrm{d}t + \sqrt{\frac{n_e b_K^2}{6\tau_i^{\mathrm{CR}}}}\,\mathrm{d}\boldsymbol{W}_i \tag{8.2.24}$$

with τ_i^{CR} being the constraint release time scale. This can be chosen from the distribution of entanglement life times during the simulation by fitting the fraction of surviving entanglements $\psi(t) = 1 - [1 - f(t)]^{n_c - 1}$ at the current time t with a function of the form $\psi(t) = \sum_i P_i \left(\frac{t}{\tau_i^{\mathrm{CR}}} + 1\right)\exp\left(-\frac{t}{\tau_i^{\mathrm{CR}}}\right)$, which for every mode (described by a single time constant) has the correct slope. In the above expressions, $f(t)$ is the fraction of entanglements that have not been abandoned by the test chain from time $t = 0$ to the current time t, n_c (≈ 5.4) is the number of chains making up an entanglement according to the Cornet criterion,[25,26] and P_i are weight factors corresponding to the constraint release time constants τ_i^{CR}. The set of values $\{P_i, \tau_i^{\mathrm{CR}}\}$ obtained from the fitting is that used in the computations, implying that every entanglement at the time of creation is assigned a constraint release time τ_i^{CR} with probability P_i. In the simulations, such a methodology requires an iterative procedure until computed and fitted $f(t)$ or $\psi(t)$ curves coincide (this typically requires two iterations).

8.2.3.3 Creation and destruction of entanglements

Entanglements are created and destroyed at the ends of the chain. The corresponding creation and destruction rates should obey the criterion of detailed balance. The transition probability for creation and destruction processes is related to the transition probability per unit time that a chain with a configuration $\boldsymbol{\Omega}$ will jump to a new configuration $\boldsymbol{\Omega}'$. Based on this, the probability for the destruction of an entanglement at a dangling end in a time step Δt was proposed by Nair–Schieber[22] to be given by the expression

$$w_{\mathrm{d}} = 1 - \exp\left(\frac{1}{\tau_{\mathrm{e}}}\sqrt{3\frac{n_{\mathrm{e}}}{n_{\mathrm{E}}}\Delta t}\right), \tag{8.2.25}$$

whereas that for the creation of an entanglement is given by

$$w_{\mathrm{c}} = 1 - \exp\left(-\frac{1}{\tau_{\mathrm{e}}}\sqrt{3\frac{n_{\mathrm{e}}-2}{n_{\mathrm{e}}n_{\mathrm{E}}}\Delta t}\right), \tag{8.2.26}$$

where n_{E} is the number of Kuhn steps in the dangling end and $\tau_{\mathrm{e}} = n_{\mathrm{e}}^2\tau_{\mathrm{K}}$.

8.2.4 The stress tensor

Based on thermodynamics arguments and considering that dangling ends do not contribute to the stress, the polymer contribution to the stress tensor is given by

$$\sigma = -\frac{N_{\mathrm{c}}}{V}\int\int\sum_{Z=3}^{\infty}\sum_{j=2}^{Z-1}\boldsymbol{R}_j\left(\frac{\partial A}{\partial \boldsymbol{R}_j}\right)_{T,\{n_i\},\{\boldsymbol{R}_{j\neq i}\}} p\left(\{\boldsymbol{R}_i, n_i\}, Z; t\right)\, \mathrm{d}\boldsymbol{R}_j \mathrm{d}n_j \tag{8.2.27}$$

with N_{c}/V being the total number of chains per unit volume.

Nair and Schieber[22] have analysed the predictive power of such a model by comparing computed complex modulus G^* and Cole–Cole plots with experimental data[27] for many polymers, such as PS, PB, and PIB. The predictions were very close to the experimental measurements at low frequencies but deviated for the glassy mode at low frequencies, which can be attributed to the inability of the model to resolve time scales smaller than the characteristic time associated with strand dynamics (i.e. relaxations more rapid than the Kuhn step shuffling). The fact that the model is strictly valid for monodisperse melts (whereas the experimental samples have some polydispersity) could be another reason for the deviations at low frequencies.

8.2.5 The Twentanglement approach

A Langevin dynamics approach for the simulation of polymer melts that incorporates tracking of topological constraints akin to slip-springs has been proposed by Padding and Briels.[16] They invoke a coarse-grained representation, wherein λ consecutive monomers are lumped into a single blob or bead, represented by its centre of mass. The value of λ should be small enough, such that the size of the blob does not exceed the entanglement tube diameter in the reptation picture. At the same time, λ should be chosen so as to allow treating the degrees of freedom projected out of the description as a Brownian bath satisfying a Markovian approximation (see Section 7.7). Clearly, large values of λ are desirable, for they make effective nonbonded interactions softer and allow the use of long time steps in integrating the equations of motion. In applying their method to polyethylene (PE) melts, Padding and Briels chose a value of $\lambda = 20$ methylene units. At this level of coarse-graining, nonbonded interactions between blobs are very soft, allowing chains to pass through each other. The uncrossability constraint must be added *by hand*. To introduce it, Padding and Briels modelled the strands between blobs as elastic bands that cannot cross each other. As soon as contact is made between two such strands (see Figure 8.1c, point $\boldsymbol{X}$), an *entanglement* is created, tending to restore the strands to a rectilinear shape. This is accomplished by defining the attractive *entropic spring* potential associated with each strand between two bonded blobs i and $i+1$ to be a function of the total path length $|\boldsymbol{R}_i - \boldsymbol{X}| + |\boldsymbol{X} - \boldsymbol{R}_{i+1}|$ of the strand, going through any intermediate entanglements. It is assumed that there is an equilibrium of forces at each entanglement point at all times.[16] Padding and Briels derived their effective potentials, frictions, and random forces directly from detailed molecular dynamics (MD) simulations. Applying their Langevin dynamics to C_{80}-C_{1000} PE melts they performed a thorough study of changes in dynamics with increasing chain length. The relaxation of Rouse modes was found to depart from a purely exponential behaviour beyond C_{120} and rather follow a stretched exponential, the stretching exponent for the first mode being 0.8. A slowing down of the dynamics (increase of the friction coefficient) at a length scale smaller than that corresponding to N_e was detected and analysed. Their simulation estimates of plateau modulus, chain self-diffusivity, and zero-shear viscosity were in good agreement with experiment. The coarse-grained simulation method of Padding and Briels, implemented in the TWENTANGLEMENT code,[16] was a very insightful development. Its only disadvantage is the computational overhead associated with tracking entanglement points between strands at all times, as a result of which very long trajectories are very computationally demanding and $G(t)$ for times beyond 10 ns becomes quite noisy.

The reader is referred to the recent reviews of simulations of entangled polymers by Masubuchi[28] and Schieber[29] for more information on slip-link models.

8.3 Slip-spring-based mesoscopic simulations of entangled polymer melt dynamics

In recent years, slip-spring models have been more popular than slip-link models in coarse-grained descriptions of the dynamics of entangled polymers. In a slip-spring model (Figure 8.1d), the topological constraint is viewed as a pair of slip-links *connected by a spring*. The primitive paths of two chains involved in an entanglement do not come together at the same point but are allowed some slack by the slip-spring, whose stiffness should be commensurate with that of intramolecular strands. The use of slip-springs seems to have been stimulated by a successful model for the nonlinear elasticity of entangled polymer networks by Rubinstein and Panyukov.[30]

Müller and collaborators[31] introduced a slip-spring model for a long-chain melt which represents entanglements by local, pairwise, translationally invariant interactions which do not alter any equilibrium properties. They invoked a bead-and-spring coarse-grained representation of chains, with successive beads along a chain connected via harmonic potentials. Slip-springs were envisioned as connecting beads on different chains. They were represented by a finitely extensible nonlinear elastic (FENE) spring potential which diverges at a finite range. The number of slip-springs was allowed to fluctuate, subject to a prescribed chemical potential, as in the work of Schieber.[21,22] In other words, the simulation was canonical with respect to the polymer chains, but grand canonical with respect to the slip-springs. A very nice idea in this work[31] was the introduction of a *compensating potential* which rigorously ensures that slip-springs have no influence on the equilibrium properties of the system. The compensating potential is a sum of contributions from all pairs of beads that can be connected by slip-springs within a given configuration. The simulation involved a hybrid dissipative particle dynamics (DPD)/Monte Carlo (MC) scheme. The DPD method is presented in Chapter 9. At each time step of this scheme, one chooses with fixed probabilities between a move that updates segment positions by a short DPD simulation and one of several MC moves that move, create, or destroy slip-springs. The attempt probabilities and acceptance criteria for the latter Monte Carlo moves are designed so as to preserve microscopic reversibility and conform to the probability density of the grand canonical ensemble that governs slip-springs. Predictions for chain dimensions, mean square displacements of segments, and chain centres of mass were in agreement with theoretical expectations. Furthermore, the model was used in the framework of reverse nonequilibrium molecular dynamics (RNEMD) to simulate steady shear flow. The viscosity was found to exhibit shear thinning behaviour with a shear thinning exponent of 0.64, and a modest decrease in the number of slip-springs with increasing shear rate was observed.[31]

Slip-spring models along similar lines to those of Ref. 31 have been presented by Uneyama and Masubuchi[32] and Langeloth et al.[33]

More recently, Vogiatzis, Sgouros et al.[34–38] have developed a bottom-up Brownian dynamics/kinetic Monte Carlo (BD/kMC) approach employing slip-springs with an eye towards quantitative prediction of melt rheological properties. This approach is implemented in the code EMSIPON (*e*ngine for *m*esoscopic *si*mulations of *po*lymer *n*etworks) whose basic features are the following:

(a) At the mesoscopic level, chains are represented as sequences of beads connected by (nonlinear) springs (see Figure 8.2a). Each bead stands for a number of Kuhn segments, chosen following the considerations described in Section 8.2.4 in connection with the TWENTANGLEMENT approach.[16] Entanglements are represented by (nonlinear) slip-springs connecting beads on different chains, as in the work of Müller and collaborators[31] (see Figure 8.2b). Slip-spring ends move by successive hops between adjacent beads along the chain to which they belong (see Figure 8.2c). These hops are viewed as infrequent events, characterized by rate constants determined as discussed later and tracked via kMC simulation. On the other hand, each bead executes Brownian motion subject to all forces exerted on it. Thus, the approach, as originally conceived, is a BD/kMC simulation. The method is usually implemented in an ensemble that is canonical with respect to the polymer chains and grand canonical with respect to the slip-springs, the density of slip-springs in the system being regulated by a constant slip-spring chemical potential. Destruction of a slip-spring occurs when one of its ends chooses to hop off a chain end (see Figure 8.2d). On the other hand, new slip-springs are generated at chain ends, which may couple with segments of other chains according to a rate constant that depends on the slip-spring chemical potential (see Figure 8.2e).

(b) The dynamics is derived from a mesoscopic Helmholtz energy function that combines particle-based contributions from bonded interactions (effective bond stretching and bond-angle bending potentials) and entanglements (slip-spring and compensating potentials) with field-based contributions from excluded volume and cohesive interactions, calculated from the local density using an equation of state (see Figure 8.2f), as outlined later in this section. Thus, the EMSIPON approach, as originally conceived, is a hybrid particle-field simulation approach that bears many similarities with Milano and Kawakatsu's hybrid particle-field molecular dynamics implemented, for example, in the OCCAM code.[39,40]

(c) The approach is *bottom up*. Parameters for the mesoscopic description of thermodynamics and dynamics are deduced from the conformational, volumetric, and dynamical properties extracted from atomistic MD simulations of short-chain melts of the chemical constitution of interest. Adjustable parameters are thus kept to a minimum.

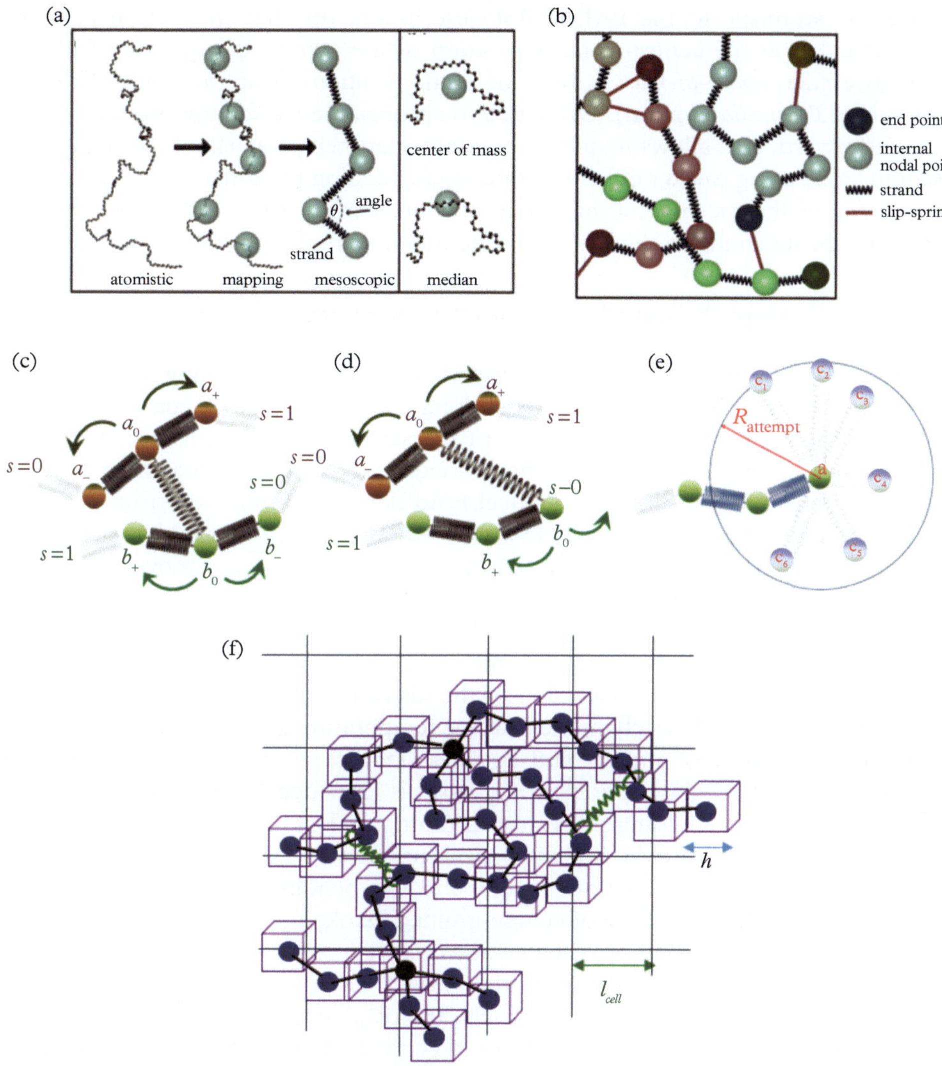

Figure 8.2 Coarse-grained representation of structure and dynamics invoked in the EMSIPON method. (a) Coarse-graining of a chain. In the implementation for PE, each bead represents a chain segment of fifty-two carbons, placed at the median position of the segment. (b) Schematic of the coarse-grained representation of a melt. The thick springs connect successive beads on the same chain; they are finitely extensible, governed by an effective potential derived from atomistic simulations. The thin springs are finitely extensible slip-springs, connecting beads on different chains. (c) Elementary hop of one end of a slip-spring along the contour of the chain on which it lies. Rate constants for these hops are determined as described in the text. (d) Destruction of a slip-spring when one of its ends hops past a chain end. (e) Creation of a slip spring by pairing a chain end with one of its neighbouring beads within a radius $R_{attempt}$ from it. (f) Spatial discretization invoked for the calculation of the nonbonded part of the free energy, shown in the case of a network with both entanglements (springs) and chemical crosslinks (tetrafunctional nodal points). The space occupied by the polymer is partitioned into cells of edge length l_{cell}. The mass of each bead is smeared uniformly within a cubic voxel of edge length h.

(a, b) Reprinted (adapted) with permission from Sgouros, A.P., Megariotis, G., and Theodorou, D.N. *Macromolecules* **50**, 4524 (2017). Copyright 2017 American Chemical Society. (c–e) Reprinted (adapted) with permission from Vogiatzis, G.G., Megariotis, G., and Theodorou, D.N. *Macromolecules* **50**, 3004 (2017). Copyright 2017 American Chemical Society.

In mapping the atomistic to the EMSIPON description, one has the option of placing the centre of a bead either at the centre-of-mass position of the atomistic sequence of monomers it represents (Figure 8.2a, *centre-of-mass* option) or at the position of the midpoint of the atomistic sequence (Figure 8.2b, *median* option). Of these two options, the *median* option has been found to be more straightforward, as it allows matching of conformational properties between the atomistic and mesoscopic levels using weaker effective bond angle bending potentials.[35]

The free energy of the mesoscopic model (potential of mean force with respect to the bead positions and slip-spring configuration) is represented as a sum of contributions:

$$A = A_{\text{strands}} + A_{\text{angles}} + A_{\text{nonbonded}} + A_{\text{entanglements}} + A_{\text{compensating}}. \tag{8.3.1}$$

The bonded contributions A_{strands} and A_{angles} comprise the effective stretching and bending potentials contributed by pairs and triplets of beads along the backbones of coarse-grained chains. They incorporate the mainly entropic elasticity of the strands. $A_{\text{nonbonded}}$ is a contribution from excluded volume and van der Waals interactions. It is expressed as a functional integral of an excess Helmholtz energy density depending on the local bead density and derived from an equation of state. It introduces the *field* part of the particle-field method and is computed by spatial discretization. $A_{\text{entanglements}}$ is the free energy of the slip-springs, while $A_{\text{compensating}}$ removes any effect of the slip-springs on the conformations of chains.

A_{strands} and A_{angles} are obtained as sums of contributions $A_{\text{strands}}(r)$ and $A_{\text{angles}}(\theta)$ from the lengths r of all strands and the angles θ between all pairs of successive strands in the system. $A_{\text{strands}}(r)$ and $A_{\text{angles}}(\theta)$ are obtained by Boltzmann inversion of the distributions r and θ obtained from equilibrium atomistic simulations of short-chain melts at the temperature and pressure of interest. A_{strands} reflects the finite extensibility of strands and deviates significantly from a quadratic potential at high extensions. An excellent description to $A_{\text{strands}}(r)$ is provided by the integral of Kröger's[41] $\mathscr{L}_{301}^{-1}$ approximant to the inverse Langevin function (compare Eq. (1.3.37); see also Problem 5.5). This analytical approximation, fitted to the numerical function derived from atomistic simulations, is employed in the mesoscopic simulations.[35] The effective bending potential $A_{\text{angles}}(\theta)$ extracted from atomistic MD is weak for the level of coarse-grained invoked (e.g. fifty-two carbon atoms per strand in PE) and can be very well approximated as a quadratic around $\theta = \pi$. $A_{\text{entanglements}}$ is a sum of contributions from all active slip-springs in the system. Each such contribution is represented as a FENE spring, following Ref. 31; alternatively, the $\mathscr{L}_{301}^{-1}$ could be used. The parameters of the slip-spring free energy are chosen[31,35,36] so as to correspond to a chain strand of length equal to the primitive path Kuhn length a_{pp}. $A_{\text{compensating}}$ is formed from the sum of Boltzmann factors of slip-spring effective potential contributions from all pairs of beads that could be bridged by a slip-spring within a given configuration times the slip-spring activity $z_{\text{activ}} = \exp\left[\mu_{\text{ss}}/(k_{\text{B}}T)\right]$, as in Ref. 31. With the bottom-up parameterization adopted in Refs. 34–37, $A_{\text{compensating}}$ can be omitted without significant effect on the unperturbed dimensions of chains.[35] $A_{\text{nonbonded}}$ is envisioned as a functional integral, over the entire domain occupied by the polymer, of an excess Helmholtz energy density $a_{\text{vol}}^{\text{ex}}\left[\rho(\boldsymbol{r}), T\right]$ that depends on the local mass density $\rho(\boldsymbol{r})$:

$$A_{\text{nonbonded}} = \int_V a_{\text{vol}}^{\text{ex}}\left[\rho(\boldsymbol{r}), T\right] \mathrm{d}\boldsymbol{r}. \tag{8.3.2}$$

Excess here is to be understood as *relative to an ideal gas of coarse-grained chains at the same temperature and density*, since intramolecular contributions are already incorporated in other terms of the Helmholtz energy function A and overall translational entropy is taken care of by the simulation approach. The excess Helmholtz energy density function $a_{\text{vol}}^{\text{ex}}\left[\rho(\boldsymbol{r}), T\right]$ can be obtained once an equation of state $P = P(\rho, T)$ for the pressure as a function of mass density ρ and temperature T is available for the melt,[37] as

$$a_{\text{vol}}^{\text{ex}}(\rho, T) = \rho \frac{RT}{M} \ln \left(\frac{Z_{\text{intra, cg chain}} \prod_i \Lambda_i^3}{Z_{\text{intra}} \Lambda_{\text{bead}}^{3n_{\text{beads/chain}}}} \right) + \rho \frac{RT}{M} \int_0^{\rho} \frac{1}{\rho'} \left(\frac{PM}{\rho' RT} - 1 \right) \mathrm{d}\rho'. \tag{8.3.3}$$

In Eq. (8.3.3) $Z_{\text{intra, cg chain}}$ is the configurational integral of a single coarse-grained chain in the ideal gas state over all but its translational degrees of freedom; Λ_{bead} is the thermal wavelength for a bead in the coarse-grained representation; $n_{\text{beads/chain}}$ is the number of coarse-grained beads per chain; Z_{intra} is the configurational integral of a detailed atomistic single chain in the ideal gas state over all but its translational degrees of freedom; Λ_i is the thermal wavelength of atom i in the atomistic representation; and M is the molar mass of chains in the melt. Note that the molecular density (chains per unit volume) is $\rho N_{\text{Avo}}/M$ with N_{Avo} being Avogadro's number and the bead density (beads per unit volume) is $\rho\, n_{\text{beads/chain}} N_{\text{Avo}}/M$.

In applications of the EMSIPON approach,[34–38] the Sanchez–Lacombe equation of state has proved useful. Its parameters can be fitted to experimental measurements or simulation estimates of the volumetric behaviour of melts of narrow molar mass distribution, preserving the *bottom-up* character of the approach. The Sanchez–Lacombe equation of state reads

$$\tilde{P} = -\tilde{\rho}^2 - \tilde{T} \left[\ln(1 - \tilde{\rho}) + \left(1 - \frac{1}{r}\right) \tilde{\rho} \right] \tag{8.3.4}$$

with the reduced mass density $\tilde{\rho}$, temperature $\tilde{T}$, and pressure $\tilde{P}$ being given in terms of the set $\{\rho^*, T^*, P^*\}$ of material-specific equation of state parameters as $\tilde{\rho} = \rho/\rho^*$, $\tilde{T} = T/T^*$, and $\tilde{P} = P/P^*$, and $r = MP^*/(\rho^* RT^*)$ being a measure of chain length. Use of Eq. (8.3.4) within Eq. (8.3.3) leads to the following expression for the excess Helmholtz energy density:

$$a_{\text{vol}}^{\text{ex}}(\rho, T) = -P^* \tilde{\rho}^2 + P^* \tilde{T} (1 - \tilde{\rho}) \ln(1 - \tilde{\rho}) + P^* \tilde{T} \tilde{\rho} \left\{ 1 + \frac{1}{r} \ln \left(\frac{Z_{\text{intra, cg chain}} \prod_i \Lambda_i^3}{Z_{\text{intra}} \Lambda_{\text{bead}}^{3n_{\text{beads/chain}}}} \right) \right\}. \tag{8.3.5}$$

Upon integration over the domain of a system with given mass, the last term in Eq. (8.3.5) (proportional to $\tilde{T}\tilde{\rho}$)—which is strictly proportional to density—yields a contribution that depends only on temperature; therefore, it does not affect the equations of motion or thermodynamic properties produced by the simulation.

To evaluate $A_{\text{nonbonded}}$ according to Eqs. (8.3.2)–(8.3.3) one can use spatial discretization of the simulation domain. One can divide the domain into orthogonal parallelepiped cells of edge length $l_{\text{cell}} = 3$ to $5\,\text{nm}$ (see Figure 8.2f) and calculate the local mass density $\rho_{\text{cell},k}$ within each cell k from the bead positions in each configuration. One way to do this[34] is to smear the mass carried by each bead uniformly within a cubic voxel of edge length $h \approx R_{g,\text{bead}}$ commensurate with the radius of gyration of the strand represented by the bead (see Figure 8.2f) and assume that the fraction of bead mass contributed to each cell equals the fraction of voxel volume contained in the cell. This simple geometric assignment leads to an analytical expression for $\rho_{\text{cell},k}$ in terms of the bead positions $\boldsymbol{r}_i$. The nonbonded free energy of Eqs. (8.3.2)–(8.3.3) is approximated as a discrete sum over the cells,

$$A_{\text{nonbonded}} = \sum_{\text{all cells } k} V_{\text{cell},k} a_{\text{vol}}^{\text{ex}} \left(\rho_{\text{cell},k}, T\right) \tag{8.3.6}$$

with $V_{\text{cell},k}$ being the volume of cell k that is accessible to polymer. As a bead i moves (bead position $\boldsymbol{r}_i$ changes), its mass gets repartitioned among the cells to which it contributes, changing the local cell densities $\rho_{\text{cell},k}$. As a consequence of Eq. (8.3.6), the nonbonded Helmholtz energy of the system changes. The negative gradient of $A_{\text{nonbonded}}$ with respect to $\boldsymbol{r}_i$ is the nonbonded force exerted on bead i in the considered configuration. In general, the force on each bead i in the EMSIPON approach is calculated as

$$\boldsymbol{F}_i = -\nabla_{\boldsymbol{r}_i} A, \tag{8.3.7}$$

with A incorporating all contributions in Eq. (8.3.1).

The stress tensor $\boldsymbol{\tau}$ is calculated through its thermodynamic definition in terms of the derivative of the free energy A with respect to the deformation gradient tensor, Eq. (3.5.23). It contains both virial theorem-type contributions from the particle-based bonded parts of A and a field contribution from the nonbonded part of A.[34–37]

The motion of beads subject to the forces generated by the Helmholtz energy A (see Eq. (8.3.7)) is tracked by Brownian dynamics simulation in the high friction limit. The following position Langevin equation (compare Eq. (7.7.57)) is used to update the position vector of each bead:

$$\boldsymbol{r}_i\left(t_n + \Delta t\right) = \boldsymbol{r}_i\left(t_n\right) + \frac{1}{\zeta_{\text{bead}}(T)}\left[\boldsymbol{F}_i\left(t_n\right)\Delta t + \frac{1}{2}\dot{\boldsymbol{F}}_i\left(t_n\right)\left(\Delta t\right)^2\right] + \boldsymbol{R}_i\left(\Delta t\right), \tag{8.3.8}$$

with the time derivative estimated by finite differences based on the previous time step and $\boldsymbol{R}_i\left(\Delta t\right)$ being a random displacement vector for bead i whose components $R_{i,a}\left(\Delta t\right)$ are independent variables following Gaussian distributions with zero mean and variance

$$\left\langle R_{i,\alpha}^2\left(\Delta t\right)\right\rangle = 2k_{\text{B}}T\frac{\Delta t}{\zeta_{\text{bead}}\left(T\right)}. \tag{8.3.9}$$

The friction factor per bead, $\zeta_{\text{bead}}(T)$, is related to the friction factor per monomer, $\zeta_0(T)$, as

$$\zeta_{\text{bead}}(T) = \frac{m_{\text{bead}}}{m_{\text{monomer}}}\zeta_0(T). \tag{8.3.10}$$

The friction factor per monomer $\zeta_0(T)$ is obtained from atomistic MD simulations or from experiment; see, for example, the friction factors extracted from united-atom MD simulations of PE melts by Harmandaris et al.[42] and those extracted from NMR measurements on polyisoprene by Klopffer et al.[43] For given chemical constitution, the monomeric friction factor $\zeta_0(T)$ generally follows a Williams–Landel–Ferry (WLF) or, equivalently, a Vogel–Fulcher–Tamman (VFT) dependence on temperature. For linear PE at 350 K a value of $\zeta_0(T) = 4.15 \times 10^{-13}$ kg/s is appropriate. An integration time step $\Delta t = 1$ ps ensures good performance of the Brownian dynamics simulation.

In parallel with the Brownian dynamics of beads, an EMSIPON simulation follows the elementary events associated with slip-springs, which change the topology and geometry of the entangled network of chains. As already mentioned, these are viewed as infrequent events, so they are tracked by kMC simulation. Elementary events involved in the simulation of a bulk melt are shown pictorially in Figures 8.2c, d, e. Each end of a slip-spring moves along the contour of the chain to which it belongs by discrete hops between neighbouring beads (Figure 8.2c). The transition state involved in such a jump is assumed to be characterized by a constant Helmholtz energy $A^{\dagger}_{\text{O}\to\text{N}}$ relative to the same configuration with the considered slip-spring absent, that is, elevated with respect to both the origin and destination states of the hop. This gives rise to a free-energy barrier for the hop. The rate constant for the hop is written as[34]

$$k_{\text{hop}} = \nu_0 \exp\left(-\frac{A^{\dagger}_{\text{O}\to\text{N}} - A_{a_0\wedge b_0}}{k_{\text{B}}T}\right) = \nu_{\text{hop}} \exp\left(\frac{A_{a_0\wedge b_0}}{k_{\text{B}}T}\right), \tag{8.3.11}$$

where $A_{a_0\wedge b_0}$ is the Helmholtz energy of the original state of the hop, measured relative to the same configuration with the considered slip-spring absent; ν_0 is an oscillation frequency along the transition coordinate in the origin state, assumed to be the same for all transitions; and $\nu_{\text{hop}} = \nu_0 \exp\left(-A^{\dagger}_{\text{O}\to\text{N}} \big/ (k_{\text{B}}T)\right)$ is a frequency obtained by multiplying ν_0 with the Boltzmann factor of the transition state free energy, also assumed to be the same for all transitions. According to Eq. (8.3.11), a strained initial state of the slip-spring (high $A_{a_0\wedge b_0}$) enhances the rate constant for transitions out of this state, while a stable initial state of the slip-spring (low $A_{a_0\wedge b_0}$) decelerates transitions out of that state. It is possible to relate the frequency factor ν_{hop} appearing in Eq. (8.3.11) to the friction factor experienced by the beads by invoking the Rouse-reptation picture of polymer dynamics. The result is[34]

$$\nu_{\text{hop}} \approx \frac{k_{\text{B}}T}{\zeta_{\text{bead}}} \frac{\langle N_{\text{ss}}\rangle}{N_{\text{beads}}} \frac{1}{n_{\text{Kuhns/bead}} b_{\text{K}}^2}, \tag{8.3.12}$$

where $n_{\text{Kuhns/bead}}$ is the number of Kuhn segments per bead, b_{K} is the Kuhn segment length, and therefore $n_{\text{Kuhns/bead}} b_{\text{K}}^2$ is the unperturbed mean squared end-to-end distance of the strand represented by a bead. $\langle N_{\text{ss}}\rangle$ is the mean number of slip-springs at equilibrium, which is controlled by the slip-spring activity z_{activ}. N_{beads} is the total number of beads in the system. Equation (8.3.12) is useful in obtaining an initial estimate for ν_{hop}. In practice, this estimate may be adjusted somewhat to achieve good agreement with experimental time scales. A value of $\nu_{\text{hop}} = 5 \times 10^5\ \text{s}^{-1}$ proved satisfactory for linear PE at 450 K.

Whenever, in the course of its hopping motion, the end of a slip-spring reaches the end of the chain to which it belongs, it may hop off of the chain, leading to destruction of the slip-spring (Figure 8.2d). On the other hand, chain ends continuously attempt to form new slip-springs by pairing with beads belonging to surrounding chains.[35] Any bead within distance R_{attempt} from the chain end, with R_{attempt} being practically equal to the maximum extension of a slip-spring, is considered as a candidate in this pairing process (Figure 8.2e). The rates of destruction of slip-springs through the process of hopping off of chain ends and of creation of new slip-springs at chain ends are required to be equal at equilibrium (microscopic reversibility). This leads to the following condition for the rate constant of formation of a new slip-spring at a chain end:[34,35]

$$k_{\mathrm{form},a_0 \wedge b_\times \to a_0 \wedge b_0} = \nu_{\mathrm{hop}} \frac{z_{\mathrm{activ}}}{n_{\mathrm{ss},a_0 \wedge b_0}}, \tag{8.3.13}$$

with z_{activ} being the activity of slip-springs and $n_{\mathrm{ss},a_0 \wedge b_0}$ the number of slip-springs connecting beads a_0 and b_0 after successful creation of a new slip-spring between them; usually, $n_{\mathrm{ss},a_0 \wedge b_0} = 1$. Slip-spring hops, destructions, and creations are implemented in a kMC simulation running in parallel to the Brownian dynamics simulation according to the rate constants of Eqs. (8.3.12) and (8.3.13). The kMC simulation proceeds as outlined in Chapter 6. Usually, one kMC move is introduced every ten BD time steps. With the resulting BD/kMC scheme, times on the order of 0.1 s can be readily simulated with modest computational means. EMSIPON affords a savings in computer time of roughly three orders of magnitude relative to atomistic MD simulations.

The EMSIPON approach can readily be implemented under nonequilibrium conditions (NEBD/kMC simulations). For example, nonequilibrium BD/kMC simulations of Couette flow under prescribed shear rates $\dot{\gamma}$ have been conducted by application of Lees–Edwards periodic boundary conditions,[35] leading to estimates of the shear viscosity (which exhibits shear thinning at high Weissenberg numbers) and of the first and second normal stress difference coefficients in good agreement with experiment. Figure 8.3 shows predictions of the frequency-dependent storage $G'(\omega)$ and loss $G''(\omega)$ moduli in a monodisperse *cis*-1,4 polyisoprene (PI) melt of molar mass 100 kg/mol at 400 K as obtained from the EMSIPON approach[34] with equilibrium BD/kMC simulations. The shear relaxation modulus $G(t)$ was obtained from the time autocorrelation function of the shear stress at equilibrium. The dynamic moduli were obtained by Fourier transformation of $G(t)$. In the mesoscopic model invoked by EMSIPON, thirty-six *cis*-1,4 PI repeat units were lumped into a single bead. The agreement of the simulation results with experimental measurements[44,45] from a melt of molar mass 90 kg/mol is generally very favourable. It is noteworthy that the dynamic mechanical moduli can be calculated for angular frequencies as low as 100 s^{-1}, which are clearly inaccessible to atomistic simulations. The quality of predictions deteriorates at high frequencies, since local segmental motions are not realistically represented in the mesoscopic model.

The EMSIPON approach has been extended to the study of polymer networks with chemical crosslinks.[36] It has also been extended to systems containing free polymer surfaces (in contact with gas phases). There, the nonbonded free energy has been augmented to include a term proportional to the square gradient of the local density, $\frac{1}{2}\frac{\kappa}{M^2}(\nabla\rho)^2$, to take care of density inhomogeneities in the system. The *influence parameter* κ can be correlated to equation of state parameters.[37] The derivation of forces and stresses can be extended to include contributions from the squared density gradient term. The nonbonded force on bead i reads

$$\boldsymbol{F}_i^{\mathrm{nonbonded}} = -\nabla_{\boldsymbol{r}_i} A_{\mathrm{nonbonded}} = -\int_V \left[\left. \frac{\partial a_{\mathrm{vol}}^{\mathrm{ex}}(\rho, T)}{\partial \rho} \right|_{\rho=\rho(\boldsymbol{r})} - \frac{\kappa}{M^2} \nabla_{\boldsymbol{r}}^2 \rho(\boldsymbol{r}) \right] \frac{\partial \rho(\boldsymbol{r})}{\partial \boldsymbol{r}_i} \mathrm{d}\boldsymbol{r}. \tag{8.3.14}$$

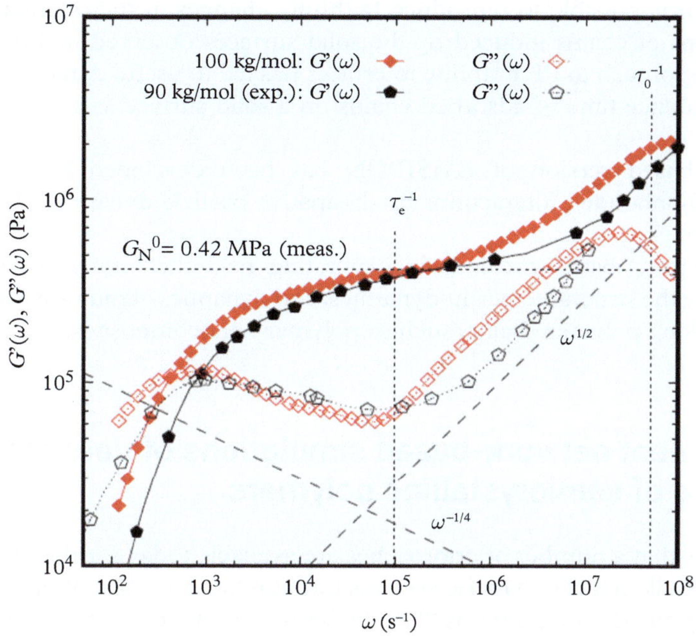

Figure 8.3 Storage $G'(\omega)$ (filled symbols) and loss $G''(\omega)$ (open symbols) moduli for monodisperse *cis*-1,4 polyisoprene (PI) melts of molar mass 100 kg/mol at 400 K as obtained from the EMSIPON approach being compared to experimental measurements[36] on a melt of similar molar mass. Predictions are shown as diamonds, experiments as pentagons. In the BD/kMC simulations, thirty-six *cis*-1,4 PI repeat units were lumped into a single bead. $G'(\omega)/\omega$ and $G''(\omega)/\omega$ were estimated as the sine and cosine Fourier transforms of $G(t)$, respectively. Also shown is the measured plateau modulus from Fetters et al.[37]

The contribution to the stress tensor from nonbonded interactions reads

$$\tau_{\alpha\beta}^{\text{nonbonded}} = \frac{1}{V}\int_V \left\{ -\frac{\kappa}{M^2}\frac{\partial\rho}{\partial r_\alpha}\frac{\partial\rho}{\partial r_\beta}\delta_{\alpha\beta}\left[a_{\text{vol}}^{\text{ex}}\left[\rho(\boldsymbol{r}), T\right] - \rho(\boldsymbol{r}) \left.\frac{\partial a_{\text{vol}}^{\text{ex}}(\rho, T)}{\partial\rho}\right|_{\rho=\rho(\boldsymbol{r})} + \frac{\kappa}{M^2}\rho(\boldsymbol{r})\,\nabla_{\boldsymbol{r}}^2\rho(\boldsymbol{r}) + \frac{1}{2}\frac{\kappa}{M^2}\left(\nabla_{\boldsymbol{r}}\rho(\boldsymbol{r})\right)^2\right]\right\} \mathrm{d}\boldsymbol{r}. \tag{8.3.15}$$

With the square gradient term incorporated in EMSIPON, the surface tension of polymer melts can be represented very well as a function of chain molar mass and cavitation phenomena can be captured.[37] The EMSIPON approach has also been extended to polymer–solid interfaces by incorporating bead adsorption to and desorption from a solid surface as elementary dynamical events, governed by atomistically derived rate constants.[38] If care is exercised in the development of the mesoscopic model, it is possible to reproduce faithfully changes in the orientational and conformational distributions of chains induced by the solid surfaces observed in atomistic simulations. Application of this approach to PE/graphite interfaces has led to useful dynamical predictions, e.g. estimates of the residence time of adsorbed chains on a solid surface as a function of melt molar mass.[38]

A fully particle-based version of EMSIPON has been developed by replacing the field-based treatment of nonbonded interactions by dissipative particle dynamics (DPD), leading to a DPD/kMC approach.[36]

In general, slip-spring-based methods incorporating field theoretic ideas offer a promising avenue for predicting the structure, thermodynamics, and dynamics of multiphase, multicomponent systems containing long polymer chains, such as polymer nanocomposites.

8.4 Entanglement network-based simulations of deformation to fracture of semicrystalline polymers

In Chapter 7 we saw that a number of approaches are available today for the precise identification of entanglements in bulk model polymeric systems through topological analysis of well-equilibrated atomistic or coarse-grained configurations.[46–50] These approaches open the way to the development of hierarchical modelling strategies going all the way from detailed atomistic to entanglement network modelling of large-scale deformation and flow. Apart from rheological properties in the melt state[51] discussed in Sections 8.1 to 8.3, the large-strain plastic deformation and failure characteristics of solid polymers composed of flexible chains in the bulk or at interfaces with other polymers or nonpolymeric solid substrates depend intimately on entanglements.

Entanglement network models have been used to study the fracture of polymer-substrate interfaces.[52–55] A typical example is the interface between a homopolymer and a solid, strengthened by grafted chains identical in composition to the homopolymer.[56–61] We mention, in particular, the interface between isotactic polypropylene (PP) and poly(amide 6) (PA6), strengthened by interposition of functionalized chains of polypropylene grafted with maleic anhydride (PP-g-MA). PP-g-MA is often referred to as a *compatibilizer*. Experimentally, such interfaces are prepared[56–61] by clamping sheets of PA6 and polypropylene-compatibilizer mixture (PP + 5% PP-g-MA) together in an airtight, Teflon-lined mould under slight pressure and then heating the mould in a temperature-controlled oven between 185 and 220°C. This temperature range is above the melting temperature of PP but below that of PA6. The functionalized PP-g-MA chains diffuse to the interface and their MA groups react with NH_2 groups on the PA6, forming an imide. Thus, PP chains grafted chemically to the PA6 surface are generated. From the experimental literature it appears that most of the MA groups of PP-g-MA are located at the ends of the chains and the NH_2 groups with which they react are mainly present at the ends of PA6 chains. Thus, the copolymer of PP and PA6 resulting from the reaction is expected to have essentially a diblock architecture,[60,61] with the compatibilizer chains that protrude from the PA6 crystal playing the role of chains that have been essentially grafted onto a *solid* surface (Figure 8.4). Their areal density is controlled by the temperature and duration of the annealing process.

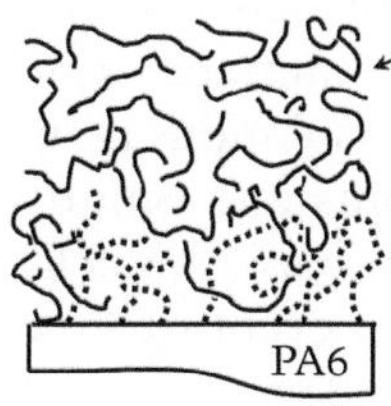

Figure 8.4 Schematic representation of the PP–PA6 interfacial system. The PA6 phase is represented by an equivalent *solid* substrate, on which the PP chains of the compatibilizer are terminally grafted, extending onto a bulk PP phase.

For such a PP–compatibilizer–PA6 interface, one can develop an appropriate model by considering the crystalline PA6 matrix as an equivalent *solid* substrate, on which the PP chains of the compatibilizer are terminally grafted, extending into a bulk PP phase. Based on the characteristics of the experimental samples, PP chains (either free or grafted) should be further taken to have a rather high polydispersity index I (significantly larger than unity). The main question to address then is how to generate a polymer entanglement network representative of the interface under these conditions. Strictly speaking, such a network should be built through an approach that starts from the atomistic level and leads safely to the generation of a network, in which the positions of all chain ends and all entanglement points have been precisely identified in the system. Since polymer systems of the molecular weights used in industrial applications are difficult to equilibrate at an atomistic level, a hierarchical equilibration approach based on a coarse-grained polymer model, as discussed in Chapter 7, is more viable. This approach may involve the following:

- reduction of the detailed atomistic model into a coarse-grained model involving fewer degrees of freedom,
- iterative Boltzmann inversion techniques to parameterize the coarse-grained model,
- implementation of state-of-the art MC methods to thoroughly equilibrate the coarse-grained model,
- topological analysis of the relaxed coarse-grained model to calculate the entanglement network, and
- use of the generated network structure as input to deformation simulations, to address mechanical properties, e.g. under conditions of constant strain rate, until fracture occurs.

Here we focus on the last item in this list, outlining kMC strategies that can be used to simulate deformation at experimentally relevant strain rates once an entanglement network is available. Such strategies have been pioneered for bulk semicrystalline polymers in a series of papers by Termonia et al.,[62–64] and rely on the Eyring chemical activation rate theory.[65,66] They have been refined and extended to three-dimensional entanglement network models of interfaces in continuous space by Terzis et al.[67] Note that the full multiscale approach outlined above (first four items in the list) for generating entanglement networks representative of the interface has not been applied in its entirety. Instead, Terzis et al. invoked an approximate, computationally much less expensive, method for generating the interfacial entanglement network. The method makes use of a self-consistent mean-field (SCF) lattice model for the interface, developed along the lines originally proposed by Scheutjens and Fleer[68–69] for polymer solutions at interfaces.

In Section 8.4.1 we discuss the work of Termonia and Smith on deformation and fracture of semicrystalline polymers as an example of a mesoscopic modelling approach that combines the picture of an entanglement network with infrequent event treatment of elementary processes that change the topology and geometry of that network. Analogies to the more sophisticated treatment of entangled melt rheology outlined in Section 8.3 are obvious. One difference is that in the much earlier model of Termonia and Smith, Brownian motion is not considered. Instead, the requirement that the network be in detailed mechanical equilibrium (i.e. at a local minimum of the mesoscopic free-energy function) at all times is invoked. This is sufficient for probing the large deformation behaviour of the solid polymers considered over experimentally relevant time scales.

8.4.1 Phenomenological entanglement model based on the Eyring chemical activation rate theory

According to Termonia et al.,[62–64] an undeformed (isotropic) polymer solid can be represented by a dense network of coiled, entangled macromolecules. A schematic representation of the undeformed polymer solid is shown in Figure 8.5a. Since the coordination number of an entanglement is only four, and to make the problem amenable to the computational resources available at the time, Termonia et al. considered only a two-dimensional model network (see Figure 8.5). From the figure, which is intended to represent PE at room temperatures, it is clear that no details of the semicrystalline morphology in the solid are specified.[70] The rationale offered is that, in a material kept together by weak van der Waals interactions, the crystal structure will be destroyed relatively quickly by pulling during the early stages of plastic deformation. Prior to deformation, chain segments between entanglements are tied together through numerous weak inter- and intramolecular bonds (e.g. van der Waals forces); these are symbolized by the dotted lines in Figure 8.5a and provide the initial stiffness to the material. In a more coarse-grained representation of the network, details of the chain configurations are omitted, implying that the individual van der Waals bonds can be replaced by *overall* bonds joining each entanglement point to its neighbours. They are shown as dotted lines in Figure 8.5b. The corresponding bold solid lines denote chain vectors between entanglements, while chain ends are indicated by dashed lines. The chain vectors in an actual, undeformed specimen are randomly oriented in 3D space; i.e. $\langle \cos^2\theta \rangle = \frac{1}{3}$, where θ is the vector's angle with the draw axis. In the simplified, regular 2D representation considered by Termonia et al.,[63] the same initial value $\theta = 54.7°$ was assumed for the orientation of all vectors along the y-axis. As also found from topological analyses presented in Chapter 7, the molecular weight between entanglements is independent of the chain length for sufficiently long chains. This implies that the number of entanglements per macromolecule increases linearly with its length.

8.4.1.1 Bond breaking

To study chain fracture phenomena, Termonia and Smith considered the network undergoing elongation along the y-axis of the coordinate system at constant rate $\dot{\varepsilon}$ and temperature T. This causes straining of the weak van der Waals (vdW) bonds, which break according to the kinetic theory of

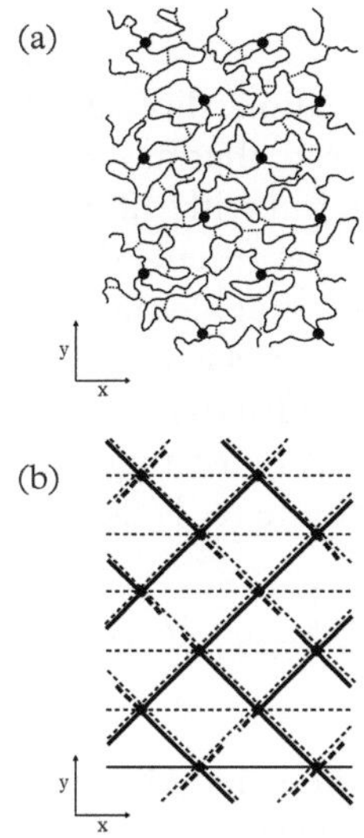

Figure 8.5 (a) Schematic representation of the undeformed polymer solid in the method of Termonia and Smith.[63–66] (b) Simplified representation of the network, in which details of the chain configurations are omitted and individual van der Waals bonds are replaced by 'overall' bonds (dotted lines) joining vectors between entanglements; chain ends are indicated by dashed lines.

fracture with a rate

$$\nu = \nu_0 \exp\left[-\frac{U^* - \sigma V^*}{k_B T}\right], \tag{8.4.1}$$

where ν_0 is the thermal vibration frequency, U^* and V^* are the activation energy and activation volume, respectively, and σ is the local stress given by $\sigma = k\varepsilon$, with k denoting the elastic constant of the bond and ε the local strain. The breaking of the vdW bonds along a chain strand between two entanglements leads to a transfer of the external load to the bonds surrounding the ends of the strand.

The simulation of vdW bond breakage events was performed with the help of a kMC scheme that breaks a bond i according to the probability

$$p_i = \frac{\nu_i}{\nu_{max}}, \tag{8.4.2}$$

where ν_{max} is the rate of breakage of the most strained bond in the array. After each visit of a bond, the time t was incremented by $\frac{1}{n(t)\,\nu_{max}}$, where $n(t)$ is the total number of intact bonds at time t. For reasons of simplicity, broken vdW bonds were assumed not to reform during the deformation process.

8.4.1.2 Chain slippage

As the load increases upon further deformation, slippage of the chain through entanglements sets in. It is assumed[52,53,66] that this occurs at a rate that has the same functional form as that used for the breakage of vdW bonds (see Eq. (8.4.1)), but with different values for the activation energy U^* and the activation volume V^*. For slippage, σ represents the difference in stress experienced by the two strands of a chain coming together at an entanglement. In order to estimate that stress, Termonia et al. turned to the classical theory of rubber elasticity. According to this theory, incorporating the effects of finite extensibility of strands as discussed in Section 1.3.4, they estimate the stress on a chain strand having vector length r and consisting of N_K statistical segments by the following approximate expression, which they attribute to Treloar:

$$\sigma = \frac{N_{\text{strands}}}{3V} k_B T N_K^{1/2} \mathscr{L}^{-1}\left(\frac{r}{N_K b_K}\right). \tag{8.4.3}$$

In Eq. (8.4.3), b_K is the statistical segment length, $\mathscr{L}^{-1}$ is the inverse Langevin function, and N_{strands}/V is the density of strands in the system.

Chain slippage leads to a change in the number of statistical chain segments between entanglements and, occasionally, to chain disentanglement.

8.4.1.3 Rupture of a strand

In general, if the rate of slippage is low, maximum elongation of the chain strands between entanglements can be reached and chain fracture occurs. A strand is assumed to rupture whenever it reaches full extension.

8.4.1.4 Network deformation

After a very small time interval Δt has elapsed, the bond-breaking and chain slippage/fracture processes are stopped and the network is strained (along the y-axis) by a small, constant amount determined by the strain rate $\dot{\varepsilon}$. Subsequently, the network is allowed to relax to its minimum-energy configuration.[71,72] During this stage, a displacement of the entanglement points along the x- and y-axes will take place. To save computer time, only displacements along the draw axis were explicitly calculated, whereas distances in the transverse x-direction were assumed to contract homogeneously by a factor $\lambda^{-1/2}$, where λ is the draw ratio $(= \varepsilon + 1)$ along the y-axis. After these relaxation steps, the axial nominal stress at one end of the network was calculated. The Monte Carlo process of bond breakages and slippages was then restarted for another time interval Δt. The procedure was continued up to fracture of the computer specimen.

Termonia et al.[62–64] applied this phenomenological model to study the effect of molecular weight on the tensile deformation of melt-crystallized PE at 75°C for various molecular weights M, and compared against experimental data using the following parameter values. For vdW bond breakage,[63] molecular weight of the chain statistical segment = 140 g/mol, $b_K \approx 10$ Å, atomic vibration frequency $v_0 = 10^{12}\,\text{s}^{-1}$, $\dot{\varepsilon} = 5\,\text{min}^{-1}$, $k = 4\,\text{MPa}$, $U^* = 30\,\text{kcal/mol}$, and $V^* = \left(26\ \text{Å}\right)^3$. For chain slippage (involving the coordinated movement of one to two statistical segments at most),[63] molecular weight between entanglements = 1900 g/mol, $U^* = 20\,\text{kcal/mol}$, and $V^* = \left(7\ \text{Å}\right)^3$.

For $M = 1900$ g/mol, brittle fracture was observed. For $M = 9500$ g/mol, a neck appeared, which then propagated along the computational specimen. For $M = 250000$ g/mol, homogeneous deformation throughout the sample was observed. At an intermediate molecular weight of $M = 20000$ g/mol, multiple necking was observed. All these results are in agreement with experimental observations, indicating that the simple model developed by Termonia and Smith can capture the salient features of terminal mechanical behaviour in a semicrystalline polyolefin.

Additional results for the effect of entanglement spacing, deformation rate, temperature, and polydispersity can be found in the original papers by Termonia, Smith, and collaborators.[62–64,71–73] Capturing temperature effects, in particular, necessitated introducing a mechanism which allowed disentangled chain ends to reentangle with neighbouring chains. This was done by introducing two processes. In the first process, two interpenetrating chain ends with the number of segments larger than the molecular weight between entanglements were allowed to re-entangle and thus to form two unstrained strands with end-to-end vector lengths of $r = \sqrt{N_K} b_K$. In the second process, a single terminal strand of N_K statistical segments was allowed to entangle with a neighbouring chain strand in its middle point, when the latter lay within the former's radius of gyration. This leads to the formation of an unstrained strand of $N'_K < N_K$ statistical segments and end-to-end vector length $r' = \sqrt{N'_K} b_K$. These moves led to a number of chain disentanglements approximately equal to the number of chain *re-entanglements*, so that the total number of entanglements remained constant. Given that PE can undergo crystallization during deformation, Termonia et al.[64] also proposed a mechanism incorporating temperature effects on orientation efficiency. To this end, they considered that every chain strand between two entanglements can be represented by a string of only two elements: *helix* elements (denoted as h) oriented along the draw axis, and *coil* or unoriented elements, denoted as u. For a chain strand i forming an angle θ with the draw axis, the respective fractions of these two types (helix and coil) of elements are $\upsilon_h = \cos\theta/(\cos\theta + \sin\theta)$ and $\upsilon_u = \sin\theta/(\cos\theta + \sin\theta)$. The corresponding expression for elastic modulus of strand i along the draw axis is subsequently written as $E_i = (\upsilon_h/E_h + \upsilon_u/E_u)^{-1}$, in which E_h and E_u are the moduli for the two elements. E_h (= 300 GPa for PE) represents the axial tensile modulus of a defect-free crystal, while E_u (= 11.6 GPa for PE) coincides with the shear modulus of the crystalline phase. Calculations with such a mechanism confirmed the experimentally observed existence of a maximum drawing temperature, above which the efficiency of the orientation process decreases, presumably due to disentangling and recoiling of the polymer chains. This maximum temperature was predicted to increase linearly with the logarithm of the weight-average molecular weight. Polydispersity, on the other hand, was found to significantly decrease the efficiency of drawing at high temperatures.

8.4.2 Extension to interfaces

Terzis et al.[67] ported ideas and concepts of the Termonia and Smith model to the interfacial geometry of Figure 8.4, in order to study deformation and fracture over experimentally relevant time scales upon imposition of uniaxial elongation normal to the interface. They invoked a fully 3D model, with grafted and free chain configurations generated based on the probabilities obtained from a self-consistent field treatment of the interfacial system. They introduced a free-energy functional of the density to describe nonbonded interactions in the spirit of Eq. (8.3.2). They considered elementary processes of chain slippage by a Kuhn length across an entanglement, chain disentanglement, chain re-entanglement, and chain rupture, which they tracked by kMC simulation, using rate constants parameterized on the basis of melt viscosity data. Strain was introduced in small steps $\Delta\varepsilon = \dot{\varepsilon}\,\Delta t$. After each incremental strain, the total free energy of the system was minimized at constant network topology and strand lengths, to impose mechanical equilibrium. Subsequently, the elementary events leading to changes in the strand lengths and network topology were allowed to occur for time Δt according to their rate constants. This was followed by a new free-energy minimization under constant topology and strand lengths, and calculations looped back to introduce the next strain step. True stress versus elongation curves were obtained for finite interfacial specimens of dimensions 45 nm × 45 nm × 70 nm at various surface grafting densities, at strain rates $\dot{\varepsilon}$ of 0.01 and 0.1 s^{-1}. An interesting finding was the existence of an optimal surface grafting density for adhesion.[67] No explicit consideration was given to the semicrystalline morphology of polypropylene.

Problems

Problem 8.1: In connection with the Schieber slip-link model discussed in Section 8.2, show that the distributions, Eqs. (8.2.14) and (8.2.15), are normalized and that $\langle Z \rangle_{eq} = N/n_e$.

References

1. Doi, M., and Edwards, S.F. *The Theory of Polymer Dynamics*. Oxford University Press, Oxford (1986).
2. Doi, M., *J. Polym. Sci.* **21**, 667 (1983).
3. des Cloizeaux, J.J. *Physique Lett.* **45**, 17 (1984).
4. Needs, R.J. *Macromolecules* **17**, 437 (1984).
5. Noolandi, J., Slater, G.W., Lim, H.A., and Viovy, J.L. *Science* **243**, 1456 (1989).
6. Levene, S.D., and Zimm, B.H. *Science* **245**, 396 (1989).
7. Richter, D., Farago, B., Fetters, L. J., Huang, J.S., Ewen, B., and Lartigue, C. *Phys. Rev. Lett.* **64**, 1389 (1990).
8. Lodge, T.P. *Phys. Rev. Lett.* **83**, 3218 (1999).
9. Watanabe, H. *Prog. Polym. Sci.* **24**, 1253 (1999).
10. McLeish, T.C.B. *Adv. Phys.* **51**, 1379 (2002).
11. Watanabe, H., Ishida, S., Matsumiya, Y., and Inoue, T. *Macromolecules* **37**, 6619 (2004).
12. Rubinstein, M., and Colby, R.H. *Polymer Physics*. Oxford University Press, Oxford (2011).
13. Doi, M., and Edwards, S.F. *J. Chem. Soc. Faraday Transactions* **2**, 74, 1802 (1978).
14. Doi, M., and Takimoto, J.-I *Phil. Trans. R. Soc. Lond. A* **361**, 641 (2003).
15. Tzoumanekas, C., and Theodorou, D.N. *Curr. Opin. Solid State Mater. Sci.* **10**, 61 (2006).
16. Padding, J. and Briels, W. *J. Chem. Phys.* **177**, 925 (2002).
17. Masubuchi, Y., Takimoto, J.-I., Koyama, K., Ianniruberto, G., Marrucci, G., and Greco, F. *J. Chem. Phys.* **115**, 4387 (2001).
18. Masubuchi, Y., Ianniruberto, G., Greco, F., and Marrucci, G. *J. Chem. Phys.* **119**, 6925 (2003).
19. Masubuchi, Y., Ianniruberto, G., Greco, F., and Marrucci, G. *J. Non-Newt. Fluid Mech.* **149**, 87 (2008).
20. Schausberger, A., Schindlauer, G., and Janeschitz-Kriegl, H. *Rheol. Acta* **24**, 220 (1985).
21. Schieber, J.D., Neergaard, J., and Gupta, S. *J. Rheology* **47**, 213 (2003).
22. Nair, D.M., and Schieber, J.D. *Macromolecules* **39**, 3386 (2006).

23. Graessley, W.W. *Adv. Polym. Sci.* **47**, 67 (1982).
24. Rubinstein, M., and Colby, R.H. *J. Chem. Phys.* **89**, 5291 (1988).
25. Fetters, L.J., Lohse, D.J., Milner, S.T., and Graessley, W.W. *Macromolecules* **32**, 6847 (1999).
26. Raoult, J. *J. Non-Newt. Fluid Mech.* **23**, 229 (1987).
27. Marin, G., and Graessley, W.W. *Rheol. Acta* **16**, 527 (1977).
28. Masubuchi, Y. *Ann. Rev. Chem. Biomol.* **5**, 11 (2014).
29. Schieber, J.D., and Andreev, M. *Annu. Rev. Chem. Biomol.* **5**, 367 (2014).
30. Rubinstein, M., and Panyukov. S. *Macromolecules* **35**, 6670 (2002).
31. Chappa, V.C., Morse, D.C., Zippelius, A., and Müller, M. *Phys. Rev. Lett.* **109**, 148302 (2012).
32. Uneyama, T., and Masubuchi, Y. *J. Chem. Phys.* **137**, 154902 (2012).
33. Langeloth, M., Masubuchi, Y., Böhm, M.C., and Müller-Plathe, F. *J. Chem. Phys.* **138**, 104907 (2013).
34. Vogiatzis, G.G., Megariotis, G., and Theodorou, D.N. *Macromolecules* **50**, 3004 (2017).
35. Sgouros, A.P., Megariotis, G., and Theodorou, D.N. *Macromolecules* **50**, 4524 (2017).
36. Megariotis, G., Vogiatzis, G.G., Sgouros, A.P., and Theodorou, D.N. *Polymers* **10**, 1156 (2018).
37. Sgouros, A.P., Lakkas, A.T., Megariotis, G., and Theodorou, D.N. *Macromolecules* **51**, 9798 (2018).
38. Sgouros, A.P., Vogiatzis, G.G., Megariotis, G., Tzoumanekas, C., and Theodorou, D.N. *Macromolecules* **52**, 7503 (2019).
39. Milano, G., and Kawakatsu, T. *J. Chem. Phys.* **130**, 214106 (2009).
40. De Nicola, A., Kawakatsu, T., and Milano, G. *J. Chem. Theory Comput.* **10**, 5651 (2014).
41. Kröger, M. *J. Non-Newton. Fluid Mech.* **223**, 77 (2015).
42. Harmandaris, V.A., Mavrantzas, V.G., and Theodorou, D.N. *Macromolecules* **31**, 7934 (1998).
43. Klopffer, M.-H., Bokobza, L., and Monnerie, L. *Polymer* **39**, 3445 (1998).
44. Auhl, D., Ramirez, J., Likhtman, A.E., Chambon, P., and Fernyhough, C. *J. Rheol.* **52**, 801 (2008).
45. Fetters, L.J., Lohse, D.J., Richter, D., Witten, T.A., and Zirkel, A. *Macromolecules* **27**, 4639 (1994).
46. Everaers, R., Sukumaran, S.K., Grest, G.S., Svaneborg, C., Sivasubramanian, A., and Kremer, K. *Science* **303**, 823 (2004).
47. Kröger, M. *Computer Physics Communications* **168**, 209 (2005).
48. Tzoumanekas, C., and Theodorou, D.N. *Macromolecules* **39**, 4592 (2006).
49. Foteinopoulou, K., Karayiannis, N.C., Mavrantzas, V.G., and Kröger, M. *Macromolecules* **39**, 4207 (2006).
50. Shanbhag, S., and Larson, R.G. *Macromolecules* **39**, 2413 (2006).
51. Ferry, J.D. *Viscoelastic Properties of Polymers* (3rd ed.). Wiley, New York (1980).
52. Treloar, L.R.G. *The Physics of Rubber Elasticity* (2nd ed.). Clarendon Press, Oxford (1958).
53. Kausch, H.H. *Polymer Fracture.* Springer Verlag, Berlin (1978).
54. Ward, I.M. *Mechanical Properties of Solid Polymers* (2nd ed.). Wiley, New York (1983).
55. Kinloch, A.J., and Young, R.J. *Fracture Behavior of Polymers.* Applied Science, London (1983).
56. Creton, C., Kramer, E. J., Hui, C.-Y., and Brown, H.R. *Macromolecules* **25**, 3075 (1992).
57. Washiyama, J., Creton, C., and Kramer, E.J. *Macromolecules* **25**, 4751 (1992).
58. Washiyama, J., Kramer, E.J., Creton, C.F., and Hui, C.-Y. *Macromolecules* **27**, 2019 (1994).
59. Cho, K., and Li, F. *Macromolecules* **31**, 7495 (1998).
60. Boucher, E., Folkers, J.P., Hervet, H., Leger, L., and Creton, C. *Macromolecules* **29**, 774 (1996).
61. Boucher, E., Folkers, J.P., Creton, C., Hervet, H., and Leger, L. *Macromolecules* **30**, 2102 (1997).
62. Termonia, Y., Meakin, P, and Smith, P. *Macromolecules* **18**, 2246 (1985); Termonia, Y., Meakin, P, and Smith, P. *Macromolecules* **19**, 154 (1986).
63. Termonia, Y., and Smith, P. *Macromolecules* **20**, 835 (1987); Termonia, Y., and Smith, P. *Macromolecules* **21**, 2184 (1988).
64. Termonia, Y., Allen, S.R., and Smith, P. *Macromolecules* **21**, 3485 (1988).

65. Glasstone, S., Laidler, K.J., and Eying, H. *The Theory of Rate Processes*. McGraw-Hill, New York (1941).
66. Ree, T., and Eyring, H. In Eirich, F.R. (ed.) *Rheology*, Vol 2, p. 83. Academic Press, New York (1958).
67. Terzis, A.F., Theodorou, D.N., and Stroeks, A. *Macromolecules* **33**, 1385 (2000); Terzis, A.F., Theodorou, D.N., and Stroeks, A. *Macromolecules* **33**, 1397 (2000); Terzis, A.F., Theodorou, D.N., and Stroeks, A. *Macromolecules* **35**, 508 (2002).
68. Scheutjens, J.M.H.M., and Fleer, G.J. *J. Phys. Chem.* **83**, 1619 (1979); Scheutjens, J.M.H.M., and Fleer, G.J. *J. Phys. Chem.* **84**, 178 (1980).
69. Scheutjens, J.M.H.M., and Fleer, G.J. *Macromolecules* **18**, 1882 (1985).
70. Meinel, G., and Peterlin, A. *J. Polym. Sci.* **9**, 67 (1971).
71. Termonia, Y., and Smith, P. *Colloid Polym. Sci.* **270**, 1085 (1992).
72. Termonia, Y., and Smith, P. *Macromolecules* **26**, 3738 (1993).
73. Termonia, Y. *Macromolecules* **29**, 4891 (1996).

9

Dissipative Particle Dynamics

Dissipative particle dynamics (DPD) is a method designed to describe the dynamics of soft particle systems where one is not interested in the complete detail of the actual motion of the fluid molecules. The particles in DPD are not real atoms but groups of atoms (or molecules) lumped together to form *fluid particles* or *beads*. During particle–particle collisions the energy is not conserved and, to deal with this, one should introduce dissipation. This is achieved by adding to the force felt by any particle not only an additional *random force*, but also a *viscous frictional force*. This results in a simulation method where, in addition to the number of particles, the total momentum of the system is conserved. That is, in contrast to the Brownian dynamics method (see Section 7.7) where the macroscopic behaviour is just diffusive, in DPD the macroscopic behaviour is hydrodynamic, since (coupled to the transport of mass) there is a transport equation for the momentum density field. On the other hand, energy is not conserved; there is no transport equation for energy in DPD. Overall, DPD is a particle-based method which (a) uses soft spheres to represent groups of atoms, (b) describes isothermal situations (although, as we will see later, there are expressions relating the temperature of the system to model parameters such as the friction coefficient and the noise amplitude), and (c) incorporates hydrodynamic behaviour via random noise which, as we will see later, is coupled to a pairwise dissipation.

From a statistical mechanics point of view, DPD is essentially a molecular dynamics (MD) method for systems of beads interacting with each other through direct conservative and dissipative forces, provided the effective potentials governing interparticle interactions are soft enough to allow for considerable overlap among particles. The method was introduced by Hoogerbrugge and Koelman[1,2] for simulating hydrodynamic behaviour and was conceived as an improvement over the conventional MD method, since it can describe complex hydrodynamic behaviour with computational efficiency. As noted by Español and Warren,[3] in particular, in MD simulations of hydrodynamic phenomena[4–7] the number of particles needed to obtain collective behaviour is extremely large, since hydrodynamic collective behaviour appears typically only for distances considerably larger than the interparticle distance and for time scales much longer than the characteristic collision time. Other techniques that can address hydrodynamic flow include the lattice-gas (LG)[8] and the more efficient lattice Boltzmann (LB) methods involving a discretization of the Boltzmann equation.[9]

A theoretical foundation for DPD has been provided by Español and Warren[3] through a formulation of the continuous stochastic differential equation that corresponds to the original algorithm proposed by Hoogerbrugge and Koelman[1,2] to derive the associated Fokker–Planck equation and

Multiscale Modelling of Polymers. Doros N. Theodorou and Vlasis G. Mavrantzas, Oxford University Press.
 DOI: 10.1093/oso/9780199226191.003.0009

study its equilibrium properties. Without such a basic Fokker–Planck equation for the microscopic dynamics, it would not be possible to derive in a rigorous way the hydrodynamic equations for DPD. In addition, this helped obtain a clear definition of the temperature and understand the significance of the relevant time step (because it is only for sufficiently small time steps that the equilibrium results can be recovered correctly).

9.1 The DPD equations of motion

According to Hoogerbrugge and Koelman,[1,2] DPD is described by the set of governing equations

$$\begin{aligned} \dot{\boldsymbol{r}}_i &= \frac{\boldsymbol{p}_i}{m_i} \\ \dot{\boldsymbol{p}}_i &= \sum_{j \neq i} \boldsymbol{F}_{ij}^{\mathrm{C}} + \sum_{j \neq i} \boldsymbol{F}_{ij}^{\mathrm{D}} + \sum_{j \neq i} \boldsymbol{F}_{ij}^{\mathrm{R}}, \end{aligned} \tag{9.1.1}$$

where $\boldsymbol{r}_i$ is the position vector of particle i, $\boldsymbol{p}_i$ its momentum, m_i its mass, and $\boldsymbol{F}_{ij}^{\mathrm{C}}$, $\boldsymbol{F}_{ij}^{\mathrm{D}}$, and $\boldsymbol{F}_{ij}^{\mathrm{R}}$ denote the conservative, dissipative, and random force, respectively, exerted on particle i from particle j. In Eq. (9.1.1), pair additivity has been assumed.

The force $\boldsymbol{F}_{ij}^{\mathrm{C}}$ has typically the form corresponding to the nonideal fluid studied by Groot and Warren,[10]

$$\boldsymbol{F}_{ij}^{\mathrm{C}} = \begin{Bmatrix} a_{ij}\left(1 - \dfrac{r_{ji}}{r_{\mathrm{c}}}\right)\hat{\boldsymbol{r}}_{ji}, & r_{ji} \le r_{\mathrm{c}} \\ 0, & r_{ji} > r_{\mathrm{c}} \end{Bmatrix}, \tag{9.1.2}$$

and its magnitude is controlled by the value of the parameter a_{ij} defining the maximum repulsion between particles i and j. In Eq. (9.1.2), r_{c} denotes the cutoff distance of the conservative force, $\hat{\boldsymbol{r}}_{ji} = \dfrac{\boldsymbol{r}_i - \boldsymbol{r}_j}{|\boldsymbol{r}_i - \boldsymbol{r}_j|}$ is the unit vector directed from particle j to particle i, and r_{ji} is the magnitude of vector $\boldsymbol{r}_{ji}\,(= \boldsymbol{r}_i - \boldsymbol{r}_j)$. a_{ij} should be sufficiently small to allow the two particles to overlap significantly. For beads adjacent along a polymer chain, in addition to the basic conservative force implied by Eq. (9.1.2), one should account also for chain connectivity. A choice often proposed for the spring force keeping two beads connected along the chain is[11–12]

$$\boldsymbol{F}_{ij,\mathrm{b}}^{\mathrm{C}} = -k_{ij}\left(r_{ji} - r_{\mathrm{eq}}\right)\hat{\boldsymbol{r}}_{ji}, \tag{9.1.3a}$$

where the subscript b is used to emphasize bonded interactions and r_{eq} is the equilibrium length of the bond connecting beads i and j. Another choice for the total spring force on particle i is[10]

$$\boldsymbol{F}_{i,\mathrm{b}}^{C} = -\sum_{j} C\boldsymbol{r}_{ji}, \tag{9.1.3b}$$

where the sum runs over all particles to which particle i is connected in the polymer chain, and C is the spring constant.

As far as the other two types of forces are concerned ($\boldsymbol{F}_{ij}^{\mathrm{D}}$ and $\boldsymbol{F}_{ij}^{\mathrm{R}}$), Galilean invariance requires that they should depend only on the differences $\boldsymbol{r}_{ji} = \boldsymbol{r}_i - \boldsymbol{r}_j$ and $\boldsymbol{v}_{ji} = \boldsymbol{v}_i - \boldsymbol{v}_j$ of the position vectors $\boldsymbol{r}_i$ and $\boldsymbol{r}_j$ of the two particles and their velocities $\boldsymbol{v}_i$ and $\boldsymbol{v}_j$, respectively.[3] A further requirement of isotropy demands that the forces should transform under rotations as vectors. Finally, it is required that the dissipative forces $\boldsymbol{F}_{ij}^{\mathrm{D}}$ be linear in the momentum and the random forces $\boldsymbol{F}_{ij}^{\mathrm{R}}$ be independent of the momentum. These requirements imply a Fokker–Planck equation for the temporal evolution of the distribution function $\rho(\boldsymbol{r}, \boldsymbol{p}; t)$ of the positions and momenta of all particles with a drift term which is linear in relative velocity and a diffusion term which is independent of relative velocity.[13] A simple form of the forces that satisfies these requirements is[3]

$$\boldsymbol{F}_{ij}^{\mathrm{D}} = -\gamma \omega_{\mathrm{D}}(r_{ji}) \left(\hat{\boldsymbol{r}}_{ji} \cdot \boldsymbol{v}_{ji} \right) \hat{\boldsymbol{r}}_{ji} \tag{9.1.4}$$

and

$$\boldsymbol{F}_{ij}^{\mathrm{R}} = \sigma\, \omega_{\mathrm{R}}(r_{ji}) \zeta_{ij} \hat{\boldsymbol{r}}_{ji}, \tag{9.1.5}$$

where ζ_{ij} is a white-noise term with the properties

$$\begin{aligned} \zeta_{ij}(t) &= \zeta_{ji}(t) \\ \langle \zeta_{ij}(t) \rangle &= 0 \\ \langle \zeta_{ij}(t)\, \zeta_{i'j'}(t') \rangle &= \left(\delta_{ii'} \delta_{jj'} + \delta_{ij'} \delta_{ji'} \right) \delta(t - t'). \end{aligned} \tag{9.1.6}$$

In the above expressions, the weight functions ω_{D} and ω_{R} provide the range of interaction for the dissipative and random forces, while γ and σ are the friction coefficient and amplitude of noise, respectively. In the original derivation by Hoogerbrugge and Koelman,[1,2] the following relationship was assumed between dissipative and random terms: $\omega_{\mathrm{D}}(r_{ji}) = \omega_{\mathrm{R}}(r_{ji})$. Español and Warren,[3] however, showed that independent expressions can be proposed for the weight functions $\omega_{\mathrm{D}}(r_{ji})$ and $\omega_{\mathrm{R}}(r_{ji})$, the friction constant γ, and the amplitude of the noise σ. In order to recover the equilibrium distribution of the canonical ensemble, in particular, the following relation should be satisfied:

$$\left[\omega_{\mathrm{R}}(r_{ji}) \right]^2 \sigma^2 = 2 \omega_{\mathrm{D}}(r_{ji}) \gamma k_{\mathrm{B}} T. \tag{9.1.7}$$

This is the fluctuation–dissipation theorem for the DPD method, which has exactly the same structure as the fluctuation–dissipation theorem in conventional Brownian motion. To satisfy this requirement in practical applications of the DPD method, the following set of equations is often employed:

$$\begin{aligned} \left[\omega_{\mathrm{R}}(r_{ji}) \right]^2 &= \omega_{\mathrm{D}}(r_{ji}) = \omega(r_{ji}) \\ \sigma^2 &= 2 \gamma k_{\mathrm{B}} T, \end{aligned} \tag{9.1.8}$$

with the weight function $\omega(r_{ji})$ setting the length scale of the interactions due to dissipative and random forces. A typical choice often made in actual simulations is

$$\omega(r_{ji}) = \begin{cases} \left(1 - \dfrac{r_{ji}}{r_c}\right)^2, & r_{ji} < r_c \\ 0, & r_{ji} \geq r_c \end{cases}, \tag{9.1.9}$$

implying that $\omega_R(r_{ji})$ is described by the same function of r_{ji} as the conservative force but without including the repulsion parameter a_{ij}. The dissipative forces have the following physical significance: if $(\hat{\boldsymbol{r}}_{ji} \cdot \boldsymbol{\nu}_{ji}) > 0$, then particle i is moving away from particle j and, therefore, it feels a viscous force towards j; in contrast, if $(\hat{\boldsymbol{r}}_{ji} \cdot \boldsymbol{\nu}_{ji}) < 0$, then particle i moves towards j and the viscous force is in the opposite direction. The random forces $\boldsymbol{F}_{ij}^{R}$, on the other hand, provide the continuous random displacements that keep the system in thermal motion. These are also radially directed and satisfy Newton's third law; thus, they conserve the total momentum.

The symmetry property $\zeta_{ij}(t) = \zeta_{ji}(t)$ and the fact that all forces are pair-wise ensure that the total momentum is conserved, i.e. $\dfrac{\mathrm{d}\left(\sum_i \boldsymbol{p}_i\right)}{\mathrm{d}t} = \boldsymbol{0}$.

An interesting consequence of the fluctuation–dissipation theorem, Eq. (9.1.7), for the relationship between $\omega_D(r_{ji})$ and $\omega_R(r_{ji})$ is that one can show that the canonical ensemble is a solution of the conservative DPD system. To prove this, and following Español and Warren,[3] we first substitute the expressions for the three forces into Newton's second law, second expression in Eq. (9.1.1), to obtain a set of Langevin equations which, in the form of stochastic differential equations, are written as

$$\begin{aligned} \mathrm{d}\boldsymbol{r}_i &= \frac{\boldsymbol{p}_i}{m_i}\mathrm{d}t \\ \mathrm{d}\boldsymbol{p}_i &= \left[\sum_{j\neq i} \boldsymbol{F}_{ij}^{C}(r_{ji}) + \sum_{j\neq i} -\gamma\omega_D(r_{ji})\left(\hat{\boldsymbol{r}}_{ji} \cdot \boldsymbol{\nu}_{ji}\right)\hat{\boldsymbol{r}}_{ji}\right]\mathrm{d}t + \sum_{j\neq i} \sigma\omega_R(r_{ji})\hat{\boldsymbol{r}}_{ji}\,\mathrm{d}W_{ij}, \end{aligned} \tag{9.1.10}$$

where $\mathrm{d}W_{ij} = \mathrm{d}W_{ji}$ denote independent increments of the Wiener process. By assuming the Itô interpretation which implies the Itô calculus rule (Chapter 7), namely[14,15]

$$\mathrm{d}W_{ij}\,\mathrm{d}W_{i'j'} = \left(\delta_{ii'}\delta_{jj'} + \delta_{ij'}\delta_{ji'}\right)\mathrm{d}t \tag{9.1.11}$$

(i.e. $\mathrm{d}W_{ij}(t)$ is an infinitesimal of order 1/2), a set of general model equations is then obtained with which one can execute a DPD simulation.

We consider next the Fokker–Planck equation that corresponds to the above stochastic differential equation (SDE),[10] by taking the differential df of an arbitrary function f to second order and substituting in the SDE.[5] By using Eq. (9.1.11) and the fact that $\mathrm{d}W_{ij}$ is an infinitesimal of order 1/2, we can obtain $\left\langle \frac{\mathrm{d}f}{\mathrm{d}t} \right\rangle$ and from that the Chandrasekhar equation governing the temporal evolution of the distribution function $\rho(\boldsymbol{r},\boldsymbol{p};t)$ of the positions and momenta of all particles as[14]

$$\frac{\partial}{\partial t}\rho(\boldsymbol{r},\boldsymbol{p};t) = L_{\mathrm{C}}\rho(\boldsymbol{r},\boldsymbol{p};t) + L_{\mathrm{D}}\rho(\boldsymbol{r},\boldsymbol{p};t), \tag{9.1.12}$$

where L_{C} and L_{D} are the operators

$$L_{\mathrm{C}}\rho(\boldsymbol{r},\boldsymbol{p};t) = -\left[\sum_i \frac{\boldsymbol{p}_i}{m_i}\frac{\partial}{\partial \boldsymbol{r}_i} + \sum_{i,j\neq i} \boldsymbol{F}_{ij}^{\mathrm{C}}\frac{\partial}{\partial \boldsymbol{p}_i}\right]\rho(\boldsymbol{r},\boldsymbol{p};t)$$

$$L_{\mathrm{D}}\rho(\boldsymbol{r},\boldsymbol{p};t) = \sum_{i,j\neq i} \hat{\boldsymbol{r}}_{ji}\frac{\partial}{\partial \boldsymbol{p}_i}\left[\gamma\omega_{\mathrm{D}}(r_{ji})\,(\hat{\boldsymbol{r}}_{ji}\cdot\boldsymbol{v}_{ji}) + \frac{\sigma^2}{2}\omega_{\mathrm{R}}^2(r_{ji})\hat{\boldsymbol{r}}_{ji}\cdot\left(\frac{\partial}{\partial \boldsymbol{p}_i} - \frac{\partial}{\partial \boldsymbol{p}_j}\right)\right]\rho(\boldsymbol{r},\boldsymbol{p};t). \tag{9.1.13}$$

L_{C} is the usual Liouville operator of a Hamiltonian system interacting with conservative forces $\boldsymbol{F}_{ij}^{\mathrm{C}}$, while L_{D} contains second derivatives and takes into account the effects of the dissipative $\boldsymbol{F}_{ij}^{\mathrm{D}}$ and random $\boldsymbol{F}_{ij}^{\mathrm{R}}$ forces. Note that the diffusion tensors accompanying the second derivatives with respect to the momentum do not depend on the momenta of the particles. This implies that the Itô and the Stratonovich interpretations (Chapter 7) provide exactly the same answers.[14]

The steady-state solution of Eqs. (9.1.12) and (9.1.13), $\rho^{\mathrm{eq}}(\boldsymbol{r},\boldsymbol{p})$, is obtained by requiring that $\frac{\partial}{\partial t}\rho(\boldsymbol{r},\boldsymbol{p};t) = 0$. And then we should consider the conditions under which $\rho^{\mathrm{eq}}(\boldsymbol{r},\boldsymbol{p})$ is the Gibbs–Boltzmann distribution of the canonical ensemble, i.e.

$$\rho^{\mathrm{eq}}(\boldsymbol{r},\boldsymbol{p}) = \frac{1}{Q}\exp\left[-\frac{\mathcal{H}(\boldsymbol{r},\boldsymbol{p})}{k_{\mathrm{B}}T}\right] = \frac{1}{Q}\exp\left[-\frac{\sum_i \frac{\boldsymbol{p}_i^2}{2m_i} + \mathcal{V}(\boldsymbol{r})}{k_{\mathrm{B}}T}\right]. \tag{9.1.14}$$

where $\mathcal{H}$ denotes the Hamiltonian of the system, $\mathcal{V}$ the potential energy function (more generally: potential of mean force) that gives rise to the conservative forces $\boldsymbol{F}_{ij}^{\mathrm{C}}$, k_{B} Boltzmann's constant, T the equilibrium temperature, and Q the partition function. We easily see that the canonical ensemble is the equilibrium solution for the conservative system, since we immediately recognize that $L_{\mathrm{C}}\rho^{\mathrm{eq}}(\boldsymbol{r},\boldsymbol{p}) = 0$. On the other hand, we can satisfy $L_{\mathrm{D}}\rho^{\mathrm{eq}}(\boldsymbol{r},\boldsymbol{p}) = 0$ by requiring that $[\omega_{\mathrm{R}}(r_{ji})]^2\sigma^2 = 2\omega_{\mathrm{D}}(r_{ji})\gamma k_B T$, which is exactly Eq. (9.1.7). With the choice $[\omega_{\mathrm{R}}(r_{ji})]^2 = \omega_{\mathrm{D}}(r_{ji})$, for example, the DPD system at temperature $k_{\mathrm{B}}T = \frac{\sigma^2}{2\gamma}$ has as an equilibrium solution the same as the conservative system (the system described by Eq. (9.1.10) with $\gamma = \sigma = 0$).

The Chandrasekhar equation, Eq. (9.1.12), has been the starting point for a formal derivation, for example by means of projection operators[16,17] (see Section 7.7), of the hydrodynamic equations that govern the mass and momentum density fields, in the same way as the

Liouville equation is the starting point in classical statistical mechanics.[18] Indeed, using the technique of projection operators, Español[19] derived the hydrodynamic equations of the mass and momentum density fields. He neglected the energy density field as a proper hydrodynamic variable because this is not conserved in DPD; its balance equation has source and dissipative terms that cannot be expressed as the divergence of a flux. Therefore, the energy is a rapidly varying function for which a proper hydrodynamic equation does not exist. On the other hand, the transport equations he found for the mass and momentum density fields have the form of the continuity and Navier–Stokes equations (as expected from the Galilean invariance and isotropy of the microscopic equations together with the conservation of mass and momentum).

9.2 DPD as a mesoscopic simulation method—parameterization

To carry out DPD simulations for actual systems, one must make specific choices for the parameters of the model, especially the noise amplitude σ (and thus also the friction factor γ), the repulsion parameters a_{ij}, and the time step Δt of the integration. Such a parameterisation is not a trivial task and remains an open issue even nowadays. A significant step forward was made by Groot and Warren,[10] who proposed matching the compressibility in the DPD model to the compressibility of the liquid under study; from this, the repulsion parameters between fluid particles in the soft model liquid can be fixed. In two-component systems, Groot and Warren[10] also made a link between these parameters and the χ parameter in the Flory–Huggins model,[20,21] since DPD can be viewed as an off-lattice simulation method for Flory–Huggins-type models. We briefly review these considerations in the following.

9.2.1 Pure fluid

For such a system, there is only one DPD interaction parameter, namely $a_{ii} = a$ for all i and j. To determine the repulsion parameter a defining the conservative force, Groot and Warren[10] required that the DPD soft sphere model should describe correctly the density fluctuations in the system, as determined by the compressibility. Thus, analogously to the Weeks–Chandler–Andersen perturbation theory of liquids,[18] the model should be chosen such that the dimensionless isothermal compressibility, defined through

$$\kappa^{-1} = \frac{m}{\rho k_B T \kappa_T} = \frac{m}{k_B T}\left(\frac{\partial p}{\partial \rho}\right)_T = \frac{1}{k_B T}\left(\frac{\partial p}{\partial n}\right)_T, \qquad (9.2.1)$$

has the correct value. In Eq. (9.2.1), m is the mass of a particle, $n = \rho/m$ is the number density of particles, and κ_T is the usual (dimensional) isothermal compressibility. For water at room temperature (300 K), $\kappa_{\text{water}}^{-1} = 15.9835$.

To find this correspondence, an equation of state should be established. By making use of the virial theorem, the pressure tensor in the DPD simulations can be obtained as[10]

$$
\begin{aligned}
p &= n k_{\mathrm{B}} T + \frac{1}{3V}\left\langle \sum_{j>i} (\boldsymbol{r}_i - \boldsymbol{r}_j)\cdot \boldsymbol{F}_i \right\rangle \\
&= n k_{\mathrm{B}} T + \frac{1}{3V}\left\langle \sum_{j>i} (\boldsymbol{r}_i - \boldsymbol{r}_j)\cdot \boldsymbol{F}_{ij}^{\mathrm{C}} \right\rangle \\
&= n k_{\mathrm{B}} T + \frac{1}{6V}\sum_{i\neq j} \left\langle \boldsymbol{r}_{ji}\cdot \boldsymbol{F}_{ij}^{\mathrm{C}}(\boldsymbol{r}_{ji}) \right\rangle \\
&= n k_{\mathrm{B}} T + \frac{1}{6V}\sum_{i\neq j} \int \mathrm{d}\boldsymbol{r} \left\langle \delta(\boldsymbol{r} - \boldsymbol{r}_{ji})\right\rangle \boldsymbol{r}_{ji}\cdot \boldsymbol{F}_{ij}^{\mathrm{C}}(\boldsymbol{r}_{ji}) \\
&= n k_{\mathrm{B}} T + \frac{2\pi}{3} n^2 \int_0^{r_c} r F^{\mathrm{C}}(r)\, g(r)\, r^2 \mathrm{d}r, \qquad (9.2.2)
\end{aligned}
$$

where $g(r)$ denotes the radial pair distribution function defined by

$$
g(r) = \frac{V}{N_{\mathrm{b}}^2} \sum_{i\neq j} \langle \delta(\boldsymbol{r} - \boldsymbol{r}_{ji})\rangle, \qquad (9.2.3)
$$

with N_{b} being the total number of interacting DPD beads.

In the following, we will work mostly with dimensionless numbers. To this end, and following Groot and Warren,[10] let us use r_c, m, and $k_{\mathrm{B}}T$ as the units for length, mass, and energy, respectively. Then, velocities are made dimensionless by dividing with $(k_{\mathrm{B}}T/m)^{\frac{1}{2}}$, times by dividing with $r_c(m/k_{\mathrm{B}}T)^{\frac{1}{2}}$, pressures by dividing with $(k_{\mathrm{B}}T/r_c^3)$, number densities by dividing by $(1/r_c^3)$, and mass densities by dividing by (m/r_c^3). Moreover, the parameters a_{ij}, γ, and σ appearing in the DPD equations of motion are made dimensionless by dividing them with $(k_{\mathrm{B}}T/r_c)$, $(mk_{\mathrm{B}}T)^{\frac{1}{2}}/r_c$, and $(mk_{\mathrm{B}}^3T^3)^{\frac{1}{4}}/r_c^{\frac{1}{2}}$, respectively. Also, following Maiti and McGrother,[22] we will use an overbar to distinguish dimensionless from the corresponding dimensional quantities.

Through test DPD simulations where the density was varied from $\bar{\rho} = 1$ to $\bar{\rho} = 8$ in steps of 0.5 using $\bar{a}_{ij} = \bar{a} = 15$ for the repulsion parameter and by studying some less extensive density variations for $\bar{a} = 25$ and $\bar{a} = 30$, Groot and Warren[10] came up with the following equation of state that holds for sufficiently high density ($\bar{\rho} > 2$) (Problem 9.1):

$$
p = n k_{\mathrm{B}} T + a c n^2 \left(c/r_c^4 = 0.101 \pm 0.001\right) \qquad (9.2.4)
$$

or, in dimensionless form,[22]

$$
\bar{p} = \bar{\rho} + \bar{c}\bar{a}\bar{\rho}^2 \qquad (9.2.5a)
$$

or, equivalently, in terms of the dimensionless number density,

$$\bar{p} = \bar{n} + \bar{c}\,\bar{a}\,\bar{n}^2. \tag{9.2.5b}$$

In Eqs. (9.2.5), $\bar{c} = c/r_c^4$ is a constant given in terms of the pair distribution function as[22]

$$\bar{c} = \frac{2\pi}{3}\int_0^1 \bar{r}^3\,(1-\bar{r})\,g(\bar{r})\,d\bar{r}, \tag{9.2.6}$$

and $g(\bar{r})$, the pair distribution function, is expressed as a function of $\bar{r}$. In general, $g(\bar{r})$, and therefore $\bar{c}$, is a function of the density $\bar{\rho}$ as well as of the interaction parameter $\bar{a}$. However, the DPD simulations of Groot and Warren[10] showed that $\bar{c} \approx 0.1$ varies very slowly with both $\bar{\rho}$ and $\bar{a}$ within the ranges of interest ($3 \leq \bar{\rho} \leq 5$ and $\bar{a} \geq 15$); thus, practically, $\bar{c} \approx 0.1$ is independent of $\bar{\rho}$ and $\bar{a}$ for sufficiently high densities (practically, $\bar{\rho} > 2$).

Based on the equation of state, Eq. (9.2.4), the Helmholtz energy density a_V in a single-component DPD fluid can be shown (Problem 9.2) to be

$$\frac{a_V}{k_B T} = n\ln\left(n\,\Lambda^3\right) - n + \frac{a\,c}{k_B T}n^2, \tag{9.2.7}$$

with Λ being the thermal wavelength of a DPD particle.

From the equation of state, Eq. (9.2.4), the dimensionless compressibility introduced in Eq. (9.2.1) is given by

$$\kappa^{-1} = 1 + 2a\,c\frac{n}{k_B T} \tag{9.2.8a}$$

or, given that $\bar{c} = c/r_c^4 \approx 0.1$,

$$\kappa^{-1} \approx 1 + 0.2\,a\,r_c^4\frac{n}{k_B T}. \tag{9.2.8b}$$

Combining this with the known compressibility of water, $\kappa_{\text{water}}^{-1} \approx 16$, Groot and Warren[10] estimated that $a\,r_c^4\dfrac{n}{k_B T} = \bar{a}\,\bar{n} \approx 75$. Thus, the repulsion parameter for water should be chosen as

$$a \approx 75\frac{k_B T}{n\,r_c^4}. \tag{9.2.9}$$

In principle, the density chosen for the DPD simulation is a free parameter, but since the number of interactions for each particle increases linearly with the density, the required CPU time per time step and per unit of volume will increase with the square of the density. For efficiency reasons, one would therefore choose the lowest possible density for which the scaling relation still holds. Overall, Groot and Warren[10] found that a reasonable choice for the dimensionless density is $\bar{n}\,(=\bar{\rho}) = 3$ (and, as mentioned earlier, to have the correct compressibility, the repulsion parameter should be chosen as $a = 75\dfrac{k_B T}{n\,r_c^4}$).

9.2.2 Two-component system: mapping onto the Flory–Huggins model

According to the Flory–Huggins lattice model,[20] the Helmholtz energy of mixing per lattice site of a polymer solution is given by

$$\frac{\Delta_{\text{mix}}A}{N_{\text{b}}k_{\text{B}}T} = \frac{\phi_A}{N_A}\ln\phi_A + \frac{\phi_B}{N_B}\ln\phi_B + \chi_{AB}\phi_A\phi_B \tag{9.2.10}$$

where ϕ_A and ϕ_B are the volume fractions of components A and B, N_A and N_B denote the numbers of segments per A or B molecule, and χ_{AB} is a dimensionless interchange energy between A and B segments per site.[20,21] Assuming that the lattice is completely filled, we have $\phi_A + \phi_B = 1$ and therefore $\phi_B = 1 - \phi_A$, leaving ϕ_A as the only degree of freedom.

We consider first a mixture of monomeric A and B particles (beads), i.e. $N_A = N_B = 1$.[22] Let the total number of A-type and B-type beads be N_{b}^A and N_{b}^B, respectively, in a total volume V. The total number of beads is $N_{\text{b}} = N_{\text{b}}^A + N_{\text{b}}^B$. Due to the assumption of full lattice coverage, N_{b} is also equal to the total number of sites constituting the lattice. We assume the same interaction cutoff r_{c}, independent of A and B. Therefore, the average number of A and B beads within a cube of side r_{c} is

$$\bar{n}_A = \frac{N_{\text{b}}^A r_{\text{c}}^3}{V} \text{ and } \bar{n}_B = \frac{N_{\text{b}}^B r_{\text{c}}^3}{V} \tag{9.2.11}$$

If $x = \dfrac{N_{\text{b}}^A}{N_{\text{b}}}$ is the volume fraction of component A and $1 - x = \dfrac{N_{\text{b}}^B}{N_{\text{b}}}$ the volume fraction of component B, then $\bar{n}_A = x\,\bar{n}$ and $\bar{n}_B = (1-x)\,\bar{n}$, where $n = \dfrac{N_{\text{b}}^A + N_{\text{b}}^B}{V}$ is the total number density of beads, $\bar{n}$ ($= \bar{\rho}$ if equal masses m for the particles of the two components are assumed) is the normalized total mass density or total number density of beads in the volume r_{c}^3, $n_A = \dfrac{N_{\text{b}}^A}{V}$ is the number density of beads of component A, and $n_B = \dfrac{N_{\text{b}}^B}{V}$ is the number density of beads of component B. Also, for the volume fractions of the two components we can write $\phi_A = x$ and $\phi_B = 1 - x$.

Then, following the same methodology as for the one-component system (Problem 9.2) and assuming the same interaction cutoff r_{c} for $\bar{a}_{AA}$, $\bar{a}_{AB}$, and $\bar{a}_{BB}$, we can find the following expression for the virial (non-ideal gas) contribution to the free-energy density:

$$a_{V,\text{virial}} = \frac{A_{\text{virial}}}{V} = \frac{\bar{c}\,k_{\text{B}}T\bar{n}^2}{r_{\text{c}}^3}\left[\bar{a}_{AA}x^2 + 2\bar{a}_{AB}x\,(1-x) + \bar{a}_{BB}(1-x)^2\right]. \tag{9.2.12}$$

Since this expression must be equivalent to Eq. (9.2.10), we conclude that the virial contribution to the total free energy ($A = Va_V$) is given by (Problem 9.3)

$$A_{\text{virial}}\,(N_{\text{b}}, x) = N_{\text{b}}\bar{c}\,\bar{n}\,k_{\text{B}}T\left[\bar{a}_{AA}x^2 + 2\bar{a}_{AB}x\,(1-x) + \bar{a}_{BB}(1-x)^2\right], \tag{9.2.13}$$

where $N_{\mathrm{b}} = \bar{n}\frac{V}{r_{\mathrm{c}}^3}$. To map onto the Flory–Huggins theory, we need the virial contribution to the free energy of mixing, namely,

$$\Delta_{\mathrm{mix}} A_{\mathrm{virial}}\left(N_{\mathrm{b}}, x\right) = A_{\mathrm{virial}}\left(N_{\mathrm{b}}, x\right) - x A_{\mathrm{virial}}\left(N_{\mathrm{b}}, 1\right) - (1-x) A_{\mathrm{virial}}\left(N_{\mathrm{b}}, 0\right). \tag{9.2.14}$$

With the help of Eq. (9.2.13), the result is

$$\Delta_{\mathrm{mix}} A_{\mathrm{virial}}\left(N_{\mathrm{b}}, x\right) = 2 N_{\mathrm{b}} \bar{c}\,\bar{n}\, k_B T \left[\bar{a}_{AB} - \frac{\bar{a}_{AA} + \bar{a}_{BB}}{2}\right] x\,(1-x). \tag{9.2.15}$$

By comparing with the interaction term in the Flory–Huggins theory, Eq. (9.2.10), according to which

$$\Delta_{\mathrm{mix}} A_{\mathrm{virial}}\left(N_{\mathrm{b}}, x\right) = N_{\mathrm{b}} k_{\mathrm{B}} T \chi_{AB}\, x\,(1-x), \tag{9.2.16}$$

we readily identify

$$\chi_{AB} = 2\bar{c}\,\bar{n}\left[\bar{a}_{AB} - \frac{\bar{a}_{AA} + \bar{a}_{BB}}{2}\right]. \tag{9.2.17}$$

This expression is fully consistent with the original definition of the χ_{AB} factor as a dimensionless interchange energy per bead,[20,21] which compares interactions between like and unlike beads.

The Flory–Huggins theory assumes that all beads have the same equilibrium volume. Together with the fact that the pairwise DPD interactions between all bead-species have the same cutoff r_{c}, this implies that $\bar{n}_A = \bar{n}_B$ for all pure components A and B.[22] A more interesting case arises when the two species A and B have a sufficiently large Flory–Huggins χ_{AB} parameter to drive segregation under equilibrium conditions. If mechanical equilibrium is assumed at the interface, then the pressure in the phases of the two pure species must be equal.[11] Using Eq. (9.2.4) for the pressure and that r_{c} and $\bar{n}$ are the same on the two sides of the interface, we find that

$$\bar{a}_{AA} = \bar{a}_{BB}, \tag{9.2.18}$$

i.e. intraspecies interactions are equal.[12] With the above result, Eq. (9.2.17) simplifies to

$$\chi_{AB} = 2\bar{c}\,\bar{n}\left[\bar{a}_{AB} - \bar{a}_{AA}\right]. \tag{9.2.19}$$

Then, by choosing $a_{AA} = a_{BB}$ and assuming that $n_A + n_B = n =$ const., one finds from Eq. (9.2.10) that

$$\frac{\Delta_{\mathrm{mix}} a_V}{\left(n_A + n_B\right) k_{\mathrm{B}} T} \approx \frac{x}{N_A}\ln x + \frac{1-x}{N_B}\ln(1-x) + \frac{2c\left(a_{AB} - a_{AA}\right)\left(n_A + n_B\right)}{k_{\mathrm{B}} T} x\,(1-x) \tag{9.2.20}$$

where $x = \dfrac{n_A}{n_A + n_B}$, and thus,

$$\chi_{AB} = \frac{2c\,(a_{AB} - a_{AA})\,(n_A + n_B)}{k_B T}. \tag{9.2.21}$$

That is, one has the correspondence $\dfrac{\Delta_{\rm mix} a_V\,(\rm DPD)}{(n_A + n_B)}$ $\left(= \dfrac{\Delta_{\rm mix} A_V\,(\rm DPD)}{(N_b^A + N_b^B)} \right) = \dfrac{\Delta_{\rm mix} A\,(\text{Flory-Huggins})}{N_b}$, with the χ_{AB}-parameter mapping given by Eq. (9.2.19). The validity of Eq. (9.2.21) has been checked by simulations of binary mixtures of both monomers and polymers.[10]

9.2.3 Two-component system: interaction parameters in terms of the solubility parameters

By using r_c as the basic length-scale of the system, we have seen that the dimensionless bead density $\bar{\rho}$ is defined through $\bar{n} = n\,r_c^3$ or, alternatively, through $\bar{n} = \dfrac{1}{\upsilon_b} r_c^3$, where υ_b is the volume of a bead. For a heterogeneous system consisting of several different species, the corresponding basic assumption is that all bead-types (each representing a single species) are of the same volume υ_b. This has been an important assumption in order to map DPD onto the Flory–Huggins chi-parameter theory. Maiti and McGrother[22] have further noted that the derivation of the previous equations on the mapping of the DPD model to the Flory–Huggins model assumes that the spatial variation of the volume fraction x is small, implying a quasi-homogeneous mixture (this is the case for small values of the χ_{AB}-parameter,[13] $\chi_{AB} < \chi_{AB}^{\rm critical}$). For a segregated mixture with a sharp interface, such an approximation breaks down. In such a case, Maiti and McGrother[22] propose computing χ_{AB} from solubility parameters using the formula[23,24]

$$\chi_{AB} = \frac{\upsilon_b}{k_B T}(\delta_A - \delta_B)^2, \tag{9.2.22}$$

where δ_A and δ_B are the solubility parameters of systems A and B, respectively, and υ_b the volume of a bead (all beads having the same average volume $r_c^3/\bar{n}$). All DPD interaction parameters can then be determined once the value of $\bar{a}_{AA}$ for the specific system has been calibrated. If water is chosen as such a system, one can equate the dimensionless isothermal density derivative of pressure to the inverse (dimensionless) isothermal compressibility κ^{-1}; see Eq. (9.2.1). As already mentioned, for water at room temperature, $\kappa_{\rm water}^{-1} \approx 16$ and $\bar{c} \approx 0.1$; also, for typically used values of $\bar{n} = 3$ and $\bar{n} = 5$, we have $\bar{a}_{AA} \approx 25$ and $\bar{a}_{AA} \approx 15$, respectively. Having determined $\bar{a}_{AA}$ and χ_{AB}, the cross-species interactions $\bar{a}_{AB}$ can be estimated by using Eq. (9.2.19) for $\chi_{AB} < \chi_{AB}^{\rm critical}$[13] or by

$$\begin{aligned} \chi_{AB} &= (0.286 \pm 0.002)\,(\bar{a}_{AB} - \bar{a}_{AA})\ (\text{for } \bar{n} = 3) \\ \chi_{AB} &= (0.689 \pm 0.002)\,(\bar{a}_{AB} - \bar{a}_{AA})\ (\text{for } \bar{n} = 5) \end{aligned} \tag{9.2.23}$$

for $\chi_{AB}^{\rm critical} < \chi_{AB}$ as suggested by the linear fits derived by Groot and Warren.[10]

9.2.4 Two-component system: mapping onto the Scatchard–Hildebrand regular solution theory

Travis et al.[25] have noted that Eq. (9.2.18), which suggests equal like–like interactions, is justified only in a few special cases and does not generally give rise to the expected phase behaviour. In particular, Travis et al.[25] have suggested deriving the magnitude of the conservative parameter from a general expression for the free energy of mixing of two DPD fluids starting from the equation of state. We illustrate the method here by restricting ourselves to binary systems, but generalization to more than two components is rather straightforward. We start with an expression for the pressure of a binary mixture of fluid components in terms of partial radial distribution functions (through a straightforward generalization of the virial equation) as follows from classical statistical mechanics of systems with pair-wise additive potentials,[25]

$$p = n k_B T - \frac{2\pi}{3} n^2 \left[x^2 \int_0^\infty g_{11}(r) \frac{du_{11}(r)}{dr} r^3 dr + 2x(1-x) \int_0^\infty g_{12}(r) \frac{du_{11}(r)}{dr} r^3 dr + (1-x)^2 \int_0^\infty g_{22}(r) \frac{du_{22}(r)}{dr} r^3 dr \right], \tag{9.2.24}$$

where x is the composition variable, g_{ij} the partial radial distribution functions, n the total number density of the mixture, and $u_{ij}(r)$ the pair potential between species i and j. We introduce the DPD conservative force expression (employing a linear weight function) supplemented by the requirement that all species have the same interaction range r_c, and we define the interaction parameters (compare Eq. (9.2.6))

$$\bar{c}_{ij} = \frac{2\pi}{3} \int_0^1 \bar{r}^3 (1-\bar{r}) g_{ij}(\bar{r}) d\bar{r}, \tag{9.2.25}$$

where $\bar{r} = \frac{r}{r_c}$. We also assume that (to a good level of approximation) the a's are independent of the total density[22] (by virtue of removing the dependence on r_c) and furthermore that $\bar{c}_{11} \approx \bar{c}_{12} \approx \bar{c}_{22} = \bar{c}$ (this last identity is not as serious an assumption as might be thought; it could easily be relaxed if the $\bar{c}_{ij}$'s were absorbed into the definition of the a_{ij} parameters). We then find that

$$\begin{aligned} p &= n k_B T + r_c^4 n^2 \bar{c} \left[x^2 a_{11} + 2x(1-x) a_{12} + (1-x)^2 a_{22} \right] \\ &= n k_B T + p_{\text{virial}} \end{aligned} \tag{9.2.26}$$

where

$$p_{\text{virial}} = r_c^4 n^2 \bar{c} \left[x^2 a_{11} + 2x(1-x) a_{12} + (1-x)^2 a_{22} \right] \tag{9.2.27}$$

is the virial contribution to the pressure. By integrating the equation of state, we can derive then the Helmholtz energy of the binary mixture through the following formula of thermodynamics[21,26]

$$A = \int_V^{\infty} \left[p - \frac{n_T RT}{V} \right] dV - RT \sum_i n_i \ln \left(\frac{V}{n_i RT} \right) + \sum_i n_i \left(u_i^0 - Ts_i^0 \right), \tag{9.2.28}$$

where the '0' superscripts refer to a pure ideal gas reference state. In Eq. (9.2.28), n_T is the total number of moles, V the volume of the system, n_i the number of moles of species i, and s and u stand for the molar entropy and molar internal energy, respectively. The total number of moles n_T is related to the total number density n through $n_T = \frac{nV}{N_{Avo}}$, where N_{Avo} stands for Avogadro's number. From Eqs. (9.2.26)–(9.2.28), and by carrying out the integrations, we obtain the following expression for the Helmholtz energy per mole of the mixture

$$\begin{aligned} \frac{A}{n_T} &= \frac{r_c^4 n_T N_{Avo}^2 \bar{c}}{V} \left[x^2 a_{11} + 2x(1-x) a_{12} + (1-x)^2 a_{22} \right] \\ &\quad - xRT \ln \left(\frac{V}{n_1 RT} \right) - (1-x) RT \ln \left(\frac{V}{n_2 RT} \right) + x \left(u_1^0 - Ts_1^0 \right) + (1-x) \left(u_2^0 - Ts_2^0 \right). \end{aligned} \tag{9.2.29}$$

From this last expression, we can also obtain the corresponding Helmholtz energy of mixing $\Delta_{mix} A$ by just subtracting the free energies of the two pure components from that of the mixture,

$$\frac{\Delta_{mix} A}{n_T} = \frac{A}{n_T} - x \frac{A_1}{n_1} - (1-x) \frac{A_2}{n_2}, \tag{9.2.30}$$

where A_1 and A_2 are the respective Helmholtz energies of pure components 1 and 2:

$$\frac{A_i}{n_i} = \frac{r_c^4 n_i N_{Avo}^2 \bar{c} a_{ii}}{V_i} - RT \ln \left(\frac{V_i}{n_i RT} \right) + \left(u_i^0 - Ts_i^0 \right), \quad i = 1, 2. \tag{9.2.31}$$

The final result is

$$\begin{aligned} \frac{\Delta_{mix} A}{n_T} &= \frac{r_c^4 N_{Avo}^2 \bar{c}}{\nu_T} \left[x^2 a_{11} + 2x(1-x) a_{12} + (1-x)^2 a_{22} \right] - x \frac{r_c^4 N_{Avo}^2 \bar{c} a_{11}}{\nu_1} - (1-x) \frac{r_c^4 N_{Avo}^2 \bar{c} a_{22}}{\nu_2} \\ &\quad + xRT \ln \left(\frac{x \nu_1}{\nu_T} \right) + (1-x) RT \ln \left(\frac{(1-x) \nu_2}{\nu_T} \right) \end{aligned} \tag{9.2.32}$$

in which we have introduced the molar volumes $\nu_i \left(= \frac{V_i}{n_i} \right)$ in place of the total volumes of the pure components as well as the molar volume $\nu_T \left(= \frac{V}{n_1 + n_2} = \frac{V}{n_T} \right)$ in place of the total system volume. The last two terms on the right-hand side of Eq. (9.2.32), having an explicit temperature dependence, may be regarded as the entropic contribution to the Helmholtz energy of mixing.

By introducing the volume fraction $\varphi_1 = \dfrac{V_1}{V} = \dfrac{x_1\nu_1}{\nu_T}\ (=\varphi)$ and by assuming that (as in regular solution theory) there is no volume change upon mixing (partial molar volumes are equal to the molar volumes), from Eq. (9.2.32) we get the following expression for the entropy of mixing per mole of the mixture:

$$\frac{\Delta_{\text{mix}} S}{n_{\text{T}} R} = -x \ln \varphi - (1-x) \ln (1-\varphi) \tag{9.2.33}$$

This is very similar to the expression for the ideal entropy of mixing, becoming exactly so when the volume fractions and mole fractions coincide, which happens for components which have equal molar volumes.

The remaining contribution to the free energy of mixing in Eq. (9.2.32) is the so-called internal energy of mixing. After some algebraic re-arrangements, this comes out to be

$$\frac{\Delta_{\text{mix}} U}{n_{\text{T}}} = -r_{\text{c}}^4\, \nu_{\text{T}}\, N_{\text{Avo}}^2\, \bar{c} \left[\frac{a_{11}}{\nu_1^2} - 2\frac{a_{12}}{\nu_1 \nu_2} + \frac{a_{22}}{\nu_2^2} \right] \varphi (1-\varphi) . \tag{9.2.34}$$

Equation (9.2.34) has almost the same form as the free energy of mixing derived by Scatchard[27] and Hildebrand and Wood[28] for systems described as regular mixtures by Hildebrand and Scott,[23] namely for mixtures for which the excess entropy S_{exc} and excess volume V_{exc} of mixing are both zero. The following equation holds for the internal energy of mixing per mole in regular solution theory (RST),[27,28]

$$\frac{\Delta_{\text{mix}} U}{n_{\text{T}}} = \nu_{\text{T}} \left(e_{\text{c},11} - 2 e_{\text{c},12} + e_{\text{c},22} \right) \varphi (1-\varphi) \equiv \frac{G_{\text{exc}}^{\text{RST}}}{n_{\text{T}}}, \tag{9.2.35}$$

where the last identity follows from $S_{\text{exc}} = V_{\text{exc}} = 0$, and the parameters $e_{\text{c},11}$ and $e_{\text{c},22}$ are the cohesive energy densities of the two pure liquids defined in general as

$$e_{\text{c}} \equiv \frac{\Delta_{\text{vap}} u}{\nu^{\text{L}}}, \tag{9.2.36}$$

in which $\Delta_{\text{vap}} u$ is the molar internal energy of vaporization and ν^{L} the molar volume of the liquid.

From Eqs. (9.2.34) and (9.2.35), it is advantageous therefore to make a correspondence between DPD and regular solution theory, since there is a direct link between the conservative interaction parameters a_{11} and a_{22} and the two cohesive energy densities $e_{\text{c},11}$ and $e_{\text{c},22}$. As far as the cross interaction is concerned, in the Hildebrand theory the corresponding cohesive energy density is given as the geometric mean of the pure component cohesive energy densities,

$$e_{\text{c},12} = \sqrt{e_{\text{c},11} e_{\text{c},22}}, \tag{9.2.37}$$

which has some physical justification for nonpolar molecules based on London's dispersion force expression, but is only approximately true. The use of Eq. (9.2.37) yields the following expression for the excess free energy of mixing per mole in RST:

$$g_{\text{exc}}^{\text{RST}} = \nu_{\text{T}} (\delta_1 - \delta_2)^2 \varphi (1-\varphi), \tag{9.2.38}$$

in which δ_1 and δ_2 are now the square roots of the pure component cohesive energy densities, or the Hildebrand solubility parameters. Comparing Eq. (9.2.38) with Eq. (9.2.34) gives the following mapping between the DPD parameters and cohesive energy densities:

$$(\delta_1 - \delta_2)^2 = -r_c^4 \, N_{\text{Avo}}^2 \, \bar{c} \left[\frac{a_{11}}{\nu_1^2} - 2\frac{a_{12}}{\nu_1 \nu_2} + \frac{a_{22}}{\nu_2^2} \right] \tag{9.2.39a}$$

or, equivalently,

$$(\delta_1 - \delta_2)^2 = -r_c^4 \, \bar{c} \left[\left(\frac{N_{\text{Avo}} \, n_1}{V_1} \right)^2 a_{11} - 2 \left(\frac{N_{\text{Avo}} \, n_1}{V_1} \right) \left(\frac{N_{\text{Avo}} \, n_2}{V_2} \right) a_{12} + \left(\frac{N_{\text{Avo}} \, n_2}{V_2} \right)^2 a_{22} \right], \tag{9.2.39b}$$

where $\frac{N_{\text{Avo}} \, n_1}{V_1}$ denotes the number density of pure component 1 and $\frac{N_{\text{Avo}} \, n_2}{V_2}$ denotes the number density of pure component 2 (recall that earlier in this section, Section 9.2.4, we denoted the number of moles of components 1 and 2 by n_1 and n_2, respectively). Given the correspondence between a_{11} and δ_1, and between a_{22} and δ_2, Eqs. (9.2.39) can be thought of as a definition of a_{12}, further suggesting that a_{12} should not be taken to be equal to the geometric mean of a_{11} and a_{22}.

9.2.5 A general methodology for deriving conservative forces

In order to elucidate the generic shape of consistent expressions for the conservative forces, first Pagonabarraga and Frenkel[29,30] and then Trofimov et al.[31] considered that, in general, the conservative force acting on particle i can be rewritten as

$$\boldsymbol{F}_i^{\text{C}} = -\sum_{j=1}^{N} \frac{\partial \psi_{\text{exc}} \left(n_j \left(\{ \boldsymbol{R} \} \right) \right)}{\partial \boldsymbol{R}_i}, \tag{9.2.40}$$

where $\psi_{\text{exc}} \left(n_j \left(\{ \boldsymbol{R} \} \right) \right)$ is the excess Helmholtz energy per particle and $n_j \left(\{ \boldsymbol{R} \} \right)$ the number density distribution of particle j throughout the system. To proceed, we can invoke a local density approximation to replace $n_j \left(\{ \boldsymbol{R} \} \right)$ by $\bar{n}_j \left(\boldsymbol{R} \right)$, i.e. by the local density of particle j at the corresponding $\boldsymbol{R}$ but averaged over a small fluid volume around its position $\boldsymbol{R}$. ψ_{exc} defines the excess free energy of the system through

$$A_{\text{exc}} = \sum_{i=1}^{N} \psi_{\text{exc}} \left(\bar{n}_i \right). \tag{9.2.41}$$

For the local density of particles i, it can be assumed that

$$\bar{n}_i = \sum_{k \neq i} w(r_{ik}), \tag{9.2.42}$$

where $w(r)$ is a weight function vanishing beyond r_c and normalized so that $\int_0^\infty 4\pi r^2 w(r)\,dr = 1$. Equation (9.2.42) implies that each DPD particle has a sphere of influence expressed by its weight function, and the local-density approximation at a point is a sum of contributions from nearby particles. A DPD method in which the definition of the conservative forces acting between the DPD particles are assumed to depend not on the interparticle separation but on the instantaneous local particle density (which, in turn, depends on the positions of all neighbouring particles) is called multibody DPD (M-DPD), since the conservative force is a many-body force. From Eq. (9.2.40), one obtains the following expression for the conservative force in M-DPD:

$$\begin{aligned}\boldsymbol{F}_i^C &= -\sum_j \frac{\partial \psi_{\text{exc}}(\bar{n}_j)}{\partial \boldsymbol{r}_i} = -\sum_j \frac{\partial \psi_{\text{exc}}(n_j(\{\boldsymbol{r}\}))}{\partial \boldsymbol{r}_i} \\ &= -\sum_{j\neq i} \left(\psi'_{\text{exc}}(\bar{n}_i) + \psi'_{\text{exc}}(\bar{n}_j)\right) w'(r_{ij})\,\hat{\boldsymbol{r}}_{ji} \\ &= \sum_{j\neq i} \boldsymbol{F}_{ij}^{\text{C}},\end{aligned} \tag{9.2.43}$$

where

$$\boldsymbol{F}_{ij}^{\text{C}} = -\left[\psi'_{\text{exc}}(\bar{n}_i) + \psi'_{\text{exc}}(\bar{n}_j)\right] w'(r_{ij})\,\hat{\boldsymbol{r}}_{ji} \tag{9.2.44}$$

and the primes denote derivatives with respect to the corresponding function arguments. Using these equations, one can easily calculate the conservative forces acting on the particles for a prescribed free energy. These forces are effectively many-body forces but are still represented as a sum of pair-wise components. Moreover, as noted by Pagonabarraga and Frenkel,[30] the fact that these forces depend on the positions of all particles in the system (through their corresponding local weighted densities) implies a smoother local structure for the fluid than when these forces are derived from a pair potential. This is a very attractive feature of the M-DPD model, since the local structure of the fluid (which is intimately related to its microscopic structure) should be smeared out at a mesoscopic, coarse-grained scale.

The conservative forces in M-DPD depend not only on the free energy but also on the form of the weight function $w(r)$, which is not specified by the method. As noted by Trofimov et al.,[31] there are no a priori arguments for the selection of a particular weight function, apart from that it should vanish beyond r_c to decrease the computational requirements by limiting the number of particle interactions in the spirit of DPD and be continuous and smooth to avoid spurious jumps in the calculated forces. A typical choice for $w(r)$ is the quadratic expression

$$w(r) = \begin{cases} \dfrac{15}{2\pi}\left(1 - \dfrac{r}{r_c}\right)^2 & (r \leq r_c) \\ 0 & (r > r_c) \end{cases}. \tag{9.2.45}$$

9.2.5.1 *The original DPD model*

This is recovered by employing Eq. (9.2.45) for the weight function together with the following expression for the free energy per particle:

$$\psi_{\text{exc}}(\bar{n}) = k_B T \frac{\pi}{30} \bar{a}\,\bar{n}. \tag{9.2.46}$$

For this model, the corresponding equation of state is given by Eq. (9.2.4).

9.2.5.2 The van der Waals fluid

The van der Waals fluid which is characterized by an equation of state of the form

$$p = \frac{n k_B T}{1 - b n} - a n^2 \tag{9.2.47}$$

is recovered by employing Eq. (9.2.45) for the weight function together with the following expression for the free energy per particle:

$$\psi_{exc}(\bar{n}) = -k_B T \ln(1 - b\bar{n}) - a\bar{n}. \tag{9.2.48}$$

The resulting expression for the pairwise conservative force is

$$\boldsymbol{F}_{ij} = \left\{ \left(k_B T \frac{b}{1 - b\bar{n}_i} - a \right) + \left(k_B T \frac{b}{1 - b\bar{n}_j} - a \right) \right\} w'(r_{ij})\, \hat{\boldsymbol{r}}_{ji}. \tag{9.2.49}$$

9.2.5.3 A multicomponent system

For a multicomponent system, let us adopt a notation where Greek letters (α, β, γ) are used to symbolize different components and Latin ones (i, j, k) to symbolize individual particles. In a multicomponent system, each particle may have neighbouring particles of different kinds. Then, Eq. (9.2.42) should be generalized by defining all partial local densities, namely,

$$\bar{n}_i^{\alpha} = \sum_{j \in \alpha, j \neq i} w(r_{ij}). \tag{9.2.50}$$

If $\psi_{exc}^{\alpha}(\{n_\beta\})$ is the equivalent of the excess free energy per particle for a single-component system so that the excess part of the Helmholtz free energy of the system is represented as

$$A_{exc} = \sum_{\alpha} N_\alpha \psi_{exc}^{\alpha}(\{n_\beta\}), \tag{9.2.51}$$

then one obtains the following generalization of Eq. (9.2.44):

$$\boldsymbol{F}_{ij}^{C} = -\left(\frac{\partial \psi_{exc}^{k(i)}(\{n_\alpha\}_i)}{\partial n_{k(j)}} + \frac{\partial \psi_{exc}^{k(j)}\left(\{n_\alpha\}_j\right)}{\partial n_{k(i)}} \right) w'(r_{ij})\, \hat{\boldsymbol{r}}_{ji}. \tag{9.2.52}$$

In Eq. (9.2.52), $k(i)$ denotes the component that particle i belongs to $(k(i) = \alpha, \text{if } i \in \alpha)$ while $\{n_\alpha\}_i$ is the set of local number densities of the different components at the position of particle i.

For a binary system, in particular, we can consider that it consists of N_α particles of component α and N_β particles of component β. Then, the corresponding two partial local density fields around the reference particle i are

$$\bar{n}_{\alpha,i} = \sum_{k \in \alpha, k \neq i} w(r_{ik})$$

$$\bar{n}_{\beta,i} = \sum_{k\in\beta,k\neq i} w(r_{ik}). \tag{9.2.53}$$

Let us also denote by n_α and n_β the continuum limit of these two discrete densities. Pagonabarraga and Frenkel[30] have suggested the following simplest form of the excess free energy for such a mixture that leads to a miscibility gap:

$$\begin{aligned} A_{\text{exc}} &= \int \left\{ 2\lambda\, n_\alpha(\boldsymbol{r})\, n_\beta(\boldsymbol{r}) + \lambda_\alpha\, n_\alpha(\boldsymbol{r})^2 + \lambda_\beta\, n_\beta(\boldsymbol{r})^2 \right\} \mathrm{d}\boldsymbol{r} \\ &= \sum_{i\in\alpha} (\lambda\, n_{\beta,i} + \lambda_\alpha\, n_{\alpha,i}) + \sum_{i\in\beta} (\lambda\, n_{\alpha,i} + \lambda_\beta\, n_{\beta,i}). \end{aligned} \tag{9.2.54}$$

In this case, the resulting conservative force on any particle i has still the form of a pair-wise additive force from all other particles j, that is $\boldsymbol{F}_i^C = \sum_{j\neq i} \boldsymbol{F}_{ij}^{\mathrm{C}}$, with

$$\boldsymbol{F}_{ij}^{\mathrm{C}} = \begin{Bmatrix} -2\lambda_{\alpha,\beta}\, w'(r_{ij})\, \hat{\boldsymbol{r}}_{ji}, & ij \text{ of the same type} \\ -2\lambda\quad w'(r_{ij})\, \hat{\boldsymbol{r}}_{ji}, & ij \text{ of different type} \end{Bmatrix}. \tag{9.2.55}$$

The M-DPD formalism has been used[31] to study the thermodynamic behaviour of the compressible Huggins fluid for monomeric species, for which the Helmholtz free energy reads (compare Eq. (9.2.10))

$$\frac{\Delta_{\text{mix}} A}{N_{\text{b}} k_B T} = \phi_\alpha \ln \phi_\alpha + \phi_\beta \ln \phi_\beta - \chi_{\alpha\beta}\phi_\alpha\phi_\beta + (1-(\phi_\alpha+\phi_\beta)) \ln (1-(\phi_\alpha+\phi_\beta)) \tag{9.2.56}$$

Compared to Eq. (9.2.10), Eq. (9.2.56) contains the last term which ensures that the system density cannot become larger than a specific value; moreover, the interaction term (the third term in the equation) enters with a negative sign to emphasize intersite attraction. Equation (9.2.56) is easily cast in the form of Eq. (9.2.51) by recognizing that it can be written as

$$\Delta_{\text{mix}} A = N_\alpha\, \psi^\alpha + N_\beta\, \psi^\beta \tag{9.2.57}$$

with

$$\frac{\psi^\alpha}{k_{\text{B}} T} = \ln \phi_\alpha - \chi_{\alpha\beta}\phi_\beta + (1-(\phi_\alpha+\phi_\beta)) \ln (1-(\phi_\alpha+\phi_\beta)) \tag{9.2.58a}$$

$$\frac{\psi^\beta}{k_{\text{B}} T} = \ln \phi_\beta - \chi_{\alpha\beta}\phi_\alpha + (1-(\phi_\alpha+\phi_\beta)) \ln (1-(\phi_\alpha+\phi_\beta)) \tag{9.2.58b}$$

and recalling that $N_\alpha = N_{\text{b}}\, \phi_\alpha$ and $N_\beta = N_{\text{b}}\, \phi_\beta$ with N_{b} denoting the total number of lattice sites. The corresponding excess functions ψ_{exc}^α and ψ_{exc}^β needed to compute the conservative forces through Eq. (9.2.52) are derived from the previous expressions by subtracting the respective ideal contributions.

9.3 Thermostats in DPD

Integrating Eq. (9.1.10) for the velocity term gives

$$\mathrm{d}\boldsymbol{\nu}_i = \frac{1}{m_i}\left[a\,\mathrm{d}t\sum_{j\neq i}\omega^{\mathrm{C}}(r_{ji})\hat{\boldsymbol{r}}_{ji} - \gamma\mathrm{d}t\sum_{j\neq i}\omega^{\mathrm{D}}(r_{ji})\left(\boldsymbol{\nu}_{ji}\cdot\hat{\boldsymbol{r}}_{ji}\right)\hat{\boldsymbol{r}}_{ji} + \sigma\sqrt{\mathrm{d}t}\sum_{j\neq i}\omega^{\mathrm{R}}(r_{ji})\xi_{ij}\hat{\boldsymbol{r}}_{ji}\right], \qquad (9.3.1)$$

which shows that, due to the stochastic nature of interactions, time reversibility is no longer guaranteed.[32,33] Moreover, the dissipative forces depend on the pairwise velocities of all pairs of particles which can cause numerical artefacts in the computed values of many of the physical properties of the system.[32–34] An algorithm that preserves hydrodynamics is the so-called *Lowe–Andersen thermostat (LAT)* in which one changes the velocities of a pair of particles, rather than the velocities of single particles. More specifically, in the LAT thermostat,[35–37] a pair of particles being a distance less than r_c apart is selected and its momentum is exchanged with frequency Γ by projecting the relative velocity on the line connecting the centres of mass of the two particles; finally, this is replaced with a value drawn from a Maxwell distribution so that the total momentum is conserved. This is achieved by transforming the particle velocities according to[38]

$$\boldsymbol{\nu}_i \leftarrow \boldsymbol{\nu}_i + m_{ij}\left[\xi_{ij}\sqrt{\frac{k_{\mathrm{B}}T_0}{m_{ij}}} - (\boldsymbol{\nu}_i - \boldsymbol{\nu}_j)\cdot\hat{\boldsymbol{r}}_{ji}\right]\frac{1}{m_i}\hat{\boldsymbol{r}}_{ji}$$

$$\boldsymbol{\nu}_j \leftarrow \boldsymbol{\nu}_j - m_{ij}\left[\xi_{ij}\sqrt{\frac{k_{\mathrm{B}}T_0}{m_{ij}}} - (\boldsymbol{\nu}_i - \boldsymbol{\nu}_j)\cdot\hat{\boldsymbol{r}}_{ji}\right]\frac{1}{m_j}\hat{\boldsymbol{r}}_{ji}. \qquad (9.3.2)$$

In the above equations ξ_{ij} is a Gaussian random variable satisfying $\langle\xi_{ij}(t)\,\xi_{km}(t')\rangle = \delta_{ik}\delta_{jm} + \delta_{im}\delta_{jk}$ when $t = t'$ and zero otherwise, and $m_{ij} = \dfrac{m_i m_j}{m_i + m_j}$ is the reduced mass of the pair of particles (i, j).

Stoyanov and Groot[38] have proposed combining the Lowe–Andersen thermostat with a local, Galilean invariant version of the *Nosé–Hoover thermostat* (*NHT*), which acts on pairs of particles rather than on single particles.[39,40] This is the so-called *Nosé–Hoover–Lowe–Andersen thermostat* (*NHLAT*). In this case, for each particle pair, one effectively chooses between NHT and LAT with probability $P = \Gamma\,\Delta t$, where Δt is the integration time step and Γ denotes the Lowe–Andersen exchange frequency, and the thermostating force acting on every pair of particles i and j within the cutoff distance r_c is

$$\boldsymbol{F}_{ij} = -\alpha\,w(r_{ij})\left(\frac{T}{T_0} - 1\right)\left[(\boldsymbol{\nu}_i - \boldsymbol{\nu}_j)\cdot\hat{\boldsymbol{r}}_{ji}\right]\hat{\boldsymbol{r}}_{ji}. \qquad (9.3.3)$$

In Eq. (9.3.3), α is a thermostat coupling parameter which is kept constant during the simulation, w a smearing function of distance chosen such that $4\pi\int_0^1 w(\bar{r})\,\bar{r}^2\mathrm{d}\bar{r} = 1$ and $w(\bar{r}) = 0$ for $\bar{r} \geq 1$, $r_{ij} = |\boldsymbol{r}_{ij}|$ the distance between particles i and j, T the instantaneous value of the system temperature, T_0 the set value of the temperature, $\boldsymbol{\nu}_i$ the velocity of particle i, and $\hat{\boldsymbol{r}}_{ij} = \dfrac{\boldsymbol{r}_{ij}}{|\boldsymbol{r}_{ij}|}$ the unit vector in the direction of $\boldsymbol{r}_{ij}$. In the above expressions, distances have been made dimensionless by dividing them with the conservative force cutoff value r_c; see Eq. (9.1.2).

The thermostating force of Eq. (9.3.3) has the same form as the dissipative force defined in Eq. (9.1.4) but, in addition, it contains the factor $\dfrac{T}{T_0} - 1$ whose sign depends on the ratio of the instantaneous temperature T over the desired temperature T_0. Thus, in contrast to Eq. (9.1.4) which describes an interparticle force that is always dissipative (it acts in the direction opposite to that of the relative velocity of the two particles), Eq. (9.3.3) describes a force which can either dissipate energy out of the system or put energy back to the system depending on the ratio $\dfrac{T}{T_0}$ of the two temperatures T and T_0 (see Problem 9.4). Moreover, since the thermostating force fluctuates due to fluctuations in mutual particle velocities, it serves as either a dissipative or a random force, but without using any random numbers. The NHLAT is completely deterministic, like the Nosé–Hoover and Berendsen ones discussed in Chapter 4.

To guarantee Galilean invariance, Stoyanov and Groot[38] noted that the instantaneous value of the system temperature should also be defined with respect to some relative velocity, by introducing a co-moving frame for each particle. They proposed two definitions. In the first, one uses an average temperature through the global scheme

$$k_B T = \frac{\sum\limits_{j}\sum\limits_{i>j} \zeta\,(\bar{r}_{ij})\, m_{ij}(\boldsymbol{\nu}_i - \boldsymbol{\nu}_j)^2}{3\sum\limits_{j}\sum\limits_{i>j} \zeta\,(\bar{r}_{ij})}, \tag{9.3.4}$$

where $\zeta\,(\bar{r})$ is a smearing function such that $\zeta = 0$ for $\bar{r} > 1$, and m_{ij} is the reduced mass of particles i and j. In the above expressions, distances have been made dimensionless by dividing them with a cutoff value which we call temperature cutoff (this, in principle, can be different from the conservative force cutoff value appearing in Eq. (9.1.2)). In the second, suitable for large and dense systems, one uses a local (particle-based) definition of the instantaneous temperature,

$$k_B T_i = \frac{\sum\limits_{j\neq i} \zeta\,(\bar{r}_{ij})\, m_{ij}(\boldsymbol{\nu}_i - \boldsymbol{\nu}_j)^2}{3\sum\limits_{j\neq i} \zeta\,(\bar{r}_{ij})}, \tag{9.3.5}$$

by also modifying slightly the thermostating force of Eq. (9.3.3) to read[38]

$$\boldsymbol{F}_{ij} = -\alpha\, w(r_{ij}) \left[\frac{1}{2}\frac{T_i + T_j}{T_0} - 1\right] \left[(\boldsymbol{\nu}_i - \boldsymbol{\nu}_j) \cdot \hat{\boldsymbol{r}}_{ji}\right] \hat{\boldsymbol{r}}_{ji}. \tag{9.3.6}$$

A flow diagram of the coupled NHLAT just described is provided in Table 9.1 based on the article of Stoyanov and Groot.[38]

Table 9.1 *A Block-Scheme of the Nosé–Hoover–Lowe–Andersen thermostat (NHLAT) for a Single Time-Step Implementation Based on the Velocity Verlet Integration Algorithm*[38]

Block 1

$\forall i$:

1a) Update coordinates: $\boldsymbol{r}_i(t+\Delta t) = \boldsymbol{r}_i(t) + \boldsymbol{\nu}_i(t)\,\Delta t + \frac{(\Delta t)^2}{2}\frac{\boldsymbol{F}_i(t)}{m_i}$

1b) Update velocities by half step: $\tilde{\boldsymbol{\nu}}_i(t) = \boldsymbol{\nu}_i(t) + \frac{\Delta t}{2}\frac{\boldsymbol{F}_i(t)}{m_i}$

1c) Put particles in cell

Block 2

$\forall\,(i<j)\,|\,(r_{ij}<r_c)$:

2a) Calculate conservative force $\boldsymbol{F}_{ij}(t+\Delta t)$ and add it to $\boldsymbol{F}_i$ and $-\boldsymbol{F}_j$

2b) Add $\mathrm{Sv2} \equiv \mathrm{Sv2} + \zeta(r_{ij})\, m_{ij}[\boldsymbol{\nu}_i(t) - \boldsymbol{\nu}_j(t)]^2$; $\mathrm{Sz} \equiv \mathrm{Sz} + \zeta(r_{ij})$

2c) • If (random $< P$) add pair ij to neighbour list of particle i

• Else calculate $\Delta\boldsymbol{F}_{ij} = \alpha\, w(r_{ij})\,[(\tilde{\boldsymbol{\nu}}_i(t) - \tilde{\boldsymbol{\nu}}_j(t))\cdot\hat{\boldsymbol{r}}_{ji}]\,\hat{\boldsymbol{r}}_{ji}$ and add $\Delta\boldsymbol{F}_i \equiv \Delta\boldsymbol{F}_i + \Delta\boldsymbol{F}_{ij}$; $\Delta\boldsymbol{F}_j \equiv \Delta\boldsymbol{F}_j - \Delta\boldsymbol{F}_{ij}$

Block 3

$\forall i$:

3a) Calculate bond forces and add them to $\boldsymbol{F}_i(t+\Delta t)$

3b) Update velocities: $\boldsymbol{\nu}_i(t+\Delta t) = \tilde{\boldsymbol{\nu}}_i(t) + \frac{\Delta t}{2}\frac{\boldsymbol{F}_i(t+\Delta t)}{m_i}$

3c) Calculate temperature: $T = \frac{1}{3}\frac{\mathrm{Sv2}}{\mathrm{Sz}}$; reset Sv2 and Sz

3d) Correct the velocities: $\boldsymbol{\nu}_i \equiv \boldsymbol{\nu}_i - \left(\frac{T}{T_0} - 1\right)\Delta t\frac{\Delta\boldsymbol{F}_i}{m_i}$

Block 4

$\forall i,j\,|\,j \in$ {neighbor list of i}:

4a) Calculate $\Delta\nu_{ji} = (\boldsymbol{\nu}_i - \boldsymbol{\nu}_j)\cdot\hat{\boldsymbol{r}}_{ji}$

4b) Calculate $\Delta_{ij} = \xi_{ij}\sqrt{\frac{k_B T}{m_{ij}}} - \Delta\nu_{ji}$ (ξ_{ij} Gaussian random)

4c) Update $\boldsymbol{\nu}_i \equiv \boldsymbol{\nu}_i + \frac{m_{ij}\Delta_{ij}}{m_i}\hat{\boldsymbol{r}}_{ji}$; $\quad \boldsymbol{\nu}_j \equiv \boldsymbol{\nu}_j - \frac{m_{ij}\Delta_{ij}}{m_j}\hat{\boldsymbol{r}}_{ji}$

9.4 Multiple time-step DPD

In Chapter 4, we saw that the classical Liouville operator iL for a system of f degrees of freedom can be decomposed into two (noncommuting) parts as $iL = iL_1 + iL_2$, where the indices '1' and '2' denote contributions from degrees of freedom of different character (e.g. short- versus long-ranged) or by degrees of freedom grouped in two different categories (e.g. based on particles of different masses or of different species). An iterative application of the Trotter theorem results in an approximate factorization of the form

$$e^{i(L_1+L_2)\Delta t} = e^{iL_1\frac{\Delta t}{2}}\left[e^{iL_2\frac{\Delta t}{n}}\right]^n e^{iL_1\frac{\Delta t}{2}}, \tag{9.4.1}$$

which is particularly useful for integrating out degrees of freedom characterized by a time-scale separation. Through such a splitting, e.g. *fast* degrees of freedom (labelled by the index 2), can be updated more frequently with a time step $\delta t = \dfrac{\Delta t}{n}$ in n *inner loops* in the simulation code, while *slow* degrees of freedom (index 1) are updated less frequently in two bracketing *outer loops* corresponding to a total time step Δt. This has formed the context for the development of the so-called reversible reference system propagator algorithms (rRESPA), which, as we saw in Chapter 4, have been applied extremely successfully in MD. In turn, each exponential in Eq. (9.4.1) can be further factorized in terms of the position and momentum parts of the iL_1 and iL_2 operators. An advantage of these factorizations is that they can be directly translated into an update scheme by applying successive propagators (e.g. from right to left) to perform corresponding translations of momenta and positions.

Based on the above scheme, for a lipid–water model system, in particular, Jakobsen et al.[41] have proposed the three different schemes, labelled as A, B, and C in Table 9.2, based on different definitions of the two Liouville operators iL_1 and iL_2, with the subscripts 'b' and 'nb' denoting bonded and nonbonded interactions, respectively, while the superscripts are used to distinguish between

Table 9.2 *Three Different rRESPA Schemes (A, B, C) as Proposed by Jakobsen et al.*[41] *for the Simulation of a Lipid–Water System Based on Different Definitions of the two Liouville Operators iL_1 and iL_2*

Scheme	Definition of iL_1 and iL_2
A	$iL_1 \equiv iL_{\mathrm{nb}}^{\mathrm{WW}} + iL_{\mathrm{nb}}^{\mathrm{LW}} + iL_{\mathrm{nb}}^{\mathrm{LL}}$
	$iL_2 \equiv iL_{\mathrm{b}}^{\mathrm{LL}}$
B	$iL_1 \equiv iL_{\mathrm{nb}}^{\mathrm{WW}} + iL_{\mathrm{nb}}^{\mathrm{LW}}$
	$iL_2 \equiv iL_{\mathrm{b}}^{\mathrm{LL}} + iL_{\mathrm{nb}}^{\mathrm{LL}}$
C	$iL_1 \equiv iL_{\mathrm{nb}}^{\mathrm{WW}}$
	$iL_2 \equiv iL_{\mathrm{b}}^{\mathrm{LL}} + iL_{\mathrm{nb}}^{\mathrm{LL}} + iL_{\mathrm{nb}}^{\mathrm{LW}}$

the three different types of interacting beads describing the lipid–water model: water beads (W) and lipid entities made up of one head (H) bead and six tail (T) beads. On the other hand, the superscript (L) denotes lipid beads, i.e. H and T beads together. For example, by LW we denote a pair of an L bead (H or T) with a water (W) bead. We also see that iL_1 and iL_2 involve a sum of partial Liouville operators, each one describing a certain type of interaction between specific pairs of beads.

In scheme A, all nonbonded interactions are computed in the outer loop, and all bonded ones in the inner loops. In scheme C, the outer loop addresses only nonbonded interactions between water beads while all other interactions are computed in the inner loops. Scheme B represents an intermediate case wherein the outer loop addresses the nonbonded interactions involving water beads (either with other water beads or with lipid beads) while nonbonded as well as bonded interactions between lipid beads are dealt with in the inner loops. Multiple time-step algorithms for DPD simulations have also been proposed by Thalmann and Farago.[42]

9.5 An application of DPD to simulate self-assembly of nanoparticle mixtures in diblock copolymers

Over the years, DPD has evolved into a friendly simulation method for a variety of systems in soft matter: polymer–clay nanocomposites, hydrodynamic interactions in dilute and concentrated polymer solutions, lipid bilayers, interfaces in incompatible A/B homopolymer blends, and block copolymers. In particular, DPD has been used[43] to build multiscale modelling approaches for the study of self-assembly at the nanometer scale of organic/inorganic hybrid materials prepared

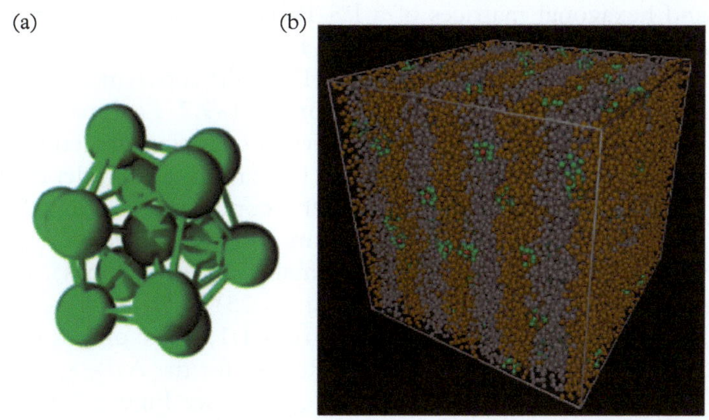

Figure 9.1 (a) Geometry and definition of the icosahedral nanoparticles considered in the work of Ref. 43: a central core DPD bead of type N is surrounded by another twelve DPD beads of type A (PS), B (PVP), or A and B, depending on the covering type considered, on each one of the icosahedron's vertices. (b) An equilibrated snapshot of the A_5B_5 system with a lamellar morphology for $\chi_{AB}N_{DPD} = 20.57$. The DPD particles of the polymer matrix of type A (PS blocks) are depicted in gold, while those of type B (PVP blocks) in white. In each nanoparticle, the central DPD bead is coloured red, while the remaining twelve DPD particles, making up the surface of the real icosahedral nanoparticles, are coloured in green.

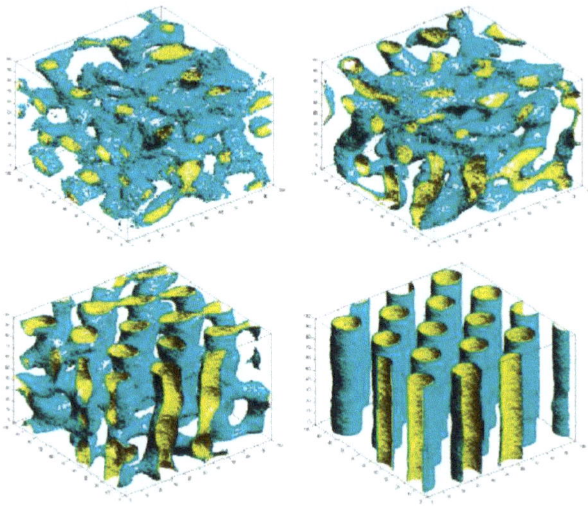

Figure 9.2 Isosurface visualization as a function of $\chi_{AB}N_{DPD}$. Top left: $\chi_{AB}N_{DPD} = 15.46$. Top right: $\chi_{AB}N_{DPD} = 17.68$. Bottom left: $\chi_{AB}N_{DPD} = 22.07$. Bottom right: $\chi_{AB}N_{DPD} = 29.67$. The polymer matrix DPD particles of type A (PS) are depicted in yellow while those of type B (PVP) in cyan.

from nanoparticles mixed with microphase-separating diblock copolymers. Emphasis was placed on the prediction of the formation of the diblock mesophases and of the nanoparticle distribution in the lamellar and hexagonal matrices of a PS–PVP diblock copolymer, in dependence of copolymer–particle and particle–particle interactions. Representative results obtained from such DPD simulations for the morphology of a number of PS–PVP copolymers corresponding to two chain architectures, A_5B_5 and A_3B_7, are shown in Figures 9.1–9.2. Each nanoparticle was considered to have an icosahedral structure and to be composed of a neutral, central DPD bead (type N) surrounded by twelve other DPD beads of type A (PS) and B (PVP), depending on the covering type considered (see Figure 9.1). The length of the copolymer was equal to 10 DPD units ($N_{DPD} = 10$). Due to its symmetry, A_5B_5 could form either a lamellar morphology (for large $\chi_{AB}N_{DPD}$ values) or a disordered morphology (for small $\chi_{AB}N_{DPD}$ values). Figure 9.1 demonstrates that for a large enough value of $\chi_{AB}N_{DPD}$ (e.g. $= 25$) A_5B_5 converges to a well-defined lamellar phase. But if the ratio A:B is changed away from 1:1, other phases should be expected; that was indeed observed in the DPD simulations of Ref. 43 for the A_3B_7 system, where by varying the parameter $\chi_{AB}N_{DPD}$ different morphologies develop (see Figure 9.2), in agreement with experimental evidence.[44]

Problems

Problem 9.1: Prove Eq. (9.2.4) for the equation of state of a single-component DPD fluid.

Problem 9.2: Prove Eq. (9.2.7) for the free-energy density of the single-component DPD fluid.

Problem 9.3: Prove Eq. (9.2.13) for the Helmholtz energy of a DPD mixture of monomeric species A and B.

Problem 9.4: Show that when the instantaneous temperature is higher than the equilibrium temperature ($T > T_0$), then the thermostating force of Eq. (9.3.3) dissipates energy from the system while when $T < T_0$ it adds energy to the system.

References

1. Hoogerbrugge, P.J., and Koelman, J.M.V.A. *Europhys. Lett.* **19**, 155 (1992).
2. Koelman, J.M.V.A., and Hoogerbrugge, P.J. *Europhys. Lett.* **21**, 363 (1993).
3. Español, P., and Warren, P.B. *Europhys. Lett.* **30**, 191 (1995).
4. Rapaport, D.C., and Clementi, E. *Phys. Rev. Lett.* **57**, 695 (1986).
5. Mareschal, M., Mansour, M.M., Puhl, A., and Kestemont, E. *Phys. Rev. Lett.* **61**, 2550 (1988).
6. Rapaport, D.C. *Phys. Rev. Lett.* **60**, 2480 (1988).
7. Zúñiga, I., and Español, P. *Phys. Rev. Lett.* **71**, 3665 (1993).
8. Frisch, U., Hasslacher, B., and Pomeau, Y. *Phys. Rev. Lett.* **56**, 1505 (1986).
9. Ladd, A.J.C. *J. Fluid Mech.* **271**, 285 (1994); Ladd, A.J.C. *J. Fluid Mech.* 271, 311 (1994).
10. Groot, R.D., and Warren, P.B. *J. Chem. Phys.* **107**, 4423 (1997).
11. Kong, Y., Manke, C.W., Madden, W.G., and Schlijper, A.G. *Int. J. Thermophys.* **15**, 1093 (1994).
12. Schlijper, A.G., Hoogerbruge, P.J., and Manke, C.E. *J. Rheol.* **39**, 567 (1995).
13. Risken, H. *The Fokker-Planck Equation.* Springer-Verlag, Berlin (1989).
14. Gardiner, C.W. *Handbook of Stochastic Methods.* Springer-Verlag, Berlin (1983).
15. Öttinger, H.C. *Stochastic Processes in Polymeric Fluids.* Springer, New York (1996).
16. Mori, H. *Prog. Theor. Phys.* **33**, 423 (1965).
17. Grabert, H. Projection operator techniques. In *Nonequilibrium Statistical Mechanics.* Springer-Verlag, Berlin (1983).
18. McQuarrie, D.A. *Statistical Mechanics.* Harper Collins, New York (1976).
19. Español, P. *Phys. Rev. E* **52**, 1734 (1995).
20. Flory, P.J. *Principles of Polymer Chemistry.* Cornell University Press, Ithaca, NY (1953).
21. Prausnitz, J.M., Lichtenthaler, R.N., and de Azevedo, E.G. *Molecular Thermodynamics of Fluid-phase Equilibria.* Prentice-Hall, Englewood Cliffs, NJ (1999).
22. Maiti, A., and McGrother, S. *J. Chem. Phys.* **120**, 1594 (2004).
23. Hildebrand, J.H., and Scott, R.L. *The Solubility of Non-electrolytes* (3rd ed.). Reinhold, New York (1950).
24. Case, F., and Honeycutt, J.D. *Trends Polym. Sci.* **2**, 259 (1994).
25. Travis, K.P., Bankhead, M., Good, K., and Owens, S.L. *J. Chem. Phys.* **127**, 014109 (2007).
26. Beattie, J.A. *Chem. Rev. (Washington, D.C.)* **44**, 141 (1949).
27. Scatchard, G. *Chem. Rev. (Washington, D.C.)* **8**, 321 (1931).
28. Hildebrand, J.H., and Wood, S.E. *J. Chem. Phys.* **1**, 817 (1933).
29. Pagonabarraga, I., and Frenkel, D. *Mol. Simul.* **25**, 167 (2000).
30. Pagonabarraga, I., and Frenkel, D. *J. Chem. Phys.* **115**, 5015 (2001).
31. Trofimov, S.Y., Nies, E.L.F., and Michels, M.A.J. *J. Chem. Phys.* **117**, 9383 (2002).
32. den Otter, W.K., and Clarke, J.H.R. *Europhys. Lett.* **53**, 426 (2001).
33. Novik, K.E., and Coveney, P.V. *J. Chem. Phys.* **109**, 7667 (1998).
34. Vattulainen, I., Karttunen, M., Besold, G., and Polson, J.M. *J. Chem. Phys.* **116**, 3967 (2002).
35. Lowe, C.P. *Europhys. Lett.* **47**, 145 (1999).

36. Lowe, C.P., and Dreischor, M.W. In Karttunen, M., Vattulainen, I., and Lukkarinen, A. (eds.), *Novel Methods in Soft Matter Simulations*, Lecture Notes in Physics. Springer, Berlin (2004), vol. 640, pp. 39–68.
37. Koopman, E.A., and Lowe, C.P. *J. Chem. Phys.* **124**, 204103 (2006).
38. Stoyanov, S.D., and Groot, R.D. *J. Chem. Phys.* **122**, 114112 (2005).
39. Groot, R.D. *J. Chem. Theory Comput.* **2**, 568 (2006).
40. Phares, D.J., and Srinivasa, A.R. *J. Phys. Chem. A* **108**, 6100 (2004).
41. Jakobsen, A.F., Mouritsen, O.G., and Besold, G. *J. Chem. Phys.* **122**, 204901 (2005).
42. Thalmann, F., and Farago, J. *J. Chem. Phys.*, **127**, 124109 (2007).
43. Maly, M., Posocco, P., Pricl, S., and Fermeglia, M. *Ind. Eng. Chem. Res.* **47**, 5023 (2008).
44. Chiu, J.J., Kim, B.J., Kramer, E.J., and Pine, D.J. *J. Am. Chem. Soc.* **127**, 5036 (2005).

10

Beyond-Equilibrium Simulations Based on Principles of Nonequilibrium Thermodynamics

In Chapters 3 and 4 we saw that, from a statistical mechanics point of view, any physical system at the atomic level is described by Hamilton's equations of motion. In Chapters 7–9, we also saw that in order to minimize the computational effort required to solve a given problem, degrees of freedom irrelevant for the understanding of specific phenomena are often projected out of the description, with appropriate modification of the equations of motion. On the other hand, the majority of natural phenomena at the macroscopic level involves transport processes of mass, momentum, and energy, as well as chemical reactions, which are accompanied by significant changes in the values of structural variables describing the system at the chosen level of analysis. We talk therefore about an *evolutionary system*. At equilibrium, the science which provides a universally accepted framework for the study of physical phenomena is *statistical thermodynamics*. This is achieved by computing thermodynamic potentials through partition functions and statistical weights and then using standard formulations to extract the values of all thermodynamic properties of the system. Ideally, one would like to use a similar systematic framework (based on universally accepted principles and procedures) to understand complex phenomena evolving under nonequilibrium or beyond equilibrium conditions, often over multiple length and time scales. Such a formulation is the subject of *nonequilibrium thermodynamics*. Nonequilibrium thermodynamics aims at exploring the unifying role of thermodynamics in the derivation of the balance equations of mass, momentum, energy, and entropy and in the formulation of additional evolution equations for the key structural variables, for coupled processes in physical, chemical, and biological systems. This can provide a bridge between science and engineering for phenomena that are inherently time-dependent or take place under an applied field that brings the system to beyond-equilibrium conditions. For complex fluids, in particular, such as polymers, how to apply thermodynamics under nonequilibrium conditions is by no means obvious. As a result, an overwhelming number of different theories and approaches have appeared in the literature: classical nonequilibrium thermodynamics with internal variables, Lagrangian methods, bracket formulations, continuum or rational thermodynamics, variational formulations, extended irreversible thermodynamics, the matrix model, network thermodynamics, and the GENERIC formalism. This chapter is devoted to a brief presentation of the key ideas underlying two of the above-mentioned formalisms which have attracted considerable interest in the past two decades, finding numerous applications in a variety of phenomena of relevance to polymers:

Multiscale Modelling of Polymers. Doros N. Theodorou and Vlasis G. Mavrantzas, Oxford University Press.
 DOI: 10.1093/oso/9780199226191.003.0010

the generalized bracket and the GENERIC formalisms. The two formalisms have been presented in detail in two different books,[1,2] so here we will restrict ourselves to a brief illustration of the basic concepts and key principles, and we will discuss their utility in the multiscale modelling of polymers.

10.1 The Poisson bracket for a system at equilibrium

At equilibrium, a classical atomistic system is fully specified by the set of generalized coordinates (positions and momenta) of all atoms in the system which obey Hamilton's equations of motion. If $F = F(\boldsymbol{q},\boldsymbol{p})$ is an arbitrary functional of the sets $\{\boldsymbol{q}_i,\ i = 1,2,\ldots,N\}$ and $\{\boldsymbol{p}_i,\ i = 1,2,\ldots,N\}$ of the generalized positions and momenta, respectively, of the N atoms in the system, then the total derivative of F with respect to time t can be obtained by the rule of implicit differentiation as

$$\frac{\mathrm{d}}{\mathrm{d}t}F(\boldsymbol{q},\boldsymbol{p}) = \sum_{i=1}^{N}\left[\frac{\partial F}{\partial \boldsymbol{q}_i}\cdot\frac{\partial \boldsymbol{q}_i}{\partial t} + \frac{\partial F}{\partial \boldsymbol{p}_i}\cdot\frac{\partial \boldsymbol{p}_i}{\partial t}\right] \tag{10.1.1}$$

(we have assumed here that F is not an explicit function of time t; thus, $\frac{\partial F}{\partial t} = 0$). Substitution of Hamilton's equations, Eqs. (1.1.33) and (1.1.34), into (10.1.1) yields

$$\frac{\mathrm{d}}{\mathrm{d}t}F(\boldsymbol{q},\boldsymbol{p}) = \sum_{i=1}^{N}\left[\frac{\partial F}{\partial \boldsymbol{q}_i}\cdot\frac{\partial \mathcal{H}}{\partial \boldsymbol{p}_i} - \frac{\partial \mathcal{H}}{\partial \boldsymbol{q}_i}\cdot\frac{\partial F}{\partial \boldsymbol{p}_i}\right], \tag{10.1.2}$$

with $\mathcal{H}(\boldsymbol{q},\boldsymbol{p})$ being the Hamiltonian. The expression on the right-hand side (Liouville operator acting on F, compare Eq. (7.7.3)) is termed the Poisson bracket (of classical mechanics).[3] In general, for two arbitrary functions $F(\boldsymbol{q},\boldsymbol{p})$ and $G(\boldsymbol{q},\boldsymbol{p})$, the Poisson bracket is denoted as $\{F, G\}$ and is expressed as

$$\{F, G\} = \sum_{i=1}^{N}\left[\frac{\partial F}{\partial \boldsymbol{q}_i}\cdot\frac{\partial G}{\partial \boldsymbol{p}_i} - \frac{\partial G}{\partial \boldsymbol{q}_i}\cdot\frac{\partial F}{\partial \boldsymbol{p}_i}\right]. \tag{10.1.3}$$

Hamilton's equations of motion can therefore be expressed alternatively in a more compact form as

$$\frac{\mathrm{d}F}{\mathrm{d}t} = \{F, H\}. \tag{10.1.4}$$

The Poisson bracket as defined by Eq. (10.1.3) has some important properties. First, it satisfies the antisymmetry property

$$\{F, G\} = -\{G, F\}, \tag{10.1.5}$$

which ensures that

$$\frac{\mathrm{d}\mathcal{H}}{\mathrm{d}t} = \{\mathcal{H}, \mathcal{H}\} = 0; \tag{10.1.6}$$

i.e. the total energy of the system is conserved. It also satisfies

$$\{F, G_1 + G_2\} = \{F, G_1\} + \{F, G_2\} \tag{10.1.7}$$

$$\{F_1 F_2, G\} = F_1 \{F_2, G\} + \{F_1, G\} F_2 \tag{10.1.8}$$

and

$$\{F, \{G, H\}\} + \{G, \{H, F\}\} + \{H, \{F, G\}\} = 0. \tag{10.1.9}$$

The latter is the well-known *Jacobi identity* which guarantees that the Poisson bracket is invariant under any arbitrary *canonical transformation* (i.e. any transformation that preserves the form of Hamilton's equations of motion, a property also known as *form invariance*). The same property also ensures that the bracket fulfils the conditions of a Lie algebra.[4–8]

In Chapter 7 we saw that the density distribution function in phase space $\rho(\boldsymbol{q}, \boldsymbol{p}, t)$ satisfies the conservation law (continuity equation in phase space)

$$\frac{\partial \rho}{\partial t} + \sum_{i=1}^{N} \left[\frac{\partial}{\partial \boldsymbol{q}_i} \cdot (\dot{\boldsymbol{q}}_i \rho) + \frac{\partial}{\partial \boldsymbol{p}_i} \cdot (\dot{\boldsymbol{p}}_i \rho) \right] = 0. \tag{10.1.10}$$

Using Hamilton's canonical equations of motion and the definition of the Poisson bracket, this is also written as

$$\frac{\partial \rho}{\partial t} + \{\rho, \mathcal{H}\} = 0, \tag{10.1.11}$$

which is the well-known *Liouville equation* of statistical mechanics.

10.2 The generalized bracket

Guided by the above ideas which are valid under equilibrium conditions, the generalized bracket formalism is an attempt to provide a unified view of the dynamical equations of motion at a more coarse-grained level of description than the atomistic one by using as the fundamental quantity the total energy function of the system, i.e. the Hamiltonian $\mathcal{H}$. For a given application, $\mathcal{H}$ should be defined as a functional in terms of the independent dynamical variables which are assumed to uniquely determine the macroscopic state of the system. Historically, the development of the Hamiltonian formalism for dissipative systems started with the pioneering works of Kaufman,[9] Morrison,[10,11] and Grmela.[12–14] This was followed by significant research activity, which at the end culminated in the two most systematic and broadest descriptions known today as the generalized bracket and the GENERIC formalisms of nonequilibrium thermodynamics.[1,2]

The key underlying assumption of the Hamiltonian approach is the existence of a partial (local) thermodynamic equilibrium, which can be described by a few internal variables (such as the second moment of the distribution function for the polymer end-to-end distance vector), in addition to the traditional and widely used ones, such as the density and the temperature. The Hamiltonian formalism has enabled the description of the dynamics of many flowing systems. In simple test cases, it has led to the same transport equations as those obtained from the application of more conventional conservation and continuum mechanics principles. In more complicated cases, it has made it possible to evaluate the thermodynamic consistency of various proposed models. In such cases, it needs to be coupled with an analysis of the microscopic physics in order to produce more valuable results. It is this unique combination of microscopic modelling in terms of the system's internal variables with the macroscopic Hamiltonian formalism that provides a powerful means for studying phenomena over many length and time scales.

In the generalized bracket formalism (but also in GENERIC), even before $\mathcal{H}$ is defined, one must decide on the vector $\boldsymbol{x}$ of state variables. A good choice for $\boldsymbol{x}$ is crucial for the success of the overall methodology. Typically, $\boldsymbol{x}$ is expressed as $\boldsymbol{x} = \{\rho(\boldsymbol{r}, t), \boldsymbol{M}(\boldsymbol{r}, t), s(\boldsymbol{r}, t), \boldsymbol{X}(\boldsymbol{r}, t)\}$, where ρ is the mass density, $\boldsymbol{M}$ the momentum density (the drift velocity is therefore given as $\boldsymbol{\nu} = \boldsymbol{M}/\rho$), s the entropy density, $\boldsymbol{X}$ an internal (structural) variable which also must be a density, $\boldsymbol{r}$ the position vector in space, and t the time. The generalized bracket formalism can address internal variables which are only scalars, vectors, or tensors (it cannot accommodate distribution functions). Thus, for polymers, a common choice for $\boldsymbol{X}$ is the dimensional conformation tensor $\boldsymbol{C}$ defined as $\boldsymbol{C} = \rho\boldsymbol{c}$, where $\boldsymbol{c}$ refers to a single chain. Formally, $\boldsymbol{c}$ is defined as the second moment of the distribution function $\psi(\boldsymbol{R}, \boldsymbol{r}, t)$ for the chain end-to-end vector $\boldsymbol{R}$ at position $\boldsymbol{r}$ and time t, i.e. $c_{\alpha\beta}(\boldsymbol{r}, t) = \langle R_\alpha R_\beta \rangle (\boldsymbol{r}, t) = \int R_\alpha R_\beta \psi(\boldsymbol{R}, \boldsymbol{r}, t) \mathrm{d}\boldsymbol{R}$, where the brackets denote a configurational average. In practice, $\boldsymbol{c}$ is often replaced by the dimensionless single-chain conformation tensor $\tilde{\boldsymbol{c}}$, defined as

$$\tilde{\boldsymbol{c}}(\boldsymbol{r}, t) \equiv \frac{3\boldsymbol{c}(\boldsymbol{r}, t)}{\langle R^2 \rangle_{\mathrm{eq}}}, \tag{10.2.1}$$

where the subscript eq denotes that the average is computed under equilibrium conditions. According to Eq. (10.2.1), at equilibrium (zero flow field), $\tilde{\boldsymbol{c}}$ coincides with the second-rank unit tensor $\boldsymbol{I}$.

The second step in the generalized bracket formalism is to define the functional form of the Hamiltonian $\mathcal{H}$ of the system. $\mathcal{H}$, being the total energy of the system, is the sum of the kinetic (E_k), potential (E_p), and internal (U) energies. Let us denote the corresponding energy densities as e_k, e_p, and u, respectively. e_k and u will be assumed to be functions of the components of the vector $\boldsymbol{x}$, while e_p is usually an explicit function of the position vector $\boldsymbol{r}$ and time t. Thus, we usually write

$$\mathcal{H}(\boldsymbol{x}) = \mathcal{H}(\rho, \boldsymbol{M}, s, \boldsymbol{C}) = E_\mathrm{k}(\rho, \boldsymbol{M}) + E_\mathrm{p} + U(\rho, s, \boldsymbol{C}) \tag{10.2.2a}$$

or, equivalently,

$$\begin{aligned}\mathcal{H}(\rho, \boldsymbol{M}, s) &= \int (e_\mathrm{k} + e_\mathrm{p} + u)\, \mathrm{d}V \\ &= \int \left[\frac{|\boldsymbol{M}(\boldsymbol{r}, t)|^2}{2\rho(\boldsymbol{r}, t)} + e_\mathrm{p}(\boldsymbol{r}, t) + u\left(\rho(\boldsymbol{r}, t), s(\boldsymbol{r}, t), \boldsymbol{C}(\boldsymbol{r}, t)\right) \right] \mathrm{d}V. \end{aligned} \tag{10.2.2b}$$

For a bulk, homogeneous system, potential energy contributions are usually absent, unless an external field is applied. They are important, however, in the case of inhomogeneous polymer systems (as e.g. is the case when a polymer flows near a solid boundary).

With the choice $\boldsymbol{x} = \{\rho(\boldsymbol{r}, t), \boldsymbol{M}(\boldsymbol{r}, t), s(\boldsymbol{r}, t), \boldsymbol{C}(\boldsymbol{r}, t)\}$ for the vector of state variables, the evolution equations will consist of the continuity equation for the total mass density ρ, the conservation equation for the total momentum density $\boldsymbol{M}$, the transport equation for the entropy s, and the evolution equation for the conformation tensor density $\boldsymbol{C}$. In the generalized bracket formalism, the dynamical equations for the components of $\boldsymbol{x}$ are derived by postulating the following master equation for the dynamics of the arbitrary functional F:

$$\frac{\mathrm{d}F}{\mathrm{d}t} = \{F, \mathcal{H}\} + [F, \mathcal{H}], \tag{10.2.3}$$

where $\mathcal{H}$ is the Hamiltonian, $\{\ldots\}$ denotes the Poisson bracket describing conservative (convective) effects, and $[\ldots]$ is the dissipative bracket accounting for nonconservative phenomena. Equation (10.2.3) constitutes a generalization of Eq. (10.1.4) in the sense that, when we are to attempt a description of the system at a coarse-grained level, we must account for the increase in entropy and dissipation (or friction) associated with the elimination of irrelevant fast degrees of freedom in favour of random fluctuations in the coarser description.[15] Based on Eq. (10.2.3), the coupled time evolution equations for the dynamical variables are obtained through a direct comparison of the master equation, Eq. (10.2.3), with the expression that results by differentiating $\frac{\mathrm{d}F}{\mathrm{d}t}$ by parts.

The two brackets appearing in Eq. (10.2.3) must satisfy certain relationships in order for the master equation to be compatible with fundamental physical laws and also to have the correct mathematical structure required for an evolution equation. In particular, the Poisson bracket:

(a) must be bilinear

(b) must satisfy the first law of thermodynamics for a conservative system; i.e. it has to be antisymmetric; see Eq. (10.1.5);

(c) must satisfy the requirement of zero rate of entropy production for a conservative system, i.e.

$$\{S, \mathcal{H}\} = 0; \tag{10.2.4}$$

and

(d) must be invariant under any arbitrary canonical transformation, i.e. satisfy the Jacobi identity; see Eq. (10.1.9).

The generalized bracket, on the other hand, must satisfy the following properties:

a) it must be linear in F;

b) it must be frame indifferent;

c) in the most general case, it is nonlinear in $\mathcal{H}$; however, close to equilibrium it may be assumed to be a linear function of $\mathcal{H}$ (in the lowest-order approximation it is always assumed to be so), in which case it can be symmetric with respect to both F and $\mathcal{H}$;

d) it must satisfy the first law of thermodynamics, i.e. the requirement of the conservation of the total energy:

$$\frac{\mathrm{d}\mathcal{H}}{\mathrm{d}t} = 0, \tag{10.2.5}$$

which translates into

$$[\mathcal{H}, \mathcal{H}] = 0; \tag{10.2.6}$$

e) It must satisfy the second law of thermodynamics, i.e. the requirement of a nonnegative rate of entropy production in an isolated system,

$$\frac{dS}{\mathrm{d}t} \geq 0, \tag{10.2.7}$$

which translates into

$$[S, \mathcal{H}] \geq 0; \tag{10.2.8}$$

f) in the particular case one is interested in describing a system where heat effects are assumed to be unimportant (as e.g. in the consideration of incompressible fluids), the generalized bracket may also be applied to the mechanical energy contribution to the total energy, $\mathcal{H}_\mathrm{m}$, to formulate equations of motion. In this case, the degradation of mechanical energy to heat requires that

$$\frac{\mathrm{d}\mathcal{H}_\mathrm{m}}{\mathrm{d}t} = [\mathcal{H}_\mathrm{m}, \mathcal{H}_\mathrm{m}] \leq 0, \tag{10.2.9}$$

where (if gravitational effects are neglected)

$$\mathcal{H}_\mathrm{m}[\boldsymbol{M}] = \int \left[\frac{\boldsymbol{M}\cdot\boldsymbol{M}}{2\rho}\right] \mathrm{d}V; \tag{10.2.10}$$

and

g) close to equilibrium, the dissipation bracket must have a symmetric structure indicative of the Casimir–Onsager reciprocal relationships between the transport coefficients relating fluxes to affinities (the derivatives or the gradient of the Hamiltonian with respect to the dynamical variables).[16–19] A nice feature of the generalized bracket formalism is that the appropriate expressions for the fluxes arise automatically from the formulation (they come out to depend only on the dissipation bracket) and should not be specified explicitly.[1]

Based on the above formulation, Beris and Edwards[1] have considered many examples of applications for complex materials and processes, such as the spatial description of the ideal fluid flow, the incompressible ideal fluid, the general isotropic fluid, the incompressible Navier–Stokes equations, the hydrodynamic equations for a polar fluid, the material description of nonlinear elasticity, the incompressible viscoelastic fluid in terms of both a single and multiple conformation tensors, the bead–spring chain model of kinetic theory, a two-fluid model for polymer solutions, surface effects on the microstructure and concentration of incompressible and isothermal viscoelastic fluids, nonconventional transport phenomena (such as relaxation in heat and mass transfer, phase transitions in inhomogeneous media, and the inertial description of incompressible viscoelastic fluids), models for the flow of liquid crystals, and multifluid transport/reaction models for multicomponent systems, as well as models for reacting systems and the dynamics of weakly ionized plasma. They have also successfully accounted for various basic principles, such as the first and second laws of thermodynamics, the Onsager–Casimir relations, the Curie principle, and the principle of material objectivity (frame indifference). We discuss some of these applications in the following sections of this chapter.

10.2.1 The Poisson part of the generalized bracket

The requirements imposed on the Poisson bracket result in an expression which, although dependent on the variables of interest, is quite standard. For a structureless system described by the set of hydrodynamic variables $\boldsymbol{x} = \{\rho(\boldsymbol{r}, t), \boldsymbol{M}(\boldsymbol{r}, t), s(\boldsymbol{r}, t)\}$, it has the form[1,20–23]

$$\begin{aligned}\{F, G\} = &-\int \left[\frac{\delta F}{\delta \rho}\nabla_\beta\left(\frac{\delta G}{\delta M_\beta}\rho\right) - \frac{\delta G}{\delta \rho}\nabla_\beta\left(\frac{\delta F}{\delta M_\beta}\rho\right)\right] \mathrm{d}V \\ &-\int \left[\frac{\delta F}{\delta M_\gamma}\nabla_\beta\left(\frac{\delta G}{\delta M_\beta}M_\gamma\right) - \frac{\delta G}{\delta M_\gamma}\nabla_\beta\left(\frac{\delta F}{\delta M_\beta}M_\gamma\right)\right] \mathrm{d}V \\ &-\int \left[\frac{\delta F}{\delta s}\nabla_\beta\left(\frac{\delta G}{\delta M_\beta}s\right) - \frac{\delta G}{\delta s}\nabla_\beta\left(\frac{\delta F}{\delta M_\beta}s\right)\right] \mathrm{d}V, \end{aligned} \qquad (10.2.11)$$

while for a system with internal microstructure described by the set $\boldsymbol{x} = \{\rho(\boldsymbol{r}, t), \boldsymbol{M}(\boldsymbol{r}, t), s(\boldsymbol{r}, t), \boldsymbol{C}(\boldsymbol{r}, t)\}$, implying that the structural variable is a tensor $\boldsymbol{C}$, a typical expression is

$$\begin{aligned}\{F, G\} = &-\int \left[\frac{\delta F}{\delta \rho}\nabla_\beta\left(\frac{\delta G}{\delta M_\beta}\rho\right) - \frac{\delta G}{\delta \rho}\nabla_\beta\left(\frac{\delta F}{\delta M_\beta}\rho\right)\right] \mathrm{d}V \\ &-\int \left[\frac{\delta F}{\delta M_\gamma}\nabla_\beta\left(\frac{\delta G}{\delta M_\beta}M_\gamma\right) - \frac{\delta G}{\delta M_\gamma}\nabla_\beta\left(\frac{\delta F}{\delta M_\beta}M_\gamma\right)\right] \mathrm{d}V \\ &-\int \left[\frac{\delta F}{\delta s}\nabla_\beta\left(\frac{\delta G}{\delta M_\beta}s\right) - \frac{\delta G}{\delta s}\nabla_\beta\left(\frac{\delta F}{\delta M_\beta}s\right)\right] \mathrm{d}V\end{aligned}$$

$$-\int\left[\frac{\delta F}{\delta C_{\alpha\beta}}\nabla_\gamma\left(\frac{\delta G}{\delta M_\gamma}C_{\alpha\beta}\right)-\frac{\delta G}{\delta C_{\alpha\beta}}\nabla_\gamma\left(\frac{\delta F}{\delta M_\gamma}C_{\alpha\beta}\right)\right]\mathrm{d}V$$
$$-\int C_{\gamma\alpha}\left[\frac{\delta G}{\delta C_{\alpha\beta}}\nabla_\gamma\left(\frac{\delta F}{\delta M_\beta}\right)-\frac{\delta F}{\delta C_{\alpha\beta}}\nabla_\gamma\left(\frac{\delta G}{\delta M_\beta}\right)\right]\mathrm{d}V$$
$$-\int C_{\gamma\beta}\left[\frac{\delta G}{\delta C_{\alpha\beta}}\nabla_\gamma\left(\frac{\delta F}{\delta M_\alpha}\right)-\frac{\delta F}{\delta C_{\alpha\beta}}\nabla_\gamma\left(\frac{\delta G}{\delta M_\alpha}\right)\right]\mathrm{d}V, \qquad (10.2.12)$$

where the last three terms have been chosen so as to give rise to a frame invariant, objective time derivative for the second-rank object $\boldsymbol{C}$ (here, of the upper-convected type).[24–28] Please note that in the above equations and in the following, Einstein's notation implying summation over repeated indices has been tacitly adopted. The above brackets are bilinear and antisymmetric and satisfy the Jacobi identity. The corresponding expressions for the Hamiltonian $\mathcal{H}$ are

$$\mathcal{H}(\rho,\boldsymbol{M},s)=\int(e_\mathrm{k}+e_\mathrm{p}+u)\,\mathrm{d}V=\int\left[\frac{|\boldsymbol{M}(\boldsymbol{r},t)|^2}{2\rho(\boldsymbol{r},t)}+e_\mathrm{p}(\boldsymbol{r},t)+u\left(\rho(\boldsymbol{r},t),s(\boldsymbol{r},t)\right)\right]\mathrm{d}V \qquad (10.2.13a)$$

for the structureless fluid, and

$$\mathcal{H}(\rho,\boldsymbol{M},s)=\int(e_\mathrm{k}+e_\mathrm{p}+u)\,\mathrm{d}V$$
$$=\int\left[\frac{|\boldsymbol{M}(\boldsymbol{r},t)|^2}{2\rho(\boldsymbol{r},t)}+e_\mathrm{p}(\boldsymbol{r},t)+u\left(\rho(\boldsymbol{r},t),s(\boldsymbol{r},t),\boldsymbol{C}(\boldsymbol{r},t)\right)\right]\mathrm{d}V \qquad (10.2.13b)$$

for the structured fluid.

Equations (10.2.11) and (10.2.12) for the Poisson bracket constitute the noncanonical, spatial equivalent of the canonical Poisson bracket in the material description (Lagrangian specification) in either discrete or continuum space and are derived by following a reduction procedure originally proposed by Marsden et al.[18] and later by Abarbanel et al.[20] (see also Chapter 5, Ref. 3).

Example 10.1

Assuming that the properties of a structureless fluid (i.e. of a fluid where no internal structural variables are considered) are described solely by the above Poisson bracket (such a fluid is called ideal or inviscid), derive the corresponding equations of motion.

Solution: The spatial form of the evolution equations for such a fluid is derived by the generalized bracket equation, Eq. (10.2.3), by (a) assuming that the set of state variables describing the system consists only of the hydrodynamic set $\boldsymbol{x}=\{\rho(\boldsymbol{r},t),\boldsymbol{M}(\boldsymbol{r},t),s(\boldsymbol{r},t)\}$, (b) neglecting the dissipative bracket, (c) using expression (10.2.11) for the Poisson bracket, (d) using expression (10.2.13a) for the Hamiltonian, (e) writing the dynamical equations for any functional $F[\rho, \boldsymbol{M}, s] = \int f(\rho, \boldsymbol{M}, s)\, \mathrm{d}V$ in the form

$$\frac{\mathrm{d}F}{\mathrm{d}t} = \int \left[\frac{\delta F}{\delta \rho} \frac{\partial \rho}{\partial t} + \frac{\delta F}{\delta M_\alpha} \frac{\delta M_\alpha}{\delta t} + \frac{\delta F}{\delta s} \frac{\delta s}{\delta t} \right] \mathrm{d}V, \tag{10.2.14}$$

and (f) comparing terms one by one. To see this, according to the generalized bracket expression (10.2.3), we have that

$$\begin{aligned}\frac{\mathrm{d}F}{\mathrm{d}t} = &- \int \left[\frac{\delta F}{\delta \rho} \nabla_\beta \left(\frac{\delta \mathcal{H}}{\delta M_\beta} \rho \right) - \frac{\delta \mathcal{H}}{\delta \rho} \nabla_\beta \left(\frac{\delta F}{\delta M_\beta} \rho \right) \right] \mathrm{d}V \\ &- \int \left[\frac{\delta F}{\delta M_\gamma} \nabla_\beta \left(\frac{\delta \mathcal{H}}{\delta M_\beta} M_\gamma \right) - \frac{\delta \mathcal{H}}{\delta M_\gamma} \nabla_\beta \left(\frac{\delta F}{\delta M_\beta} M_\gamma \right) \right] \mathrm{d}V \\ &- \int \left[\frac{\delta F}{\delta s} \nabla_\beta \left(\frac{\delta \mathcal{H}}{\delta M_\beta} s \right) - \frac{\delta \mathcal{H}}{\delta s} \nabla_\beta \left(\frac{\delta F}{\delta M_\beta} s \right) \right] \mathrm{d}V. \end{aligned} \tag{10.2.15}$$

The second term in each of the above integrals can be integrated by parts which (upon neglecting the surface contribution terms) helps us write the equation for $\frac{\mathrm{d}F}{\mathrm{d}t}$ as

$$\begin{aligned}\frac{\mathrm{d}F}{\mathrm{d}t} = &- \int \left[\frac{\delta F}{\delta \rho} \nabla_\beta \left(\frac{\delta \mathcal{H}}{\delta M_\beta} \rho \right) + \frac{\delta F}{\delta M_\beta} \rho \nabla_\beta \left(\frac{\delta \mathcal{H}}{\delta \rho} \right) \right] \mathrm{d}V \\ &- \int \left[\frac{\delta F}{\delta M_\gamma} \nabla_\beta \left(\frac{\delta \mathcal{H}}{\delta M_\beta} M_\gamma \right) + \frac{\delta F}{\delta M_\beta} M_\gamma \nabla_\beta \left(\frac{\delta \mathcal{H}}{\delta M_\gamma} \right) \right] \mathrm{d}V \\ &- \int \left[\frac{\delta F}{\delta s} \nabla_\beta \left(\frac{\delta \mathcal{H}}{\delta M_\beta} s \right) + \frac{\delta F}{\delta M_\beta} s \nabla_\beta \left(\frac{\delta \mathcal{H}}{\delta s} \right) \right] \mathrm{d}V \end{aligned} \tag{10.2.16}$$

which, after collecting similar terms together, takes the form

$$\begin{aligned}\frac{\mathrm{d}F}{\mathrm{d}t} = &- \int \left[\frac{\delta F}{\delta \rho} \nabla_\beta \left(\frac{\delta \mathcal{H}}{\delta M_\beta} \rho \right) \right] \mathrm{d}V \\ &- \int \left[\frac{\delta F}{\delta M_\beta} \left(\rho \nabla_\beta \left(\frac{\delta \mathcal{H}}{\delta \rho} \right) + \nabla_\gamma \left(\frac{\delta \mathcal{H}}{\delta M_\gamma} M_\beta \right) + M_\gamma \nabla_\beta \left(\frac{\delta \mathcal{H}}{\delta M_\gamma} \right) + s \nabla_\beta \left(\frac{\delta \mathcal{H}}{\delta s} \right) \right) \right] \mathrm{d}V \\ &- \int \left[\frac{\delta F}{\delta s} \nabla_\beta \left(\frac{\delta \mathcal{H}}{\delta M_\beta} s \right) \right] \mathrm{d}V. \end{aligned} \tag{10.2.17}$$

continued

Example 10.1 *continued*

By comparing the two different expressions for the time evolution of F, Eqs. (10.2.14) and (10.2.17), we obtain the following equations of motion for the inviscid fluid:

$$\frac{\partial \rho}{\partial t} = -\nabla_\beta \left(\frac{\delta \mathcal{H}}{\delta M_\beta} \rho \right) \tag{10.2.18a}$$

$$\frac{\partial M_\alpha}{\partial t} = -\left[\rho \nabla_\alpha \frac{\delta \mathcal{H}}{\delta \rho} + \nabla_\beta \left(\frac{\delta \mathcal{H}}{\delta M_\beta} M_\alpha \right) + M_\beta \nabla_\alpha \frac{\delta \mathcal{H}}{\delta M_\beta} + s \nabla_\alpha \frac{\delta \mathcal{H}}{\delta s} \right] \tag{10.2.18b}$$

$$\frac{\partial s}{\partial t} = -\nabla_\beta \left(\frac{\delta \mathcal{H}}{\delta M_\beta} s \right). \tag{10.2.18c}$$

The first is the conservation equation for the mass (the continuity equation), the second is the conservation equation for the momentum (the momentum balance equation), and the third is the balance equation for the entropy. The three balance equations can be written in a more familiar form by computing the necessary functional derivatives of the Hamiltonian based on expression (10.2.13a) for $\mathcal{H}$:

$$\frac{\delta \mathcal{H}}{\delta M_\beta} = \frac{M_\beta}{\rho} = \boldsymbol{\nu}_\beta \tag{10.2.19a}$$

$$\frac{\delta \mathcal{H}}{\delta \rho} = -\frac{M_\alpha M_\alpha}{2\rho^2} + \frac{e_{\mathrm{p}}}{\rho} + \frac{\partial u}{\partial \rho} \tag{10.2.19b}$$

$$\frac{\delta \mathcal{H}}{\delta s} = \frac{\partial u}{\partial s} = T. \tag{10.2.19c}$$

Substituting back into Eqs. (10.2.18), defining the pressure P through

$$P = \rho \left.\frac{\partial u}{\partial \rho}\right|_s + s \left.\frac{\partial u}{\partial s}\right|_\rho - u(\rho, s), \tag{10.2.20}$$

rearranging terms in the momentum equation, and making use also of the continuity equation results in the following set of transport equations for an inviscid fluid:

$$\frac{\partial \rho}{\partial t} = -\nabla_\beta (\boldsymbol{\nu}_\beta \rho) \tag{10.2.21a}$$

$$\rho \frac{\partial \boldsymbol{\nu}_\alpha}{\partial t} = -\rho \boldsymbol{\nu}_\beta \nabla_\beta \boldsymbol{\nu}_\alpha - \rho \nabla_\alpha \frac{e_{\mathrm{p}}}{\rho} - \nabla_\alpha P \tag{10.2.21b}$$

$$\frac{\partial s}{\partial t} = -\nabla_\beta (\boldsymbol{\nu}_\beta s). \tag{10.2.21c}$$

Example 10.2

Write the evolution equation describing the conservation of energy for an incompressible ideal fluid.

Solution: It is easy to show that

$$\frac{\partial u}{\partial t} = -\nabla_\alpha \left(\boldsymbol{\nu}_\alpha u\right) - p\nabla_\alpha \boldsymbol{\nu}_\alpha \tag{10.2.22}$$

and

$$\frac{\partial e}{\partial t} = -\nabla_\alpha \left(\boldsymbol{\nu}_\alpha e\right) - \nabla_\alpha \left(\boldsymbol{\nu}_\alpha P\right) - \rho\boldsymbol{\nu}_\alpha \nabla_\alpha \frac{e_{\mathrm{p}}}{\rho}, \tag{10.2.23}$$

where $e = e_{\mathrm{k}} + u = \frac{1}{2}\rho v^2 + u$. The last equation can also be found by applying the general evolution equation (10.2.3) for an arbitrary functional F for the case $F = E_{\mathrm{k}}$, where $E_{\mathrm{k}} = \int \frac{M_\alpha M_\alpha}{2\rho} \mathrm{d}V$, i.e. by using that $\frac{\mathrm{d}E_{\mathrm{k}}}{\mathrm{d}t} = \{E_{\mathrm{k}}, \mathcal{H}\}$.

Example 10.3

Assuming that the properties of a complex fluid (i.e. of a fluid with internal structural variables) are described solely by the Poisson bracket given by Eq. (10.2.12) (such a fluid is called a nonlinear elastic fluid), derive the corresponding equations of motion.

Solution: Following exactly the same procedure but for the set $\boldsymbol{x} = \{\rho(\boldsymbol{r}, t), \boldsymbol{M}(\boldsymbol{r}, t), s(\boldsymbol{r}, t), \boldsymbol{C}(\boldsymbol{r}, t)\}$, implying that the dynamical equations for any functional $F[\rho, \boldsymbol{M}, s, \boldsymbol{C}] = \int f(\rho, \boldsymbol{M}, s, \boldsymbol{C})\, \mathrm{d}V$ should be written in the form

$$\frac{\mathrm{d}F}{\mathrm{d}t} = \int \left[\frac{\delta F}{\delta \rho}\frac{\partial \rho}{\partial t} + \frac{\delta F}{\delta M_\alpha}\frac{\delta M_\alpha}{\delta t} + \frac{\delta F}{\delta s}\frac{\delta s}{\delta t} + \frac{\delta F}{\delta C_{\alpha\beta}}\frac{\delta C_{\alpha\beta}}{\delta t}\right] \mathrm{d}V,$$

and making use of $\frac{\delta H}{\delta C_{\alpha\beta}} = \frac{\partial u}{\partial C_{\alpha\beta}}$ from Eq. (10.2.13b), we are led to the following dynamical equations for the components of $\boldsymbol{x}$:

$$\frac{\partial \rho}{\partial t} = -\nabla_\beta \left(\boldsymbol{\nu}_\beta \rho\right) \tag{10.2.24a}$$

$$\rho\frac{\partial \boldsymbol{\nu}_\alpha}{\partial t} = -\rho\boldsymbol{\nu}_\beta \nabla_\beta \boldsymbol{\nu}_\alpha - \rho\nabla_\alpha \frac{e_p}{\rho} - \nabla_\alpha P + \nabla_\gamma \left(2C_{\gamma\beta}\frac{\partial u}{\partial C_{\alpha\beta}}\right) \tag{10.2.24b}$$

$$\frac{\partial s}{\partial t} = -\nabla_\beta \left(\boldsymbol{\nu}_\beta s\right) \tag{10.2.24c}$$

continued

Example 10.3 *continued*

and

$$\frac{\partial C_{\alpha\beta}}{\partial t} = -\nabla_\gamma \left(\nu_\gamma C_{\alpha\beta}\right) + C_{\alpha\gamma}\nabla_\gamma \boldsymbol{\nu}_\beta + C_{\gamma\beta}\nabla_\gamma \boldsymbol{\nu}_\alpha, \tag{10.2.24d}$$

where the pressure P has now been defined as

$$P = \rho \left.\frac{\partial u}{\partial \rho}\right|_{s,C} + s \left.\frac{\partial u}{\partial s}\right|_{\rho,C} + \boldsymbol{C} : \left.\frac{\partial u}{\partial \boldsymbol{C}}\right|_{\rho,s} - u\left(\rho, s, \boldsymbol{C}\right). \tag{10.2.25}$$

Equation (10.2.24d) for the evolution of $\boldsymbol{C}$ is usually written as

$$\dot{C}_{\alpha\beta,[1]} = 0, \tag{10.2.26}$$

where

$$\dot{C}_{\alpha\beta,[1]} \equiv \frac{\partial C_{\alpha\beta}}{\partial t} + \nabla_\gamma \left(\nu_\gamma C_{\alpha\beta}\right) - C_{\alpha\gamma}\nabla_\gamma \boldsymbol{\nu}_\beta - C_{\gamma\beta}\nabla_\gamma \boldsymbol{\nu}_\alpha \tag{10.2.27}$$

denotes the upper-convected Maxwell derivative introduced by Oldroyd. This is a materially objective derivative (a contravariant derivative) and arises here because $\boldsymbol{C}$ is a second-rank tensor, so it has to transform as an absolute, second-rank contravariant tensor. If one works with an absolute, second-rank tensor which is covariant or mixed, then other materially objective Oldroyd derivatives arise. For a brief introduction to covariant or contravariant or mixed tensors, the interested reader is referred to Appendix A8 of Ref. 26 or to Ref. 29. Equation (10.2.24a) is the usual conservation equation for the mass, Eq. (10.2.24b) is the conservation equation for the momentum containing now an extra term due to the material elasticity, Eq. (10.2.24c) is the entropy equation, and Eq. (10.2.24d) is the evolution equation for the conformation tensor for a purely elastic material with nonlinear (in general) elasticity. The extra term of the form $2C_{\beta\gamma}\frac{\partial u}{\partial C_{\gamma\alpha}}$ appearing in Eq. (10.2.24b) for the momentum is called the extra stress term or the polymer contribution to the stress and is symbolized as $\boldsymbol{\tau}$. In the most general case, this is defined as

$$\tau_{\alpha\beta} = 2C_{\beta\gamma}\frac{\delta \mathcal{H}}{\delta C_{\gamma\alpha}}. \tag{10.2.28}$$

Thus, we usually write Eq. (10.2.24b) as

$$\rho\frac{\partial \boldsymbol{\nu}_\alpha}{\partial t} = -\rho\boldsymbol{\nu}_\beta\nabla_\beta\boldsymbol{\nu}_\alpha - \rho\nabla_\alpha\frac{e_p}{\rho} - \nabla_\alpha P + \nabla_\beta\tau_{\beta\alpha}. \tag{10.2.29}$$

In the literature, authors[30,31] have also discussed reducing the above equations when the conformation tensor is constrained to having a unit trace or a unit determinant.

10.2.2 The dissipative part of the generalized bracket

For a system described by the set $\boldsymbol{x} = \{\rho(\boldsymbol{r}, t), \boldsymbol{M}(\boldsymbol{r}, t), s(\boldsymbol{r}, t)\}$ of hydrodynamic variables, the simplified expression of a bilinear form for the dissipative bracket is

$$\begin{aligned}
[F, G] = &-\int Q_{\alpha\beta\gamma\varepsilon} \nabla_\alpha \frac{\delta F}{\delta M_\beta} \nabla_\gamma \frac{\delta G}{\delta M_\varepsilon} \mathrm{d}V \\
&-\int A_{\alpha\beta} \nabla_\alpha \frac{\delta F}{\delta s} \nabla_\beta \frac{\delta G}{\delta s} \mathrm{d}V \\
&+\int \frac{1}{T}\frac{\delta F}{\delta s} Q_{\alpha\beta\gamma\varepsilon} \nabla_\alpha \frac{\delta G}{\delta M_\beta} \nabla_\gamma \frac{\delta G}{\delta M_\varepsilon} \mathrm{d}V \\
&+\int \frac{1}{T}\frac{\delta F}{\delta s} A_{\alpha\beta} \nabla_\alpha \frac{\delta G}{\delta s} \nabla_\beta \frac{\delta G}{\delta s} \mathrm{d}V,
\end{aligned} \tag{10.2.30}$$

where the various phenomenological coefficients should satisfy the following constraints due to the Onsager–Casimir reciprocity principles:[16–19]

$$\begin{aligned}
A_{\gamma\beta} &= A_{\beta\gamma}, \quad \forall \gamma, \beta \\
Q_{\alpha\beta\gamma\varepsilon} &= Q_{\gamma\varepsilon\alpha\beta}, \quad \forall \alpha, \beta, \gamma, \varepsilon.
\end{aligned} \tag{10.2.31}$$

For an isotropic fluid (a material whose properties are independent of space direction), the two phenomenological coefficients should have the form[32–37]

$$\begin{aligned}
A_{\gamma\beta} &= A_{00}\delta_{\gamma\beta}, \quad \forall \gamma, \beta \\
Q_{\alpha\beta\gamma\varepsilon} &= \mu\left(\delta_{\alpha\gamma}\delta_{\beta\varepsilon} + \delta_{\alpha\varepsilon}\delta_{\beta\gamma}\right) + \eta\left(\delta_{\alpha\gamma}\delta_{\beta\varepsilon} - \delta_{\alpha\varepsilon}\delta_{\beta\gamma}\right) + \kappa'\delta_{\alpha\beta}\delta_{\gamma\varepsilon}, \quad \forall \alpha, \beta, \gamma, \varepsilon,
\end{aligned} \tag{10.2.32}$$

with the three numerical coefficients μ, η, and κ' having units of viscosity. For an isotropic, structureless, viscous medium, in particular, it can be shown that $\eta = 0$.[1] In the above expressions, μ is the shear viscosity, κ' is related to the bulk viscosity κ through $\kappa = \kappa' + \frac{2}{3}\mu$, and the thermal conductivity k is defined as $k = A_{00}T$.[1]

On the other hand, for a compressible, nonisothermal viscoelastic melt described by the set $\boldsymbol{x} = \{\rho(\boldsymbol{r}, t), \boldsymbol{M}(\boldsymbol{r}, t), s(\boldsymbol{r}, t), \boldsymbol{C}(\boldsymbol{r}, t)\}$ which includes also a structural variable (the conformation tensor $\boldsymbol{C}$), a typical expression postulated for the dissipation bracket (a bilinear form with respect to F and G implying that nonlinear terms are neglected) is

$$\begin{aligned}
[F, G] = &-\int Q_{\alpha\beta\gamma\varepsilon} \nabla_\alpha \frac{\delta F}{\delta M_\beta} \nabla_\gamma \frac{\delta G}{\delta M_\varepsilon} \mathrm{d}V \\
&-\int A_{\alpha\beta} \nabla_\alpha \frac{\delta F}{\delta s} \nabla_\beta \frac{\delta G}{\delta s} \mathrm{d}V \\
&-\int \Lambda_{\alpha\beta\gamma\varepsilon} \frac{\delta F}{\delta C_{\alpha\beta}} \frac{\delta G}{\delta C_{\gamma\varepsilon}} \mathrm{d}V
\end{aligned}$$

$$
\begin{aligned}
& - \int B_{\alpha\beta\gamma\varepsilon\eta\nu} \nabla_\gamma \frac{\delta F}{\delta C_{\alpha\beta}} \nabla_\nu \frac{\delta G}{\delta C_{\varepsilon\eta}} \mathrm{d}V \\
& - \int L_{\alpha\beta\gamma\varepsilon} \left(\nabla_\alpha \frac{\delta F}{\delta M_\beta} \frac{\delta G}{\delta C_{\gamma\varepsilon}} - \nabla_\alpha \frac{\delta G}{\delta M_\beta} \frac{\delta F}{\delta C_{\gamma\varepsilon}} \right) \mathrm{d}V \\
& + \int \frac{1}{T} \frac{\delta F}{\delta s} Q_{\alpha\beta\gamma\varepsilon} \nabla_\alpha \frac{\delta G}{\delta M_\beta} \nabla_\gamma \frac{\delta G}{\delta M_\varepsilon} \mathrm{d}V \\
& + \int \frac{1}{T} \frac{\delta F}{\delta s} A_{\alpha\beta} \nabla_\alpha \frac{\delta G}{\delta s} \nabla_\beta \frac{\delta G}{\delta s} \mathrm{d}V \\
& + \int \frac{1}{T} \frac{\delta F}{\delta s} \Lambda_{\alpha\beta\gamma\varepsilon} \frac{\delta G}{\delta C_{\alpha\beta}} \frac{\delta G}{\delta C_{\gamma\varepsilon}} \mathrm{d}V \\
& + \int \frac{1}{T} \frac{\delta F}{\delta s} B_{\alpha\beta\gamma\varepsilon\eta\nu} \nabla_\gamma \frac{\delta G}{\delta C_{\alpha\beta}} \nabla_\nu \frac{\delta G}{\delta C_{\varepsilon\eta}} \mathrm{d}V.
\end{aligned} \tag{10.2.33}
$$

In this equation, $\boldsymbol{\Lambda}$ is a fourth-rank semi-definite symmetric matrix expressing relaxation effects (it is inversely proportional to a characteristic relaxation time λ_0 for the system), $\boldsymbol{B}$ a sixth-rank tensor representing inhomogeneous translational diffusivity which is seldom considered in viscoelastic modelling, $\boldsymbol{Q}$ a fourth-rank tensor representing solvent effects (it is exactly zero for polymer melts), and $\boldsymbol{L}$ a fourth-rank tensor describing nonaffine motion.

For the relaxation matrix $\boldsymbol{\Lambda}$, Beris and Edwards report that most of the existing viscoelastic models are special choices of the more general construction[1]

$$
\begin{aligned}
\Lambda_{\alpha\beta\gamma\varepsilon}(\tilde{\mathbf{c}}) = {} & f_1(I_1)\left(\tilde{c}_{\alpha\gamma}\delta_{\beta\varepsilon} + \tilde{c}_{\alpha\varepsilon}\delta_{\beta\gamma} + \tilde{c}_{\beta\gamma}\delta_{\alpha\varepsilon} + \tilde{c}_{\beta\varepsilon}\delta_{\alpha\gamma}\right) + 2f_2(I_1)\left(\tilde{c}_{\alpha\gamma}\tilde{c}_{\beta\varepsilon} + \tilde{c}_{\alpha\varepsilon}\tilde{c}_{\beta\gamma}\right) \\
& + f_3(I_1)\left(\tilde{c}_{\alpha\beta}\delta_{\gamma\varepsilon} + \tilde{c}_{\gamma\varepsilon}\delta_{\alpha\beta}\right) + f_4(I_1)\left(\tilde{c}_{\alpha\beta}\tilde{c}_{\gamma\varepsilon}\right),
\end{aligned} \tag{10.2.34}
$$

where I_1 is the first invariant of $\tilde{\mathbf{c}}$ (the trace of $\tilde{\mathbf{c}}$), $\boldsymbol{\delta}$ is the Kronecker symbol, and f_1, f_2, f_3 and f_4 are arbitrary functions of I_1, which, however, should not violate the second law of thermodynamics. A few examples are reported in Table 10.1, which summarizes choices of $\boldsymbol{\Lambda}$ that reproduce well-known viscoelastic models such as the upper-convected Maxwell (UCM), the Giesekus, the Phan-Tien/Tanner (PTT), the finitely extensible nonlinear elastic (FENE), the hybrid FENE/PTT/Giesekus, the modified Bird-DeAguiar, the extended White-Metzner, and a recent generalization of the Giesekus model.[38]

For the matrix $\boldsymbol{L}$ describing nonaffine effects, Beris–Edwards report that this is usually of the form[1]

$$
L_{\alpha\beta\gamma\varepsilon}(\mathbf{c}) = -\frac{\xi}{2}\left(c_{\alpha\gamma}\delta_{\beta\varepsilon} + c_{\alpha\varepsilon}\delta_{\beta\gamma} + c_{\beta\gamma}\delta_{\alpha\varepsilon} + c_{\beta\varepsilon}\delta_{\alpha\gamma}\right), \tag{10.2.35}
$$

where ξ denotes the slip or nonaffine parameter with values such that $0 \leq \xi < 1$ in order for the model to be thermodynamically admissible (i.e. not to violate the second law of thermodynamics).[38] For $\xi = 0$, the model reduces to the upper-convected Maxwell derivative, for $\xi = 1$ to the co-rotational derivative, and for $\xi = 2$ to the lower-convected Maxwell derivative.

Table 10.1 *Mathematical Form of the Functions $f_1(I_1)$, $f_2(I_1)$, $f_3(I_1)$, and $f_4(I_1)$ Appearing in the General Expression, Eq. (10.2.34), of the Relaxation Matrix Λ for Several Viscoelastic Models*

Model	$f_1(I_1)$	$f_2(I_1)$	$f_3(I_1)$	$f_4(I_1)$
UCM, FENE	$\frac{1}{2\lambda_0 nk_BT}$	0	0	0
Giesekus	$\frac{1}{2\lambda_0 nk_BT}(1-\alpha)$	$\frac{1}{2\lambda_0 nk_BT}\alpha$	0	0
PTT	$\frac{1}{2\lambda_0 nk_BT}\exp\left[(1-\xi)\,\varepsilon\,(I_1-3)\right]$	0	0	0
Extended White/Metzner	$\frac{1}{2\lambda_0 nk_BT}\left(\frac{I_1}{3}\right)^{-k}$	0	0	0
Modified Bird-DeAguiar	$\frac{1}{2\lambda_0 nk_BT}\sigma$	0	0	$\frac{2}{\lambda_0 nk_BT}\frac{1-\sigma}{I_1}$
Leonov	0	$\frac{1}{2\lambda_0 nk_BT}$	0	$-\frac{2}{3\lambda_0 nk_BT}$

Note: λ_0 denotes the chain longest characteristic time, n the chain number density, ξ the slip coefficient, and α, σ, k, and ε are parameters of the various models.

The tensor $\boldsymbol{Q}$, on the other hand, is taken to be equal to

$$Q_{\alpha\beta\gamma\varepsilon} = \mu\left(\delta_{\alpha\gamma}\delta_{\beta\varepsilon} + \delta_{\alpha\varepsilon}\delta_{\beta\gamma}\right) + \eta\left(\delta_{\alpha\gamma}\delta_{\beta\varepsilon} - \delta_{\alpha\varepsilon}\delta_{\beta\gamma}\right) + \kappa'\delta_{\alpha\beta}\delta_{\gamma\varepsilon}, \quad \forall\alpha, \beta, \gamma, \varepsilon. \tag{10.2.36}$$

In actual applications, $\boldsymbol{Q}$-dependent terms show up in the momentum equation for the total mass density as the gradient of a stress tensor term and in the entropy equation as a bulk entropy production term, again through the stress tensor.

In the next examples, we discuss applications of the generalized bracket formalism in the modelling of Newtonian and non-Newtonian fluids. A typical example of the first class is the Navier–Stokes equations for structureless fluids. A typical example of the second class is polymeric fluids (melts and solutions). Polymer melts are usually modelled as incompressible viscoelastic fluids; in polymer solutions, on the other hand, one often must account for effects related to gradients in the polymer concentration even in the bulk of the flow (due to inhomegeneities in the flow field).

Example 10.4

Derive the equations of motion for an isotropic structureless fluid (known also as the general Navier–Stokes equations).

Solution: The equations of motion for a structureless fluid described by the vector $\boldsymbol{x} = \{\rho(\boldsymbol{r}, t), \boldsymbol{M}(\boldsymbol{r}, t), s(\boldsymbol{r}, t)\}$ of hydrodynamic variables are derived by applying the general evolution equation (10.2.3) for a Poisson bracket given by expression (10.2.11), a dissipation bracket given by expression (10.2.30) and a Hamiltonian described by Eq. (10.2.13a). The result is the set of equations

$$\frac{\partial\rho}{\partial t} = -\nabla_\beta\left(\boldsymbol{\nu}_\beta\rho\right) \tag{10.2.37a}$$

continued

Example 10.4 *continued*

$$\rho\frac{\partial \boldsymbol{\nu}_\alpha}{\partial t} = -\rho\boldsymbol{\nu}_\beta\nabla_\beta\boldsymbol{\nu}_\alpha - \rho\nabla_\alpha\frac{e_{\mathrm{p}}}{\rho} - \nabla_\alpha P + \nabla_\gamma\tau_{\gamma\alpha} \tag{10.2.37b}$$

$$\frac{\partial s}{\partial t} = -\nabla_\beta\left(\boldsymbol{\nu}_\beta s\right) + \frac{1}{T}\nabla_\gamma\left(k\nabla_\gamma T\right) + \frac{1}{T}\tau_{\gamma\alpha}\nabla_\alpha\boldsymbol{\nu}_\gamma, \tag{10.2.37c}$$

where the stress tensor $\boldsymbol{\tau}$ is given by

$$\tau_{\gamma\alpha} = \mu\left(\nabla_\gamma\boldsymbol{\nu}_\alpha + \nabla_\alpha\nu_\gamma\right) + \kappa'\delta_{\gamma\alpha}\nabla_\varepsilon\nu_\varepsilon \tag{10.2.38}$$

and, as before, $k = A_{00}T$ denotes the thermal conductivity.

The corresponding expression for the energy conservation is

$$\frac{\partial e}{\partial t} = -\nabla_\alpha\left(\boldsymbol{\nu}_\alpha e\right) - \nabla_\alpha\left(\boldsymbol{\nu}_\alpha P\right) - \rho\boldsymbol{\nu}_\alpha\nabla_\alpha\frac{e_{\mathrm{p}}}{\rho} + \nabla_\alpha\left(k\nabla_\alpha T\right) + \nabla_\gamma\left(\tau_{\gamma\alpha}\boldsymbol{\nu}_\alpha\right), \tag{10.2.39}$$

where $e = e_{\mathrm{k}} + u = \frac{1}{2}\rho v^2 + u$. The last equation can also be found by applying the general evolution equation (10.2.3) for an arbitrary functional F, for the case $F = E_{\mathrm{k}}$ where $E_{\mathrm{k}} = \int \frac{M_\alpha M_\alpha}{2\rho}\mathrm{d}V$, i.e. by using that $\frac{\mathrm{d}E_{\mathrm{k}}}{\mathrm{d}t} = \{E_{\mathrm{k}}, H\} + [E_{\mathrm{k}}, H]$. The corresponding expression for the entropy production σ is

$$\sigma = \frac{1}{T}\left[Q_{\alpha\beta\gamma\varepsilon}\left(\nabla_\alpha\boldsymbol{\nu}_\beta\right)\left(\nabla_\gamma\nu_\varepsilon\right) + A_{00}^{\gamma\beta}\left(\nabla_\gamma T\right)\left(\nabla_\beta T\right)\right] \tag{10.2.40}$$

and according to the second law of thermodynamics, it should be nonnegative. Using Eq. (10.2.36) corresponding to an isotropic fluid, the requirement $\sigma \geq 0$ is equivalent to the following requirement (also known as the Clausius inequality):[39]

$$\sigma = \frac{1}{T}\left[2\mu\left(\nabla_\alpha\boldsymbol{\nu}_\beta + \nabla_\beta\boldsymbol{\nu}_\alpha\right) + \kappa'\left(\nabla_\alpha\boldsymbol{\nu}_\alpha + \nabla_\beta\boldsymbol{\nu}_\beta\right) + A_{00}\left(\nabla_\gamma T\right)\left(\nabla_\beta T\right)\right] \geq 0. \tag{10.2.41}$$

Given that $T > 0$, in order for this inequality to hold under general conditions, Beris–Edwards[1] showed (by breaking the first two terms on the right-hand-side of the equation into independent terms) that the following three inequalities should be satisfied:

$$\kappa' + \frac{2}{3}\mu \geq 0 \quad \text{and} \quad \mu \geq 0 \quad \text{and} \quad A_{00} \geq 0, \tag{10.2.42}$$

which are known as the Stokes relations. Due to the first inequality in the above set of constraints, we naturally define as a coefficient of viscosity the bulk viscosity κ through $\kappa \equiv \kappa' + \frac{2}{3}\mu$; thus, $\kappa \geq 0$.

Example 10.5

Derive the general form of the equations of motion for a compressible, nonisothermal viscoelastic melt described by a single conformation tensor.

Solution: This is the most general case where one accounts explicitly for temperature effects by including the entropy density s among the state variables. The Poisson bracket is given by Eq. (10.2.12) while the dissipation bracket can be assumed to be given by Eq. (10.2.33). The resulting evolution equations read

$$\frac{\partial \rho}{\partial t} = -\nabla_\beta (\boldsymbol{\nu}_\beta \rho) \tag{10.2.43a}$$

$$\rho \frac{\partial \boldsymbol{\nu}_\alpha}{\partial t} = -\rho \boldsymbol{\nu}_\beta \nabla_\beta \boldsymbol{\nu}_\alpha - \nabla_\alpha P + \nabla_\beta \tau_{\alpha\beta} \tag{10.2.43b}$$

$$\begin{aligned}\frac{\partial s}{\partial t} &= -\nabla_\beta (\boldsymbol{\nu}_\beta s) + \frac{1}{T} Q_{\alpha\beta\gamma\varepsilon} \nabla_\alpha \boldsymbol{\nu}_\beta \nabla_\gamma \nu_\varepsilon + \frac{1}{T} \nabla_\alpha (A_{\alpha\beta} T \nabla_\beta T) \\ &\quad + \frac{1}{T} \Lambda_{\alpha\beta\gamma\varepsilon} \frac{\delta \mathcal{H}}{\delta C_{\alpha\beta}} \frac{\delta \mathcal{H}}{\delta C_{\gamma\varepsilon}} + \frac{1}{T} B_{\alpha\beta\gamma\varepsilon\zeta\eta} \nabla_\gamma \left(\frac{\delta \mathcal{H}}{\delta C_{\alpha\beta}} \right) \nabla_\eta \left(\frac{\delta \mathcal{H}}{\delta C_{\varepsilon\zeta}} \right)\end{aligned} \tag{10.2.43c}$$

and

$$\begin{aligned}\frac{\partial C_{\alpha\beta}}{\partial t} &= -\nabla_\gamma (\nu_\gamma C_{\alpha\beta}) + C_{\alpha\gamma} \nabla_\gamma \boldsymbol{\nu}_\beta + C_{\gamma\beta} \nabla_\gamma \boldsymbol{\nu}_\alpha - \Lambda_{\alpha\beta\gamma\varepsilon} \frac{\delta \mathcal{H}}{\delta C_{\gamma\varepsilon}} \\ &\quad + L_{\alpha\beta\gamma\varepsilon} \nabla_\gamma \nu_\varepsilon + \nabla_\gamma \left(B_{\alpha\beta\gamma\varepsilon\zeta\eta} \nabla_\eta \frac{\delta \mathcal{H}}{\delta C_{\varepsilon\zeta}} \right).\end{aligned} \tag{10.2.43d}$$

In the above expressions, P is the pressure defined according to Eq. (10.2.25), $\boldsymbol{\tau}$ is the polymer contribution to the stress tensor (i.e. the extra stress tensor) which is now given by

$$\tau_{\alpha\beta} = 2C_{\beta\gamma} \frac{\delta \mathcal{H}}{\delta C_{\gamma\alpha}} + Q_{\alpha\beta\gamma\varepsilon} \nabla_\gamma \nu_\varepsilon + 2L_{\alpha\beta\gamma\varepsilon} \frac{\delta \mathcal{H}}{\delta C_{\gamma\varepsilon}}, \tag{10.2.44}$$

and in Eq. (10.2.43d) we recognize the combination of terms which leads to the appearance of the upper-convected Maxwell derivative of $\boldsymbol{C}$. As far as the tensor of thermal conductivity $\boldsymbol{A}$ is concerned, usually this is assumed to depend on the structure of the fluid. Using the Cayley–Hamilton theorem,[27] such a dependence is mathematically captured through an expression of the form[1]

$$A_{\alpha\beta} = \kappa_1 \delta_{\alpha\beta} + \kappa_2 C_{\alpha\beta} + \kappa_3 C_{\alpha\gamma} C_{\gamma\beta}, \tag{10.2.45}$$

where the coefficients κ_i, $i = 1, 2, 3$ are in general functions of the invariants of $\boldsymbol{C}$ (and possibly of the rest of dynamic variables). By checking the entropy inequality, Beris–Edwards have found the following restrictions on these constants:

$$\kappa_1 \geq 0 \quad \text{and} \quad \kappa_2 + \kappa_3 \geq 0. \tag{10.2.46}$$

10.2.3 The generalized bracket formalism for incompressible and/or isothermal systems

In practical applications, of great interest are the cases (a) of incompressible fluids, and (b) of isothermal fluids. How the general equations in Sections 10.2.1 and 10.2.2 are reduced in these two cases is discussed just below.

For a fluid that maintains a constant density, the corresponding velocity field is divergence free:

$$\left.\begin{array}{l}\dfrac{\partial \rho}{\partial t} = -\nabla_\beta \left(\boldsymbol{\nu}_\beta \rho\right) \\ \text{and} \\ \rho = \text{const.}\end{array}\right\} \Rightarrow \nabla_\alpha \boldsymbol{\nu}_\alpha = 0. \tag{10.2.47}$$

The divergence-free velocity plays the role of an internal constraint for the dynamical equations. It is taken into account[40–43] by correcting the functional derivative $\dfrac{\delta F}{\delta \boldsymbol{M}}$ where $\boldsymbol{M} = \rho\boldsymbol{\nu}$ as[44]

$$\frac{\delta F}{\delta \boldsymbol{M}} = \boldsymbol{I}\left(\frac{\partial F}{\partial \boldsymbol{M}}\right) \in \mathrm{P} = \frac{\partial F}{\partial \boldsymbol{M}} - \nabla P, \tag{10.2.48}$$

where $\boldsymbol{I}\left(\dfrac{\partial F}{\partial \boldsymbol{M}}\right) \in \mathrm{P}$ means a projection operator on the divergence-free space P and P is a scalar (yet undetermined) function (the pressure) to be eventually computed by taking the divergence of the momentum equation for the total density ρ resulting in an elliptic (Poisson) equation subject to flux (von Neumann) boundary conditions, which makes it possible to calculate P up to an unknown numerical constant.

Example 10.6

Write down the Navier–Stokes equations for an incompressible fluid.

Solution: These are similar to those given by Eqs. (10.2.37), but in the following form now:

$$\rho = \text{const.} \Leftrightarrow \boldsymbol{\nabla} \cdot \mathbf{v} = 0 \tag{10.2.49a}$$

$$\rho \frac{\partial \boldsymbol{\nu}_\alpha}{\partial t} = -\rho \boldsymbol{\nu}_\beta \nabla_\beta \boldsymbol{\nu}_\alpha - \nabla_\alpha e_p - \nabla_\alpha P + \nabla_\gamma \tau_{\gamma\alpha} \tag{10.2.49b}$$

$$s = \text{const.}, \tag{10.2.50c}$$

where $\boldsymbol{\tau}$ denotes the extra stress given now by

$$\tau_{\gamma\alpha} = \mu\left(\nabla_\gamma \boldsymbol{\nu}_\alpha + \nabla_\alpha \nu_\gamma\right) \tag{10.2.51}$$

and P a scalar (the pressure) which, as explained earlier, will be determined by taking the divergence of Eq. (10.2.49b) to formulate a Poisson equation for P with appropriate boundary conditions.

On the other hand, if the system is isothermal, we can forget about the direct dependence of the potential energy on entropy. The vector of state variables will therefore include the mass density ρ, the momentum density $\boldsymbol{M}$ (defined as $\boldsymbol{M} = \rho\boldsymbol{\nu}$ where $\boldsymbol{\nu}$ is the fluid velocity), and the conformation tensor density $\boldsymbol{C} = \rho\mathbf{c}$, where $\mathbf{c}$ is the single-chain conformation tensor; that is, $\boldsymbol{x} = \{\rho(\boldsymbol{r}, t), \boldsymbol{M}(\boldsymbol{r}, t), \boldsymbol{C}(\boldsymbol{r}, t)\}$. The governing equations will therefore consist only of the continuity equation for the total mass density ρ, the conservation equation for the total momentum density $\boldsymbol{M}$, and the evolution equation for the conformation density $\boldsymbol{C}$. We can also write the Hamiltonian $\mathcal{H}$ as

$$\begin{aligned}
&\mathcal{H} = \mathcal{H}(\rho, \boldsymbol{M}, s, \boldsymbol{C}) = K(\rho, \boldsymbol{M}) + U(\rho, s, \boldsymbol{C}) = K(\rho, \boldsymbol{M}) + U(\rho, s, \boldsymbol{C}) - TS + TS \\
&\Rightarrow \\
&\mathcal{H} = K(\rho, \boldsymbol{M}) + A(\rho, T, \boldsymbol{C}) + TS = \mathcal{H}(\rho, T, \boldsymbol{C}) + \text{const.}
\end{aligned} \tag{10.2.52}$$

where U denotes the potential energy, S the entropy, and T the temperature defined as the Volterra derivative of $\mathcal{H}$ with respect to the entropy density s, $T \equiv \dfrac{\delta \mathcal{H}}{\delta s}$. According to Eq. (10.2.52), $\mathcal{H}$ is defined as the sum of the kinetic energy K, of the Helmholtz free energy A defined as $A = A(\rho, T, \boldsymbol{C}) = U(\rho, \boldsymbol{C}) - TS(\rho, \boldsymbol{C})$, and of the (constant) term TS. Since for an isothermal system, this term TS does not contribute, all that matters is the sum of the first two terms which has the meaning of an extended Helmholtz free energy.

Example 10.7

Write down the dynamical equations for an isothermal viscoelastic fluid described by the conformation tensor $\boldsymbol{C}$.

Solution: The expression for the Poisson bracket is still given by Eq. (10.2.12). The corresponding expression for dissipative bracket is also given by Eq. (10.2.33), but we should not consider the derivatives with respect to the entropy variable s; thus,

$$\begin{aligned}
[F, G] = &-\int Q_{\alpha\beta\gamma\varepsilon} \nabla_\alpha \frac{\delta F}{\delta M_\beta} \nabla_\gamma \frac{\delta G}{\delta M_\varepsilon} \mathrm{d}V \\
&-\int \Lambda_{\alpha\beta\gamma\varepsilon} \frac{\delta F}{\delta C_{\alpha\beta}} \frac{\delta G}{\delta C_{\gamma\varepsilon}} \mathrm{d}V \\
&-\int B_{\alpha\beta\gamma\varepsilon\eta\nu} \nabla_\gamma \frac{\delta F}{\delta C_{\alpha\beta}} \nabla_\nu \frac{\delta G}{\delta C_{\varepsilon\eta}} \mathrm{d}V \\
&-\int L_{\alpha\beta\gamma\varepsilon} \left(\nabla_\alpha \frac{\delta F}{\delta M_\beta} \frac{\delta G}{\delta C_{\gamma\varepsilon}} - \nabla_\alpha \frac{\delta G}{\delta M_\beta} \frac{\delta F}{\delta C_{\gamma\varepsilon}} \right) \mathrm{d}V.
\end{aligned} \tag{10.2.53}$$

Inserting the two expressions in the fundamental evolution equation, Eq. (10.2.3), and making use of Eq. (10.2.52) for the Hamiltonian $\mathcal{H}$ leads to the following general evolution equations (for an incompressible system, the density ρ is constant; therefore, the final governing equations for the momentum and the conformation tensor are written directly in terms of the velocity $\boldsymbol{\nu}$ and the single-chain conformation tensor $\mathbf{c}$):

continued

Example 10.7 *continued*

$$\frac{\partial \rho}{\partial t} = -\nabla_\beta \left(\boldsymbol{\nu}_\beta \rho \right) \tag{10.2.54a}$$

$$\rho \frac{\partial \boldsymbol{\nu}_\alpha}{\partial t} = -\rho \boldsymbol{\nu}_\beta \nabla_\beta \boldsymbol{\nu}_\alpha - \nabla_\alpha P + \nabla_\beta \tau_{\alpha\beta} \tag{10.2.54b}$$

and

$$\frac{\partial C_{\alpha\beta}}{\partial t} = -\nabla_\gamma \left(\nu_\gamma C_{\alpha\beta} \right) + C_{\alpha\gamma} \nabla_\gamma \boldsymbol{\nu}_\beta + C_{\gamma\beta} \nabla_\gamma \boldsymbol{\nu}_\alpha - \Lambda_{\alpha\beta\gamma\varepsilon} \frac{\delta \mathcal{H}}{\delta C_{\gamma\varepsilon}}$$
$$+ L_{\alpha\beta\gamma\varepsilon} \nabla_\gamma \nu_\varepsilon + \nabla_\gamma \left(B_{\alpha\beta\gamma\varepsilon\zeta\eta} \nabla_\eta \frac{\delta \mathcal{H}}{\delta C_{\varepsilon\zeta}} \right), \tag{10.2.54c}$$

where the equations defining the pressure P and the extra stress tensor $\boldsymbol{\tau}$ are identical to Eqs. (10.2.25) and (10.2.44), respectively, for the nonisothermal case. The set of transport equations (10.2.54) is identical to that of Eqs. (10.2.43) except that there is no evolution equation for the entropy.

An interesting case arises if we assume that the temperature (and not the entropy) appears as a natural variable. Then, if $\mathcal{H}(\rho, \boldsymbol{M}, s, \boldsymbol{C}) = K(\rho, \boldsymbol{M}) + U(\rho, s, \boldsymbol{C})$ is the Hamiltonian in the $(\rho, \boldsymbol{M}, s, \boldsymbol{C})$ representation, a representation in the $(\rho, \boldsymbol{M}, T, \boldsymbol{C})$ space is obtained by utilizing a Legendre transform of the form

$$A(\rho, \boldsymbol{M}, T, \boldsymbol{C}) = \mathcal{H}(\rho, \boldsymbol{M}, s, \boldsymbol{C}) - S \left. \frac{\delta \mathcal{H}}{\delta s} \right|_{\rho, \boldsymbol{M}, \boldsymbol{C}} = \mathcal{H}(\rho, \boldsymbol{M}, s, \boldsymbol{C}) - ST$$

$$\Rightarrow$$

$$\mathcal{H}(\rho, \boldsymbol{M}, s, \boldsymbol{C}) = A(\rho, \boldsymbol{M}, T, \boldsymbol{C}) + ST, \tag{10.2.55}$$

where by $A(\rho, \boldsymbol{M}, T, \boldsymbol{C})$ we denote the extended Helmholtz free energy of the system including the kinetic energy. This implies the following derivative tranfromations:[1]

(a)
$$\left. \frac{\delta \mathcal{H}}{\delta \boldsymbol{C}} \right|_{\rho, \boldsymbol{M}, s} = \left. \frac{\delta A}{\delta \boldsymbol{C}} \right|_{\rho, \boldsymbol{M}, T} \tag{10.2.56a}$$

(b)
$$\left. \frac{\delta \mathcal{H}}{\delta \rho} \right|_{s, \boldsymbol{M}, \boldsymbol{C}} = \left. \frac{\delta A}{\delta \rho} \right|_{T, \boldsymbol{M}, \boldsymbol{C}} \tag{10.2.56b}$$

(c)
$$\left. \frac{\delta \mathcal{H}}{\delta \boldsymbol{M}} \right|_{s, \rho, \boldsymbol{C}} = \left. \frac{\delta A}{\delta \boldsymbol{M}} \right|_{T, \rho, \boldsymbol{C}} \tag{10.2.56c}$$

(d)
$$\left. \frac{\delta \mathcal{H}}{\delta s} \right|_{\rho, \boldsymbol{M}, \boldsymbol{C}} = T. \tag{10.2.56d}$$

In the same way, one can transform the equation defining the pressure from a representation in terms of the internal energy to a representation in terms of the extended Helmholtz free energy. In more detail, starting from

$$P = -\left.\frac{\partial A}{\partial V}\right|_{T,\boldsymbol{C}} \tag{10.2.57}$$

one can first show (by invoking the corresponding Euler expression for a thermodynamic description of the nonequilibrium system in terms of the internal energy) that

$$P = -u + \rho\left.\frac{\partial u}{\partial \rho}\right|_{s,\boldsymbol{C}} + s\left.\frac{\partial u}{\partial s}\right|_{\rho,\boldsymbol{C}} + \boldsymbol{C} : \left.\frac{\partial u}{\partial \boldsymbol{C}}\right|_{\rho,s}, \tag{10.2.58}$$

where u is the internal energy density, and then prove that

$$P = -a + \rho\left.\frac{\partial a}{\partial \rho}\right|_{T,\boldsymbol{C}} + \boldsymbol{C} : \left.\frac{\partial a}{\partial \boldsymbol{C}}\right|_{\rho,T}, \tag{10.2.59}$$

where a is the Helmholtz free energy density (but without including the kinetic energy density), given as $a = u - Ts$. By substituting these expressions into the dynamical evolution equations as derived by starting from a representation in terms of the set of variables $\boldsymbol{x} = \{\rho(\boldsymbol{r},t), \boldsymbol{M}(\boldsymbol{r},t), s(\boldsymbol{r},t), \boldsymbol{C}(\boldsymbol{r},t)\}$, we arrive at the coupled set of evolution equations:

$$\frac{\partial \rho}{\partial t} = -\nabla_\beta\left(\boldsymbol{\nu}_\beta \rho\right) \tag{10.2.60a}$$

$$\rho\frac{\partial \boldsymbol{\nu}_\alpha}{\partial t} = -\rho\boldsymbol{\nu}_\beta\nabla_\beta\boldsymbol{\nu}_\alpha - \nabla_\alpha P + \nabla_\beta\tau_{\alpha\beta} \tag{10.2.60b}$$

$$\begin{aligned}\frac{\partial s}{\partial t} &= -\nabla_\beta\left(\boldsymbol{\nu}_\beta s\right) + \frac{1}{T}Q_{\alpha\beta\gamma\varepsilon}\nabla_\alpha\boldsymbol{\nu}_\beta\nabla_\gamma\nu_\varepsilon + \frac{1}{T}\nabla_\alpha\left(A_{\alpha\beta}T\nabla_\beta T\right)\\ &\quad + \frac{1}{T}\Lambda_{\alpha\beta\gamma\varepsilon}\frac{\delta A}{\delta C_{\alpha\beta}}\frac{\delta A}{\delta C_{\gamma\varepsilon}} + \frac{1}{T}B_{\alpha\beta\gamma\varepsilon\zeta\eta}\nabla_\gamma\left(\frac{\delta A}{\delta C_{\alpha\beta}}\right)\nabla_\eta\left(\frac{\delta A}{\delta C_{\varepsilon\zeta}}\right)\end{aligned} \tag{10.2.60c}$$

and

$$\begin{aligned}\frac{\partial C_{\alpha\beta}}{\partial t} &= -\nabla_\gamma\left(\nu_\gamma C_{\alpha\beta}\right) + C_{\alpha\gamma}\nabla_\gamma\boldsymbol{\nu}_\beta + C_{\gamma\beta}\nabla_\gamma\boldsymbol{\nu}_\alpha - \Lambda_{\alpha\beta\gamma\varepsilon}\frac{\delta A}{\delta C_{\gamma\varepsilon}}\\ &\quad + L_{\alpha\beta\gamma\varepsilon}\nabla_\gamma\nu_\varepsilon + \nabla_\gamma\left(B_{\alpha\beta\gamma\varepsilon\zeta\eta}\nabla_\eta\frac{\delta A}{\delta C_{\varepsilon\zeta}}\right).\end{aligned} \tag{10.2.60d}$$

These are identical to the set of Eqs. (10.2.43) but they are expressed in terms of the extended Helmholtz free energy $A(\rho, \boldsymbol{M}, T, \boldsymbol{C})$, and not in terms of the Hamiltonian $\mathcal{H} = \mathcal{H}(\rho, \boldsymbol{M}, s, \boldsymbol{C})$, signalling a representation in terms of the set $\boldsymbol{x} = \{\rho(\boldsymbol{r},t), \boldsymbol{M}(\boldsymbol{r},t), T(\boldsymbol{r},t), \boldsymbol{C}(\boldsymbol{r},t)\}$.

10.2.4 The extended Helmholtz free energy *A* and the spring force-law constant

We consider a system containing N_{ch} polymer chains in volume V and temperature T. We can make its Helmholtz free energy dimensionless by dividing it with $N_{ch}k_BT$ where k_B is the Boltzmann constant. Equivalently, we can make its Helmholtz free-energy density a dimensionless by dividing it with nk_BT, where $n = \frac{N_{ch}}{V}$ denotes the chain number density. To account for the fact that the entropy of the melt should remain finite at high deformation rates,[45] the dimensionless Helmholtz free energy $\tilde{A}$ is typically assumed to be of the form

$$\tilde{A}(\tilde{\mathbf{c}}) = b\frac{\tilde{A}_0(\tilde{\mathbf{c}})}{b+\tilde{A}_0(\tilde{\mathbf{c}})}, \quad (10.2.61)$$

where

$$\tilde{A}_0 = \frac{1}{2}\int\left[\tilde{\phi}\left(\mathrm{tr}\left(\tilde{\mathbf{c}}-\boldsymbol{I}\right)\right) - \ln\left(\det\tilde{\mathbf{c}}\right)\right]\mathrm{d}V \quad (10.2.62)$$

denotes the unbounded dimensionless free energy; the corresponding unbounded free-energy density is therefore $\tilde{a}_0 = \frac{1}{2}\left[\tilde{\phi}\left(\mathrm{tr}\left(\tilde{\mathbf{c}}-\boldsymbol{I}\right)\right) - \ln\left(\det\tilde{\mathbf{c}}\right)\right]$.[46–47] In the above expressions, $\tilde{\mathbf{c}}$ denotes the dimensionless single-chain conformation tensor defined such that at equilibrium it is equal to the unit tensor $\boldsymbol{I}$, and b is the extensibility parameter (the dimensionless maximum length of the chain) defined as $b \equiv 3L^2/\langle R^2\rangle_{eq}$, where L is the maximum length of the chain and $\langle R^2\rangle_{eq}$ the mean square chain end-to-end distance at equilibrium. The dimensionless potential term $\tilde{\phi}\left(\mathrm{tr}\left(\tilde{\mathbf{c}}-\boldsymbol{I}\right)\right)$ on the right-hand-side of Eq. (10.2.62) accounts for contributions associated with molecular stretching (a typical and simple choice is $\tilde{\phi} = \mathrm{tr}\left(\tilde{\mathbf{c}}-\boldsymbol{I}\right)$) while the term $-\ln\left(\det\tilde{\mathbf{c}}\right)$ accounts for intrachain entropic contribution due to deformation. Of importance is the first term, since its partial derivative with respect to the first invariant (the trace) of the conformation tensor defines the (dimensionless) effective spring constant $\tilde{h}$ through[2]

$$\tilde{h}\left(\mathrm{tr}\left(\tilde{\mathbf{c}}-\boldsymbol{I}\right)\right) = 2\frac{\delta\tilde{A}(\tilde{\mathbf{c}})}{\delta\,\mathrm{tr}\left(\tilde{\mathbf{c}}-\boldsymbol{I}\right)} = 2B(\tilde{\mathrm{c}})\frac{\partial\tilde{\phi}\left(\mathrm{tr}\left(\tilde{\mathbf{c}}-\boldsymbol{I}\right)\right)}{\partial\,\mathrm{tr}\left(\tilde{\mathbf{c}}-\boldsymbol{I}\right)}, \quad (10.2.63a)$$

i.e.

$$\tilde{h}\left(\mathrm{tr}\left(\tilde{\mathbf{c}}-\boldsymbol{I}\right)\right) = B(\tilde{\mathbf{c}})\,h_0\left(\mathrm{tr}\left(\tilde{\mathbf{c}}-\boldsymbol{I}\right)\right), \quad (10.2.63b)$$

where

$$B(\tilde{\mathbf{c}}) \equiv \left[\frac{b}{b+\tilde{A}_0(\tilde{\mathbf{c}})}\right]^2. \quad (10.2.64)$$

Here, $0 \leq B(\tilde{\mathbf{c}}) \leq 1$ while $\tilde{h}_0$ is the spring constant that corresponds to the unbounded free energy. Equation (10.2.64) for $B(\tilde{\mathbf{c}})$ guarantees that, consistent with the constraint of finite chain extensibility, the Helmholtz free energy of the deformed polymer chains remains finite at high deformation rates (since the entropy is finite for a finite value of the extensibility parameter b). With the definition of the parameter $B(\tilde{\mathbf{c}})$, the Volterra derivative of $\tilde{A}$ with respect to $\tilde{\mathbf{c}}$ needed in the general evolution equations (see e.g. the term $\frac{\delta \mathcal{H}}{\delta \boldsymbol{C}}$ in Eq. (10.2.44)) is expressed as

$$\frac{\delta \tilde{A}(\tilde{\mathbf{c}})}{\delta \tilde{\mathbf{c}}} = \frac{1}{2} B(\tilde{\mathbf{c}}) \left[\tilde{h}_0 \left(\operatorname{tr}(\tilde{\mathbf{c}} - \boldsymbol{I}) \right) \boldsymbol{I} - \tilde{\mathbf{c}}^{-1} \right]. \tag{10.2.65}$$

If we adopt Peterlin's preaveraging approximation,[48] we can take the spring constant to be of the form

$$\tilde{h}_0^{\mathrm{W}} \left(\operatorname{tr}(\tilde{\mathbf{c}} - \boldsymbol{I}) \right) = \frac{1}{1 - \frac{1}{b-3} \operatorname{tr}(\tilde{\mathbf{c}} - \boldsymbol{I})} = \frac{b-3}{b - \operatorname{tr}(\tilde{\mathbf{c}})} \tag{10.2.66a}$$

corresponding to Warner's approximation (W for Warner)[49] to the inverse Langevin function, or of the form

$$\tilde{h}_0^{\mathrm{C}} (\operatorname{tr}(\tilde{\mathbf{c}} - \boldsymbol{I})) = \frac{3(b-2) - \operatorname{tr}(\tilde{\mathbf{c}})}{3(b - \operatorname{tr}(\tilde{\mathbf{c}}))}, \tag{10.2.66b}$$

corresponding to Cohen's approximation (C for Cohen)[50] to the inverse Langevin function. For some simple but accurate approximants to the inverse Langevin (and Brillouin) functions, of relevance to large polymer deformation under strong flows, the interested reader is referred to Ref. 51. The potential energy terms generating Eqs. (10.2.66a) and (10.2.66b) can be found by simple integration and turn out to be

$$\tilde{\phi}^{\mathrm{P}} = -(b-3) \ln \left(1 - \frac{\operatorname{tr}(\tilde{\mathbf{c}})}{b} \right) \tag{10.2.67a}$$

and

$$\tilde{\phi}^{\mathrm{C}} = \frac{1}{3} \left[\operatorname{tr}(\tilde{\mathbf{c}} - \boldsymbol{I}) - 2(b-3) \ln \left(\frac{b - \operatorname{tr}(\tilde{\mathbf{c}})}{b-3} \right) \right], \tag{10.4.67b}$$

respectively. The extensibility parameter b appearing in the above expressions can be directly evaluated from the molecular characteristics of the polymer.[52,53]

10.3 Using the generalized bracket formalism to model concentration variations in an isothermal polymer solution through a two-fluid hydrodynamic model

We consider a polymer solution under flow, and we wish to derive a thermodynamically consistent set of transport equations that can account for possible concentration variations due to variations in the stress field. We can adopt a description in terms of two fluids, hence a two-fluid model,[54–57] in the sense that we can view the polymer solution as consisting of two interpenetrating continua, the polymer chains corresponding to medium 1 and the solvent molecules corresponding to medium 2. We further assume that the two media are in thermal equilibrium and interact with each other both thermodynamically and hydrodynamically.

Guided from the previous analysis, we can restrict ourselves to isothermal conditions so that the relevant set of state variables to consider includes the mass density ρ_1 of the polymer, the momentum density $\boldsymbol{M}_1$ $(=\rho_1\boldsymbol{\nu}_1)$ of the polymer, the conformation tensor density of the polymer $\boldsymbol{C}\,(=\rho_1\,\mathbf{c})$, the mass density ρ_2 of the solvent, and the momentum density $\boldsymbol{M}_2$ $(=\rho_2\boldsymbol{\nu}_2)$ of the solvent. The vector of state variables is therefore $\boldsymbol{x}=\{\rho_1,\boldsymbol{M}_1,\rho_2,\boldsymbol{M}_2,\boldsymbol{C}\}$. The corresponding Hamiltonian is practically replaced by the extended Helmholtz free energy of the system; i.e. it is written in the form

$$\mathcal{H}=E_{\mathrm{k}}+\mathcal{H}_i=\int\left[\frac{\boldsymbol{M}_1\cdot\boldsymbol{M}_1}{2\rho_1}+\frac{\boldsymbol{M}_2\cdot\boldsymbol{M}_2}{2\rho_2}\right]\mathrm{d}V+\int h_i\left(\rho_1,\boldsymbol{C}\right)\mathrm{d}V+\text{const.},\tag{10.3.1}$$

emphasizing the contributions due to the kinetic energy of the two species and the internal energy due to (a) polymer–solvent mixing described (e.g.) through the Flory–Huggins theory and (b) polymer microstructure (and its variation under flow) for which we can use the expressions discussed in Section (10.2.4), while the numerical constant accounts for the neglected entropy effects. For such a system, the Poisson bracket involves (a) the direct sum of the terms corresponding to a single incompressible fluid written for the two species 1 and 2, i.e. for the pairs $(\rho_1,\boldsymbol{M}_1)$ and $(\rho_2,\boldsymbol{M}_2)$, and (b) the terms involving $\boldsymbol{C}$ coupled with its own momentum density $\boldsymbol{M}_1$. Mathematically,

$$\{F,G\}=\{F,G\}_1+\{F,G\}_2,\tag{10.3.2}$$

where

$$\begin{aligned}
\{F,G\}_1=&-\int\left[\frac{\delta F}{\delta\rho_1}\nabla_\beta\left(\frac{\delta G}{\delta M_{1,\beta}}\rho_1\right)-\frac{\delta G}{\delta\rho_1}\nabla_\beta\left(\frac{\delta F}{\delta M_{1,\beta}}\rho_1\right)\right]\mathrm{d}V\\
&-\int\left[\frac{\delta F}{\delta M_{1,\gamma}}\nabla_\beta\left(\frac{\delta G}{\delta M_{1,\beta}}M_{1,\gamma}\right)-\frac{\delta G}{\delta M_{1,\gamma}}\nabla_\beta\left(\frac{\delta F}{\delta M_{1,\beta}}M_{1,\gamma}\right)\right]\mathrm{d}V\\
&-\int\left[\frac{\delta F}{\delta C_{\alpha\beta}}\nabla_\gamma\left(\frac{\delta G}{\delta M_{1,\gamma}}C_{\alpha\beta}\right)-\frac{\delta G}{\delta C_{\alpha\beta}}\nabla_\gamma\left(\frac{\delta F}{\delta M_{1,\gamma}}C_{\alpha\beta}\right)\right]\mathrm{d}V\\
&-\int C_{\gamma\alpha}\left[\frac{\delta G}{\delta C_{\alpha\beta}}\nabla_\gamma\left(\frac{\delta F}{\delta M_{1,\beta}}\right)-\frac{\delta F}{\delta C_{\alpha\beta}}\nabla_\gamma\left(\frac{\delta G}{\delta M_{1,\beta}}\right)\right]\mathrm{d}V\\
&-\int C_{\gamma\beta}\left[\frac{\delta G}{\delta C_{\alpha\beta}}\nabla_\gamma\left(\frac{\delta F}{\delta M_{1,\alpha}}\right)-\frac{\delta F}{\delta C_{\alpha\beta}}\nabla_\gamma\left(\frac{\delta G}{\delta M_{1,\alpha}}\right)\right]\mathrm{d}V
\end{aligned}\tag{10.3.3}$$

and

$$\{F,G\}_2 = -\int\left[\frac{\delta F}{\delta\rho_2}\nabla_\beta\left(\frac{\delta G}{\delta M_{2,\beta}}\rho_2\right) - \frac{\delta G}{\delta\rho_2}\nabla_\beta\left(\frac{\delta F}{\delta M_{2,\beta}}\rho_2\right)\right]\mathrm{d}V$$
$$-\int\left[\frac{\delta F}{\delta M_{2,\gamma}}\nabla_\beta\left(\frac{\delta G}{\delta M_{2,\beta}}M_{2,\gamma}\right) - \frac{\delta G}{\delta M_{2,\gamma}}\nabla_\beta\left(\frac{\delta F}{\delta M_{2,\beta}}M_{2,\gamma}\right)\right]\mathrm{d}V. \qquad (10.3.4)$$

If no dissipation (and thus no dissipative bracket) is considered, through such a consideration one derives the continuity and momentum balance equations for the two uncoupled, individual species and the evolution equation for the conformation tensor $\boldsymbol{C}$:

$$\frac{\partial\rho_1}{\partial t} = -\nabla_\beta\left(\nu_{1\beta}\rho_1\right) \qquad (10.3.5a)$$

$$\frac{\partial\rho_2}{\partial t} = -\nabla_\beta\left(\nu_{2\beta}\rho_2\right) \qquad (10.3.5b)$$

$$\frac{\partial M_{1\alpha}}{\partial t} = -\nabla_\beta\left(\nu_{1\beta}M_{1\alpha}\right) - \rho_1\nabla_\alpha\frac{\delta\mathcal{H}}{\delta\rho_1} - M_{1\beta}\nabla_\alpha\nu_{1\beta} - C_{\gamma\beta}\nabla_\alpha\frac{\delta\mathcal{H}}{\delta C_{\gamma\beta}} + \nabla_\gamma\left(2C_{\gamma\beta}\frac{\delta\mathcal{H}}{\delta C_{\beta\alpha}}\right) \qquad (10.3.5c)$$

$$\frac{\partial M_{2\alpha}}{\partial t} = -\nabla_\beta\left(\nu_{2\beta}M_{2\alpha}\right) - \rho_2\nabla_\alpha\frac{\delta\mathcal{H}}{\delta\rho_2} - M_{2\beta}\nabla_\alpha\nu_{2\beta} \qquad (10.3.5d)$$

and

$$\frac{\partial C_{\alpha\beta}}{\partial t} = -\nabla_\gamma\left(\nu_{1\gamma}C_{\alpha\beta}\right) + C_{\alpha\gamma}\nabla_\gamma\nu_{1\beta} + C_{\gamma\beta}\nabla_\gamma\nu_{1\alpha}. \qquad (10.3.5e)$$

To impose the incompressibility constraint for the mass average velocity $\boldsymbol{\nu}$, we introduce two new sets of variables: the total mass density ρ with its corresponding momentum density $\boldsymbol{M}$, and the reduced mass density ρ_- with its corresponding momentum density $\boldsymbol{M}_-$. That is, we introduce the sets $(\rho, \boldsymbol{M})$ and $(\rho_-, \boldsymbol{M}_-)$ defined as

$$\rho = \rho_1 + \rho_2, \quad \boldsymbol{M} = \boldsymbol{M}_1 + \boldsymbol{M}_2$$
$$\rho_- = \rho_1, \quad \boldsymbol{M}_- = \frac{\rho_2}{\rho_1+\rho_2}\boldsymbol{M}_1 - \frac{\rho_1}{\rho_1+\rho_2}\boldsymbol{M}_2. \qquad (10.3.6)$$

The velocities $\boldsymbol{\nu}$ and $\Delta\boldsymbol{\nu}$ of the two new momentum densities are given by

$$\boldsymbol{\nu} \equiv \frac{\boldsymbol{M}}{\rho} = \frac{\rho_1}{\rho_1+\rho_2}\boldsymbol{\nu}_1 + \frac{\rho_2}{\rho_1+\rho_2}\boldsymbol{\nu}_2$$
$$\Delta\boldsymbol{\nu} \equiv \frac{\boldsymbol{M}_-}{\rho_-} = \boldsymbol{\nu}_1 - \boldsymbol{\nu}_2, \qquad (10.3.7)$$

respectively. In this way, the incompressibility constraint is easily taken into account by imposing

$$\nabla \cdot \boldsymbol{M} = 0 \Leftrightarrow \nabla \cdot \boldsymbol{\nu} = 0. \tag{10.3.8}$$

By using the chain rule of differentiation to re-express the generalized brackets of Eqs. (10.3.3) and (10.3.4) in terms of the new variables, the following conservation equations are obtained for the new sets of variables:

$$\frac{\partial \rho_-}{\partial t} = -\boldsymbol{\nu}_\beta \nabla_\beta \rho_- - \nabla_\beta \left[\rho_- (1-\phi)\, \Delta \boldsymbol{\nu}_\beta\right] \tag{10.3.9a}$$

$$\rho = \text{const.} \tag{10.3.9b}$$

$$\begin{aligned} \frac{\partial M_\alpha}{\partial t} &= -\nabla_\beta \left(\nu_{1\beta} M_{1\alpha} + \nu_{2\beta} M_{2\alpha}\right) - \left(M_{1\beta} \nabla_\alpha \nu_{1\beta} + M_{2\beta} \nabla_\alpha \nu_{2\beta}\right) \\ &\quad - \rho_1 \nabla_\alpha \frac{\delta \mathcal{H}_i}{\delta \rho_1} - \nabla_\alpha P - C_{\gamma\beta} \nabla_\alpha \frac{\delta \mathcal{H}_i}{\delta C_{\gamma\beta}} + \nabla_\gamma \left(2 C_{\gamma\beta} \frac{\delta \mathcal{H}}{\delta C_{\beta\alpha}}\right) \end{aligned} \tag{10.3.9c}$$

$$\begin{aligned} \frac{\partial M_{-\alpha}}{\partial t} &= -(1-\phi) \nabla_\beta \left(\nu_{1\beta} M_{1\alpha}\right) + \phi \nabla_\beta \left(\nu_{2\beta} M_{2\alpha}\right) - (1-\phi) M_{1\beta} \nabla_\alpha \nu_{1\beta} \\ &\quad + \phi M_{2\beta} \nabla_\alpha \nu_{2\beta} - (1-\phi)\, \rho_1 \nabla_\alpha \frac{\delta \mathcal{H}_i}{\delta \rho_1} \end{aligned} \tag{10.3.9d}$$

and

$$\frac{\partial C_{\alpha\beta}}{\partial t} = -\nabla_\gamma \left(\nu_{1\gamma} C_{\alpha\beta}\right) + C_{\alpha\gamma} \nabla_{1\gamma} \boldsymbol{\nu}_\beta + C_{\gamma\beta} \nabla_\gamma \nu_{1\alpha}. \tag{10.3.9e}$$

In the above, $\phi = \dfrac{\rho_1}{\rho}$ is the mass fraction of the polymer component and P denotes the scalar pressure which can be obtained by solving the Poisson equation that results by taking the divergence of Eq. (10.3.9c) describing the conservation of the total momentum for the system.

The above equations contain only conservative terms, as they have been derived by making use only of the Poisson bracket. Additional dissipative effects can be included by providing a suitable expression for the dissipative bracket $[F, G]$. For the problem at hand, to the lowest approximation, this can be considered to involve three terms accounting for (a) solvent viscous effects, (b) drag force effects as the two components are moving in general with different velocities, and (c) chain relaxation due to polymer viscoelasticity. The corresponding entropy correction terms are neglected since we restrict our analysis to isothermal conditions. Mathematically,[57]

$$
\begin{aligned}
[F, G] = & -\int \frac{\eta_s}{2}\left(\nabla_\alpha \frac{\delta F}{\delta M_\beta} + \nabla_\beta \frac{\delta F}{\delta M_\alpha}\right)\left(\nabla_\alpha \frac{\delta G}{\delta M_\beta} + \nabla_\beta \frac{\delta G}{\delta M_\alpha}\right) \mathrm{d}V \\
& - \int Z_{\alpha\beta} \frac{\delta F}{\delta M_{-\alpha}} \frac{\delta G}{\delta M_{-\beta}} \mathrm{d}V \\
& - \int \Lambda_{\alpha\beta\gamma\varepsilon} \frac{\delta F}{\delta C_{\alpha\beta}} \frac{\delta G}{\delta C_{\gamma\varepsilon}} \mathrm{d}V,
\end{aligned}
\tag{10.3.10}
$$

where $\boldsymbol{Z}$ is the friction or drag coefficient tensor (a symmetric and nonnegative definite 3×3 matrix). To be able to apply the resulting set of equations to confined systems (e.g. to a polymer solution flowing past a solid boundary in order to study wall effects on fluid dynamics and concentration), we also consider that there is an additional direct dependence of the internal component of the Helmholtz free energy h_i on the local position vector $\boldsymbol{r}\,(= (x_1, x_2, x_3))$; i.e. we assume that

$$
\mathcal{H} = E_k + \mathcal{H}_i = \int \left[\frac{\boldsymbol{M}_1 \cdot \boldsymbol{M}_1}{2\rho_1} + \frac{\boldsymbol{M}_2 \cdot \boldsymbol{M}_2}{2\rho_2}\right] \mathrm{d}V + \int h_i\left(\rho_1, \boldsymbol{C}, \boldsymbol{r}\right) \mathrm{d}V + \text{const.} \tag{10.3.11}
$$

The resulting equations then read

$$
\frac{\partial \rho_-}{\partial t} = -\boldsymbol{\nu}_\beta \nabla_\beta \rho_- - \nabla_\beta \left[\rho_- (1-\phi)\, \Delta \boldsymbol{\nu}_\beta\right] \tag{10.3.12a}
$$

$$
\begin{aligned}
\frac{\partial M_\alpha}{\partial t} = & -\nabla_\beta \left(\nu_{1\beta} M_{1\alpha} + \nu_{2\beta} M_{2\alpha}\right) - M_{1\beta} \nabla_\alpha \nu_{1\beta} - M_{2\beta} \nabla_\alpha \nu_{2\beta} - \nabla_\alpha P \\
& - \nabla_\alpha \Pi - \frac{\partial h_i}{\partial x_\alpha} + \nabla_\beta \tau_{\beta\alpha}
\end{aligned}
\tag{10.3.12b}
$$

$$
\begin{aligned}
\frac{\partial M_{-\alpha}}{\partial t} = & -(1-\phi)\nabla_\beta \left(\nu_{1\beta} M_{1\alpha}\right) + \phi \nabla_\beta \left(\nu_{2\beta} M_{2\alpha}\right) - (1-\phi) M_{1\beta} \nabla_\alpha \nu_{1\beta} \\
& + \phi M_{2\beta} \nabla_\alpha \nu_{2\beta} - Z_{\alpha\beta} \Delta \boldsymbol{\nu}_\beta - (1-\phi)\left[\nabla_\alpha \Pi + \frac{\partial h_i}{\partial x_\alpha} - \nabla_\beta \tau_{\mathrm{p},\beta\alpha}\right]
\end{aligned}
\tag{10.3.12c}
$$

$$
\frac{\partial C_{\alpha\beta}}{\partial t} = -\nabla_\gamma \left(\nu_{1\gamma} C_{\alpha\beta}\right) + C_{\alpha\gamma} \nabla_\gamma \nu_{1\beta} + C_{\gamma\beta} \nabla_\gamma \nu_{1\alpha} - \Lambda_{\alpha\beta\gamma\varepsilon} \frac{\delta \mathcal{H}}{\delta C_{\varepsilon\gamma}}. \tag{10.3.12d}
$$

In the above expressions, we have introduced the polymer contribution $\boldsymbol{\tau}_\mathrm{p}$ to the stress tensor through $\boldsymbol{\tau}_{\mathrm{p},\alpha\beta} \equiv 2C_{\alpha\gamma} \dfrac{\delta \mathcal{H}}{\delta C_{\gamma\beta}}$, the total (solvent + polymer) stress tensor $\boldsymbol{\tau}$ through $\boldsymbol{\tau}_{\alpha\beta} \equiv 2C_{\alpha\gamma} \dfrac{\delta \mathcal{H}}{\delta C_{\gamma\beta}} + \eta_s \left[\nabla_\alpha \boldsymbol{\nu}_\beta + \nabla_\beta \boldsymbol{\nu}_\alpha\right]$, and the osmotic pressure Π through

$$
\Pi = \rho_1 \frac{\delta \mathcal{H}_i}{\delta \rho_1} + C_{\alpha\beta} \frac{\delta \mathcal{H}}{\delta C_{\beta\alpha}} - h_i. \tag{10.3.13}
$$

Quite often, the inertial terms in Eq. (10.3.12c) can be neglected. Then, Eq. (10.3.12c) can be solved to provide the following expression for the differential velocity of the two fluids:

$$\Delta \boldsymbol{\nu}_\alpha = -Z_{\alpha\beta}^{-1}(1-\phi)\left[\nabla_\beta \Pi + \frac{\partial h_{\mathrm{i}}}{\partial x_\beta} - \nabla_\gamma \boldsymbol{\tau}_{\mathrm{p},\gamma\beta}\right]. \tag{10.3.14}$$

Inserting this into Eq. (10.3.12a) helps one obtain the following concentration or *diffusion equation* for the polymer component, which can be considered as the one-fluid model approximation to the two-fluid model invoked here for the analysis of transport phenomena taking place in an inhomogeneous polymer solution:

$$\frac{\partial \rho_-}{\partial t} = -\nabla_\beta(\boldsymbol{\nu}_\beta \rho_-) + \nabla_\alpha\left\{Y_{\alpha\beta}\left[\nabla_\beta \Pi + \frac{\partial h_{\mathrm{i}}}{\partial x_\beta} - \nabla_\gamma \tau_{\mathrm{p},\gamma\beta}\right]\right\}. \tag{10.3.15}$$

The tensor $\boldsymbol{Y}$ showing up in this equation has been introduced as

$$\boldsymbol{Y} = \rho_-(1-\phi)^2 \boldsymbol{Z}^{-1} \tag{10.3.16}$$

and has the meaning of a diffusivity tensor.

The final set of transport equations (defining a one-fluid approximation of the two-fluid model) consists therefore of the following dynamical equations:

(a) the diffusion equation for the polymer species,

$$\frac{\partial \rho_1}{\partial t} = -\boldsymbol{\nu}_\beta \nabla_\beta \rho_1 + \nabla_\alpha\left\{Y_{\alpha\beta}\left[\nabla_\beta \Pi + \frac{\partial h_{\mathrm{i}}}{\partial x_\beta} - \nabla_\gamma \boldsymbol{\tau}_{\mathrm{p},\gamma\beta}\right]\right\}; \tag{10.3.17a}$$

(b) the momentum equation for the total density or mass-averaged velocity,

$$\frac{\partial \boldsymbol{\nu}_\alpha}{\partial t} = -\boldsymbol{\nu}_\beta \nabla_\beta \boldsymbol{\nu}_\alpha - \nabla_\alpha P + \eta_s \nabla^2 \boldsymbol{\nu}_\alpha + \nabla_\beta \tau_{\mathrm{p},\beta\alpha} - \nabla_\alpha \Pi - \frac{\partial h_{\mathrm{i}}}{\partial r_\alpha}; \tag{10.3.17b}$$

and

(c) the constitutive equation for the polymer conformation,

$$\dot{C}_{\alpha\beta,[1]}\left(\equiv \frac{\partial C_{\alpha\beta}}{\partial t} + \nu_{1,\gamma}\nabla_\gamma C_{\alpha\beta} - C_{\alpha\gamma}\nabla_\gamma \nu_{1,\beta} - C_{\gamma\beta}\nabla_\gamma \nu_{1,\alpha}\right) = -\Lambda_{\alpha\beta\gamma\delta}\frac{\delta \mathcal{H}}{\delta C_{\delta\gamma}}. \tag{10.3.17c}$$

Equation (10.3.17a) for the polymer concentration deserves some more discussion. According to it, gradients in the osmotic pressure Π and/or polymer stress tensor $\boldsymbol{\tau}_{\mathrm{p}}$ will lead to gradients in the concentration of the polymer solution, as has been discussed extensively in the literature.[54–71] This is a remarkable finding, since such an effect has indeed been observed experimentally.[72–75] On the other hand, the momentum equation for the total velocity field, Eq. (10.3.17b), involves (in addition to the usual contributions due to the velocity gradient, the pressure gradient, the viscous solvent, and the extra or elastic stress tensor $\boldsymbol{\tau}_{\mathrm{p}}$) a term related to the gradient of the osmotic pressure Π, plus a local term due to the spatial dependence of the internal free energy of the solution. It is also important to note that in the context of the two-fluid model introduced in this section, the upper convected derivative in Eq. (10.3.17c) for the evolution of the conformation tensor is based on the polymer-phase velocity $\boldsymbol{v}_1$.

Example 10.8

How do the evolution Eqs. (10.3.17) and the expression for the elastic or polymer contribution to the stress tensor reduce in the bulk of the flow where $\boldsymbol{Y}$ can be assumed to be constant?

Solution: Let us make use of the chain number density n in place of the mass density ρ_1. Then, the corresponding evolution equations (for the chain number density, the total momentum density, and the conformation tensor density) take the following form:

(a) The diffusion equation for the polymer concentration becomes

$$\frac{Dn}{Dt} = D_{\mathrm{tr}}\nabla^2 n - \frac{D_{\mathrm{tr}}}{k_{\mathrm{B}}T}\boldsymbol{\nabla\nabla} : \boldsymbol{\tau}_{\mathrm{p}}, \tag{10.3.18}$$

where

$$\frac{D}{Dt} = \frac{\partial}{\partial t} + \boldsymbol{v}\cdot\boldsymbol{\nabla} \tag{10.3.19}$$

denotes the substantial derivative, $\boldsymbol{\tau}_{\mathrm{p}}$ is the polymer contribution to the stress, and D_{tr} is the translational polymer diffusivity.

(b) The momentum equation becomes

$$\rho\frac{D\boldsymbol{v}}{Dt} = -\boldsymbol{\nabla}P + \boldsymbol{\nabla}\cdot\boldsymbol{\tau}_{\mathrm{p}} + \eta_{\mathrm{s}}\nabla^2\boldsymbol{v}, \tag{10.3.20}$$

where ρ is the (constant) mass density of the solution, $\boldsymbol{v}$ the mass average velocity of the solution, P the pressure, and η_{s} the solvent contribution to the total viscosity η of the solution, $\eta = \eta_{\mathrm{s}} + \eta_{\mathrm{p}}$, where η_{p} is the polymer contribution. According to the linear viscoelastic theory,[76] this is proportional to the chain density of the solution, i.e. $\eta_{\mathrm{p}} = n\,\lambda_0 k_{\mathrm{B}}T$, where λ_0 is the characteristic (longest) polymer chain relaxation time.

(C) The constitutive equation for the chain-weighted conformation tensor $\boldsymbol{C}(= n\boldsymbol{c}$, where $\boldsymbol{c}$ is the single-chain conformation tensor) is

$$\lambda_0 k\,\dot{\boldsymbol{C}}_{[1]} + k\boldsymbol{C} - nk_{\mathrm{B}}T\boldsymbol{I} = \boldsymbol{0}, \tag{10.3.21}$$

continued

Example 10.8 *continued*

where $\boldsymbol{I}$ is the unit tensor of rank two, k denotes the Hookean spring constant and the subscript [1] in the constitutive equation stands for the upper-convected Maxwell time derivative defined as

$$\dot{\boldsymbol{C}}_{[1]} = \frac{\partial \boldsymbol{C}}{\partial t} + \boldsymbol{\nu}_1 \cdot \nabla \boldsymbol{C} - \boldsymbol{C} \cdot \nabla \boldsymbol{\nu}_1 - (\nabla \boldsymbol{\nu}_1)^T \cdot \boldsymbol{C}. \tag{10.3.22}$$

The polymer-phase velocity $\boldsymbol{v}_1$ that enters in the above equations is

$$\boldsymbol{\nu}_1 = \boldsymbol{\nu} + (1 - w)\, \Delta\boldsymbol{\nu}, \tag{10.3.23}$$

with w being the polymer mass fraction and $\Delta\boldsymbol{\nu}$ the differential velocity (the difference in the velocities of the two components of the fluid); the latter is also given by

$$\Delta\boldsymbol{\nu} = +\frac{\boldsymbol{J}}{n}, \tag{10.3.24}$$

where, from Eq. (10.3.18),

$$\boldsymbol{J} = -D_{\text{tr}} \nabla n + \frac{D_{\text{tr}}}{k_B T} \nabla \cdot \boldsymbol{\tau}_{\text{p}} \tag{10.3.25}$$

denotes the total (Fickian+stress induced) chain flux.

(d) The conformation tensor is connected to the polymer stress tensor through the formula

$$\boldsymbol{\tau}_{\text{p}} = k_B T \left[\frac{k}{k_B T} \boldsymbol{C} - n \boldsymbol{I} \right]. \tag{10.3.26}$$

According to Eq. (10.3.25), nonuniform stress gradients will always produce a concentration gradient; this justifies the term *stress gradient-induced migration*. Interestingly enough, concentration changes will also develop even if the stresses are uniform but in a curvilinear system of coordinates. This happens because of the structure of the term $\nabla \cdot \boldsymbol{\tau}_{\text{p}}$, which, in a cylindrical system of coordinates, for example, involves terms of the form $\tau_{\text{p},\alpha\beta}/r$.[63] In such a case, a polymer concentration profile should be expected to develop. As already mentioned, these stress gradient-induced polymer migration phenomena have been indeed observed experimentally.[72–75] Note that the diffusion equation itself does not specify the direction of this flux, nor its magnitude. To do so one needs to solve (in most cases numerically)[67,71] the full coupled set of transport equations.

10.4 The GENERIC formalism

The GENERIC formalism differs from the generalized bracket formalism in that it employs two separate generators of the dynamics, the energy E and entropy S (as opposed to a single one used in the generalized bracket, the Hamiltonian $\mathcal{H}$). Then, the vector $\boldsymbol{x}$ of the set of independent state variables required for a complete description of the nonequilibrium system obeys the fundamental evolution equation [2,77,78]

$$\frac{\partial \boldsymbol{x}}{\partial t} = \boldsymbol{L}(\boldsymbol{x}) \cdot \frac{\delta E(\boldsymbol{x})}{\delta \boldsymbol{x}} + \boldsymbol{M}(\boldsymbol{x}) \cdot \frac{\delta S(\boldsymbol{x})}{\delta \boldsymbol{x}}, \tag{10.4.1}$$

where the functionals $E(\boldsymbol{x})$ and $S(\boldsymbol{x})$ represent the total energy and total entropy of the system expressed in terms of the state variables $\boldsymbol{x}$, and $\boldsymbol{L}(\boldsymbol{x})$ and $\boldsymbol{M}(\boldsymbol{x})$ are the Poisson and friction matrices representing the reversible and dissipative contributions, respectively, to the dynamics in terms of linear operators. $\boldsymbol{L}(\boldsymbol{x})$ turns energy gradients into reversible dynamics, and $\boldsymbol{M}(\boldsymbol{x})$ turns entropy gradients into irreversible dynamics.[2]

The Poisson matrix $\boldsymbol{L}$:

- is always antisymmetric;
- satisfies the degeneracy requirement

$$\boldsymbol{L}(\boldsymbol{x}) \cdot \frac{\delta S(\boldsymbol{x})}{\delta \boldsymbol{x}} = 0, \tag{10.4.2}$$

  which expresses the conservation of entropy for any reversible process; and
- satisfies the Jacobi identity; that is, if we consider the bracket

$$\{A, B\} = \frac{\delta A(\boldsymbol{x})}{\delta \boldsymbol{x}} \cdot \boldsymbol{L}(\boldsymbol{x}) \cdot \frac{\delta B(\boldsymbol{x})}{\delta \boldsymbol{x}} \tag{10.4.3}$$

  then this should satisfy

$$\{A, \{B, C\}\} + \{B, \{C, A\}\} + \{C, \{A, B\}\} = 0. \tag{10.4.4}$$

On the other hand, the friction matrix $\boldsymbol{M}$:

- is usually symmetric, expressing the Onsager–Casimir symmetry of linear irreversible thermodynamics;
- is semi-positive definite (a consequence of the second law of thermodynamics); and
- satisfies the degeneracy condition

$$\boldsymbol{M}(\boldsymbol{x}) \cdot \frac{\delta E(\boldsymbol{x})}{\delta \boldsymbol{x}} = 0, \tag{10.4.5}$$

expressing the conservation of energy for an isolated system even in the presence of dissipation.

10.4.1 The energy and the entropy

Assuming that most of the external-field effects on the overall structure of the system are absorbed into the entropy functional (i.e. the energetic effect is relatively negligible), E and S are commonly expressed as

$$E(\boldsymbol{x}) = \int \left[\frac{\boldsymbol{M}(\boldsymbol{r})^2}{2\rho(\boldsymbol{r})} + \varepsilon(\boldsymbol{r}) \right] \, \mathrm{d}\boldsymbol{r} \tag{10.4.6}$$

and

$$S(\boldsymbol{x}) = \int s\left(\rho(\boldsymbol{r}), \varepsilon(\boldsymbol{r}), \boldsymbol{X}(\boldsymbol{r})\right) \, \mathrm{d}\boldsymbol{r}. \tag{10.4.7}$$

Typical choices for $\boldsymbol{X}$ include the distribution function $\Psi(\boldsymbol{R}, \boldsymbol{r}, t)$ for the chain end-to-end vector $\boldsymbol{R}$ at position $\boldsymbol{r}$ and time t; the conformation tensor $\boldsymbol{C}$ defined as the second moment of the distribution function $\Psi(\boldsymbol{R}, \boldsymbol{r}, t)$ for $\boldsymbol{R}$, i.e. $C_{\alpha\beta}(\boldsymbol{r}, t) = \langle R_\alpha R_\beta \rangle (\boldsymbol{r}, t) = \int R_\alpha R_\beta \Psi(\boldsymbol{R}, \boldsymbol{r}, t) \mathrm{d}\boldsymbol{R}$, where the brackets denote a configurational average; the distribution function $\Psi\ (s, \boldsymbol{u}, \boldsymbol{r}, t)$ for the chain tangent vector at point s along the chain contour in a melt of entangled polymer chains to be within $\mathrm{d}\boldsymbol{u}$ around $\boldsymbol{u}$ at position $\boldsymbol{r}$ and time t; and a few others.

10.4.2 The Poisson operator

Like the Poisson bracket in the generalized bracket formalism, the matrix $\boldsymbol{L}$ that generates the conservative contribution to the dynamics depends on the type of coarse-grained variables chosen (scalar, scalar density, vector, vector density, tensor, tensor density, or distribution function) but for each one of them its form is quite standard. In general, there exist two representations of the various components of $\boldsymbol{L}$. The first is the operator form, in which the matrix $\boldsymbol{L}$ is explicitly given in terms of the vector of state variables $\boldsymbol{x}$. The second assumes that $\boldsymbol{L}$ is labelled by two position vectors, thus $\boldsymbol{L}(\boldsymbol{r}, \boldsymbol{r}')$, and to get $\boldsymbol{L}(\boldsymbol{r})$ we should integrate over $\boldsymbol{r}'$:

$$\boldsymbol{L}(\boldsymbol{r}) \cdot \frac{\delta E}{\delta \boldsymbol{x}(\boldsymbol{r})} = \int \boldsymbol{L}(\boldsymbol{r}, \boldsymbol{r}') \cdot \left(\frac{\delta E}{\delta \boldsymbol{x}(\boldsymbol{r}')} \right)^{\mathrm{T}} \mathrm{d}\boldsymbol{r}'. \tag{10.4.8}$$

This second representation is easier to use in practice because one can easily accommodate in it the antisymmetry property of $\boldsymbol{L}$.

For a system described by a set $\boldsymbol{x}$ of state variables that includes the mass density $\rho(\boldsymbol{r})$, the momentum density $\boldsymbol{M}(\boldsymbol{r})$, the internal energy density $\varepsilon(\boldsymbol{r})$, a conformation tensor $\mathbf{c}$ (or a conformation tensor density $\boldsymbol{C}$) which exhibits lower- or upper-convected dynamics, and a scalar λ, the matrix $\boldsymbol{L}$ has the general form

$$\boldsymbol{L} = \begin{pmatrix} 0 & L_{\rho\boldsymbol{M}} & 0 & 0 & 0 \\ L_{\boldsymbol{M}\rho} & L_{\boldsymbol{M}\boldsymbol{M}} & L_{\boldsymbol{M}\varepsilon} & L_{\boldsymbol{M}\mathbf{c}} & L_{\boldsymbol{M}\lambda} \\ 0 & L_{\varepsilon\boldsymbol{M}} & 0 & 0 & 0 \\ 0 & L_{\mathbf{c}\boldsymbol{M}} & 0 & 0 & 0 \\ 0 & L_{\lambda\boldsymbol{M}} & 0 & 0 & 0 \end{pmatrix}. \tag{10.4.9}$$

In general, for a variable of a given type (scalar, scalar density, tensor, tensor density, etc.), several general expressions for the nonzero components of $\boldsymbol{L}$ are reported in Ref. 2.

10.4.3 The friction operator

Of interest is the friction matrix $\boldsymbol{M}$. For the simple set of hydrodynamic variables described by the vector $\boldsymbol{x} = (\rho(\boldsymbol{r}), \boldsymbol{M}(\boldsymbol{r}), \varepsilon(\boldsymbol{r}), \boldsymbol{c}(\boldsymbol{r}))$, a typical expression for $\boldsymbol{M}$ is

$$\boldsymbol{M} = \begin{pmatrix} 0 & 0 & 0 & 0 \\ 0 & M_{\boldsymbol{MM}} & M_{\boldsymbol{M}\varepsilon} & 0 \\ 0 & M_{\varepsilon\boldsymbol{M}} & M_{\varepsilon\varepsilon} & 0 \\ 0 & 0 & 0 & M_{\boldsymbol{cc}} \end{pmatrix}, \tag{10.4.10}$$

with expressions for its nonzero components as reported in Ref. 78.

A very attractive feature of the GENERIC formalism is that one can derive analytical expressions for the full set $\{E, S, \boldsymbol{L}, \boldsymbol{M}\}$ of its building blocks for systems for which there is a clear separation of time scales, by employing Grabert's projection operator method.[79] The approach is outlined in detail by Öttinger[2,80] and starts by considering the mapping $\Pi(\boldsymbol{z})$ from the atomistic-level microstate $\boldsymbol{z}$ to the coarse-grained-level macrostate $\boldsymbol{x}$. In particular, it is assumed that the values of the state variables $\boldsymbol{x}$ are the averages of $\Pi(\boldsymbol{z})$, and these are evaluated with the probability density $\rho_{\boldsymbol{x}}(\boldsymbol{z})$ to find a microstate $\boldsymbol{z}$ for the given value of the coarser variable $\boldsymbol{x}$:

$$\boldsymbol{x} = \langle \Pi(\boldsymbol{z}) \rangle_{\boldsymbol{x}} = \int \rho_{\boldsymbol{x}}(\boldsymbol{z}) \Pi(\boldsymbol{z}) \mathrm{d}\boldsymbol{z}. \tag{10.4.11}$$

Clearly, $\boldsymbol{z}(t)$ satisfies a purely reversible dynamics, namely,

$$\frac{\partial \boldsymbol{z}}{\partial t} = \boldsymbol{L}_0 \cdot \frac{\delta E_0(\boldsymbol{z})}{\delta \boldsymbol{z}} \tag{10.4.12}$$

with $\boldsymbol{z}(0) = \boldsymbol{z}_0$ and the antisymmetric matrix $\boldsymbol{L}_0$ defined (according to Hamilton's equations of motion) as

$$\boldsymbol{L}_0 = \begin{pmatrix} 0 & 1 \\ -1 & 0 \end{pmatrix} \tag{10.4.13}$$

for each degree of freedom. Then, for any arbitrary phase-space function $A = A(\boldsymbol{z}(t))$, we have:

$$\frac{\mathrm{d}A}{\mathrm{d}t} = \frac{\partial A(\boldsymbol{z})}{\partial \boldsymbol{z}} \cdot \frac{\mathrm{d}\boldsymbol{z}}{\mathrm{d}t} = \frac{\partial A(\boldsymbol{z})}{\partial \boldsymbol{z}} \cdot \boldsymbol{L}_0 \cdot \frac{\partial E_0}{\partial \boldsymbol{z}} = \{A, E_0\}, \tag{10.4.14}$$

where we have made use of the well-known Poisson bracket of classical mechanics already defined in Eq. (10.1.3). Equation (10.4.14) can also be written as an operator equation, namely,

$$\frac{\mathrm{d}A(\boldsymbol{z})}{\mathrm{d}t} = iLA(\boldsymbol{z}), \tag{10.4.15}$$

where we have introduced the Liouville operator iL where i is the imaginary unit. The reason why the Liouville operator incorporates also the imaginary unit is in order to make it a self-adjoint (or, in the language of physics, a Hermitian; see Chapter 7) operator:[2]

$$L^{\dagger} = L \tag{10.4.16}$$

By viewing Eq. (10.4.15) as a homogeneous linear differential equation, its formal solution can be written as

$$A(t) = e^{iLt} A(0) \tag{10.4.17}$$

To proceed, it is important to make a choice of a generalized ensemble in order to specify the probability density $\rho_{\boldsymbol{x}}(\boldsymbol{z})$. Öttinger discusses two options,[2] the case of a *generalized microcanonical* ensemble and the case of a *generalized canonical* ensemble, and has also derived the corresponding expressions for the entropy building block of GENERIC, S, as natural generalizations of the corresponding expressions of the respective equilibrium ensembles. Having chosen a generalized ensemble, the next step is to consider the fundamental projection operator P (compare with Section 7.7) which singles out slow variables (and thus puts the emphasis on the rapid fluctuations of the dynamics around the imposed values of the slowly evolving coarse-grained variables) through the following action on an arbitrary observable $\mathcal{X}$:

$$P(\boldsymbol{x})\mathcal{X} = \int \rho_{\boldsymbol{x}}(\boldsymbol{z}')\,\mathcal{X}(\boldsymbol{z}')\,\mathrm{d}\boldsymbol{z}' + \sum_k (\Pi_k - x_k) \int \mathcal{X}(\boldsymbol{z}') \frac{\delta \rho_{\boldsymbol{x}}(\boldsymbol{z}')}{\delta x_k} \mathrm{d}\boldsymbol{z}'. \tag{10.4.18}$$

For the energy building block of GENERIC, in particular, for which there are typically only macroscopic contributions (and no rapidly fluctuating parts), we have that $P(\boldsymbol{x})E_0 = E_0$, which implies that

$$E(\boldsymbol{x}) = E(\Pi(\boldsymbol{z})) = E_0(\boldsymbol{z}) \tag{10.4.19}$$

and therefore

$$\frac{\partial E_0}{\partial \boldsymbol{z}} = \frac{\partial (P(\boldsymbol{x})E_0)}{\partial \boldsymbol{z}} = \sum_k \frac{\partial \Pi_k}{\partial \boldsymbol{z}} \frac{\delta}{\delta x_k} \int \rho_{\boldsymbol{x}}(\boldsymbol{z}')\, E_0(\boldsymbol{z}')\,\mathrm{d}\boldsymbol{z}'. \tag{10.4.20}$$

Equation (10.4.19) can be considered as the mapping of the energy from its microstate value $E_0(\boldsymbol{z})$ to its macroscale value $E(\boldsymbol{x})$ for a given value of $\boldsymbol{x}$.

To define the corresponding mappings for the $\boldsymbol{L}$ and $\boldsymbol{M}$ building blocks of GENERIC, it is useful to consider also the orthogonal operator Q defined as (see also Eq. (7.7.14)) $Q = 1 - P$. Then, the following identity can be used to formulate the entire projection-operator formalism,[2,79]

$$e^{iLt} = e^{iLt}P(t) + Q(0)\,G(0,t) + \int_0^t e^{iLu}P(u)\left[iL - \dot{P}(u)\right] Q(u)\, G(u,t)\mathrm{d}u, \tag{10.4.21}$$

where L denotes the Liouville operator defined above, the dot in $\dot{P}(u)$ indicates differentiation, and $G(u,t)$ is the time-ordered exponential $G(u,t) = \exp\left\{\int_u^t Q(s)\,iLQ(s)\,\mathrm{d}s\right\}$ with operators ordered from left to right as time increases. We have also made use of the shorthand notation $P(\boldsymbol{x}(t)) = P(t)$.

Example 10.9

Prove Eq. (10.4.21).

Solution: First, we note that by definition $P(t) + Q(t) = 1$ and $P(t)Q(t) = 0$. One can also show that $P(\boldsymbol{x}(t'))\,P(\boldsymbol{x}(t)) = P(\boldsymbol{x}(t))$ or in shorthand notation $P(t')\,P(t) = P(t)$, from which we get (by differentiating with respect to t, and then applying the result for $t' = t$) $P(t)\dot{P}(t) = \dot{P}(t)$ and $\dot{P}(t)P(t) = 0$, thus also $\dot{P}(t)Q(t) = \dot{P}(t)$. For a more detailed discussion about these equalities, interested readers are referred to Chapter 6 of Ref. 2. Next, we define the quantity

$$F(t) = e^{iLt}Q(t), \tag{10.4.22}$$

from which, after differentiating with respect to time and using the above identities, we get

$$\begin{aligned}
\dot{F}(t) &= e^{iLt}iLQ(t) + e^{iLt}\dot{Q}(t) \\
&= e^{iLt}iLQ(t) - e^{iLt}\dot{P}(t) \\
&= e^{iLt}\left(P(t) + Q(t)\right) iLQ(t) - e^{iLt}\left(P(t)\dot{P}(t)\right) \\
&= e^{iLt}P(t)iLQ(t) + e^{iLt}Q(t)iLQ(t) - e^{iLt}\left(P(t)\dot{P}(t)Q(t)\right) \\
&= e^{iLt}Q(t)iLQ(t) + e^{iLt}P(t)iLQ(t) - e^{iLt}P(t)\dot{P}(t)Q(t) \\
&= e^{iLt}Q(t)iLQ(t) + e^{iLt}P(t)\left[iL - \dot{P}(t)\right]Q(t).
\end{aligned} \tag{10.4.23}$$

We also have that

$$F(t)Q(t)iLQ(t) = e^{iLt}Q(t)Q(t)iLQ(t) = e^{iLt}\left(1 - P(t)\right)Q(t)iLQ(t) = e^{iLt}Q(t)iLQ(t),$$

i.e.

$$e^{iLt}Q(t)iLQ(t) = F(t)Q(t)iLQ(t).$$

By replacing this last equality into Eq. (10.4.23), we finally get

$$\dot{F}(t) = F(t)Q(t)iLQ(t) + e^{iLt}P(t)\left[iL - \dot{P}(t)\right]Q(t). \tag{10.4.24}$$

This has the form of an inhomogeneous linear differential equation whose formal solution is

$$F(t) = F(0)\,G(0,t) + \int_0^t e^{iLu}P(u)\left[iL - \dot{P}(u)\right]Q(u)\,G(u,t)\mathrm{d}u. \tag{10.4.25}$$

continued

Example 10.9 *continued*

By definition, $e^{iLt} = e^{iLt}(P(t) + Q(t))$. Thus,

$$e^{iLt} = e^{iLt}P(t) + F(t). \tag{10.4.26}$$

Substituting Eq. (10.4.25) into Eq. (10.4.26) yields the fundamental equality that we wanted to prove, Eq. (10.4.21).

By acting on $iL\Pi_j$ with both sides of Eq. (10.4.21) and averaging over a generalized canonical initial ensemble,[2] one obtains (see Ref. 2 for all intermediate steps)

$$\begin{aligned}\frac{\mathrm{d}x_j}{\mathrm{d}t} &= \int \rho_{\boldsymbol{x}(t)}(\boldsymbol{z})iL\Pi_j(\boldsymbol{z})\mathrm{d}\boldsymbol{z} \\ &+ \int_0^t \int \rho_{\boldsymbol{x}(u)}(\boldsymbol{z})iLQ(u)\,G(u,t)iL\Pi_j(\boldsymbol{z})\mathrm{d}\boldsymbol{z}\;\mathrm{d}u.\end{aligned} \tag{10.4.27}$$

Nicely, the first term on the right-hand-side of Eq. (10.4.27) expresses the reversible contribution to the dynamics,

$$\left.\frac{\mathrm{d}x_j}{\mathrm{d}t}\right|_{\mathrm{rev}} = \int \rho_{\boldsymbol{x}(t)}(\boldsymbol{z})iL\Pi_j(\boldsymbol{z})\mathrm{d}\boldsymbol{z}, \tag{10.4.28}$$

and the second term the irreversible one:

$$\left.\frac{\mathrm{d}x_j}{\mathrm{d}t}\right|_{\mathrm{irrev}} = \int_0^t \int \rho_{\boldsymbol{x}(u)}(\boldsymbol{z})iLQ(u)\,G(u,t)iL\Pi_j(\boldsymbol{z})\mathrm{d}\boldsymbol{z}\,\mathrm{d}u. \tag{10.4.29}$$

Clearly, the reversible contribution must have the form of Eq. (10.4.15), namely $\frac{\mathrm{d}A}{\mathrm{d}t} = iLA(\boldsymbol{z})$. Indeed, if we choose

$$L_{jk}(\boldsymbol{x}) = \int \rho_{\boldsymbol{x}}(\boldsymbol{z})\,\{\Pi_j, \Pi_k\}\,\mathrm{d}\boldsymbol{z}, \tag{10.4.30}$$

then we see that

$$\left.\frac{\mathrm{d}x_j}{\mathrm{d}t}\right|_{\mathrm{rev}} = \sum_k L_{jk}(\boldsymbol{x}(t))\,\frac{\delta E(\boldsymbol{x}(t))}{\delta x_k}, \tag{10.4.31}$$

which is exactly of the form of Eq. (10.4.15) describing the time evolution of a general function of phase space; it is also compatible with the reversible part of the fundamental GENERIC dynamics,

Eq. (10.4.1). Equation (10.4.30) is a very important outcome of the projection-operator formalism just outlined because it offers a means of computing the Poisson matrix $\boldsymbol{L}$ from atomistic-level simulations, using a probability distribution function that is compatible with the imposed macrostate at the particular time.

Similarly, for the term expressing the irreversible contribution, Eq. (10.4.29), one can carry out first an integration by parts (to allow the first Liouville operator to act on the distribution function), then apply the Markovian approximation (to approximate all slow variables at times u in the interval $[t-\tau,\tau]$ with their value at time t), and then use a self-adjointness property of the projection operator $P(\boldsymbol{x})$,[2] namely that $\int \rho_{\boldsymbol{x}}(\boldsymbol{z})A(\boldsymbol{z})P(\boldsymbol{x})B(\boldsymbol{z})\mathrm{d}\boldsymbol{z} = \int \rho_{\boldsymbol{x}}(\boldsymbol{z})B(\boldsymbol{z})P(\boldsymbol{x})A(\boldsymbol{z})\mathrm{d}\boldsymbol{z}$, to finally bring it to the form[2]

$$\left.\frac{\mathrm{d}x_j}{\mathrm{d}t}\right|_{\mathrm{irrev}} = \sum_k M_{jk}\left(\boldsymbol{x}(t)\right)\frac{\delta S\left(\boldsymbol{x}(t)\right)}{\delta x_k}, \tag{10.4.32}$$

where

$$M_{jk}(\boldsymbol{x}) = \frac{1}{k_{\mathrm{B}}}\int_0^{\tau}\int \rho_{\boldsymbol{x}}(\boldsymbol{z})\left[Q(\boldsymbol{x})iL\Pi_k(\boldsymbol{z})\right]e^{Q(\boldsymbol{x})iLQ(\boldsymbol{x})u}\left[Q(\boldsymbol{x})iL\Pi_j(\boldsymbol{z})\right]\mathrm{d}\boldsymbol{z}\,\mathrm{d}u \tag{10.4.33}$$

or, equivalently, as

$$M_{jk}(\boldsymbol{x}) = \frac{1}{k_{\mathrm{B}}}\int_0^{\tau}\left\langle\hat{\Pi}_k(\boldsymbol{z})e^{i\hat{L}t}\hat{\Pi}_j(\boldsymbol{z})\right\rangle_{\boldsymbol{x}}\mathrm{d}t, \tag{10.4.34}$$

where we have defined

$$\hat{\Pi}_k(\boldsymbol{z}) \equiv Q(\boldsymbol{x})iL\Pi_k(\boldsymbol{z}),\quad \hat{L} \equiv Q(\boldsymbol{x})LQ\left(\boldsymbol{x}.\right) \tag{10.4.35}$$

Equation (10.4.34) has the form of a Green–Kubo equation and is one of the most important outcomes of the projection-operator approach. It nicely demonstrates that the magnitude of the components of the friction matrix $\boldsymbol{M}(\boldsymbol{x})$ is governed by the amplitude and relaxation time of the fluctuations that exist on top of the much slower changes that the coarse-grained (i.e. the slowly evolving) degrees of freedom undergo. In the projection-operator approach just described, these coarse-grained variables vary over times much longer than τ, and they are singled out by the action of $P(\boldsymbol{x})$.

A detailed comparison between the GENERIC and the generalized bracket formalisms[81] shows that the two approaches are intimately coupled, and that the single generator formalism can be considered as a particular realization of the more general double-generator formalism. Overall, GENERIC appears to be more structured and more powerful in the sense that (a) it allows for a more general choice of variables, (b) it has more predictive power and a more symmetric form than the bracket formalism, (c) one can derive microscopic expressions for the set of building blocks

$\{E, S, \boldsymbol{L}, \boldsymbol{M}\}$ by projection-operator techniques, and (d) it can be used to derive the Boltzmann equation,[82] a case where neither the entropy density nor the temperature appear in the vector of state variables.

A nice feature of the two formalisms is that both can be applied to derive a set of Lorentz-covariant equations for an imperfect fluid with viscosity, dilatational viscosity, and anisotropic thermal conductivity.[83,84] A general covariant version of the GENERIC structure for constructing transport equations for self-gravitating imperfect fluids has also been presented.[85]

10.5 Beyond-equilibrium polymer simulations inspired by the generalized bracket and GENERIC formalisms

The two formalisms provide an excellent framework for the design of beyond-equilibrium simulations in expanded statistical mechanics ensembles. Indeed, we can write the general thermodynamic equation for the entropy density s in GENERIC in the form[2,86]

$$s(\boldsymbol{x}) = k_{\mathrm{B}} \sum_k \lambda_k x_k \tag{10.5.1}$$

or, equivalently, as

$$\mathrm{d}s = k_{\mathrm{B}} \sum_k \lambda_k \mathrm{d}x_k \tag{10.5.2}$$

by making use of the Lagrange multipliers or conjugate fields λ_k defined as

$$\frac{\delta s(\boldsymbol{x})}{\delta x_k} = k_{\mathrm{B}} \lambda_k \tag{10.5.3}$$

to pass from the generalized microcanonical to the generalized canonical ensemble. By taking the total derivative of Eq. (10.5.1) and by comparing with Eq. (10.5.2), we obtain the generalized Gibbs–Duhem relation

$$\sum_k x_k \, \mathrm{d}\lambda_k = 0. \tag{10.5.4}$$

Then, based on the fundamental statistical principle of *the equal a priori probability for every phase-space point on a constant energy surface*,[87,88] the probability density function in a generalized canonical ensemble corresponding to the above macroscopic thermodynamic relations involves the extensive thermodynamic variables Π_k (whose ensemble average corresponds to the thermodynamic state variable x_k) paired with their conjugate thermodynamic fields (the Lagrange multipliers λ_k):[2]

$$\rho_{[\lambda]}(\boldsymbol{z}) = \frac{1}{Q(\boldsymbol{\lambda})} \exp\left(-\sum_k \lambda_k \Pi_k(\boldsymbol{z})\right); \quad Q(\boldsymbol{\lambda}) = \int \exp\left(-\sum_k \lambda_k \Pi_k(\boldsymbol{z})\right) \mathrm{d}\boldsymbol{z}. \tag{10.5.5}$$

The generalized probability function of Eq. (10.5.5) provides the basis for designing simulations to sample nonequilibrium states of a system characterized by a structural or conformational property that evolves very slowly in time compared to the rest of the variables, upon applying an external field. Let us explain this by considering the case of an unentangled polymer melt coarse-grained to the level of the chain conformation tensor $\mathbf{c}$ and subjected to a flow field. By separating temperature effects (based on the well-known notion of purely entropic elasticity[89–94]) in the Lagrange multiplier, we can write the following expression for the internal energy function of this system under the applied flow field,[95]

$$U = TS - PV + \mu N_{\mathrm{ch}} + N_{\mathrm{ch}} k_{\mathrm{B}} T \boldsymbol{\alpha} : \tilde{\mathbf{c}}, \tag{10.5.6}$$

implying that

$$\mathrm{d}U(S, V, N_{\mathrm{ch}}, N_{\mathrm{ch}}\tilde{\mathbf{c}}) = T\mathrm{d}S - P\mathrm{d}V + \mu \mathrm{d}N_{\mathrm{ch}} + k_B T \boldsymbol{\alpha} : \mathrm{d}\left(N_{\mathrm{ch}}\tilde{\mathbf{c}}\right). \tag{10.5.7}$$

From these equations, all other thermodynamic functions can be derived[95–99] through appropriate Legendre transforms.[100] For example, the extended Helmholtz energy A and Gibbs energy G of the nonequilibrium system will be given through

$$\mathrm{d}A(T, V, N_{\mathrm{ch}}, N_{\mathrm{ch}}\tilde{\mathbf{c}}) = -S\mathrm{d}T - P\mathrm{d}V + \mu \mathrm{d}N_{\mathrm{ch}} + k_{\mathrm{B}} T \boldsymbol{\alpha} : \mathrm{d}\left(N_{\mathrm{ch}}\tilde{\mathbf{c}}\right) \tag{10.5.8}$$

and

$$\mathrm{d}G(T, P, N_{\mathrm{ch}}, \boldsymbol{\alpha}) = -S\mathrm{d}T + V\mathrm{d}P + \mu\,\mathrm{d}N_{\mathrm{ch}} - N_{\mathrm{ch}} k_{\mathrm{B}} T \tilde{\mathbf{c}} : \mathrm{d}\boldsymbol{\alpha}, \tag{10.5.9}$$

respectively. The differential of the Helmholtz free-energy density can alternatively be written as

$$\mathrm{d}\left(\frac{A}{V}\right) = -\frac{S}{V}\mathrm{d}T + \mu \mathrm{d}\left(\frac{N_{\mathrm{ch}}}{V}\right) + \frac{N_{\mathrm{ch}}}{V} k_{\mathrm{B}} T \boldsymbol{\alpha} : \mathrm{d}\tilde{\mathbf{c}}, \tag{10.5.10}$$

from which we recognize that

$$\boldsymbol{\lambda}_{\tilde{\mathbf{c}}} \equiv \frac{1}{k_{\mathrm{B}}} \frac{\delta S}{\delta \tilde{\mathbf{c}}} = -\frac{1}{k_{\mathrm{B}} T} \frac{\delta A(\tilde{\mathbf{c}})}{\delta \tilde{\mathbf{c}}} = -\frac{N_{\mathrm{ch}}}{V} \boldsymbol{\alpha}. \tag{10.5.11}$$

The above expressions define the framework for designing beyond-equilibrium MC simulations by expanding the equilibrium statistical ensemble to include the synthetic tensorial field $\boldsymbol{\alpha}$.[95] We can denote such an expanded ensemble as $\{N_{\mathrm{ch}} NPT \boldsymbol{\mu}^* \boldsymbol{\alpha}\}$ because the following variables should be specified (held fixed): the number of chains N_{ch}, the average number of atoms per chain N, the pressure P, the temperature T, the spectrum of chain relative chemical potentials $\boldsymbol{\mu}^*$ controlling the distribution of chain lengths in the system (and thus also the system polydispersity),[101] and the tensorial field $\boldsymbol{\alpha}$ accounting indirectly for flow effects through its direct coupling to the conformation tensor $\tilde{\mathbf{c}}$, thereby driving polymer configurations away from their mean spherical shape representative of a quiescent (no flow applied) system. The corresponding probability density function is[101]

$$\rho^{N_{\mathrm{ch}}NPT\boldsymbol{\mu}^*\boldsymbol{\alpha}}(\boldsymbol{r}_1,\boldsymbol{r}_2,\ldots,\boldsymbol{r}_n,V)$$
$$\propto \exp\left[-\beta\left(\mathcal{V}(\boldsymbol{r}_1,\boldsymbol{r}_2,\ldots,\boldsymbol{r}_n,V)+PV-\sum_{k=1}^{N_{\mathrm{ch}}}\mu_k^* N_k - k_{\mathrm{B}}T\boldsymbol{\alpha}:\sum_{k=1}^{N_{\mathrm{ch}}}\tilde{\mathbf{c}}_k\right)\right], \tag{10.5.12a}$$

from which we understand that system configurations can be sampled with a Monte Carlo algorithm similar to the end-bridging one described in Chapter 5 on the basis of the Metropolis acceptance criterion

$$p_{\mathrm{acc}}^{N_{\mathrm{ch}}NPT\boldsymbol{\mu}^*\boldsymbol{\alpha}} = \min\left\{1,\exp\left[-\beta\left(\Delta\mathcal{V}+P\Delta V-\sum_{k=1}^{N_{\mathrm{ch}}}\Delta\left(\mu_k^* N_k\right)-k_{\mathrm{B}}T\boldsymbol{\alpha}:\sum_{k=1}^{N_{\mathrm{ch}}}\Delta\tilde{\mathbf{c}}_k\right)\right]\right\}, \tag{10.5.12b}$$

or, for the case of a system simulated under conditions of constant volume V (i.e. constant density ρ) according to

$$p_{\mathrm{acc}}^{N_{\mathrm{ch}}NVT\boldsymbol{\mu}^*\boldsymbol{\alpha}} \min\left\{1,\exp\left[-\beta\left(\Delta\mathcal{V}-\sum_{k=1}^{N_{\mathrm{ch}}}\Delta\left(\mu_k^* N_k\right)-k_B T\boldsymbol{\alpha}:\sum_{k=1}^{N_{\mathrm{ch}}}\Delta\tilde{\mathbf{c}}_k\right)\right]\right\}, \tag{10.5.12c}$$

where $\beta \equiv \dfrac{1}{k_{\mathrm{B}}T}$. In the above equations, $n\,(=N_{\mathrm{ch}}\times N)$ is the total number of atoms in the system, $\{\boldsymbol{r}\}=\{\boldsymbol{r}_1,\boldsymbol{r}_2,\ldots,\boldsymbol{r}_n\}$ denotes the space of their position vectors, V the volume, $\mathcal{V}$ the potential energy, $\mu_k^*\ (1\le k\le N_{\mathrm{ch}})$ the relative chemical potential of the kth chain, and $\tilde{\mathbf{c}}_k(1\le k\le N_{\mathrm{ch}})$ its conformation tensor. With appropriate input data for the set $\{N_{\mathrm{ch}}NPT\boldsymbol{\mu}^*\boldsymbol{\alpha}\}$, such a method allows one to sample phase points of a polymer melt beyond equilibrium, by assigning nonzero values to $\boldsymbol{\alpha}$.

Despite its fundamental rooting on nonequilibrium thermodynamics, the above modelling scheme has the drawback that defining the field $\boldsymbol{\alpha}$ for a given flow kinematics (i.e. a given velocity gradient) is not straightforward. According to the GENERIC or generalized bracket formalisms, the evolution equation for the structural variable $\tilde{\mathbf{c}}$ reads (see e.g. Eq. (10.3.12d))

$$\dot{\tilde{\mathbf{c}}}_{[1]} = -\boldsymbol{\Lambda}:\frac{\partial A}{\partial\tilde{\mathbf{c}}}, \tag{10.5.13}$$

where on the left hand-side we recognize the upper-convected Maxwell derivative of $\tilde{\mathbf{c}}$ and on the right hand-side the appearance of the fourth-rank relaxation matrix $\boldsymbol{\Lambda}$ and the free-energy functional

A. Then, by identifying the partial derivative $\dfrac{\partial A}{\partial \tilde{\boldsymbol{c}}}$ with $-k_{\mathrm{B}}T\boldsymbol{\lambda}_{\tilde{\boldsymbol{c}}}$ (see Eq. (10.5.11)), we can connect the elements of the tensor $\boldsymbol{\alpha}$ with the elements of the velocity gradient $\nabla\boldsymbol{\nu}$ hidden inside the objective derivative $\overset{\nabla}{\tilde{\boldsymbol{c}}}_{[1]}$ under steady-state flow conditions. More formally, according to the GENERIC framework, Eq. (10.4.1), the Lagrange multipliers have a clear kinematic interpretation, since under steady-state conditions

$$\boldsymbol{\lambda} = -\frac{1}{k_{\mathrm{B}}}\boldsymbol{M}^{-1}\cdot\boldsymbol{L}(\boldsymbol{x})\cdot\frac{\delta E(\boldsymbol{x})}{\delta \boldsymbol{x}}. \qquad (10.5.14)$$

This implies that the set $\boldsymbol{\lambda}$ conveys information related to the dissipative matrix $\boldsymbol{M}$. More specifically, for a given flow, nonzero values to $\boldsymbol{\alpha}$ can be assigned through Eq. (10.5.13) if we know the matrix $\boldsymbol{M}$. The matrix $\boldsymbol{M}$, however, is known today only approximately through the various proposed viscoelastic models. Thus, any proposition for $\boldsymbol{\alpha}$ would be *model-dependent*, and thus only approximately correct. To understand this better, let us (e.g.) restrict ourselves to the case of steady shear flow, which is described by the velocity gradient tensor

$$\nabla\boldsymbol{\nu} = \begin{pmatrix} 0 & 0 & 0 \\ \dot{\gamma} & 0 & 0 \\ 0 & 0 & 0 \end{pmatrix}. \qquad (10.5.15)$$

With the choice $\boldsymbol{X} = \tilde{\boldsymbol{c}}$, the constitutive equation for the viscoelastic material takes the general form of Eq. (10.5.13). As we discussed in Section 10.2, the relaxation matrix $\boldsymbol{\Lambda}$ (see Eq. (10.2.34)), must satisfy Onsager's symmetry property and the thermodynamic admissibility criteria (i.e. the second law of thermodynamics). The extended Helmholtz free-energy density A, on the other hand, is usually taken to be a function of the invariants of the conformation tensor (see Section (10.2.4)), and with suitable choices of $\boldsymbol{\Lambda}$ and A, a number of well-established and widely used viscoelastic models are reproduced.[1,38] We further recall that, in order for the model to satisfy the second law of thermodynamics (such as Eq. (10.2.9) for an isothermal system), the four functions $f_1(I_1), f_2(I_1)$, $f_3(I_1)$ and $f_4(I_1)$ in expression (10.2.34) must be constrained. For certain special cases considered in the literature, one finds that $f_i(I_1) \geq 0, \forall i$. Then, according (e.g.) to the simple upper-convected Maxwell (UCM) viscoelastic model characterized by a tensor $\boldsymbol{M}$ or, equivalently, $\boldsymbol{\Lambda}$ of the form (10.2.34) with $f_2 = 0, f_3 = 0, f_4 = 0$ (see Table 10.1), the tensor $\boldsymbol{\alpha}$ comes out to be

$$\boldsymbol{\alpha} = \begin{bmatrix} \dfrac{1}{2}\dfrac{(\lambda_0\dot{\gamma})^2}{1+(\lambda_0\dot{\gamma})^2} & \dfrac{1}{2}\dfrac{\lambda_0\dot{\gamma}}{1+(\lambda_0\dot{\gamma})^2} & 0 \\ \dfrac{1}{2}\dfrac{\lambda_0\dot{\gamma}}{1+(\lambda_0\dot{\gamma})^2} & -\dfrac{1}{2}\dfrac{(\lambda_0\dot{\gamma})^2}{1+(\lambda_0\dot{\gamma})^2} & 0 \\ 0 & 0 & 0 \end{bmatrix}, \qquad (10.5.16)$$

where λ_0 is the longest relaxation time of the polymer. Similarly, according to the more refined and widely used Giesekus model, the elements of $\boldsymbol{\alpha}$ (again for the case of steady shear) are obtained from the solution to the set of algebraic equations

$$\begin{aligned}
&(1-2a)\,\tilde{c}_{yy} + a\left(\tilde{c}_{xy}^2 + \tilde{c}_{yy}^2\right) - 1 + a = 0 \\
&(1-2a)\,\tilde{c}_{xy} + a\left(\tilde{c}_{xx}\tilde{c}_{xy} + \tilde{c}_{xy}\tilde{c}_{yy}\right) - (\lambda_0\dot{\gamma})\,\tilde{c}_{yy} = 0 \\
&(1-2a)\,\tilde{c}_{xx} + a\left(\tilde{c}_{xx}^2 + \tilde{c}_{xy}^2\right) - 2\,(\lambda_0\dot{\gamma})\,\tilde{c}_{xy} - 1 + a = 0 \\
&\tilde{c}_{yy} - 1 = 0
\end{aligned} \tag{10.5.17a}$$

together with

$$\boldsymbol{\alpha} = \frac{1}{2}\lambda_0[\boldsymbol{I} + a\,(\tilde{\boldsymbol{c}} - \boldsymbol{I})]^{-1}\cdot\left[\nabla\boldsymbol{\nu} + \tilde{\boldsymbol{c}}^{-1}\cdot\nabla\boldsymbol{\nu}^T\cdot\tilde{\boldsymbol{c}}\right], \tag{10.5.17b}$$

where a denotes the Giesekus parameter (it usually takes values between 0 and 0.5), $\tilde{\boldsymbol{c}}^{-1}$ denotes the inverse of $\tilde{\boldsymbol{c}}$, and $\nabla\boldsymbol{\nu}^{\mathrm{T}}$ denotes the transpose of $\nabla\boldsymbol{\nu}$. Similar expressions are found for the other models reported in Table 10.1 or in the literature.

To overcome this obstacle and compute *model-independent* values of $\boldsymbol{\alpha}$ one should run GENERIC Monte Carlo simulations in parallel with direct nonequilibrium molecular dynamics (NEMD) simulations for the specified value of shear rate $\dot{\gamma}$ considered, using (e.g.) the most general form of $\boldsymbol{\alpha}$ possible, and then attempt to match the resulting chain conformation tensors from the two methods.[96,97] For an unentangled polymer melt described by its conformation tensor and subjected to a shear flow, the tensors $\tilde{\boldsymbol{c}}$ and $\boldsymbol{\alpha}$ have only four independent, nonzero components, the sets $\{\tilde{c}_{xx}, \tilde{c}_{xy} = \tilde{c}_{yx}, \tilde{c}_{yy}, \tilde{c}_{zz}\}$ and $\{\alpha_{xx}, \alpha_{xy} = \alpha_{yx}, \alpha_{yy}, \alpha_{zz}\}$, respectively. Thus, the *most general form* of the tensor $\boldsymbol{\alpha}$ which should be used in the GENERIC Monte Carlo solver in order to generate a true shear flow should be

$$\boldsymbol{\alpha} = \begin{pmatrix} \alpha_{xx} & \alpha_{xy} & 0 \\ \alpha_{xy} & \alpha_{yy} & 0 \\ 0 & 0 & \alpha_{zz} \end{pmatrix}, \tag{10.5.18a}$$

but leaving the exact dependence of the set of the four nonzero elements $\{\alpha_{xx}, \alpha_{xy} = \alpha_{yx}, \alpha_{yy}, \alpha_{zz}\}$ on the imposed shear rate $\dot{\gamma}$ undetermined. This implies that a state point in the NEMD simulation corresponding to a given value of $\dot{\gamma}$ is represented in the GENERIC Monte Carlo simulation by the set $\{\alpha_{xx}, \alpha_{xy} = \alpha_{yx}, \alpha_{yy}, \alpha_{zz}\}$. We term the NEMD simulation the solver for the micro-model (the *micro-solver*), and the GENERIC Monte Carlo simulation the solver for the macroscopic model (the *macro-solver*).

In fact, for the majority of the members of the family of viscoelastic models reported in Table 10.1, the coefficients f_3 and f_4 are usually assumed to be zero; then, it can be shown that

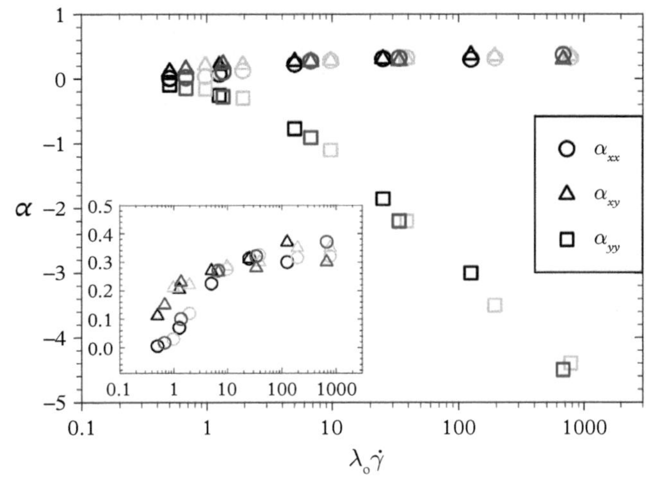

Figure 10.1 Variation of the thermodynamic force field (Lagrange multiplier) α components with the applied Weissenberg number (dimensionless shear rate) for three simulated PE melts: $C_{50}H_{102}$ (black symbols), $C_{78}H_{158}$ (light grey symbols), and $C_{128}H_{258}$ (dark grey symbols) at $T = 450K$ and P=1 atm. The inset displays a more detailed comparison of α_{xx} and α_{xy} values.

$\alpha_{zz} = 0$,[97] and thus the tensor α assumes the form

$$\alpha = \begin{pmatrix} \alpha_{xx} & \alpha_{xy} & 0 \\ \alpha_{xy} & \alpha_{yy} & 0 \\ 0 & 0 & 0 \end{pmatrix}, \tag{10.5.18b}$$

from which we understand that the number of the unknown components to be determined for a given value of the applied shear rate $\dot{\gamma}$ decreases to 3. Such a methodology has been implemented to study the response to a flow field of a number of unentangled linear polyethylene (PE) melts and obtain their nonequilibrium thermodynamic properties.[95–98] A typical result is given in Figure 10.1 showing the (model-independent) values of the components of the tensor α which for a wide range of shear rates examined (covering both the linear and the nonlinear viscoelastic regimes) bring results for the overall polymer conformation of three such PE melts ($C_{50}H_{102}$, $C_{78}H_{158}$, $C_{128}H_{258}$) from the two models (NEMD or micro-solver and GENERIC Monte Carlo or macro-solver) on top of each other. The shear rate in the figure is shown in dimensionless units and is known as the Weissenberg (Wi) number; it is equal to the product of the applied shear rate $\dot{\gamma}$ and the longest relaxation time λ_0 of the polymer melt at equilibrium, the latter being estimated by monitoring the time it takes the end-to-end unit vector time autocorrelation function to drop to 0.

An interesting feature of the entire approach is that, by comparing the values of α computed from the two solvers with those suggested by the macroscopic model, one can validate the model or deduce information on how to modify it so that model predictions and simulation data for the same flow superimpose. It defines therefore a powerful framework for improving available macroscopic

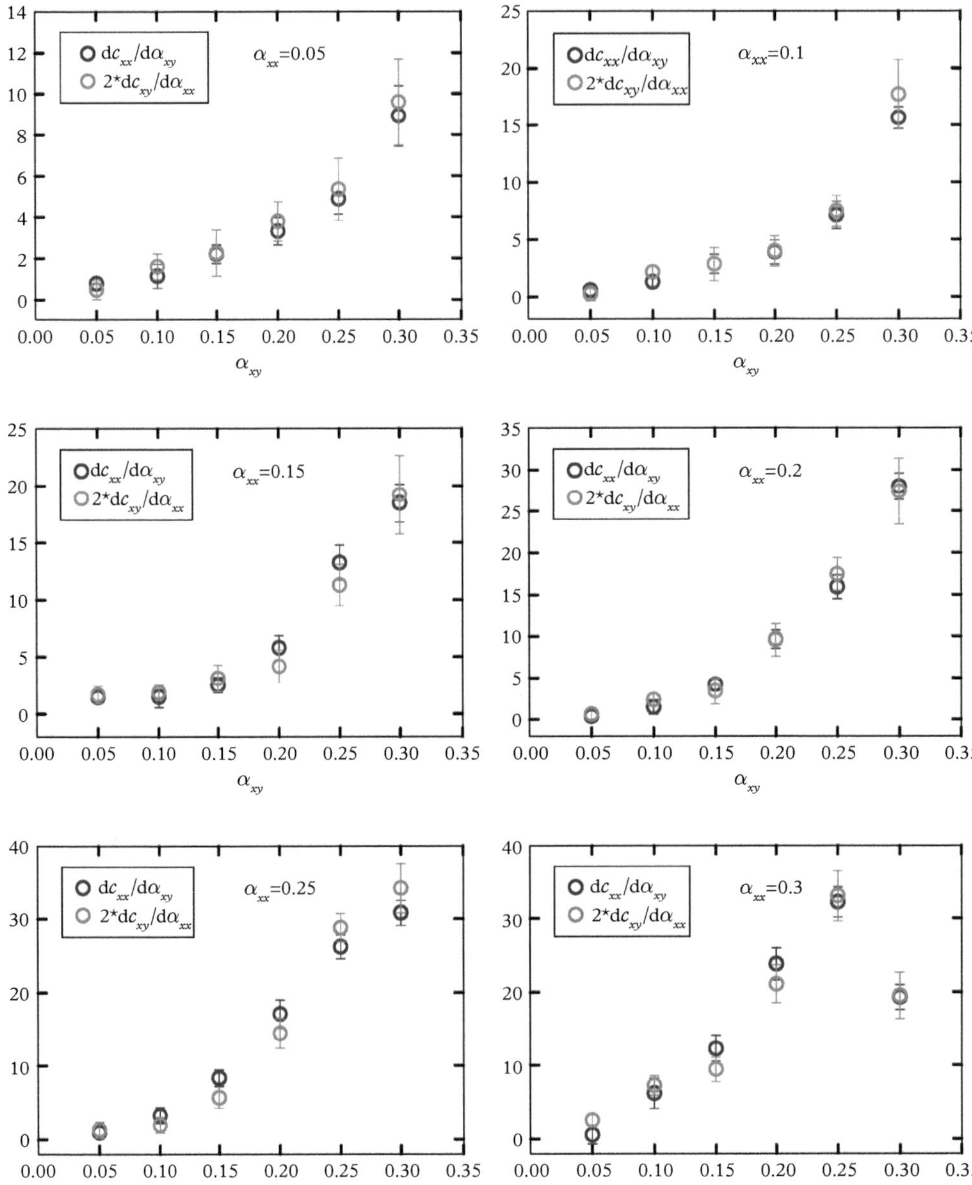

Figure 10.2 Validity of the generalized Maxwell equation, Eq. (10.5.19), from GENERIC Monte Carlo simulations with an unentangled $C_{78}H_{158}$ PE melt at 450 K over a wide range of values of the field α.
Reprinted (adapted) with permission from Baig, C., Mavrantzas, V.G., and Öttinger, H.C. *Macromolecules* **44**, 640 (2011).

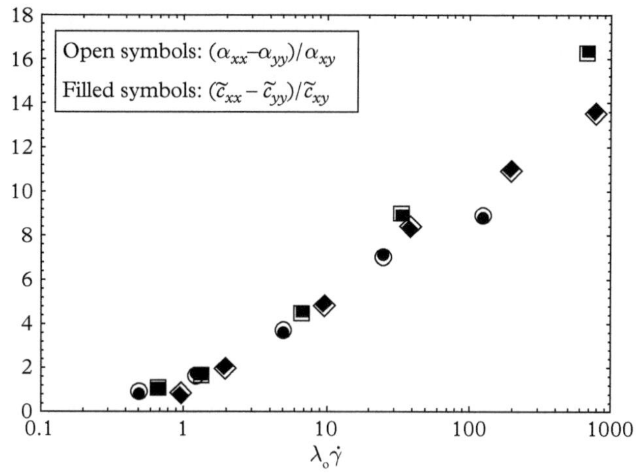

Figure 10.3 Validity of the symmetry equation, Eq. (10.5.20), from GENERIC Monte Carlo simulations with an unentangled $C_{78}H_{158}$ PE melt at 450 K over a wide range of values of the field $\boldsymbol{\alpha}$.

models for the viscoelasticity of polymer melts. It can also be used to test fundamental thermodynamic equations far away from equilibrium. For example, the form of $\boldsymbol{\alpha}$ provided by Eq. (10.5.18b) implies the generalized Maxwell relation

$$\left(\frac{\partial c_{xx}}{\partial \alpha_{xy}}\right)_{T,V,N_{\mathrm{ch}},\alpha_{xx}} = 2\left(\frac{\partial c_{xy}}{\partial \alpha_{xx}}\right)_{T,V,N_{\mathrm{ch}},\alpha_{xy}}. \tag{10.5.19}$$

The GENERIC Monte Carlo simulations[98] provide direct evidence for the validity of such an equation over a wide range of shear flows (Figure 10.2).

An additional important relationship between $\boldsymbol{\alpha}$ and $\tilde{\mathbf{c}}$ that can be derived based on the symmetry of the elastic stress is

$$\frac{\alpha_{xx}-\alpha_{yy}}{\alpha_{xy}} = \frac{\tilde{c}_{xx}-\tilde{c}_{yy}}{\tilde{c}_{xy}}, \tag{10.5.20}$$

which, as Figure 10.3 nicely demonstrates, is also confirmed by the GENERIC Monte Carlo simulations.[96,97]

The GENERIC Monte Carlo simulations presented in this chapter are extremely powerful. In addition to providing a path for improved constitutive modelling,[38,97] they can be used (a) to calculate the free energy of the deformed polymer melt through thermodynamic integration by performing a series of simulations over the thermodynamic state points where only one component of $\boldsymbol{\alpha}$ is varied and the rest are fixed;[98] (b) to partition the free energy into an energetic and an entropic component and investigate whether the melt response to the applied flow is purely entropic or if significant enthalpic interactions develop;[95,98,102] (c) to study when the flow leaves the intrinsic shape of the chains practically unaltered or when it distorts it significantly;[95] and (d) to compute the birefringence of the deformed polymer melt and check the validity of the stress optical law.[103]

Problems

Problem 10.1: Prove Eq. (10.1.11).

References

1. Beris, A.N., and Edwards, B.J. *Thermodynamics of Flowing Systems with Internal Microstructure*. Oxford University Press, Oxford (1994).
2. Öttinger, H.C. *Beyond Equilibrium Thermodynamics*. Wiley-Interscience, Hoboken, NJ (2004).
3. Poisson, S.D. *J. de l' École Polytechnique* **8**, 266 (1809).
4. Lanczos, C. *The Variational Principles of Mechanics* (4th ed.). University of Toronto Press, Toronto (1970). Also available in paperback (Dover, New York, 1986).
5. Lanczos, C. In Salam, A., and Wigner, E.P. (eds.), *Aspects of Quantum Theory*. Cambridge University Press, Cambridge (1972), pp. 169–178.
6. Truesdell C., and Noll W. In Flügge, S., (ed.), *Handbook der Physik*, vol. III. Springer-Verlag, Berlin (1965), pp. 1–154.
7. Truesdell C. *The Mechanical Foundations of Elasticity and Fluid Dynamics*. Gordon and Breach, New York (1966).
8. Truesdell, C. *Rational Thermodynamics* (2nd ed.). Springer-Verlag, New York (1984).
9. Kaufman, A.N. *Phys. Lett. A* **100**, 419 (1984).
10. Morrison, P.J. *Phys. Lett. A* **100**, 423 (1984).
11. Morrison, P.J. *Physica D* **18**, 410 (1986).
12. Grmela, M. *Phys. Lett. A* **102**, 355 (1984); Grmela, M. *Phys. Lett. A* **111**, 36 (1985); Grmela, M. *Phys. Lett. A* **130**, 81 (1988).
13. Grmela, M. *Physica D* **21**, 179 (1986).
14. Grmela, M. *J. Phys. A: Math. Gen.* **22**, 4375 (1989).
15. Öttinger, H.C. *MRS Bull.* **32**, 936 (2007).
16. Onsager, L. *Phys. Rev.* **37** 405, 1931; Onsager, L. *Phys. Rev.* **38**, 2265 (1931).
17. de Groot, S.R. In Domingos, J.J.D., Nina, M.N.R., and Whitelaw, J.H. (eds.), *Foundations of Continuum Thermodynamics*. Macmillan Press, London (1974), pp. 159–183.
18. de Groot, S.R., and Mazur, P. *Non-equilibrium Thermodynamics*. North-Holland, Amsterdam (1962). Also available in paperback (Dover, New York, 1984).
19. Marsden, J.E., Ratiu, T., and Weinstein, A. *Trans. Amer. Math. Soc.* **281**, 147 (1984).
20. Abarbanel, H.D.I., Brown, R., and Yang, Y.M. *Phys. Fluids* **31**, 2802 (1988).
21. Morrison, P.J., and Greene, J.M. *Phys. Rev. Lett.* **45**, 790 (1980); Morrison, P.J., and Greene, J.M. *Phys. Rev. Lett.* **48**, 569 (1982).
22. Morrison, P.J., and Hazeltime, R.D. *Phys. Fluids* **27**, 886 (1984).
23. Edwards, B.J., and Beris, A.N. *J. Phys. A: Math. Gen.* **24**, 2641 (1991).
24. Oldroyd, J.G. *Proc. Roy. Soc. Lond. A* **200**, 523 (1950).
25. Aris, R. *Vectors, Tensors, and the Basic equations of Fluid Mechanics*. Prentice-Hall, Englewood-Cliffs, NJ (1962).

26. Bird, R.B., Armstrong, R.C., and Hassager, O. *Dynamics of Polymeric Liquids, Fluid Mechanics* (2nd ed.), Vol. 1. John Wiley & Sons, New York (1987).
27. Bird, R.B., Curtiss, C.F., Armstrong, R.C., and Hassager, O. *Dynamics of Polymeric Liquids, Kinetic Theory* (2nd ed.), Vol. 2. John Wiley & Sons, New York (1987).
28. Larson, R.G. *Constitutive Equations for Polymer Melts and Solutions*. Butterworth-Heinemann, London (1988).
29. Fleisch, D.A. *A Student's Guide to Vectors and Tensors* (Student's Guides, 1st ed.). Cambridge University Press, Cambridge (2012).
30. Edwards, B.J., Beris, A.N., and Grmela, M. *J. Non-Newt. Fluid Mech.* **35**, 51 (1990).
31. Mwasame, P.M., Wagner, N.J, and Beris, A.N. *J. Fluid Mech.* **831**, 433 (2017).
32. Müller, I. *Arch. Rat. Mech. Anal.* **45**, 241 (1972).
33. Edelen, D.G.B. *Int. J. Eng. Sci.* **10**, 481 (1972).
34. Edelen, D.G.B., and McLennan, J.A. *Int. J. Eng. Sci.* **11**, 813 (1973).
35. Wang, C.C. *Arch. Rat. Mech. Anal.* **58**, 381 (1975).
36. Ryskin, G. *Phys. Rev. A* **32**, 1239 (1985); Ryskin, G. *Phys. Rev. A* **36**, 4526 (1987).
37. Spezialle, C.G. *Phys. Rev. A* **36**, 4522 (1987).
38. Stephanou, P.S., Baig, C., and Mavrantzas V.G. *J. Rheol.* **53**, 309 (2009).
39. Woods, L.C. *The Thermodynamics of Fluid Systems*. Oxford University Press, Oxford (1975).
40. Arnold, V.I. *J. Appl. Math. Mech.* **29**, 1002 (1965).
41. Arnold, V.I. *Mathematical Methods of Classical Mechanics*. Springer-Verlag, New York (1978).
42. Marsden, J.E., and A. Weinstein, A. *Physica D* **7**, 305 (1983).
43. Holm, D.D., Marsden, J.E., Ratiu, T., and Weinstein A. *Phys. Rep.* **123**, 1 (1985).
44. Beris, A.N., and Edwards, B.J. *J. Rheol.* **34**, 55 (1990).
45. Housiadas, K.D., and Beris, A.N. *J. Non-Newt. Fluid Mech.* **140**, 41 (2006).
46. Booij, H.C. *J. Chem. Phys.* **80**, 4571 (1984).
47. Flory, P.J. *Principles of Polymer Chemistry*. Cornell University Press, Ithaca, NY (1953).
48. Peterlin, A. *J. Polym. Sci.: Polym. Lett.* **4**, 287 (1966).
49. Warner Jr., H.R. *Ind. Eng. Chem. Fund.* **11**, 379 (1972).
50. Cohen, A. *Rheol. Acta* **30**, 270 (1991).
51. Kröger, M. *J. non-Newt. Fluid Mech.* **223**, 77 (2015).
52. Stephanou, P.S., Mavrantzas, V.G., and Georgiou, G.C. *Macromolecules* **47**, 4493 (2014).
53. Stephanou, P.S., Tsimouri, I.Ch., and Mavrantzas, V.G., *Macromolecules* **49**, 3161 (2016).
54. Onuki, A. *Phys. Rev. Lett.* **62**, 2472 (1989).
55. Milner, S.T. *Phys. Rev. Lett.* **66**, 1477 (1991).
56. Mavrantzas, V.G., and Beris, A.N. *Phys. Rev. Lett.* **69**, 273 (1992); Mavrantzas, V.G., and Beris, A.N. *Phys. Rev. Lett.* **70**, 2659 (1993).
57. Mavrantzas, V.G. *Surface effects on the conformation and rheology of polymer solutions*. Ph.D. Thesis, University of Delaware (1994).
58. Aubert, J.H., and Tirrell, M. *J. Chem. Phys.* **72**, 2694 (1980); Aubert, J.H., and Tirrell, M. *J. Chem. Phys.* **77**, 553 (1982).
59. Aubert, J.H., Prager, S., and Tirrell, M. *J. Chem. Phys.* **73**, 4103 (1980).
60. Brunn, P.O. *J. Chem. Phys.* **80**, 3420 (1984).
61. Helfand, E., and Fredrickson, G.H. *Phys. Rev. Lett.* **62**, 2468 (1989).
62. Onuki, A. *J. Phys. Soc. Japan* **59**, 3423 (1990).
63. Doi, M. In Onuki, A., and Kawasaki, K. (eds.), *Dynamics and Patterns in Complex Fluids: New Aspects of Physics and Chemistry of Interfaces*. Springer-Verlag, New York (1990), pp. 100–112.
64. Bhave, A.V., Armstrong, R.C., and Brown, R.A. *J. Chem. Phys.* **80**, 2988 (1991).

65. Öttinger, H.C. *Rheol. Acta* **31**, 14 (1992).
66. Larson, R.G. *Rheol. Acta* **31**, 497 (1992).
67. Mavrantzas, V.G., Souvaliotis, A., and Beris, A.N. *Theor. Comp. Fluid Dynam.* **5**, 3 (1993).
68. Beris, A.N., and Mavrantzas, V.G. *J. Rheol.* **38**, 1235 (1995).
69. Mavrantzas, V.G., and Beris, A.N. *J. Chem. Phys.***110** 616 (1999); Mavrantzas, V.G., and Beris, A.N. *J. Chem. Phys.* **110**, 628 (1999).
70. Curtiss, C.F., and Bird, R.B. *J. Chem. Phys.* **111**, 10362 (1999).
71. Apostolakis, M., Mavrantzas, V.G., and Beris, A.N. *J. non-Newt. Fluid Mech.* **102**, 409 (2002).
72. Dill, K.A., and Zimm, B.H. *Nucleic Acids Res.* **7**, 647 (1979).
73. Baumert, B.M., and Muller, S.J. *Rheol. Acta* **34**, 147 (1995).
74. Baumert, B.M., and Muller, S.J. *Phys. Fluids* **9**, 566 (1997).
75. MacDonald, M.J., and Muller, S.J. *J. Rheol.* **40**, 259 (1999).
76. Doi, M., and Edwards, S.F. *The Theory of Polymer Dynamics*. Oxford University Press, New York (1986).
77. Grmela, M., and Öttinger, H.C. *Phys. Rev. E* **56**, 6620 (1997).
78. Öttinger, H.C., and Grmela, M. *Phys. Rev. E* **56**, 6633 (1997).
79. Grabert, H. *Projection Operator Techniques in Nonequilibrium Statistical Mechanics*. Springer, Berlin (1982).
80. Öttinger, H.C. *Phys. Rev. E* **57**, 1416 (1998).
81. Edwards, B.J. *J. Non-Equil. Thermodyn.* **23**, 301 (1998).
82. Öttinger, H.C. *J. Non-Equil. Thermodyn.* **22**, 386 (1997); Öttinger, H.C. *J. Non-Equil. Thermodyn.* **27**, 105 (2002).
83. Öttinger, H.C. *Physica A* **259**, 24 (1998); Öttinger, H.C. *Physica A* **254**, 433 (1998).
84. Mavrantzas, V.G. *J. Non-Equil. Thermodyn.* **49**, 89 (2024).
85. Öttinger, H.C. *Phys. Rev. D* **60**, 103507 (1999).
86. Mavrantzas, V.G., and Öttinger, H.C. *Macromolecules* **35**, 960 (2002).
87. Tolman, R.C. *The Principles of Statistical Mechanics*. Oxford University Press, London (1938).
88. Landau, L.D., and Lifshitz, E.M. Statistical Physics, Part 1, Vol. 5. In *Course of Theoretical Physics* (3rd ed.). Butterworth-Heinemann, London (1976).
89. Mark, J.E., and Erman, B. *Rubberlike Elasticity: A Molecular Primer*. John Wiley, New York (1988).
90. Treloar, L.R.G. *The Physics of Rubber Elasticity* (3rd ed.). Clarendon Press, Oxford (1975).
91. Astarita, G., and Marrucci, M. *Principles of Non-Newtonian Fluid Mechanics*. McGraw Hill, London (1974).
92. Weiner, J.H. *Statistical Mechanics of Elasticity*. Wiley, New York (1983).
93. Tanner, R.I. *Engineering Rheology*. Clarendon Press, Oxford (1985).
94. Morrison, F.A. *Understanding Rheology*. Oxford University Press, New York (2001).
95. Mavrantzas, V.G., and Theodorou, D.N. *Macromolecules* **31**, 6310 (1998).
96. Baig, C., and Mavrantzas, V.G. *Phys. Rev. Lett.* **99**, 257801 (2007).
97. Baig, C., and Mavrantzas, V.G. *Phys. Rev. B* **79**, 144302 (2009).
98. Baig, C., Mavrantzas, V.G., and Öttinger, H.C. *Macromolecules* **44**, 640 (2011).
99. Mavrantzas, V.G. *Front. Phys.* **9**, 661367 (2021).
100. Callen, H.B. *Thermodynamics and an Introduction to Thermostatistics* (2nd ed.). John Wiley & Sons, New York (1985).
101. Pant, P.V.K., and Theodorou, D.N. *Macromolecules* **28**, 7224 (1995).
102. Ionescu, T.C., Edwards, B.J., Keffer, D.J., and Mavrantzas, V.G. *J. Rheol.* **52**, 105 (2008); Ionescu, T.C., Edwards, B.J., Keffer, D.J., and Mavrantzas, V.G. *J. Rheol.* **52**, 567 (2008).
103. Mavrantzas, V.G., and Theodorou, D.N. *Theor. & Comp. Polym. Sci.* **10**, 1 (2000).

Name Index

General Index

F

I

J

K

L

M

N

O